Chemistry in Context

Third Edition

A Project of the American Chemical Society

Chemistry in Context

Applying Chemistry to Society

Third Edition

Conrad L. Stanitski
University of Central Arkansas

Lucy Pryde Eubanks
Clemson University

Catherine H. Middlecamp
University of Wisconsin-Madison

Wilmer J. Stratton
Earlham College

A Project of the American Chemical Society

Boston   Burr Ridge, IL   Dubuque, IA   Madison, WI
New York   San Francisco   St. Louis
Bangkok   Bogotá   Caracas   Lisbon   London   Madrid   Mexico City
Milan   New Delhi   Seoul   Singapore   Sydney   Taipei   Toronto

McGraw-Hill Higher Education

A Division of The McGraw-Hill Companies

CHEMISTRY IN CONTEXT: APPLYING CHEMISTRY TO SOCIETY, THIRD EDITION

 This book is printed on recycled, acid-free paper containing 10% postconsumer waste.

1 2 3 4 5 6 7 8 9 0 QPD/QPD 9 0 9 8 7 6 5 4 3 2 1 0

ISBN 0–697–36024–5

Vice president and editorial director: *Kevin T. Kane*
Publisher: *James M. Smith*
Sponsoring editor: *Kent A. Peterson*
Developmental editor: *Margaret B. Horn*
Editorial assistant: *Jennifer L. Bensink*
Senior marketing manager: *Martin J. Lange*
Project manager: *Vicki Krug*
Senior production supervisor: *Sandra Hahn*
Design manager: *Stuart D. Paterson*
Senior photo research coordinator: *Carrie K. Burger*
Supplement coordinator: *Tammy Juran*
Compositor: *York Graphic Services, Inc.*
Typeface: *10/12 Times Roman*
Printer: *Quebecor Printing Book Group/Dubuque, IA*

Cover/interior designer: *Maureen McCutcheon*
Cover images: *Tony Stone Images*
Photo research: *Connie Gardner*

The credits section for this book begins on page 521 and is considered an extension of the copyright page.

Library of Congress Cataloging-in-Publication Data

Chemistry in context: applying chemistry to society/Conrad L.
 Stanitski, editor-in-chief . . . [et al.].—3rd ed.
  p.   cm.
 Includes index.
 ISBN 0–697–36024–5
 1. Biochemistry. 2. Environmental chemistry. 3. Geochemistry.
I. Stanitski, Conrad L.
QD415.C482   2000
540—dc21   99–40273
                                                      CIP

www.mhhe.com

Brief Contents

Contents

CHAPTER 4

Energy, Chemistry, and Society　137

CHAPTER 5

Take A Drink: The Wonder of Safe Drinking Water　183

CHAPTER 6

Neutralizing the Threat of Acid Rain　229

CHAPTER 11

Nutrition: Food for Thought 415

CHAPTER 12

Genetic Engineering and the Chemistry of Heredity 455

APPENDIX 1

Measure for Measure: Conversion Factors and Constants 493

APPENDIX 2

The Power of Exponents 495

APPENDIX 3

Clearing the Logjam 497

APPENDIX 4

Answers to Your Turn Questions Not Answered in Text 499

APPENDIX 5

Answers to Selected End-of-Chapter Questions 505

Foreword

The Chinese characters representing "chemistry" literally signify "the study of change." Change is what chemistry is all about: the transformation of one substance into another through the rearrangement of atoms. Chemical education is also undergoing change, and *Chemistry in Context* has proved to be a potent catalyst for altering the way in which chemistry is taught. The first edition of the book (1994) was the first college text to use contemporary issues in a consistent and concerted fashion to introduce chemical concepts, and to develop those concepts, *as needed*, to inform an understanding of the issues. The second edition (1997) built on the success of the first and further refined its pedagogical techniques. The book you hold in your hands continues the tradition and further improves upon it.

A text built on current events requires frequent updating, and that has most certainly been done for this edition. Moreover, the many Web and internet-based activities will ensure that instructors and students have access to the latest information. Many new Your Turn and Consider This activities have been added, and, at the request of many users, the end-of-chapter questions have been significantly expanded.

The writing team has also undergone some changes, with Conrad Stanitski, a veteran of the first two editions now serving as senior author and editor-in-chief. New contributors have brought new skills and former members have moved on to the Advisory Board. But the vision of *Chemistry in Context* remains clear and unaltered. It is a vision that has not only served the intended audience well, but has also influenced other courses, other syllabi, and other textbooks for non-science majors and for science majors as well. If imitation is the sincerest form of flattery, then *Chemistry in Context* has been frequently flattered. Through this book and many other innovative curriculum projects, the American Chemical Society continues to be a major agent for change in the teaching of chemistry.

A. Truman Schwartz
Advisory Board Chair

Preface

The Conclusion section of the last chapter of this textbook begins with a question related to genetic engineering: How can we balance the great potential benefits of modern chemical sciences and technology and the risks that seem inevitably to be part of the Faustian bargain that brought us knowledge? (Faust is a literary figure, an old philosopher who sells his soul to the devil in exchange for knowledge and power.)

It might seem unusual, perhaps even odd, to begin the Preface of a book with a question from the last chapter of the work. Yet, the compelling question given above shapes and motivates much of what the third edition of *Chemistry in Context* and its two previous editions are about—establishing chemical principles within a contextual framework of significant societal-technological issues. Thus, the basic philosophy of the third edition is unchanged from that of its antecedents. The wide acceptance of *Chemistry in Context* affirms the validity of its approach to develop the chemical content on a need-to-know basis so that students can more fully appreciate its contextual relationships to the intricate web of topics addressed in the book. Indeed, for three editions, a web is featured on the cover of the book because the word "context" is derived from the Latin word meaning "to weave"; this book weaves chemistry into the web of issues developed herein. By using this approach, we believe that students develop critical thinking ability, the competence to better assess risks and benefits, and the skills that lead them to be able to make informed and thus, reasoned decisions about technology-based matters.

With this third edition, the concept of the web takes on an expanded, contemporary meaning—the use of the internet worldwide web to seek answers to in-chapter and end-of-chapter questions. On the web, students can: (1) apply the Toxics Release Inventory information and other local environmental data to link environmental quality issues to their own city or state; (2) use real-time data, such as that provided by satellites about stratospheric ozone levels; (3) get up-to-date information concerning rapidly changing topics such as electric cars, nuclear waste, new pharmaceuticals, and genetic engineering; and (4) evaluate controversies using the web sites provided by experts, would-be experts, and skeptics on controversial topics such as global warming, food irradiation, and nuclear energy. Thus, the web presents students with the opportunity *and the responsibility* to critically evaluate web information among web sites of widely differing quality and validity. *Chemistry in Context* has its own web site at www.mhhe.com/cic with

password-protected items exclusively for faculty members as well as those available to students.

Changes in this edition reflect comments from users and reviewers of the second edition, consultants, as well as the opportunities presented by technological developments, such as the worldwide web, not widely available for the first two editions. There is material in a book such as this one that is time sensitive, and efforts have been made to have the tables, figures, and new information as up-to-date as possible using a printed format. Even so, the pace of chemical discoveries and their applications outstrips that of the printed word. Using the worldwide web helps to offset this imbalance. We also urge professors to supplement the text material with topical and current items from national, regional, and local media (print and non-print forms). Supplemental readings are listed on the *Chemistry in Context* web site rather than in the text so that updates can be provided in a more timely way.

In this edition, the number of chapters has been reduced from thirteen to twelve by incorporating a number of items from Onondaga Lake: A Case Study, Chapter 7 of the earlier editions, into other chapters. With this exception, the sequence of chapters from the second edition remains unchanged. The focus of the water chapter (5) has shifted from water quantity to water quality. The shift also brings into play three brief case studies of selected water quality issues, each coupled to the analytical chemical technique used to identify the particular substance of concern. As in prior editions, the first six chapters have an environmental focus. The next two bring a non–fossil fuel energy orientation and the remaining ones feature aspects of organic and biochemistry. Thus, one-third of the book has an organic/biochemistry flavor. We continue to suggest that the first four to six chapters serve as a core, with additional chapters selected as time and interests dictate. Most users teach from seven to nine chapters in a typical one-semester course.

In addition to the web-based activities, several other items are new to the third edition. The green leaf inset on the cover is emblematic of green chemistry, which makes its debut in this edition. Green chemistry, an important and rapidly emerging field, is introduced in Chapter 1 and then developed further with mini–case studies of its applications throughout the rest of the book. Risk-benefit analysis treatment has been expanded so that it appears as an aspect of additional issues. Other new items or expanded coverage include sun tanning (Chapter 2), low-level nuclear waste (Chapter 7), photovoltaics, fuel cells, and electric vehicles

(Chapter 8), generic and pioneer drugs (Chapter 10), Olestra as well as the use of irradiation to preserve food (Chapter 11), and mammalian cloning (Chapter 12). The coverage of covalent bonding, molecular shapes, and bond energies has been expanded a bit to provide a more leisurely discussion of those topics. The art program has increased the number of illustrations and photographs.

The **Your Turn**, **Consider This**, and **Sceptical Chymist** in-chapter activities remain integral and unique features of this book. They and the end-of-chapter questions have been thoroughly revised and expanded. Each chapter has 50 end-of-chapter questions. Interpretive questions related to molecular-scale interpretations of macroscale phenomena, graphical presentation and interpretation of data, and web-based assignments have been incorporated. As with the first two editions, a new edition of the detailed *Instructor's Resource Guide* is available, compiled by Joseph Bieron. For those who have a laboratory program with this course, a new edition of the *Laboratory Manual* is available, which includes experiments using microscale equipment (wellplates and Beral-type pipets) and common materials. Instructors are urged to have their students read the introductory section "Tracing the Web: A Reader's Guide to *Chemistry in Context*," which describes the purposes of the various pedagogical devices that are an integral part of the text.

A work of this type is not done in isolation; it depends on the assistance of others. Truman Schwartz, the Editor-in-Chief of the first two editions of *Chemistry in Context,* chaired the Advisory Board for this edition with his typical erudition and keen insights. Robert Silberman, Diane Bunce, and Arden Zipp were members of the author team for the first edition, the Editorial Team for the second edition, and served on the Advisory Board for the third edition. Advisory Board members contributed technical expertise as well as close reading of the manuscript as it developed. Our gratitude goes to the ACS Division of Education and International Activities staff, lead by Sylvia Ware, whose vision and professionalism continue to advance the agenda for quality chemistry education from kindergarten through graduate school by this project and other curricular initiatives. No other scientific society matches such efforts, which are reflections of the commitment of the leadership and membership of the American Chemical Society to curricular enhancement. The unstinting support of Kent Peterson (Chemistry Editor), Margaret Horn (Developmental Editor), and Vicki Krug (Project Manager), along with members of the production staff at McGraw-Hill is gratefully acknowledged. Sincere thanks also go to the reviewers, users, and consultants who provided comments that kept the manuscript on course. Errors in the Acknowledgments list are those of omission, not commission.

Another important third edition change to be noted is the revised editorial team. Two new members, Catherine Middlecamp and Lucy Pryde Eubanks, join Wilmer Stratton and me, veterans of the first two editions of this textbook. The third edition of *Chemistry in Context* maintains the outlook, philosophy, and strengths of its previous editions while bringing new and exciting dimensions to the textbook that has become the leader in its field. We are excited about this new edition and look forward to your comments.

Conrad L. Stanitski
Senior Author and Editor-in-Chief
conrads@mail.uca.edu
September 1999

Acknowledgements

Advisory Board (Third Edition)

A. Truman Schwartz, *Macalester College, Chair*
William Beranek, Jr., *Indianapolis Environmental Institute, Inc.*
Joseph F. Bieron, *Canisius College*
Joe Breen, *The Green Chemistry Institute*
Diane M. Bunce, *The Catholic University of America*
Al Campion, *University of Texas at Austin*
Joan Lebsack, *Fullerton College*
Robert Silberman, *State University of New York College at Cortland*
Arden P. Zipp, *State University of New York College at Cortland*

Advisory Board (Previous Editions)

Ronald D. Archer, *University of Massachusetts, Chair*
William Beranek, Jr., *Indianapolis Center for Advanced Research*
Glenn A. Crosby, *Washington State University*
Alice J. Cunningham, *Agnes Scott College*
Joseph N. Gayles, *Jon Mon Associates*
Ned D. Heindel, *Lehigh University*
Glenn L. Taylor, *Consultant*

Reviewers and Consultants (Third Edition)

Elisabeth Bell-Loncella, *University of Pittsburgh at Johnstown*
Paul H. Benoit, *University of Arkansas*
James Beres, *Shippensburg University*
David Bergbreiter, *Texas A&M University*
Joseph F. Bieron, *Canisius College*
Ron Caple, *University of Minnesota, Duluth*
Gregory Choppin, *Florida State University*
S. Todd Deal, *Georgia Southern University*
William Ellis, *Maine Maritime Academy*
Lance Funderburk, *Washington & Jefferson College*
Ana Gaillat, *Greenfield Community College*
John George, *Mary Washington College*
Marcia Gillette, *Indiana University-Kokomo*
Lynn Hartshorn, *University of St. Thomas*
Benjamin Huddle, *Roanoke College*
Mark Jackson, *Florida Atlantic University*
Angela Glisan King, *Wake Forest University*
Joseph Kirsch, *Butler University*
Cindy Klevickis, *James Madison University*
Howard Knachel, *University of Dayton*
Andrew Koch, *St. Mary's College of Maryland*
Martha Kurtz, *Central Washington University*
Robert Libby, *Truman State University*

James Long, *University of Oregon*
John Matacheck, *Hamline University*
Brent May, *Widener University*
Joseph Morse, *Western Washington University*
Gary Mort, *Lane Community College*
Ruth Nalliah, *Huntington College*
Deborah Otis, *Virginia Wesleyan College*
David Redfield, *Northwest Nazarene College*
Jeffrey Smiley, *Youngstown University*
Janice Smith, *York College*
Larry Smith, *Southern Methodist University*
Daniel Sullivan, *University of Nebraska-Omaha*
Glenn Taylor, *Consultant*
John Warner, *University of Massachusetts-Boston*
Bruce Winkler, *University of Tampa*

Web Site Reviewers

Elisabeth Bell-Loncella, *University of Pittsburgh at Johnstown*
Mark Jackson, *Florida Atlantic University*
Robert Libby, *Truman State University*

Reviewers and Consultants (Previous Editions)

Kyle D. Bayes, *University of California, Los Angeles*
Thomas Benzig, *James Madison University*
Paul P. Blanchette, *St. Mary's College of Maryland*
Stacey Lowery Bretz, *University of Michigan–Dearborn*
Colleen Byron, *Ripon College*
William F. Coleman, *Wellesley College*
Wendall H. Cross, *Georgia Institute of Technology*
Julie Cullen, *Bates College*
Paul M. Dullea, *North Atlantic Energy Service Corporation*
Steven W. Effler, *Upstate Freshwater Institute (NY)*
Lucy Pryde Eubanks, *Clemson University*
Linda Farber, *Sacred Heart University*
Newton C. Fawcett, *University of Southern Mississippi*
John E. Frederick, *University of Chicago*
Marcia L. Gillette, *Indiana University at Kokomo*
Ramaswamy Gnanasekaran, *Mansfield University of Pennsylvania*
Kenneth A. Goldsby, *Florida State University*
Mary L. Good, *Allied-Signal, Inc.*
Tiffany A. Grant, *Texas A&M University*
B. Welling Hall, *Earlham College*
Galen Hansen, *Fairmount State University*
Ronny Harris, *Texas A&M University*
Shirley Holden Helberg, *Baltimore, MD*
John Hogg, *Texas A&M University*

Thomas A. Holme, *University of Wisconsin–Madison*
Michael Imhoff, *Austin College*
A. H. Johnstone, *University of Glasgow*
Evelyn Kennedy, *Texas A&M University*
Joan Lebsack, *Fullerton College*
Robert A. Libby, *Truman State University*
Stephen Lindberg, *Oak Ridge National Laboratory*
Richard F. Malm, *Medina, WA*
Lilly Ng, *Cleveland State University*
Kevin W. O'Connor, *Office of Technology Assessment*
Michael Ogawa, *Bowling Green State University*
Larry M. Peck, *Texas A&M University*
Sarah Penhale, *Earlham College*
Robert Place, *Otterbein College*
Paul S. Poskozim, *Northeastern Illinois University*
Neil H. Potter, *Susquehanna University*
Ronald O. Ragsdale, *University of Utah*
James Rahn, *Concordia University*
Gulnar Rawji, *Southwestern University*
Jill Rawlings, *Auburn University of Montgomery*
Jeffrey Robertson, *Texas A&M University*
Billye Ross, *Pueblo Community College*
Martin St. Clair, *Coe College*
Steven M. Schildcrout, *Youngstown State University*
Irene Slagle, *The Catholic University of America*
W. Ross Stevens, III, *E. I. du Pont de Nemours*
Mary Stratton, *Earlham College*
Steven H. Strauss, *Colorado State University*
Jerry P. Suits, *McNeese State University*
Ronald S. Tjeerdema, *University of California, Santa Cruz*
Sheila Tobias, *University of Arizona*
Chad A. Tolman, *E. I. du Pont de Nemours*
Carl C. Wamser, *Portland State University*
Karen Warren, *Macalester College*
Martha J. M. Wells, *Tennessee Technological University*
Jay L. Wile, *Pathologists Associated*
Bruce Winkler, *University of Tampa*
William H. Zuber, Jr., *The University of Memphis*

George Gilbert, *Denison University*
William Hendrickson, *University of Dallas*
Judith Kelley, *University of Lowell*
H. Graden Kirksey, *The University of Memphis*
Werner Kolln, *Simpson College*
Ken Latham, *Lakewood Community College*
Joan Lebsack, *Fullerton College*
Marsh I. Lester, *University of Pennsylvania*
R. Bruce Martin, *University of Virginia*
Clifford Meints, *Simpson College*
Alan Pribula, *Towson State University*
Jeanne Robinson, *Seminole Community College*
Richard G. Scamehorn, *Ripon College*
Keith Schray, *Lehigh University*
A. Truman Schwartz, *Macalester College*
Robert G. Silberman, *State University of New York College at Cortland*
Charles Spink, *State University of New York College at Cortland*
Conrad L. Stanitski, *Mount Union College and University of Central Arkansas*
Wilmer J. Stratton, *Earlham College*
Dean Van Galen, *Northeast Missouri State University*
Arden P. Zipp, *State University of New York College at Cortland*
William H. Zuber, Jr., *The University of Memphis*

Evaluators (Trial Edition)

Mary B. Nakhleh, *Purdue University*
Jack E. Rossmann, *Macalester College*
Timm Thorsen, *Alma College*

American Chemical Society Division of Education and International Activities

Sylvia A. Ware, *Director, Division of Education and International Activities*
Janet M. Boese, *Assistant Director for Academic Programs*
T. L. Nally, *Department Head, Higher Education*
Christine Brennan, *Chemistry in Context Project Manager*

Chemistry in Context Examination Committee

ACS-DivCHED Examinations Institute, *Clemson University*
Diane M. Bunce, *The Catholic University of America*
Mary D. B. Dillingham, *Lander College*
Anne G. Glenn, *Guilford College*
Thomas A. Holme (Chair), *University of Wisconsin-Milwaukee*
Jeffrey L. Seela, *Abraham Baldwin Agricultural College*
Julianne M. Smist, *Springfield College*
Conrad L. Stanitski, *University of Central Arkansas*
Daniel M. Sullivan, *University of Nebraska-Omaha*

Student Aides (First Edition)

Melissa Dovi, *State University of New York College at Cortland*
Steven Longenecker, *Earlham College*
Holly Vande Wall, *Macalester College*
Maria Verbrugge, *Earlham College*
Joe Ziegelbauer, *Earlham College*

Test Site Directors/Instructors (Trial Edition)

Joseph F. Bieron, *Canisius College*
Diane M. Bunce, *The Catholic University of America*
Colleen Byron, *Ripon College*
Timothy D. Champion, *Johnson C. Smith University*
Charles E. Eaker, *University of Dallas*
Gordon J. Ewing, *New Mexico State University, Las Cruces*

Tracing the Web:
A Reader's Guide to *Chemistry in Context*

The symbol selected for *Chemistry in Context* is a spider web, a remarkably strong and flexible structure created by elegant chemistry within a spider's body. The web also represents the complex connections that exist between chemistry and society. Appropriately, the word "context" derives from a Latin word meaning "to weave." Thus, the title says it all: a major purpose of this book is to weave chemistry into its social, political, economic, and ethical context. This introduction is intended to help you find your way through the web without getting trapped.

If you have already had a chemistry course, the first thing you likely will notice is that the **Table of Contents** of this text is quite different from that of a more traditional chemistry book. In *Chemistry in Context,* the chapter titles reflect today's headlines and the issues surrounding them. Ozone depletion, global warming, new energy sources, nutrition, genetic engineering, and the other topics treated in this text are all closely connected to chemistry. For you to understand and respond thoughtfully in an informed manner to these vitally important issues, you must know something about the chemical principles that underlie the socio-technological issues. This book presents those principles, when and where they are needed, in a manner intended to better prepare you to be a well-informed citizen.

We have introduced a number of features that we hope will be helpful to you, and we encourage you to make use of them. The most traditional features are the **Your Turn** activities. They provide opportunities for you to practice a skill or calculation that has, in most cases, already been illustrated in the text, usually just before the Your Turn activity. These activities test your understanding and are good practice for chapter-end questions or examinations. Even if your instructor does not assign all of the Your Turn activities, we urge you to attempt them. Hints or complete solutions are often provided, and answers appear either immediately following the question or in Appendix 4.

Many modern problems involving chemistry are a good deal more complicated than those in the Your Turn category, and for that reason we have constructed **Consider This** questions. Here you may be asked to engage in risk-benefit analysis, consider opposing viewpoints, speculate on the consequences of a particular action, or formulate and defend a personal position. Some of these questions invoke the use of the worldwide web to gather data and update information in the textbook. These decision-making activities can be used in a wide variety of ways, and your instructor will provide detailed assignments related to them that might involve library research, writing, group work, discussion, debate, or role playing. A typical Consider This activity, like many aspects of life, may not have a single "right" answer. However, the issues raised demand correct information, critical thinking, sound reasoning, and clear communication to form and present an informed opinion. Although no course will be able to address all of the Consider This questions, we encourage you to read them all, because they provide much food for thought.

The third special feature, **The Sceptical Chymist**, takes its title (and peculiar spelling) from an influential book published in 1661 written by Robert Boyle, an early investigator of the properties of air. In his book, Boyle observed that scientific truth would be more solidly established "if men would more carefully distinguish those things that they know from those that they ignore or do but think." The popular press is full of statements and stories that seem to confuse what is known and what is thought. Modern "chymists" and students of "chymistry" are advised to develop the critical habit of mind that leads them to doubt and question what they read or hear. The Sceptical Chymist gives you an opportunity to sharpen those analytical skills by responding to a variety of statements and assertions. Often we provide some guidance; sometimes we leave you to your own devices.

Each chapter begins with a **Chapter Overview** that indicates what is to follow and ends with a **Conclusion** that draws together the major themes. The **Chapter Summary**, with its "Having studied this chapter, you should be able to:" heading calls attention to the most important skills and applications introduced and developed in the chapter. **Marginal Notes** are used throughout the book to succinctly summarize and emphasize key points, or to point out linkages with sections in other chapters.

The end-of-chapter study materials also include **Questions**, some of which will likely be assigned by your instructor. The Questions are divided into three categories. *Emphasizing Essentials* are questions on which you practice and sharpen the chemistry skills developed in the chapter. *Concentrating on Concepts* questions have you focus on chemical concepts and their relationships to the socio-technological topics under discussion. Questions in the *Exploring Extensions* category present a challenge to go beyond the textbook material by providing an opportunity to extend and integrate skills, concepts, and communication. The latter two categories of questions also incorporate the use of the worldwide web as a source of data and opinions. The questions with colored numbers are answered in Appendix 5.

Among the in-chapter questions and the end-of-chapter questions are questions that have you use the worldwide web to find data and commentary on a variety of topics to answer the questions. An important resource in this regard is the *Chemistry in Context* web site at www.mhhe.com/cic.

The end of the book includes **Appendices** on conversion factors and constants (Appendix 1), exponents (Appendix 2), and logarithms, including an expanded treatment of pH (Appendix 3); answers to Your Turn questions (Appendix 4), and selected end-of-chapter questions (Appendix 5); a **Glossary** that defines the major terms used in the text and provides references to the pages where the terms are explained in context; and an **Index**.

We recognize that the great majority of the readers of *Chemistry in Context* will not become chemists. But, we are convinced that the areas where chemistry has an impact on society are far too important to be left to chemists alone, or to politicians, for that matter. We obviously cannot include all the current or potential socio-technological problems that involve chemistry. Nevertheless, we hope that the issues selected, the facts and principles presented, and the habits of mind you and your fellow readers develop as you study in this course will assist you to live responsibly in a future in which chemistry will continue to be the science that is central to better understand our world.

Conrad L. Stanitski
Lucy Pryde Eubanks
Catherine H. Middlecamp
Wilmer J. Stratton

Guided Tour

5.5 How Pure is Drinking Water?

Regardless of its source, drinking water is rarely, if ever, just pure H_2O. You can be assured that almost certainly it contains other substances. For example, a label on Evian bottled water includes the following information:

Mineral Composition, mg/liter			
Calcium	78	Bicarbonates	357
Magnesium	24	Sulfates	10
Silica	14	Chlorides	4
		Nitrates (N)	1

Of these seven items for Evian water, all but one of them will be discussed in this chapter. We will also find out that the items listed on the label are themselves not chemical compounds, but are just parts of compounds. The number given with each item indicates how much of that substance (in milligrams) is present in one liter of Evian water. This raises a reasonable question: Should we be concerned about any of these substances and its amount? Calcium, for example, has a definite health benefit in producing stronger bones. Milk, not Evian water, is the preferred source for calcium; you would have to drink 4 liters of Evian water to get the same amount of calcium as that in one 8-oz glass of milk. In contrast, nitrate can be a health problem, especially for infants, depending on its concentration. The other listed substances in Evian bottled water are not likely to be a health problem. Elsewhere on the label it is noted that sodium, a health issue for some people, is present at less than 5 mg per bottle.

Such composition information can also be obtained for tap water supplied by municipal water companies. A typ...
the authors of this text reveale...

1 liter = 1.00 quart.

Min...	
Calci...	
Magn...	
Sodiu...	

5.7 Your Turn

One 500-mL bottle of E...
requirement of calcium...

a. Use the label inform...
   that is recommende...
b. How many 500-mL b...

1.14 Breathing Lessons—Indoor Air

The Consider This 1.1 exercise had you "take a breath." That one breath adds up to between 1×10^4 to 2×10^4 L (10–20 m^3) of air each of us breathes daily. As we have learned from previous sections, higher air quality standards have decreased the allowable concentrations of various air pollutants by controlling emissions from automobiles and industries. But air quality depends on where we are. Ironically, such standards have been established for outdoor air, but not for indoor air. Yet, most of us sleep, work, study, and play indoors, spending up to 95% of our time in our dorm rooms, classrooms, offices, and residences. Consequently, we should be concerned not just with outdoor air quality, but that of indoor air as well. Figure 1.11 indicates the concentrations of common pollutants in indoor and outdoor air. Notice from Figure 1.11 that, with the exception of sulfur dioxide, the average indoor concentrations of the pollutants exceed their average outdoor concentrations.

Indoor air is a complex mixture; typically nearly 1000 substances are detectable in it at the ppb level or higher. However, indoor air sources are limited to either the outside air that enters buildings, or air that comes from within buildings. Tobacco smoke, cooking by-products, substances emitted from rugs, furniture, construction materials, and office products are some of the many materials that can degrade indoor air quality. Table 1.10 lists some of the sources of indoor air pollutants.

How quickly indoor air pollutants build up depends on the rates at which outdoor air moves inside and indoor air moves out, as well as on how rapidly pollutants are generated indoors. An insufficient exchange of outside air can cause the concentration of indoor air pollutants to build up to troublesome levels. Consider the risk-benefit trade-off in which buildings constructed within the past two decades have been more airtight to minimize drafts and increase energy efficiency. Although enhanced energy efficiency has been achieved, it has been at the cost of decreasing the flow of outside air into the building to replace the indoor air. When this reduction in air exchange occurs, it allows the concentration of indoor air pollutants to increase. Therefore, initially what was a benefit (better energy efficiency) can turn into an increased risk (increased pollutants concentration). Construction of some large office buildings has been so highly energy efficient that little exchange of outside air occurs within them. In some of these cases, the reduced air exchange has allowed indoor air pollutants to reach levels hazardous to the health of some individuals, creating a condition known as "sick building syndrome."

Figure 1.11
Indoor and outdoor concentrations of common pollutants

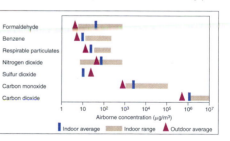

2.4 Waves of Light

To better understand how stratospheric ozone screens out much of the Sun's harmful radiation requires knowing the molecular structure of ozone as well as some fundamental properties of light. The interaction of sunlight with matter is important, such as in photosynthesis or the damage high-energy solar radiation can cause in living organisms. Therefore, we turn now to develop an understanding of light.

Every second, five million tons of the Sun's matter are converted into energy, which is radiated into space. The fact that we detect color indicates that the radiation that reaches us is not all identical. Prisms and raindrops break sunlight into a spectrum of colors. Each of these colors can be identified by the numerical value of its wavelength. The word correctly suggests that light behaves rather like a wave in the ocean. The **wavelength** is the distance between successive peaks (Figure 2.1). It is expressed in units of length and symbolized by the Greek letter lambda (λ).

It is both interesting and humbling to realize that out of the vast array of radiant energies, our eyes are sensitive to light only in a very tiny portion of the electromagnetic spectrum—wavelengths between about 700×10^{-9} m (corresponding to red) and 400×10^{-9} m (corresponding to violet). These lengths are very short, so we typically express them in nanometers. One nanometer (nm) is defined as one one-billionth of a meter (m). In symbols,

$$1 \text{ nm} = \frac{1}{1,000,000,000} \text{ m} = \frac{1}{10^9} \text{ m} = 1 \times 10^{-9} \text{ m}$$

We can use this equivalency to convert meters to nanometers, for example, to convert 700×10^{-9} meters to nanometers.

$$\text{wavelength} = \lambda = 700 \times 10^{-9} \text{ m} \times \frac{1 \text{ nm}}{1 \times 10^{-9} \text{ m}} = 700 \text{ nm}$$

The meter units cancel and we are left with nanometers.

2.9 Your Turn

Violet light in the visible part of the spectrum has a wavelength of 408 nm. Express this wavelength in meters.

Ans. 4.08×10^{-7} m

Figure 2.1
Wave motion.

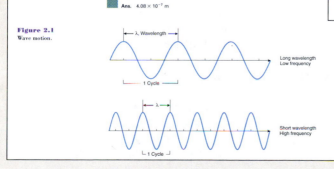

4.10 Your Turn

Use the bond energies in Table 4.1 to calculate the heat of combustion of propane, C_3H_8 (LP or "bottled gas"). Report your answer both in kJ/mole of C_3H_8 and kJ/g C_3H_8. This is the equation for the reaction, written with structural formulas.

$$H-\overset{\overset{\displaystyle H}{|}}{\underset{\underset{\displaystyle H}{|}}{C}}-\overset{\overset{\displaystyle H}{|}}{\underset{\underset{\displaystyle H}{|}}{C}}-\overset{\overset{\displaystyle H}{|}}{\underset{\underset{\displaystyle H}{|}}{C}}-H + 5\,\ddot{O}{=}\ddot{O} \longrightarrow 3\,\ddot{O}{=}C{=}\ddot{O} + 4\,H{-}\ddot{O}{-}H$$

Hint: Note that there are 8 moles of C—H bonds, 2 moles of C—C bonds, 5 moles of O=O bonds, 6 moles of C=O bonds and 8 moles of O—H bonds involved in the reaction.

Ans. Energy change = −2016 kJ/mole C_3H_8 or −45.8 kJ/g C_3H_8
Heat of combustion = 2016 kJ/mole C_3H_8 or 45.8 kJ/g C_3H_8.

YOUR TURN

These activities allow students to practice a skill or calculation that has, in most cases, previously been illustrated in the text, typically just before the Your Turn activity. Excellent as reinforcement for chapter-end questions and examinations. Hints and complete solutions are often provided immediately following the questions.

CONSIDER THIS

Develop your students' critical thinking skills with these more complicated problems. These decision-making activities can be used as assignments involving library research, writing, group work, discussion, debate, or role playing.

1.1 Consider This: Take a Breath

One breath does not use much air, but what total volume of air do you exhale in a typical day? One approach to this question is simply to guess, but an educated guess is more reliable than a wild one. By designing and executing a simple experiment, you can come up with a reasonably accurate answer. You will need to determine how much air you exhale in a single breath and how many breaths you take per minute. Once you establish this information, determine how much air you exhale in a day (24 hours). Describe the experiment you performed, the data you obtained, and any factors that you can identify that affect the accuracy of your answer.

3.23 The Sceptical Chymist: Cooler Heads

Has the greenhouse effect been amplified by human activities? **a.** Find some web links to several organizations that don't believe so. Try the Global Warming Skeptics web page, or that from the Greening Earth Society. Newspapers, television, and radio-related web pages may contain editorials that criticize global warming as a scientific concept. **b.** Is there scientific merit to the arguments of those opposing global warming? Cite specific cases.

SCEPTICAL CHYMIST

Students are challenged to use their critical habits of mind to doubt and question chemistry-related statements and stories made by the popular press. Each Sceptical Chymist exercise contains a variety of questionable statements and assertions, providing students an opportunity to sharpen their analytical skills.

GREEN CHEMISTRY

Green Chemistry principles are introduced and used repeatedly to demonstrate creative contemporary chemical approaches that minimize and prevent technological difficulties.

 An obvious way to reduce pollution is not to have the pollutants form in the first place. Over the past decade, an important initiative known as "green chemistry," the use of chemistry to prevent pollution, has taken place. **Green chemistry**—designing chemical products and processes that reduce or eliminate the use and/or generation of hazardous substances. Begun under the EPA's Design for the Environment Program, green chemistry reduces pollution through fundamental chemical breakthroughs in designing and re-designing chemical processes, with an eye toward making them environmentally friendly, that is "benign by design." In this regard, Dr. Barry Trost, a Stanford University chemist, advocates an "atom economy" approach to the synthesis of commercial chemical products such as pharmaceuticals, plastics, or pesticides. Such syntheses would be designed so that all reactant atoms end up as desired products, not

Halons are greenhouse gas compounds composed of carbon, fluorine, and bromine atoms used in fire fighting and other applications. As a greenhouse gas, halons are also scheduled to be phased out under the modified Montreal Protocol. Using a green chemistry approach, Pyrocool Technologies has synthesized a halon substitute. The product, called Pyrocool FEF, is an environmentally benign foam that is more effective than halons in fighting fires, even large-scale fires such as those on oil tankers and jet airplanes (Figure 2.20). Pyrocool Technologies won a 1998 Presidential Green Chemistry Challenge award for this development.

The phaseout of CFCs is not without major economic considerations. At its peak, the annual worldwide market for CFCs reached $2 billion, but that was only the tip of a very large financial iceberg. In the United States alone, chlorofluorocarbons were used in or used to produce goods valued at about $28 billion per year. Even today, over $100 billion worth of equipment, probably including your refrigerator and automobile air conditioner, rely on CFCs. Although the conversion to CFC replacements has had some additional costs associated with it, the overall effect on the U.S. economy has been minimal. Companies that produce refrigerators, air conditioners, insulating plastics, and other goods have adapted to using the new compounds. Some substitutes for CFC refrigerants are less energy efficient, hence increasing energy consumption by several percent. But the conversions will provide a market opportunity for innovative syntheses using green chemistry to produce environmentally benign substances.

One of the most interesting efforts to eliminate CFCs has been the Super Efficient Refrigerator Program (SERP). Historically, refrigerators have been major users of electrical

GREEN CHEMISTRY ICONS

Icons identify areas within the text where Green Chemistry principles are discussed.

WEB-BASED ACTIVITIES

Web-based activities are built into the in-chapter and chapter-end questions. Students can explore a variety of issues, including evaluating the accuracy of various web sites.

2.1 Consider This: Ozone Levels Above Your House

What are the current ozone levels in the stratosphere? The National Aeronautics and Space Administration (NASA) can provide you with values. In fact, if a satellite is sending back data as you read this, you may be able to get today's ozone level in the stratosphere right above where you live.

a. Where do you live? Determine your latitude and longitude from a map, from a nearby airport, or by using the link provided at the *Chemistry in Context* web site.

b. Use the NASA link at the *Chemistry in Context* web site to access satellite data. Enter your latitude and longitude from part **a** to find the total column ozone amount at your location for today. Request an earlier date if today's is not available. As you will see later in the chapter, 320 Dobson units is the average ozone level over the northern U.S. How does your value compare with the average?

c. Again using the NASA link, obtain ozone values for some other parts of the world.

47. How do researchers determine whether the negative effects of acid deposition on aquatic life are a direct consequence of low pH or the result of Al^{3+} released from rocks and soil? Find at least one web article that gives the details of such a study. In your own words write a summary of the experimental plan and its results.

WEB ICON

A web icon alerts students to web-based activities within the text.

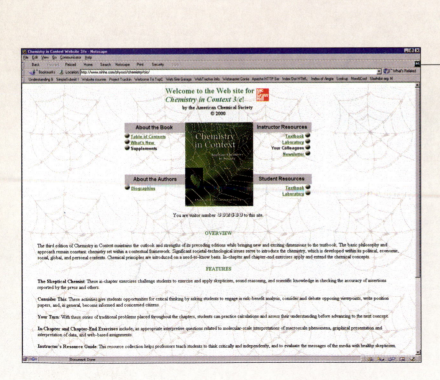

ON-LINE LEARNING CENTER

Log on at—

www.mhhe.com/cic

Also new to the third edition is a *Chemistry in Context* (CIC) web site for both students and instructors. This site includes web activities from the text, information to help students evaluate the quality of web sites, and searching tools. Instructors can find links to documents that are useful source material for lectures and classroom discussions. Also available on-line will be parts of the Instructor's Resource Guide (IRG), such as the answers to end-of-chapter questions.

Chapter Summary

Note: The numbers that follow indicate the sections in which the topics are introduced and explained.

Having studied this chapter, you should be able to:

- Recognize the composition of air and reasons for local and regional variations of it (1.1 – 1.3);
- Understand factors behind air quality and the chief components of air pollution (1.3, 1.11, 1.12);
- Evaluate conditions significant in risk-benefit analysis (1.4);
- Identify the general regions of the atmosphere with respect to altitude and the relationship of air pressure to altitude (1.5);
- Interpret air quality data in terms of concentration units (ppm, ppb) and pollution levels, including unreasonableness of "pollution free" levels (1.2 – 1.3, 1.12, 1.14, 1.15);
- Differentiate among mixtures, elements, and compounds (1.6);
- Understand the differences between atoms and molecules, between symbols for elements and formulas for chemical compounds (1.7);

- Name selected chemical elements and compounds (1.7);
- Write and interpret chemical formulas (1.8);
- Balance chemical equations, including using sphere equation representations (1.9 – 1.10);
- Discuss the green chemistry initiative (1.11);
- Describe the nature of air quality policies in this country and abroad in terms of their effectiveness in controlling air pollution (1.12 – 1.13);
- Identify the sources and nature of indoor air pollution (1.14);
- Interpret the nature of air at the molecular level (1.15);
- Use scientific notation and significant figures in performing basic calculations (1.4 and 1.15, respectively).

Interpretive questions related to molecular-scale interpretations of macroscale phenomena, graphical presentation and interpretation of data, and web-based assignments are incorporated throughout the text.

Questions

The questions in thi
chapters, are divided

- **Emphasizing Ess**
  nity to practice t
  chapter. This set
  Turns in the chap
  questions whose

- **Concentrating o**
  on the chemical c
  lationships to the
  grate and to apply

28. In Consider This 1.3, you considered how life on Earth would change if the concentration of oxygen were doubled. Now consider the opposite case; discuss how life on Earth would change if the concentration of O_2 were only 10%. Give some specific examples of how burning, rusting, and most metabolic processes in humans and plants would be affected.

29. Explain why the concentrations of some components in the atmosphere are expressed in percent (parts per hundred) and others are given in ppm (parts per million).

30. Consider the following table of data from the U.S. Environmental Protection Agency, Office of Air Quality Planning and Standards. The data indicate the number of days metropolitan statistical areas failed to meet acceptable air-quality standards (Pollutant Standards Index rating over 100).

Air Quality of Selected U.S. Metropolitan Areas, 1986–94

Metropolitan statistical area	1986	1988	1990	1992	1994
Atlanta, GA	18	21	17	5	4
Bakersfield, CA	54	85	48	16	45

36. In these diagrams, the larger circles represent one kind of atom and the smaller circles represent a different kind. Characterize each of the samples as an element, compound, or mixture and give reasons for your answers.

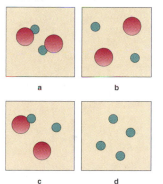

a   b

c   d

There are 50 questions at the end of each chapter, an increase of more than 25%.

INSTRUCTOR'S RESOURCE GUIDE AND LABORATORY MANUAL

As with the first two editions of **Chemistry in Context,** a new edition of the detailed *Instructor's Resource Guide* and the *Laboratory Manual* are available.

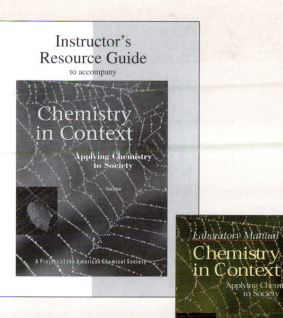

The Air We Breathe

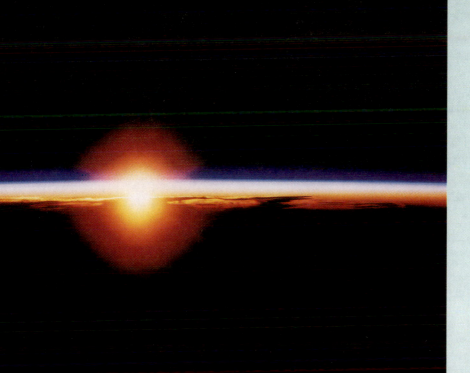

The Earth's fragile atmosphere is seen in this photo of a sunrise over the West Indies taken by the crew of the Space Shuttle Discovery (July, 1995). The thin blue band is the stratosphere and the red-orange band is the troposphere. A cloud layer (white) is seen in the upper troposphere.

"Finally it shrank to the size of a marble, the most beautiful marble anyone can imagine."

Astronaut James Erwin, May 1969

Spoken more than three decades ago, the striking words of James Erwin compel us to consider the awesome spectacle of our home planet. Only a few women and men have actually observed what James Erwin saw in May 1969, but most of us have seen the spectacular photographs of the Earth taken from outer space. From that vantage point, our planet looks magnificent — a blue and white ball compounded of water, earth, air, and fire. It is where thousands upon thousands of species of plants and animals live, all interrelated in a global community. Nearly six billion of us belong to one particular species with special responsibilities for the protection of our beautiful marble.

The "blue marble," our Earth, as seen from outer space.

Mount St. Helen's volcano, Washington photo from the Space Shuttle (September 1994). The volcano erupted in May, 1980 blowing out its northern wall causing the largest avalanche ever to occur in the western hemisphere. The eruption sent clouds of ash and smoke 70,000 feet into the stratosphere and the blast and avalanche destroyed over 270 square miles of forest in five seconds.

A closer view from a satellite reveals more detail. The landforms visible in computer-enhanced photographs remind us of the great geological diversity of our planet and the many biological changes that have occurred in adaptation to these varied environments. We see mountains, forests, deserts, prairies, glaciers, jungles, lakes, and rivers. In some cases, these natural features are boundaries that, for better or worse, separate our kind into national communities.

But the communities that we know best are those closest to home—the cities, towns, ranches, and farms where we live, work, play, and sleep. Our neighbors, friends, and families are here. The people, customs, habits, and laws that create and constitute these regional environments shape each one of us.

As individuals, we simultaneously inhabit these concentric communities. Our personal lives are imbedded in our immediate surroundings, the countries we live in, and the entire globe. Changes in any of these environments affect us, and we, in turn, have obligations at each level of community. This book is about some of those responsibilities and the ways in which a knowledge of chemistry can help us meet them with intelligence, understanding, and wisdom.

Chapter Overview

To be an informed (and healthy) citizen, you should know about the air you breathe. What is in it and its quality are essential for your existence, and what might endanger it. Because air is a complex mixture of substances, such knowledge requires

A crowd celebrating at the River Festival in Little Rock on the banks of the Arkansas River.

familiarity with certain fundamental chemical concepts and facts. Therefore, this chapter begins by considering the chemical composition of the air, its major components and minor constituents including pollutants, and how their presence is expressed. The next part looks at the matter of air quality, which raises a theme that will frequently recur in this text—the difficult challenge of risk assessment. In addition, you will learn about the structure of the atmosphere, and encounter the ways in which chemists organize matter into elements and compounds and the symbols and formulas used to represent these substances. A consideration of atoms and molecules then provides a submicroscopic view of matter that helps further our comprehension. These elements and compounds, atoms and molecules undergo an amazing range of transformations. Such chemical reactions are at the very heart of chemistry, and we will consider a few of the more important reactions that occur in the atmosphere. These, in turn, will provide an opportunity to explore the powerful shorthand of the chemist—chemical equations. Thus armed, we will examine reasons for the dramatic improvement in U.S. air quality over the past 30 years and consider the sources of the most important air pollutants in the United States and abroad. Also included is a brief discussion of indoor air quality. The chapter ends by reexamining at the molecular level the breath that initiated it.

1.1 Take a Breath

We begin by asking you to do something you do automatically and unconsciously thousands of times each day— take a breath. You certainly do not need textbook authors to tell you to breathe! A doctor may have encouraged your first breath with a well-placed slap, but from then on nature took over. Childish threats to hold your breath forever probably did not worry your parents. They knew that within a minute or two you would involuntarily gasp a lungful of that invisible stuff we call air. Indeed, you could not survive more than 5–10 minutes without a fresh supply of air.

1.1 Consider This: Take a Breath

One breath does not use much air, but what total volume of air do you exhale in a typical day? One approach to this question is simply to guess, but an educated guess is more reliable than a wild one. By designing and executing a simple experiment, you can come up with a reasonably accurate answer. You will need to determine how much air you exhale in a single breath and how many breaths you take per minute. Once you establish this information, determine how much air you exhale in a day (24 hours). Describe the experiment you performed, the data you obtained, and any factors that you can identify that affect the accuracy of your answer.

Consider This 1.1 addresses how much air you breathe, but not the equally important topic of *what* you breathe, and whether it might be harmful. For a commentary on air quality we turn to this statement from a Shakespearean play about a troubled young man, Hamlet: "This most excellent canopy, the air, look you, this brave o'erhanging firmament, this majestical roof fretted with golden fire, why, it appears no other thing to me but a foul and pestilent congregation of vapors." To be sure, the speaker had a lot on his mind, especially the allegation that his uncle killed Hamlet's father before marrying his mother. Hamlet spends the rest of the play trying to decide what to do about his dysfunctional family, and no further reference to air pollution is made—except perhaps in the observation that "Something is rotten in the state of Denmark."

Actually, the air is probably worse in Los Angeles, Mexico City, or Bangkok than it is in Elsinore (Hamlet's home) or nearby Copenhagen. But wherever you live, there is a good chance that the lungful of air you just inhaled contains some substances that, depending on their amount, could be harmful to your health. The health threat can be so serious that laws are passed in an effort to limit pollution by curtailing some of the ways we normally do things. Passing such legislation is not without controversy, as illustrated by the following newspaper headlines.

AUTO AND OIL COMPANIES DISAGREE ON HOW TO CURB SULFUR POLLUTION (*Wall Street Journal;* May 13, 1998)

LIGHT TRUCKS INCREASE PROFITS BUT FOUL AIR MORE THAN CARS (*New York Times*; November 30, 1997)

EASTERN STATES BACK MEASURES TO REDUCE SMOG-CAUSING EMISSIONS (*New York Times;* June 7, 1997)

STATE HEARS BOTH SIDES ON STRENGTHENING AIR POLLUTION STANDARDS (*Houston Chronicle*; January 16, 1997)

EPA STUDY CALLS FOR TOUGHER CONTROLS ON EMISSIONS OF AUTOMOBILES BY 2004 (*Wall Street Journal;* April 23, 1998)

The chief points of the story behind the last headline are:

- A U.S. Environmental Protection Agency (EPA) study calls for tougher auto-emission controls by 2004 by seeking better ways to clean up exhaust from auto diesel engines, light trucks, and sport-utility vehicles.
- Four pollutants are cited in the study: soot particles, carbon monoxide, volatile organic compounds, and nitrogen oxides. The EPA claims that exhaust cleaning technologies are available, at an additional cost of up to $161 per vehicle.
- Although new cars are almost 97% cleaner than 1970 models, the exhausts of larger vehicles such as sport-utility vehicles, minivans, and small pickup trucks are not as clean as those from automobiles.
- The constant growth in the volume of traffic and in the number of vehicles could degrade the nation's air quality, especially in cities and urban areas such as New York,

Atlanta, Baltimore-Washington, Dallas-Fort Worth, Houston, Chicago, Pittsburgh, and Cincinnati.

- Auto makers see reducing gasoline's sulfur content as a key technological step to cleaner emissions, an issue not addressed by the EPA study.
- The EPA might be willing to allow greater flexibility in how the new standards are met by having different guidelines for different categories, a system now used in California.

In this chapter and those that follow, news headlines and reports will be used to illustrate chemically significant issues, and you will be asked to analyze the news articles with a critical eye— for scientific accuracy, bias, timeliness, and other criteria.

1.2 Consider This: A Visit to the EPA

The U.S. EPA maintains an extensive web site providing resources for both scientists and the general public. The EPA, like many government agencies, can be found by entering www.agency.gov, substituting the name or abbreviation for the agency. Thus, a likely place to find the EPA is www.epa.gov.*

Go to the EPA web site and explore it for a few minutes. Periodically use your back button (or Go button) to keep track of where you are. If you get lost, start again at the EPA's home page. As you browse, you may come across the EPA's Office of Air and Radiation (OAR) that also contains a number of "consumer-friendly" documents.

The EPA offers many useful documents on air quality. Find one and provide the following information:

Name of the EPA document

URL

Summary of information provided

Last updated (if provided)

Notable features (good or bad)

Something interesting that you learned

Get directions from your instructor if you are to share the information you found with your classmates, and how to do so.

*You also can use the search button on your browser to locate the EPA.

Unfortunately, it is nearly impossible to completely avoid polluted air or to remove all pollutants from it. But actions or proposed actions such as those described in the earlier news articles and the natural regenerative properties of the atmosphere have improved air quality in many parts of the world, including the United States, over the past two decades.

1.2 What's in a Breath? The Composition of Air

The air we breathe is a mixture of several substances. For the moment we will focus on only five: oxygen, nitrogen, argon, carbon dioxide, and water. The first four normally exist as gases. Although we usually think of water as a liquid, it can also be a gas— in which case we often call it "water vapor" to distinguish between the two physical states. The concentration of water vapor varies widely; it can be close to 0% in very dry desert air or 5–6% in a tropical rain forest. Because of this variability, reference tables typically list the composition of dry air. The normal composition of dry air is 78% nitrogen, 21% oxygen, and 1% other gases by volume. Figure 1.1 displays this information in the form of a pie chart and a bar graph. Both of these are important, widely used methods for displaying numerical information and we will use each at various places in this text. The pie chart emphasizes the fractions of the total, whereas the bar graph emphasizes the relative sizes of each. Regardless of how we present the data, notice that 99% of the total is made up of only two substances, nitrogen and oxygen.

Figure 1.1
The composition of dry air,
by volume.

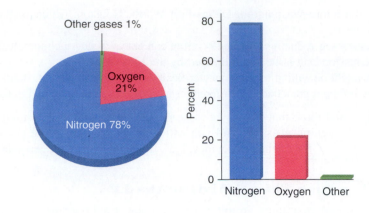

The 21% oxygen is what is most immediately essential for sustaining life. Oxygen is absorbed into our blood via the lungs, and reacts with the foods we eat to release the energy needed for all life processes within our bodies. Life on Earth bears the stamp of oxygen. Indeed, it is difficult to conceive of life on any planet without this remarkable chemical. Oxygen is also a participant in burning, in rusting, and in other corrosion reactions. Because it is a constituent of water and many rocks, oxygen is the most abundant element in the Earth's crust and the human body. Given this broad distribution and high reactivity, it is somewhat surprising that oxygen was not isolated as a pure substance until the 1770s. But once oxygen was isolated and purified, it proved to be of great significance in establishing the principles of the young science of chemistry.

| You can think about these percentages by considering that if you had 100 quarts of air, you would have 78 quarts of nitrogen, 21 quarts of oxygen, and 1 quart of other gases.

| Table 12.1 gives the elementary composition of the human body.

1.3 Consider This: Increasing the Oxygen in the Atmosphere

Humans are accustomed to living in a world with an atmosphere containing 21% oxygen. In such a world, a match from a paper matchbook burns completely in less than a minute and a log with a diameter of 3 inches is consumed in about 20 minutes in a fireplace fire. You have already determined how many times a minute you exhale air. Burning, rusting, and most metabolic processes in humans, plants, and animals are dependent on oxygen. How would life on earth be different if the oxygen content in the atmosphere were doubled to 40%, or higher? List at least four effects such a change would have on life as we know it.

Nitrogen is the most abundant substance in the air and constitutes over three-fourths of the air we inhale. However, it is much less reactive than oxygen, and it is exhaled from our lungs unchanged. Although nitrogen is essential for life and is a part of all living things, most plants and animals obtain the nitrogen they require from mineral sources, not directly from the atmosphere.

The remaining 1% of air is mostly argon, a substance so unreactive that it is said to be "chemically inert." This inertness is recognized in the name *argon,* which means "lazy." Because argon refuses to form a chemical combination with anything (even itself), it does not normally make its presence known. Hence, it was not discovered until 1894.

| Conversion factors relating units are found in Appendix 1.

It is important to remember that the percentages we have been using to describe the composition of the atmosphere are based on volume. Thus, a mixture of 78 liters of nitrogen (1 liter = 1 L = 1.06 quart), 21 L oxygen, and 1 L argon would yield 100 L of a mixture of gases very closely approximating the composition of dry air. But because the volume of a gas sample increases with increasing temperature and decreases with increasing pressure, all gas volumes must be measured at the same temperature and pressure.

An equivalent way of representing composition is in terms of the relative amounts of the particles of the various components present in the mixture. In a total of 100 particles of air, 78 are nitrogen, 21 are oxygen, and 1 is argon. These two ways of expressing composition give identical results because equal volumes of gases at the same temperature and pressure contain equal numbers of particles. Per cent means "parts per hundred (pph)." In this case, the parts are particles. Thus, 21% oxygen means 21 particles of oxygen per 100 air particles = 21 particles oxygen/100 particles air. Hence, a percentage based on gas volumes is the same as a percentage based on the number of particles present. We will soon be more specific about the identity of those particles.

Carbon dioxide is a very important constituent of the atmosphere, although its concentration is only 0.036%. Thus 0.036% carbon dioxide can be written as a fraction or ratio: 0.036 parts carbon dioxide/100 parts air. Because it is difficult to envision less than one part, in the case of carbon dioxide 0.036 parts, we scale up the measurement from parts per hundred to **parts per million** (abbreviated as **ppm**); 0.036 parts per hundred equals 360 parts per million (360 ppm). It is fairly simple to do this conversion. We set up a pair of ratios, each representing the concentration, and set the ratios equal to each other.

$$\frac{0.036 \text{ particles carbon dioxide}}{100 \text{ particles air}} = \frac{\text{number of carbon dioxide particles}}{1,000,000 \text{ air particles}}$$

The next step is to solve the equation for the number of carbon dioxide particles for every million air particles.

$$(100) (\text{number of carbon dioxide particles}) = (0.036) \times (1,000,000)$$

$$\text{number of carbon dioxide particles} = \frac{0.036 \times 1,000,000}{100} = (0.036 \times 10,000)$$

to yield the following result: number of carbon dioxide particles = 360. This means that a sample consisting of 1,000,000 particles of air will contain 360 particles of carbon dioxide. Hence, the carbon dioxide concentration is 360 ppm, which is equivalent to 0.036%.

1.4 Your Turn

a. The EPA permissible limit for carbon monoxide is 9 ppm. Express this concentration as a percentage.

b. Exhaled air is typically 75% nitrogen gas. Express this concentration in parts per million.

Ans. **a.** 0.0009% **b.** 750,000 ppm

We human beings and our fellow members of the animal kingdom add carbon dioxide to the atmosphere every time we exhale. Table 1.1 indicates the difference in composition between inhaled dry air and exhaled air. Clearly some changes have taken place that use up oxygen and give off both carbon dioxide and water. Not surprisingly, chemistry is involved. In the biological process of metabolism, oxygen reacts with foods to yield carbon dioxide and water. However, most of the water in exhaled air is simply the result of evaporation from the moist surfaces within the lungs. Note that even exhaled air still contains 16% oxygen. Some people mistakenly think that in respiration most of the oxygen is replaced with carbon dioxide. But if this were true, mouth-to-mouth resuscitation would not work.

Table 1.1 **Typical Composition of Inhaled and Exhaled Air**

Substance	Inhaled Air (%)	Exhaled Air (%)
Nitrogen	78	75
Oxygen	21	16
Argon	0.9	0.9
Carbon dioxide	0.04	4
Water	0	4

Changing from percent to ppm or vice versa is a matter of moving the decimal point four places:

move decimal point
four places to the **right**

parts per $\longrightarrow$ **parts per**
hundred (%) $\longleftarrow$ **million (ppm)**

move decimal point
four places to the **left**

One part per million corresponds to: one second in nearly 12 days; one step in a 568-mile journey; one penny out of $10,000; a pinch of salt on 20 pounds of potato chips.

You will learn a good deal more about carbon dioxide in Chapter 3

1.3 What Else Is in a Breath? Minor Components

The air is obviously different in a pine forest, a bakery, an Italian restaurant, a locker room, or a barnyard. Even blindfolded, we can *smell* where we are. Pine needles, fresh bread, garlic, sweat, and manure all have distinctive odors. These odors are carried by matter. Hence, air must contain trace quantities of substances not included among the five substances listed in Table 1.1. The major components of air are odorless (at least to our noses), but many other substances do have pronounced odors. In fact, the human nose is an extremely sensitive odor detector. In some cases, only a minute trace of a substance is needed to trigger the olfactory receptors. Thus, tiny amounts of substances can have a powerful effect on our noses—as well as on our emotions.

Some of the trace substances that we can smell are pleasant and quite harmless; others can be very dangerous depending on their concentrations and our time of exposure to them. Our noses warn us to avoid certain places because of the odors. But some of the most dangerous air pollutants have no odor at all, and others are dangerous at concentrations low enough that they cannot be detected by smell. As a result, it is often necessary to rely on specialized scientific equipment to measure the presence and concentrations of such substances in the air.

In this chapter we will concentrate on four gases that contribute to air pollution at the surface of the Earth. One of these gases, carbon monoxide, is odorless; ozone, sulfur dioxide, and nitrogen oxides have distinctly unpleasant smells. With sufficient exposure to them, all of these substances are hazardous to health, even at concentrations well below 1 ppm. Together they represent the most serious air pollutants at the Earth's surface. But there is good news, because the concentrations of three of these four gases have decreased significantly over the last 15 years.

Table 1.2 lists concentrations for the four major gaseous air pollutants as measured over 12 cities across the United States. Notice that the concentrations are all expressed as parts per million, though the concentrations of sulfur oxides and nitrogen oxides are sufficiently low that they could conveniently be reported in parts per billion, ppb (sulfur oxides 30 ppb; nitrogen oxides 53 ppb). We will first consider briefly the behavior and damaging effects of these pollutants. Then we will examine the data in the table.

> To convert ppm to ppb, move the decimal point three places to the right. For example, 0.0075 ppm equals 7.5 ppb; 6 ppm = 6000 ppb.

Table 1.2 Major Gaseous Air Pollutants for Selected Cities in the United States, 1996

City	Carbon Monoxide* (ppm)	Ozone** (ppm)	Sulfur Oxides*** (ppm)	Nitrogen Oxides**** (ppm)
(Permissible Limits)	9 ppm	0.12 ppm‡	0.030 ppm	0.053 ppm
New York City	6	0.12	0.055	0.042
Atlanta	4	0.14	0.022	0.027
Boston	5	0.11	0.037	0.031
Chicago	5	0.13	0.032	0.032
Los Angeles	7	0.18	0.046	0.023
Pittsburgh	4	0.11	0.070	0.030
St. Louis	6	0.13	0.102	0.025
San Francisco	5	0.10	0.007	0.022
Detroit	6	0.11	0.079	0.021
Houston	15	0.20	0.011	0.045
New Orleans	4	0.11	0.035	0.018
Indianapolis	3	0.12	0.041	0.018

Source: EPA Annual report, 1996

*Second highest 8-hour average
**Second highest 1-hour average
***Second highest 24-hour average
****Yearly average
‡Changed to 0.08 ppm for an 8-hour average

The first substance, carbon monoxide, enters the bloodstream and disrupts the delivery of oxygen throughout the body. In extreme cases (as in long exposure in a confined space to auto exhaust or a faulty furnace) it can lead to death. The health threat from carbon monoxide is especially serious for individuals suffering from cardiovascular disease, but healthy individuals are also affected. Visual perception, manual dexterity, and learning ability can all suffer.

Ozone is a special form of oxygen. It has a characteristic sharp odor that is frequently detected around electric motors, transformers, and welding torches. Unlike normal oxygen, ozone is very toxic. It affects the respiratory system and even very low concentrations will produce reduced lung function in normal, healthy people during periods of exercise. Symptoms include chest pain, coughing, sneezing, and pulmonary congestion. At the Earth's surface, ozone is definitely a bad actor, but you will see in Chapter 2 that it plays an essential role at high altitudes (10–30 miles).

Sulfur oxides and nitrogen oxides are respiratory irritants that can affect breathing and lower resistance to respiratory infections. People most susceptible include the elderly, young children, and individuals with emphysema or asthma. A particularly severe example of these effects was the London fog of December 1952. The fog lasted five days and led to approximately 4000 deaths. The oxides of sulfur and nitrogen also contribute to acid precipitation—a subject to be explored in considerable depth in Chapter 6.

The numbers in the top row of Table 1.2 are the air quality standards established by the U.S. EPA. Based on scientific studies, these are the maximum concentrations considered to be safe for the general population. Notice that the numbers are quite different: 9 ppm per eight hours for carbon monoxide, but only 0.12 ppm per one hour for ozone, and even less for the sulfur and nitrogen oxides based on a second highest 24-hour average. According to these numbers, ozone is about 100 times more hazardous to breathe than carbon monoxide, and sulfur oxides are four times as hazardous as ozone. You should also be aware that the measured values reported in Table 1.2 are not the same sorts of averages for all four gases. For carbon monoxide and ozone, the data represent high values measured over relatively brief periods (hours). Sulfur oxides and nitrogen oxides are reported as second highest 24-hour averages. As you can see from the table, there is a good deal of variability in the pollution levels in these cities. Measured values of nitrogen oxides are generally well below the permissible limits. Most of the carbon monoxide concentrations are below the limit of 9 ppm per eight hours, but five of the 12 cities report ozone levels greater than the permissible value, and nine cities exceed the limit for sulfur oxides.

1.5 Consider This: Gaseous Pollutant Levels for Selected Cities

In Table 1.2, the top line that appears in boldface type lists the EPA accepted limits for different pollutants. By examining the pollutant levels for the cities listed, it is obvious that ozone and sulfur oxides are the pollutants whose levels are most often exceeded. Use both the data you gain from the table and your knowledge of chemistry to answer these questions.

a. List the cities that violate both the ozone and sulfur oxides EPA accepted limits.
b. The only city listed that is close to the nitrogen oxides level limit in addition to violating the carbon monoxide and ozone limits is Houston. What factors are primarily responsible for the numerous pollutant violations in this city?
c. By examining the data provided, which city has the best overall pollutant record? Justify your choice by discussing the data that led to your decision.

1.6 Consider This: Ozone Across the Country

Table 1.2 lists peak ozone levels in 12 U.S. cities. As you can see from the table, ozone is the pollutant that most often exceeds the air quality standards set by the EPA. What are the current ozone levels in these cities? To answer this question, use the AIRNOW ozone maps provided by the EPA. You can either search for "AIRNOW" on the web or use the direct link provided at the *Chemistry in Context* web site.

Estimate the ozone levels for one of the cities listed in Table 1.2. First select the appropriate state or region and then see what you can find about its ozone values by using the color-coded data provided. Notice that the data are reported in several different ways. Compare what you find on the web with the values listed in Table 1.2. What factors might be contributing to any differences you observe?

1.4 Taking and Assessing Risks

Air quality provides an opportunity for our first look at the subject of risk, an important topic and one to which we will return repeatedly throughout this text. Indeed, it is an issue that is central to life itself, because everything we do carries a certain level of risk, although these levels vary greatly. We are often presented with warnings about certain activities that are believed to carry high risk. For example, the law requires cigarette packages to carry the message "WARNING: Smoking cigarettes may be dangerous to your health." Still other practices have been declared illegal because the level of risk is judged unacceptable to society. On the other hand, there are many other activities that carry no warning, presumably because the degree of risk is quite low, the risk is obvious or unavoidable, or the benefits of the activity far outweigh the risk.

One of the features of such warnings is a characteristic of risk itself. The warnings do not say that a specific individual *will* be affected by a particular activity. They only indicate the statistical probability or chance that an individual will be affected. For example, if the odds of dying from an accident while traveling 300 miles in a car are one in a million, this means that, on average, one person out of every million people traveling 300 miles by car would be killed in an accident. Such predictions are not simply guesses, but are the result of evaluating scientific data and making predictions in an organized manner about the probabilities of an occurrence. These studies are referred to as **risk assessment.**

For air pollutants, the assessment of risk requires knowledge of two factors: **toxicity** and **exposure.** In other words, it is necessary to consider the intrinsic hazard of a substance (the toxicity) and the amount of the substance encountered (the exposure). Exposure is the easier factor to evaluate; it depends simply on the concentration of the substance in the air, the length of time a person is exposed, and the amount of air inhaled into the lungs in a given time. As you already know, the latter depends on lung size and breathing rate. Concentrations in air are usually expressed either as parts per million (ppm) or as micrograms per cubic meter ($\mu g/m^3$). (One microgram is one one-millionth of a gram, $1\ \mu g = 0.000001$ g; one cubic meter is 1000 liters, $1\ m^3 = 1000$ L.)

Consider the risk related to carbon monoxide, one of the four major gaseous air pollutants. There are billions of tons of carbon monoxide spread throughout the atmosphere. Yet, by itself, this prodigious amount does not tell the true story of risk regarding this substance. Because carbon monoxide is not evenly distributed atmospherically, its concentration varies widely by location. In some places the concentration is so low, far less than the air quality standards, that it is not really considered a pollutant. In others, the carbon monoxide concentration is sufficiently high to create a hazardous situation. Therefore, to assess our risk to this possible pollutant, we need to consider our exposure to it as well as its potential toxicity.

Depending on our size and other factors, each of us breathes 10,000 to 20,000 L of air daily. If we were to breathe 10,000 L (10 m^3) of air of moderate quality that contains carbon monoxide at a concentration of 10,000 μg carbon monoxide/m^3 of air, we would be exposed to 100,000 μg of carbon monoxide in a day.

$$10 \text{ m}^3 \text{ air} \times \frac{10,000 \ \mu g \text{ carbon monoxide}}{\text{m}^3 \text{ air}} = 100,000 \ \mu g \text{ CO in a day}$$

This text does not make extensive use of calculations, but it is important for you to be able to do and understand calculations of the type shown here.

Someone who breathes twice as much of this air in a day obviously would be exposed to twice as much carbon monoxide (200,000 μg).

For large numbers such as 100,000 μg CO per day and even larger values, we will use **scientific notation** to avoid turning the text into strings of zeros. This particular number is written in scientific notation as 1×10^5. The easy way to make this conversion is to simply count the number of zeros to the right of the initial 1. There are five of them, and 5 becomes the exponent of 10. The number 1 is then multiplied by 10^5 to obtain the number of micrograms of CO per day, 1×10^5. Consider the number of molecules in a typical breath, a much larger number where the usefulness of scientific notation will be more readily apparent. There are more than 20,000,000,000,000,000,000,000 of them, a number large enough to take your breath away! In scientific notation this particular number is written as 2×10^{22}. Why 20,000,000,000,000,000,000,000 equals 2×10^{22} takes a bit more explaining. Remember that 10^{22} means 10 multiplied by itself 22 times. This is simply another instance in the following series:

$$10^1 = 10$$
$$10^2 = 10 \times 10 - 100$$
$$10^3 = 10 \times 10 \times 10 = 1000$$

Note that 10^1 is 1 followed by 1 zero, that is, ten; 10^2 is 1 followed by 2 zeros; and 10^3 is 1 followed by 3 zeros. Continuing this pattern, 10^{22} is 1 followed by 22 zeros. Therefore, 2×10^{22} must equal $2 \times 10,000,000,000,000,000,000,000$ or 20,000,000,000,000,000,000,000. If exponents and scientific notation are new to you, help is available in Appendix 1.

The 1×10^5 μg CO/day might seem like a large amount of carbon monoxide. But keep in mind that this amount, when ingested over a 24-hour period, is not toxic because it is less than the amount determined as hazardous. For carbon monoxide to be considered hazardous over 24 hours, we would need to be exposed to about 5×10^5 to $9 \times 10^5 \mu g$ of it in air, nearly 5 to 10 times the amount found in air of moderate quality, and about 100 times more than in air of good quality. The important point here is that when relating exposure with toxicity, it is necessary to compare the exposure level to the minimum amount required for the pollutant to become a public health risk or to be present at toxic levels. Even though a pollutant may be present, it is not considered hazardous unless it exceeds the toxic threshold, the amount that causes harmful effects.

Specific chemicals can be beneficial or harmful, depending on their location and concentration.

Toxicity is more difficult to know with accuracy, in part because it is considered unethical to do controlled experiments with human subjects. This leaves scientists with three choices: human population studies, animal studies, and bacterial studies. Population studies involve collecting data on affected groups of people. For example, a researcher may determine what percentage of people who smoke one pack of cigarettes per day get lung cancer. Such studies are necessarily limited and may require many years of observation to obtain results that are statistically significant and reflect accurately the long-term risk. For this reason, animal studies have been a widely used substitute for those involving human subjects. Animals are given controlled doses of the substance being tested and observed for harmful effects. Aside from questions of animal rights, the problem here is that we do not know with certainty whether specific animal species respond the same as humans. There is a growing awareness among scientists that animal studies must be interpreted with great caution. A more recent area of toxicity measurements relies on studies with bacteria. An important advantage of bacteria is that they grow and reproduce very rapidly, thus many studies can be done quickly and inexpensively.

Even if data are available to calculate the risks from a given pollutant, we still have to ask what level of risk is acceptable and for what groups of people. Various government agencies are charged with establishing safe limits of exposure for the major air pollutants. Table 1.3 gives current outdoor air quality standards established by the U.S. EPA for the pollutants discussed in this chapter. Some states, including California and Oregon, have their own, stricter, standards.

1.7 Consider This: Toxicity vs. Exposure

Consider this graph showing the effects of carbon monoxide inhalation on humans.

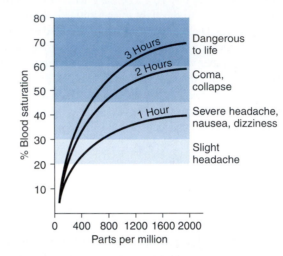

a. Use this graph to illustrate that both the avenue of exposure and the duration of exposure have an effect on CO toxicity in humans. Explain your reasoning.
b. Use the information in this graph to write a paragraph intended to inform purchasers of a home carbon monoxide detection kit about the potential health hazards of carbon monoxide gas.

1.8 Consider This: Communicating Air Pollution Levels

Newspapers often provide a color-coded "Pollution Index" to give the public a qualitative indication of the level of air pollution expected in the region. Typically, this is based on computer-based models predicting anticipated levels of ground-level ozone. Green indicates low levels, with no cautionary action required. Yellow indicates moderate levels and often leads to warnings that outdoor activity should be limited, particularly for children and adults with respiratory problems. Red levels indicate that ozone is likely to exceed federal standards and outdoor activity should be avoided.

a. What are the advantages to using a color-coded qualitative index rather than simply reporting the predicted parts per million of ozone?
b. If a "red alert" is forecast for tomorrow, what can you do to help? Prepare a list of at least five actions that would help reduce air pollution tomorrow, particularly if everyone were to follow your suggestions.
c. Check the weather page of your local newspaper to see if an air pollution index is included. You may also want to check an appropriate state or regional web site for this information. Describe the pollution indices you have located and comment on their effectiveness in communicating with the public.

An important factor in dealing with risks is not only the actual risk, but people's perception of a particular risk. For example, the majority of people surveyed were willing to accept a 1 in 10,000 (1 in 1×10^4) to 1 in 100,000 (1 in 1×10^5) risk of cancer from eating peanut butter that contains a naturally occurring carcinogen (cancer-causing ma-

Table 1.3 National Ambient Air Quality Standards (1999)

Pollutant	Limit
Carbon monoxide	
8-hr average	9 ppm
1-hr average	35 ppm
Nitrogen dioxide	
Annual arithmetic mean	0.053 ppm
Ozone	
1-hr average	0.12 ppm
8-hr average	0.08 ppm
Lead	
Quarterly average	1.5 $\mu g/m^3$
Particulates	
$< 10\ \mu m$, 24-hr average	150 $\mu g/m^3$
$< 2.5\ \mu m$, 24-hr average	65 $\mu g/m^3$
Sulfur dioxide	
Annual arithmetic mean	0.03 ppm
24-hr average	0.14 ppm
3-hr average	0.50 ppm

terial). Yet, these same people found it unacceptable to use a synthetic material with a 1 in 1,000,000 (1 in 1×10^6) risk factor, a case whose cancer risk is 10 to 100 times *less* than from eating peanut butter. The perception was that a natural material had to be less risky than a synthetic one, a common misconception.

1.9 Consider This: Risk Analysis

"The general public is uncomfortable with uncertainties. Too often we think in terms of absolutes and demand that scientists and decision makers be held accountable for their risk decisions."*

Do you agree or disagree with these statements? Support your opinion with reasonable arguments, giving a specific example from your personal experience in considering a risk of importance to you.

*Chemical Risk: A Primer. Department of Government Relations and Science Policy, American Chemical Society, 1996, p. 11.

1.5 The Atmosphere: Our Blanket of Air

The most familiar kinds of air pollution and their influences on air quality occur in the **troposphere,** the part of the atmosphere that lies directly at the surface of the Earth. Figure 1.2 provides the names of regions of the atmosphere and some reference points in relation to altitude. As one rises in the troposphere, the temperature decreases until it reaches about −40°C (also −40°F). That temperature roughly marks the beginning of the **stratosphere,** which includes the ozone layer, the subject of Chapter 2. The temperature of the stratosphere increases from about −40°C at 20 km to 0°C (32°F) at 50 km. Above that altitude, the temperature of the atmosphere again begins to decrease on passing through the **mesosphere.** The issues we will study in the first three chapters of this book will take us to these various regions, which differ in atmospheric properties and phenomena. Bear in mind that there are no sharp physical boundaries that separate these layers. The atmosphere is a continuum with gradually changing composition, concentrations, pressure, and temperature. In fact, temperature changes account for the organization of the atmosphere.

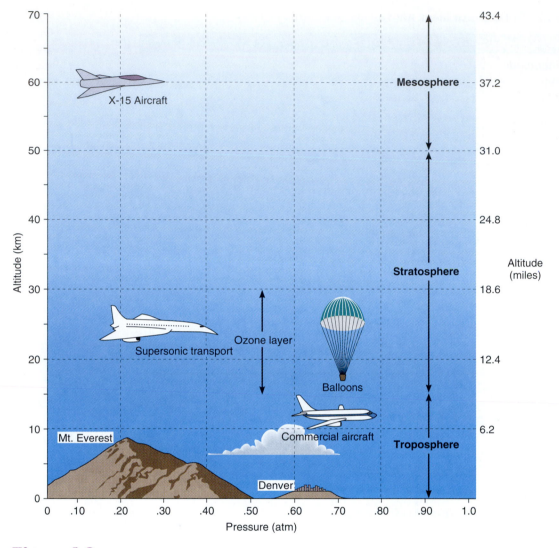

Figure 1.2
The regions of the atmosphere at various altitudes.

The relative concentrations of the major components of the atmosphere are nearly constant at all altitudes. In other words, the concentration of oxygen remains about 21% and that of nitrogen is 78%. However, you might know from reading about it, from the experience of hiking in the mountains, or flying in a jet plane that the air gets "thinner" with increasing altitude. As you climb up into the blanket of air, there is less of it— fewer particles in a given volume. Moreover, as you move up through the atmosphere, the mass of air above you decreases. Therefore, the **atmospheric pressure,** the force with which the atmosphere presses down on a given area, decreases with increasing altitude as shown in Figure 1.3. Atmospheric pressure is measured with a device called a barometer. At sea level, such as in Boston or Los Angeles, the barometric pressure is 14.7 pounds per square inch. A pressure of this magnitude is defined as 1 atmosphere (1 atm). In Denver, the "mile high city," the pressure is 12.0 pounds per square inch— about 0.8 atmosphere. You will note from Figure 1.3 that the plot of pressure versus altitude is not a straight line. Below about 20 kilometers (km) the pressure drops very sharply with increasing altitude. In this region, the pressure decreases by about 50% for every 5-kilometer increase in altitude (1 km = 0.62 mi, 5 km = 3.1 mi). At higher altitudes, the pressure decreases more gradually. Somewhere above 100 km, the atmosphere simply fades into the almost perfect vacuum of outer space.

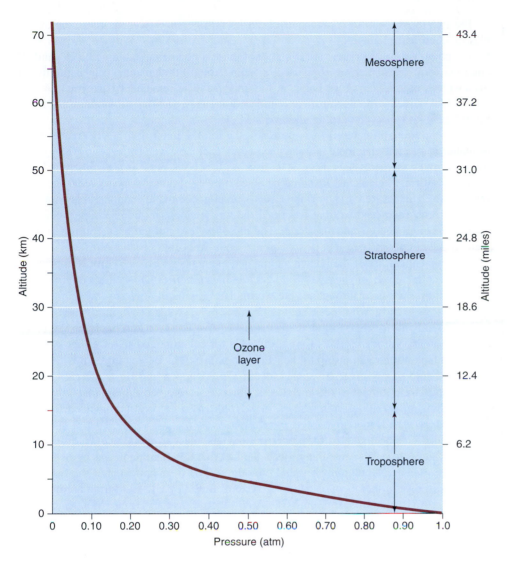

Figure 1.3
The pressure of the atmo-
sphere changes with altitude.
An increase in altitude is accompanied by a drop in air
pressure.

1.6 Classifying Matter: Mixtures, Elements, and Compounds

In describing the atmosphere and air quality, we have referred to a number of substances and used a bit of chemical terminology. Therefore, before proceeding further, some clarification is probably necessary. For one thing, we need some understanding of the way chemists describe the composition of different types of matter. Chemists classify matter as mixtures and pure substances; the pure substances are elements or compounds. A breath of air is a **mixture**—a physical combination of two or more substances that may be present in variable amounts. Pure air and polluted air differ in composition, and we have seen that exhaled air differs from inhaled air. As composition varies, so do many of the properties of mixtures. Much of the matter we encounter in everyday life is in the form of mixtures. The fuels we burn, the foods we eat, the beverages we drink, our bodies themselves are all complex mixtures of individual substances. By appropriate experiments, these individual substances can be isolated and shown to have reproducible, characteristic properties and fixed composition.

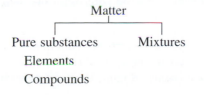

Matter can be classified as a mixture
or a pure substance. Elements and
compounds are pure substances;
mixtures are not.

The two most plentiful components of air are nitrogen and oxygen. These are examples of **chemical elements**—substances that cannot be broken down into simpler stuff by any **chemical** means. There are over 100 of these fundamental building blocks and all common forms of matter are composed of them. About 90 elements occur naturally on planet Earth and, as far as we know, in the universe. The remainder have been created from other elements through artificially induced nuclear reactions. (Plutonium is probably the best known of the artificially produced elements, although it does occur in very low concentrations in nature.) In some cases, the total amount of a newly created form of matter is so small that there is some uncertainty (perhaps even some controversy) over just how many elements have been identified. As this book goes to press, there are 112 known (confirmed) elements. It should be noted, however, that in January 1999 the Joint Institute for Nuclear Research at Dubna, Russia announced the formation of element 114. The research group reported that it formed the new element by bombarding plutonium atoms with high-energy calcium atoms. The discovery has not yet been confirmed. Element 113 had not yet been reported when this book went into production.

An alphabetical list of the known and named elements and their **symbols** appears in the inside back cover of the text. The symbols have been established by international agreement and are used throughout the world. The origins of some of the symbols are quite obvious to those who speak English. For example, oxygen is O, nitrogen is N, carbon is C, and sulfur is S. Most symbols consist of two letters, again based on the name of the element: Ni for nickel, Cl for chlorine, Ca for calcium, and so on. Other symbols appear to have little relationship to their English names. Thus, Fe is iron, Pb is lead, Au stands for gold, Ag is silver, Sn stands for tin, Cu is copper, and Hg is mercury. All of these metals were known to the ancients, and hence were given Latin names long ago. The symbols reflect those Latin names, for example, *ferrum* for iron, *plumbum* for lead, and *hydrargyrum* for mercury.

Elements have been named for properties, planets, places, and people. Hydrogen (H) means "water former," a name that reflects the fact that this flammable gas burns in oxygen to form water. Neptunium (Np) and plutonium (Pu) were named after the two most recently discovered members of our solar system. Berkeley and California are honored in berkelium (Bk) and californium (Cf). And Albert Einstein, Dmitri Mendeleev, and more recently Lise Meitner (co-discoverer of nuclear fission) have attained elementary immortality in einsteinium (Es), mendelevium (Md), and meitnerium (Mt).

The formation of elements 116 and 118 in Berkeley California was reported in June 1999.

1.10 Consider This: Adopt an Element

Periodic tables are available on the web that list the properties of elements, their date of discovery, their naturally occurring isotopes, and much more. Thus, the web can give you quick access to information that it might take you hours to find using reference books.

Use a search engine to bring up a list of periodic tables. Go to one of the periodic tables to find out more about an element of your choice. You probably will obtain more complete information if you select an element with atomic number of 94 (plutonium) or less.

Find out what year your element was discovered; whether it occurs naturally as a solid, liquid, or gas; its appearance; where it is found; and any two other facts, such as toxicity, cost, uses, etc.

Following the directions given by your instructor, get together with other students in your class to answer questions such as: Are most elements gases, solids, or liquids? Which elements were discovered first? Last? Are most elements found "free" in nature, that is, not combined with any other element? Do the elements chosen combine with other elements to form compounds?

It is particularly appropriate that Mendeleev should have his own element, because the most common way of arranging the elements reflects the periodic system developed by this 19th century Russian chemist. Figure 1.4 is the **periodic table.** We will explain

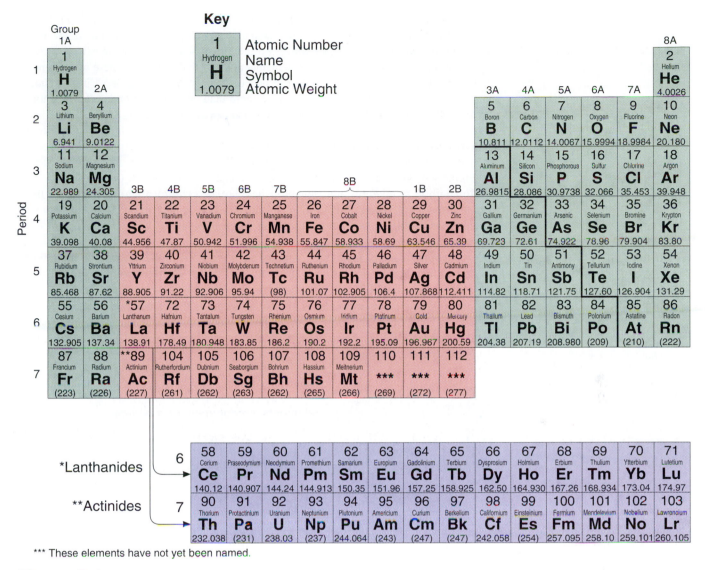

Figure 1.4
The periodic table of the elements.

*** These elements have not yet been named.

the significance of the numbers and the elements' order in Chapter 2. For the moment it is sufficient to note that about the time of the American Civil War, Mendeleev arranged the then-known elements so that the elements with similar chemical and physical properties fell in families or groups designated by vertical columns. Thus, the members of Group 1A include lithium (Li), sodium (Na), potassium (K), and three other very reactive metals. Similarly, Group 7A consists of very reactive nonmetals, including fluorine (F), chlorine (Cl), bromine (Br), and iodine (I). Nitrogen and oxygen, the two most common elements in the atmosphere, are side by side in Groups 5A and 6A. Some helpful generalizations about elements come from the table; the vast majority of elements are solids, some are gases, and only two—bromine and mercury—are liquids at room temperature and pressure. Note also that most elements are metals; fewer of them are nonmetals. The Group 8A elements, such as neon and radon, are known as the *noble gases*. Some of them, such as helium, neon, and argon do not combine chemically with any elements.

The periodic table is a very handy data base, an amazingly useful way of organizing the stuff of the universe, and we will return to it throughout the text. The periodically repeating properties can be beautifully explained by our knowledge of atomic structure. Moreover, the table continues to grow as new elements are made.

Radioactivity is an important topic in Chapter 7.

By tradition, the discoverers of elements have been given the right to name them. Things get a bit complicated with recently discovered elements, especially when only a few atoms of the new element exist, and then for only a fraction of a second before they radioactively decompose. This is the case with the elements numbered 104 to 109. These new forms of matter were variously created in particle accelerators or "atom smashers" in the United States, Germany, and Russia. In some cases more than one research group claims priority of discovery. The International Union of Pure and Applied Chemistry (IUPAC) is the body that formally approves the names of elements, but its preliminary recommendations for elements 104–109 were controversial. After more than a decade of debate, IUPAC made a final decision in 1997 regarding these elements. Their proper names and symbols appear in Figure 1.4 and in the table of elements in the inside back cover of the text.

In addition to mixtures and elements, there are also **chemical compounds,** the third class of matter. A compound is a pure substance made up of two or more elements in a fixed, characteristic chemical composition and combination. Although there are only about one hundred elements, over fifteen million chemical compounds have been isolated, identified, and characterized. Among these are some of the most familiar naturally occurring substances, including water, salt, and sugar. But most of the known compounds do not exist in nature; they have been synthesized by chemists. The motivation for making new forms of matter is almost as varied as the compounds themselves—to make synthetic fibers and plastics, to find drugs to cure AIDS or cancer, or just for the intellectual fun of it. In later chapters we will look at some examples of how chemists synthesize new compounds.

See especially Chapters 9 and 10.

1.11 Your Turn

Classify the following as elements, compounds, or mixtures.

a. water   **b.** nickel   **c.** U.S. nickel coin
d. diamond   **e.** sulfur dioxide   **f.** lemonade

Ans. **a.** compound   **b.** element   **c.** mixture

There is no residual pure oxygen or carbon uncombined in carbon dioxide. In CO_2 the two elements are chemically combined and are no longer in their elemental forms.

1 gram = 0.00220 pound = 0.0352 ounce. A gram is approximately the mass of a peanut.

Two important compounds in the atmosphere have already been mentioned: carbon dioxide and water. As its name implies, carbon dioxide is made up of chemically combined carbon and oxygen. All pure samples of carbon dioxide contain 27% carbon and 73% oxygen by **weight** (or **mass**). A 100 gram (g) sample of carbon dioxide will always consist of 27 g of carbon and 73 g of oxygen, chemically combined to form this particular compound. These values never vary, no matter what the source of the carbon dioxide. This is simply one example of the fact that every compound exhibits a constant characteristic chemical composition. This is reflected in the fact that although carbon monoxide is also a compound of carbon and oxygen, the combination results in a compound with 43% carbon and 57% oxygen, by weight. Thus, 100 g of carbon monoxide contain 43 g of carbon and 57 g of oxygen, a much different composition than that of carbon dioxide, but not surprising because CO and CO_2 are two different compounds.

Moreover, the composition of a compound is constant, as are its physical properties (such as boiling point) and its chemical reactivity. Consider water, a compound consisting of 11% hydrogen and 89% oxygen, by weight. At room temperature, water is a colorless, tasteless liquid. It boils at 100°C and it freezes at 0°C. It is an excellent solvent and participates in many chemical reactions. Like any compound, water can be broken down into its constituent elements. A 100 gram sample of water will yield 11 grams of hydrogen and 89 grams of oxygen.

1.7 Atoms and Molecules

The definitions of elements and compounds given earlier are valid, in spite of the fact that no assumptions were made about the physical structure of matter. But modern insights into the organization of matter help us to better understand matter, the "stuff" of the universe. It is now well established that elements are made up of **atoms.** An atom is

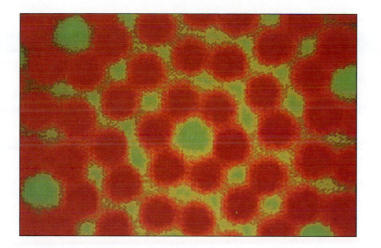

Figure 1.5
Silicon atoms (orange) viewed with a scanning tunneling microscope.

the smallest unit of an element that can exist as a stable, independent entity. The word *atom* comes from the Greek for "uncuttable." Today we know that atoms consist of smaller particles, and that atoms can be "split" by high-energy processes. However, atoms remain indivisible to chemical or mechanical means. Atoms are extremely small — many billions of times smaller than anything we can detect with our senses. Because of this small size, there must be huge numbers of atoms in any sample of matter that we can see or touch or weigh by conventional means. But as Figure 1.5 reveals, scanning tunneling microscopy has recently made the invisible visible.

The existence of atoms provides a means of refining our earlier definitions of elements and compounds. Each element has a different kind of atom, but within a sample of any given element all the atoms are chemically the same. By contrast, compounds are made up of the atoms of two or more elements. For example, the compound carbon dioxide has a ratio of one carbon atom for every two oxygen atoms and is symbolized as CO_2. In a similar manner, water contains two hydrogen atoms for each oxygen atom and is represented as H_2O. CO_2 and H_2O are examples of **chemical formulas,** which are symbolic representations of the elementary composition of chemical compounds. Note that when an atom is used once in a formula, such as oxygen in water, it carries no subscript of "1."

Carbon dioxide, water, and millions of other compounds exist as **molecules.** A molecule is a combination of a fixed number of atoms held together in a certain geometric arrangement. Thus, one molecule of carbon dioxide consists of one carbon atom bonded to two oxygen atoms. Like the compound itself, the molecule is represented by the formula CO_2. Similarly, a molecule of water, H_2O, contains exactly two hydrogen atoms and one oxygen atom, never any other combination. The constant ratio of atoms that characterizes each compound explains why compounds always have fixed composition. Elements can exist either as molecules or as single atoms. Thus, the nitrogen and oxygen of the atmosphere are *diatomic* (two atoms per molecule) made up of N_2 and O_2 *molecules,* whereas argon consists of individual Ar *atoms.*

Section 2.2 includes information about atomic structure and Section 7.2 discusses atom "splitting."

The number of atoms in a single drop of water is huge—about 5×10^{21} atoms. This is about a trillion times greater than the approximately six billion people on Earth, enough to give each person a trillion atoms from that water drop.

Information about molecular structure is found in Chapters 2 and 3.

1.12 Your Turn

Identify the elements that are in each of these compounds.

a. sulfur dioxide, SO_2 **b.** carbon tetrachloride, CCl_4
c. hydrogen peroxide, H_2O_2 **d.** sucrose, $C_{12}H_{22}O_{11}$

Ans. **a.** sulfur, oxygen   **b.** carbon, chlorine

Table 1.4 describes elements, compounds, and mixtures in two ways: the behavior we can observe experimentally, and the theory or model scientists use to explain what is happening at the incredibly small atomic level. Both are correct, and they complement

Table 1.4   Classification of Matter

	Observable Properties	Atomic Theory
Element	Cannot be broken down into simpler substances	Only one kind of atom
Compound	Fixed composition, but capable of being broken down into elements	Two or more atoms in fixed combination
Mixture	Variable composition of elements and/or compounds	Variable assortment of atoms and/or molecules

each other. We can now apply these concepts to the atmosphere. Air is a mixture, which means that its composition can vary with time and place. Some of its components, such as nitrogen, oxygen, and argon, are individual elements; others, notably carbon dioxide and water, are compounds. All of the compounds and some of the elements are present as molecules (for example, CO_2 and O_2), but some elements exist as uncombined atoms (for example, Ar). Furthermore, we can give a more precise definition of the concentrations of gases in air. Recall that earlier we referred to "particles" of nitrogen and "particles" of air. These particles are the smallest units in which the constituents of air normally exist in their stable form—N_2 molecules, O_2 molecules, Ar atoms, CO_2 molecules, and so on. When we say that air is 78% nitrogen, we mean that in every 100 particles of air, 78 of them will be N_2 molecules. Similarly, 100 particles of air will include 21 O_2 molecules and 1 Ar atom. A concentration of 360 ppm carbon dioxide means that there will be 360 CO_2 molecules in 1×10^6 molecules and atoms of air.

1.8 Formulas and Names: The Vocabulary of Chemistry

If elementary symbols are the alphabet of chemistry, then chemical formulas are the words. And, the language of chemistry, like any other language, has rules of spelling and syntax. The symbols of the elements must be combined in ways that correctly correspond to the composition of the compounds in question. Although this system of chemical symbolism and nomenclature is logical, precise, and extremely useful (at least to chemists) it does present somewhat of a barrier to others studying the discipline. This, of course, is true of any specialized vocabulary, which often sounds like jargon to the uninitiated. This book does not seek to make its readers expert in all aspects of chemical nomenclature; however, some familiarity with the rules of assigning formulas and names to simple chemical compounds will be helpful. In this chapter we will consider only the simplest compounds, those consisting of two elements.

The chemical formula of a compound reveals the elements that make it up, and the atom ratio of those elements. The name usually conveys similar information. For example, the name "magnesium oxide" indicates a compound consisting of magnesium (Mg) and oxygen (O). Because the magnesium and oxygen combine in a one-to-one atomic ratio, the formula of magnesium oxide is MgO. Similarly, "sodium chloride" indicates a compound composed of the elements sodium (Na) and chlorine (Cl) with a formula of NaCl. The rule for naming such two-element compounds is simple: The name of the more metallic element comes first, followed by the name of the less metallic element, modified to end in "ide." So let's try to name the compound composed of potassium (K) and iodine (I). First we need to determine which of these two elements is more metallic, and for this we turn to the periodic table. As noted from the table, an important generalization is that the metallic elements are on the left side of the periodic table and the nonmetallic elements on the right. From this it follows that potassium must be more metallic than iodine. Applying the rule, potassium is first and the compound is named potassium iod*ide*. The formula turns out to be KI.

1.13 Your Turn

Name the compounds that contain each pair of elements.

a. magnesium, bromine **b.** barium, oxygen
c. hydrogen, chlorine **d.** sodium, sulfur

Ans. **a.** magnesium bromide **b.** barium oxide

1.14 Your Turn

Name the compounds that have these formulas.

a. $ZnCl_2$ **b.** Al_2O_3
c. CaS **d.** Li_3N

Ans. **a.** zinc chloride **b.** aluminum oxide

Unfortunately, assignment of formulas is a bit more complicated than we have just suggested. Not all compounds exhibit one-to-one atomic ratios such as MgO and KI. For example, in calcium chloride there are two atoms of chlorine for each atom of calcium and the formula is therefore $CaCl_2$. Someone familiar with the periodic table and atomic structure should be able to predict the atomic ratio and the formula of just about any two-element compound and to name it. But we have not yet provided you with enough information to develop that skill. At present it is probably sufficient if you can name a two-element compound when presented with its formula. Thus, the formula H_2S represents a compound called hydrogen sulfide, a gas with the unmistakable smell of rotten eggs.

One of the complications associated with the rules that we have been applying is that the name of the compound does not always unambiguously reveal its formula. A way of eliminating the ambiguity is to use prefixes that indicate the number of atoms of an element specified by the formula. A good example is one of the atmospheric compounds we have already introduced—carbon dioxide. *Di-* means "two" and thus the name carbon *di*oxide implies that each molecule of the compound includes two oxygen atoms. The corresponding formula, CO_2, indicates the two oxygen atoms with a subscript 2 on the symbol O. The use of the prefixes listed in Table 1.5 makes it possible to distinguish between two or more compounds consisting of the same elements, but in different atomic ratios. Thus, carbon *mon*oxide also consists of carbon and oxygen, but the prefix *mon-* or *mono-* reveals that there is only one oxygen atom in each molecule of this compound. It follows that the formula of carbon monoxide is CO. Using the same logic and the same set of prefixes, the chemical name for water, H_2O, is dihydrogen oxide. Note that in most compounds in which a formula contains only one atom of an element, the *mono-* prefix is omitted. Carbon monoxide is an exception; the mono- prefix is used to avoid possible confusion with carbon dioxide.

The reason why the formula for calcium chloride is $CaCl_2$ will become clear in Chapter 5.

You will read more about atomic ratios and formulas in Chapter 2, Section 2.3.

Table 1.5 Prefixes Used in Naming Compounds

Prefix	Meaning	Prefix	Meaning
Mono-	One	Hexa-	Six
Di- or bi-	Two	Hepta-	Seven
Tri-	Three	Octa-	Eight
Tetra-	Four	Nona-	Nine
Penta-	Five	Deca-	Ten

1.15 Your Turn

What information does each formula convey about these substances? Each is an important atmospheric component.

a. NO_2 **b.** SO_2 **c.** SO_3 **d.** O_3

Ans. **a.** A molecule of the compound represented by the formula NO_2 consists of one atom of the element nitrogen combined with two atoms of the element oxygen.

1.16 Your Turn

Use appropriate prefixes to name the compounds whose formulas are given in **1.15 Your Turn**.

Ans. **a.** nitrogen dioxide   **b.** sulfur dioxide

1.9 Chemical Change: Reactions and Equations; Oxygen's Role in Burning

The first pollutant listed in Table 1.2 is carbon monoxide, CO, whereas all air, polluted or unpolluted, contains carbon dioxide, CO_2. Carbon monoxide and carbon dioxide can both arise from the same source: combustion. **Combustion** (or burning) is the rapid combination of oxygen with another material. When carbon or carbon-containing compounds burn, oxygen combines with the carbon to form CO_2 and/or CO. Similarly, combustion reactions produce water (H_2O) and sulfur dioxide (SO_2) by the burning of hydrogen and sulfur, respectively.

Combustion is a major type of **chemical reaction** or a **chemical change.** A chemical reaction is a process whereby substances described as **reactants** are transformed into different substances called **products.** The process can be represented by an expression called a chemical equation. Chemical equations are the sentences in the language of chemistry. They are made up of elementary symbols (corresponding to letters), which are often combined in the formulas of compounds (the "words" of chemistry). Like a sentence, a chemical equation conveys information, in this case about the chemical change taking place. But a chemical equation must also obey some of the same constraints that apply to a mathematical equation.

At its most fundamental level, a chemical equation is very simple indeed. It is a qualitative description of the reaction.

$$\text{Reactant(s)} \longrightarrow \text{Product(s)}$$

By convention, the reactants are always written on the left and the products on the right. The arrow represents a chemical transformation and is read as "is converted to" or "yields." Thus, reactants are converted to products in the sense that the reaction creates products whose properties are different from those of the starting materials, the reactants.

The combustion of carbon to produce carbon dioxide (for example, the burning of charcoal in air, Figure 1.6) can be represented in several ways. One way is by a "word equation":

$$\text{carbon} + \text{oxygen} \longrightarrow \text{carbon dioxide}$$

It is much more common to use chemical symbols and formulas for the elements and compounds involved.

$$C + O_2 \longrightarrow CO_2 \tag{1.1}$$

This compact symbolic statement conveys a good deal of information to a chemist. A translation of equation 1.1 into words might read something like this: "One atom of the element carbon reacts with one molecule of the element oxygen (consisting of two oxygen atoms joined to each other) to yield one molecule of carbon dioxide, a compound consisting of one carbon atom linked to two oxygen atoms."

If we use a black sphere to represent a carbon atom and a red sphere to represent an oxygen atom, the rearrangement of atoms by this reaction looks something like this:

Likewise, the burning of sulfur to produce the air pollutant sulfur dioxide can be represented by a chemical equation:

$$S + O_2 \longrightarrow SO_2 \qquad (1.2)$$

or symbolically with spheres, where a yellow sphere represents a sulfur atom and red spheres represent oxygen atoms.

It is possible to pack more information into an equation by specifying the physical states of the reactants and products. A solid is designated by (s) following the symbol or formula; a liquid is designated by (l); and a gas is indicated by (g). Because carbon and sulfur are solids, and oxygen, carbon dioxide, and sulfur dioxide are gases at ordinary temperatures and pressures, equations 1.1 and 1.2 become:

$$C(s) + O_2(g) \longrightarrow CO_2(g)$$
$$S(s) + O_2(g) \longrightarrow SO_2(g)$$

In this text we will designate the physical states of the substances participating in a reaction when that information is particularly important, but in many cases we will omit it for simplicity.

You will note that equation 1.1 has some of the characteristics of a mathematical equation—in this case, the number and kinds of atoms on the left equal those on the right. One carbon atom and two oxygen atoms are on the left side of the arrow and one carbon atom and two oxygen atoms are on the right. This is the test of a correctly balanced equation: the number and kinds (which element) of the atoms on the reactant side of the arrow must equal the number and identity of atoms on the product side. Atoms are neither created nor destroyed in a chemical reaction and the elements present do not change when converted from reactants to products. This relationship is called the **law of conservation of matter and mass:** In a chemical reaction, matter and mass are conserved. The mass of the reactants consumed equals the mass of the products formed. The total mass does not change, because no matter is created or destroyed.

Of course, atoms are rearranged during a chemical reaction. That is what a chemical change is all about: the atoms in the products are in a different arrangement than they were as reactants. Therefore, there is no requirement that the number of molecules must be the same on both sides of the arrow. In fact, the number of molecules changes during many reactions. In equation 1.1, one atom of carbon plus one molecule of oxygen yields one molecule of carbon dioxide. This looks suspiciously like $1 + 1 = 1$. This is not a cause for alarm; a chemical equation is not exactly the same as a mathematical equation. Remember, a chemical equation represents a transformation, not a simple equality. In a correctly balanced chemical equation, some things must be equal, others need not be. Table 1.6 summarizes these peculiarities.

Equation 1.1 describes the combustion of pure carbon in an ample supply of oxygen. However, if the oxygen supply is limited, the product is carbon monoxide, CO, not carbon dioxide. Like any reaction, this one can be expressed in equation form. First, we write down the symbols and formulas of the reactants and products:

$$C + O_2 \longrightarrow CO \qquad \text{(unbalanced equation)}$$

Figure 1.6
The burning of charcoal in air.

Equation 1.2 is balanced because an equal number of sulfur atoms are in the reactants and products and the numbers of oxygen atoms in the reactants and products are also equal.

Consider the following analogy in terms of the reorganization of reactants to form products in a chemical reaction. Building materials (reactants) used to construct an apartment building can be disassembled and rearranged to build three houses and a garage (products).

Table 1.6   **Characteristics of Chemical Equations**

Always Conserved
Number of atoms in reactants = Number of atoms in products
Identity of atoms in reactants = Identity of atoms in products
Total mass of reactants = Total mass of products

Not Necessarily Conserved
Number of molecules of reactants may or may not equal number of molecules of products
Volume of reactants may or may not equal volume of products

A quick count of atoms on either side of the arrow reveals that the expression does not balance. There are two oxygen atoms on the left and one on the right. We cannot balance the equation by simply adding an additional oxygen atom to the product side. *Once we write the **correct** symbols and formulas for the reactants and products, we cannot change them, including their subscripts!* That would imply a different reaction. All we can do is to place whole-number coefficients before the various symbols and formulas. In simple cases like this, the coefficients can be found quite easily by inspection or simple trial and error. If we place a 2 to the left of the symbol CO, it signifies two molecules of carbon monoxide. This corresponds to a total of two carbon atoms and two oxygen atoms. Because there are also two oxygen atoms on the left side of the arrow, the oxygen atoms have been equalized.

> A subscript follows a chemical symbol, as in O_2 or CO_2; a coefficient precedes a symbol or a formula, for example 2 C or 2 CO.

$$C + O_2 \longrightarrow 2\ CO\ \text{(but, equation still not balanced)}$$

But now the carbon atoms are out of balance. There are two on the right and only one on the left. Fortunately, this is easily corrected by placing a 2 in front of the C:

$$2\ C + O_2 \longrightarrow 2\ CO\ \text{(balanced equation)} \tag{1.3}$$

The balanced equation can also be represented by colored spheres—black for carbon and red for oxygen atoms:

Note that the balanced equation includes quantitative and qualitative information. It tells us qualitatively what atoms are present, and quantitatively how many carbon atoms and how many oxygen molecules react to form carbon monoxide. It is evident from comparing equations 1.1 and 1.3 that, relatively speaking, more oxygen is required to form CO_2 from carbon than is needed to form CO.

The same kind of reasoning can be applied to balance the equation for the formation of another air pollutant, nitrogen monoxide (commonly called nitric oxide), created by the reaction of nitrogen with oxygen. We begin by writing down the unbalanced equation giving the correct symbols and formulas of the reactants and products:

$$N_2 + O_2 \longrightarrow NO\ \text{(unbalanced equation)}$$

We notice that the equation is not balanced because there are two oxygen atoms on the left, and only one on the right. The same is true for nitrogen atoms. Placing a two to the left of the formula NO supplies two nitrogen *and* two oxygen atoms to that side, the same number that are to the left of the arrow; the equation is now balanced

$$N_2 + O_2 \longrightarrow 2\ NO$$

> Nitrogen atoms are represented by blue spheres.

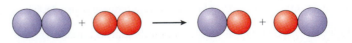

1.17 Your Turn

Balance each of these equations and then draw a representation of each equation using spheres.

a. $H_2 + O_2 \longrightarrow H_2O$
b. $N_2 + O_2 \longrightarrow NO_2$

Ans.   **a.** Balanced equation: $2 H_2 + O_2 \longrightarrow 2 H_2O$

1.10 Fire and Fuel: Air Quality and Burning Hydrocarbons

We saw in the previous section that the combustion of some elements — carbon, nitrogen, and sulfur—produces air pollutants. The combustion of fuels can also produce carbon monoxide and carbon dioxide. Many fuels, including those obtained from petroleum, are **hydrocarbons**, compounds of hydrogen and carbon. The simplest of these is methane, CH_4, the primary component of natural gas. When a hydrocarbon burns completely, all of the carbon combines with oxygen to form carbon dioxide and all of the hydrogen combines with oxygen to form water. We can use this reaction to again illustrate the process of balancing equations. First we write formulas that qualitatively represent the combustion of methane:

$$CH_4 + O_2 \longrightarrow CO_2 + H_2O$$

The expression is already balanced with respect to carbon; there is one C, indicating one carbon atom, on each side of the arrow. But the equation is not balanced with respect to hydrogen and oxygen. It is easier to start with hydrogen because that element is present in only one substance on each side of the arrow: CH_4 on the left and H_2O on the right. Oxygen, on the other hand, winds up in both CO_2 and H_2O. There are currently four hydrogen atoms on the left of the expression (in CH_4) and two hydrogen atoms on the right (in H_2O). To bring the hydrogen atoms into balance, we place the number 2 in front of H_2O.

$$CH_4 + O_2 \longrightarrow CO_2 + 2 H_2O$$

The coefficient 2 is multiplied through the substance to the right of it, in this case water. Thus, it signifies 4 H atoms and 2 O atoms. Because a CO_2 molecule contains 2 O atoms, there are now a total of 4 O atoms on the right of the equation and 2 O atoms on the left. We equalize the number of O atoms by placing a 2 before O_2.

$$CH_4 + 2 O_2 \longrightarrow CO_2 + 2 H_2O \qquad (1.4)$$

That should balance the equation, but it is always a good idea to check. The nice thing about writing balanced equations is that you can always tell if you are correct by counting and comparing atoms on either side of the arrow:

Carbon: Left: 1 CH_4 molecule $\times$ 1 C atom/CH_4 molecule = 1 C atom
             Right: 1 CO_2 molecule $\times$ 1 C atom/CO_2 molecule = 1 C atom
Hydrogen: Left: 1 CH_4 molecule $\times$ 4 H atoms/CH_4 molecule = 4 H atoms
             Right: 2 H_2O molecules $\times$ 2 H atoms/H_2O molecule = 4 H atoms
Oxygen: Left: 2 O_2 molecules $\times$ 2 O atoms/O_2 molecule = 4 O atoms
             Right: 1 CO_2 molecule $\times$ 2 O atoms/CO_2 molecule = 2 O atoms
                    +2 H_2O molecules $\times$ 1 O atom/H_2O molecule = 2 O atoms

This tabulation confirms that the equation is indeed balanced.

Chapter 4 contains many examples of the burning of fuels.

1.18 Your Turn

a. "Bottled gas" or "liquid petroleum gas" (LPG) is mostly propane, C_3H_8. Balance this equation that represents the burning of propane.

$$C_3H_8 + O_2 \longrightarrow CO_2 + H_2O$$

b. Cigarette lighters burn butane, C_4H_{10}. Write a balanced equation for this combustion reaction.

Ans. a. $C_3H_8 + 5\ O_2 \longrightarrow 3\ CO_2 + 4\ H_2O$

One of the most widely used hydrocarbon fuels in automobiles is gasoline, a mixture of dozens of different compounds. One of the compounds is octane, C_8H_{18}. If a sufficient supply of oxygen is delivered to the auto engine when the octane burns, only carbon dioxide and water are formed:

$$2\ C_8H_{18} + 25\ O_2 \longrightarrow 16\ CO_2 + 18\ H_2O \qquad (1.5)$$

In practice, however, not all of the carbon is converted to carbon dioxide. The amount of oxygen present and amount of time available for reaction (before the materials are ejected in the exhaust) are insufficient for the reaction represented by equation 1.5 to occur entirely. Instead, some CO is formed. An extreme situation is represented by equation 1.6, in which all of the carbon in the octane is converted to carbon monoxide.

$$2\ C_8H_{18} + 17\ O_2 \longrightarrow 16\ CO + 18\ H_2O \qquad (1.6)$$

Note that the coefficient of O_2 in equation 1.5 is 25, whereas the corresponding coefficient in equation 1.6 is 17. This indicates that less oxygen is used in the latter reaction.

What really happens in a car's engine is a combination of the two reactions. Most of the carbon released in automobile exhaust is in the form of CO_2, although there is also some CO. The relative amounts of these two gases indicate how efficiently the car burns the fuel, which is evidence of how well tuned the engine is. States that monitor auto emissions check for this by sampling exhaust emissions using a probe that detects CO. The measured CO concentrations are compared to established standards (1.20% in the state of Minnesota). A car whose CO emissions exceed the standard must be serviced so that it complies.

1.19 Your Turn

Demonstrate that equations 1.5 and 1.6 are balanced by counting the number of atoms of each element on either side of the arrow.

Ans. In equation 1.5, there are 16 C, 36 H, and 50 O on each side of the equation.

1.11 Air Quality: Some Good News

Bad news always seems to get more attention than good news. Newspaper headlines and newscasts dwell on atmospheric pollutants, not the components of the atmosphere that are essential for life. And because pollutants such as carbon monoxide and sulfur dioxide are compounds generated by chemical processes, chemistry often gets blamed for pollution. Of course, oxygen, nitrogen, and carbon dioxide are also chemicals, and we would not be here without them.

Obviously, it is essential that science and society be concerned with hazardous components in the air we breathe. Chemists help monitor the concentrations of these pollutants and work to reduce them. Figure 1.7 indicates some of the results they have obtained recently. The data, gathered by the EPA, reveal how the total amounts (millions of tons/year) of several important air pollutants have changed from 1970 to 1997. Included in the graph are three gaseous pollutants listed in Table 1.2: sulfur dioxide, nitrogen oxides, and carbon monoxide. Because nitrogen forms several polluting com-

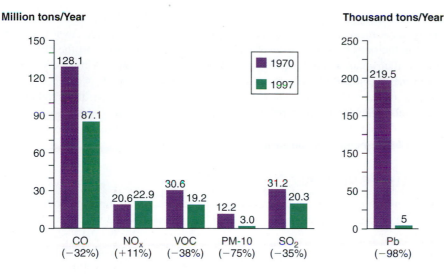

Figure 1.7

Changes in the average amounts of air pollutants in the United States, 1970 to 1997.

(Source: Data from *National Air Quality and Emissions Trends Report,* 1998, U.S. EPA.)

VOC means volatile organic compounds. PM-10 means particulate matter $< 10\ \mu$m.

pounds with oxygen, these oxides are often referred to collectively as NO_x, where X can take several different integer values, as for example one and two in NO and NO_2, respectively. In addition, Figure 1.7 includes data for particulate matter (PM) and lead. The same information is also presented in Table 1.7.

Looking at the graph and Table 1.7, it is easy to see that the concentrations of most of these pollutants have decreased markedly in the United States. The most dramatic changes have been in lead, sulfur dioxide, and carbon monoxide. The concentrations of carbon monoxide and particulates have decreased by 32 and 75%, respectively, but nitrogen oxides have increased by 11%.

An obvious way to reduce pollution is not to have the pollutants form in the first place. Over the past decade, an important initiative known as "green chemistry," the use of chemistry to prevent pollution, has taken place. **Green chemistry—**designing chemical products and processes that reduce or eliminate the use and/or generation of hazardous substances. Begun under the EPA's Design for the Environment Program, green chemistry reduces pollution through fundamental chemical breakthroughs in designing and re-designing chemical processes, with an eye toward making them environmentally friendly, that is "benign by design." In this regard, Dr. Barry Trost, a Stanford University chemist, advocates an "atom economy" approach to the synthesis of commercial chemical products such as pharmaceuticals, plastics, or pesticides. Such syntheses would be designed so that all reactant atoms end up as desired products, not as wasteful by-products. This approach would save money, as well as materials; undesired products would not be produced as waste, which requires disposal.

Dr. Lynn R. Goldman, an EPA administrator, says "Green chemistry is preventative medicine for the environment." Innovative "green" chemical methods already have made an impact on a wide variety of chemical manufacturing processes by decreasing or eliminating the use or creation of toxic substances. For example, cheaper, less wasteful, and less toxic methods have been developed to produce ibuprofen, pesticides, new materials for disposable diapers and contact lenses, new dry cleaning methods, and recycled silicon

Table 1.7   **Changes in Average Concentrations of Air Pollutants in the United States, 1970–1997, millions of tons/year**

Pollutant	1970	1997	Change
Sulfur dioxide	31.2	20.3	35% decrease
Nitrogen oxides	20.6	22.9	11% increase
Carbon monoxide	128.1	87.1	32% decrease
Particulates	12.2	3.0	75% decrease
Lead	0.220	0.005	98% decrease

wafers for integrated circuits. The developers of these and other green chemistry approaches have received the Presidential Green Chemistry Challenge Award. Begun in 1995, it is the only Presidential-level award recognizing chemists and the chemical industry for their innovations for a less polluted world; its theme is "Chemistry is not the problem, it's the solution." At various places throughout this book, we will discuss applications of green chemistry. They will be designated by an icon, the symbol of the Presidential Green Chemistry Challenge.

This is the icon for the Presidential Green Chemistry Challenge:

To better understand how pollutant levels can be reduced, we need to know some things about the pollutants and their sources.

1.12 Air Pollutants: Sources

By now you should recognize that the atmospheric concentration of air pollutants such as carbon monoxide, the oxides of sulfur and nitrogen, and ozone is far less than that of nitrogen and oxygen, the major atmospheric components. But by their presence, these minor components can compromise the quality of the air we breathe, even at parts per million or parts per billion levels. Much of the carbon monoxide, sulfur dioxide, nitrogen oxides, and ozone in air result from modern society's demands for energy. Most of the energy is used in generating electricity and in transportation, for which gasoline powers millions of cars and trucks. The burning of coal is the major U.S. source of electric power and of SO_2. Coal is mostly carbon and hydrogen, and thus the major products of its combustion are carbon dioxide and water. But coal is a complex mixture of variable composition, not merely carbon and hydrogen. Most coals contain rocklike minerals and 1–3% sulfur. When coal is burned, the sulfur is converted into gaseous sulfur dioxide, and the minerals are converted into fine ash particles. If they are not removed, the particles and the sulfur dioxide gas go up the smokestack.

You will learn in Chapter 3 that the CO_2 released by combustion is believed to contribute to global warming.

Sulfur dioxide can react with more oxygen to form sulfur trioxide, SO_3.

$$2\,SO_2 + O_2 \longrightarrow 2\,SO_3 \tag{1.7}$$

This reaction is normally quite slow, but it is much faster in the presence of small ash particles. The ash particles also aid another process: if there is high humidity in the air, they promote the conversion of water vapor into an **aerosol** of tiny water droplets, which we call fog. An aerosol is a form of liquid in which the droplets are so small that they stay suspended in the air rather than settling out. Once sulfur trioxide is formed, it dissolves readily in the water droplets to form an aerosol of sulfuric acid, H_2SO_4.

$$H_2O + SO_3 \longrightarrow H_2SO_4 \tag{1.8}$$

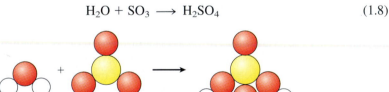

When inhaled, the sulfuric acid aerosol droplets are small enough to be trapped in the lung tissue where they cause severe damage. Moreover, the sulfur oxides and sulfuric acid are also major contributors to acid precipitation, the topic of an entire chapter in this book (Chapter 6).

Atmospheric levels of SO_2 have been decreasing slowly as a result of the Clean Air Act of 1970, which mandated reductions in power plant emissions. More stringent regulations were established in the Clean Air Act Amendments of 1990. Progress will not come easily or cheaply. More information about the strategies and technologies available to reduce atmospheric SO_2 and their economic and political costs is included in Chapter 6.

The vast majority of cars are powered by internal combustion engines that run on gasoline. Because gasoline contains only very small quantities of sulfur, the automobile is not a significant source of sulfur dioxide. However, the ubiquitous motor car does contribute to atmospheric concentrations of carbon monoxide, lead, nitrogen oxides, ozone, and a number of other unhealthful substances. The problem is particularly acute in America because the United States has more automobiles per capita than any other nation. In 1998 we had over 200 million cars, more than one for every two Americans. In some cities, such as Denver, Houston, and Los Angeles, 90% of the working population commutes to work by car—often with only one person per vehicle.

The combustion reaction of octane, a component of gasoline, has already been discussed. Ideally, from an energy and environmental standpoint, the only combustion products would be CO_2 and H_2O (equation 1.5). But a modern high-performance automobile, capable of operating at high speeds and with fast acceleration, is a source of carbon monoxide and some incompletely burned fragments of gasoline molecules called *volatile organic compounds* or VOCs. This incomplete combustion is caused by either insufficient oxygen or insufficient time in the engine cylinders for all the hydrocarbons to be burned to carbon dioxide and water. The problem of automobile carbon monoxide emissions has been partially solved by the installation of catalytic converters, devices that accelerate the conversion of CO in the exhaust stream to CO_2. The dramatic reduction in CO emissions shown in Figure 1.7 has occurred even though the number of cars has doubled in the past 25 years. The CO decrease is due to several factors: better engine design; computerized sensors that better adjust the fuel/oxygen mixture; and most importantly, all new cars since the mid-1970s have catalytic converters (Figure 1.8).

1.20 Consider This: Electric Cars

Many people believe that the only true solution to the pollution caused by gasoline-powered cars is to promote widespread development and use of electric cars. Such cars are no longer just a hope for the future, but are currently available in some areas. What are the criteria that you would use when deciding whether to buy an electric car?

There is even more good news. For more than 50 years, a compound named tetraethyl lead was added to gasoline to make it burn more smoothly by eliminating premature explosion or "knocking." It worked beautifully, but unfortunately the lead was released through the tailpipe to the atmosphere and ultimately into water supplies. Lead is a highly toxic element—a cumulative poison that can cause a wide variety of neurological problems, especially if ingested by children. Moreover, lead can also destroy the effectiveness of catalytic converters. With the advent of catalytic converters it became

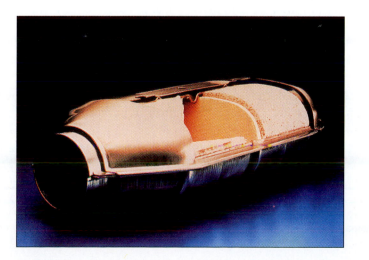

Figure 1.8

A cut-away view of an automobile catalytic converter. Semi-precious metals such as platinum and rhodium are coated on the surface of ceramic beads. The metals catalyze the combustion of CO to CO_2. Other catalysts in the converter accelerate the conversion of nitrogen oxides to N_2 and O_2.

Figure 1.9

Annual lead usage in gasoline and the variations in lead concentration in human blood during that time (selected U.S. cities). Notice the sharp decline in average blood lead levels corresponding to when unleaded gasoline began to be required.

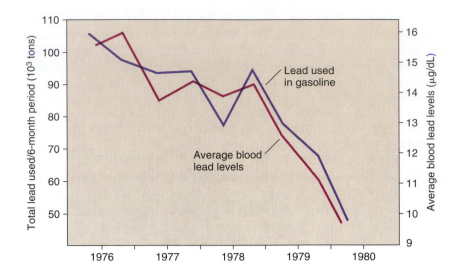

necessary to formulate gasoline without lead. Since 1976, all new cars and trucks sold in the United States have been designed to use only unleaded fuel. The result has been a dramatic decrease in lead emissions—from more than 219,000 tons in 1970 to 5,000 tons in 1997 (see Figure 1.7 and Table 1.7). Leaded fuel was banned by law in the United States in 1997. The switch to non-leaded fuels caused a dramatic drop in blood lead levels (Figure 1.9).

The U.S. has had less success curbing the emission of nitrogen oxides. Cars are also a major source of these noxious gases. Nitrogen and oxygen are always present wherever there is air. And when this mixture is subjected to high temperatures, as in an internal combustion engine, the following reaction occurs.

$$N_2 + O_2 \longrightarrow 2 \, NO \qquad \qquad (1.9)$$

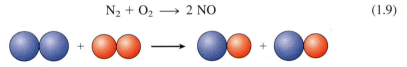

Subsequent reactions with atmospheric oxygen can generate other oxides of nitrogen, including nitrogen dioxide, NO_2. These oxides can be highly toxic and contribute to acid rain, the subject of Chapter 6.

The quantity of nitrogen oxides released to the atmosphere increased steadily up to 1980 and has only slightly decreased since then. But the situation would have been much worse (given the increased number of autos and miles driven) if emission controls had not been mandated starting in 1970. The Clean Air Act of 1970 set tailpipe emission standards, to go into effect in 1975, and these were subsequently revised (Table 1.8). Despite early claims from the auto industry that it would be impossible, or too costly, to meet the standards, the industry has, in fact, achieved these goals by using improved catalytic converters, engine designs, and gasoline formulations.

Table 1.8 National Tailpipe Emission Standards, Passenger Cars*
All values in grams/mile; (California standards in parentheses)
Based on 5 years/50,000 miles

NMHC**	CO	NO$_x$
0.25 (0.25)	3.4 (3.4)	0.4 (0.4)

* New standards began with 1996 models. If the EPA determines by 1999 that lower standards are necessary, the standards will be cut in half beginning with the 2004 model year vehicles.
** Non-methane hydrocarbons, that is, all hydrocarbons except CH_4.

1.13 Air Quality at Home and Abroad

Anyone who reads the newspapers or listens to broadcast news analysis knows that the control of automobile emissions is not a matter of technology alone. Economics and politics are also part of this complex web. For example, the U.S. automobile industry did not act to reduce emissions until forced to do so by federal legislation in 1970. There were no significant changes in the standards during the 1980s, when the national administration advocated less governmental regulation of private industry. The Clean Air Act signed by former President George Bush in November 1990 was the first major new clean air legislation in 20 years, but it did not come about easily. There was a great deal of political infighting as Congress and the administration struggled to find acceptable compromises. The controversy continues. Many members elected to Congress in the mid-1990s have expressed their opposition to government regulations, and a number of environmental policies continue to be under threat.

Clean air legislation is revisited in Section 6.14.

Air pollution is primarily an urban problem, and more than fifty percent of all Americans live in cities with populations over 500,000. Many of these cities fail to meet the national air quality standards, at least during periodic pollution alerts. In spite of recent improvement in air quality we still have difficulties, especially with nitrogen oxides and ozone. Children and adults with chronic respiratory problems or heart disease are most at risk from exposure to these pollutants, and the risk is increased by vigorous physical activity. Furthermore there is evidence that the present air quality standards provide little margin of safety in protecting public health. The American Lung Association estimates that $50 billion in health benefits could be realized annually in the United States if air quality standards were met.

Increasingly, U.S. industries have been held accountable for their waste emissions. Annually since 1987, the EPA has published the Toxics Release Inventory (TRI), a national directive requiring companies to make available to the public data on the amounts of certain chemicals they have released into the air, water, and land. TRI is part of the Community Right To Know Law, and a kind of snapshot for a given time of the state of pollution from a list of pollutants as reported by U.S. industries. Vice-President Gore praised the program noting: "It has spurred innovation to help business work smarter and cleaner and become more profitable."

The law has had a significant impact; toxic emissions from all industries declined by 46% since the law took effect. The EPA data indicate that the U.S. chemical industry has decreased its emissions 51%, the largest reduction by any manufacturing industry, in the years since the law began. In 1996, half of the total chemical emissions were released into the air (Figure 1.10.) Just seven compounds make up over 65% of the emissions from the chemical industry, an indication of how directly these substances are associated with synthesizing the compounds produced by that industry. The compounds are ammonia (NH_3), methanol (CH_4O), carbon disulfide (CS_2), hydrochloric acid (HCl), and three compounds important in plastics production—ethylene (C_2H_4), propylene (C_3H_6), and toluene (C_7H_8).

TRI data are available for each state. The following exercise gives the opportunity to check on the toxic emissions in your state or locale, and the progress made in reducing those emissions.

1.21 Consider This: TRI and You

Go to the EPA web site and find the Toxic Release Inventory (TRI) for your state and for your locale.

a. Compare the current levels of toxic emissions with what they were one or two years ago.

b. Which emissions (if any) have decreased and which have remained the same or increased? Determine, if you can, a reason for the changes.

Figure 1.10

Emission of chemicals in the United States. The 1996 releases amounted to 785 million pounds.

(Source: Data from the U.S. EPA).

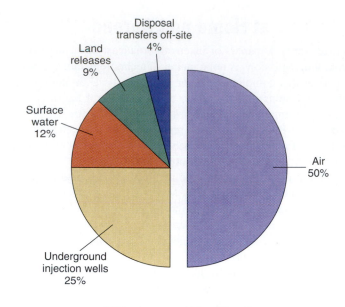

1996 releases = 785 million lb

We face difficult political and economic choices. Are we willing to spend the money that would be needed to really clean up the air we breathe? What would happen if regulations were dropped or relaxed, as some have proposed? Would the supposed boon to the American economy compensate for the hidden costs? In considering the risks with the benefits, a tightening of regulations could mean a boon to health and a significant reduction in health care costs associated primarily with respiratory conditions. The improved air quality that we have enjoyed could be short-lived. Our environment is a fragile system that could quickly revert to a status of severe air pollution.

1.22 Consider This: Growing Interest in Air Pollution

Air pollution has not occurred overnight. It has been a growing problem since at least the time of the Industrial Revolution. Why have we as a nation and a world community become so concerned with it lately? Through discussion and/or library research, identify at least four factors that have combined to make air pollution an important issue for the present.

International comparisons indicate that air pollution is no respecter of nations (Table 1.9). It has the potential to be a major problem anywhere the defining characteristics of modern industrial development—electrical power generation and many automobiles—exist. Air pollution problems in the United States pale compared to those in other parts of the world. Many countries have few or no controls on pollutant emissions. In some cases this is because of the political system. Regulatory laws have been passed, but enforcement of them has been lax, often because of the enormous pressures to put economic development before environmental quality. Germany has 10 times as many automobiles per square mile as the United States, yet it has fewer emission controls on cars (or on electric power plants). As a result, parts of Germany face much more serious air pollution problems than we have in this country.

The situation in Eastern Europe is especially bad because heavy industrialization has occurred without the constraints of pollution controls. Portions of the region have become nearly uninhabitable. A very low grade of coal, called "brown coal," is widely used; large lead smelters release huge amounts of lead into the atmosphere. In China,

28 northern cities have SO_2 and particulate concentrations that are three to eight times higher than the guideline limits set by the World Health Organization (WHO). And the 20 million residents of Mexico City breathe ozone levels that are more than 50% above WHO guidelines for most of the year. Among the steps that have been taken there is a mandated certified analysis of exhaust gases each six months. Analyses classify cars into three groups, identified by a windshield tag. Category Zero includes those cars manufactured since 1992 and those with exhaust within legal limits for pollutants. Such cars can be driven daily. Category One cars have a higher, but not excessive, level of exhaust pollutants. These cars can be driven six days a week with the "off" days distributed evenly across the week. Category 2 cars produce high levels of exhaust pollutants; driving these cars is limited during an "air quality alert" to either day during the first two days of such an alert. No cars in this category can be driven from the third day onward until the air quality alert is lifted.

Table 1.9 **International Air Quality**
(a) World Heath Organization (WHO) Air Quality Guidelines (1993–98)

Pollutant	Guideline, $\mu g/m^3$
SO_2	5–400
Particulates	Standard being revised
Lead	0.01–2
CO	500–7000
NO_2	10–150
O_3	10–100

(b) Air Quality In Megacities Around the World (1992 Data) Data from the Atmospheric Research and Information Centre, Manchester Metropolitan University, Manchester, England

City	SO_2	Particulate Matter	Lead	CO	NO_2	O_3
Bangkok, Thailand	*	***	**	*	*	*
Beijing, China	***	***	*	–	*	**
Cairo, Egypt	–	***	***	**	–	–
Jakarta, Indonesia	*	***	**	**	*	**
London, England	*	*	*	**	*	*
Mexico City, Mexico	***	***	**	***	**	***
Moscow, Russia	–	**	*	**	**	–
Rio de Janeiro, Brazil	**	**	*	*	–	–
Sao Paulo, Brazil	*	**	*	**	**	***
Tokyo, Japan	*	*	–	*	*	***

* Low pollution—normally meets WHO guidelines (on occasion may exceed guidelines short-term)
** Moderate to heavy pollution—WHO guidelines exceeded up to twice as much (short-term guidelines regularly exceeded at certain locations)
*** Serious pollution problem—WHO guidelines exceeded at greater than twice as much.
— Data not available

1.23 Consider This: Air Quality Around the World

Based on the data in Table 1.9:

a. Which city has the worst overall air quality?
b. Of the 10 cities, which one has the best air quality?
c. In which cities do automobiles contribute significantly to air pollution? Give a rationale for your answer.

1.14 Breathing Lessons—Indoor Air

The Consider This 1.1 exercise had you "take a breath." That one breath adds up to between 1×10^4 to 2×10^4 L (10–20 m³) of air each of us breathes daily. As we have learned from previous sections, higher air quality standards have decreased the allowable concentrations of various air pollutants by controlling emissions from automobiles and industries. But air quality depends on where we are. Ironically, such standards have been established for outdoor air, but not for indoor air. Yet, most of us sleep, work, study, and play indoors, spending up to 95% of our time in our dorm rooms, classrooms, offices, and residences. Consequently, we should be concerned not just with outdoor air quality, but that of indoor air as well. Figure 1.11 indicates the concentrations of common pollutants in indoor and outdoor air. Notice from Figure 1.11 that, with the exception of sulfur dioxide, the average indoor concentrations of the pollutants exceed their average outdoor concentrations.

Indoor air is a complex mixture; typically nearly 1000 substances are detectable in it at the ppb level or higher. However, indoor air sources are limited to either the outside air that enters buildings, or air that comes from within buildings. Tobacco smoke, cooking by-products, substances emitted from rugs, furniture, construction materials, and office products are some of the many materials that can degrade indoor air quality. Table 1.10 lists some of the sources of indoor air pollutants.

How quickly indoor air pollutants build up depends on the rates at which outdoor air moves inside and indoor air moves out, as well as on how rapidly pollutants are generated indoors. An insufficient exchange of outside air can cause the concentration of indoor air pollutants to build up to troublesome levels. Consider the risk-benefit trade-off in which buildings constructed within the past two decades have been more airtight to minimize drafts and increase energy efficiency. Although enhanced energy efficiency has been achieved, it has been at the cost of decreasing the flow of outside air into the building to replace the indoor air. When this reduction in air exchange occurs, it allows the concentration of indoor air pollutants to increase. Therefore, initially what was a benefit (better energy efficiency) can turn into an increased risk (increased pollutants concentration). Construction of some large office buildings has been so highly energy efficient that little exchange of outside air occurs within them. In some of these cases, the reduced air exchange has allowed indoor air pollutants to reach levels hazardous to the health of some individuals, creating a condition known as "sick building syndrome."

Figure 1.11

Indoor and outdoor concentrations of common pollutants

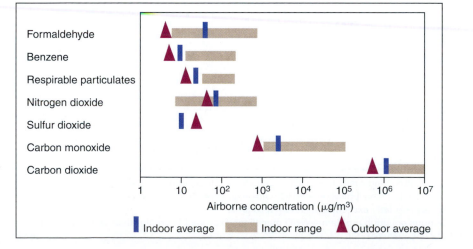

Table 1.10 Indoor Air Pollutants and Their Sources

Phases of Matter	Source	Pollutant
Solid/particulate	Floor tile	Asbestos
	Pets	Pet dander, dust
	Plants	Molds, mildew, bacteria, viruses
Liquid/gases	Carpet	Styrene
	Cigarette smoke	Formaldehyde, carbon monoxide
	Clothes	Drycleaning fluid, moth balls
	Electric arcing	Ozone
	Faulty furnace or space heater	Carbon monoxide
	Furniture	Formaldehyde
	Glues and solvents	Acetone, toluene
	Paint and paint thinners	Methanol, methylene chloride
	Soil and rocks under house	Radon

Radon, a colorless, odorless, tasteless, but radioactive gas, is a possible indoor air pollutant, depending on your location. Radon is generated naturally by the nuclear decomposition of uranium in the rocks and soil on which buildings rest. In some regions of the country, sufficient radon gas is present to seep into buildings through cracks in the foundations. Extended inhalation of radon gas can result in lung cancer. But the overall health effects of indoor radon, and the exposure required to create health problems are controversial. Home radon testing kits are commercially available.

> Radioactivity is discussed in detail in Chapter 7.

1.24 Consider This: Rating Radon

As a public service, local and national agencies provide documents on the web about radon. Search to bring up a list about radon. In addition to searching just for "radon," you might want to add the terms "detection," air quality," and/or "EPA."

a. Find two web sites about radon provided by government agencies. For each, list the title, the source, and the URL.

b. How can you measure the radon levels in your home? Search the web for a company that sells radon test kits. Describe the kit, including its price. If you don't find anything, switch to another search engine.

c. Is information from commercial sources about radon any different in its objectivity from that provided by agencies as a public service? If so, discuss the differences and reasons for them.

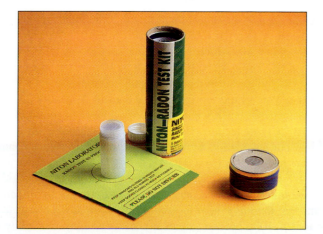

A home radon test

Whether it is indoor or outdoor air we breathe, we inhale (and exhale) a truly prodigious number of molecules and atoms during a lifetime. On a molecular and atomic level, these particles have some fascinating properties, ones that we will consider next.

1.15 Back to the Breath—at the Molecular Level

The maximum concentrations of pollutants specified in Table 1.2 seem very small — and they are. Nine CO molecules out of one million particles of the mixture called air is a tiny fraction. But, as we will soon calculate, at this CO concentration a breath of air contains a staggering number of carbon monoxide molecules. This apparent contradiction is a consequence of the minuscule size of molecules and the immense numbers of them. Recall 1.1 Consider This. If you are an average-sized adult in good physical condition, the total capacity of your two lungs is approximately 1 liter (about a quart). Determining the number of molecules in this volume of air is no easy task, but it can be done. As a result of experiments (as well as theories) we know that a typical breath contains more than 2×10^{22} particles—molecules such as N_2, O_2, CO, and individual atoms like Ar.

Using the number of particles in a breath, we can calculate the number of CO molecules in the breath you just inhaled. We will assume the breath contained 2×10^{22} molecules, and that the CO concentration in the air was the national ambient air quality standard of 9 ppm. This means that out of every million (1×10^6) molecules of air, nine will be CO molecules. To compute the number of CO molecules in the breath we multiply the total number of air molecules by the fraction of them that are carbon monoxide molecules.

$$\text{number of CO molecules} = 2 \times 10^{22} \, \text{air particles} \times \frac{9 \text{ CO molecules}}{1 \times 10^6 \text{ air particles}}$$

$$= \frac{18 \times 10^{22}}{1 \times 10^6} \text{ CO molecules} = \frac{18}{1} \times \frac{10^{22}}{10^6} \text{ CO molecules}$$

Note that in writing out this problem, we have retained the labels on the numbers. This is a reminder of the physical entities involved, but it also provides a guide for setting up the problem correctly. The labels "air particles" cancel each other, and we are left with what we want—CO molecules.

However, we need to divide 10^{22} by 10^6 to convert the answer into a simpler number. To *divide* powers of ten, you simply *subtract* the exponents. In this case,

$$\frac{10^{22}}{10^6} = 10^{(22-6)} = 10^{16}$$

So, this means that there are $\frac{18}{1} \times \frac{10^{22}}{10^6}$ CO molecules $= 18 \times 10^{16}$ CO molecules in the breath (Figure 1.12).

The above answer is mathematically correct, but in scientific notation it is customary to have only one digit to the left of the decimal point. Here we have two: 18. Therefore, our last step will be to rewrite 18×10^{16} as 1.8×10^{17}. We can make this conversion because $18 = 1.8 \times 10$, which is the same as 1.8×10^1. We *add* exponents to *multiply* powers of 10. Thus, $18 \times 10^{16} = (1.8 \times 10^1) \times 10^{16}$ CO molecules, which equals 1.8×10^{17} CO molecules in that last breath you exhaled. (If all of this use of exponents is coming at you a little too fast, please consult Appendix 2.)

It may sound surprising, but it would be more accurate to round off the answer and report it as 2×10^{17} CO molecules. Certainly 1.8×10^{17} looks more accurate, but the data that went into our calculation were not very exact. The breath contains *about* 2×10^{22} molecules, but it might be 1.6×10^{22}, 2.3×10^{22}, or some other number. The jargon is that 2×10^{22} expresses a physically based property to "one **significant figure**." Only one digit, the initial 2, is used, and so there is only one significant figure in 2×10^{22}. That means that the number of molecules in the breath is closer to 2×10^{22} than to 1×10^{22} or to 3×10^{22}, but we cannot say with certainty much beyond that. Simi-

Figure 1.12
Carbon monoxide monitors
are commercially available to
be used in residences, as well
as offices and business sites.

larly, unless the analytical data are very good, the concentration of carbon monoxide is also known to only one significant figure, 9 ppm. The product 2×9 equals 18. That is certainly correct mathematically, but our question about CO is based on physical data. The answer, 1.8×10^{17} CO molecules includes two significant figures, the 1 and the 8. It implies a level of knowledge that is not justified. The accuracy of a calculation is limited by the *least accurate* piece of data that goes into it. In this case, both the concentration of CO and the number of air particles in the breath were each known to only one significant figure (9 and 2, respectively); two significant figures in the answer are unjustified. The rule is that you cannot improve the accuracy of experimental measurements by ordinary mathematical manipulations like multiplying and dividing. Therefore, the answer must also contain only one significant figure, hence 2×10^{17}.

1.25 Your Turn

The local news has just reported that today's ground-level ozone readings are right at the acceptable standard, 0.12 ppm. How many molecules of ozone, O_3, are in each breath of this air, assuming there are 2×10^{22} molecules of air in each breath?

Ans. 2×10^{22} air particles $\times \dfrac{0.12 \ O_3 \text{ molecules}}{10^6 \text{ air particles}} = 2 \times 10^{15} \ O_3$ molecules in each breath

You may well question the significance of all of this talk about significant figures, but it is very important in interpreting numbers associated with physical quantities. It has been observed that "figures don't lie, but liars can figure." Numbers often lend an air of authenticity to newspaper or television stories, so the popular press is full of numbers. Some are meaningful and some are not, and the informed citizen must be able to discriminate between the two types. For example, the assertion that the concentration of carbon dioxide in the atmosphere is 358.5537 ppm should be taken with a rather large grain of sodium chloride (salt). The estimate of 360 ppm (three significant figures) is reasonable; the previous assertion with seven significant figures is not valid.

There are other ways in which numbers sometimes introduce ambiguity. You have just encountered some conflicting information. The concentration of CO in air is very small, 9 parts per million. Nevertheless, the number of CO molecules in a breath is almost unimaginably large, about 2×10^{17}. Both statements are true. The consequence of these numbers is that it is *impossible* to completely remove pollutant molecules from the air. "Zero pollutants" is an unattainable goal; using the most sophisticated detection

In expressing the presence or lack of a substance in a sample, absence of evidence is not the same as evidence of absence.

methods you still could not even determine whether it had been achieved. At present, our most sensitive methods of chemical analysis are capable of detecting one target molecule out of a trillion. One part per trillion corresponds to: moving six inches in the 93 million–mile trip to the sun; a single second in 320 centuries; or a pinch of salt in 10,000 tons of potato chips. And yet, a chemical could be undetectable at this level, and a breath might still include 1×10^5 molecules of the substance.

1.26 Your Turn

To help you comprehend the magnitude of the 2×10^{17} CO molecules in just one of your breaths, assume that they were equally distributed among the five billion (5×10^9) human inhabitants of the Earth. Calculate each person's share of the 2×10^{17} CO molecules you just inhaled.

Hint: You are trying to distribute the huge number of molecules in a breath among all the human inhabitants of the Earth. Each person's share can be found by dividing the total number of CO molecules by the total number of humans.

$$\text{Each person's share is } \frac{2 \times 10^{17} \text{ CO molecules}}{5 \times 10^9 \text{ people}}$$

Now see if you can demonstrate that each person's share is 4×10^7 or 40,000,000 molecules of CO per person.

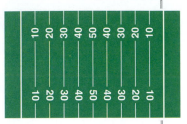

pph = 1 yd out of length of one field (100 yards)

ppthousand = 1 yd out of length of ten fields

ppm = 1 yd out of length of 10^4 fields

ppb = 1 yd out of length of 10^7 fields

Parts per hundred, parts per million (ppm), and parts per billion (ppb).

A breath of air typically contains molecules of hundreds—perhaps thousands—of different compounds, most in minuscule concentrations (Figure 1.13). For almost all of these substances, it is impossible to say whether the origin is natural or artificial. Indeed, many trace components, including the oxides of sulfur and nitrogen, come from both natural sources and those related to human activity. And, as with all chemicals, "natural" is not necessarily good and "human-made" is not necessarily bad. As you read in Section 1.4, what matters is exposure, toxicity, and the assessment of risk.

In addition to being extremely small, the particles in your breath possess other remarkable characteristics. In the first place, they are in constant motion. At room temperature and pressure, a nitrogen molecule travels at about 1000 feet per second and ex-

Figure 1.13

A spirometer is an instrument used for measuring an individual's breathing capacity.

periences approximately 400 billion collisions with other molecules in that time interval. Nevertheless, relatively speaking, the molecules are quite far apart. The actual volume of the molecules making up the air is only about 1/1000th of the total volume of the gas. If the particles in your one liter breath were all squeezed together, their volume would be about 1 milliliter (1 mL)—about one-third of a teaspoon. Sometimes people mistakenly think that air is empty space. It's 99.9% empty space, but the matter that is in it is literally a matter of life and death!

Moreover, it is matter that we continuously exchange with other living things. The carbon dioxide that we exhale is used by plants to make the food we eat, and the oxygen that plants release is essential for our existence. Our lives are linked together by the elusive medium of air. With every breath we exchange millions of molecules with each other. As you read this, your lungs contain 4×10^{19} molecules that have been previously breathed by other human beings, and 6×10^8 molecules that have been breathed by some *particular* person—say Julius Caesar, Marie Curie, or Martin Luther King, Jr. Pick your favorite hero or heroine—your body almost certainly contains atoms that were once in his or her body. In fact, the odds are very good that right now your lungs contain one molecule that was in Caesar's *last* breath. The consequences are breathtaking!

1.27 The Skeptical Chymist: Caesar's Last Breath

We just claimed that your lungs currently contain one molecule that was in Caesar's last breath. That assertion is based on some assumptions and a calculation. We are not asking you to reproduce the calculation, but rather to identify some of the assumptions and arguments that we might have used. Are they reasonable?

Hint: Here is a start. The calculation assumes that all of the molecules in Caesar's last breath have been uniformly distributed throughout the atmosphere.

In starting this chapter with "taking a breath," we began a pattern for a kind of activity that we will ask you to reflect on in several places in this text. The next "take a…" activity will come up in Chapter 5 where you will be asked to "take a drink" of water. Several of the themes that arose in Chapter 1 regarding air will also appear in that later chapter dealing with water—the quality of a natural resource essential to life, its sources, its special properties, and the risks associated with its quality, or lack of it.

Conclusion

The air we breathe has a personal and immediate effect on our health. Our very existence depends on having a large supply of relatively clean, unpolluted air with its essentials for life—the elements, oxygen and nitrogen, and two compounds, water and carbon dioxide. But air can be polluted with potentially toxic substances such as carbon monoxide, ozone, sulfur oxides, and nitrogen oxides. This is true especially in the urban environments of our large cities, the very places where the majority of Americans live. The major pollutants are, for the most part, relatively simple chemical substances. Carbon monoxide and the oxides of sulfur and nitrogen are compounds that exist as molecules made from atoms of their constituent elements. These compounds are formed by chemical reactions, often as unavoidable consequences of our dependence on coal for energy production in power plants and gasoline in internal combustion engines. Over the past thirty years, governmental regulations (such as TRI), industrial participation, modern technology, and green chemistry have resulted in large reductions in many pollutants. But it is not possible to reduce pollutant concentrations to zero because of the minuscule size of atoms and molecules and their immense numbers. Rather we must ask what the risk is from a given level of pollutant and then what level of risk is acceptable for various population groups.

The oxygen-laden air we breathe, whether indoors or out, is, of course, very close to the surface of the Earth. But the Earth's atmosphere extends upward for considerable distance and contains other substances that are also essential for life on this planet. In the next two chapters we will consider two of these substances and how they are changing, perhaps as a result of human activities.

Chapter Summary

Note: The numbers that follow indicate the sections in which the topics are introduced and explained.

Having studied this chapter, you should be able to:

- Recognize the composition of air and reasons for local and regional variations of it (1.1 – 1.3);
- Understand factors behind air quality and the chief components of air pollution (1.3, 1.11, 1.12);
- Evaluate conditions significant in risk-benefit analysis (1.4);
- Identify the general regions of the atmosphere with respect to altitude and the relationship of air pressure to altitude (1.5);
- Interpret air quality data in terms of concentration units (ppm, ppb) and pollution levels, including unreasonableness of "pollution free" levels (1.2 – 1.3, 1.12, 1.14, 1.15);
- Differentiate among mixtures, elements, and compounds (1.6);
- Understand the differences between atoms and molecules, between symbols for elements and formulas for chemical compounds (1.7);

- Name selected chemical elements and compounds (1.7);
- Write and interpret chemical formulas (1.8);
- Balance chemical equations, including using sphere equation representations (1.9 – 1.10);
- Discuss the green chemistry initiative (1.11);
- Describe the nature of air quality policies in this country and abroad in terms of their effectiveness in controlling air pollution (1.12 – 1.13);
- Identify the sources and nature of indoor air pollution (1.14);
- Interpret the nature of air at the molecular level (1.15);
- Use scientific notation and significant figures in performing basic calculations (1.4 and 1.15, respectively).

Questions

The questions in this chapter, as well as those in the remaining chapters, are divided into three categories:

- **Emphasizing Essentials** These questions give you the opportunity to practice the fundamental skills to be developed in the chapter. This set of questions relates most closely to the *Your Turns* in the chapter. Answers are provided in Appendix 4 for questions whose numbers are in color.

- **Concentrating on Concepts** These questions ask you to focus on the chemical concepts developed in the chapter and their relationships to the topics under discussion. They serve to integrate and to apply chemical concepts. This set of questions most closely resembles *Consider This* activities you have been engaged with throughout the chapter. Answers are provided in Appendix 5 for questions whose numbers are in color.

- **Exploring Extensions** These questions challenge you to go beyond the information presented in the text. They provide an opportunity for extending and integrating the skills, concepts, and communication abilities practiced in the chapter. Some extension questions are closely related to the type of analysis practiced in the *Sceptical Chymist* activities in the chapter. Questions marked with the web icon require using the worldwide web to obtain further information.

Emphasizing Essentials

1. Calculate the volume of air that a person exhales in an 8-hr day. Assume that each breath has a volume of about 1 L and that the person exhales 15 times a minute.

2. Given that air is 78% nitrogen by volume, how many liters of nitrogen are in 500 L of dry air?

3. A 5.0-L mixture of gases is prepared for photosynthesis studies by combining 0.75 L of oxygen, 4.0 L of nitrogen, and 0.25 L of carbon dioxide. Compare the percentage of carbon dioxide gas with that normally found in the atmosphere.

4. Air contains 9000 ppm (parts per million) argon. Express this value as a percentage.

5. The smoke inhaled from a cigarette contains about 0.04 % carbon monoxide, CO. Express this concentration in ppm.

6. The concentration of water vapor in the atmosphere of a tropical rain forest may reach 50,000 ppm. Express this value as a percentage.

7. According to Table 1.1, the percentage of carbon dioxide in inhaled air is *lower* than it is in exhaled air, but the percentage of oxygen in inhaled air is *higher* than in exhaled air. How can you account for these relationships?

8. The permissible limit for ozone for a 1-hr average is 0.12 ppm. If Greenville, South Carolina registers a reading of 0.15 ppm for 1 hour, by what percent is Greenville over the limit for atmospheric ozone?

9. Express each of these numbers in scientific notation.
   a. 1500 m, the distance of a foot race
   b. 0.0000000000958 m, the distance between O and H atoms in water

c. 0.0000075 m, the diameter of a red blood cell

d. 150,000 mg of CO, the approximate amount breathed daily

10. Write each of these values in non–scientific notation.

    a. 8.5×10^4 g, the mass of air in an average room

    b. 1.0×10^7 gallons, the volume of crude oil spilled by the Exxon Valdez

    c. $5.0 \times 10^{-3}\%$, the concentration of CO in the air of a city street

    d. 1×10^{-5} g, the recommended daily allowance of vitamin D

11. Express each of these numbers in scientific notation.

    a. 72000000 cigarettes; the number of cigarettes smoked per hour in the United States

    b. 15000°C; the temperature near the spark plug in an automobile engine

    c. 0.000000003 g; the number of grams of the insecticide DDT that dissolves in 1 g of water

    d. 0.00022 g; the number of grams of NO_2 that can be detected by smell in 1 m^3 of air

12. Use Figure 1.3 to verify this statement: "Below about 20 km . . . the pressure decreases by about 50% for every 5 km increase in altitude." Does this relationship hold throughout the troposphere?

13. Consider this periodic table.

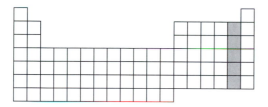

    a. What is the group number indicated by the shading on this periodic table?

    b. What elements make up this group?

    c. What is a general characteristic of the elements in this group?

14. Consider this periodic table.

    a. What is the group number indicated by the shading on this periodic table?

    b. What elements make up this group?

    c. What is a general characteristic of the elements in this group?

15. Classify each of these substances as an element, compound, or mixture.

    a. a sample of "laughing gas" (dinitrogen monoxide)

    b. steam coming from a pan of boiling water

    c. a bar of deodorant soap

    d. a sample of graphite

    e. a cup of mayonnaise

    f. the helium filling a balloon

16. Name the compounds that contain these pairs of elements.

    a. potassium and oxygen

    b. aluminum and chlorine

    c. sodium and iodine

    d. magnesium and bromine

17. Give the correct formula for each of these substances.

    a. "laughing gas," chemically named dinitrogen monoxide

    b. ozone, more properly named trioxygen

    c. sodium fluoride, an anti-cavity ingredient in toothpaste

    d. carbon tetrachloride, formerly used as a dry cleaning agent

18. What information does each formula convey about its compound, each a trace component of the atmosphere?

    a. CH_2O, formaldehyde

    b. H_2O_2, hydrogen peroxide

    c. CH_3Br, methyl bromide

19. Write balanced chemical equations to represent these reactions.

    a. nitrogen (N_2) reacting with oxygen gas (O_2) to form nitric oxide (NO)

    b. ozone (O_3) decomposing into oxygen gas (O_2) and atomic oxygen (O)

    c. sulfur (S) reacting with oxygen gas (O_2) to form sulfur trioxide (SO_3)

20. Write balanced sphere equations to represent each of the reactions in question 19.

21. Balance these equations, all of which involve the reaction of ethylene gas, C_2H_4, with oxygen gas, O_2.

    a. $C_2H_4(g) + O_2(g) \longrightarrow C(s) + H_2O(g)$

    b. $C_2H_4(g) + O_2(g) \longrightarrow CO(g) + H_2O(g)$

    c. $C_2H_4(g) + O_2(g) \longrightarrow CO_2(g) + H_2O(g)$

22. Consider the three equations you balanced in question 21. Compare the coefficient for oxygen gas in these equations. How is it related to the products formed in each case?

23. Demonstrate that each of these equations is balanced by counting atoms of each element on either side of the arrow.

    a. $2\ C_3H_8(g) + 7\ O_2(g) \longrightarrow 6\ CO(g) + 8\ H_2O(l)$

    b. $2\ C_8H_{18}(g) + 25\ O_2(g) \longrightarrow 16\ CO_2(g) + 18\ H_2O(l)$

24. Platinum, palladium, and rhodium are used in automobile catalytic converters.

    a. What is the symbol for each of these metals?

    b. Where are these metals located on the periodic table?

    c. What can you infer about the properties of these metals, given that they are useful in this application?

25. If a room is 6 m long, 5 m wide, and 3 m high, how many milligrams of formaldehyde must be present if the concentration is reported as 40 ppm?

Concentrating on Concepts

26. In Section 1.1, air was referred to as ". . . that invisible stuff . . . ." Is this always true? What factors influence if air appears "invisible" or if you can "see" it?

27. In Consider This 1.1, you calculated the volume of air exhaled in a day. How does this volume compare with the volume of air in your chemistry classroom? Show your calculations. *Hint:* Think ahead about the most convenient unit to use for measuring or estimating the dimensions of your classroom.

28. In Consider This 1.3, you considered how life on Earth would change if the concentration of oxygen were doubled. Now consider the opposite case; discuss how life on Earth would change if the concentration of O_2 were only 10%. Give some specific examples of how burning, rusting, and most metabolic processes in humans and plants would be affected.

29. Explain why the concentrations of some components in the atmosphere are expressed in percent (parts per hundred) and others are given in ppm (parts per million).

30. Consider the following table of data from the U.S. Environmental Protection Agency, Office of Air Quality Planning and Standards. The data indicate the number of days metropolitan statistical areas failed to meet acceptable air-quality standards (Pollutant Standards Index rating over 100).

Air Quality of Selected U.S. Metropolitan Areas, 1986–94

Metropolitan statistical area	1986	1988	1990	1992	1994
Atlanta, GA	18	21	17	5	4
Bakersfield, CA	54	85	48	16	45

   a. Prepare a visual representation of these data. Use any type of representation that you feel will best convey the information to the general public.

   b. Use your representation to discuss the trends in these two cities from 1986–94.

   c. Can you use your representation to *predict* the air quality in Atlanta and Bakersfield in 1993? Discuss your reasoning.

31. A certain city has an ozone reading of 0.13 ppm for one hour, and the permissible limit is 0.12 for that time. You have the choice of reporting that the city has exceeded the ozone limit by 0.01 ppm or saying that it has exceeded the limit by 8%. What are the advantages of each method?

32. a. Arrange these measurements in order of increasing size: 1 m, 3.0×10^2 m, 5.0×10^{-3} m.

   b. Draw an analogy between these three length measurements and time. Let one year be equal to 1 m. How long will the other two measurements be in terms of time expressed in years?

33. Air quality reports are often published in local newspapers, but rarely reported during televised weather reports, unless there is a dangerously high level of air pollution. Why do you think this is the case?

34. If risk is related to public perception, what is your feeling about the relative risks associated with each of these items? Rank them in order of your perception of the *most* risky to the *least* risky. Be prepared to explain your choices to class members.

   smoking, using roller blades, eating beef, driving to work, getting a suntan, taking aspirin, drinking tap water, breathing polluted air

35. The cabins of commercial airliners flying at 30,000 feet are pressurized. Buildings in Denver, the "mile-high city," do not need to be pressurized. What is the best explanation for these observations? Figure 1.3 might be helpful in answering this question.

36. In these diagrams, the larger circles represent one kind of atom and the smaller circles represent a different kind. Characterize each of the samples as an element, compound, or mixture and give reasons for your answers.

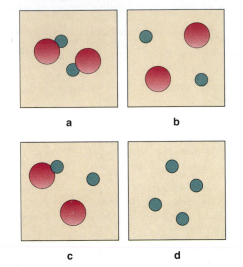

a   b

c   d

37. Consider this representation of the reaction between nitrogen and hydrogen to form ammonia.

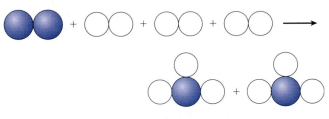

Use it to comment on these questions.

   a. Are the mass of reactants and products the same?

   b. Are the number of molecules of reactants and products the same?

   c. Are the number of atoms of reactants and products the same?

38. Consider the information in Figure 1.7.

   a. Control measures have not uniformly decreased all air pollutants during the period 1970 to 1997. Rank the pollutants from the one that shows the *greatest* percent reduction to the one having the *smallest* percentage reduction during this period.

   b. What factors help explain the differences in the percent reduction?

39. Consider Figure 1.9. Is there a direct correlation between the annual variation in lead concentrations in human blood and the lead compounds in gasoline in selected U.S. cities? Explain your reasoning.

40. Young adults in Beijing, China are heading to bars after work, not for glasses of beer or wine, but for fresh air. These "oxygen bars" provide one half-hour of deep breathing for the equivalent of $6.

a. What does this tell you about air pollution in Beijing?

b. Consider the information in Table 1.9. If you wanted to set up "oxygen bars" in other cities of the world, which would be your first likely markets?

41. Air quality in Santiago, Chile has become such a major problem that driving private cars has been severely restricted. Special decals indicate which days a particular car can be driven. Some citizens have purchased a second car and have obtained a decal for that car. However, the increase in the total number of cars may make the pollution problem even worse. Write a letter to a friend in Santiago suggesting a possible solution to this problem and defending your suggestion.

42. The concentration of formaldehyde in *outdoor* air is typically about 0.01 ppm in urban areas, unless conditions are right for smog formation. The level of formaldehyde *indoors* can average 0.1 ppm, which is the level at which most people can smell its pungent odor. What factors can lead to formaldehyde accumulation indoors?

Exploring Extensions

43. The percentage of oxygen gas in the atmosphere (21%) is usually expressed as the volume of oxygen gas relative to the total volume of the atmosphere being considered. The percentage can also be reported as the mass of oxygen gas relative to the total mass of the atmosphere being considered; in this case it is 23%. Offer a possible explanation for why these two values are not the same.

44. 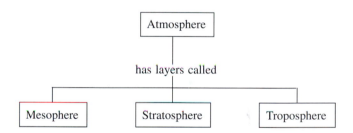 The EPA oversees the Presidential Green Chemistry Challenge awards. Use the EPA web site to find when the program started and to find the list of the most recent winners of the Presidential Green Chemistry Challenge awards. Pick one winner and summarize in your own words the green chemistry advance that merited the award.

45. Recreational scuba divers usually use compressed air that has the same composition as normal air. A new mixture being used is called Nitrox®. What is its composition and why is it being used?

46. Here are some data from the U.S. EPA, Office of Air Quality Planning and Standards. Data indicate the number of days metropolitan statistical areas failed to meet acceptable air-quality standards (Pollutant Standards Index rating over 100).

Air Quality of Selected U.S. Metropolitan Areas, 1986–94

Metropolitan statistical area	1986	1988	1990	1992	1994
Boston, MA	2	15	1	1	1
Denver, CO	49	19	9	7	2
Houston, TX	55	61	61	31	29

a. Do these values show the same types of trends shown in question 30?

b. What factors influence these values? Offer some reasonable explanations, based on your research or knowledge about these cities.

47. An article in *USA Today* on January 12, 1999, is called "Taking technology from here to the infinitesimal." By the year 2020, the article predicts that ". . . the age of atomic engineering . . . , called nanotechnology, will dawn." What does this

term imply? What kinds of applications will be possible that are not now part of our technology?

48. Consider this graph showing the dependence of hydrocarbon and ozone concentrations with time for a major metropolitan area.

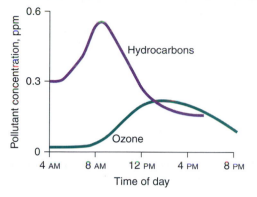

a. Interpret the two curves, explaining what they imply about air pollution in an urban area.

b. Where do you think the curve for NO would fit on this graph?

c. What type of health effects would be felt in this major metropolitan area?

49. A concept web or concept map is a convenient way to represent knowledge and connection among ideas. Concept webs are constructed by joining a word or expression to another one by means of linking words. For example, the atmosphere has three layers, the mesosphere, stratosphere, and troposphere.

```
                    ┌────────────┐
                    │ Atmosphere │
                    └────────────┘
                          │
                   has layers called
                          │
        ┌─────────────────┼─────────────────┐
  ┌───────────┐    ┌──────────────┐    ┌──────────────┐
  │ Mesophere │    │ Stratosphere │    │ Troposphere  │
  └───────────┘    └──────────────┘    └──────────────┘
```

What advantages or disadvantages does this representation have compared with Figure 1.2? Explain your reasoning.

50. Consider these statements given in the text or found in Figure 1.10.

- There was a 51% reduction in the release of toxics by chemical industries from 1987 to 1996.

- 785 million pounds of toxics were released by all industries in 1996.

- Half of all toxics released by all industries in 1996 went into the air.

- Seven compounds make up 65% of all toxics released by chemical industries.

A newspaper used these facts to write an article about the effects of the Toxics Release Inventory. The headline reads: "Success of TRI for Chemical Industries." The opening of the story reads: "Releases of Seven Compounds Released by Chemical Industries to the Air Down From 500 Million Pounds in 1987 to 255 Million Pounds in 1996." Comment on the accuracy of both the headline and the opening to the story. Explain your reasoning.

Protecting the Ozone Layer

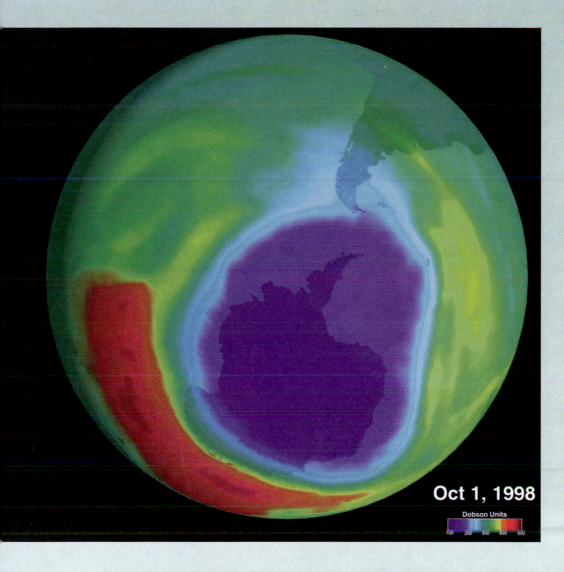

Oct 1, 1998

Dobson Units

A view of the 1998 Antarctic stratospheric ozone hole (inside the purple color) taken from a National Oceanic and Atmospheric Administration (NOAA) satellite. The 1998 hole is the largest one yet, extending over about 26 million square kilometers, an area larger than North America.

"Orbiting above the Earth, an astronaut can look down on our home and see the thin blue ribbon that rims our planet. That transparent blanket— our atmosphere— makes life possible. It provides the air we breathe and regulates our global temperature. And it contains a special ingredient called ozone that filters deadly solar radiation. Life as we know it is possible because of the protection afforded by the ozone layer. Gradually it has become clear to scientists and to

> governments alike that human activities
> are threatening our ozone shield. Behind
> this environmental problem lies a tale of
> twin challenges: the scientific quest to
> understand our ozone shield and the
> debate among governments about how
> best to protect it..."
>
> Daniel Albritton *et al*

So begins *Our Ozone Shield,* a 1992 Report to the Nation On Our Changing Planet, a publication of the University Corporation for Atmospheric Research in conjunction with the National Oceanic and Atmospheric Administration. Written nearly a decade ago, the challenges raised in the opening statement nevertheless are still with us. Essential to life on Earth, ozone in the stratosphere remains a subject of close scrutiny by scientists, politicians, and diplomats alike. Although its concentration can vary by season, ozone in the stratosphere plays a vital role in protecting the Earth's surface and those who live there from damaging solar radiation. Where you live also plays a role in the amount of stratospheric ozone overhead and how well it provides its protective effects.

2.1 Consider This: Ozone Levels Above Your House

What are the current ozone levels in the stratosphere? The National Aeronautics and Space Administration (NASA) can provide you with values. In fact, if a satellite is sending back data as you read this, you may be able to get today's ozone level in the stratosphere right above where you live.

a. Where do you live? Determine your latitude and longitude from a map, from a nearby airport, or by using the link provided at the *Chemistry in Context* web site.

b. Use the NASA link at the *Chemistry in Context* web site to access satellite data. Enter your latitude and longitude from part **a** to find the total column ozone amount at your location for today. Request an earlier date if today's is not available. As you will see later in the chapter, 320 Dobson units is the average ozone level over the northern U.S. How does your value compare with the average?

c. Again using the NASA link, obtain ozone values for some other parts of the world.

CFCs are no longer produced in the U.S. They are replaced by a class of compounds known as HCFCs such as HCFC-22.

There is now ample experimental evidence that the concentration of ozone in the stratosphere has diminished beyond that expected seasonally from natural causes. A class of compounds called **chlorofluorocarbons (CFCs)**, used in air conditioners, refrigerators, and other applications, has been implicated in the ozone reduction. The story of the stratospheric ozone layer and the role of these chemicals in its depletion is a superb example of chemistry in context. As you will discover in the pages that follow, the "chemistry" includes topics such as atomic and molecular structure, chemical reactions, bond breaking and bond making, the interaction of radiation and matter, the effect of radiant energy on living things, and the laboratory search for CFC substitutes. The "context" is created by the global environmental impact of stratospheric ozone depletion, the widespread use of chlorofluorocarbons, and the economic and political issues associated with achieving international agreement on the phaseout of compounds that attack stratospheric ozone.

The 1995 Nobel Prize in chemistry was awarded to Dr. F. Sherwood Rowland, Dr. Mario Molina, and Dr. Paul Cruzen for their ground-breaking research on ozone depletion. A *New York Times* editorial called the award "A Nobel Prize With Political Punch" in reference to the fact that Rowland, Molina, and Cruzen had to win over international scientists and government officials who were skeptical of the award-winning research that affirmed the stratospheric ozone problem as real. Yet unanswered questions still remain: Just how serious is the ozone depletion that has already occurred? How dangerous is the associated increase in exposure to ultraviolet radiation? What can be done to halt or correct the problem? Will the proposed measures work, and how much will they cost?

Chapter Overview

In this chapter, we attempt to address the questions raised in the introduction by considering both scientific and societal issues. We begin by investigating the properties of ozone, and we soon find that an understanding of how it acts in the stratosphere to filter the Sun's harmful radiation requires some knowledge of its molecular structure and the nature of light. This leads to the next section, which describes some fundamental properties of atomic structure. These ideas are then used to predict the molecular structures of a number of substances, including ozone. A discussion of sunlight in particular and radiation in general follows. You will find that light is strangely schizophrenic—it can behave both like waves and like little particles of energy. The particulate properties are especially useful in describing how ozone and oxygen molecules absorb ultraviolet radiation and how radiation damages biological materials. The formation and fate of ozone and its distribution in the atmosphere are considered and we then turn to an analysis of the various mechanisms for the depletion of stratospheric ozone, some involving naturally occurring chemicals and others involving synthetic compounds. Among the latter are the chlorofluorocarbons, whose properties and uses are discussed at length. The following section describes the stratospheric interaction of chlorofluorocarbons and ozone. After a discussion of the Antarctic ozone hole, the chapter concludes with a consideration of the social and technical problems associated with reducing chlorofluorocarbon emissions and developing substitutes.

2.1 Ozone: What Is It?

The central substance in this chapter is ozone, an atmospheric gas. If you have ever been near a sparking electric motor or an arc welding machine, or in a severe lightning storm, you have probably smelled ozone. Its odor is unmistakable, but hard to describe. One can smell concentrations as low as 10 parts per billion (ppb)—10 molecules out of one billion. Appropriately enough, the name "ozone" comes from a Greek word meaning "to smell."

Ozone is oxygen that has undergone rearrangement from the normal diatomic molecule, O_2, to a triatomic form, O_3. A simple chemical equation summarizes the reaction:

$$\text{Energy} + 3\,O_2 \longrightarrow 2\,O_3 \tag{2.1}$$

We have inserted a reminder that energy must be absorbed for this reaction to occur, which accounts for the fact that ozone forms when oxygen is subjected to electrical discharge, whether from an electric spark or from lightning.

Ozone is called an **allotrope** or **allotropic form** of oxygen. Allotropes are two or more forms of the same element that differ in their molecular or crystal structure, and hence in their properties. The familiar allotropes of carbon—diamond and graphite— have different crystal structures, as do fullerenes (buckyballs), the recently discovered, but less common, carbon allotrope (see Chapter 9). Diatomic oxygen, O_2, and triatomic ozone, O_3, obviously differ in molecular structure. This variance is responsible for slight

The structures of diamond and graphite are given in Figure 9.3.

differences in the physical and chemical properties of the two allotropes. For example, ordinary oxygen (O_2) is odorless. It condenses and changes from a colorless gas to a light blue liquid at $-183°C$ and a pressure of one atmosphere. Ozone is more easily liquefied, changing its physical state from gas to a dark blue liquid at $-112°C$. Because ozone is chemically more reactive than oxygen, O_3 is used in the purification of water and the bleaching of paper pulp and fabrics. At one time it was even advocated as a deodorant for air in crowded interiors.

In the troposphere, the region of the atmosphere in which we live, ozone forms in photochemical smog and other kinds of air pollution. But what is detrimental in one region of the atmosphere may be essential in another. In the stratosphere, at an altitude of 20 to 30 km where its concentration is the highest, ozone performs most of its filtering function on ultraviolet light from the Sun. That process involves the interaction of matter and radiant energy, and to understand it requires knowledge about both of these fundamental topics. We turn first to a submicroscopic view of matter.

2.2 Atomic Structure and Elementary Periodicity

The chemical and physical properties of oxygen and ozone and the interaction of these allotropes with sunlight are intimately related to the structure of the O_2 and O_3 molecules. Before we can speak about molecular structure, we must consider the atoms from which molecules are formed. You will recall from Chapter 1 or from your previous study that each element consists of its own distinctive, characteristic atoms. During the twentieth century, chemists and other scientists made great progress in discovering details about the structure of atoms and the particles that make them up. The physicists have been almost too successful; they have found more than 200 subatomic particles. Fortunately, most chemistry can be explained with only three.

We now know that every atom has at its center a minuscule **nucleus.** This nucleus is composed of particles called **protons** and **neutrons.** Protons are positively charged and neutrons are electrically neutral, but both have almost exactly the same mass. Indeed, the protons and neutrons in the nucleus account for almost all of an atom's mass. Well beyond the nucleus are the **electrons** that define the outer boundary of the atom. An electron has a mass equal to only about 1/2000th the mass of a proton or neutron. Moreover, an electron has a negative electrical charge that is equal in magnitude to that of a proton but opposite in sign. The charge and mass properties of these particles are summarized in Table 2.1.

In any electrically neutral atom, the number of electrons equals the number of protons. This number of protons is called the **atomic number.** The atomic number is very important because it determines the elementary identity of the atom. Each element has its own characteristic atomic number. For example, the simplest atom is hydrogen, and each hydrogen atom contains one proton, and thus has an atomic number of 1. Helium (He) has an atomic number of 2, hence each atom of this element contains two protons. With each successive element, the atomic number increases, right up through element 112, whose atoms contain 112 protons.

The atomic number also indicates the number of electrons in a *neutral* atom. To be electrically neutral, an atom must have the same number of protons (positive charge) as electrons (negative charge). Correspondingly, a neutral hydrogen atom contains 1 proton and 1 electron; a neutral helium atom has 2 protons and 2 electrons.

Table 2.1 **Properties of Subatomic Particles**

	Relative Mass	Relative Charge
Proton	1	+1
Neutron	1	0
Electron	1/1838	−1

2.2 Your Turn

Using the periodic table as a guide, specify the number of protons and electrons in a neutral atom of each of these elements.

a. aluminum (Al) **b.** bromine (Br)
c. gold (Au)   **d.** chromium (Cr)

Ans. **a.** 13 protons, 13 electrons   **b.** 35 protons, 35 electrons

We wish that we could include a drawing of a typical atom. However, atoms defy such representation, and any of the depictions in textbooks are at best great oversimplifications. Electrons are often pictured as moving in orbits about the nucleus, but reality is a good deal more complicated and abstract. For one thing, the relative size of the nucleus and the atom create serious problems for the illustrator. If the nucleus of a hydrogen atom were the size of a period on this page, the atom's single electron would most likely be found at a distance of about 10 feet from that period. An atom is thus mostly empty space. Moreover, electrons do not really follow specific circular orbits. In spite of what one reads, an atom is really not very much like a miniature solar system. Rather, the distribution of electrons in an atom is best represented by probability and statistics. The nucleus is surrounded by sort of a fuzzy cloud in which one is more or less likely to find the electrons at various points.

If this sounds rather vague to you, you are not alone. Common sense and our experience of ordinary things are not particularly helpful in our efforts to visualize the interior of an atom. Instead, we are forced to resort to mathematics and metaphors. The mathematics required (a field called quantum mechanics) can be formidable. Chemistry majors do not normally encounter it until rather late in their undergraduate study. We cannot fully share with you the strange beauties of the peculiar quantum world of the atom, although we can provide some useful generalizations.

In the periodic table (see Figure 1.4), the elements are arranged in order of increasing atomic number. The periodic table also organizes elements so that those with similar chemical and physical properties fall in the same columns (groups). Hence, the properties of the elements vary in a regular way with increasing atomic number, repeating themselves periodically. Thus, lithium (Li, atomic number 3), sodium (Na, 11), potassium (K, 19), rubidium (Rb, 37), and cesium (Cs, 55) must share something besides their behavior as highly reactive metals. What they also have in common is the fundamental feature that accounts for these similar properties.

Today we know that the periodicity of properties is chiefly the consequence of the number and distribution of electrons in the atoms of the elements. Because the atomic number represents the number of protons in each atom (or electrons in a neutral atom) of each particular element, properties vary with atomic number. And when properties repeat themselves, it signals a repeat in electronic arrangement.

It can be demonstrated by experiment and calculation that the electrons are arranged in levels (sometimes called shells) about the nucleus. The electrons in the innermost level are the most strongly attracted by the positively charged nucleus. The greater the distance between an electron and the nucleus, the weaker the attraction between them. We say that the more distant electron is in a higher energy level, which means that the electron itself possesses more energy.

An important feature of these energy levels is the fact that they have maximum electron capacities and are particularly stable when they are fully occupied. The innermost level, corresponding to the lowest energy, can hold only two electrons. The second level has a maximum capacity of eight, and the higher levels are also particularly stable when they contain eight electrons.

Because it is difficult to picture an atom accurately, Table 2.2 is used to represent the arrangement of electrons in the atoms of the first 18 elements. The total number of electrons in each atom (the atomic number) is printed in blue; the number of outer electrons is printed in red. The number of electrons in fully filled inner or lower energy levels is printed in parentheses. The number of outer electrons is particularly important because

Table 2.2   Electronic Arrangements in Atoms of the First 18 Elements

Group 1A	2A	3A	4A	5A	6A	7A	Noble Gases 8A
1							2
H							He
1							2
3	4	5	6	7	8	9	10
Li	Be	B	C	N	O	F	Ne
(2) + 1	2	3	4	5	6	7	8
11	12	13	14	15	16	17	18
Na	Mg	Al	Si	P	S	Cl	Ar
(2) + (8) + 1	2	3	4	5	6	7	8

Number of electrons in atom (atomic number)
Number of outer electrons
( ) indicates fully filled electron energy levels
Note that the family designation, as it appears above and in some periodic tables, corresponds to the number of outer electrons.

these electrons account for many of the chemical and physical properties of the corresponding elements.

Note, for example, that in spite of the fact that they have different total numbers of electrons, lithium and sodium atoms both have one outer electron per atom. This common features explains much of the chemistry that these two alkali metals have in common. This fact places them in Group 1A of the periodic table (the 1 indicates one outer electron). Moreover, we would be correct in assuming that potassium, rubidium, and the other elements in column 1A of the periodic table also have a single outer electron in each of their atoms. They are all soft metals that react readily with oxygen, water, and a wide range of other chemicals. In fact, chemical reactivity and the bonding that holds atoms together to form molecules and crystals are largely consequences of the number of outer electrons in any element.

The periodic table is a useful guide to electron arrangement in the various elements. In the elementary families or groups marked "A" the number that heads the column indicates the number of outer electrons in each atom. You have already seen that Group 1A elements are characterized by one outer electron. Similarly, the atoms of the Group 2A elements (the "alkaline earths") all have two outer electrons. Seven outer electrons characterize the atoms of the "halogens" that make up Group 7A—fluorine (F), chlorine (Cl), bromine (Br), and iodine (I). The situation gets a bit more complicated with the B families, but the next two exercises provide some practice with elements in the A series.

Lithium

Sodium

Potassium

Rubidium

Group 1A elements.

2.3 Your Turn

Predict the number of outer electrons in neutral atoms of each of these elements.

a. oxygen (O)   **b.** silicon (Si)
c. nitrogen (N)   **d.** calcium (Ca)

Ans. **a.** 6 (Group 6)   **b.** 4 (Group 4)

2.4 Your Turn

What feature of atomic structure is shared by oxygen (O), sulfur (S), selenium (Se), and tellurium (Te)?

In addition to electrons and protons, essentially all atoms also contain neutrons. The exception is ordinary hydrogen, which consists of only one electron and one proton. But even in pure hydrogen, one atom out of 6700 also has a neutron in its nucleus. Recall our earlier statement that most of the mass of any atom is associated with its nucleus. Because both the proton and the neutron have relative masses of almost exactly 1, the relative mass of an atom of this "heavy hydrogen" very nearly equals 2. This form of hydrogen is also called deuterium. It is an example of a naturally occurring **isotope** of hydrogen. Isotopes are two or more forms of the same element (same number of protons) whose atoms differ in number of neutrons and hence in mass.

Isotopes are identified by their **mass numbers**—the sum of the number of protons and the number of neutrons in an atom. Thus, the mass number of ordinary hydrogen is 1, reflecting the fact that the nucleus contains only one proton. On the other hand, the nucleus of an atom of deuterium contains one proton plus one neutron and is therefore assigned a mass number of 2. In identifying isotopes, the mass number follows the name or symbol of the element. Thus, deuterium is designated as hydrogen-2, H-2; it can also be represented as ^2H where the mass number is indicated by a superscript. There is also a third isotope of hydrogen, called tritium, whose atoms consist of two neutrons in addition to the one proton and one electron characteristic of all hydrogen atoms. Tritium, a radioactive isotope that is rare in nature, thus has a mass number of 3 (1 proton + 2 neutrons). It is represented as hydrogen-3, H-3, or ^3H. Although the concept of atomic mass is an important one, we do not require it at this time. Following our general rule of introducing information only as needed, we will defer a discussion of atomic masses to Chapter 3.

Hydrogen isotopes	Protons	Neutrons
H-1, ^1_1H	1	0
H-2, ^2_1H	1	1
H-3, ^3_1H	1	2

Atomic masses are discussed in Section 3.8.

2.5 Your Turn

Specify the number of protons, electrons, and neutrons in a neutral atom of these isotopes.

a. carbon-14 (C-14; ^{14}C)   **b.** gold-198 (Au-198; ^{198}Au)
c. copper-64 (Cu-64; ^{64}Cu)   **d.** uranium-235 (U-235; ^{235}U)

Ans. **a.** 6 protons, 6 electrons, 8 neutrons   **b.** 79 protons, 79 electrons, 119 neutrons

2.3 Molecules and Models

After this excursion into the atom, we come to our primary motivation for studying atoms, which is molecular structure. The stability of filled electron shells can be invoked to explain why atoms bond to each other to form molecules. The simplest case is H_2, a diatomic molecule we encountered in Chapter 1. A hydrogen atom has only one electron, but if two hydrogen atoms come together, the two electrons become common property. Each atom effectively has a share in both electrons. The resulting H_2 molecule has a lower energy than two individual H atoms, and consequently the molecule with its bonded atoms is more stable than the separate atoms. The two shared electrons constitute what is called a **covalent bond**. Appropriately, the name "covalent" implies "shared strength."

If we represent each atom by its symbol and each electron by a dot, the two individual hydrogen atoms might look something like this:

H· and ·H

Bringing the two atoms together yields a molecule that can be represented thus:

H : H

This is called a dot or **Lewis structure,** after Gilbert Newton Lewis (1875–1946), an American chemist who pioneered its use. Lewis structures can be predicted for any molecule by following a few simple steps. We will first illustrate the procedure with hydrogen fluoride, HF, a very reactive compound used to etch glass.

1. **Starting with the chemical formula of the compound, note the number of outer electrons contributed by each of the atoms.** (Remember that the periodic table is a useful guide.)

$$1 \text{ H} \cdot \text{ atom} \times 1 \text{ outer electron per atom} = 1 \text{ outer electron}$$
$$1 \cdot \ddot{\underset{..}{\text{F}}} : \text{ atom} \times 7 \text{ outer electrons per atom} = 7 \text{ outer electrons}$$

2. **Add the outer electrons contributed by the individual atoms to obtain the total number of outer electrons available.**

$$1 + 7 = 8 \text{ outer electrons}$$

3. **Arrange the outer electrons in pairs. Then distribute them in such a way as to maximize stability by giving each atom a share in enough electrons to fully fill its outer shell—2 electrons in the case of hydrogen, 8 electrons for most other atoms.**

$$\text{H} : \ddot{\underset{..}{\text{F}}} :$$

We have surrounded the F with eight dots, organized into four pairs. The pair of dots between the H and the F represent the electron pair that forms the bond uniting the hydrogen and fluorine atoms. The other three pairs of dots are the three pairs of electrons that are not shared with other atoms and hence not involved in bonding. As such, they are called "nonbonding" electrons or "lone pairs."

When only one pair of electrons is involved in a covalent bone (as it is in HF) the linkage is called a **single bond.** Single covalent bonds are often indicated by a horizontal line connecting the symbols for the two atoms.

$$\text{H} - \ddot{\underset{..}{\text{F}}} :$$

Remember that the single line represents one pair of shared electrons. These two electrons plus the six electrons in the three nonbonding pairs mean that the fluorine atom is associated with a total of eight outer electrons.

The fact that in many molecules electrons are arranged so that every atom (except hydrogen) shares in eight electrons is called the **octet rule.** This generalization is a useful guide for predicting Lewis structures and the formulas of compounds. Consider the Cl_2 molecule, the diatomic form of elemental chlorine. From the periodic table, we can see that chlorine, like fluorine, is in Group 7A, which means that its atoms each have seven outer electrons. Using the scheme given for HF earlier, we first count and add up the outer electrons for Cl_2.

$$2 \cdot \ddot{\underset{..}{\text{Cl}}} : \text{ atoms} \times 7 \text{ outer electrons per Cl atom} = 14 \text{ outer electrons}$$

For Cl_2 to exist, there must be a bond between the two atoms, which we show by a single line designating a shared electron pair, a single covalent bond. The remaining twelve electrons constitute six nonbonding pairs, distributed in such a way as to give each chlorine atom eight electrons (two bonding and six nonbonding). This meets the octet rule. Accordingly, the Lewis structure for Cl_2 is

$$: \ddot{\underset{..}{\text{Cl}}} - \ddot{\underset{..}{\text{Cl}}} :$$

In the compound ICl, each atom has seven outer electrons (I and Cl are Group 7A elements), like each chlorine has in Cl_2. Reasoning by analogy, the Lewis structure for ICl must be like that of Cl_2, with each atom having a bonding pair and three nonbonding pairs to achieve an overall octet.

$$: \ddot{\underset{..}{\text{I}}} - \ddot{\underset{..}{\text{Cl}}} :$$

Sometimes nonbonding electrons are not shown in a Lewis structure.

2.6 Your Turn

Use the procedure just outlined to draw the Lewis structures for each of these substances; both species obey the octet rule.

a. HBr   **b.** I_2

Ans. **a.**

H· 1 H atom × 1 outer electron per atom = 1 outer electron

·B̈r: 1 Br atom × 7 outer electrons per atom = 7 outer electrons

The total number of electrons is 8. The Lewis structures are

H:B̈r: or   H—B̈r:

So far we have dealt only with molecules having just two atoms. But there are many compounds whose molecules contain more than two atoms. The octet rule is a generalization that applies to many of these compounds as well. Another generalization is that in most molecules where there is only one atom of one element bonded to two or more atoms of another element (or elements), the single atom goes in the center of the Lewis structure. There are exceptions to these generalizations, but this is a good place to begin to apply the generalizations. We start with a water molecule, H_2O, as an example.

Following the scheme described above, we first count and add up the outer electrons.

2 H· atoms × 1 outer electron per H atom = 2 outer electrons

1 ·Ö· atom × 6 outer electrons per O atom = 6 outer electrons

Total = 8 outer electrons

We place the O representing the oxygen atom in the center and distribute the eight electrons (dots) around the O, in conformity with the octet rule. Each of the hydrogen atoms is bonded to the oxygen atom with a pair of electrons. The remaining four electrons are also placed on the oxygen, but as two nonbonding pairs. The result is:

H : Ö : H

Each hydrogen forms only one bond (two shared electrons). Oxygen, because it can form two bonds, is the central atom.

A quick count confirms that the O is surrounded by eight dots, representing the eight electrons predicted by the octet rule. Alternatively, we could symbolize the water molecule with lines for the single bonds.

H—O̤—H

These Lewis representations provide more information than the chemical formula, H_2O, does because they indicate how the atoms are connected to each other. On the other hand, Lewis structures do not directly reveal the shape of a molecule. From the structure given here, it might appear that the atoms of the water molecule all fall in a straight line. In fact, the molecule is bent. It looks something like this:

```
      ..
      O           O
     / \   or    / \
    H   H       H   H
```

Section 3.4 in Chapter 3 will describe how the Lewis structure can lead to the prediction of this bent structure.

Another example of a polyatomic molecule is methane, CH_4, a gas found in trace amounts in the atmosphere, as we will see in Chapter 3. Using the rules and generalizations given earlier, and recognizing that carbon is in Group 4A, we can write the Lewis structure of methane.

4 H· atoms × 1 outer electron per H atom = 4 outer electrons

1 ·C· atom × 4 outer electrons per C atom = 4 outer electrons

Total = 8 outer electrons

The C representing a carbon atom goes in the center and is surrounded by the eight electrons, so as to give carbon an octet of electrons. Each of the four hydrogen atoms use two of the electrons to form a shared pair with carbon, for a total of four single covalent bonds. This gives us the Lewis structure of methane.

The geometry of the methane molecule will be described in Section 3.4

$$H : \overset{..}{\underset{..}{C}} : H \quad \text{or} \quad H-\overset{\overset{\displaystyle H}{|}}{\underset{\underset{\displaystyle H}{|}}{C}}-H$$

Checking the methane structure we see that the carbon atom has a share in eight electrons, complying with the octet rule; each hydrogen atom is satisfied by having a share in a pair of electrons (remember that H can only accommodate a pair of electrons).

2.7 Your Turn

Use the procedure just outlined to draw the Lewis structures for each of these compounds. Both species obey the octet rule.

a. hydrogen sulfide (H_2S)   **b.** dichlorodifluoromethane (CCl_2F_2)

Ans.   a.

H ·   2 H atoms × 1 outer electron per atom = 2 outer electrons

· $\overset{..}{\underset{..}{S}}$ ·   1 S atom × 6 outer electrons per atom = 6 outer electrons

The total number of electrons is 8. The Lewis structures are

$$H : \overset{..}{\underset{..}{S}} : H \quad \text{or} \quad H-\overset{..}{\underset{..}{S}}-H$$

The octet rule often leads to the correct conclusions, but not always. It turns out that the O_2 molecule is a case where the rule is slightly misleading. Here we have 12 outer electrons to distribute, six from each of the Group 6A oxygen atoms. There are not enough electrons to give each of the atoms a share in eight electrons if only one pair is held in common. However, the octet rule can be satisfied if the two atoms share four electrons (two pairs). A covalent bond consisting of two pairs of shared electrons is called a **double bond** and it is represented by four dots or two lines:

$$\overset{..}{\underset{..}{O}} :: \overset{..}{\underset{..}{O}} \quad \text{or} \quad \overset{..}{\underset{..}{O}} = \overset{..}{\underset{..}{O}}$$

Double bonds are shorter, stronger, and harder to break than single bonds involving the same atoms. The length and strength of the bond in the O_2 molecule corresponds to a double bond. However, oxygen has a peculiar property that is not fully consistent with the Lewis structure drawn above. When liquid oxygen is poured between the poles of a strong magnet, it sticks there like iron filings. Such magnetic behavior implies that the electrons are not as neatly paired as the octet rule would suggest. But this little dis-

Liquid oxygen is attracted to the poles of a magnet.

crepancy is hardly a reason to discard a useful generalization. After all, simple scientific models seldom if ever explain all phenomena, but they can be helpful approximations.

For completeness, we need to note that a **triple bond** is a covalent linkage made up of three pairs of shared electrons. The bond in the nitrogen molecule, N_2, is an example of a triple bond. Each Group 5A nitrogen atom contributes five outer electrons for a total of ten. These ten electrons can be distributed in accordance with the octet rule if six of them (three pairs) are shared between the two atoms, leaving four of them to form two nonbonding pairs, one on each nitrogen.

$$: N ::: N : \quad \text{or} \quad : N \equiv N :$$

The ozone molecule introduces another structural feature. We again start with the octet rule. Each of the three oxygen atoms contributes six outer electrons for a total of 18. These 18 electrons can be arranged in two ways, each of which give each atom a share in eight outer electrons:

$$\overset{..}{O} :: \overset{..}{\underset{..}{O}} : \overset{..}{\underset{..}{O}} : \quad \text{or} \quad : \overset{..}{\underset{..}{O}} : \overset{..}{\underset{..}{O}} :: \overset{..}{O}$$
$$\text{a.} \qquad\qquad\qquad\qquad \text{b.}$$

There are experimental techniques for determining molecular structure, and it turns out that neither of the above versions is exactly correct. Structures *a.* and *b.* predict that the molecule should contain one single bond and one double bond. In *a.* the double bond is to the left of the central atom; in *b.* it is to the right. But experiment reveals that the two bonds in the O_3 molecule are identical in length and strength, being somewhere between a single bond and a double bond. Structures *a.* and *b.* are called **resonance forms.** They represent hypothetical extremes of electron arrangements that do not exist. The actual structure of the ozone molecule is sort of an average or hybrid of the two resonance forms. The phenomenon of **resonance** is represented by a double-headed arrow linking the resonance forms.

$$\overset{..}{O} :: \overset{..}{\underset{..}{O}} : \overset{..}{\underset{..}{O}} : \longleftrightarrow : \overset{..}{\underset{..}{O}} : \overset{..}{\underset{..}{O}} :: \overset{..}{O}$$

This representation and the word "resonance" seem to imply that the electrons are jumping back and forth between the two arrangements, but in fact this does not happen. Resonance is just another useful concept invented by chemists to describe the complex microworld of molecules.

A closer experimental inspection of that microworld reveals that the O_3 molecule is bent, not linear as the simple Lewis structures would suggest. Thus a more accurate picture of the resonance forms is:

An explanation of why the O_3 molecule is bent will have to wait until Chapter 3. At this point we are more concerned about how bonding in O_2 and O_3 influences their interaction with sunlight.

2.8 Your Turn

Use the octet rule to draw the Lewis structures for each of these compounds.

a. carbon monoxide (CO)   **b.** sulfur dioxide (SO_2)

Ans.   a.

$\cdot \overset{..}{C} \cdot$ 1 C atom × 4 outer electrons per atom = 4 outer electrons

$\cdot \overset{..}{O} \cdot$ 1 O atom × 6 outer electrons per atom = 6 outer electrons

The total number of electrons is 10. The Lewis structures are

$$: C ::: O : \quad \text{or} \quad : C \equiv O :$$

Note that the number of electrons, 10, is the same as the number in the N_2 molecule, which also has a triple bond.

2.4 Waves of Light

To better understand how stratospheric ozone screens out much of the Sun's harmful radiation requires knowing the molecular structure of ozone as well as some fundamental properties of light. The interaction of sunlight with matter is important, such as in photosynthesis or the damage high-energy solar radiation can cause in living organisms. Therefore, we turn now to develop an understanding of light.

Every second, five million tons of the Sun's matter are converted into energy, which is radiated into space. The fact that we detect color indicates that the radiation that reaches us is not all identical. Prisms and raindrops break sunlight into a spectrum of colors. Each of these colors can be identified by the numerical value of its wavelength. The word correctly suggests that light behaves rather like a wave in the ocean. The **wavelength** is the distance between successive peaks (Figure 2.1). It is expressed in units of length and symbolized by the Greek letter lambda (λ).

It is both interesting and humbling to realize that out of the vast array of radiant energies, our eyes are sensitive to light only in a very tiny portion of the electromagnetic spectrum—wavelengths between about 700×10^{-9} m (corresponding to red) and 400×10^{-9} m (corresponding to violet). These lengths are very short, so we typically express them in nanometers. One nanometer (nm) is defined as one one-billionth of a meter (m). In symbols,

$$1 \text{ nm} = \frac{1}{1,000,000,000} \text{ m} = \frac{1}{10^9} \text{ m} = 1 \times 10^{-9} \text{ m}$$

We can use this equivalence to convert meters to nanometers; for example, to convert 700×10^{-9} meters to nanometers.

$$\text{wavelength} = \lambda = 700 \times 10^{-9} \text{ m} \times \frac{1 \text{ nm}}{1 \times 10^{-9} \text{ m}} = 700 \text{ nm}$$

The meter units cancel and we are left with nanometers.

2.9 Your Turn

Violet light in the visible part of the spectrum has a wavelength of 408 nm. Express this wavelength in meters.

Ans. 4.08×10^{-7} m

Figure 2.1

Wave motion.

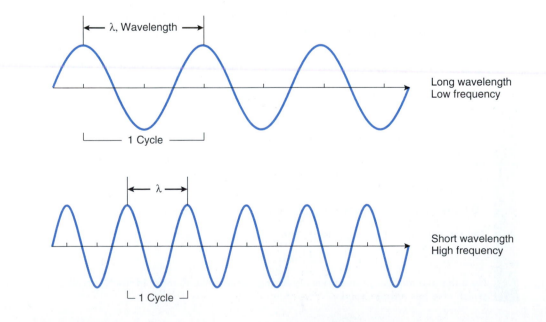

Another way of quantifying color is to express it in terms of frequency. If you were watching waves on the surface of a lake or the ocean, you could measure the distance between successive crests (the wavelength). But you could also determine how often the crests passed your point of observation by counting the number in a particular time interval. That would give you the frequency of the waves. The same idea applies to radiation. The **frequency of light** is the number of waves passing a fixed point in one second. Frequency and wavelength are related; the shorter the wavelength, the higher the frequency, that is, the greater the number of waves that pass the observer in one second. We have symbolized this relationship in the margin. The arrows signify that frequency increases as wavelength decreases; their values change in opposite directions.

As wavelength ↓, frequency ↑.

Instead of reporting frequency as "waves per second," the units are shortened to "per second" and written as 1/s or s^{-1}. This unit is also referred to as hertz (Hz), a term that may be familiar to you from radio station frequencies. A companion unit is a megahertz (mHz), one million hertz, $1 \times 10^6 \ s^{-1}$. Frequency is symbolized by the Greek letter nu (ν).

The relationship between frequency and wavelength that we have just described in words can be summarized in a simple equation in which ν is the frequency and c represents the constant speed at which light and other forms of radiation travel.

$$\text{Frequency} = \nu = \frac{c}{\lambda} \qquad (2.2)$$

In metric units, the speed of light is 3.00×10^8 meters/second or $3.00 \times 10^8 \ m \cdot s^{-1}$. You may be more familiar with the speed of light as 186,000 miles/second. The form of equation 2.2 indicates that wavelength and frequency are *inversely* related: as the value for λ decreases, the value for ν increases, and vice versa. Red light, which has a wavelength of 700 nm or 700×10^{-9} m has a frequency of $4.29 \times 10^{14} \ s^{-1}$.

$$\text{Frequency} = \nu = \frac{c}{\lambda} = \frac{3.00 \times 10^8 \ m \cdot s^{-1}}{700 \times 10^{-9} \ m} = 4.29 \times 10^{14} \ s^{-1}$$

Violet light has a shorter wavelength (400 nm) and hence a higher frequency ($7.50 \times 10^{14} \ s^{-1}$) than red light.

2.10 Your Turn

The violet light mentioned in 2.9 Your Turn has a wavelength of 408 nm. Calculate its frequency.

Ans. $7.35 \times 10^{14} \ s^{-1}$

The light that we can see directly represents only a narrow band in the electromagnetic spectrum. The **electromagnetic spectrum** is the entire range of radiant energy, and it is a very wide range indeed. Figure 2.2 indicates that the spectrum extends in both directions from the visible region. Scientists have devised a variety of detectors that are sensitive to the radiation in various parts of this broad band. As a consequence, we can speak with confidence about the regions of the spectrum that are invisible to our eyes.

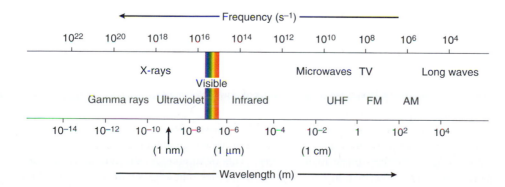

Figure 2.2

The elecromagnetic spectrum.

Figure 2.3

The distribution of energy in
solar radiation above the
Earth's atmosphere.

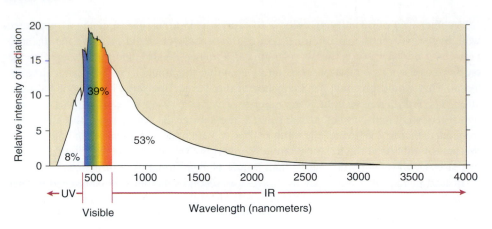

At wavelengths longer than red one first encounters infrared (IR) or heat rays, which we cannot see, but can certainly feel. The microwaves used in radar and to cook food quickly have wavelengths of about 1 centimeter (cm) (1 cm = 1×10^{-2} m). At still longer wavelengths (1 m to 1000 m) are the regions of the spectrum used to transmit your favorite AM and FM radio and television programs.

In this chapter we are most concerned with the **ultraviolet (UV)** region, which lies at wavelengths shorter than those of violet. At still shorter wavelengths are the **X-rays** used in medical diagnosis and the determination of crystal structure and **gamma rays** that are given off in certain radioactive processes. Some gamma rays can have wavelengths as short as 10^{-13} m.

2.11 Your Turn

Arrange these types of radiation in order of *increasing* wavelength.

a. microwaves   **b.** ultraviolet   **c.** visible   **d.** infrared

Ans. ultraviolet, visible, infrared, microwaves

See Section 3.5.

Our local star, the Sun, emits infrared, visible, ultraviolet, and cosmic radiation, but not all with equal intensity. This is evident from Figure 2.3, a plot of the relative intensity of solar radiation as a function of wavelength. The curve represents the spectrum as measured *above* the atmosphere, before there has been opportunity for interaction of radiation with the molecules of the air. The peak indicating the greatest intensity is in the visible region. However, infrared radiation is spread over a much wider wavelength range, with the result that 53% of the total energy emitted by the Sun is radiated to Earth as infrared radiation. This is the major source of heat for the planet. Approximately 39% of the energy comes to us as visible light and only about 8% as ultraviolet. (The areas under the curve give an indication of these percentages.) But in spite of its small percentage, the Sun's UV radiation is potentially the most damaging to living things. To understand why, we must look at radiation in a different light, this time in terms of its energy.

2.5 "Particles" of Energy

The idea that radiation can be described in terms of wave-like character is well established and very useful. However, around the beginning of the twentieth century, scientists found a number of phenomena that seemed to contradict this model. In 1990, a German physicist named Max Planck (1858–1947) argued that the shape of the energy distribution curve pictured in Figure 2.3 could only be explained if the energy of the radiating body were the sum of many energy levels of minute but discrete size. In other words, the energy distribution is not really continuous, but consists of many individual steps. The energy is said to be **quantized.** An often-used analogy is that the quantized energy of a radiating body is like steps on a staircase, which are quantized (no partial steps allowed), not like a ramp, which allows any sized stride. Five years later, in the work that won him his Nobel Prize, Albert Einstein (1879–1955) suggested that radiation itself should be viewed as constituted of individual bundles of energy called **photons.** One can regard these photons as "particles of light," but they are definitely not particles in the usual sense. For example, they have no mass.

The atomization of energy by quantum theory did not displace the utility of the wave model. Both are valid descriptions of radiation. This dual nature of radiant energy seems to defy common sense. How can light be two different things at the same time, both waves and particles? There is no obvious answer to that very reasonable question—that's just the way nature is. The two views are linked in a simple relationship that is one of the most important equations in modern science. It is also an equation that is very relevant to the role of ozone in the atmosphere.

> Planck and Einstein were amateur violinists who played duets together.

$$\text{Energy, } E = h\nu = \frac{hc}{\lambda} \qquad (2.3)$$

Here E represents the energy of a single photon. It is *directly* proportional to ν, the frequency of radiation, and *inversely* proportional to the wavelength, λ. Consequently, as the wavelength of radiation gets shorter, its energy increases; as the energy decreases, the wavelength increases. On the other hand, as the frequency of the radiation increases, so does its energy. These qualitative relationships we represent by arrows in the margin and ask you to use them in the next exercise. The symbol h indicates **Planck's constant,** which has a value of 6.63×10^{-34} joule · second (J · s). Note the important fact that a joule is a unit of energy.

> As wavelength ↓, energy ↑, frequency ↑

2.12 Your Turn

Arrange these colors of the visible spectrum in order of *increasing* energy per photon.

a. green **b.** red **c.** yellow **d.** violet

Ans. red, yellow, green, violet

We can also introduce numbers into equation 2.3 and use them to compare the energy of different photons. For example, the energy of a photon of ultraviolet light with a wavelength of 300 nm and a frequency of 1.00×10^{15} s^{-1} can be shown to have an energy of 6.63×10^{-19} joule (J), which is a very tiny amount of energy. (One joule is approximately equal to the energy required for one beat of a human heart.)

$$E = h\nu = (6.63 \times 10^{-34} \text{ J} \cdot \text{s}) \times (1.00 \times 10^{15} \text{ s}^{-1}) = 6.63 \times 10^{-19} \text{ J}$$

By contrast, the photon of a 100 mHz FM radio signal with a wavelength of 300×10^7 nm has an energy of only 6.63×10^{-26} J. Although these energies are very small, there is a significant difference in the energies of the photon of UV radiation and the photon of the radio signal.

$$\text{UV radiation: } 6.63 \times 10^{-19} \text{ J per photon}$$
$$\text{Radio signal: } 6.63 \times 10^{-26} \text{ J per photon}$$

The energy of a photon of UV radiation is 10^7 or ten million times larger than the energy of a photon emitted by your favorite radio station. Remember that as wavelength decreases (from radio waves to ultraviolet radiation), the energy per photon of radiation increases.

One consequence of this great difference in energy is the fact that you cannot get a tan from listening to the radio—unless you happen to be outside in the sunlight. Whether or not your radio is turned on, you are continuously bombarded by radio waves. Your body cannot detect them, but your radio can. As we have just seen, the energy associated with each of the radio photons is very low—about 7×10^{-26} J. That is not enough energy to result in a local increase in the concentration of the skin pigment, melanin, to cause tanning. That process involves a quantum jump, an electronic transition that requires approximately 7×10^{-19} J, far more than radio wave photons can supply.

Your body cannot store the 10 million low-energy photons of radio frequency that would be necessary to equal the energy required for the tanning reaction. It is an either/or situation: either a photon has enough energy to cause a specific chemical change or it does not. Photons of ultraviolet radiation of 300 nm or shorter do have sufficient energy to bring about the changes that result in tanning, burning, or in some cases, skin cancer.

All of this may seem quite trivial. But it was essentially this line of reasoning, on a different system, that won Einstein his Nobel Prize in 1905. Moreover, this same logic extends to any interaction of radiation and matter.

2.13 Your Turn

Return once more to the violet light of 2.9 and 2.10 Your Turns. Calculate the energy of a photon of this radiation, expressing your answer in joules.

Ans. 4.87×10^{-19} J

2.6 Radiation and Matter

The Sun bombards the Earth with countless photons—indivisible packages of energy. Many of these photons are absorbed by the atmosphere, the surface of the planet, and its living things. Radiation in the infrared region of the spectrum warms the Earth and its oceans, causing molecules to move, rotate, and vibrate. The cells of our retinas are tuned to the wavelengths of visible light. Photons of the proper wavelength are absorbed and the energy is used to "excite" electrons in biological molecules. The electrons jump to higher energy levels, triggering a series of complex chemical reactions that ultimately lead to sight. Green plants capture photons in an even narrower region of the visible spectrum (corresponding to red light) and use the energy to convert carbon dioxide and water into food, fuel, and oxygen in the process of photosynthesis.

As the wavelength of light decreases, the energy carried by each photon increases. Consequently, the interaction of shorter wavelength radiation and matter becomes more violent. Photons in the UV region of the spectrum are sufficiently energetic to eject electrons from atoms and molecules, converting them into positively charged species. Even shorter UV wavelength photons break bonds causing molecules to come apart. In living things, such changes disrupt cells, and create the potential for genetic defects and cancer.

It is part of the fascinating symmetry of nature that this interaction of radiation with matter explains both the damage ultraviolet radiation can cause and the atmospheric mechanism that protects us from it. We turn first to the shield of oxygen and ozone.

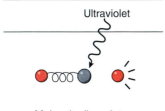

Ultraviolet

Molecule dissociates

Ultraviolet radiation can break chemical bonds.

2.7 The Oxygen/Ozone Screen

The presence of oxygen and ozone in the Earth's stratosphere means that the radiation that reaches the surface of the planet is different from that emitted by the Sun in some important respects. Much of the ultraviolet light is blocked out by these two allotropes of the same element (Figure 2.4).

As we noted in Chapter 1, approximately 21% of the atmosphere consists of ordinary diatomic oxygen. The forms of life that inhabit our planet are absolutely dependent upon the chemical properties of this gas and on its interaction with ultraviolet radiation. The strong covalent bond holding the two oxygen atoms together in the O_2 molecule can be broken by the absorption of a photon of the proper radiant energy. The photon excites a bonding electron to a higher energy level, causing the atoms to come apart. But the bond will be broken and the molecule will dissociate only if the photon has energy cor-

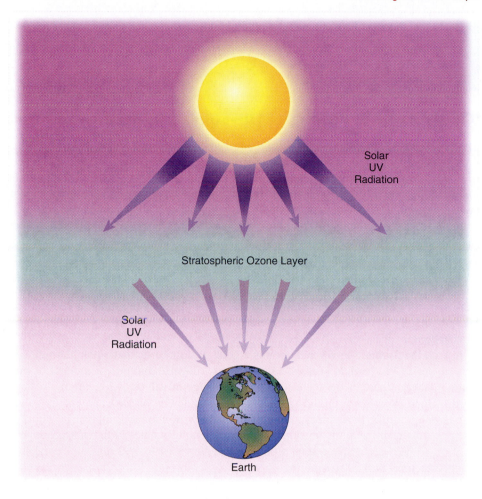

Figure 2.4
Solar UV radiation is greatly
diminished by passing through
the stratospheric ozone layer.

responding to a wavelength of 242 nm or less ($\lambda \leq 242$ nm). This energy is in the ultraviolet region of the spectrum.

$$O_2 + \text{photon} \longrightarrow 2\,O \qquad\qquad (2.4)$$
$$\lambda \leq 242 \text{ nm}$$

Because of this reaction, stratospheric oxygen shields the surface of the Earth from high-energy radiation. As green plants flourished on the young planet, they released oxygen to the atmosphere. The increasing oxygen concentration led to more effective interception of ultraviolet radiation. Consequently, forms of life evolved that were less resistant to UV radiation than they would have been otherwise.

Ordinary diatomic oxygen screens out radiation with wavelengths shorter than 242 nm. However, if O_2 were the only UV absorber in the atmosphere, the surface of the Earth and the creatures that live on it would still be subjected to potentially damaging radiation in the 242–320 nm range. It is here that ozone plays its protective role. The fact that ozone is more reactive than diatomic oxygen suggests that the O_3 molecule is more easily broken apart than O_2. Recall that the atoms in the O_2 molecule are connected with a strong double bond. Each of the bonds in O_3 is somewhere between a single and double bond in length and in strength. This makes the bonds in O_3 weaker than the double bonds in O_2. Therefore, photons of a lower energy (longer wavelength) should be sufficient to separate the atoms in O_3. This is in fact the case: radiation of wavelength 320 nm or less will induce the following reaction.

$$O_3 + \text{photon} \longrightarrow O_2 + O \qquad\qquad (2.5)$$
$$\lambda \leq 320 \text{ nm}$$

Because of this reaction and that represented by equation 2.4, only a relatively small fraction of the Sun's UV radiation reaches the surface of the Earth. However, what does arrive can do damage.

2.8 Biological Effects of Ultraviolet Radiation

The consequences of ultraviolet radiation on plants and animals depend on two factors: the intensity of UV radiation and the sensitivity of organisms to that radiation. The vertical scale of Figure 2.5 indicates the quantity of ultraviolet solar energy (expressed in joules) falling on a surface one square meter in area in one second. The graph shows how this energy varies with wavelength. For any wavelength, the total amount of energy is the product of the number of photons striking the surface and the energy per photon. The rather flat upper curve reveals that the energy input above the atmosphere does not depend significantly upon wavelength. If the Earth's atmosphere did not exist, the surface of the planet and the creatures on it would be subjected to these punishingly high levels of radiant energy. However, the lower curve indicates that the energy reaching the surface of the earth varies markedly with wavelength. It starts dropping at 330 nm and falls off sharply as the wavelength decreases, due to the absorption of UV radiation by stratospheric ozone.

In fact, the decrease in UV radiation is a good deal more dramatic than the figure at first suggests. The vertical scale in Figure 2.5 is a logarithmic one, a method of presenting data that permits the inclusion of a wide range of values. A logarithmic scale is one in which every mark on the axis represents a value (in this case, an energy value) that is one-tenth of that corresponding to the mark immediately above it. Thus, at 320 nm, where ozone starts absorbing, the energy input to the Earth's surface is 1×10^{-1} or 0.1 joules per square meter per second ($J/m^2 \cdot s$). At 300 nm, the energy has dropped to 10^{-4} or 0.0001 $J/m^2 \cdot s$.

Highly energetic photons can excite electrons and break bonds in biological molecules, rearranging them and altering their properties. Solar radiation at wavelengths below 300 nm is almost completely screened out by O_2 and O_3 in the stratosphere. This is most fortunate, because radiation in this region of the spectrum is particularly damaging to living things. This relationship is evident from Figure 2.6, where biological sensitivity is plotted versus wavelength. As defined here, biological sensitivity is based on experiments in which the damage to deoxyribonucleic acid (DNA), the chemical basis

A discussion of DNA appears in Chapter 12.

Figure 2.5

Variation of solar energy with wavelength of UV radiation above and below the atmosphere.

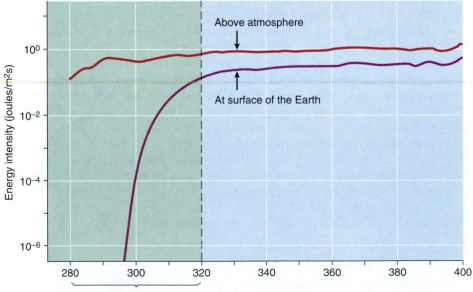

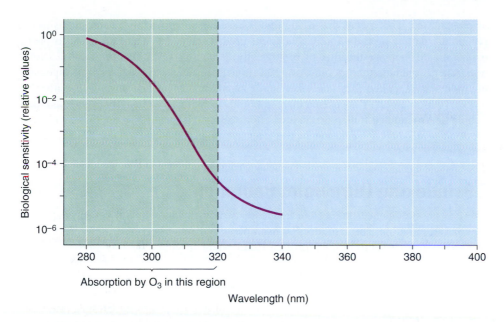

Figure 2.6
Variation of biological sensitivity of DNA with wavelength of UV radiation.

of heredity, is measured at various wavelengths. In the figure, the biological sensitivity is expressed in relative units, once more on a logarithmic scale. Biological sensitivity at 320 nm is about 10^{-5} or 0.00001 unit. But at 280 nm, the sensitivity is 10^0 or 1 unit. This means that radiation at 280 nm is 10^5 or 100,000 times more damaging than radiation at 320 nm. As we have seen, this is because the energy per photon and the potential for biological damage increase as the wavelength decreases.

Wavelength ↓
Energy ↑
DNA damage ↑

2.15 Your Turn

Arrange these types of radiation in order of *decreasing* energy.

a. microwaves   **b.** ultraviolet   **c.** visible   **d.** cosmic rays

Ans. cosmic rays, ultraviolet, visible, microwaves

Figure 2.7 presents evidence that the average ozone concentration in the most heavily populated region of the Earth has dropped by 6% since 1979. Consequently, the ability of the atmosphere to screen out UV radiation with wavelengths below 320 nm has

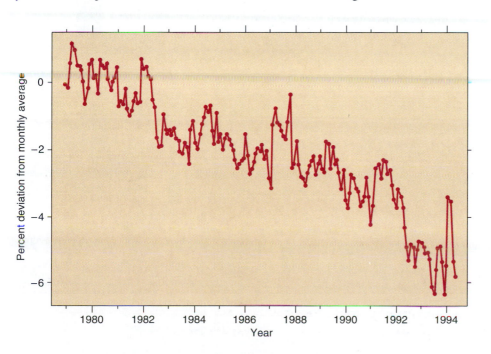

Figure 2.7
Change in stratospheric ozone concentration (60 degrees south to 60 degrees north) expressed as percent deviation from the monthly average (1979 to 1995).

(Source: Data from *Scientific Assessment of Ozone Depletion, 1994. Executive Summary, World Meteorological Organization Global Ozone Research and Monitoring Project,* Report No. 37, Geneva, 1995.)

also decreased. This means that living things have been exposed to greater intensities of potentially damaging radiation. Calculations predict that a 6% decrease in stratospheric ozone should increase the flux of biologically damaging UV radiation by 12%. Some researchers conclude this should also result in a 12% increase in skin cancer, especially the more easily treated form, non-melanoma skin cancer. This condition is considerably more common among Caucasians than among those with more heavily pigmented skin. People of African and Indian origin are better equipped by natural selection to withstand the high levels of UV radiation in the intense sunlight that strikes the Earth near the equator.

2.9 Indexing Ultraviolet Radiation

There is good evidence linking the incidence of non-melanoma skin cancer, the intensity of UV radiation, and latitude. For example, the disease becomes more prevalent as one moves further south in the Northern Hemisphere (Figure 2.8).

2.16 Consider This: Geography of Skin Cancer

Many generations of immigrants have come to the United States. There are many fair-skinned Northern Europeans, for example, who have settled in the area around San Antonio, Tex; there are many other equally fair-skinned immigrants who have settled in the area around Seattle, Wash. Based on Figure 2.8, compare their relative risks of developing skin cancer. Identify several other factors that may affect the risk for any individual in these two populations.

The UV radiation coming from the sun can be categorized by its wavelength as UV-A (320–400 nm), UV-B (280–320 nm), and UV-C (< 280 nm). Although very dangerous because of its very short wavelength (high energy), UV-C thankfully is completely

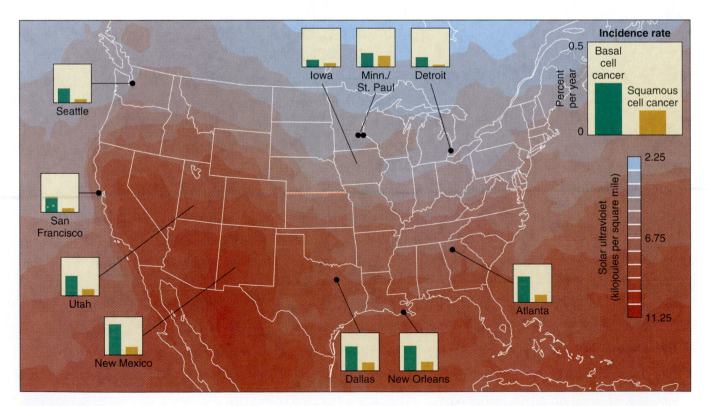

Figure 2.8
Skin cancer risks.

Table 2.3   UV Index

UV Index	Minutes for some damage to occur in light-skinned people who "never tan"	Minutes for some damage to occur in dark-skinned people who "never tan"
0–2 Minimal	30	>120
3–4 Low	15–20	75–90
5–6 Moderate	10–12	50–60
7–9 High	7–8.5	33–40
10–15 Very High	<4–6	20–30

screened by stratospheric ozone and oxygen. The ozone layer protects against most of the UV-B radiation, although some UV-B still reaches the Earth's surface. UV-B is especially effective at damaging DNA and has been linked to weakening the human immune system; eye damage; skin cancers, including non-melanoma basal cell cancer, as well as damage to crops and marine organisms. Recent research has linked non-melanoma skin cancers specifically to mutations of the *p*53 gene in skin cells. UV-B photons are sufficiently energetic to directly alter the DNA structure in those cells, causing the mutations. Unlike UV-B or UV-C, the longer wavelength UV-A is not screened by stratospheric ozone. UV-A is suspected in the premature wrinkling and aging of skin, as well as skin cancer, including melanomas that can be fatal. There is also mounting evidence that tanning beds, which emit UV-B radiation, may increase the risk of skin cancer, even melanomas.

Because of the damage that can be caused by unhealthy exposure to UV radiation, the National Weather Service daily issues an ultraviolet index forecast that appears nationally in newscasts, in newspapers, and on the web. UV Index values range from 0 to 15 based on how long it takes for skin damage to occur (Table 2.3).

2.17 Consider This: UV Index Forecasts

The UV Index indicates the amount of UV radiation reaching the earth's surface at solar noon (which may be closer to 1 P.M., if you are on daylight time). It is generated daily for more than 50 cities in the United States.

a. The UV Index depends on latitude, the day of the year, the time of day, the amount of ozone above the city, the elevation, and the predicted cloud cover. How is the UV index affected by each of these?

b. The UV Index Forecast is available on the web, compliments of a satellite launched by NOAA, the National Oceanographic and Atmospheric Administration. Search either for "Current UV Index Forecast, NOAA" or go to the *Chemistry in Context* web site for a direct link. Account for the range of values that you see on today's map of the United States.

c. Surfaces such as snow, sand, and water all can intensify your exposure to UV radiation, because they can reflect it back at you. What outdoor activities might increase your risk from exposure?

Tanning is a risk/benefit activity. Many sunworshippers consider a golden tan a cosmetic benefit worth taking a risk for. The risk undoubtedly exists, as evidenced by the fact that about one million new cases of skin cancer occur each year in the United States,

Figure 2.9
Increases in melanoma skin
cancer in the United States,
1973–1995.

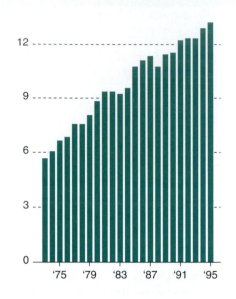

Melanoma Skin Cancer on the Rise
New Cases Diagnosed per 100,000
Population (United States, 1973-1995)

Figure 2.9
Increases in melanoma skin
cancer in the United States,
1973–1995.

*Sunscreen lotions contain
varying amounts of compounds
that absorb some UV radiation.*

almost as many as the total number of cases of all other cancers. Skin cancers can develop many years after repeated, excessive exposure has stopped. Fair-haired, fair-skinned individuals are at the highest risk to develop skin cancer from UV-B radiation; fair-skinned whites in Australia have the highest skin cancer rates in the world. But it is also important to note that all skin types face greater risk with increased exposure to the sun (Figure 2.9).

Some protection against harmful UV rays is offered by sunscreens, products that contain compounds that absorb UV-B to some extent; some sunscreens also contain compounds that absorb UV-A. The American Academy of Dermatology recommends a sunscreen with a skin protection factor (SPF) of 15 to 30. But wearing a sunscreen does not mean that you are without risk from the Sun's UV rays. Because sunscreens let you be exposed for a longer time without burning, they may ultimately cause greater UV damage. "Sunscreen alone does not prevent melanoma" according to epidemiologist Dr. Marianne Berwick of the Sloan-Kettering Cancer Center (*Newsweek,* March 2, 1998, p. 61).

The potential danger of UV-B exposure is not just a summer phenomenon. Because snow reflects UV-B radiation, harm can also occur in winter, especially at high altitudes, creating a condition called snow blindness, which can lead to eye damage. Cataracts, a clouding of the lens of the eye, can be caused by excessive exposure to UV-B radiation. It has been estimated that a 10% decrease in the ozone layer could create up to two million new cataract cases globally.

2.18 Consider This: Sunscreens and You

Even though sunblocks with skin protection factors (SPF) ratings of 25 or greater are the fastest-growing segment of the $400 million market in the United States, there is a countering trend. Coppertone, the company that helped pioneer SPF ratings and the UV Index, has introduced Coppertone Gold®, a product line low in protection. Banana Boat is competing with Tan Express® and Hawaiian Tropic is launching a suntan oil called Total Exposure®. Identify some possible explanations for these actions.

Those who endure the long nights and short days of northern winters are compensated by a level of non-melanoma skin cancer that is only about half that of those who enjoy year-round Florida sunshine. In fact, the geographical effect on radiation intensity and skin cancer is, at least to date, much greater than that caused by ozone depletion.

That seemingly reassuring statement should not be interpreted as suggesting that the problem of ozone depletion can be ignored. Human beings are, after all, not the only creatures on the globe; our existence is inextricably linked to the entire ecosystem. Plant growth is suppressed by UV radiation, and experiments have measured the negative impact that increased UV-B radiation has on phytoplankton. These photosynthetic microorganisms live in the oceans where they occupy a fundamental niche in the food chain. The phytoplankton ultimately supply the food for all the animal life in the oceans, and any significant decrease in their number could have a major effect globally. Moreover, these tiny plant-like organisms play an important role in the carbon dioxide balance of the planet by absorbing approximately 80% of the atmospheric CO_2 created by human activities. Thus, it is possible that ozone depletion may influence another atmospheric problem—the greenhouse effect, which is the topic of Chapter 3. There is also experimental evidence of DNA damage in the eggs of Antarctic icefish.

Most would agree that the disappearing stratospheric ozone and potential consequences of this disappearance are cause for concern and action. But action requires knowledge of the chemistry that occurs 15 miles above the surface of the earth.

2.10 Stratospheric Ozone: Its Formation and Fate

Every day, 300,000,000 (3×10^8) tons of stratospheric ozone are formed and an equal mass is destroyed. Of course, matter is not really created or destroyed. As in any chemical or physical change, matter merely undergoes changes in its chemical or physical form. In this particular case, the overall concentration of ozone remains constant. The process is an example of a **steady state,** a condition in which a dynamic system is in balance so that there is no net change in concentration of the major species involved.

A steady state arises when a number of chemical reactions, typically competing reactions, balance each other. In the case of stratospheric ozone, the steady state is the net result of four reactions that constitute the **Chapman cycle,** named after Sydney Chapman, a physicist who first proposed it in 1929 and 1930. Equations for the reactions in the four steps of the cycle are:

Step 1: Monatomic oxygen formation (decomposition of O_2):

$$O_2 + \text{UV photon} \longrightarrow 2\ O$$

Step 2: Ozone formation (O_2 and O consumption):

$$O_2 + O \longrightarrow O_3$$

Step 3: Ozone decomposition (O_2 and O formation; opposite of Step 2):

$$O_3 + \text{UV photon} \longrightarrow O_2 + O$$

Step 4: Diatomic oxygen formation (O_3 and O conversion):

$$O_3 + O \longrightarrow 2\ O_2 \quad \text{(slow)}$$

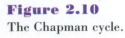

Figure 2.10

The Chapman cycle.

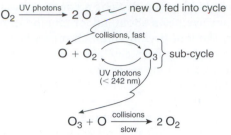

This natural process is, in effect, a cycle related to ozone formation (green) and decomposition (blue). Ozone forms from O_2 via steps 1 and 2, and decomposes back to O_2, the material from which it originates, in steps 3 and 4, thus completing the cycle (Figure 2.10).

You have already encountered the reaction identified as step 1 (equation 2.4). It is the process by which oxygen molecules absorb photons of UV radiation and dissociate into individual oxygen atoms. These reactive atoms tend to combine readily with other atoms and molecules. One such reaction is step 2, which occurs when a monatomic oxygen atom strikes an oxygen molecule (O_2) to generate an ozone molecule, O_3. As we can see from step 3, once an O_3 molecule is generated, it can absorb a photon of UV radiation, causing the molecule to dissociate and regenerate O_2 and O. It is, of course, by means of the reaction in step 3 that ozone screens out UV radiation. Most of the oxygen molecules and atoms formed in step 3 recombine to form ozone molecules via step 2. Occasionally, however, an O_3 molecule collides with an O atom to form two O_2 molecules (step 4). This slow reaction removes the "odd oxygen" species O and O_3 from the cycle. The "lifetime" of an ozone molecule depends strongly on altitude, ranging from days to years. In the center of the ozone layer, an O_3 molecule can persist for several months before it dissociates into O_2 and O.

The four reactions of the Chapman cycle constitute a steady state in which the rate of O_3 formation equals the rate of O_3 destruction. Although reactions are going on, no net change in the concentrations of the reactants or products is observed. The balance point depends upon the details of the system. In this particular case, the steady-state concentration of ozone depends on such factors as the intensity of the UV radiation, the concentration of O_2 and other reacting species, temperature, and the rates and efficiencies of the individual steps in the cycle. To further complicate things, all of these factors vary with altitude. When these variables are properly evaluated and included, it becomes apparent that the Chapman cycle does not tell the whole story. It is fundamentally correct in its description of a natural process, but the steady-state concentration of O_3 is lower than that predicted by this simple model. The real world is inevitably more complicated than such idealized constructions.

2.11 Distribution of Ozone in the Atmosphere

One key to understanding ozone depletion is reliable information about atmospheric ozone concentrations. The total amount of ozone in a vertical column of air of known volume can be determined with relative ease. It is done from the surface of the Earth by measuring the amount of UV radiation reaching the detector: the lower the intensity of the radiation, the greater the amount of ozone. Such data have been collected since the late 1930s. Since the 1970s, measurements of total ozone have also been made from the top of the atmosphere. Satellite-mounted detectors record the intensity of the ultraviolet radiation that is scattered by the upper atmosphere. The results are then related to the amount of O_3 present.

Measuring the ozone concentration at various altitudes is a good deal more difficult. Detectors are carried aloft on airplanes, rockets, balloons, and satellites. The results of some of these measurements are summarized in Figure 2.11. This plot includes a good deal of important information and is worth some careful study. First of all, note that the concentration scale on the horizontal axis has units of O_3 molecules per cubic meter. Now look at the numbers along this axis. Here we are once more using a logarithmic scale to include a wide range of data. (If a typical linear graph were used to present these data, it would be inconveniently long.) There is a tenfold increase in concentration between two successive markers. Thus, 10^{15} is ten times larger than 10^{14}. Although the distances between successive numbers on the horizontal axis are equal, the concentration intervals are not. The difference between 10^{16} and 10^{15} is ten times the difference between 10^{15} and 10^{14}.

To put these concentrations into perspective, recall the Chapter 1 (1.4) calculation of the concentration of CO molecules in a breath. There we concluded that the average 1-L breath would contain 2×10^{17} CO molecules. This corresponds to 2×10^{20} CO molecules per cubic meter (1 m^3 = 1000 L). Note that according to Figure 2.11, the highest concentration of ozone in the stratosphere is between 10^{18} and 10^{19} O_3 molecules per cubic meter. This means that the maximum stratospheric ozone concentration is about one percent of the maximum tropospheric carbon monoxide concentration.

The horizontal lines that cross the curve in Figure 2.11 are called uncertainty bars or error bars. Each bar is a composite of the individual concentration values obtained in a number of separate measurements. Not all measurements give identical results, and the length of the bar indicates the variability of the data. In some cases, especially at low and at high altitudes, the variation is considerable. The cause of this variability may be limitations in instrument sensitivity, experimental error, or genuine variation in concentration. It is impossible to determine the source of variability from the graph. In this case, most of the scatter is due to seasonal and geographical variation.

The message is that when evaluating experimental results, whether published in a daily newspaper, on the web, or in a scientific journal, it is important to know the limitations of the data. Just how good are the measurements and how many were performed? Are the results statistically significant? How great is the uncertainty? How much confidence can you have in the conclusions? A political poll reporting that 51% ± 5% of those surveyed favored one candidate and 49% ± 5% favored the other

Exponents are discussed in Appendix 2.

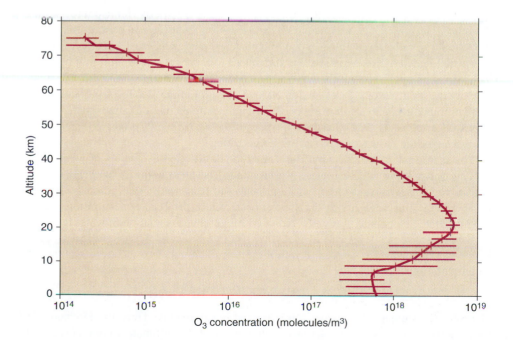

Figure 2.11

Ozone concentrations at various altitudes.

(Source: Data from *United States Standard Atmosphere,* 1976, U.S. EPA.)

candidate would be of little value. The potential errors ($\pm 5\%$) are greater than the difference between the percentages for the two candidates. Healthy and informed skepticism is a useful attribute in analyzing any experimental evidence or statistical summary.

In Figure 2.11 the curve is drawn through the midpoints of the uncertainty bars—points that represent the averages of the measurements. Even allowing for uncertainties, it is clear that the highest concentration of ozone occurs between 10 and 30 km, with a maximum around 20 km. Roughly 91% of the Earth's ozone is found in the stratosphere, between the altitudes of 10 and 50 km.

Because this 40-km band is so broad, the concept of an ozone layer can be a little misleading. At the altitudes of the maximum ozone concentration, the atmosphere is very thin so the total amount of ozone is surprisingly small. If all of the O_3 in the atmosphere could be isolated and brought to the average pressure and temperature at the surface of the Earth (1.0 atm and 15°C), the resulting layer of gas would be just 0.30 cm thick, or about $^1/_8$ of an inch. On a global scale, this is a minute amount of matter. Yet, this fragile shield protects the surface of the Earth and its inhabitants from the harmful effects of ultraviolet radiation. Because ozone is present in a small and finite quantity, it is important that we protect and preserve it.

2.12 Natural Pathways for Ozone Destruction

Stratospheric ozone concentrations have been measured over the past 80 years at ground experimental stations spread over the planet, and for more than 20 years by satellite-mounted detectors. The results are clear: the concentration levels are lower than those predicted using the simple Chapman cycle mechanism. That in itself is neither cause for alarm nor proof that the lower ozone concentrations are the consequence of human intervention. In fact, many of the factors influencing the ozone layer are natural in origin. We know that the processes establishing the steady-state concentration of ozone are more complicated than originally believed.

For one thing, the natural concentration of stratospheric ozone is not uniform over all parts of the globe. On average, the total O_3 concentration increases the closer one gets to either pole (with the exception of the "hole" over the Antarctic). The formation of ozone via steps 1 and 2 of the Chapman cycle is triggered when an O_2 molecule absorbs a photon of ultraviolet light. Therefore, ozone production increases with the intensity of the radiation striking the stratosphere, and that intensity is not constant. It obviously varies with the seasons, reaching its maximum (in the Northern Hemisphere) in March and its minimum in October (just the reverse of the Southern Hemisphere). Consequently, stratospheric ozone concentrations also follow this seasonal pattern. In addition, the amount of radiation emitted by the Sun changes over an 11- to 12-year cycle related to sunspot activity. This variation also influences O_3 concentrations, but only by 1–2%. The winds blowing through the atmosphere cause other variations in ozone concentrations, some on a seasonal basis and others over a 28-month cycle. To further complicate matters, random fluctuations often occur in large samples like the atmosphere. Finally, it is well established that other gases, some of them artificially produced, are also responsible for the destruction of stratospheric ozone.

> The Antarctic ozone hole is the subject of Sections 2.13 and 2.16.

> Free radicals are also discussed in Chapter 9 in conjunction with plastics formation.

The major naturally occurring cause of ozone destruction is a series of reactions involving water vapor and its breakdown products. The great majority of the H_2O molecules that evaporate from the oceans and lakes fall back to the surface of the Earth as rain or snow. But a few reach the stratosphere, where the H_2O concentration is about 5 ppm. There, photons of ultraviolet radiation trigger the dissociation of water molecules into hydrogen atoms (H ·) and hydroxyl (· OH) free radicals. A **free radical** is an unstable chemical species with an unpaired electron. The unpaired electron is often indicated with a dot, as it is here.

$$H_2O + UV \text{ photon} \longrightarrow H \cdot + \cdot OH$$

Because of its unpaired electron, a free radical reacts readily. Thus, the H · and · OH radicals participate in many reactions, including some that ultimately convert O_3 to O_2.

It turns out that this is the most efficient mechanism for destroying ozone at altitudes greater than 50 km.

Water molecules and their breakdown products are not the only agents responsible for natural ozone destruction. Another is NO (nitrogen oxide). Most of the NO in the stratosphere is of natural origin. It is formed when nitrous oxide, N_2O, reacts with oxygen atoms. The N_2O is produced in the soil and oceans by microorganisms and gradually drifts up to the stratosphere. There is really little that can or should be done to control this process. It is part of a cycle involving compounds of nitrogen and living things.

However, not all the nitric oxide in the atmosphere is of natural origin; human activities can alter steady-state concentrations. That is why, in the 1970s, chemists became concerned about the increase in NO that would result from developing and deploying a fleet of supersonic transport (SST) airplanes. These planes were designed to fly at altitudes of 15 to 20 km, the region of the ozone layer. The scientists calculated that much additional NO would be generated by the direct combination of nitrogen and oxygen:

$$\text{Energy} + N_2 + O_2 \longrightarrow 2\ NO$$

The equation emphasizes that this reaction requires large amounts of energy, which can be supplied by lightning or the high-temperature jet engines of the SSTs. To evaluate this risk/benefit situation, many experiments and calculations were carried out, leading to predictions about the net effect of a fleet of SSTs. The conclusion was that the potential risks outweighed the benefits, and the decision was made, partly on scientific grounds, not to build an American fleet. The Anglo-French Concorde is the only commercial plane that currently operates at this altitude.

Subsequent research has indicated that atmospheric reactions involving NO and other oxides of nitrogen are more numerous and more complicated than originally thought. The National Aeronautics and Space Administration (NASA) is currently sponsoring a project to investigate effects of another generation of high altitude supersonic aircraft. As you already know from your reading of Chapter 1, a more serious pollution problem involving the oxides of nitrogen occurs at ground level.

NO has an uneven number of electrons, 11 (5 from N and 6 from O). This makes an NO molecule a free radical because of its unpaired electron. As a free radical, NO is chemically reactive and reacts with additional oxygen to form NO_2 and N_2O_4, other oxides of nitrogen.

Nitrogen oxides are also prominently featured in Chapter 6 dealing with acid rain.

2.13 The Case of the Absent Ozone

Even when the effects of water, nitrogen oxides, and other naturally occurring compounds are included in stratospheric models, the measured ozone concentration is still lower than predicted. Moreover, these measurements indicate that the ozone concentration has been decreasing over the past 20 years. Figure 2.12 is a plot of the percent

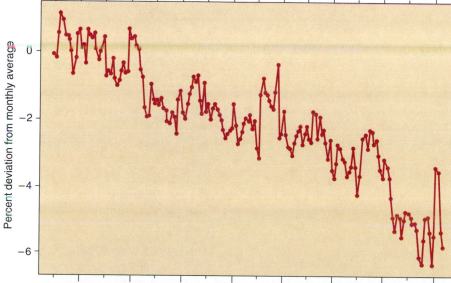

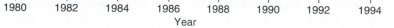

Figure 2.12

Change in stratospheric ozone concentration (60 degrees south to 60 degrees north) expressed as percent deviation from the monthly average (1979 to 1995).

(Source: Data from *Scientific Assessment of Ozone Depletion, 1994. Executive Summary, World Meteorological Organization Global Ozone Research and Monitoring Project,* Report No. 37, Geneva, 1995.)

change in ozone concentration from 1979 to 1995. There is a good deal of fluctuation in the data, but the trend is clear. Over this 16-year period, the stratospheric ozone concentration at midlatitudes (60° south to 60° north) has decreased by about 6%. The change cannot be correlated with changes in the intensity of solar radiation.

2.20 Consider This: Up and Down the Latitudes

In an earlier exercise, you used the web to get stratospheric ozone data at a location of your choice (presumably above your head). Now go to NASA's archive of satellite data on stratospheric ozone levels to find out how the values have varied between 1979 and 1992 over the lower northern hemisphere latitudes. You may wish to coordinate your efforts with other students, so that together you cover a range of years.

a. Obtain values of stratospheric ozone levels at latitudes from +45 degrees north to +0 (the equator) for the year of your choice. Enter −90 degrees west (the middle of the United States) as the longitude and use the satellite Nimbus-7 and a June 15th date. Obtain readings 5 degrees apart. Make a table of the stratospheric ozone values and compute the average.

b. Compare with others in your class data over these 13 years. Note that you may not always be able to use the average as a meaningful comparison, because satellite data may be missing at some latitudes.

The decrease over the South Pole has been far more dramatic. Indeed, it is so pronounced that when the British monitoring team at Halley Bay in Antarctica first observed it in 1985 they thought their instruments were malfunctioning. Some of their subsequent results are reproduced in Figure 2.13, along with data from earlier ground-based studies. The graph reports total ozone levels above the surface of the Earth in Dobson units (DU). A Dobson unit corresponds to about one ozone molecule for every billion molecules of air. For our purposes, the precise definition of a Dobson unit is less im-

$$\frac{DU \ rating}{100} = mm \ O_3 \ at \ Earth's \ surface$$

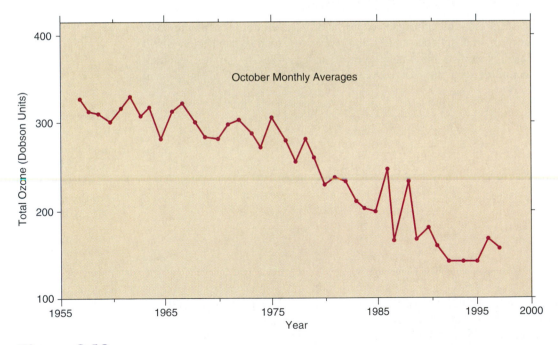

Figure 2.13

Total monthly (October) averages of ozone levels over Halley Bay, Antarctica (1956–1997).

(Source: Data from *Scientific Assessment of Ozone Depletion, 1994. Executive Summary, World Meterological Organization Global Ozone Research and Monitoring Project*, Report No. 37, Geneva, 1995 and ICD [1933–1997].)

portant than the fact that a value of 320 DU represents the average O_3 level over the northern United States. A value of 250 DU is typical at the equator. The data in Figure 2.13 were collected in October of every year from 1956 through 1997. Over this period, the total ozone over Halley Bay decreased from about 320 DU to below 120 DU. In early October 1993, the total ozone there dropped to 91 DU, the lowest recorded anywhere in the world in 36 years of measurement, less than 30% of the 1956 value. The 1998 ozone depletion almost matched this record, with a value of only 92 DU. The 1998 hole, the largest ozone hole observed since such measurements began in 1970, covered 27.3 million km^2 (10.5 million mi^2), an area larger than North America. Total ozone destruction in the hole occurred from an altitude of 15 km to 21 km, the deepest that it has ever been. Keep in mind that there has always been a seasonal variation in ozone concentration over the South Pole, with a minimum in October, the Antarctic spring. What is unprecedented is the dramatic decrease in this minimum that has been observed over the 40 years.

Additional computerized images at the NASA web site show ozone concentrations over the South Pole since 1978. The images were acquired by an analytical instrument known as TOMS (Total Ozone Mapping Spectrometer) carried by a series of satellites for the past twenty years. The latest TOMS was launched in 1996 and still provides daily data at the time this text went to press.

The TOMS images, are color coded to show ozone concentrations in Dobson units. The violet and purple regions are those where the greatest destruction of O_3 occurred, that is, lowest ozone concentrations. If you examined a series of these images, each taken annually in October, you would see a dramatic reduction in ozone concentration over Antarctica. Since the early to mid-1990s, the size of the depleted ozone region annually equals nearly the total area of the North American continent, in some cases exceeding it.

2.21 Consider This: Purple Octobers

NASA satellites provide stratospheric ozone data over time that can be tabulated in a number of ways including global images, Antarctica ozone minima, and size of the ozone hole. All three are provided at the *Chemistry in Context* web site.

a. Using the web site, look first at the global images that are centered on Antarctica. Describe what is happening with the passage of time.
b. Now look at the graphs that show the minimum ozone levels and the size of the region affected. What information does each plot give you?
c. Use the information from all three views to write a description of what is meant by the term "ozone hole." In your statement, include references to the region of the globe, area affected, amount of ozone, and time.

2.14 Chlorofluorocarbons: Properties and Uses

Fortunately, a major cause of stratospheric ozone depletion has been found through the masterful scientific sleuthing of 1995 Nobel laureates F. Sherwood Rowland and Mario Molina and of other chemists and physicists. Vast quantities of atmospheric data have been collected and analyzed, hundreds of chemical reactions have been studied, and complicated computer programs have been written to identify the chemical culprit. As with most scientific results, some uncertainties remain, but there is now compelling evidence implicating an unlikely group of compounds—the **chlorofluorocarbons (CFCs).**

As the name implies, chlorofluorocarbons are compounds composed of the elements chlorine, fluorine, and carbon. Fluorine (F) and the more familiar chlorine (Cl) are members of the same chemical family, the halogens. The other halogens are bromine (Br) and iodine (I). These elements appear in a column labeled Group 7A in the periodic table. At ordinary temperature and pressures, fluorine and chlorine exist as gases made

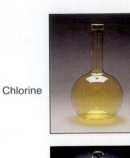

Chlorine

Bromine

Iodine

Some group 7A elements.

Room temperature is typically 20–25°C.

up of diatomic molecules, F_2 and Cl_2. Bromine and iodine also form diatomic molecules, but the former is a liquid and the latter a solid at room temperature. Fluorine is the most reactive element known. It combines with many other elements to form a wide variety of compounds, including those in "Teflon" (a trademark of the DuPont Company) and other synthetic materials. Chlorine is best known as a water purifier, but it is also a very important starting material in the chemical industry. The element itself ranks tenth among all chemicals in total production in the United States. Moreover, four chlorine-containing compounds are among the top 50 industrial chemicals.

Chlorofluorocarbons do not occur in nature; they are artificially produced (an important verification point in the CFC/ozone depletion debate.) Two of the most widely used have the formulas CF_2Cl_2 and $CFCl_3$. They are commonly known as CFC-12 and CFC-11, respectively, following a scheme developed in the 1930s by chemists at DuPont. Their scientific names and Lewis structures are given in Table 2.4. Note that the scientific names for these two compounds are based on methane, CH_4. The prefixes *di-* and *tri-* specify the number of halogen atoms that substitute for hydrogen atoms. ("Freon" is a trademark of the DuPont Company.)

The introduction of CFC-12 (Freon 12) as a refrigerant in the 1930s was rightly hailed as a great triumph of chemistry and an important advance in consumer safety and environmental protection. This synthetic substance replaced ammonia or sulfur dioxide, two naturally occurring toxic and corrosive compounds that made leaks in refrigeration systems extremely hazardous. In many respects, CFC-12 was (and is) an ideal substitute. It has a boiling point that is in the right range and it is almost completely inert. It is not poisonous, it does not burn, and the CCl_2F_2 molecule is so stable that it does not react with much of anything.

This might be an appropriate place to offer a little information about how refrigerators and air conditioners work. They are effectively heat pumps, designed to pump heat from a region of colder temperature (the freezing compartment, for example) to a region of warmer temperature (the room). As you well know, this is an unnatural process. Heat will not flow by itself from a colder to a hotter region. The spontaneous transfer of heat is always in the other direction: from hot to cold. (Chapter 4 goes into more detail about this important generalization, which is known as the second law of thermodynamics). But if you are willing to plug in the refrigerator and pay the electric bill, you can reverse nature and keep your food and drinks cold.

Figure 2.14 is a schematic drawing of a refrigerator, and we suggest you refer to it as you read on. The electricity is used to run a compressor that transforms the refrigerant from a low pressure gas into a high pressure gas, which then flows into the condenser coil. Typically, the refrigerant has a normal boiling point somewhat below room temperature, for example −30°C for CFC-12. This means it is quite easily liquefied in the condenser. As the gas is condensed into a liquid, heat is evolved, usually at the back or bottom of the refrigerator. The liquefied refrigerant then flows into the evaporator coils of the freezing compartment where it converts back into a gas. This change in physical state absorbs heat from the surroundings, including the contents of the compartment. It is the same phenomenon that is responsible for the cooling effect when water evapo-

Table 2.4 Two Important Chlorofluorocarbons

CFC-11	CFC-12
Freon 11	Freon 12
$CFCl_3$	CF_2Cl_2
trichlorofluoromethane	dichlorodifluoromethane
Cl—C—Cl with F up, Cl down	Cl—C—F with F up, Cl down

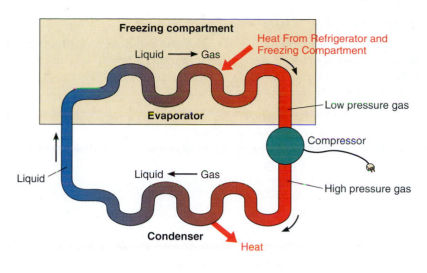

Figure 2.14
Schematic diagram of a refrigerator (see text for explanation).

rates from a wet bathing suit or after-shave lotion or cologne evaporate from the skin. The gas then returns to the compressor and the cycle is repeated. A well-designed refrigeration system is completely sealed and should run for many years without leaking refrigerant into the surroundings. Incidentally, it turns out that the amount of heat released to the room from the back of the refrigerator is always greater than the heat removed from the freezing compartment. The difference corresponds to the work done by the compressor. Consequently, you cannot cool your kitchen by leaving the refrigerator door open. Air conditioners cool rooms because the heat is pumped outside.

The many desirable properties of CFCs soon led to other uses: as propellants in aerosol spray cans, as the gases blown into polymer mixtures to make expanded plastic foams, as solvents for oil and grease, and as sterilizers for surgical instruments. Similar compounds in which bromine replaces some of the chlorine or fluorine, have proved to be very effective fire extinguishers. These halons are used to protect property that would be especially vulnerable to water and other conventional fire-fighting chemicals. Thus they have found applications in electronic and computer installations, chemical storerooms, aircraft, and rare book rooms.

Before we consider the key issue of how CFCs interact with stratospheric ozone, it is important to put into perspective the impact of their usefulness. Because they were nontoxic, nonflammable, cheap, and widely available, CFCs revolutionized air conditioning, making it readily accessible in the United States for homes, office buildings, shops, schools, even automobiles. Oppressive summer heat and humidity became manageable with the use of low-cost CFCs as coolants. Throughout the American South beginning in the 1960s and '70s, air conditioning and CFCs helped to spur on the booming growth of cities such as Atlanta, Dallas-Fort Worth, and Houston, to be followed by others such as San Antonio, Austin, Charlotte, Phoenix, Memphis, Orlando, and Tampa. Some of these are now among our nation's most populated metropolitan areas. In effect, a major sociological shift occurred because of CFCs-based technology that transformed the economy and business potential of an entire region of the country.

By 1985, the combined annual international production of CFC-11 and CFC-12 was approximately 850,000 tons. Some of this material was released into the atmosphere by venting of refrigerators and air conditioners, evaporation of solvents, and escape during polymer foaming. In 1985, the ground-level atmospheric concentration of CFCs was about six molecules out of every 10 billion (0.6 ppb)—a value that has been increasing by about 4% per year. Of course, the fact that ozone levels have been decreasing while CFC levels have been increasing does not prove that the two are causally related. However, other evidence suggests that there is a connection. Ironically, it is the very property that makes CFCs so ideal for so many applications—inertness—that may pose a threat to the environment.

2.15 Interaction of CFCs with Ozone

Chlorofluorocarbons represent a classic case where an apparent virtue becomes a liability. Many of the uses of these compounds capitalize on their low reactivity. The carbon-chlorine and carbon-fluorine bonds in the CFCs are so strong that the molecules can remain intact for long periods. For example, it has been estimated that an average CF_2Cl_2 molecule will persist in the atmosphere for 120 years before it is destroyed. In a much shorter time, typically about five years, many CFC molecules penetrate to the stratosphere with their structures intact.

In 1973, Rowland and Molina, motivated largely by intellectual curiosity, set out to study the fate of these stratospheric CFC molecules. They knew that as altitude increases and the concentrations of oxygen and ozone decrease, the intensity of ultraviolet radiation increases. Therefore, they reasoned that in the stratosphere high-energy photons, such as UV-C, corresponding to wavelengths of 220 nm or less, can break carbon-chlorine bonds. This reaction releases chlorine atoms.

$$CCl_2F_2 + photon \longrightarrow \cdot CClF_2 + \cdot Cl \qquad (2.6)$$
$$\lambda \leq 220 \text{ nm}$$

Similar reactions occur with other CFCs.

Atomic chlorine, $\cdot Cl$, is a very reactive free radical. A Cl atom has seven outer electrons, six of them paired and one unpaired. In equation 2.6, we emphasize the unpaired electron by writing the atom as $\cdot Cl$. The point is that an unbonded chlorine atom exhibits a strong tendency to achieve a stable octet by combining and sharing electrons with another atom. Rowland and Molina, and subsequent researchers hypothesized that this reactivity would result in the following chain of reactions.

First, the chlorine atom pulls an oxygen atom away from the O_3 molecule, forming chlorine monoxide, $\cdot ClO$, and leaving an O_2 molecule. Equation 2.7 shows this for two $\cdot Cl$ atoms reacting with two ozone molecules to form two O_2 molecules and two chlorine monoxide molecules, key reaction intermediates in ozone depletion.

$$2 \cdot Cl + 2 O_3 \longrightarrow 2 \cdot ClO + 2 O_2 \qquad (2.7)$$

Recent experimental evidence indicates that 75–80% of stratospheric ozone depletion involves $\cdot ClO$ joining to form Cl_2O_2, which is subsequently broken down by UV photons to produce $\cdot Cl$ and O_2 as shown in equations 2.8–2.9b. The $\cdot ClO$ molecule is another free radical; it has 13 outer electrons $(7 + 6)$. It readily reacts with another $\cdot ClO$ molecule to form $ClOOCl$ (Cl_2O_2), as noted in equation 2.8.

$$\cdot ClO + \cdot ClO \longrightarrow ClOOCl \qquad (2.8)$$

The ClOOCl decomposes in a two-step sequence as shown in equations 2.9a and 2.9b.

$$ClOOCl + UV \text{ photon} \longrightarrow \cdot ClOO + \cdot Cl \qquad (2.9a)$$

$$\cdot ClOO + UV \text{ photon} \longrightarrow \cdot Cl + O_2 \qquad (2.9b)$$

Recall from Section 2.12 that free radicals have an unpaired electron.

Rowland's and Molina's initial hypothesis was that $\cdot Cl$ reacted with O_3 to form $\cdot ClO$ and O_2:

$$\cdot Cl + O_3 \longrightarrow \cdot ClO + O_2;$$

the $\cdot ClO$ then reacted with oxygen atoms, O, to form O_2 and regenerate $\cdot Cl$

$$\cdot ClO + O \longrightarrow \cdot Cl + O_2$$

The $\cdot Cl$ then recycled to convert other O_3 molecules to O_2.

Note that $\cdot Cl$, $\cdot ClO$, and $ClOOCl$ appear on both sides of equation sequence 2.7–2.9b. We again treat this series of chemical equations as if they were a mathematical equation. We subtract the $\cdot Cls$, the $\cdot ClOs$, and the $ClOOCls$ from both sides of the chemical equations, or, if you prefer, we cancel the $\cdot Cls$, the $\cdot ClOs$, and the $ClOOCls$. What remains is the net equation that indicates the conversion of ozone into O_2.

$$\text{Net equation: } 2\,O_3 \longrightarrow 3\,O_2 \tag{2.10}$$

Interaction of ozone with atomic chlorine thus provides a pathway for the destruction of ozone.

The fact that $\cdot Cl$ appears as a reactant (equation 2.7) and a product (equations 2.9a and 2.9b). is an important one. This indicates that $\cdot Cl$ is both consumed and regenerated in the cycle, so there is no net change in its concentration. Such behavior is characteristic of a **catalyst,** a chemical substance that participates in a chemical reaction and influences its speed without undergoing permanent change. Again and again, atomic chlorine acts catalytically by being regenerated and recycled to remove more ozone molecules. On average, a single $\cdot Cl$ atom can catalyze the destruction of as many as 1×10^5 O_3 molecules before it is carried back to the lower atmosphere by winds.

Atomic chlorine can also become incorporated in stable compounds that do not react to destroy ozone. Hydrogen chloride, HCl, and chlorine nitrate, $ClONO_2$, are two of these "safe" compounds that are quite readily formed at altitudes below 30 km. Thus, chlorine atoms are fairly effectively removed from the region of highest ozone concentration (about 20 km). Maximum ozone destruction by chlorine atoms appears to occur at about 40 km, where the normal ozone concentration is quite low.

Rowland, a professor at the University of California at Irvine, and Molina, then a post-doctoral fellow in Rowland's laboratory, published their first paper (two pages) on chlorofluorocarbons and ozone depletion in 1974 in the scientific journal *Nature*. At about the same time, other scientists were obtaining the first experimental evidence of stratospheric ozone depletion. Since then, the correctness of the Rowland-Molina hypothesis has been well established. Perhaps the most compelling evidence for the involvement of chlorine and chlorine monoxide in the destruction of stratospheric ozone is presented in Figure 2.15. These two graphs contain two plots of Antarctic data: one of O_3 concentration and the other of $\cdot ClO$ concentration. Both are plotted versus the latitude of the sampling airplane when the measurement was made. As stratospheric O_3 concentration decreases, the $\cdot ClO$ concentration increases; the two curves mirror each other almost perfectly. The major effect is a decrease in ozone and an increase in chlorine monoxide as the plane approaches the South Pole. Because ClO, Cl, and O_3 are linked by equation 2.7, the conclusion is compelling. Figure 2.15 is sometimes described as the "smoking gun," the clinching evidence. James G. Anderson of Harvard University, who conducted these measurements, has recently refined his instrumentation so that he can detect pollutants in concentrations as low as one part in 10 trillion, the equivalent of the area of a postage stamp in an area 10 times the size of Texas.

Although most of the attention has focused on chlorofluorocarbons, it is important to recognize that not all of the chlorine implicated in ozone destruction comes from CFCs. The atmosphere currently contains about 4 ppb of chlorinated carbon compounds capable of reaching the ozone layer. The comparable concentration in 1950 was 0.8 ppb, at least implying human origin. On the other hand, some critics, including radio talk-show hosts and some politicians, have argued that essentially all of the chlorine in the stratosphere comes from natural sources such as seawater and volcanoes. However, the majority of atmospheric scientists agree that most chlorine from natural sources is in water-soluble forms. Therefore, any natural chlorine-containing substances are washed out of the atmosphere by rainfall, long before they would reach the stratosphere. In 1994 NASA completed three years of measurements that unambiguously identified CFCs and their decomposition products in the stratosphere, including HCl and HF. Of particular significance are the data that high concentrations of HCl and HF (hydrogen fluoride) always occur together. Although some of the HCl might conceivably arise from a variety of natural sources, the only reasonable origin of stratospheric HF is CFCs. The conclusion is that most of the HCl comes from the same synthetic source.

Figure 2.15
Antarctic O_3 and ClO concentrations.

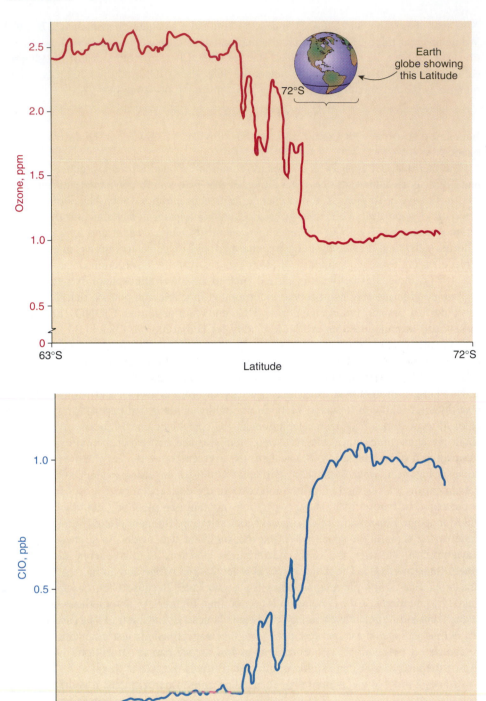

2.23 Consider This: Talk Radio Opinion

"...And if prehistoric man merely got a sunburn, how is it that we are going to destroy the ozone layer with our air conditioners and underarm deodorants and cause everybody to get cancer? Obviously we're not ... and we can't ... and it's a hoax. Evidence is mounting all the time that ozone depletion, if occurring at all, is not doing so at an alarming rate."* Consider the first thing you would ask this talk-show host about these statements. Remember that you need to formulate a short and focused question to get any air time!

*Limbaugh, R. 1993. *See, I told you so.* New York: Pocket Books.

2.24 Consider This: Graffiti with a Message

a. What was the source of humor in this cartoon when it originated in the mid-1970s?

b. Is this cartoon still relevant to the problem of ozone depletion today? Explain your reasoning.

c. If you are so inclined, create your own cartoon dealing with the issue of ozone layer depletion. Be sure that the chemistry is correct!

2.16 The Antarctic Ozone Hole

A particularly intriguing question is why the greatest losses of stratospheric ozone have occurred over Antarctica. Evidence suggests that a special mechanism is operative in that region. This mechanism is related to the fact that the lower stratosphere over the South Pole is the coldest spot on Earth. From June to September, during the Antarctic winter, circulator winds blowing around the South Pole prevent warmer air from entering the region. Temperatures get as low as $-90°C$. Under these conditions, the small amount of water vapor present freezes into thin stratospheric clouds, called polar stratospheric clouds, of ice crystals. The clouds have also been found to contain sulfate particles and droplets or crystals of nitric acid. Atmospheric scientists believe that chemical reactions occurring on the surface of these cloud particles convert otherwise safe, that is, non–ozone depleting, molecules like $ClONO_2$ and HCl, to more reactive species such as $HOCl$ and Cl_2. When the Sun comes out in late September–early October to end the long Antarctic night, the radiation breaks down the $HOCl$ and Cl_2, releasing Cl atoms. The destruction of ozone, which is catalyzed by these atoms, accounts for the hole. Notice the conditions needed for the hole to form: extreme cold and no wind for an extended period to permit ice crystals to provide a surface for the reactions; darkness followed by rapidly increasing levels of sunlight.

The seasonal variations of the ozone hole (1955–1995) over Antarctica are shown in Figure 2.16. Note the rapid ozone decline during spring at the South Pole (September–early November) compared to the summer (January–March). As the sunlight warms the stratosphere, the ice clouds evaporate, halting the chemistry that occurs on the ice crystals. Moreover, air from lower latitudes flows into the polar regions, replenishing the depleted ozone levels. Thus, by the end of November, the hole is largely refilled. However, annual repetitions of this process could result in a general lowering of ozone concentrations in the Antarctic region and eventually across the entire atmosphere.

Although the deepest decrease in the ozone layer over Antarctica occurs during the spring, creating the ozone hole, recent discoveries by British Antarctic Survey researchers, based on improved detection instruments, indicate that the ozone depletion may begin earlier, as early as midwinter at the edges of the Antarctic, including over populated southern areas of South America.

There is already evidence that the ozone reduction over the southern hemisphere is greater than would be predicted solely on the basis of the midlatitude chlorine cycle. Australian scientists believe that wheat, sorghum, and pea production have already been lowered as a result of increased ultraviolet radiation. Health officials there have also observed significant increases in skin cancers. Ultraviolet alerts have even been issued in Australia. These may be among the first manifestations of the consequences of ozone destruction over the South Pole.

A chemist, Dr. Susan Solomon, when she was just 30 years old, headed the team that gathered the stratospheric ClO and ozone data over Antarctica that solidified the connection between CFCs and the ozone hole.

Decreased stratospheric ozone over South Pole $\longrightarrow$ increased UV-B levels $\longrightarrow$ increased skin cancer rates in Australia.

Figure 2.16

Figure 2.16
1955–1995 deviations from average ozone concentration over Antarctica: Pre-ozone hole season (summer—Jan., Feb., Mar.) and ozone hole season (spring—Sep., Oct., Nov.) Note the rapid decline in ozone during the spring and the lesser decline during the austral summer.

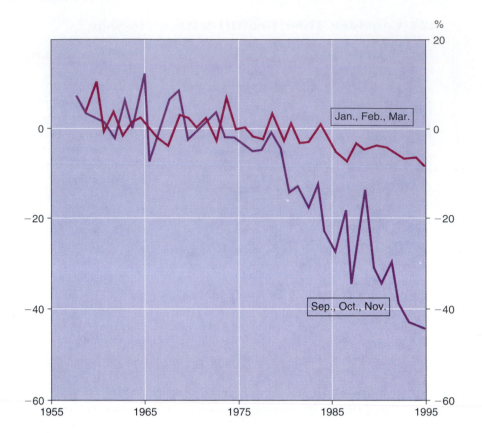

2.17 A Northern Hemisphere Ozone Hole?

The dramatic seasonal loss of stratospheric ozone near the South Pole raises the logical question of whether such a phenomenon also occurs at the North Pole. The effect, if it occurs, should be observed in the North during March/April rather than September/October because of the reversal of the seasons in the two hemispheres. Because a far larger fraction of the Earth's people live north of 50 degrees north latitude than live south of 50 degrees south latitude, the possibility of a northern ozone hole has important implications.

Recent measurements have indicated that some of the same ozone-depleting conditions also apply over the North Pole, where potentially destructive chemical species such as ClO have been detected. In fact, the highest stratospheric ClO concentration ever observed (1.5 ppb) was measured in January 1992 by the Second Airborne Arctic Stratospheric Expedition. More recent information gathered by the Finnish Meteorological Institute indicates that Arctic ozone reached a record low during the winter of 1994–95. More evidence of Arctic ozone depletion in 1994–95 was obtained from satellite measurements conducted by the United States National Oceanic and Atmospheric Administration. Ozone concentrations over Siberia dropped as much as 35% below the levels measured there in 1979. Over the same period, O_3 levels above the United States fell by 10–15%. A map of stratospheric ozone over the Northern Hemisphere from February 13, 1999 is shown in Figure 2.17.

Thus far, scientists have not classified the ozone depletion over the North Pole as a hole. The atmosphere above the North Pole is not as cold as that over its Southern Hemisphere counterpart. Although polar stratospheric clouds have been observed in the Arctic, the air trapped over the Arctic generally begins to diffuse out of the region before the Sun gets bright enough to trigger as much ozone destruction as has been observed in Antarctica. The fact that the stratosphere above the North Pole reached record low temperatures in 1994–95 may have been a factor in the uncommonly high ozone depletion. Whatever the causes, scientists are giving the situation their close attention.

Total Ozone (D.U.) for 14 February 1999

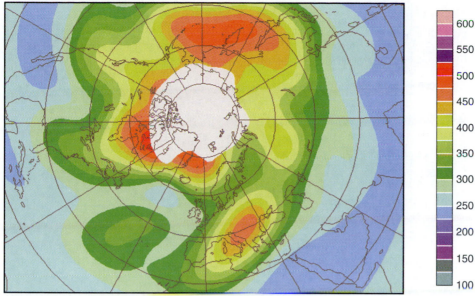

WMO Daily Ozone Maps   LAP-AUTH-GR 1999

Figure 2.17

A map of stratospheric ozone over the Northern Hemisphere.

(World Meteorological Organization web site: www.wmo.ch/web/arep/nhoz.html.)

2.25 Consider This: Comparison of Northern Hemisphere Ozone Maps

Figure 2.17 gives the ozone map of the Northern Hemisphere as of February 13, 1999. Visit the *Chemistry in Context* web site for direct access to the World Meteorological Organization site and obtain the latest map. Compare these two maps and offer some possible explanations for differences observed.

2.18 Response to a Global Crisis

About 82% of stratospheric chlorine comes from synthetic compounds, such as CFCs, which are implicated in enlarging the ozone hole. The response to the threat of ozone destruction has been surprisingly rapid and reasonably effective. The first steps were taken by individual countries. For example, the use of chlorofluorocarbons in spray cans was banned in North America in 1978, and their use as foaming agents for plastics was discontinued in 1990. The problem, however, is a global one, and it requires international cooperation.

In 1985, in response to the first experimental evidence of an ozone hole, a number of world governments participated in the Vienna Convention on the Protection of the Ozone Layer. Through action taken at the convention, these nations committed themselves to protecting the ozone layer and to conduct scientific research to better understand atmospheric processes. A major breakthrough came with the signing, in 1987, of the Montreal Protocol on Substances That Deplete the Ozone Layer. The participating nations agreed to reduce CFC production to one half of the 1986 levels by 1998. But with growing knowledge of the cause of the ozone hole and the potential for global ozone depletion, atmospheric scientists, environmentalists, chemical manufacturers, and government officials soon agreed that the original Montreal Protocol was not sufficiently stringent. In 1990, representatives of approximately 100 nations met in London and decided to ban the production of CFCs by the year 2000. Even that phaseout time was further accelerated in the amendments enacted in Copenhagen in 1992, and again in Montreal in 1997. The production of other ozone-depleting compounds, including halons (carbon-fluorine-bromine compounds), carbon tetrachloride (CCl_4), and methyl bromide (CH_3Br), a widely used agricultural fumigant, were identified for elimination between 2000 and 2005. Developing nations have until 2015 to discontinue using methyl bromide.

Figure 2.18
Estimated global consumption of CFCs (1961–1991).

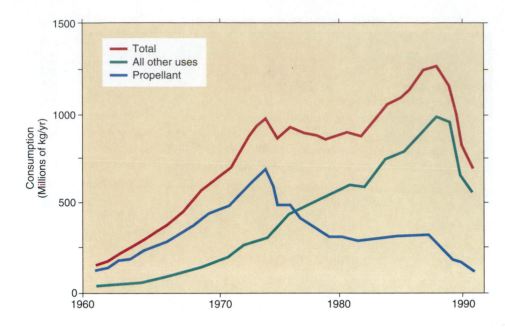

CFCs used as propellants in medical inhalers, such as those used by asthmatics, are exempt from the ban on CFCs for other uses.

The United States and 140 other countries agreed to a complete halt in CFC manufacture after December 31, 1995. Figure 2.18 indicates that the decline in global CFC consumption has been dramatic. Between 1986 and 1992, global use of ozone-depleting substances declined by 40%. It is estimated that without the international action brought to bear by the Montreal Protocol, stratospheric abundances of chlorine would have tripled by the middle of the twenty-first century (2050).

Although these protocols set dates for the halt of CFC production, the sale of existing stockpiles of CFCs will remain legal. One reason for this is the fact that in the United States alone, 140 million car air conditioners and the majority of home air conditioners are designed to use these compounds. Nevertheless, the U.S. government is promoting conversion to less harmful substitute refrigerants by imposing a tax of $5.35 per pound on chlorofluorocarbons. As a result, the price of CFCs has risen from $1 per pound in 1989 to over $15. This situation has tempted bootleggers to smuggle CFCs into the United States, largely from Eastern Europe and Mexico. Russia, although a signator of the Montreal Protocol, currently is responsible for 45% of the world's CFC production. According to law enforcement officers, CFCs are second only to illicit drugs as the most lucrative illegal import. In 1997, "freon busts" by the U.S. Justice Department led to the confiscation of nearly 12 million pounds of illegal CFCs, bringing in fines of $38 million.

2.26 Consider This: CFC Questionnaire

Now that CFC production has halted, chemical substitutes and new technology must be used to fill the gap. How has the removal of CFCs affected you and others? To help answer this question, use the following questionnaire, or one of your own design, to survey public opinion. Focus on one age group of your choosing for your survey. Compile the results of your survey with those of the rest of the class and see what differences you have uncovered in the answers to your questions.

1. Before CFC production was halted, what products did you or your parents use that contained CFCs?
2. To the best of your knowledge, how many years ago were CFCs first removed from products?
3. What changes in your lifestyle can you identify that are a result of the CFC ban?

4. There have been costs associated with the phaseout of CFCs, and many of these costs have been passed on to the consumer. Identify some of the costs that you and your family have paid.

5. Knowing what you know about the environmental impact of using CFCs, do you think the additional cost to consumers is justified? Please explain.

6. When a ban such as this occurs, how long do you think it takes society to adjust to the associated changes in lifestyle, if any?

7. Is this period associated with change longer or shorter than you would expect? Explain.

2.19 Substitutes for the Future

Where do we go from here? Atmospheric scientists feel that it is very likely that the atmosphere will eventually rid itself of the ozone-destroying agents, but it will take a long time, given the 100-year lifetime of many CFCs. Analysis of trends in atmospheric chlorine levels indicates that stratospheric chlorine will peak at 4.1 ppb around the year 2000 and then diminish slowly. This is taken as evidence that the Montreal Protocol and its amendments have slowed the release of CFCs and related ozone-depleting materials. But we are not completely in the clear. Scientists estimate that even under the most stringent international controls on the use of ozone-depleting chemicals, the stratospheric chlorine concentration would not drop to 2 ppb until 2075. This value is significant because the Antarctic ozone hole first appeared when chlorine levels reached 2 ppb.

To be sure, some have proposed attempting to scrub the atmosphere of chlorine and some chlorine-containing compounds by intentionally introducing other chemicals. For example, a researcher at the University of California, Los Angeles, suggested injecting electrons into the stratosphere. He postulated that these electrons would react with chlorine atoms, converting them into negatively charged chloride ions (Cl^-). But a recent report has argued that this approach would not be effective. Another proposal advocated adding propane (C_3H_8) to the stratosphere, reasoning that the hydrocarbon would react with chlorine atoms and remove them from the ozone-depletion cycle. The author of the scheme estimated that 50,000 tons of C_3H_8 would have to be added each year, at an annual cost of $100 million. This strategy has also been discredited.

Nor can we adopt the solution proposed by the industrialist in Sidney Harris's cartoon (Figure 2.19). The authors of this book calculated that it would take about 17 million planeloads of ozone to replenish 10% of the stratospheric ozone. Sherwood Rowland has estimated that "the energy that would be needed to move the ozone up [to the stratosphere] is about $2\frac{1}{2}$ times all of our current global power use." Even if we could temporarily replace the lost ozone, the steady-state cycle would soon re-establish itself.

Although we cannot undo what already has been done, we can stop doing it. Key to the process will be the success of chemists in finding replacements for CFCs. No one advocates the return to ammonia and sulfur dioxide in refrigeration units. In designing replacement molecules, chemists are concentrating on compounds similar to the CFCs. The assumption is that the substitute molecules will include one or two carbon atoms, at least one hydrogen atom, and several chlorine and/or fluorine atoms. The rules of molecular structure limit the options. For example, in the molecules under consideration each carbon atom forms single bonds to four other atoms.

The chemical and physical properties of all compounds depend upon elementary composition and molecular structure. In synthesizing substitutes for CFCs, chemists must weigh three potentially undesirable properties—toxicity, flammability, and extreme stability—and attempt to achieve the most suitable compromise. Compounds containing only carbon and fluorine (fluorocarbons) are neither toxic nor flammable, and they are not decomposed by ultraviolet radiation, even in the stratosphere. Consequently, they would not catalyze the destruction of ozone. This would be ideal, were it not for the fact that the undecomposed fluorocarbons would eventually build up in the atmosphere and contribute to the greenhouse effect by absorbing infrared radiation.

The greenhouse effect is taken up in detail in Chapter 3.

Figure 2.19

A solution to ozone depletion?

(© Sidney Harris. Reprinted by permission.)

"OH, FOR PETE'S SAKE, LET'S JUST GET SOME OZONE AND SEND IT BACK UP THERE!"

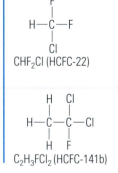

CHF$_2$Cl (HCFC-22)

C$_2$H$_3$FCl$_2$ (HCFC-141b)

Introducing hydrogen atoms in place of one or more halogen atoms reduces molecular stability and promotes their destruction at low altitudes, long before they enter the ozone-rich regions of the atmosphere. However, too many hydrogen atoms increase flammability and too many chlorine atoms seem to increase toxicity. For these reasons, chloroform, CHCl$_3$, would not be a good substitute. Moreover, when a halogen atom is replaced by a much lighter hydrogen atom, the total mass of the molecule is decreased. This results in a decrease in boiling point—a trend observed in many families of similar compounds. A boiling point in the -10 to $-30°C$ range is an important property for a refrigerant. Therefore, the relationship between composition, molecular structure, boiling point, and proposed use must be considered along with toxicity, flammability, and stability.

Fortunately, chemists already know a good deal about how these variables are related, and they have used this knowledge to synthesize some promising replacements for CFCs. The current substitutes are the hydrochlorofluorocarbons (HCFCs). These compounds of hydrogen, chlorine, fluorine, and carbon decompose more readily than CFCs, and hence do not accumulate to the same extent in the stratosphere. HCFC-22 (CHF$_2$Cl) is the most widely used HCFC, being used in air conditioners and in the production of foamed fast-food containers. Its ozone-depleting potential is about 5% that of CFC-12 and its estimated atmospheric lifetime is only 20 years, compared to 111 years for CFC-12. HCFC-141b (C$_2$H$_3$FCl$_2$) is also used to form foam insulation and more than 250 million pounds of this HCFC were produced worldwide in 1996 for this purpose. But because HCFCs have some adverse effects on the ozone layer, they are regarded as only an interim solution to the problem. The 1992 Copenhagen amendments to the Montreal Protocol call for a halt in the manufacture of these compounds by 2030.

In the long run, hydrofluorocarbons (HFCs), compounds of hydrogen, fluorine, and carbon, may be more suitable. HFC-134a (CF$_3$CH$_2$F), with a boiling point of $-26°C$, could prove to be the substitute of choice for CFC-12. HFC-134a has no chlorine atoms to interact with ozone, and its two hydrogen atoms facilitate its decomposition in the lower atmosphere without making it flammable under normal conditions. Other potential replacement refrigerants include hydrocarbons such as propane (C$_3$H$_8$) and isobutane (C$_4$H$_{10}$). Hydrocarbons do not destroy ozone, but they are flammable and can contribute to global warming.

2.27 Consider This: Air Conditioning Your Car

What information do automobile manufacturers provide you about a car's air conditioning system?

a. Locate and compare the relevant information in an owner's manual for a pre-1994 motor vehicle and a post-1994 vehicle.

b. If you need to have your air conditioning system's refrigerant changed, can it be replaced with the same material?

c. Do you consider that the manufacturers are providing sufficient explanation about the air cooling system being used? What other information about the refrigerant would you like included in the vehicle's owner's manual?

Halons are greenhouse gas compounds composed of carbon, fluorine, and bromine atoms used in fire fighting and other applications. As a greenhouse gas, halons are also scheduled to be phased out under the modified Montreal Protocol. Using a green chemistry approach, Pyrocool Technologies has synthesized a halon substitute. The product, called Pyrocool FEF, is an environmentally benign foam that is more effective than halons in fighting fires, even large-scale fires such as those on oil tankers and jet airplanes (Figure 2.20). Pyrocool Technologies won a 1998 Presidential Green Chemistry Challenge award for this development.

The phaseout of CFCs is not without major economic considerations. At its peak, the annual worldwide market for CFCs reached $2 billion, but that was only the tip of a very large financial iceberg. In the United States alone, chlorofluorocarbons were used in or used to produce goods valued at about $28 billion per year. Even today, over $100 billion worth of equipment, probably including your refrigerator and automobile air conditioner, rely on CFCs. Although the conversion to CFC replacements has had some additional costs associated with it, the overall effect on the U.S. economy has been minimal. Companies that produce refrigerators, air conditioners, insulating plastics, and other goods have adapted to using the new compounds. Some substitutes for CFC refrigerants are less energy efficient, hence increasing energy consumption by several percent. But the conversions will provide a market opportunity for innovative syntheses using green chemistry to produce environmentally benign substances.

One of the most interesting efforts to eliminate CFCs has been the Super Efficient Refrigerator Program (SERP). Historically, refrigerators have been major users of electrical

Figure 2.20
Pyrocool FEF being used in a test to extinguish a military jet fuel engine fire. This test duplicated an actual jet crash fire that took hours to extinguish. Using Pyrocool FEF, the test fire was extinguished in 34 seconds.

power and heavy consumers of CFCs, both as refrigerants and as blowing agents for polymer foam insulation. SERP was created by 24 U.S. utilities companies, the Natural Resources Defense Council, the EPA, and a number of other organizations. In 1992, these groups announced a competition for a new CFC-free refrigerator design that would cut energy consumption 30% below the then-existing federal standards. The "golden carrot" that motivated this Great Refrigerator Race is $30 million, to be awarded when 300,000 to 500,000 of the environmentally friendly refrigerators have been sold to customers of the participating electric utilities. In June 1993, Whirlpool was selected as the winner, out of an initial field of 14 contenders. The Golden Carrot, as the winning model has been named, uses HFC-134a as a refrigerant and HFC-14lb as a blower for its foam insulation. Whirlpool started shipping the new "green" refrigerators in 1993, but the company is already encountering competition from Frigidaire, runner-up in the Great Refrigerator Race. The profit motive may yet be turned to the benefit of the environment.

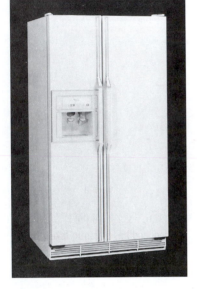

The winner of the Great Refrigerator Race contest uses HFC-134a,

2.28 Consider This: HFCs and the Great Refrigerator Race

Two different CFC substitutes were used to win the Great Refrigerator Race. The structure of HFC-134a, shown next to the photo of the winning refrigerator, was used as the refrigerant. HCFC-141b was used as the blower for its foam insulation. Its structural formula is given on page 84. Explain why neither of these qualifies as a CFC.

On a domestic level, the political dimension of CFC regulation raises many issues. What agency establishes the limits? Where is the legislation enacted? Is this a national, state, or local affair? Who will enforce the regulations? What limitations and what time constraints are reasonable, responsible, possible? How much testing is necessary before replacement compounds can be introduced? How can the country be confident that those making the political, legal, and economic decisions are getting the best scientific advice and interpreting it correctly? There are no easy answers to these questions, maybe not even any right or wrong answers, but 2.29 Consider This gives you an opportunity to struggle with some of them.

2.29 Consider This: Environmental Legislation and States' Rights

In 1995, the Arizona legislature, in defiance of the federal government, passed a law permitting the manufacture and use of CFCs within the state. In defining the law, state representative Robert N. Blendu made the following statement: "Before we ask people to spend millions and millions of dollars in Arizona to replace the Freon in their equipment, we need proof [CFCs] are harmful. We heard testimony on both sides of the issue, and it's only a matter of opinion that CFCs are bad." The Arizona law is a symbolic protest, because federal law banning the manufacture of CFCs supersedes it. On the other hand, the U.S. EPA has ruled that its federal regulations governing CFCs do not preempt the rights of states and cities to enact legislation with stricter, more stringent controls. At first sight, this appears to be an unequal application of states' rights. Write an essay in which you present arguments either for or against a system that prohibits state laws that are more lenient than federal environmental regulations, but allows states to set stricter limits.

Developing countries face another set of economic problems and priorities. Chlorofluorocarbons have played an important role in improving the quality of life in the industrialized nations. Few would be willing to give up the convenience and health benefits of refrigeration or the comfort of air conditioning. It is understandable that millions of people over the globe aspire to the lifestyle of the industrialized nations. As an example, over the past decade, the annual production of refrigerators in China has increased from 500,000 to eight million. But if the developing nations are banned from using the relatively cheap CFC-based technology, they may not be able to afford alternatives. "Our development strategies cannot be sacrificed for the destruction of the environment caused by the West," asserts Ashish Kothari, a member of an Indian environmental group.

In recognition of these legitimate expectations, the Montreal Protocol and its amendments have established a more lenient timetable for developing countries to phase out ozone-depleting substances. These nations will not be expected to begin cutting back on the use of CFCs and halons until 1999, and a complete halt on production is not required until 2010.

As a result of these international agreements, CFC consumption by industrialized countries dropped between 1986 and 1998. But during the same interval, use of these compounds by the developing world increased. This suggests that the use of CFCs by developing nations will continue to grow during the next decade. Without further restrictions, total CFC emissions by developing countries might easily equal one million tons by 2010. Although the result may be industrial and economic progress for one segment of the world's population, it hardly represents progress in the protection of the environment.

Recently, environmentalists have urged that the phaseout schedule for developing countries should be accelerated, but such action is rife with political complications and may well require the infusion of funds from the industrialized nations. There is a precedent. Both India and China refused to sign the original Montreal Protocol because they felt that it discriminated against developing countries. To gain the participation of these highly populated nations, the industrially developed nations created a special fund in 1990.

Developing nations apply to this special fund for grants to underwrite specific projects that lead to discontinuation of CFC use. The United States has contributed nearly one quarter of the money to this fund, which totals more than a half million dollars. But this amount is insufficient to cover the costs of conversion and phaseout in many nations. To make matters worse, a number of industrialized nations are behind in their payments to the fund, and the United States has not always been supportive of foreign aid. Without financial assistance, the developing nations may not be able or willing to meet a more stringent timetable for discontinuing their use of substances that deplete the ozone layer. Clearly, a knowledge of chemistry is necessary to protect the ozone layer, but it is not sufficient. And thus, the second part of the twin challenge—the debate among governments about how best to protect the ozone layer—continues in the global political arena.

In 1991, China signed the Montreal Protocol and has begun to take steps to phase out the production and use of CFCs. Starting in 1998, it banned the industrial use of CFCs as aerosol propellants.

2.30 Consider This: Equity and National Rights

Think back to when the United States itself was a developing nation. How do you think that our economic growth might have been different if we had not had access to CFCs? If we had the luxury of using CFCs with no concern for the long-range consequences, do we have the right to deny this privilege to countries that are presently starting to develop? Write an essay in which you address these issues.

Conclusion

Chemistry is intimately entwined with the story of ozone depletion. Chemists created the chlorofluorocarbons whose near-perfect properties only recently revealed their dark side as predators of stratospheric ozone. Chemists worked internationally to discover the mechanism by which CFCs destroy this ozone and warned of the dangers of increasing ultraviolet radiation. And chemists will continue to synthesize the substitutes necessary to ultimately replace CFCs. But the issues involve more than just chemistry. Philip Elmer-DeWitt said it well in the article in *Time* on February 17, 1992, that provided some of the quotations used in this chapter: "Chlorofluorocarbons have worked their way deep into the machinery of what much of the world thinks of as modern life—air-conditioned homes and offices, climate-controlled shopping malls, refrigerated grocery stores, squeaky-clean computer chips. Extricating the planet from the chemical burden of that high-tech lifestyle—for both those who enjoy it and those who aspire to it—will require not just technical ingenuity but extraordinary diplomatic skill."

Chapter Summary

Having studied this chapter, you should be able to:

- Describe the chemical nature of ozone, the ozone layer, and factors affecting its existence (2.1, 2.10)
- Understand how the ozone layer protects against harmful ultraviolet radiation (2.7–2.9)
- Differentiate among the energies and biological effects of UV-A, UV-B, and UV-C radiation (2.8)
- Evaluate news articles on ozone depletion with an enhanced confidence in your ability to interpret their accuracy and conclusion (2.1–2.18)
- Appreciate the complexities of collecting accurate data for stratospheric ozone depletion and interpreting them correctly and unambiguously (2.10–2.16)
- Apply the basics of atomic structure, that is, protons, neutrons, and electrons to particular elements (2.2)
- Translate an element's atomic number to its position in the periodic table (2.2)
- Write electron distributions, by levels, for a given element (2.2)
- Differentiate atomic number from mass number; apply the latter to isotopes (2.2)

- Write Lewis dot structures using the octet rule; interpret such structures in terms of the nature of their bonds (2.3)
- Describe the electromagnetic spectrum in terms of frequency, wavelength, and energy, and use appropriate calculations to determine these quantities (2.4–2.5)
- Discuss the interaction of radiation with matter and changes caused by such interactions, including biological sensitivity and the use of the UV index (2.6–2.9)
- Differentiate among energies of UV-A, UV-B, and UV-C radiation
- Understand the Chapman cycle (2.10) and the role of nature (2.12) and the role of CFCs (2.13–2.16) in stratospheric ozone depletion, that is, ozone hole formation (2.13, 2.16)
- Interpret graphs related to wavelength and energy, radiation and biological damage, and ozone depletion (2.4, 2.8, 2.16)
- Summarize the political dimensions of CFC regulation and the Montreal Protocol and its amendments (2.18–2.19)
- Evaluate articles on green chemistry alternatives to stratospheric ozone-depleting compounds and recognize that market forces will determine the success of these innovations (2.19).

Questions

Emphasizing Essentials

1. The text states that ozone can be detected in concentrations as low as 10 ppb. Will you be able to detect the odor of ozone in any of these air samples?

   a. 0.118 ppm ozone, a concentration reached in the troposphere

   b. 25 ppm ozone, a concentration reached in the stratosphere

2. Which of these pairs are allotropes?

   a. diamond and graphite

   b. water, H_2O, and hydrogen peroxide, H_2O_2

   c. white phosphorus, P_4, and red phosphorus, P_8

3. Using the periodic table as a guide, specify the number of protons and electrons in a neutral atom of each of these elements.

   a. oxygen (O)   b. nitrogen (N)

   c. magnesium (Mg)   d. sulfur (S)

4. Consider this periodic table.

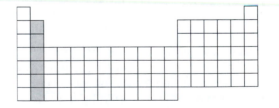

   a. What is the group number indicated by the shading?

   b. What elements make up this group?

   c. What is the number of electrons for each element in this group?

   d. What is the number of outer electrons for each element of this group?

5. Using the periodic table as a guide, give the name and symbol of the element that has the given number of protons in the nucleus of its atoms.

   a. 2

   b. 19

   c. 29

6. Give the number of protons, neutrons, and electrons in each of these isotopes.

   a. oxygen-18 (O-18, $^{18}_{8}O$)

   b. sulfur-35 (S-35, $^{35}_{16}S$)

   c. uranium-238 (U-238, $^{238}_{92}U$)

7. Give the symbol showing the atomic number and the mass number for the element that has

   a. 9 protons and 10 neutrons (an isotope used in nuclear medicine)

   b. 26 protons and 30 neutrons (the most stable isotope of this element)

   c. 86 protons and 136 neutrons (the radioactive gas found in many homes)

8. Give the number of protons, neutrons, and electrons in each of these isotopes.

   a. bromine-82 (Br-82, $^{82}_{35}Br$)

   b. neon-19 (Ne-19, $^{19}_{10}Ne$)

   c. radium-226 (Ra-226, $^{226}_{88}Ra$)

9. Write the electron dot structures for each of these elements.

   a. calcium

   b. nitrogen

   c. chlorine

   d. helium

10. Assuming that the octet rule applies, write Lewis structures for each of these compounds. Start by counting the number of available outer electrons. (a) Write the complete electron dot structure; (b) then write the structure representing shared pairs with a dash, showing unshared electrons as dots.

    a. CCl_4 (carbon tetrachloride, a substance formerly used as a cleaning agent)

    b. H_2O_2 (hydrogen peroxide, a mild disinfectant; the atoms are bonded in this order: H to O to O to H)

    c. H_2S (hydrogen sulfide, a gas with the unpleasant odor of rotten eggs)

11. Assuming that the octet rule applies, write Lewis structures for each of these compounds. Start by counting the number of available outer electrons. (a) Write the complete electron dot structure; (b) then write the structure representing shared pairs with a dash, showing unshared electrons as dots.

    a. N_2 (nitrogen gas, the major component of the atmosphere)

    b. HCN (hydrogen cyanide, a molecule found in space and used in some "death" chambers)

    c. N_2O (nitrous oxide, "laughing gas"; the atoms are bonded in this order: N to N to O)

    d. CS_2 (carbon disulfide, used as a rodenticide; the atoms are bonded in this order: S to C to S)

12. Consider these two waves representing different parts of the electromagnetic spectrum.

Wave 1   Wave 2

   a. How do these two waves compare in wavelength?

   b. How do these two waves compare in frequency?

   c. How do these two waves compare in forward speed?

13. Use Figure 2.2 to specify the region of the electromagnetic spectrum where radiation of each wavelength is found. *Hint:* change each wavelength to meters before making the comparison.

   a. 2 cm

   b. 400 nm

   c. 50 μm

   d. 150 mm

14. Calculate the frequency that corresponds to each of the wavelengths in question 13.

15. Calculate the energy of a photon for each wavelength in question 13. Which wavelength possesses the most energetic photons?

16. Arrange these types of radiation in order of *increasing* energy per photon.

   gamma rays   infrared radiation   radio waves   visible light

17. The microwaves in home microwave ovens have a frequency of $2.45 \times 10^9 \ s^{-1}$. What is the wavelength of these waves in meters?

18. Ultraviolet radiation coming from the sun is categorized by wavelength into three bands. These are UV-A, UV-B, and UV-C. Arrange these three bands in order of their increasing:

   a. wavelengths

   b. energies

   c. potential for biological damage

19. Consider the Chapman cycle in Figure 2.10. Rewrite each step of the cycle using a sphere equation to illustrate the changes that take place.

20. Cl, NO_2, ClO, and HO are all free radicals that catalyze atmospheric ozone depletion.

   a. Count the number of outer electrons available and then draw a Lewis structure for each of these species.

   b. What characteristic is shared by these free radicals that makes them so reactive?

21. In Chapter 1, the role of nitric oxide, NO, in forming photochemical smog was discussed. What role does NO play, if any, in ozone depletion? Are the sources of the NO the same in the troposphere and in the stratosphere?

22. Which graph shows how measured increases in UV-B radiation correlate with percent reduction in the concentration of ozone in the stratosphere over the South Pole? All data were

taken between February 1991, and December 1992, and reported in *Science* in 1993.

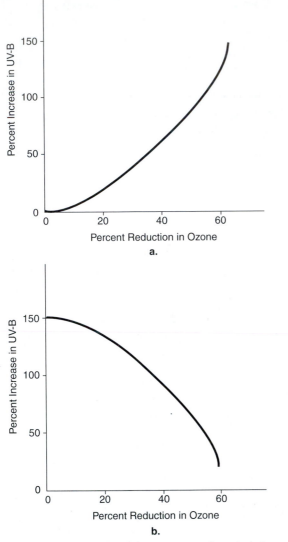

a.

b.

23. **a.** Most CFCs that have been used are based on methane, CH_4, or ethane, C_2H_6. Use structural formulas to represent these two compounds.

   b. Substituting chlorine and/or fluorine for hydrogen atoms, how many different CFCs can be formed from methane?

   c. Which of the compounds you wrote in part **b** have been most successful?

   d. Why weren't all of these compounds equally successful?

24. **a.** How were the original measurements of increases in chlorine monoxide and the stratospheric ozone depletion over the Antarctic gathered?

   b. How are these measurements made today?

Concentrating on Concepts

25. The allotropes oxygen and ozone differ in their molecular structures. What differences does this produce in their properties, uses, and significance?

26. Why is it possible to detect the pungent odor of ozone after a lightning storm or around electrical transformers?

27. How do *allotropes* of oxygen and *isotopes* of oxygen differ? Explain your reasoning.

28. Consider the Lewis structure for SO_2. How is it similar to or different from the Lewis structure for ozone?

29. It is possible to write three resonance structures for ozone, not just the two shown in the text. Verify that all three structures satisfy the octet rule and offer an explanation as to why the triangular structure is not used.

$$:\overset{..}{O}:\overset{..}{O}::\overset{..}{O}: \leftrightarrow :\overset{..}{O}::\overset{..}{O}:\overset{..}{O}: \longleftrightarrow :\overset{..}{O}-\overset{..}{O}:$$

30. The average bond length of an oxygen-to-oxygen single bond is 132 pm. The average bond length of an oxygen-to-oxygen double bond is 121 pm. What do you predict the oxygen-to-oxygen bond lengths will be in ozone? Will they all be the same? Explain your predictions.

31. The formula $E = \dfrac{h\nu}{\lambda}$ indicates the relationships among energy, frequency, and wavelength for electromagnetic radiation. Which variables are directly related and which are inversely related?

32. Which of these forms of electromagnetic radiation from the Sun has the lowest energy and therefore the least potential for damage to biological systems?

   infrared radiation, ultraviolet radiation, visible radiation, radio waves

33. Why is it that you cannot get a suntan from standing in front of your radio in your living room or dorm room?

34. The morning newspaper reports a UV index of 6.5. What should that mean to you as you plan your daily activities?

35. UV-C has the shortest wavelengths of all UV radiation, and therefore the highest energies. All of the reports of the damage caused by UV radiation focus on UV-A and UV-B radiation. Why isn't the focus of attention on the damaging effects that UV-C radiation can have on our skin?

36. If all the 3×10^8 tons of stratospheric ozone that are formed every day are also destroyed every day, how is it possible for stratospheric ozone to offer any protection from ultraviolet radiation?

37. Explain how the small changes in ClO concentrations in Figure 2.15 (measured in parts per billion) can cause the much larger changes in O_3 concentrations (measured in parts per million).

38. Development of the stratospheric ozone hole has been most dramatic over Antarctica. What set of conditions exist over Antarctica that help to explain why this area is well-suited to studying changes in the stratospheric concentration of ozone?

39. The free radical $CF_3O \cdot$ is produced during the decomposition of HFC-134a.

   a. Propose a Lewis structure for this free radical.

   b. Offer a possible reason why this free radical does not cause ozone depletion.

40. Consider this graph from the World Meteorological Organization.

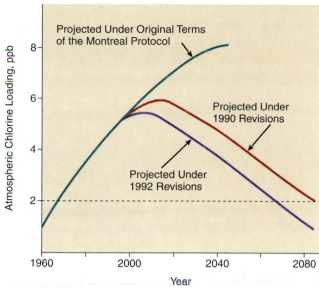

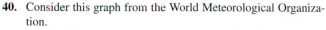

Note: The original Montreal Protocol was signed in 1987 but was amended in 1990 in London, in 1992 in Copenhagen, and again in 1997 in Montreal.

a. Scientists from the World Meteorological Organization think that if atmospheric chlorine loading in the atmosphere were reduced to 2 ppb, shown with the dashed line on the graph, the ozone hole in the Antarctic would disappear. According to the predictions based on the 1992 revisions to the Montreal Protocol, what is the approximate total time span (in years) that the Antarctic ozone hole will be in existence?

b. Imagine that a local newspaper published this graph with the headline "Montreal Protocol Right on Target—Ozone Depletion Problem Solved by the Year 2000." Does this headline correctly convey the information in the graph? Explain your reasoning.

Exploring Extensions

41. Consider this periodic table.

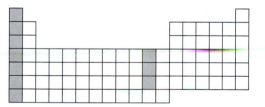

Two groups are highlighted; the first is Group 1A and the second is Group 1B. The text states that although **A** groups have very regular patterns, the "...situation gets a bit more complicated with the **B** families...." Researching in other resources, find out which of these predictions becomes more complicated.

a. predicted number of electrons for each element in each group

b. predicted number of outer electrons for each element of each group

c. predicted formula when each element joins with chlorine

42. Resonance structures can be used to explain the bonding in charged groups of atoms as well as in neutral molecules, such as ozone. The nitrate ion, NO_3^-, has one additional electron beyond the outer electrons contributed by nitrogen and oxygen atoms. That extra electron gives the ion its charge. Draw the resonance structures, verifying that each obeys the octet rule.

43. Although oxygen exists as O_2 and O_3, nitrogen only exists as N_2. Propose an explanation for these facts. *Hint:* Try drawing a Lewis structure for N_3.

44. Consider Figure 2.11 showing ozone concentrations at various altitudes.

a. What does this graph tell you about the concentration of ozone as you travel outward from the surface of the earth? Write a brief description of the trends shown in the graph, describing the location of the ozone layer.

b. The y-axis in this graph starts at zero. Why doesn't the x-axis start at zero?

45. It has been suggested that the term "ozone screen" would be a better descriptor than the "ozone layer" to describe ozone in the stratosphere. What are the advantages and disadvantages to each term?

46. The effect a chemical substance has on the ozone layer is measured by its ozone depleting potential, ODP. This is a numerical scale that estimates the lifetime potential stratospheric ozone that could be destroyed by a given mass of the substance. All values are relative to CFC-11, which has an ODP defined as being equal to 1.0. Use those facts to consider these questions.

a. What factors do you think will influence the ODP value for a chemical? Why?

b. Most CFCs have ODP values ranging from 0.6 to 1.0. What range do you expect for HCFCs? Explain your reasoning.

c. What ODP values do you expect for HFCs? Explain your reasoning.

47. Recent experimental evidence indicates that OCl· initially reacts to form Cl_2O_2.

a. Predict a reasonable Lewis structure for this molecule.

b. What impact does this evidence have on understanding the mechanism for the catalytic destruction of ozone by ClO?

48. The chemical formulas for individual CFCs, such as CFC-11 ($CFCl_3$), can be figured out from their code numbers. A quick way to interpret the code number for CFCs is to add 90 to the number. In this case, $90 + 11 = 101$. The first number in this sum is the number of carbon atoms, the second is the number of hydrogen atoms, and the third is the number of fluorine atoms. $CFCl_3$ has one carbon, no hydrogens, and one fluorine atom. All remaining bonds are assumed to be chlorine until carbon has the required four single covalent bonds to satisfy the octet rule.

a. What is the chemical formula for CFC-12?

b. What is the code number of CCl_4?

c. Will this "90 +" method work for HCFCs? Use HCFC-22, which is CHF_2Cl, to explain your answer.

49. As halons are being phased out under the conditions of the Montreal Protocol, are halon substitutes ready to take their place? Halon substitutes such as Pyrocool FEF, are effective and environmentally benign. Are they also economically successful? Use the resources of the web to assess the scientific and economic success of halon substitutes.

50. One of the outcomes of regulations regarding ozone-depleting substances has been the development of a black market for CFCs.

a. Why is this a significant problem for the United States?

b. Has anyone ever been penalized by the United States for purchasing or possessing illegal CFCs?

The Chemistry of Global Warming

A Landsat computer-enhanced photo of deforestation in the Brazilian rain forest. In this photo, the dark green is the natural forest and the pale green and brown areas are areas where the forest has been cut. Deforestation creates environmental impact in two ways: (1) By using carbon dioxide in photosynthesis, trees help to maintain atmospheric levels of CO_2; and (2) burning the trees after they are cut adds CO_2 to the atmosphere, a factor in enhancing global warming.

Who's Afraid of CO₂?

"...To hear the catastrophists tell it, a rise in the Earth's temperature caused by an accumulation of greenhouse gases—mostly carbon dioxide and water— threatens an environmental doomsday. Unless we curtail our use of fossil fuels, the oceans will rise, disease will spread, the polar caps will melt, wildlife will die out, and millions of human beings will starve. Global warming, they say, is the worst disaster our planet has ever faced.

No one sounds this alarm more urgently than Al Gore, who for years has championed a steep "carbon tax" to drastically force down American energy con- sumption. In his book *Earth in the Balance,* Gore writes that our "dysfunctional civilization" is "killing the atmosphere with carbon dioxide and other pollutants."

Pollutants? Carbon dioxide (CO_2) is a pollutant like water is a poison. It is one of the mainstays of life on earth. Plants cannot survive without it, just as we cannot survive without the oxygen plants exhale. Of course, too much of a good thing can be a bad thing, but how much CO_2 is too much? Federal mine safety regulations warn that carbon dioxide in the air becomes toxic at 5,000 parts per million (ppm). Currently, CO_2 concentrations in the earth's atmosphere clock in at roughly 360 ppm. Don't rush for the gas masks yet.

True, at 360 ppm, we have more CO_2 in the air than we used to; in 1940 the level was about 300 ppm. It's also true that the increase in the last 40 years is due to human activity—the vast industrial infrastructure on which modern civilization rests. Technology requires fuel, and as coal, oil, and natural gas are consumed, CO_2 is released.

But it is far from clear that human-generated CO_2 can affect the world's climate. Gobal temperatures fluctuate. Much of the 17th century was so cold it is known as the "Little Ice Age." For three centuries, the world has been recovering from that global cold snap; only in the last few decades has the warming finally tapered off. That's right, tapered off. The period of greatest emissions of CO_2 has coincided with a *cessation* of global warming. Hard to believe? Only if you've been listening to the alarmists...." THE BOSTON GLOBE   Jeff Jacoby

March 12, 1998

Jeff Jacoby, a *Boston Globe* columnist, serves up a hot dish of issues regarding the controversial subject of global warming. What is global warming? What are the controversies surrounding it? Are legitimate answers available to questions about it? Why this fuss about carbon dioxide, an essential component of the atmosphere—a gas that all animals exhale and all green plants absorb?

3.1 Consider This: Right on Target or Missing the Mark?

How do you react to the information presented in Jacoby's column? To help in analyzing the factual content of it, make a list of the assertions made. For each statement, categorize your response as: **a.** having enough information to judge it is a valid statement; **b.** having enough information to judge it is not a valid statement; or **c.** needing more information about the assertion before reaching a decision. Bring your list to class for further discussion throughout this chapter.

Jacoby's comments came shortly after the Kyoto Conference, a meeting held in December 1997, and attended by nearly 10,000 delegates from around the world. At the conference, intense, even round-the-clock diplomatic negotiations created proposals and counterproposals to establish an agreement among nations to meet the objective of the conference: "...to achieve...stabilization of greenhouse gas concentrations in the atmosphere at a level that would prevent dangerous anthropogenic interference with the climate system."

Why is the global warming issue, dismissed almost flippantly by Jacoby, of enough importance that so many attended the multinational Kyoto Conference? The contrast between Jacoby and the Kyoto delegates could hardly be more pro-

Raul Estrada-Oyuela, chair of the 1997 Kyoto Conference.

nounced. There seems clearly to be a serious misunderstanding of the issue, but on whose part? Can reasonable explanations be offered to support the tone and scepticism of Jacoby's article? Are conflicting national self-interests at work here? Are other motives involved? Do atmospheric scientists agree on the major issues related to global warming? Do economists agree about those issues and their impact on the world economy? We need to develop answers to such questions and to the items identified for further study in 3.1 Consider This.

Which is the defensible or supportable position? Jacoby's? The Kyoto Conference signators? Perhaps others? This chapter responds to "Who's Afraid of CO_2?" and the Kyoto Conference by examining two major issues central to understanding the underlying issues: (1) the way in which carbon dioxide and other gases, generated in part by human activity, contribute to global warming; and (2) to what extent global warming has occurred.

The official title of the Kyoto Conference was the Third Session of the Parties to the United Nations Framework Convention on Climate Change.

Chapter Overview

The two sections that immediately follow this overview provide a general description of the greenhouse effect and its relationship to the evolution of the Earth and its atmosphere. Central to the issue of global warming is the Earth's energy balance and the molecular mechanism by which carbon dioxide and other compounds absorb the infrared radiation emitted by the planet. Some knowledge of molecular structure and shape is necessary to understand this mechanism. Therefore, in the next two sections we develop a general method for predicting molecular geometry and then relate it to infrared-induced vibrations. The two sections that follow make it clear that most of the CO_2 in the atmosphere is of natural origin, but increased human contributions are the chief cause of the current concern about the greenhouse effect. These concerns have a significant quantitative component; we need numbers to help assess the seriousness of the situation. That need justifies several sections in which we introduce and illustrate some fundamental chemical concepts, including atomic and molecular mass, Avogadro's number, and the mole concept. Examples and exercises demonstrate how the important ideas of mass and moles are related. Thus armed, we return to a brief look at methane and several other greenhouse gases. A discussion of predictions based on computer modeling of the climate leads to an assessment of the current situation. The chapter ends with some suggested answers to the all-important question about global warming: "What can we do?" Recognizing that global warming has international implications, should we and can we, as nations, collectively address global warming through initiatives such as the Kyoto Conference protocols?

3.1 In the Greenhouse

The brightest and most beautiful body in the night sky, after our own moon, is Venus. It is ironic that the planet named for the goddess of love is a most unlovely place. Spacecraft launched by the United States and the former Soviet Union have revealed a desolate, eroded surface with an average temperature of about 450°C (840°F). The Venusian atmosphere has a pressure 90 times greater than that of the Earth, and it is 96% carbon dioxide, with clouds of sulfuric acid. It makes the worst smog-bound southern California day seem like a breath of country air. The beautiful blue-green ball we inhabit has an average annual temperature of 15°C (59°F). The point of this little astronomical digression is that both Venus and Earth are warmer than one would expect solely on the basis of their distances from the Sun and the amount of solar radiation they receive. If distance were the only determining factor, the temperature of Venus would average approximately 100°C, the boiling point of water. The Earth, on the other hand, would have an average temperature of −18°C (0°F), and the oceans would be frozen year-round.

3.2 Consider This: Science Fiction Story

A number of successful writers of science fiction began their careers as science majors. Their best work reveals a sound understanding of scientific phenomena and principles. Often a good science fiction story assumes a slightly different scientific reality than the one we know. For example, *Dune,* by Frank Herbert, takes place on a desert planet.

Here is an opportunity to exercise your imagination in a different climate. Suppose the planet had an average temperature of $-18°C$ (0°F)? What would human life be like? Write a brief description of a day on a frozen planet. (Residents of Northern climates should have a great advantage in this exercise.)

The composition of the atmosphere is central to understanding why our planet is 33°C warmer than we would expect, considering the amount of solar energy reaching its surface. The moderating effect is primarily due to two of the minor constituents of the atmosphere: water vapor and carbon dioxide. There is a sort of wonderfully harmonious symmetry in the fact that the two compounds that keep our planet warm enough to sustain life are also among the essential ingredients of all living things.

The idea that atmospheric gases might somehow be involved in trapping some of the Sun's heat was first proposed around 1800 by the French mathematician and physicist, Jean-Baptiste Joseph Fourier (1768–1830). Fourier compared the function of the atmosphere to that of the glass in a "hothouse" (his term) or **greenhouse.** Although he did not understand the mechanism or know the identity of the gases responsible for the effect, his metaphor has persisted. Some 60 years later, John Tyndall (1820–1893) in England experimentally demonstrated that carbon dioxide and water vapor absorb heat radiation. In addition, he calculated the warming effect that would result from the presence of these two compounds in the atmosphere. Given the perspective of time and additional research to augment Fourier's and Tyndall's work, we now know three things: carbon dioxide absorbs heat; the concentration of CO_2 in the atmosphere has increased over the past 150 years; and the Earth's average temperature has not remained constant. The major question is: Are these three items related?

3.2 The Testimony of Time

In the 4.5 billion years that our planet has existed, its atmosphere and climate have varied widely. Evidence from the composition of volcanic gases suggests the concentration of carbon dioxide in the early atmosphere of the Earth was perhaps 1000 times what it is today. Much of the CO_2 dissolved in the oceans became incorporated in rocks such as limestone, which is calcium carbonate, $CaCO_3$, but the high concentration of carbon dioxide also made possible the most significant event in the history of our planet. Although the Sun's energy output was 25–30% less than it is today, the ability of CO_2 to trap heat kept the Earth sufficiently warm to permit the development of life. As early as three billion years ago, the oceans were filled with primitive plants such as cyanobacter. Like their more sophisticated descendants, these simple plants were capable of **photosynthesis.** They were able to use chlorophyll to capture sunlight and use its energy to combine carbon dioxide gas and water to form more complex molecules such as glucose.

$$6\,CO_2 + 6\,H_2O \xrightarrow{\text{chlorophyll}} \underset{\text{glucose}}{C_6H_{12}O_6} + 6\,O_2$$

Photosynthesis not only dramatically reduced the CO_2 concentration of the atmosphere, it increased the amount of O_2 present. The microbiologist, Lynn Margulis, has called this "the greatest pollution crisis the Earth has ever endured." We and our kin are its beneficiaries. The increase in oxygen concentration made possible the evolution of animals. But even in the time of the dinosaurs, 100 million years ago, well before humans walked the Earth, the average temperature is estimated to have been 10–15°C warmer than it is today and the CO_2 concentration is assumed to have been considerably higher.

Reasonably reliable evidence is available about temperature fluctuations during the past 200,000 years—only yesterday in geological terms. Deeply drilled cores from the ocean floor give us a slice through time. The number and nature of the microorganisms present at any particular level indicate the temperature at which they lived. Supplementing this, the alignment of the magnetic field in particles in the sediment provides an independent measure of time.

Other relevant information comes from the analysis of ice cores. The Soviet drilling project at the Vostok Station in Antarctica has yielded over a mile of ice cores taken from the snows of 160 millennia. The ratio of deuterium, ^2H, to hydrogen, ^1H, in the ice can be measured and used to estimate the temperature at the time the snow fell. Water molecules containing atoms of ordinary hydrogen (mass number 1) are lighter than molecules of "heavy water," which contain hydrogen of mass number 2. The lighter H_2O molecules evaporate more readily than the heavier ones. This means that there is relatively more ordinary hydrogen and less deuterium in the water vapor of the atmosphere than in the oceans. The rain or snow that condenses from atmospheric water vapor will also reflect this enrichment of ^2H. The $^2H/^1H$ ratio in precipitation also varies with average temperature. Higher temperatures tend to increase the deuterium/hydrogen ratio in the rain or snow. This is the key to estimating ancient temperatures by the analysis of the isotopic composition of ice cores. In addition, the bubbles of air trapped in the ice can be analyzed for carbon dioxide and other gases. Both sorts of data are incorporated in Figure 3.1. The upper curve (corresponding to the scale on the left) is a plot of parts per million of carbon dioxide in the atmosphere versus time over a span of 160,000 years. The lower plot and the right-hand scale indicate how the average global temperature has varied over the same period. For example, the figure shows that 20,000 years ago, during the last ice age, the average temperature of the Earth was about 9°C below the 1950–1980 average. At the other extreme, a maximum temperature (just over 16°C) occurred approximately 130,000 years ago.

What is particularly striking about Figure 3.1 is that temperature and carbon dioxide concentration parallel each other. When the CO_2 concentration was high, the temperature was high. Other measurements show that periods of high temperature have also

Isotopes of hydrogen were discussed in Section 2.2.

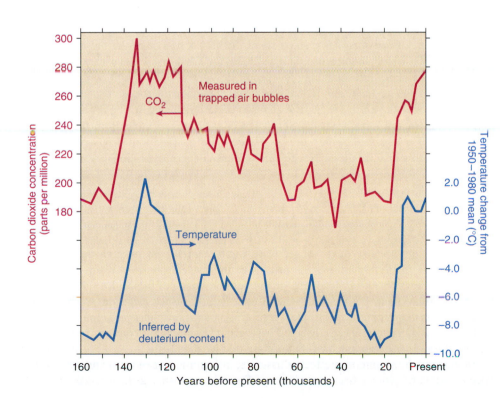

Figure 3.1

Atmospheric CO_2 concentration and average global temperature over 160,000 years (ice core data).

been characterized by high atmospheric concentrations of methane (CH_4). Such correlations do not necessarily prove that elevated atmospheric concentrations of CO_2 and CH_4 caused the temperature increases. Presumably, the converse could have taken place. But these compounds are known to trap heat, and there is no doubt that they can and do contribute to global warming.

To be sure, other mechanisms are also involved in the periodic fluctuations of global temperature. Temperature maxima seem to come at roughly 100,000-year intervals, with interspersed major and minor ice ages. Over the past million years, the Earth has experienced 10 major periods of glaciation and 40 minor ones. Some of this temperature variation is probably caused by minor changes in the Earth's orbit, which affect the distance of the Earth to the Sun and the angle with which sunlight strikes the planet. However, this hypothesis cannot fully explain the observed temperature fluctuations. It is likely that the orbital effects are coupled with terrestrial events such as changes in reflectivity, cloud cover, airborne dust, and carbon dioxide and methane concentration. These factors can diminish or enhance the orbital-induced climatic changes. The feedback mechanism is complicated and not well understood. One thing is clear: the Earth is a far different place in the 1990s than it was at the time of our last temperature maximum 130,000 years ago. Our ancestors had discovered fire by then, but they had not learned to exploit it as we have.

3.3 The Earth's Energy Balance

The major source of the Earth's energy is the Sun. About half of the radiant energy that strikes our atmosphere is either reflected or absorbed by the molecules that make up this envelope of air. You know from your study of Chapter 2 that oxygen and ozone intercept much of the ultraviolet radiation. The rays that do reach the surface of the planet are largely in the visible and infrared (heat) regions of the spectrum. This radiation is absorbed by the Earth, and as a result, the continents and oceans are warmed. The current average temperature of the planet, about 15°C (59°F), is much higher than the −270°C of outer space. Consequently, the Earth acts like a global radiator, radiating heat to its frigid surroundings.

Figure 3.2 is a schematic representation of the Earth's energy balance. The widths of the arrows are roughly proportional to energy flow. Thus, the rate at which energy escapes the surface of the Earth is over twice the rate at which the planet directly absorbs energy from the Sun. This means that the Earth acts a little like an extravagant person, spending money faster than he or she earns it. If this process were the only one occurring, the person would soon be deeply in debt and the Earth would be a very cold place. Fortunately, the planet has something few people do—a sort of built-in forced savings account. Almost 84% of the heat it radiates is absorbed by gases in the atmosphere and then reradiated back to the Earth's surface. As a result of this exchange, the books are balanced and the total energy input from the Sun balances the energy output from the Earth. A steady state is established, with more or less constant average terrestrial temperatures.

The "more or less" is the reason for the current concern over global warming. It is the return of 84% of the energy radiated from the surface of the Earth that has been termed the **global greenhouse effect.** You likely experienced a similar greenhouse effect in your automobile when the car's windows were left unopened on a sunny day. The windows allow UV and visible light in sunlight to pass through them into the car, where some of the radiation is absorbed. Some of it is also re-radiated in the car as infrared radiation, which, because of its longer wavelength, cannot pass through the windows. The infrared radiation (heat) gets trapped inside the car, causing it to warm up.

In sunny climates, temperatures in a closed car can quickly exceed 120°F. For this reason, small children or pets should not be left in cars under these conditions.

The "windows" of the atmospheric greenhouse are made of molecules that are transparent to visible light, but absorb in the infrared region of the spectrum. Carbon dioxide, water, methane, and the molecules of several other atmospheric compounds act this way, and are called *greenhouse gases.* They permit the radiation coming from the Sun to pass through, but trap much of the heat emitted by the Earth.

Obviously, the greenhouse effect is essential in keeping our planet habitable for the species that have evolved here. But if some CO_2 in the atmosphere is a good thing, more

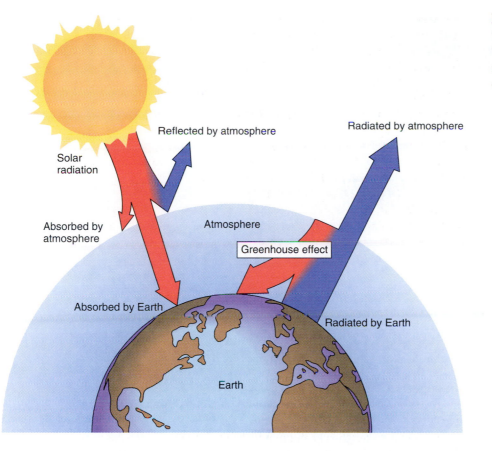

Figure 3.2
The Earth's energy balance.
The width of the arrows is
roughly proportional to energy flow.

is not necessarily better. An increase in the concentration of this infrared absorber will
very likely mean that more than 84% of the radiated energy will be returned to the
Earth's surface, with an attendant increase in average temperature. Back in 1896 the
Swedish chemist, Svante Arrhenius (1859–1927), estimated the extent of this effect. He
calculated that doubling the concentration of CO_2 would result in an increase of 5–6°C
in the average temperature of the planet's surface. Writing in the *London, Edinburgh,
and Dublin Philosophical Magazine* to announce his findings, Arrhenius colorfully described the phenomenon: "We are evaporating our coal mines into the air." At the end
of the 19th century, the Industrial Revolution was already well under way in Europe and
America, and it was "picking up steam" as well as generating it (and CO_2 also).

3.3 Consider This: Evaporating Coal Mines

Although the Arrhenius statement about "...evaporating our coal mines into
the air..." certainly was effective in grabbing attention in 1896, what process
do you think he really was referring to in discussing the amounts of CO_2 being added to the air? What is your reasoning? *Hint:* Consult Figure 3.10 on
p. 109 in this chapter.

   Key to assessing the current and future status of the greenhouse effect are recent
trends in atmospheric carbon dioxide and in average global temperatures. There is compelling evidence that CO_2 concentrations have risen by about 25% in the past century
(Figure 3.3). The best data are those acquired at Mauna Loa in Hawaii. Figure 3.3 presents values from 1957 through 1998. The zig-zag line is a consequence of seasonal variation, but the increase in average annual values from 315 ppm to about 360 ppm is clear.
Projections such as those in Figure 3.4 are computer-based and will be discussed in
greater detail in Section 3.12. Later in this chapter we will learn why scientists believe
that much of the added carbon dioxide has come from the burning of fossil fuels.

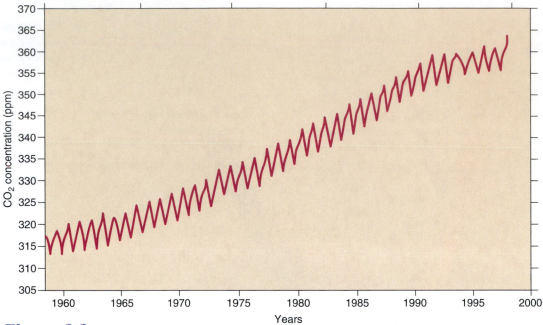

Figure 3.3

Atmospheric CO_2 concentration in ppm as measured at Mauna Loa, Hawaii from 1957 to 1998.

Source: (Data from WHO Statement on the *Status of the Global Climate in 1993*, World Meteorological Organization, WMO No. 809 and also from U.S. EPA.)

3.4 Consider This: The Cycles of Mauna Loa

The text states the zig-zag pattern observable in Figure 3.3 is due to "seasonal variation." **a.** What is the trend within each year? **b.** Propose some specific reason(s) to explain why seasonal changes cause the ebb and flow of atmospheric CO_2 concentrations in Hawaii.

Other measurements indicate that during the past nearly 120 years, the average temperature of the planet has increased by somewhere between 0.4 and 0.8°C. Figure 3.5 is a graph of the changes in the air temperature at the Earth's surface from 1880 to 1996; zero represents the average temperature from 1951 to 1980. Some scientists have correctly pointed out that a century or two is an instant in the 4.5 billion–year history of our planet. They have cautioned restraint in reading too much into what may be short-term temperature fluctuations. Nevertheless, most researchers agree that there seems to be a trend; from Figure 3.5 we see that the average temperature of the Earth is about 0.6°C higher than it was in 1880. Whether this temperature increase is a consequence of the increased CO_2 concentration cannot be concluded with absolute certainty. Never-

Figure 3.4

Atmospheric CO_2 concentrations.

(Source: Environmental Protection Agency Web: www.epa.gov/globalwarming/reports/slides/cc&i/c-concen.html)

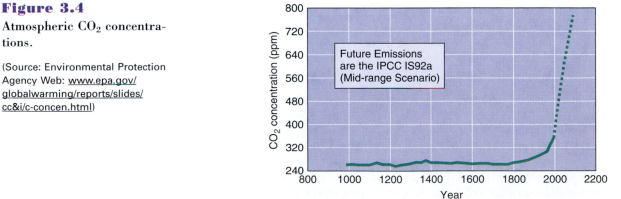

Future Emissions are the IPCC IS92a (Mid-range Scenario)

Derived from ice-core measurements (Siple and South Pole) and direct Observation (Mauna Loa, Hawaii) Source: Based on IPCC (1995)

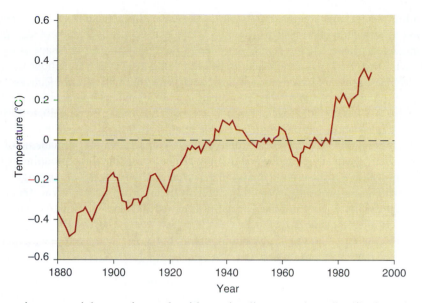

Figure 3.5
Temperature changes on the Earth's surface. Zero is the average temperature for 1951–1980 as a frame of reference.

theless, circumstantial, experimental evidence implicates carbon dioxide from human-related sources as a cause of recent global warming.

> ### 3.5 Consider This: Revisiting Global Warming Trends
>
> One of Jacoby's assertions in the chapter opener was that "...only in the last few decades has the warming finally tapered off. That's right, tapered off. The period of greatest emissions of CO_2 has coincided with a cessation of global warming...." Use information from Figures 3.3, 3.4, and 3.5 to comment on the validity of his assertion.

When temperature measurements are extrapolated into the future, Arrhenius's predictions must be revised downward. Current estimates are that doubling the CO_2 concentration will result in a temperature increase of between 1.0 and 3.5°C. If and when that doubling will occur depends, to a considerable extent, on the human beings who inhabit this planet. We are a long way from the out-of-control hothouse of Venus, but we face difficult decisions. These decisions may not be made easier, but they will be better informed with an understanding of the mechanism by which greenhouse gases interact with radiation to create the greenhouse effect. For that we must again assume a submicroscopic view of matter.

> ### 3.6 Consider This: Adding to the CO_2 Level
>
> Generally speaking, as more CO_2 is produced and released into our atmosphere, more heat will be trapped, and our planet will become warmer unless there are counterbalancing measures. There are several ways that CO_2 is produced and removed from our atmosphere. As members of the global community, we add to the production of CO_2 by our activities and we also hinder the removal of CO_2 from the atmosphere by cutting down trees that use CO_2. Review your typical activities and decide which have the potential to contribute to a net increase in the amount of CO_2 in the atmosphere.

3.4 Molecules: How They Shape Up

Methane, water, and carbon dioxide are greenhouse gases; nitrogen (N_2) and oxygen (O_2) are not. The obvious question is "Why?" The not-so-obvious answer has to do with molecular structure and shape. When you encountered molecular structure in Chapter 2, it was at the level of Lewis structures. The octet rule provides a generally reliable method for predicting bonding in molecules. Moreover, in the case of diatomic molecules, it also predicts molecular geometry. In molecules such as O_2 and N_2, shape is unambiguous; the atoms can only be in a straight line.

The Lewis structures of N_2 and O_2 are:

$:N{\equiv}N:$ and $\ddot{O}{=}\ddot{O}$

With molecules of three or more atoms, differences in molecular geometry become possible. Fortunately, knowing where the outer electrons are located provides insight into molecular shape. Therefore, the first step in predicting molecular shape is to write the Lewis structure for the molecule. If the octet rule is obeyed throughout the molecule, each atom (except hydrogen) will be associated with four pairs of electrons. Some molecules include nonbonding lone-pair electrons, but all molecules contain some bonding electrons. These bonding electrons can be grouped in one or more pairs to form one or more single bonds. In other molecules, the bonding electrons are involved in double bonds consisting of two pairs of electrons or in triple bonds, which are made up of three pairs of electrons. In any case, the key point to remember is that these groupings of negatively charged electrons will repel each other. Hence, **the most stable arrangement is one in which the mutually repelling electron groups are as far away from each other as possible.** This electronic arrangement determines the atomic arrangement and the shape of the molecule.

We illustrate this step-wise procedure for predicting molecular structure with methane, CH_4, one of the greenhouse gases.

1. **Determine the number of outer electrons associated with each atom in the molecule.** The carbon atom (atomic number 6, Group 4A) has four outer electrons; each of the four hydrogen atoms contributes one electron. Thus, there is a total of $4 + (4 \times 1)$ or 8 outer electrons.

2. **Arrange the outer electrons and the atoms in pairs in such a way as to satisfy the octet rule. This may require single, double, and/or triple bonds.** The eight outer electrons in a CH_4 molecule are arranged around the central carbon atom in four bonding pairs, each pair connecting the carbon atom to a hydrogen atom. Thus, the Lewis structure has this appearance:

$$\begin{array}{c} H \\ \cdot\cdot \\ H : C : H \\ \cdot\cdot \\ H \end{array}$$

This structure seems to imply that the CH_4 molecule is flat or planar. But here we are restricted to the two dimensions of a sheet of paper. The architecture of molecules is three dimensional, and we must look into that third dimension.

3. **Assume that the most stable molecular shape is the one in which the bonding or nonbonding electron groups attached to any atom are as far from each other as possible, within the constraints of bonding.**

The four electron pairs around the carbon atom in CH_4 repel each other, and in their most stable arrangement they will be as far from each other as they can be and still form C—H bonds. Furthermore, because a hydrogen atom is attached to each pair of electrons, the four hydrogen atoms will also be as far from each other as possible. This means that the shape of a CH_4 molecule is similar to the base of a folding music stand. The four C—H bonds correspond to the three evenly spaced legs and the vertical shaft of the stand. The angle between each pair of bonds is 109.5°. This shape is said to be tetrahedral, because the hydrogen atoms correspond to the corners of a **tetrahedron,** a four-cornered figure with four equal triangular sides. This shape has been experimentally confirmed. Indeed, the tetrahedral structure is one of the most common atomic arrangements in nature, particularly in carbon-containing molecules.

The legs and shaft of a folding music stand form a tetrahedron.

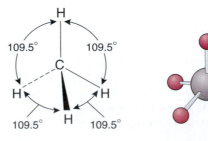

The drawing on the left is an attempt to convey this structure. The wedge-shaped line represents a bond that is coming out of the paper at an angle but generally toward the reader, the dashed line represents a bond pointing away from the reader, and the solid lines are assumed to be in the plane of the paper. This is an improvement, but the best way to visualize molecules is with wooden or plastic models, as in the picture on the right. Your instructor will certainly demonstrate such models, and may give you an opportunity to use them yourself.

We can apply the same set of steps with $CFCl_3$, a CFC that can act as a greenhouse gas. Using step 1, we determine from the molecular formula and the periodic table that there are a total of 32 outer electrons: four for carbon (Group IVA), seven for fluorine (Group VIIA), and seven for *each* chlorine (Group VIIA). Applying step 2 reveals that carbon is the central atom and the other four atoms are bonded to it by four single covalent bonds (8 electrons; four shared electron pairs). This satisfies the octet rule for carbon. The remaining 24 electrons serve as nonbonding (unshared; lone) electron pairs on the other bonded atoms, thus achieving an octet around each one.

CFCs were discussed in Section 2.14.

Like the four shared electron pairs around carbon in CH_4, the four shared pairs in $CFCl_3$ also repel each other so as to be deflected as far apart from each other as possible. The resulting molecular shape is tetrahedral, with the fluorine and chlorine atoms at the corners of a tetrahedron.

In some molecules, the central atom is surrounded by an electron octet, but not all of the electrons are bonding electron pairs. Some are nonbonding (lone) pairs, as in ammonia, NH_3, a refrigerant gas replaced by CFCs.

The nonbonding pair occupies greater space than a bonding pair. Consequently, it repels the bonding pairs somewhat more strongly than the bonding pairs repel each other. This stronger repulsion forces the bonding pairs closer to each other creating an H—N—H angle slightly less than the 109.5° of a regular tetrahedron. The experimental value is close to this, 107.5°, indicating that our model is reasonably reliable.

In describing the shape of a molecule, we do it in terms of the atoms, not the electrons. The hydrogen atoms form a triangle with the nitrogen atom above them at the top of the pyramid. Thus, ammonia is said to be a triangular pyramid; it has a triangular pyramidal structure.

107.5°

Water, another naturally occurring greenhouse gas, illustrates yet another type of molecular shape. There are eight outer electrons: one from each hydrogen atom plus six from the oxygen (Group VIA). Its Lewis structure, given below, discloses how the eight electrons on the central oxygen atom are distributed—two pairs involved in bonding and two lone pairs.

If these four pairs of electrons are arranged so that they are as far apart as possible, the distribution will be similar to that in methane, and we might predict water to have a tetrahedral shape, with a 109.5° H—O—H bond angle. But unlike the four bonding pairs in methane, water has two bonding pairs and two nonbonding pairs. The repulsion between the two nonbonding pairs and their repulsion of the bonding pairs cause the bond angle to be less than 109.5°. Experiments indicate a value of approximately 105°. Thus, a water molecule is said to have a **bent** shape.

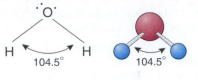

3.7 Your Turn

Using the strategies just described, predict and sketch the shape of each of these molecules.

a. CCl_4 (carbon tetrachloride)
b. CCl_2F_2 (Freon 12; dichlorodifluoromethane)
c. H_2S (hydrogen sulfide)

Ans. **a.** The first step in predicting a structure is to write the correct Lewis structure. Each of the four chlorine atoms has seven outer electrons, and the carbon atom has four. The chlorine atoms bond to the central carbon atom, forming four single bonds. Each bond is a shared pair of electrons. Thus, each atom is surrounded by eight electrons.

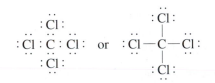

The bonding electron pairs and the attached chlorine atoms will arrange themselves so that their separation is a maximum. It follows that the shape of a carbon tetrachloride molecule is tetrahedral—the same as a methane molecule.

Carbon dioxide is another example of a greenhouse gas. A count of outer electrons reveals a total of 16: four contributed by the carbon atom and six from each of the two oxygen atoms. If only single bonds are involved, there are not enough electrons to provide eight electrons for each atom. That would require 20 electrons. However, the octet rule will be obeyed if the central carbon atom shares four electrons (a double bond) with each of the oxygen atoms. This means that two double bonds are formed.

In the CO_2 molecule, there are thus two groups of four electrons each associated with the central atom. These groups of electrons will again repel each other, and the most stable configuration will provide the furthest separation of the negative charges. This will occur when the angle between them is 180° and the molecule is **linear.** The model predicts that all three atoms in a CO_2 molecule will be in a straight line. This is, in fact, the case. Here is an instance where the simple Lewis structure reveals the correct molecular geometry.

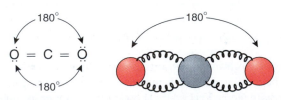

We have applied the idea of electron pair repulsion to molecules in which there are four groups of electrons (CH_4, $CFCl_3$, NH_3, and H_2O) and two groups of electrons (CO_2).

Electron pair repulsion also applies reasonably well to molecules that include three, five, or six groups of electrons. In most molecules, the electrons and atoms are arranged to keep the separation of the electrons at a maximum. This logic accounts for the bent shape we associated with the ozone molecule in Chapter 2. Remember that according to the octet rule, the O_3 molecule (18 total outer electrons) contains a single bond and a double bond. But the central oxygen atom also carries a nonbonding lone pair of electrons. Thus, there are three groups of electrons on this central atom: the pair that make up the single bond, the two pair that constitute the double bond, and the lone pair. These three groups of negatively charged particles repel each other, and the minimum energy of the molecule will correspond to the furthest separation of these electron groups. This will occur when the electron groups are all in the same plane and at an angle of about 120° from each other. Hence, we would predict that the O_3 molecule should be bent, and the angle made by the three atoms should be approximately 120°. The fact that experiment shows the angle to be 117° is confirmation of the general utility of this method.

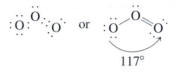

117°

The 3.8 Your Turn activity that follows gives you an opportunity to apply the method to two more molecules of atmospheric importance.

3.8 Your Turn

Using the strategies just described, predict and sketch the shapes of the following molecules.

a. SO_2 (sulfur dioxide)
b. SO_3 (sulfur trioxide)

Hint: Note the periodic table family resemblance between SO_2 and O_3.

3.5 Vibrating Molecules and the Greenhouse Effect

Now that we know the molecular shapes of some consequential greenhouse gases, we can turn to the important phenomenon of how these molecules interact with infrared radiation. When a molecule absorbs a photon, it responds to the added energy. You already learned in Chapter 2 that if the photon corresponds to the UV region of the spectrum, it can have sufficient energy to promote electrons to higher levels within the molecule. This can cause covalent bonds to break, as in the dissociation of O_2 and O_3 by UV-B and UV-C radiation.

Radiation in the infrared region of the spectrum is not sufficiently energetic to cause such molecular disruption. However, a photon of IR radiation can enhance the vibrations in a molecule. The covalent bonds holding atoms together can be thought of as springs, and the atoms can move back and forth. Depending on the molecular structure, only certain vibrations are permitted, and each of these vibrations has a characteristic set of permissible energy levels. The energy of the photon must correspond exactly to the vibrational energy of the molecule for the photon to be absorbed. This means that different molecules absorb IR radiation at different wavelengths and thus vibrate at different energies.

We illustrate these ideas with the carbon dioxide molecule, representing the atoms as balls and the bonds as springs. A CO_2 molecule can vibrate in the four ways pictured on page 106. The arrows indicate the direction of motion. Vibrations A and B are called stretching vibrations. In the vibration labeled A, the central carbon atom is stationary and the oxygen atoms move back and forth (stretch) in opposite directions. Alternatively, the oxygen atoms can move in the same direction and the carbon atom in the opposite direction (vibration B). Vibrations C and D look very much alike. In both cases, the molecule bends from its normal linear shape. The bending counts as two vibrations because it can occur in either of two planes.

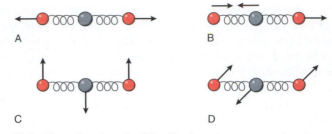

Molecular vibrations in CO₂. Each spring represents a C=O double bond.

In any molecule, the amount of energy required to cause vibration will depend on the nature of the motion, the "stiffness" and strength of the bonds, and the masses of the atoms that move. If you have ever examined a spring or played with a "Slinky" you have probably observed that more energy is required to stretch a spring than to bend it. Similarly, more energy is required to stretch a CO_2 molecule than to bend it. This means that more energetic photons—corresponding to shorter wavelengths—are needed to excite stretching vibrations A or B than those for bending vibrations C or D. The two bending motions (C and D) are both stimulated when the molecule absorbs IR radiation with a wavelength of 15.000 micrometers (μm). (A micrometer is equal to one-millionth of a meter: $1 \ \mu m = 1 \times 10^{-6}$ m.) Vibration B requires more energy; it will occur only if radiation of wavelength of 4.257 μm is absorbed. Vibrations B, C, and D account for the greenhouse properties of carbon dioxide. It turns out that vibration A cannot be triggered by the direct absorption of IR radiation. For such absorption to occur, the overall electrical charge distribution in the molecule must change during the vibration. In a CO_2 molecule, the average concentration of electrons is greater on the oxygen atoms than on the carbon atom. This means that the oxygen atoms are negatively charged relative to the carbon atom. As the bonds stretch, this charge distribution alters. But because of the linear shape and symmetry of the molecule and of vibration A, the changes in charge distribution cancel each other and no infrared absorption occurs.

The infrared energies absorbed or transmitted by molecules can be measured with an instrument called an infrared spectrometer. Heat radiation from a glowing filament is passed through a sample of the compound to be studied, in this case gaseous carbon dioxide. A detector measures the amount of radiation, at various frequencies, that is transmitted by the sample. High transmission means low absorbance, and vice versa. This information is recorded on a chart, where radiation intensity is plotted versus wavelength. The result is called the infrared spectrum of the compound. Figure 3.6 is the in-

Figure 3.6

Infrared spectrum of carbon dioxide.

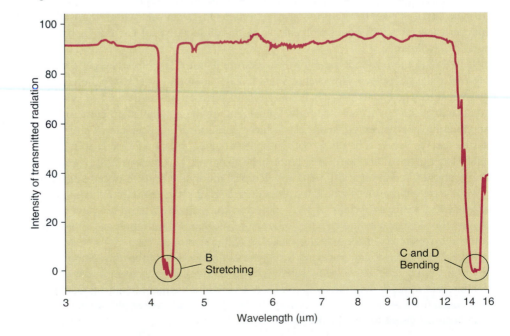

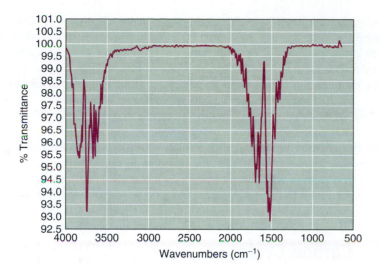

Figure 3.7
IR spectrum of water vapor.

The wavenumbers associated with the two major peaks in this IR spectrum correspond to 2.5 μm and 6.5 μm.

frared spectrum of CO_2, obtained in just this way. There are two steep valleys where the intensity of the transmitted radiation drops almost to zero. This means that most of the radiation is absorbed by the CO_2 molecules. Note that these absorbencies occur at approximately 4.26 and 15.00 μm, as predicted above. The same phenomenon occurs in the atmosphere. Carbon dioxide molecules absorb infrared energy at these wavelengths. They vibrate for a while and then re-emit the energy and return to their normal unexcited or "ground" state. **It is by this means that carbon dioxide captures and returns the infrared radiation coming from the surface of the Earth.** This is what makes carbon dioxide a greenhouse gas.

Any molecule that can vibrate in response to the absorption of infrared radiation is potentially a greenhouse gas. There are many such substances. Carbon dioxide and water are the most important in maintaining the temperature of the Earth. However, methane (CH_4), nitrous oxide (N_2O), ozone (O_3), and chlorofluorocarbons (such as $CFCl_3$) are among the other substances that help retain planetary heat. Figure 3.7, the infrared spectrum of water vapor, shows that H_2O molecules absorb infrared radiation equivalent to 2.5 and 6.5 μm. Diatomic N_2 and O_2 are not greenhouse gases. Although molecules consisting of two identical atoms do vibrate, the overall electrical charge distribution does not change during these vibrations. Hence, these molecules do not absorb infrared radiation.

You have encountered two responses of molecules to radiation. Highly energetic photons with high frequencies and short wavelengths (such as UV radiation) can break up molecules. The less energetic photons of infrared light cause many molecules to vibrate. Both these processes are depicted in Figure 3.8, but the figure also includes another response of molecules to radiant energy that is probably a good deal more familiar to you. It happens in a microwave oven. The radiation generated in such a device is of relatively long wavelength, about a centimeter. This means that the energy per photon is quite low. This energy is insufficient to cause a molecule to vibrate or dissociate, but it is enough to set the molecule spinning. Microwave ovens are tuned to generate radiation that causes water molecules to rotate. As the H_2O molecules absorb the photons and spin

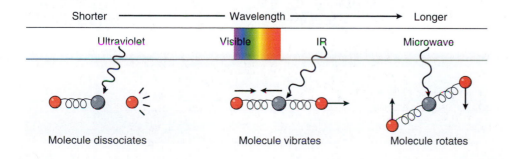

Figure 3.8
Molecular response to types of radiation.

more rapidly, the resulting friction warms up the leftovers. The same region of the spectrum is used for radar. Beams of microwave radiation are sent out from a generator. When they strike an object, such as an airplane, the microwaves bounce back and are detected by a sensor.

The practical consequences of the interaction of radiation and matter are immense, but there is another application of great significance in our understanding of nature. Spectroscopy provides a means of studying atomic and molecular structure. Electronic, vibrational, and rotational energy are all quantized: only certain energy levels are permitted. No matter what region of the spectrum is employed, spectroscopy reveals differences between energy levels. Using the appropriate mathematical model, scientists can translate these energy differences into information about bond lengths, bond strengths, and bond angles. The assurance with which chemists describe the invisible is a consequence of looking through a spectroscopic window into atoms and molecules.

3.6 The Carbon Cycle

In a book entitled *The Periodic Table,* the late Primo Levi, chemist, author, and concentration camp survivor, wrote eloquently about carbon dioxide: "This gas which constitutes the raw material of life, the permanent store upon which all that grows draws, and the ultimate destiny of all flesh, is not one of the principal components of air but rather a ridiculous remnant, an 'impurity' thirty times less abundant than argon, which nobody even notices.... [F]rom this ever renewed impurity of the air we come, we animals and we plants, and we the human species, with our four billion discordant opinions, our millenniums of history, our wars and shames, nobility and pride."

In the essay from which this quotation is taken, Levi traces a brief portion of the life history of a carbon atom from a piece of limestone (calcium carbonate, $CaCO_3$) where it lies "congealed in an eternal present," to a CO_2 molecule, to a molecule of glucose in a leaf, and ultimately to the brain of the author. And yet, that is not the final destination. "The death of atoms, unlike our own," writes Levi, "is never irrevocable." That carbon atom, already billions of years old, will continue to persist into the unimagined future.

This marvelous continuity of matter, a consequence of its conservation, is beautifully illustrated by the carbon cycle. Even without Primo Levi's poetic gifts, the story is a fascinating one. It is summarized in Figure 3.9. Approximately 200 billion metric tons (bmt) of carbon are removed from the atmosphere in the form of CO_2 each year (1 metric ton = 1000 kg or 2200 lb). Slightly over half of it, 110 bmt, is "fixed" by photo-

An elephant weighs between five and six metric tons; 1 metric ton = 2200 lb.

Figure 3.9

Carbon cycle. Carbon is exchanged between the atmosphere and the land and water reservoirs on Earth. The numbers give the approximate annual transfers of carbon (in the form of CO_2) in billions of metric tons (bmt). The existing cycles remove about as much carbon from the atmosphere as they add, but human activity is currently increasing atmospheric carbon by about 3 bmt per year.

(Source: Data based on work of Bert Bolin, University of Stockholm.)

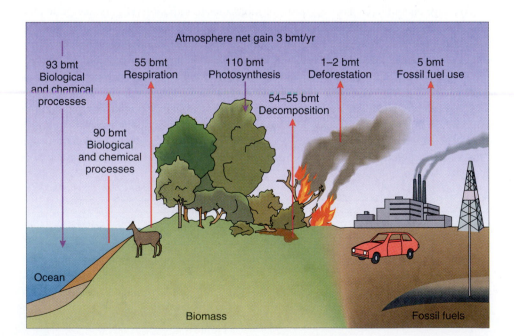

Atmosphere net gain 3 bmt/yr

93 bmt Biological and chemical processes

55 bmt Respiration

110 bmt Photosynthesis

1–2 bmt Deforestation

5 bmt Fossil fuel use

54–55 bmt Decomposition

90 bmt Biological and chemical processes

Ocean

Biomass

Fossil fuels

synthesis and incorporated into plant tissue. Most of the rest dissolves in the oceans, concentrates in coral and sea shells, and ultimately finds its way into limestone and other rocks. The Earth thus serves as a vast reservoir for carbon dioxide.

This is a dynamic, steady-state system that returns as much CO_2 to the atmosphere as is extracted. Plants die and decay, releasing CO_2. Other plants enter the food chain where their complex molecules are broken down into CO_2, H_2O, and other simple substances. Animals exhale CO_2, carbonate rocks decompose, and carbon dioxide escapes through the vents of volcanoes. And the cycle goes on and on. Michael B. McElroy of Harvard University has estimated, "The average carbon atom has made the cycle from sediments through the more mobile compartments of the Earth back to sediments, some 20 times over the course of Earth's history."

As the carbon moves in its many kinds of molecules from gaseous to liquid to solid environments, from vegetable to animal to mineral, from living to dead and back again, it encompasses much of chemistry. The subdiscipline of **biochemistry** deals with the chemical processes in living things; carbon-containing compounds that typically have their origins in living things are studied in **organic chemistry**; and substances derived from the mineral realm are the subjects of **inorganic chemistry**. Concentrations of carbon and any other element or compound are measured by applying the principles and techniques developed in **analytical chemistry.** Finally, **physical chemistry** seeks to elucidate the structure of matter and discover the general principles governing its transformation. But nature knows no such boundaries. Calcium carbonate ($CaCO_3$) and carbon dioxide may be the province of the inorganic chemist, but calcium carbonate derives from sea shells and carbon dioxide is the ultimate source of all organic matter.

3.7 Human Contributions to Atmospheric Carbon Dioxide

As members of the animal kingdom, we *Homo sapiens* participate in the carbon cycle along with our fellow creatures. But we do more than our share; we do more than simply inhale and exhale, ingest and excrete, live and die. We have acquired skills that permit us to perturb the system. Because of our involvement, the cycle is out of balance. Every year, humans release into the atmosphere more carbon dioxide than is removed from it by natural processes. We do so principally by spending sunlight that has been stored for eons in carbon-rich petroleum and coal; we burn fossil fuels.

The Industrial Revolution, which began in Europe in the late eighteenth century, was fueled largely by coal. The coal was used to power steam engines in mines, factories, locomotives, ships, and later, electrical generators. The subsequent discovery and exploitation of vast deposits of petroleum made possible the development of automobiles. To a very considerable extent, the Industrial Revolution was a revolution in energy sources and energy transfer.

As the generation of energy and the consumption of fossil fuels increased, so did the quantity of combustion products such as carbon dioxide released to the atmosphere. Since 1860, the CO_2 concentration has increased from 290 ppm to 360 ppm, and the current rate of increase is about 1.5 ppm per year. At the present level, fossil fuels containing five billion metric tons of carbon are burned annually, but not necessarily for the same purpose. Most fossil fuel–based CO_2 comes from power utilities (35%) and transportation (31%); much less, for example, comes from home and commercial heating (Figure 3.10).

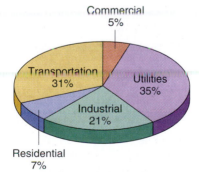

Figure 3.10

CO_2 emission sources.

(Source: Data from DOS/EPA.)

> ### 3.9 Consider This: The CO_2 Emissions—Implications for Policy
>
> Figure 3.10 gives the sources of carbon dioxide emissions from fossil fuel consumption, which also has implications for setting control policies. Do you think that national priorities for controlling CO_2 emissions are set based on the rank order of percentages in this figure? Why or why not? Explain your hypothesis.

In addition, deforestation by burning releases 1–2 bmt of carbon (in the form of CO_2) to the atmosphere each year. It is estimated that globally 150,000 km^2 of rain forest, an area equal to the combined size of Switzerland and the Netherlands, are cut down or burned annually. As a result, trees, which are very efficient absorbers of carbon dioxide, are removed from the cycle. If the wood is burned, vast quantities of CO_2 are generated; if it is left to decay, that process also releases carbon dioxide, but more slowly. Even if the lumber is harvested for construction purposes and the land is replanted in cultivated crops, the loss in CO_2 absorbing capacity may approach 80%.

The total quantity of carbon dioxide released by the human activities of deforestation and burning fossil fuels is 6–7 billion metric tons per year. About half of this is recycled into the oceans and the biosphere. The remainder stays in the atmosphere as CO_2, adding 3 bmt of carbon per year to the existing base of 740 bmt. We are primarily concerned with the increase in carbon dioxide, because this compound is implicated in global warming. Therefore, it would be useful to know the mass (bmt) of CO_2 added to the atmosphere each year. In other words, what is the mass of CO_2 that contains 3 billion metric tons of carbon? Other books might just give the answer, 11 bmt. But here the urge to work through a thoughtful explanation of the 11 bmt is too great, rather than make the assertion and leave it at that. To demonstrate where that number comes from will require a rather lengthy, but scenic, detour.

3.8 Weighing the Unweighable

To solve the problem just posed, we need to know the mass fraction or mass percent of carbon in carbon dioxide. This information can be obtained experimentally by burning a weighed sample of carbon in oxygen and capturing and weighing the carbon dioxide formed. Alternatively, we could decompose a known mass of CO_2 and weigh the carbon and oxygen formed. A third approach, the one we will use, is to calculate an answer based on the formula of the compound. In doing this calculation we will illustrate some fundamental chemical concepts, including the conservation of matter. Regardless of the source of carbon dioxide, its formula is stubbornly the same, CO_2. Thus the mass percent of carbon in CO_2 is also unwavering. As you work through Sections 3.8–3.10, keep in mind that we are seeking a value for that percentage.

The procedure requires the use of the atomic masses or atomic weights of the elements involved. But this raises an important question: How much does an individual atom weigh? Recall from Chapter 2 that most of the mass of an atom is attributable to the neutrons and protons in the nucleus. Thus, elements differ in atomic mass because their atoms differ in composition. Rather than use absolute masses of individual atoms, chemists have found it convenient to employ relative atomic masses—in other words, to relate all atomic masses to some convenient standard. The internationally accepted atomic mass standard is carbon-12, the isotope that makes up 98.9% of all carbon atoms. Carbon-12 has a mass number of 12 because each atom has a nucleus consisting of six protons and six neutrons plus six electrons outside the nucleus. The mass of one of these atoms is arbitrarily assigned a value of exactly 12 atomic mass units (amu). We can thus define the **atomic mass** of an element as the average mass of an atom of that element as compared to an atomic mass of exactly 12 amu for carbon-12. Because atoms are so very small, an atomic mass unit is an extremely small unit: 1 amu = 1.66×10^{-24} g.

You will note that in the periodic table in your text, the atomic mass of carbon is reported as 12.011, not 12.000. This is not an error; it reflects the fact that carbon exists naturally as three isotopes. Although C-12 predominates, 1.1% of carbon is C-13, with six protons and *seven* neutrons per atom and an isotopic mass very close to 13. In addition, natural carbon contains a trace of C-14, whose nuclei consist of six protons and *eight* neutrons. The tabulated atomic mass value of 12.011 is a weighted average that takes into consideration the masses of the three isotopes of carbon and their natural abundances. This isotopic distribution and this average atomic mass will characterize carbon obtained from any chemical source—a graphite ("lead") pencil, a tank of gasoline, a loaf of bread, or a lump of limestone.

Carbon-14 is radioactive, and you will read about its use in dating objects in Chapter 7. This isotope plays a key role in determining the origin of the increasing atmospheric carbon dioxide. In all living things, one out of 10^{12} carbon atoms is a C-14 atom. The plant or animal constantly exchanges CO_2 with the environment and this maintains the C-14 concentration in the organism at a constant level. However, when the organism dies the carbon-14 is no longer replenished as it undergoes radioactive decay to form nitrogen-14. This means that the concentration of C-14 decreases with time. Coal and oil are the fossilized remains of plant life that died millions of years ago. Hence, the level of C-14 is vanishingly low in fossil fuels and in the carbon dioxide released when fossil fuels burn. Experiments show that there has been a recent decrease in the concentration of C-14 in atmospheric CO_2. This strongly suggests that the origin of the added carbon dioxide is indeed the burning of fossil fuels, a decidedly human activity.

See Section 7.9

Most of the atomic masses listed in the periodic table are experimentally obtained values that correspond to averages reflecting individual isotopic masses and the natural distribution of those isotopes. Even so, the atomic masses of many elements closely approximate whole numbers. There are two reasons for this. In the first place, the nucleus of any atom consists of a whole number of particles. These neutrons and protons each have a relative mass of very nearly one atomic mass unit. Second, for many elements one isotope dominates and is by far the most plentiful. Therefore, the experimentally determined atomic mass will be close to the sum of protons and neutrons in the most plentiful isotope. For example, this is the case with nitrogen (N). The most abundant isotope of nitrogen is N-14, with seven protons and seven neutrons in each atomic nucleus. The mass number of this isotope is thus 14, a number that corresponds closely to the average atomic mass of 14.0067 for the naturally occurring isotopic mixture. Other elements with atomic masses close to whole numbers are oxygen (O, atomic mass 15.9994), argon (Ar, 39.948), uranium (U, 238.0289), and, as we have seen, carbon. When the tabulated atomic mass differs significantly from a whole number, as it does for chlorine (Cl, 35.453) or copper (Cu, 63.546) it indicates that the natural distribution involves sizeable concentrations of two or more isotopes. For some of the artificially produced heavy elements at the end of the periodic table, the atomic mass reported is the mass number of the most plentiful isotope.

Mass number = number of protons + number of neutrons (See Section 2.2).

3.10 Your Turn

Gold (Au) has an atomic mass of 196.967 and an atomic number of 79.

a. What is the number of protons, neutrons, and electrons in a neutral atom of the most common isotope, Au-197?
b. Gold-198 is a naturally radioactive isotope that is medically useful in diagnosing liver function. How does the number of protons, neutrons, and electrons in a neutral atom of Au-198 compare with that of Au-197?

3.11 Your Turn

Lithium (Li) has an atomic mass of 6.939. The isotope Li-7 makes up 92.5% of naturally occurring lithium. Lithium has only one other naturally occuring isotope. Predict the mass number of this isotope. Explain your answer.

With that bit of digression about isotopes and their relation to atomic weights as background information, we return to the matter at hand—the masses of atoms and particularly the atoms in CO_2. Not surprisingly, it is impossible to weigh a single atom. A typical laboratory balance can detect a minimum mass of 0.1 mg, and that corresponds to 5×10^{18} or 5,000,000,000,000,000,000 carbon atoms. An atomic mass unit is far too small to measure in a conventional chemistry laboratory. The gram is the chemist's mass unit of choice. Therefore, scientists use exactly 12 g of carbon-12 as the reference for the atomic masses of all the elements. **Atomic mass (or atomic weight)** can thus be alternately defined as the mass (in grams) of the same number of atoms that are found in exactly 12 g of carbon-12. The number of atoms in exactly 12 g of C-12 is called Avogadro's number after an Italian scientist with the impressive name of Lorenzo Romano Amadeo Carlo Avogadro di Quaregna e di Ceretto (1776–1856). (His friends called him Amadeo.) You will note that Avogadro's name consists of 52 letters; his number requires 24 numerals: 602,000,000,000,000,000,000,000. It is more compactly written in scientific notation as 6.02×10^{23}.

Avogadro's number may be shorter than his name, but it is so large that about the only way to hope to comprehend it is through analogies. For example, one Avogadro's number of regular-sized marshmallows, 6.02×10^{23} of them, would cover the surface of the United States to a depth of 650 miles. Or, if you are more impressed by money than marshmallows, assume 6.02×10^{23} pennies were distributed evenly among the 5.5 billion inhabitants of the Earth. Every man, woman, and child could spend one million dollars every hour, day and night, and half of the pennies would still be left unspent at death. And remember, this is the number of atoms in 12 g of carbon—a tablespoonful of soot.

Because one Avogadro's number of carbon-12 atoms has a mass of exactly 12 g, then the mass of an equal number of oxygen atoms should correspond to the atomic mass of oxygen. All we need to do is count out 6.02×10^{23} atoms and weigh them. We could use some help with this assignment, so suppose we enlist all the human beings currently alive and set them counting—one atom per second per person, 24 hours a day, 365 days a year. Even at that rate, things do not look good. It would take almost four million years for all of us to complete the job. Fortunately, we do not need to count the atoms. There are fairly accurate ways of estimating the number. And remember, we can be off by 5×10^{18} atoms and never notice the difference in mass on a laboratory balance! Without going into the experimental details, the result is that one Avogadro's number of oxygen atoms weighs 15.9994 g. That means that the atomic mass of oxygen is 15.9994. This value is consistent with the structure of the oxygen atom. By far the most common isotope of oxygen is O-16, with atoms consisting of 8 electrons, 8 protons, and 8 neutrons.

A knowledge of Avogadro's number and the atomic mass of any element permits us to calculate the average mass of an atom of that element. Thus, the mass of 6.02×10^{23} oxygen atoms is 15.9994 g. To find the mass of one oxygen atom, we simply divide.

$$15.9994 \text{ g oxygen}/6.02 \times 10^{23} \text{ oxygen atoms} = 2.66 \times 10^{-23} \text{ g oxygen/atom}$$

With few exceptions, chemists never work with individual atoms and molecules. We manipulate trillions at a time. Therefore, practitioners of this art need to measure matter with a sort of chemist's dozen—a very large one, indeed. To learn about it, read on.

3.12 Your Turn

a. Calculate the average mass (in grams) of an individual atom of nitrogen.

b. Calculate the mass (in grams) of five trillion nitrogen atoms.

c. Calculate the mass (in grams) of 6×10^{15} nitrogen atoms.

Ans.   a. 2.33×10^{-23} g/atom

3.9 Of Molecules and Moles

An Avogadro's number, 6.02 3 10^{23} of anything, is called a **mole,** a term derived from the Latin word to "heap" or "pile up." It is a chemist's way of counting. Usually the mole is used to count atoms, molecules, electrons, or other small particles. Thus, one mole of carbon atoms consists of 6.02×10^{23} C atoms, one mole of oxygen gas (O_2) is 6.02×10^{23} O_2 molecules, and one mole of carbon dioxide molecules corresponds to 6.02×10^{23} CO_2 molecules.

Moles are fundamental to chemistry because chemistry involves the interaction of individual atoms and molecules. As you already know from Chapter 1, chemical formulas and equations are written in terms of atoms and molecules. For example, reconsider the equation for the reaction of carbon and oxygen.

$$C + O_2 \longrightarrow CO_2$$

In Chapter 1 we interpreted this expression as stating that one atom of carbon combines with one molecule of diatomic oxygen to yield one molecule of CO_2. The equation reflects the ratio in which the particles interact. Thus, it would be equally correct to say that 10 carbon atoms react with 10 oxygen molecules (20 oxygen atoms) to form 10 carbon dioxide molecules. Or, putting the reaction on a grander scale for that matter, we could say 6.02×10^{23} C atoms combine with 6.02×10^{23} O_2 molecules (12.0×10^{23} oxygen atoms) to yield 6.02×10^{23} CO_2 molecules. The latter statement is equivalent to saying: "one *mole* of carbon plus one *mole* of diatomic oxygen gas yields one *mole* of carbon dioxide." The point is that the numbers of atoms and molecules taking part in a reaction are proportional to the numbers of moles of the same substances. The atomic ratio reflected in a chemical formula or the molecular ratio in a molecular equation is identical to the molar ratio. Accordingly, just as there are two oxygen *atoms* for each carbon *atom* in a CO_2 *molecule,* there are also two *moles* of oxygen atoms for each *mole* of carbon atoms in a *mole* of carbon dioxide. The ratio of two oxygens to one carbon remains the same regardless of the number of carbon dioxide molecules, as summarized in the following table.

# carbon atoms	# oxygen atoms	# carbon dioxide molecules
1	2	1
2	4	2
5	10	5
10	20	10
100	200	100
1000	2000	1000
6.02×10^{23}	$2 (6.02 \times 10^{23})$	6.02×10^{23}
1 mole	2 moles	1 mole

In the laboratory and the factory, the quantity of matter required for a reaction is usually measured by mass or weight. The mole concept is a way to simplify matters (and matter) by relating number of particles and mass. Central to this approach is the **molar mass,** defined as the mass of one Avogadro's number of whatever particles are specified. In chemistry, molar masses are almost always expressed in grams. Thus, the mass of a mole of carbon atoms, rounded to the nearest tenth of a gram, is 12.0 g. Similarly, a mole of oxygen atoms has a mass of 16.0 g. But we can also speak of a mole of O_2 molecules. Because there are two oxygen atoms in each oxygen molecule, there are two moles of oxygen atoms in each mole of molecular oxygen, O_2. Consequently, the molar mass of O_2 is 32.0 g—twice the molar mass of O. Some books refer to this as the molecular mass or molecular weight of O_2, emphasizing its similarity to atomic mass or atomic weight.

The same logic for the molar mass of O_2 applies to compounds of two or more elements, which brings us, at last, to the composition of carbon dioxide. The formula, CO_2, and the molecular structure reveal that each molecule contains one carbon atom and two oxygen atoms. Scaling up by 6.02×10^{23}, we can say that each mole of CO_2 consists of one mole of C and two moles of O atoms (see the table above). But remember that we are interested in the mass composition of carbon dioxide—the number of grams of carbon per gram of CO_2. This requires the molar mass of carbon dioxide, which we obtain by adding the molar mass of carbon to twice the molar mass of oxygen.

$$\text{Molar Mass } CO_2 = 1 \times \text{molar mass C} + (2 \times \text{molar mass O})$$

Like atoms of different elements, the masses of a golf ball and a tennis ball differ. Six tennis balls have a greater mass than six golf balls.

Substituting numerical values for the molar masses of the elements gives the desired result.

$$1 \text{ mole C} \times 12.0 \text{ g C/mole C} = 12.0 \text{ g C}$$
$$+ 2 \text{ mole O} \times 16.0 \text{ g O/mole O} = 32.0 \text{ g O}$$
$$\text{molar mass CO}_2 = 44.0 \text{ g CO}_2$$

This procedure is routinely used in chemical calculations, where molar mass is an important property. Some examples are included in the next activity. In every case, you count the number of moles of the constituent elements in one mole of the compound, multiply the number of moles of each element by the corresponding elementary molar mass (the atomic mass in grams), and add the result.

3.13 Your Turn

Calculate the molar mass of each of these substances important in atmospheric chemistry.

a. O_3 (ozone)
b. NO (nitrogen monoxide or nitric oxide)
c. $CFCl_3$ (Freon 11; trichlorofluoromethane)

Ans   a. $\dfrac{16.0 \text{ g O}}{1 \text{ mol O}} \times \dfrac{3 \text{ mol O}}{1 \text{ mol O}_3} = \dfrac{48.0 \text{ g O}}{1 \text{ mol O}_3}$

3.10 Manipulating Moles and Mass with Math

You may recall that several pages ago we set out to calculate the mass of carbon dioxide that includes 3 billion metric tons of carbon. We at last have all the pieces necessary to solve the problem. To do so, we use the quantitative compositional information implicit in the formula of a compound, in this case CO_2. Because 44.0 g CO_2 contain 12.0 g C, we can easily find the ratio (by mass) of carbon to carbon dioxide. It is 12.0 g C/44.0 g CO_2. Out of every 44.0 g CO_2, 12.0 g are C (the remaining 32.0 g is oxygen). This mass ratio holds for all samples of carbon dioxide, and we can use it to calculate the mass of carbon in any known mass of carbon dioxide. For example, we could compute the number of grams of C in 100.0 g CO_2 by setting up a proportion in the following manner.

$$\frac{12.00 \text{ g C}}{44.0 \text{ g } CO_2} = \frac{\text{grams C}}{100.0 \text{ g } CO_2}$$

The equation is then rearranged by "cross multiplication."

$$\text{grams C} \times 44.0 \text{ g } CO_2 = 100.0 \text{ g } CO_2 \times 12.0 \text{ g C}$$

Solving for grams C yields the desired result.

$$\text{grams C} = 100.0 \text{ g } CO_2 \times \frac{12.0 \text{ g C}}{44.0 \text{ g } CO_2} = 27.3 \text{ g C in } 100.0 \text{ g } CO_2$$

Note that carrying along the labels, "g CO_2" and "g C," helps you do the calculation correctly. In the center part of the above expression, "g CO_2" appears in the top (numerator) and the bottom (denominator). Hence, they can be canceled, and you are left with the desired label, "g C." This is a useful strategy in solving many problems. The fact that there are 27.3 grams of carbon in 100.0 grams of carbon dioxide is equivalent to saying that the mass percent of C in CO_2 is 27.3%. Alternatively, CO_2 is 27.3% C by mass.

To find the mass of carbon dioxide that contains 3 billion metric tons (bmt) of carbon, we use the same mass ratio and a similar approach. We could convert 3 bmt to grams, but it is really not necessary. As long as we use the same units for the mass of C and the mass of CO_2, the same numerical ratio holds: 12.0/44.0. But there is one important difference, this time we are solving for the mass of CO_2, not the mass of C. Here y stands for the mass (bmt) of CO_2.

$$\frac{12.0 \text{ bmt C}}{44.0 \text{ bmt } CO_2} = \frac{3 \text{ bmt C}}{y}$$

$$y \times 12.0 \text{ bmt C} = 3 \text{ bmt C} \times 44.0 \text{ bmt } CO_2$$

$$y = 3 \text{ bmt C} \times \frac{44.0 \text{ bmt } CO_2}{12.0 \text{ bmt C}} = 11.0 \text{ bmt } CO_2$$

Once again the labels cancel and the answer comes out in the desired form, bmt CO_2.

Our innocent question, "What is the mass of carbon dioxide added to the atmosphere each year?" has finally been answered: 11 billion metric tons. Of course, our not-so-hidden agenda was to demonstrate the problem-solving power of chemistry and to introduce five of its most important ideas: atomic mass, molecular mass, Avogadro's number, mole, and molar mass. The next few activities provide opportunities to practice your skill with these concepts and manipulations.

3.14 Your Turn

a. Calculate the mass ratio of sulfur (S) to sulfur dioxide (SO_2). Also find the mass percent of S to SO_2.
b. Calculate the mass ratio and the mass percent of N in N_2O.

Ans. a.

The mass ratio is found by comparing the number of grams of sulfur to the molar mass of SO_2.

$$\frac{32.1 \text{ g S}}{64.1 \text{ g } SO_2} = \frac{0.501 \text{ g S}}{1.00 \text{ g } SO_2} \text{ is the mass ratio.}$$

To find the mass percent, multiply the mass ratio by 100.

$$\frac{0.501 \text{ g S}}{1.00 \text{ g } SO_2} = 50.1\% \text{ Sulfur}$$

3.15 Your Turn

a. It is estimated that globally volcanoes release, on average, 19 million metric tons of SO_2 per year. Calculate the mass of sulfur in this amount of SO_2.

b. If 142 million metric tons of SO_2 are released per year by fossil fuel combustion, calculate the mass of sulfur released.

Ans. **a.** The mass ratio of S to SO_2 is known from 3.14 Your Turn, so it is the easiest way to approach this problem.

$$\frac{32.1 \text{ g S}}{64.1 \text{ g } SO_2} \times 19 \text{ million metric tons } SO_2 = 9.52 \text{ million metric tons S}$$

Note: There is no need to change the unit of SO_2 from million metric tons to grams or any other unit. The mass ratio expressed in grams to grams is in the same proportion as if expressed in million metric tons to million metric tons.

If you know how to apply these ideas, you have gained the ability to evaluate critically media reports about releases of carbon or carbon dioxide (and other substances as well) and judge their accuracy. For example, William McKibben seeks to personalize carbon dioxide production in "The End of Nature," an article in *The New Yorker* September 11, 1989. He reports that "the average American car driven the average American distance—ten thousand miles—in an average American year releases its own weight in carbon into the atmosphere." Elsewhere in the same article, McKibben writes about the carbon emitted per gallon of gasoline consumed by stating that "A clean-burning engine . . . will emit about five and a half pounds of carbon in the form of carbon dioxide for every gallon of gasoline it consumes."

One can either take such statements on faith or check their accuracy by applying mathematics to the relevant chemical concepts. Obviously, there is insufficient time to check every assertion, but we hope that readers develop questioning and critical attitudes toward all statements about chemistry and society, even those found in this book.

McKibben's statement about "A clean burning engine . . ." appears deceptively simple. But, it contains a lot of chemistry, which we can check by making some reasonable assumptions and applying a few chemical principles. In the first place, we note that the gasoline is the source of the carbon that is emitted as CO_2. In Chapter 1, gasoline was described as a mixture of hydrocarbons, with octane, C_8H_{18}, as one of the major components. Therefore, we will use C_8H_{18} to represent gasoline.

In a clean-burning engine, such as described by McKibben, all of the hydrogen in the gasoline combines with atmospheric oxygen to form water; all of the carbon reacts with oxygen to create carbon dioxide. Therefore, if we knew the mass of carbon in one gallon of gasoline, that amount would equal the mass of C released as CO_2. To determine the mass of C we have to find the mass of a gallon of octane, C_8H_{18}. To find this information we turn to the *Handbook of Chemistry and Physics* or a similar reference source. Although the *Handbook* does not list the mass of a gallon of C_8H_{18}, it does provide its density, 0.692 g of octane/mL and the fact that 1 gallon is the equivalent volume of 3790 milliliters (mL). With these two pieces of information, we know the number of milliliters in a gallon, and the mass per milliliter. To find the mass of this volume of octane, we multiply the volume of octane by its density:

> Density is the ratio of the mass of a substance divided by its volume, for example, grams/mL. The concept of density is developed more fully in Chapter 5.

$$\text{mass } C_8H_{18} = 3790 \text{ mL} \times \frac{0.692 \text{ g octane}}{1 \text{ mL octane}} = (3790 \times 0.692) \text{ g} = 2620 \text{ g octane}$$

The next question is "How much of this 2620 g of octane is due to carbon?" Fortunately, the formula C_8H_{18} enables us to calculate the mass ratio of carbon in octane. As we did with carbon dioxide in Section 3.9, we first find the molar mass of C_8H_{18}, using the molar masses of carbon and hydrogen.

Molar Mass C_8H_{18} = (8 mole C $\times$ 12.0 g C/1 mole C) + (18 mole H $\times$ 1.0 g H/1 mole H)

$$= 96.0 \text{ g C} + 18.0 \text{ g H} = 114.0 \text{ g } C_8H_{18}$$

This means that in every 114 g C_8H_{18}, there are 96.0 g of carbon and 18.0 g of hydrogen. Therefore, the carbon-to-octane ratio is 96.0 g C/114.0 g octane. Multiplying this ratio by the mass of one gallon of octane gives us the mass of carbon in that amount of octane.

$$\text{mass C} = 2620 \text{ g } C_8H_{18} \times \frac{96.0 \text{ g C}}{114.0 \text{ g } C_8H_{18}} = 2200 \text{ g C}$$

Because McKibben's statement is given in terms of pounds of carbon, the last step in the calculation is to convert grams to pounds by using the fact that 1 lb = 454 g.

$$\text{mass C} = 2200 \text{ g C} \times \frac{1 \text{ lb C}}{454 \text{ g C}} = 4.85 \text{ lb C}$$

Note that in every step of the calculation, labels (units) were carried along with numbers to provide a check for the correctness of the expression.

3.16 The Sceptical Chymist: Is William McKibben Close Enough?

William McKibben is quoted in the discussion on p. 116 as reporting that 5.5 lb of carbon are released for each gallon of gasoline burned. We calculate the value to be 4.85 lb of carbon/gallon of gasoline. This represents a 13% deviation from McKibben's calculation. McKibben uses his determination of 5.5 lb of carbon released/gallon of gasoline to calculate the total amount of carbon released into the atmosphere by a single car in a year. To help decide whether the percent error in this calculation is worth worrying about, answer these questions.

a. Identify the assumptions we made in our calculations and determine if each assumption would increase or decrease our final answer of 4.85 lb of carbon for each gallon of gasoline.
b. Do you agree with McKibben's use of the value 5.5 lb of carbon released/gallon of gasoline or do you think another value should be used? Explain your answer.

3.17 Your Turn

One assumption made in the discussion on p. 116 was that gasoline could be represented by octane, C_8H_{18}. Another common component of gasoline is toluene, C_7H_8, which has a density of 0.867 g/mL. Would the amount of carbon produced (as CO_2) per gallon of toluene be greater or lesser than the result found above? Write down your prediction and your reasons, and then carry out a calculation to support your hypothesis.

3.18 The Sceptical Chymist: Checking Another McKibben Assumption

Check McKibben's statement that "The average American car driven the average American distance—ten thousand miles—in an average American year releases its own weight in carbon into the atmosphere." Note the assumptions you make in solving this problem and compare your assumptions, and your answer, with those of your classmates.

3.11 Methane and Other Greenhouse Gases

Recent estimates suggest that about half of global warming may be attributable to compounds other than carbon dioxide. Methane, CH_4, is approximately 30 times more effective than CO_2 in its infrared trapping characteristics. The atmospheric concentration of methane is relatively low, but its current level of 1.7 ppm is more than twice that before the Industrial Revolution. Data gathered since 1979 indicate an annual increase of about 1% (Table 3.1).

Methane comes from a wide variety of sources. Most of them are natural, but they have been magnified by human activities. For example, because methane is a major component of natural gas, some has always leaked into the atmosphere from rock fissures. But the exploitation of these deposits and the refining of petroleum have led to increased emissions. Similarly, CH_4 has always been released by decaying vegetable matter. Its early name, "marsh gas," reflects this origin. Any human activity that contributes to similar conditions leads to increased methane release. Thus, the decaying organic matter in landfills and from the residue of cleared forests generates CH_4. Methane formed in the main New York City landfill is used for residential heating, but at most landfills it simply escapes into the atmosphere. Another major source of CH_4 is cultivated rice paddies.

Agriculture has also contributed additional methane as the number of cattle and sheep has increased. The digestive systems of these ruminants contain bacteria that break down cellulose. In the process, CH_4 is formed and released through belching and flatulence—about 500 L per cow per day. According to Bill McKibben, a staggering 73 million metric tons of methane are released by the ruminants of the Earth each year. Even termites, who carry on similar chemistry in their guts, generate methane. And there is more than half a ton of termites for every man, woman, and child on the planet.

There is a possibility that global warming may exacerbate the release of methane from ocean muds, bogs, peatlands, and the permafrost of northern latitudes. In these areas, a substantial amount of CH_4 appears to be trapped in "cages" made of water molecules. As the temperature increases, the escape of CH_4 becomes more likely. Fortunately, CH_4 is quite readily converted to less harmful chemical species. It has a relatively short average atmospheric lifetime of 12 years, compared to about 50–200 years for CO_2. The details of the generation and fate of methane are sufficiently complex that it is difficult to speak with a high degree of certainty about its future effect on the average temperature of the planet.

The compound N_2O has the common name nitrous oxide, a name given to it before compounds were named systematically. Its proper (systematic) name is dinitrogen monoxide.

Nitrous oxide, N_2O, also known as "laughing gas," is used as an inhaled anesthetic for dental and medical purposes. In the atmosphere, it is less useful. There, a typical N_2O molecule will persist for about 120 years, absorbing and emitting infrared radiation. Over the past decade, atmospheric concentrations of the compound have shown a slow but steady rise. Major sources are synthetic fertilizers and the burning of biomass. In addition to its role in the greenhouse effect, nitrous oxide contributes to stratospheric ozone depletion. Near the surface of the Earth, however, the reactions of nitrogen oxides and hydrocarbons (like methane) lead to the production of ozone. Ozone can also act like a greenhouse gas, but its efficiency depends very much upon altitude. It appears to have its maximum warming effect in the upper troposphere (around 10 km). Depletion of ozone has a cooling effect in the stratosphere and it may also promote slight cooling at the surface of the Earth. Chlorofluorocarbons (CFCs), already implicated in the destruction of stratospheric ozone, also absorb infrared radiation.

Table 3.1 Changes in Greenhouse Gases Since Pre-Industrial Times

	CO_2	CH_4	N_2O
Pre-industrial conc.	280 ppm	0.70 ppm	0.28 ppm
1994 conc.	358 ppm	1.7 ppm	0.31 ppm
Rate of conc. change	1.5 ppm/yr	0.010 ppm/yr	0.0008 ppm/yr
Atmospheric lifetime (yr)	50–200	12	120

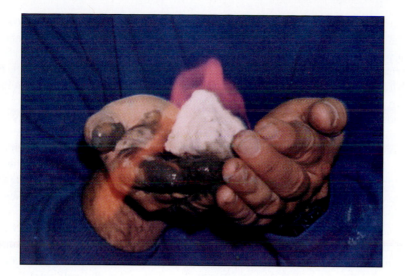

Icy methane; methane is trapped in undersea deposits of ice at depths of greater than 500 m.

One important consideration in all of this is that not all greenhouse gases are equally effective in absorbing infrared radiation. This effectiveness is quantified by the greenhouse factor, a number that represents the relative contribution of a molecule of the indicated substance to global warming. Values of the greenhouse factor for seven common atmospheric trace gases and their average concentrations in the troposphere are given in Table 3.2. Note that according to the table, one molecule of the chlorofluorocarbon CCl_2F_2 has the same global warming effect as 25,000 CO_2 molecules. Fortunately, the tropospheric abundances of most highly effective absorbers of IR radiation are very low.

3.12 Climatic Modeling

The previous paragraphs have merely hinted at the complexity of atmospheric chemistry. To accurately model global climate, one must also include a number of often poorly understood astronomical, meteorological, geological, and biological factors. Among these are variations in the intensity of the Sun's radiation as a consequence of sunspot activity, winds and air circulation patterns, cloud cover, volcanic activity, dust and soot, aerosols, ice sheets, the oceans, and the extent and nature of living things, especially human beings. Moreover, the situation is further complicated by a variety of mechanisms that make these variables interrelated.

For example, we know from the solubility properties of most gases that increasing the temperature of the oceans will decrease the solubility of CO_2, releasing more of it into the atmosphere. An increase in the temperature of the oceans may promote the growth of tiny photosynthetic plants called phytoplankton, and hence increase CO_2 absorption. But the result could be just the opposite. Water in a warmer ocean will not circulate as well as it does now, which may inhibit plankton growth and CO_2 fixing.

Table 3.2 **Greenhouse Factors for Some Common Atmospheric Trace Constituents**

Substance	Greenhouse Factor	Tropospheric Abundance (%)
CO_2	1 (assigned value)	3.6×10^{-2}
CH_4	30	1.7×10^{-3}
N_2O	160	3×10^{-4}
H_2O	0.1	1
O_3	2000	4×10^{-6}
CCl_3F	21,000	2.8×10^{-8}
CCl_2F_2	25,000	4.8×10^{-8}

Decreased snow and ice cover, which would attend global warming, would lower the amount of sunlight reflected from the Earth's surface. The resultant increase in absorbed radiation would promote a further increase in temperature.

A warmer Earth would presumably mean that the tree line would move north, bringing with it added CO_2 absorbing capacity. Countering this, related reductions in rainfall might turn areas that are currently covered by vegetation into deserts, thus reducing carbon dioxide absorption. Global warming would also cause more water to evaporate, increasing the average relative humidity and thus adding to the greenhouse effect. More clouds would form, but their influence cannot be generalized. It seems that high clouds contribute little to the greenhouse effect and reflect sufficient sunlight so that they have a net cooling effect on the surface of the Earth. Low clouds have a net warming effect.

In spite of such formidable problems and sometimes countervailing effects, scientists are developing computer programs to model the Earth's climate. As supercomputers have become more powerful, models have become more sophisticated. The oceans are represented as a multilayer circulating system and the model atmosphere is assumed to contain ten or more interacting layers. Typically, the surface of the planet is divided into about 10,000 cells, not enough to provide detailed predictions, but sufficient to include general patterns of weather development. One test of these simulations of global climate is how well they predict the 0.5°C temperature increase observed over the last century when CO_2 concentrations increased by 25%. Most models estimate a temperature increase of about twice that actually measured. This suggests that certain relevant factors may have been omitted or that some variables may have been incorrectly weighted.

3.19 Consider This: Climate Questions

If you visit climate-modeling sites on the web, you may be deluged with technical terms and numerical analyses. A good place to begin your understanding of climate modeling is to visit the National Climatic Data Center (NCDC), billed as "the world's largest active archive of weather data." A direct link to the NCDC is provided at the CiC web site. What types of data does the NCDC provide? Propose two or three questions that you might like to investigate using data provided at the NCDC.

One group of researchers, led by Benjamin Santer of Lawrence Livermore National Laboratory, has found that predictions agree more closely with observations if the model includes the cooling effect of atmospheric aerosols. These aerosols consist primarily of tiny particles of ammonium sulfate that form from sulfur dioxide released by natural or artificial sources. These particles promote global cooling by reflecting and scattering sunlight. In addition, they serve as nuclei for the condensation of water droplets and hence cloud formation. Thus, aerosols counter the effects of greenhouse gases. The temporary drop in average global temperature that followed the eruption of Mount Pinatubo in 1991 may well have been the consequence of the large volume of sulfur dioxide released by the volcano. Santer and his colleagues argue in a 1995 report by the Intergovernmental Panel on Climate Change that the evidence strongly supports the position that human activity is the cause of the increase in average global temperature observed over the last century.

It is possible that some climatic models may underestimate the amount of heat absorbed by the oceans. Much of the heat radiated by the greenhouse gases may be going into the oceans, which are acting as a thermal buffer. But although the oceans are very important in moderating the temperature of the planet, there are limits to their capacity to do so. It is also instructive that some climatologists have found that if greenhouse gases are *omitted* from their computer models, predictions *underestimate* observed temperature increases.

Given the complexity of the global system, there is considerable uncertainty associated with extrapolating the climate and the weather into the future. There is no wonder, then, that experts sometimes disagree. First of all, there is the matter of projected levels

| Ammonium sulfate is $(NH_4)_2SO_4$.

of greenhouse gases. The **rate** of their emission is currently increasing by about 1.5% per year. This is largely a consequence of growing global population, agricultural production, and industrialization. The population of the planet has tripled in this century and it is expected to double or triple again before reaching a plateau sometime in the next century. Industrial production is 50 times what it was 100 years ago. In the next 50 years, it will probably grow to 5 or 10 times what it is today. Most of this growth has been powered by the combustion of fossil fuels. Every year, 2–3% more energy is generated than in the previous one, and most of it comes from the burning of coal. If these rates of fuel consumption continue, the atmospheric concentration of CO_2 will be double its 1860 level sometime between the years 2030 and 2050.

All models predict that this doubling will result in an increase in the average global temperature, but the magnitude of that increase is variously estimated between 1.5 and 4.5°C. Many predictions, including the most recent one of the prestigious Intergovernment Panel on Climate Change (IPCC), fall in the 1.0 to 3.5°C (2–6°F) range by the year 2100. A 1.0 to 3.5°C (2–6°F) increase in average global temperature might seem trivial. But it is not; a temperature drop of 5 to 9°F (2.8–5.0°C) is the difference between the current average global temperature and that of a much chillier epoch, the last ice age, about 20,000 years ago.

Figure 3.11 is a graph developed by the IPCC depicting atmospheric carbon dioxide concentrations since 1990 and three projections of CO_2 concentrations. The projections are based on differing scenarios, each assuming that no dedicated efforts are made to diminish greenhouse gas emissions. Because increased numbers of people translate into increased energy use and greater greenhouse gas emissions through burning fossil fuels, the scenarios incorporate population growth rates. A significant uncertainty in any such projections is trying to predict whether the rate of population growth will stabilize during the next hundred years; during the twentieth century worldwide population increased from 2 billion people to approximately 5.9 billion today. The low-end scenario assumes a world population in the year 2100 of 6.4 billion and an annual economic growth rate of 1.2%. The mid-range projection is based on 11.3 billion people with a 2.3% annual

1992a low-end: 1992c mid-range; 1992e high-end

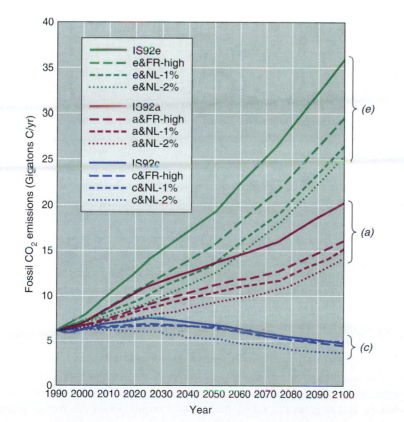

Figure 3.11

Three 1992 intergovernmental scenarios of global fossil CO_2 emissions (gigatons/yr). IS92a assumes intermediate population and economic growth, and limited fuel constraints. IS92c assumes low population growth, low economic growth, and severe constraints on fossil fuels. IS92e is based on intermediate population growth, high economic growth, and plentiful fossil fuels. Also included within each scenario are projections involving CO_2 limitation proposals by France (FR) and two by the Netherlands (NL).

economic growth rate, nearly twice that of the low-end scenario. In the high-end projection, the year 2100 population is 11.3 billion as in the mid-range calculation, but annual economic growth would occur at 3.0%.

Some indication of what we might expect, even under these best of conditions, can be gained by looking at the geological evidence for what the world was like 130,000 years ago. The average temperature was about 16°C, but very likely the poles were considerably warmer than they are today. As a result, the polar ice caps were smaller and the oceans were approximately 5 m (15 ft) higher than they are at present. A comparable increase in sea level would inundate the Netherlands, many islands, and the land where half of the 100 million people of Bangladesh currently live. An even worse catastrophe would occur if all or part of the East Antarctic ice sheet were to break loose and slip into the sea. An article in the May 2, 1995 *New York Times* reported on a conference at which scientists predicted that shedding of one-third of the ice sheet would raise the oceans by over 150 feet. Even if rises in sea level were significantly smaller— say 15 to 95 cm (6–37 in) as predicted by the IPCC—they would endanger New York, New Orleans, Miami, Venice, Bangkok, Taipei, and many other coastal cities. Millions of people might be made homeless. It is far from certain that major increases in sea level will in fact occur. Even if they do, they will take place over many years, providing considerable time for preparation and protection.

There is even more uncertainty associated with the regional weather patterns predicted by the various models. One of the more controversial forecasters is James Hansen of NASA. Hansen has estimated that doubling the concentration of greenhouse gases would mean that New York City could expect 48 days a year with temperatures above 90°F instead of the current 15. In Dallas, the number of days per year with temperatures above 100°F would increase from 19 to 87. It is important to note that many scientists have questioned Hansen's estimates.

Some climatologists also feel that an increase in the average temperature of the oceans could cause more weather extremes including storms, floods, and droughts. In the Northern Hemisphere, summers are expected to be drier and winters to be wetter. The regions of greatest agricultural productivity would probably change. Drought and high temperatures could reduce crop yields in the American Midwest, but the growing range might extend farther into Canada. It is also possible that some of what is now desert could get sufficient rain to become arable. One region's loss may well become another locale's gain.

But in other respects, we may all be losers in a warmer world. Recently, physicians and epidemiologists have attempted to assess the costs of global warming in terms of public health. An increase in average temperatures might increase the geographical range of mosquitoes, tsetse flies, and other insects. The result could be a significant increase in diseases such as malaria, yellow fever, and sleeping sickness in new areas, including Asia, Europe, and the United States (Figure 3.12). Indeed, it has been suggested that the deadly 1991 outbreak of cholera in South America is attributable to a warmer Pacific Ocean. The bacteria that cause cholera thrive in plankton, and the growth of both the plankton and the bacteria is stimulated by higher temperatures.

3.20 Consider This: Winners and Losers

If significant global warming occurs, some countries will probably be winners and some losers. Identify three nations that would most likely benefit from a warmer Earth and three that would face serious problems. State the reasons for your selections, including the gains and losses that you anticipate.

3.13 Has the Greenhouse Effect Already Started?

The answer to this question is most definitely "yes." Recall that without the greenhouse effect the average temperature of the Earth would be about −18°C. Under such circumstances we would not be here, or perhaps more correctly, we would be very different creatures. But as generally asked, the question implies "Has the average temperature

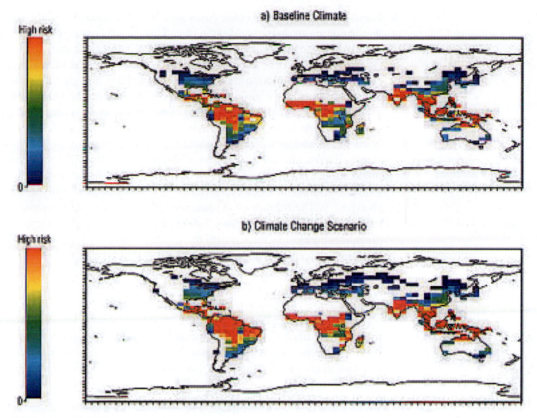

Figure 3.12

Climate change and risk due to global warming.

of the planet increased as a consequence of human activity?" That is, is global warming, the amplification of the greenhouse effect by human activities, occurring? Here the answer is somewhat less certain. Therefore, this is an excellent opportunity for the Sceptical Chymist to exhibit his or her skills of inquiry.

To better address the issues, it would be well for us to review the status of our knowledge. We do so by making some statements and then attempting to assess their accuracy. In this, we are guided by the work of J.D. Mahlman, an atmospheric scientist at Princeton University and the National Oceanic and Atmospheric Administration, as reported in "Uncertainties in Projections of Human-Caused Climate Warming" in the journal *Science* (November 1997.)

1. *Carbon dioxide contributes to an elevated global temperature.* Definitely true and supported by much experimental evidence, including the average temperatures of the Earth and Venus. The mechanism for global warming, the absorption of infrared radiation by vibrating molecules, is well understood and widely accepted.

2. *The concentration of carbon dioxide in the atmosphere has been increasing over the past century.* Definitely true. Analytical data strongly support this statement.

3. *The increase in atmospheric carbon dioxide over the past century is a consequence of human activity.* Virtually certain. Carbon isotope ratio measurements strongly suggest that at least part of the increase is attributable to human activities such as increased burning of fossil fuels and cutting of forests.

4. *There has been an increase in average global temperature during the past century.* Probably true. The data are consistent with this interpretation, indicating a rise of about 0.5°C ($\pm$0.2°C). Computer models as well as experimental data from ice core samples, tree ring measurements, and coral growth rates point to the twentieth century as the warmest since 1400. Measurement methods have changed during this period, which might call the conclusion into some question. Also, a century may be too short a time to reveal genuine temperature trends.

5. *The carbon dioxide and other gases generated by human activity are responsible for this temperature increase.* May be true. This causal connection is not unambiguously established. It is possible that there are other, natural causes for the measured temperature increase. Thus far, climate scientists have created increasingly sophisticated computer models linking the temperature increase and the increase in the concentration of greenhouse gases. The evidence implicating CO_2 from human sources is growing, but remains circumstantial. One of the most influential groups studying climate change is the Intergovernmental Panel on Climate Change, a collection of 2500 world-leading international scientists and technical experts whose mission is to assess climate change research in a balanced way through peer review. In its 1995 publication *Climate Change 1995,* the IPCC does not equivocate on the matter of global warming; it takes the position that humans have caused untoward climatic changes: "...The balance of evidence suggests a discernible human influence on global climate." It goes on to say "...With the growth in atmospheric concentrations of greenhouse gases, influence with the climate system will grow in magnitude, and the likelihood of adverse impacts from climate change that could be judged dangerous will become greater."

6. *The average global temperature will continue to increase as anthropogenic emissions of greenhouse gases increase.* Uncertain. This statement assumes that the (probable) increase in average global temperature observed over the last century has been caused by the (very likely) increase in human-generated CO_2 and other gases. As we have seen, this cause-and-effect relationship has not been unambiguously established. Extrapolations into the future are even more uncertain because of the complexities of the global system.

All of us have a strong temptation to address issues such as these by resorting to anecdote and personal experience. But such arguments can be misleading. It does not necessarily follow that the widespread North American drought of 1988, floods of 1993, or the summer heat waves of the 1990s were evidence of a global warming trend. Fluctuations in temperature and precipitation occurring over short periods of time are common and correspond to variations in *weather* patterns. They may not signal large-scale and long-range changes that shape the *climate.* On the other hand, the fact that ten of the 12 years between 1986 and 1998 have been the warmest on record may be significant and a predictor of things to come; 1998 has been the warmest to date since records have been kept (Figure 3.13).

3.21 Consider This: Deciding Whom to Believe

Given the rate at which new information about environmental effects is being generated, some parts of this book will be out of date before it comes to press. Consult the web to find two documents on global warming that were published in the last calendar year. For each, give the title, author or source, URL, and the date last updated. Summarize the new information that you found. If this information is different from your textbook, cite the differences.

Figure 3.13

Post-1980 global warming as seen from average global temperature change relative to a 1961–1990 average used as zero.

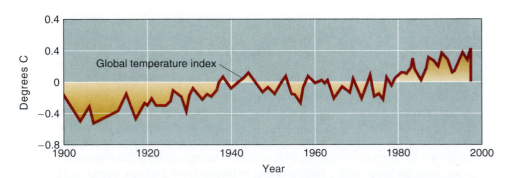

Reprinted with permission from Richard A. Kerr in *Science,* Vol. 279, January 16, 1998. Source: NCDC/NOAA. Copyright © 1998 American Association for the Advancement of Science.

3.22 The Sceptical Chymist: Global Warming Skeptics

Some people believe that human activities have amplified the greenhouse effect; others do not. Find out what such skeptics have to say about the topic. You can locate organizations that take a lukewarm view of global warming by searching for "global warming" and "skeptics." Many search engines allow "wildcards" such as *. Thus, by typing skeptic*, you will bring up sites that include related words such as skeptics, skeptical, or skepticism.

Locate a global warming skeptics web site and write a short description about the information found there. Include the URL, the title of the site, its sponsor, and the date the web site was last updated. Summarize three points made on the web site in opposition to the concept of global warming.

3.14 What Can We Do? What Should We Do?

Given these uncertainties, what can and should we do about the possibility of global warming? Climate change is a paradox, one described as "too serious to panic about" in a lengthy article in *The Economist* (June 18, 1994). One thing is clear: if the models are reasonably accurate, we will start seeing significant climatic changes within a decade or so. But can we prudently wait that long, or is prompt action essential? Whether or not to act, and how to act are not simply scientific issues. Bette Hileman in *Chemical & Engineering News,* March 13, 1989 wrote "This is a value question involving perceptions of risk and uncertainty."

Those who dare to answer that question can be segregated (somewhat arbitrarily and unfairly) into three very different camps: (1) must act now; (2) study it more; and (3) don't act because it's inevitable. In the "act now" category, Dr. Stephen Schneider, a Stanford University biologist, notes that the twentieth century has produced a distinct global warming trend accompanied by precipitation pattern changes that are difficult to explain without invoking anthropogenic generation of greenhouse gases. In the December 1, 1997 *New York Times,* Dr. Schneider expresses his support to take action now: "Is this nature being perverse or is it us? The only way to prove it for sure is hang around 10, 20, or 30 more years, when the evidence would be overwhelming. But in the meantime, we're conducting a global experiment. And we're all in the test tube."

There are those who advocate delaying because further study is needed. They argue that the uncertainties in our predictive powers and climate models are so great that it would waste money and effort to undertake preventive or ameliorative action at this time; without more knowledge we run the risk of making more mistakes. Such is the attitude of William O'Keefe, CEO of the American Petroleum Institute, who declared: "I do not agree that it's a foregone conclusion that there's going to be significant climate change if we don't take dramatic action now. We also have to reduce the scientific uncertainties. We should not base policy on worst-case scenarios." Yet, it is this attitude that Dr. Michael Oppenheimer, chief scientist of the Environmental Defense Fund rebuts: "Scientific uncertainties, which are substantial, of course, [regarding global warming] are not a reason to put off action. In fact, we only have one Earth to experiment on."

In marked contrast to the "wait and study" and "it's inevitable" schools of thought are the "do anything but do something fast" activists. The danger with such an approach is that some of the proposed cures may be worse than the disease. A number of emergency responses have been suggested, and most of them have been discredited. For example, some have proposed that we follow the example of Mount Pinatubo and release sulfur-containing compounds that would generate cooling sulfate aerosols in the atmosphere. But critics have pointed out that the equivalent of 300 Pinatubo eruptions would be required each year to counteract the projected temperature increases. The not-so-desirable associated effects would include global acid rain, serious ozone loss, and severe air pollution. Another approach suggested fertilizing southern oceans with iron to increase the growth of phytoplankton. According to the plan, these primitive green plants would thrive, absorbing vast quantities of carbon dioxide from the atmosphere.

This strategy was even tested, and found wanting. Iron did increase phytoplankton growth, but the growth of zooplankton, the animals that eat the phytoplankton, kept pace; there was no significant change in net CO_2 absorption.

Others suggest a very different approach involving various ways of putting the carbon dioxide back into the Earth. MIT scientists propose capturing the CO_2 and pumping it into the ocean depths, pointing out that much of the carbon dioxide now emitted eventually finds its way into the oceans anyway. Opponents respond, citing recent experimental evidence, that increased oceanic carbon dioxide could damage coral reefs. Underground sequestering is another approach to preventing the release of CO_2 to the atmosphere. This method uses to advantage the fact that carbon dioxide is a sometimes frustrating contaminant found in underground natural gas deposits in conjunction with crude oil. To rid themselves of this nuisance, oil companies propose separating the CO_2 from the natural gas, and pumping the CO_2 directly back underground either into nearby depleted natural gas fields, or use it to flush out residual crude oil from shrinking reservoirs. This method could be especially useful in the case of a huge Indonesian natural gas deposit. The deposit is principally CO_2 which, if released, would equal about 0.5% of all carbon dioxide emitted globally from fossil fuel burning. This approach or those previously suggested must be evaluated thoroughly because, although some beneficial and benign forms of climate engineering may be possible, we would be well advised to proceed with caution.

Still others look at the magnitude of the potential problems associated with global warming and conclude that there is nothing that humanly can be done to halt or reverse the process. Their outlook is that the generation of energy, and with it, carbon dioxide, is an essential feature of modern industrial life, and we must therefore learn to live with its consequences and begin adapting to our warm new world.

You are encouraged to debate and discuss these and other options. Clearly, your opinions and the evidence you marshal to support them are important. However, the authors of this text hope that there will be at least some advocates who develop a compromise strategy out of the best characteristics of the extreme positions just described. There is, after all, an element of truth in each position.

3.23 The Sceptical Chymist: Cooler Heads

Has the greenhouse effect been amplified by human activities? **a.** Find some web links to several organizations that don't believe so. Try the Global Warming Skeptics web page, or that from the Greening Earth Society. Newspapers, television, and radio-related web pages may contain editorials that criticize global warming as a scientific concept. **b.** Is there scientific merit to the arguments of those opposing global warming? Cite specific cases.

We take it as a given that no matter what happens, careful study is essential to improve our options. It is a response that carries little risk and the potential for great benefit. Extensive environmental monitoring is necessary to provide more reliable climatic and meteorological data. Appropriate technology must be developed and transferred to those who need it. And the public must be educated and informed. We also believe, however, that complete evidence and absolute certainty will never be ours, and that intervention, based on the best available scientific and technical information, is called for. Some of this activity can and should be designed to reduce the extent of global warming; some should be designed to mitigate the temperature increases that will very likely occur, no matter how earnest our efforts.

The most obvious strategy for dealing with global warming would seem to be to reduce reliance on fossil fuels. Such action is incredibly difficult, not only because this energy source is so important to our modern economy, but because of its international dimensions. Although the developing countries may well become the major producers of carbon dioxide and other greenhouse gases in the future, the developed countries have a prodigious lead. The annual carbon output (in the form of CO_2) of the United States is over five metric tons for each of its inhabitants compared to a worldwide average only

1 metric ton = 2200 lb.

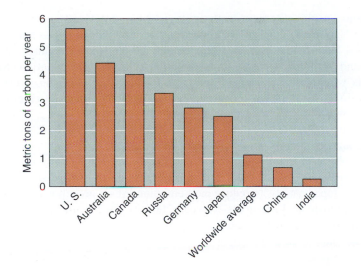

Figure 3.14

Per capita CO_2 emissions (measured as metric tons of carbon/year) for selected nations.

(Source: Data from Oak Ridge National Laboratory)

one-fifth as great (Figure 3.14). On a per capita basis the United States leads the world. Comparable per capita values for China and India are about 0.7 and 0.3 tons, respectively. Even so, the Peoples' Republic of China ranks second behind the United States in *total* carbon dioxide emissions from fossil fuels. If China were to succeed in raising its per capita gross national product to only 15% of the U.S. figure, the increase in CO_2 production would approximately equal the current American annual emissions from coal. The IPCC has estimated that by 2010, the developing countries plus the former Soviet Union nations will produce more than half of the world's CO_2 emissions (Figure 3.15). It is unrealistic for the developed countries to expect the nations of the Third World to abandon their hopes for economic growth and become good, non-polluting global citizens.

3.24 Consider This: The Top 20

It is no secret which countries are emitting the highest amounts of CO_2. You can access a list of the Top 20 fossil fuel CO_2 emitters right on the web. This list is provided by the Carbon Dioxide Information Analysis Center and gives carbon dioxide emissions both per country and per capita (person) as of a year or two ago. Three sets of data are available: regional, national, or global basis.

a. Your textbook cites the United States and Peoples' Republic of China as the leaders in total CO_2 emissions from fossil fuels. What countries rank 3rd, 4th, and 5th? Whose emissions are relatively low? Access the national database to answer these questions.

b. Now look at the emissions per capita. Which countries have the highest emissions per person? Do the top five differ per capita by very much?

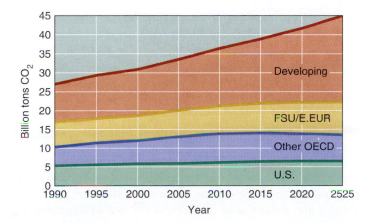

Figure 3.15

Total CO_2 emission projections from various global regions: IPCC IS92 Scenario output. (FSU/E.EUR): Former Soviet Union/Eastern Europe; (Other OECD): Organization for Economic Cooperation and Development—Western Europe and North America.

(Source: Data from IPCC/EPA Report 1992)

In response to this dilemma, anthropologist Margaret Mead and William W. Kellogg of the National Council on Atmospheric Research proposed as early as 1976 an international "law of the air" in which worldwide emission standards would be established and enforced by assigning "polluting rights" to each nation. Such global agreements would be extremely difficult to achieve. Thus far, the international response to global warming has been notably less effective than the reaction to ozone depletion.

Attempts have been made to respond to global warming. Nearly 10,000 participants from 159 countries at the Kyoto Conference in December 1997 established goals to stabilize atmospheric greenhouse gas concentrations at environmentally more responsible levels. To meet these goals, binding emission targets based on five-year averages were set for 38 developed nations to reduce their emissions of six greenhouse gases from 1990 levels. Accomplishing these goals between 2008 to 2012 would potentially decrease emissions from industrialized nations overall by about 5%. Under the Kyoto Conference protocol, the United States will be expected to reduce emissions to 7% below its 1990 levels, the European Union nations 8%, and Japan 6%. No new binding emission targets were established for developing countries, a contentious issue. Developed nations are permitted to trade emission credits to meet their targets. That is, countries that have emissions lower than their targets can sell the residual amounts to countries exceeding their targets. Developed nations also can receive further credits, as early as 2000, for investments and projects to help developing countries reduce their emission of greenhouse gases through better technologies.

Although there is no deadline for its ratification, countries are expected to sign the Kyoto agreement (treaty) by March 15, 1999. The treaty does not go into effect until 90 days after at least 55 countries that account for at least 55% of total 1990 CO_2 emissions have ratified the treaty. President Clinton has made it clear that he will not submit the treaty to the U.S. Senate for ratification until key developing countries agree to limit their greenhouse gas emissions. This step is important because the rates of greenhouse gas emissions of developing nations are increasing more than those of industrialized countries, and expected to grow even faster.

Writing in *Scientific American* (June 1995), after conferences leading up to the Kyoto conference, Tim Beardsley had this wry comment: "Just as St. Augustine prayed for chastity—'but not yet!'—parties at the climate convention meeting...expressed an earnest desire to do something about releases of greenhouse gases, chiefly carbon dioxide—but not yet." The Kyoto Conference has some "not yet" aspects about it. Because the wheels of governments move slowly, nations need ample lead time to affect changes. But the conference has also brought forth agreements from major developed nations to attempt to do something to diminish the release of greenhouse gases, and established targets to do so. Americans as well seem to agree that something needs to be done. In a Harris poll conducted in the United States shortly after the Kyoto Conference, 75% of those polled supported the Kyoto accord. But, as Bette Hileman cautioned, global warming is a value question involving perceptions of risk and uncertainty. Whether the U.S. Senate will ratify the Kyoto Protocol is problematic, with the confrontation cast along party lines, leading to a highly politicized debate. Democrats, in general, support the treaty whereas most Republicans oppose it, objecting that ratification would be economically irresponsible, especially in light of the uncertainties surrounding global warming.

3.25 Consider This: Three Reactions to Global Warming

The text identifies three extreme reactions to the problem of global warming: continue to study it, act to prevent it, or prepare for it. The most effective response will take place in the shortest amount of time if human and financial resources are directed toward a single unified plan. Carefully review the three positions and their consequences. Discuss the situation with others and prepare to state and defend your position to the rest of the class.

3.26 Consider This: Kyoto Conference Humor

What is the humor in this cartoon? Would everyone find it amusing? Explain your reaction to this cartoon, including whether or not you feel it is trying to communicate a certain point of view.

© Tribune Media Services, Inc. All Rights Reserved. Reprinted with permission.

3.27 The Sceptical Chymist: A Bit of "Spin" About the Kyoto Conference

An advertisement that was headlined "The Only Thing This Treaty Cools Down Is America's Economy" appeared in the December 1, 1997 *New York Times.* The ad was in the same issue as a feature article on the Kyoto Conference. Find a copy of this advertisement in your library or on the web. To help in analyzing the factual content of this ad, make a list of the assertions made. For each statement, categorize your response as having enough information to judge it is a valid statement, having enough information to judge it is not a valid statement, or needing more information about the assertion before reaching a decision. Bring your list to class for further discussion.

3.28 Consider This: Joint Implementation of Greenhouse Gas Guidelines

In December 1997, when the international community met in Kyoto to discuss the implementation of the Rio de Janeiro Accord of 1992 to ban the use of greenhouse gases worldwide by the year 2000, discord dominated the discussion. Developing nations were unwilling to accept limits that might imperil their economic growth, whereas industrialized nations were unwilling to bear the burden of acting alone in cutting greenhouse gas emissions. Both sides did agree to a policy of "joint implementation" that permits industrialized nations to exceed their targeted reduced production of greenhouse gases, if they financially support projects to reduce production of greenhouse gases in developing countries. This general approach to pollution control has been used successfully in other contexts.

a. List the scientific advantages and disadvantages of "joint implementation."
b. List the societal advantages and disadvantages of "joint implementation."

John Topping, president of the Climate Institute, points out the major difficulties in drafting international proposals for climate change: "The vast majority of industrialized countries are failing to meet their own targets, while developing countries' emissions are exploding. The implications are—barring the introduction of radical, new, and cost-effective technologies to replace fossil fuels—carbon dioxide emissions are going to go through the roof. We need to focus on developing a serious, international public-private partnership to bring on new clean-energy technologies. We have to transform the discussion by bringing to the table engineers and innovators from industry. If we don't, all the haggling over this regulation or that program is beside the point." There is a suggestion that the international debate on global warming has contributed little else than more hot air to an already warm planet. Richard Linzen, an MIT meteorology professor and persistent global warming sceptic, said in the *New York Times* (December 1, 1997) that "...negotiations in Kyoto were mostly focused on bolstering resumes of diplomats... Climate always changes, whether man does anything about it or not. Now any changes will be attributed to policy, not nature."

To be sure, it is technically possible to trap the CO_2 released by burning fossil fuels, but the cost in money and energy would be large. A more feasible solution would be to shift to energy sources that release little or no CO_2. For example, replacing coal with natural gas would significantly reduce carbon dioxide emissions. Nuclear power generates no greenhouse gases at all, but this alternative is not without risks, and its costs and benefits will be analyzed in Chapter 7. Renewable energy sources and wind, solar, hydroelectric, and geothermal power have been proposed. Clearly, the development of alternate energy sources must be a high global priority. Deposits of coal, natural gas, and petroleum are not only finite, they are too valuable as raw materials to be consumed totally by combustion. Conserving energy by making power plants, furnaces, factories, and automobiles more efficient is a strategy worth pursuing under any circumstances, but there are limits imposed by laws of nature as well as by engineering skill and the properties of materials.

Humans can also intervene by rectifying their assaults on the planet. Deforested areas can be replanted, converting a CO_2 source back to a CO_2 sink. And while promoting the growth of other species, we must be aware of the fact that many of our environmental problems are related to the expanding human population. Unfortunately, it is unlikely that these preventive measures, though necessary, will be sufficient to avert a temperature increase. We must also be prepared to meet and mitigate future climate changes. This includes protecting arable soil, improving water management, prudently using agricultural technology and agricultural chemicals, maintaining global food reserves, and establishing an effective mechanism for disaster relief.

3.15 Global Warming and Ozone Depletion

Global warming and ozone depletion are important environmental issues that involve the atmosphere, and both are much in the news. There are enough apparent similarities between the two phenomena that the casual reader of newspaper accounts may mix them up. Sometimes the authors of the articles get confused. One aim of this text is to avoid such mix-ups, and for that reason it is probably a good idea to conclude this chapter by summarizing some of the important differences between the greenhouse effect and ozone destruction. We do so in Table 3.3. Such a tabulation is an invitation to oversimplification, but it can be a useful reminder of some of the important aspects of these two environmental problems. 3.29 Consider This provides you with an opportunity to express your informed opinion about their relative significance.

3.29 Consider This: Air Quality, Ozone Depletion, or Global Warming

Now that you have studied air quality (Chapter 1), ozone depletion (Chapter 2), and global warming (Chapter 3), which do you believe poses the most serious problem? Which is most easily solved? Discuss your reasons with others and draft a one-page report on this question.

Table 3.3 Global Warming and Ozone Depletion: Some Characteristics

	Global Warming	Ozone Depletion
Region of atmosphere involved	Mostly troposphere	Stratosphere
Major substances involved	CO_2, CH_4, N_2O	O_3, O_2, CFCs
Interaction with radiation	IR radiation absorbed by molecules, which vibrate and remit energy to Earth	UV radiation absorbed by molecules, which are dissociated into smaller fragments
Nature of problem	Increasing concentrations of greenhouse gases are apparently increasing average global temperature	Decreasing concentration of O_3 is apparently increasing exposure to UV radiation
Source of problem	Release of CO_2 from burning fossil fuels, deforestation; CH_4 from agriculture	Release of CFCs from refrigeration, foaming agents, solvents; CFCs release Cl, which destroys O_3
Possible consequences	Altered climate and agricultural productivity, increased sea level	Increased incidence of skin cancer, damage to phytoplankton
Possible responses	Decrease use of fossil fuels and discontinue deforestation	Eliminate use of CFCs and find suitable replacements

Conclusion

"For the first time in my life I saw the horizon as a curved line. It was accentuated by a thin seam of dark blue light—our atmosphere. Obviously this was not the ocean of air I had been told it was so many times in my life. I was terrified by its fragile appearance."

Ulf Merbold

This chapter and the two that preceded it have disclosed that our atmosphere is more robust than it appeared to the German astronaut, Ulf Merbold (Figure 3.16). Nevertheless, during the twentieth century, the air upon which our very existence depends has been subjected to repeated assaults. The fact that most of these environmental insults were unintentional and, in some cases, the unexpected consequences of social progress, does not alter the problems we face. We have only recently recognized the potential harm that air pollution, ozone depletion, and global warming can bring to our personal, regional, national, and global communities. To reverse the damage already done and to prevent more, all of these

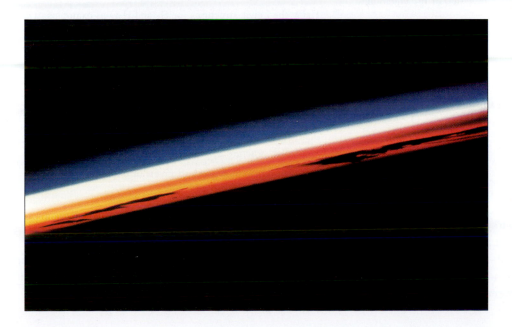

Figure 3.16

An astronaut's view of the Earth's atmosphere from an altitude of 128 miles. The black regions at the top and bottom of the picture are outer space and the Earth, respectively. The outermost layer of the atmosphere appears blue. The reddish streak immediately below the white band, at an altitude of 20–27 km, is caused by ash and sulfuric acid particles from the eruption of Mount Pinatubo in 1991. The region closest to the surface of the Earth is dark red because the sunlight is scattered by dust, smoke, and water vapor.

communities must respond with intelligence, compassion, commitment, and wisdom. It is instructive that even in the absence of threats such as global warming, much of what has been advocated in the preceding section would be sound, prudent, and responsible stewardship of our planet.

Chapter Summary

Having completed this chapter you should be able to:

- Understand the mechanism by which global warming occurs (amplified greenhouse effect), the chief role played by carbon dioxide in it, and the nature of the carbon cycle (3.3–3.6)
- Relate Lewis structures and molecular geometry to absorption of particular radiation, for example, infrared (3.4–3.5)
- Know which molecular species are greenhouse gases because of their molecular structures; write such Lewis structures (3.4)
- Summarize the contributions human activities make to the carbon cycle and through it, to global warming (3.7)
- Apply atomic weight and Avogadro's number data to molar mass and mole calculations (3.8–3.10)
- Recognize the contributions made by computer models of climatic change and the current limitations to such models (3.12)

- Consider the national and global implications of a 1.5–4.5°C rise in the Earth's average temperature (3.12–3.13)
- Describe ways in which carbon dioxide emissions can be reduced (3.14)
- Assess the factors involved in global warming (3.1, 3.3, 3.4, 3.11–3.14)
- Read and hear news articles on global warming with some measure of confidence in your ability to interpret the accuracy and conclusions of such reports
- Take an informed position with respect to issues surrounding global warming

Questions

Emphasizing Essentials

1. Concentrations of CO_2 in the early atmosphere of the earth were much higher than today. What happened to this CO_2?

2. **a.** Write and balance an equation for the photosynthetic conversion of CO_2 and H_2O to form glucose, $C_6H_{12}O_6$, and O_2. Show all states for reactants and products, and include any necessary conditions for the reaction to take place.

   b. Demonstrate that the equation is balanced by counting atoms of each element on either side of the arrow.

   c. Is the number of molecules on either side of the equation the same? Why or why not?

3. Using the analogy of a greenhouse to understand the energy radiated by the Earth, what are the "windows" of the greenhouse made of?

4. Consider Figure 3.1.

   a. How does the concentration of CO_2 in the atmosphere at present compare with the concentration of CO_2 20,000 years ago? How does the present concentration of CO_2 compare with the concentration 120,000 years ago?

   b. Compared with the 1950–1980 mean temperature of the atmosphere, how does the temperature at present compare? How does the temperature 20,000 years ago compare? How does each of these values compare with the average temperature 120,000 years ago?

   c. Do your answers to parts **a** and **b** indicate causation, correlation, or no relation? Explain.

5. Understanding the Earth's energy balance is essential to understanding the issue of global warming. For example, the solar energy striking the surface of the earth averages 169 watts per square meter, but the energy leaving the surface of the earth averages 390 watts per square meter. Why isn't the Earth cooling rapidly?

6. Explain each of the observations.

   a. The inside of a car left in the sun may become hot enough to endanger the lives of pets or small children.

   b. Clear winter nights tend to be colder than cloudy ones.

   c. There is a much wider daily temperature variation in the desert than in a moist environment.

7. Using the Lewis structures for H_2 and H_2O as examples, why is it that the Lewis structure for H_2 allows an unambiguous prediction of molecular geometry whereas this is not the case for H_2O?

8. Use a molecular model kit to make a methane molecule, CH_4. (If a kit is not available, this model can be made using Styrofoam balls or gumdrops to represent the atoms and toothpicks to represent the bonds.) Prove that the hydrogen atoms are farther from each other in a tetrahedron than they would be if the molecular structure of methane were square planar.

9. Use the step-wise procedure given in the text to predict the shape of each of these molecules.

   a. CH_2Cl_2

   b. CO

   c. HCN

   d. PH_3

10. **a.** Write the Lewis structure for H_3COH, which is methanol or wood alcohol.

b. Based on this structure, predict the hydrogen-to-carbon-to-hydrogen bond angle. Explain the reason for your prediction.

c. Based on this structure, predict the hydrogen-to-oxygen-to-carbon bond angle. Explain the reason for your prediction.

11. a. Write the Lewis structure for H_2CCH_2, ethene, a simple hydrocarbon.

b. Based on this structure, predict the hydrogen-to-carbon-to-hydrogen bond angle. Explain the reason for your prediction.

c. Sketch the molecule showing the predicted bond angles.

12. The text states that a UV photon can break chemical bonds but that an IR photon can only cause vibration in the bonds.

a. Calculate the energy associated with each of these processes by assuming a wavelength of 320 nm for the UV photon and a wavelength of 5000 nm for the IR photon.

b. What is the ratio of the energy that breaks bonds to the energy that causes vibration?

13. Here are three different modes of vibration of a water molecule. Imagine the atoms being connected by bonds that act as springs. Each vibration can be seen by moving each atom in the direction indicated and then back again.

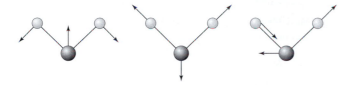

Can any of these modes of vibration contribute to the greenhouse effect? Explain your reasoning.

14. If a carbon dioxide molecule interacts with certain photons in the IR region, the molecule will vibrate. For CO_2, the major wavelengths of absorption occur at 4.26 μm and 1500 μm.

a. What is the energy for each of these IR photons?

b. What happens to the energy in the vibrating CO_2 species?

15. What effect would each of these changes have on global warming?

a. volcanic eruptions

b. CFCs in the troposphere

c. CFCs in the stratosphere

16. One of the biochemical processes that releases carbon dioxide to the atmosphere is the fermentation of sugar to produce alcohol. For example, when glucose, $C_6H_{12}O_6$, undergoes fermentation, ethanol (C_2H_5OH) and CO_2 are produced. Yeast is used to catalyze this conversion. Write a balanced chemical equation for this reaction.

17. Consider Figure 3.10.

a. What is the major source of CO_2 emission from fossil fuel combustion?

b. What are some of the alternatives for each of the major contributors to CO_2 emissions?

18. Silver (Ag) has an atomic mass of 107.870 and an atomic number of 47.

a. What is the number of protons, neutrons, and electrons in a neutral atom of the most common isotope, Ag-107?

b. How do the numbers of protons, neutrons, and electrons in a neutral atom of Ag-109 compare with those of Ag-107?

19. There are just two naturally occurring isotopes of silver. If silver-107 accounts for 52% of natural silver, what is the mass number of the other isotope of silver?

20. a. Calculate the average mass (in grams) of an individual atom of silver.

b. Calculate the mass (in grams) of ten trillion silver atoms.

c. Calculate the mass (in grams) of 5×10^{45} silver atoms.

21. Calculate the molar mass of each of these substances important in atmospheric chemistry.

a. H_2O

b. CF_2Cl_2 (Freon 12)

c. CO

22. a. Calculate the mass ratio and mass percent of oxygen in H_2O.

b. Calculate the mass ratio and the mass percent of fluorine in CF_2Cl_2.

c. Calculate the mass ratio and mass percent of carbon in CO.

23. The total mass of carbon in living systems is estimated to be 7.5×10^{17} g. Given that the total mass of carbon on the Earth is estimated to be 7.5×10^{22} g, what is the concentration of carbon atoms on the earth? Report your answer in percent and in ppm.

24. Consider the information presented in this graph.

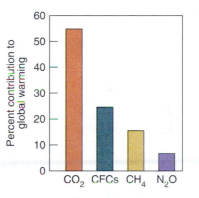

a. Which substance makes the largest contribution to global warming?

b. Use these percentages together with the greenhouse factors given in Table 3.2 to calculate the net effect of each of these gases on global warming. Which gas has the largest net effect?

25. Use the data in Table 3.2 to rank these gases in order of decreasing Greenhouse Factor. CH_4, CO_2, CCl_3F, N_2O, H_2O, CCl_2F_2, and O_3

26. Consider the information presented in Table 3.1. Calculate these changes.

a. What is the percent increase in CO_2 when comparing 1994 concentrations with pre-industrial concentrations?

b. Considering CO_2, CH_4, and N_2O, which has shown the greatest percentage increase when comparing 1994 concentrations with pre-industrial concentrations?

Concentrating on Concepts

27. The text makes a distinction between the *correlation* of two events and the *causation* of one by another. Identify each of these pairs as an example of correlation, causation, or having no relationship. Explain your reasoning.

    a. metric tons of coal burned   metric tons of CO_2 emitted

    b. national per capita income   per capita emission of CO_2

    c. number of cigarettes smoked per day   incidence of lung cancer

    d. number of bonds between two oxygen atoms   length of oxygen-to-oxygen bond

    e. building a greenhouse   raising beautiful tropical plants

    f. buying a pair of roller blades   breaking your leg

28. Why do people sometimes bring living plants, rather than cut flowers, to help a friend along the road to recovery from an illness?

29. Given that direct measurements of the Earth's atmospheric temperature over the last several thousands of years are not available, how can scientists know the fluctuations in the temperature in the past?

30. Consider Figure 3.3 showing atmospheric concentrations of CO_2 at Mauna Loa, and Figure 3.5, showing temperature changes on the Earth's surface. Why is the pattern with the trend so regular in Figure 3.3, but not as regular in Figure 3.5?

31. Consider this graph showing sources of CO_2 emissions in the United States.

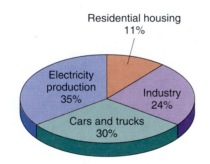

Residential housing
11%

Electricity production
35%

Industry
24%

Cars and trucks
30%

    a. If you want to cut down your personal contribution to CO_2 emissions, what changes can you make that will be most effective? Explain your reasoning.

    b. Will the entire quantity of CO_2 emitted end up in the atmosphere? Why or why not?

32. Why is it said that Lewis dot structures show linkages (what is hooked to what) but they do not show shape? Explain what is meant by this statement, using the molecule H_2S as an example.

33. The molecule BF_3 is triangular planar with 120° fluorine-to-boron-to-fluorine bond angles, whereas the NF_3 molecule is triangular pyramidal with fluorine-to-nitrogen-to-fluorine bond angles of about 103°. Account for the different geometries and bond angles. *Hint:* Boron is too small to obey the octet rule and will be stable with only six electrons around it in the Lewis structure.

34. Carbon dioxide gas and water vapor both absorb IR radiation. Do they also absorb visible radiation? Offer some evidence based on your everyday experiences to help explain your answer.

35. How do you think that the energy required to cause IR-absorbing vibrations in CO_2 would change if the carbon and oxygen atoms were bonded with single bonds, rather than with double bonds?

36. Explain why food placed in a plastic container is quickly warmed in a microwave oven, while the container warms much more slowly.

37. Consider Figure 3.9. Where is a carbon atom most likely to be found? Rank living systems, the oceans, and the Earth in order of increasing amounts of carbon they contain, giving reasons for your ranking.

38. Why is the atmospheric lifetime of a potential greenhouse gas important?

39. CO_2 gas has a greater density than N_2 gas. Why don't these gases settle out into layers in the atmosphere?

40. One of the first radar devices developed during World War II used microwave radiation of a specific wavelength that triggers the rotation of water molecules. Why was this design not successful?

41. McKibben suggests that 73 million metric tons of CH_4 are produced by the Earth's ruminants, such as cattle and sheep, each year. How many metric tons of carbon are present in this mass of CH_4?

42. **a.** The January 16, 1998 issue of *Science* magazine reported that 1997 was the warmest year of this century. Does this prove that theories about global warming are correct? Explain your reasoning.

    b. Evidence has been reported that certain butterflies have recently changed their migration patterns and are moving farther north. Does this prove that theories about global warming are correct? Explain your reasoning.

Exploring Extensions

43. Figure 3.5 shows the temperature increase on the Earth's surface from 1880 to 1996. Imagine you are in charge of extending this graph to include the present year. What kind of information would you need and how might you gather such data? *Hint:* Consider the source of the data in Figure 3.5.

44. If a water molecule interacts with certain photons in the IR region, the molecule will vibrate. This equation can be used to represent that interaction, in which the species with the * represents the vibrating molecule.

$$H_2O + hv \longrightarrow H_2O*$$

    a. The maximum absorbance in the IR for water vapor occurs at a wavelength of $6.6 * 10^{-6}$ m. Calculate the energy associated with each IR photon represented by this equation.

    b. Which absorption peak represents higher energy?

    c. Using the * notation, write an equation representing a vibrating H_2O molecule passing its energy to an H_2O molecule that is not vibrating.

45. Data taken over time show an increase in CO_2 in the atmosphere. The large increase in the combustion of hydrocarbons since the Industrial Revolution is often cited as a reason for the increasing levels of CO_2. However, an increase in water vapor has *not* been observed during the same time period. Remembering the general equation for the combustion of a hydrocarbon, does the difference in these two trends *disprove* any connection between man's activities and global warming?

46. **a.** Assuming that only single bonds are present, how many outer electrons are needed to form a three-atom molecule formed from atoms of **X, Y,** and **Z?** Assume that **X, Y,** and **Z** follow the octet rule.

    b. What is the shape of the molecule formed from **X, Y,** and **Z** in part **a?** What is the **X**-to-**Y**-to-**Z** bond angle? Explain your reasoning.

    c. How will the answers to parts **a** and **b** change if you remove the restriction on using only single bonds? Explain.

47. Consider this information about three greenhouse gases; these data are from the United Nations Intergovernmental Panel on Climate Change.

    See table below.

    Use these data to write a commentary for your local newspaper, explaining which gases are experiencing the greatest percent increases. Explain some of the reasons for the observed increases, and explain why knowing the atmospheric lifetime of a greenhouse gas is an important piece of information for setting control strategies.

48. The per capita CO_2 emissions for several countries are shown in Figure 3.14. This table gives the per capita income for some of those countries.

U.S.	$22,470
Australia	$18,054
Canada	$19,400
Japan	$19,100
China	$360
India	$380

    a. What is the relationship between the per capita CO_2 emissions and the per capita income? Offer some reasonable explanation for the relationship.

    b. Considering the relationship between per capita income and per capita CO_2 emissions, what are the policy implications for implementing the Kyoto agreement?

49. The CNN business news web site reported this headline on November 9, 1998. "Some U.S. businesses warming to global emissions treaty." What type of businesses do you predict are supporting the Kyoto agreement? Which are most likely not to be supportive? What changes in business support for the Kyoto agreement have taken place in the last year? Research this aspect of U.S. support for the Kyoto agreement, and write a report summarizing your findings.

50. The atmospheric problems described in Chapters 2 and 3 have stimulated different responses. The evidence for ozone-depletion resulted in the Montreal Protocol, which included a schedule for decreasing the production of ozone-depleting chemicals. The evidence for global warming led to the Kyoto agreement, which calls for targeted reductions in greenhouse gases. Suggest some reasons why the question of ozone depletion was dealt with by the world community before taking up the question of global warming.

Gas	Pre-Industrial Concentration	1994 Concentration	Atmospheric Lifetime
CO_2	280 ppm	358 ppm	50–200 years
N_2O	275 ppb	311 ppb	120 years
CH_4	700 ppb	1721 ppb	12 years

Energy, Chemistry, and Society

Oil wells and a pumping derrick in Huntington Beach, CA. We are a society that is dependent on oil. Because of this, the search for oil and bringing it to the surface is non-ending.

Gasoline, automobiles, and the U.S. lifestyle are inextricably bound to each other. We value our vehicles and the freedom of movement they provide, motion fueled by gasoline, itself extracted from crude oil (petroleum). Although gasoline has been extracted from petroleum since the mid-1800s, it became the most important and valuable crude oil component only with the advent of the automobile and the internal combustion engine early in the twentieth century. That fuel-engine partnership has led to our seemingly insatiable appetite for gasoline. In 1998, almost 120 billion gallons of gasoline were burned in more than 200 million American automobiles. Our capacity to consume gasoline has far outstripped our ability to produce it from crude oil extracted in this country. By the 1970s, the United States was producing less than half of the crude oil it required to power its automobiles and factories, heat its homes, and lubricate its machines (Figure 4.1a). Our fragile dependency on oil from abroad continues today. The Persian Gulf region has been the world's major petroleum producer for more than five decades (Figure 4.1b). The 1991 Gulf War with Iraq over issues relating to energy sources and fuel supplies is a reminder of this dependency.

Figure 4.1a

U.S. oil production and oil imports. At present, more than 50% of the total oil used in the United States is imported, and projections show oil imports will continue to increase.

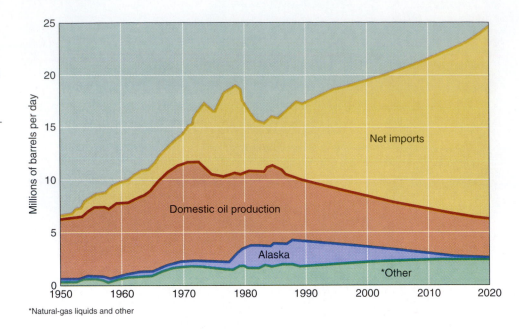

*Natural-gas liquids and other

4.1 Consider This: Global Oil Production

Figure 4.1b shows a downturn in Persian Gulf oil production in 1974. Speculate on reasons why Persian Gulf oil production decreased then, and why this caused a U.S. energy crisis.

4.2 Consider This: Oil Production Around the World

Imagine that you are living in one of the other countries for which oil production figures are given in Figure 4.1b. Assume that you must have a car to travel to school or work each day and that you will live under conditions similar to your own home. How will the production of oil influence your lifestyle in that country? Identify some of the factors that would influence your answer.

Figure 4.1b

Global oil production

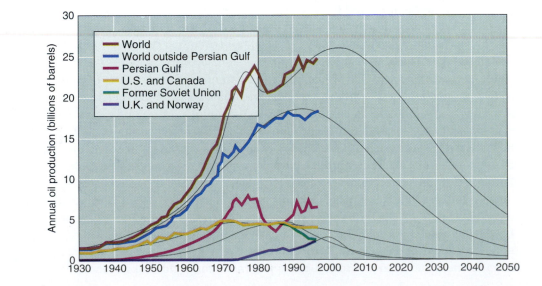

As the horsepower of automobile engines increased, higher quality gasoline was needed to operate efficiently in the higher performance engines. To meet this demand, additives that could boost gasoline performance (increase its "octane") were developed. One such octane-enhancing additive was tetraethyl lead which, since 1997, has been banned from motor fuels because of the lead it released into the atmosphere (Section 1.12). During the 1990s, **reformulated gasolines**, those containing methanol (methyl alcohol), ethanol (ethyl alcohol), or other oxygen-containing compounds, were developed to meet the mandate of the Clean Air Act of 1990 and its amendments. One of the goals of the Act is to improve air quality by reducing the amount of carbon monoxide pollution formed during combustion of automobile fuels. Because they already contain oxygen, reformulated gasolines (RFGs) burn more cleanly by producing less carbon monoxide than non-oxygenated fuels.

Ethanol (C_2H_5OH or CH_3CH_2OH) has been used as a gasoline additive for many years, including its use in gasohol, a mixture of ethanol and gasoline (10% ethanol and 90% gasoline). Two common methods are used to produce ethanol:

Reaction of water (steam) with ethylene, C_2H_4;

$$CH_2CH_2 + H_2O \longrightarrow CH_3CH_2OH$$

or by a method known since ancient times, the fermentation of the starch in grains such as corn

$$C_6H_{12}O_6 \longrightarrow 2\ CH_3CH_2OH + 2\ CO_2$$

When the first method is used to produce ethanol for oxygenated fuels, any residual water must scrupulously be removed so that it will not create problems in an automobile engine. Making ethanol by fermenting grains creates a fuel which, unlike gasoline, is renewable because grains can continue to be grown from seeds. Thus, the source of the fuel can be replaced.

Ethanol has a high octane value; it works well in automobile engines. In fact, nearly one-third of Brazil's motor vehicles are fueled by pure ethanol. However, ethanol is not without its drawbacks and critics. There are those who point to the fact that a gram of ethanol does not produce as much energy as a gram of gasoline. In addition, those opposed to ethanol as a fuel question whether valuable farmland, normally used to grow crops such as corn that feed people and animals, should be used to produce grain for ethanol. Others claim that ethanol burns incompletely to form acetaldehyde, CH_3CHO, a component of urban smog, and a compound that can cause allergic reactions in some people.

In these statements about ethanol, we see some of the basic issues to be considered in this chapter: an energy-intensive society such as the United States seeks ways to achieve what to some may be incompatible goals — improving air quality while continuing to consume increasing amounts of fossil fuels to provide relatively low-cost energy for transportation and other uses. But the days of plentiful (and consequently cheap) fuel and energy are obviously limited. The planet has a finite supply of coal, oil, and gas. Colin Campbell and Jean Laherrère, in their 1998 *Scientific American* essay "The End of Cheap Oil," suggest that world production of conventional oil will peak within the next decade to be followed by a permanent decline. The fact that previous forecasts of major energy crises have not developed thus far is by no means reason for complacency. Quite obviously energy, chemistry, and society are closely intertwined. This chapter is an attempt to untangle them.

4.3 Consider This: New Fuels for Your Car

Consider the magazine advertisement for reformulated gasoline shown in Figure 4.2. **a.** How might who sponsors the advertisement influence its content and emphasis? **b.** List the ideas presented in the ad. **c.** Then review each item in your list and decide whether it is a scientific fact, an opinion, or if you need additional information before making this judgment. **d.** What factors will affect your attitude about using reformulated fuels in your car?

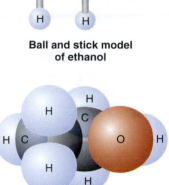

Ball and stick model of ethanol

Space-filling model of ethanol

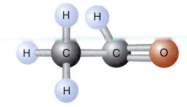

Ball and stick model of acetaldehyde

Space-filling model of acetaldehyde

Gasohol is a blend of gasoline and ethanol.

Figure 4.2

Advertisement promoting the use of oxygenated fuels to make automobile fuels burn more efficiently.

4.4 Consider This: The Corn Connection

In 4.3 Consider This, the magazine advertisement "Driving to Cleaner Air," which promotes oxygenates such as ethanol, was considered. Others advocate ethanol as well, including those who grow corn. You can locate web sites sponsored by corn growers by searching for "ethanol," "gasoline," and "additive." Alternatively, you may want to visit the ones provided at the *Chemistry in Context* web site. Select a site (or two). **a.** Who sponsored it? **b.** What is the connection between corn and ethanol? **c.** What advantages of ethanol are cited? **d.** What are some disadvantages to using corn to produce ethanol?

Chapter Overview

Energy is a common thread that runs through the first three chapters of this book. The Earth's finite energy reserves create the probability of energy shortages, a compelling reason to attend to issues related to energy. Our lives are intimately connected to this familiar yet elusive concept. The time has obviously come to take a closer look at energy.

In order to do so, we need agreement on the meanings of some fundamental concepts such as energy, work, and heat. We begin with these essential ideas and energy units. The first law of thermodynamics, a generalization that describes some of the constraints governing the generation and use of energy, links energy, work, and heat. With this foundation we turn to a consideration of energy sources and uses in the past, the present, and the future. Insight into the source of energy requires a look at what happens at the molecular level when a chemical reaction occurs. We describe these energetic transformations in qualitative terms and then use bond energies to calculate the energy changes associated with typical combustion reactions. Most of the energy currently used in home and industry comes from fossil fuels, so the topics of coal and petroleum are explored in some depth and detail. But the supplies of these fuels are limited, so the text then turns to discuss substitutes and additives from various sources, including renewable biological sources such as corn.

Mechanisms for transforming energy come under scrutiny next. Such mechanisms are subject to the second law of thermodynamics, another natural constraint. We introduce the second law to explain the inescapable inefficiencies of energy transformation. Along the way we develop the concept of entropy, a measure of disorder and an indication of the directionality of natural change. The chapter concludes with some observations on the importance of conserving energy and fuels.

4.1 Energy: Hard Work and Hot Stuff

Energy is one of those words that everyone uses, but whose precise meaning is not well understood. Unfortunately, the dictionary definition, the capacity to do work, doesn't help much because work is another common but poorly defined concept. To a scientist, work is done when movement occurs against a restraining force, and it is equal to the force multiplied by the distance over which the motion occurs. Thus, when you lift a book against the force of gravity, you are doing work. When you read a book without moving it, you are not, strictly speaking, working. On the other hand, you are again doing something that will not happen by itself, and that does require energy.

The source of much of the work done on our planet is another familiar form of energy—heat. The formal definition sounds a little strange: **heat** is that which flows from a hotter to a colder body. But a child once burned knows the meaning of heat. Our understanding of temperature is also based on experience, and we know that temperature and heat are not the same thing. But a standard definition of temperature sounds awkward and circular: **temperature** is a property that determines the direction of heat flow. When two bodies are in contact, heat always flows from the object at the higher temperature to that at the lower temperature. Consider that a burning match and a bonfire are at the same high temperature. But, as experience tells us, the bonfire produces far more heat than the match. Heat is a consequence of motion at the molecular level. When matter, for example liquid water in a pan, absorbs heat, its molecules move more rapidly. Temperature is a statistical measure of the average speed of that motion. Hence, temperature rises as the amount of heat energy in a body increases.

Consider two containers of water, each at room temperature (25°C). One contains 100 mL of water; the other holds 200 mL of water. Because the two water samples are at the same temperature, the average speed of their water molecules is the same, but their heat content is not the same. Starting at the same lower temperature, it takes twice as much heat energy to raise the temperature of 200 mL of water to 25°C than it does to reach that temperature for 100 mL. Therefore, the 200 mL of water has twice the heat energy than the smaller volume of water.

Before we turn to matters of energy demand, we need a unit in which to express it. Historically, there have been many, but recently there has been an international agreement to make the common unit the **joule.** One joule (1 J) is approximately equal to the energy required to raise a 1 kg (2 lb) book 10 cm (4 in) against the force of gravity. On a more personal basis, each beat of the human heart requires about 1 J of energy. As the name implies, one kilojoule (1 kJ) is equal to 1000 J.

Much of the published data relating to energy are reported in calories, not in joules. The **calorie** was introduced with the metric system in the late eighteenth century as a measure of heat. Originally, the calorie was defined as the amount of heat necessary to raise the temperature of exactly one gram of water by one degree Celsius. It has been redefined as exactly 4.184 J. Calories are perhaps most familiar when used to express the energy released when foods are metabolized. The values tabulated on package labels and in diet books are, in fact, kilocalories (1 kcal = 1000 cal). When Calorie is written with a capital C, it generally means kilocalorie. Thus, the energetic equivalent of a doughnut is 425 Cal (425 kcal, 425,000 calories). For most purposes we will use joules and kilojoules in this chapter, but when it seems more appropriate or more easily understandable, we will express energy in calories or kilocalories. We will not worry about British Thermal Units (BTU), ergs, or foot-pounds, but you are cautioned that the world also expresses energy in these terms.

The **calorie** is more specifically defined as the amount of heat required to raise the temperature of one gram of pure liquid water from 14.5°C to 15.5°C.

Food labels have the energy given in calories or kilojoules.

4.5 Your Turn

a. Convert to kilojoules the 425 kcal (425 food Calories) released when a donut is metabolized.

b. Calculate the number of 2-lb books you could lift to a shelf 6 ft off the floor with the amount of energy from metabolizing one doughnut.

c. A 12-oz can of a soft drink has an energy equivalent of 92 kcal. Convert the energy released when metabolizing the soft drink, expressing the answer in kJ.

d. Assume that you use this energy to lift concrete blocks that weigh 22 lb (10 kg) each. How many of these blocks could you lift to a height of 4 ft with this quantity of energy?

Ans. **a.** Consult this page to find the relationship necessary for this calculation. You will find 1 kcal is equivalent to 4.184 kJ.

$$425 \text{ kcal} \times \frac{4.184 \text{ kJ}}{1 \text{ kcal}} = 1.78 \times 10^3 \text{ kJ}$$

b. The text states on p. 142 that 1 J is approximately equal to the energy required to raise a 2-lb book a distance of 4 inches against Earth's gravity. We can use this information to find the number of 2-lb books that could be lifted 6 feet. This will clearly take more energy, but how much more? The energy required to lift each book to a height of 6 ft can be calculated using the appropriate relationships.

$$\frac{6 \text{ ft}}{1 \text{ book}} \times \frac{12 \text{ in}}{1 \text{ ft}} \times \frac{1 \text{ J}}{4 \text{ in}} = \frac{18 \text{ J}}{1 \text{ book}}$$

This value allows the final calculation to be made.

$$1.78 \times 10^3 \text{ kJ} \times \frac{1000 \text{ J}}{1 \text{ kJ}} \times \frac{1 \text{ book}}{18 \text{ J}} = 9.9 \times 10^4 \text{ books}$$

In round numbers, about 100,000 books! Lots of exercise is required to work off one donut.

4.6 The Sceptical Chymist

A simplifying (and erroneous) assumption was made in doing the calculations in parts **b** and **d** of the previous Your Turn. What was the assumption, and is it reasonable? Is the answer based on this assumption too large or too small? Give a reason for your answer.

4.2 Energy Conservation and Consumption

Strictly speaking, energy is not consumed. The **first law of thermodynamics,** also called the **law of conservation of energy,** states that energy is neither created nor destroyed. Energy is often transformed as it is transferred, but the energy of the universe is constant. However, energy sources such as coal, oil, and natural gas are consumed. In the United States we burn a prodigious quantity of these fossil fuels to generate a huge amount of energy. Figure 4.3 compares our annual per capita energy use with that of 16 other countries and the African continent. The energy comes from many different sources, but in the bar graph it is expressed as if it were all generated from oil. Thus, in 1997 the energy share of the average North American was 7.65 tons of oil, a quantity that yields about 95 million kcal. In India, the amount of energy derived from commercial fuel sources corresponded to about 0.30 tons of oil per person per year, or less than 3% of the values for the United States and Canada. Perhaps an equal quantity of India's energy was derived from traditional fuels such as wood, grass, or animal dung, but the international energy imbalance remains staggering. It is no coincidence that the nations at the top of Figure 4.3 are industrialized and wealthy, and that those on the bottom are struggling with poverty. Energy appears to drive industrial and economic progress, and gross national product correlates well with energy production and use. So do life expectancy, infant mortality, and literacy.

The great burst of energy consumption is of relatively recent origin. Two million years ago, before our primitive ancestors learned to use fire, the sole source of energy

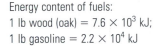

Energy content of fuels:
1 lb wood (oak) = 7.6×10^3 kJ;
1 lb gasoline = 2.2×10^4 kJ

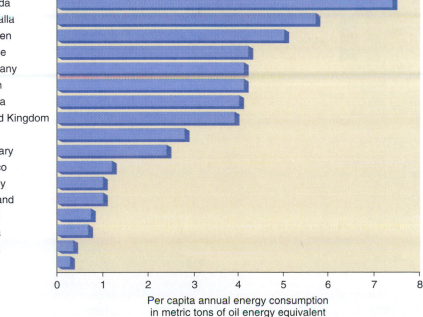

Per capita annual energy consumption in metric tons of oil energy equivalent

Figure 4.3

Annual per capita energy consumption (1997).

(Source: Data from *BP Statistical Review of World Energy,* June 1998, British Petroleum Company.)

available to an individual was that of his or her own body. Earliest hominids probably consumed the equivalent of 2000 kcal per day and expended most of it finding food. This roughly corresponds to the energy output of a 100-watt light bulb burning for 24 hours. The discovery of fire and the domestication of beasts of burden increased the energy available to an individual about six times. Hence, we estimate that about 2000 years ago a farmer with an ox or donkey had roughly 12,000 kcal at his or her disposal each day. The Industrial Revolution brought another five-or six-fold increase in the energy supply, most of it from coal via steam engines. Yet another energy jump occurred during the twentieth century. By the end of the twentieth century, the total energy used in the United States (from all sources and for all purposes) corresponds to about 650,000 kcal per person per day. This translates to an annual equivalence of 65 barrels of oil or 16 tons of coal for each American. Or, to put it in human terms, the energy available to each resident of the United States would require the physical labor of 130 workers. Yet, there are still people on the planet whose energy use and lifestyle closely approximate those of 2000 years ago.

The history of increasing energy consumption is closely related to changing energy sources and the development of devices for extracting and transforming that energy. Figure 4.4 displays the average American energy consumption from a variety of sources over a 150-year period. The data start in 1850 and are projected to 2000. The graph indicates that wood was originally the major energy source in the United States, and it continued to be until around 1890, when it was surpassed by coal. Coal provided more than 50% of the nation's energy from then until about 1940. By 1950, oil and gas were the source of more than half of the energy used in this country. Falling water has long been used to power mills and, more recently, to generate electricity, but it provides only a small percentage of our total energy output. Nuclear fission, once hailed as an almost limitless source of energy, has not achieved its full potential for a variety of reasons.

Nuclear fission is discussed in Chapter 7.

4.7 Consider This: U.S. Sources of Energy Over Time

Many changes in our sources of energy are represented in Figure 4.4. Concentrating on the period from 1950 to 2000 shown in the figure, which sources of energy have shown steady growth and which have not? Propose some possible reasons for the observed trends.

Figure 4.4

Annual United States energy consumption from various sources (1850–2000).

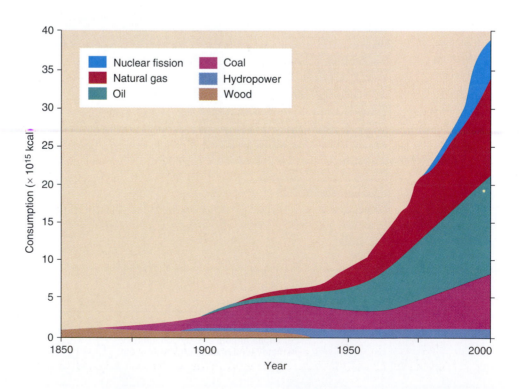

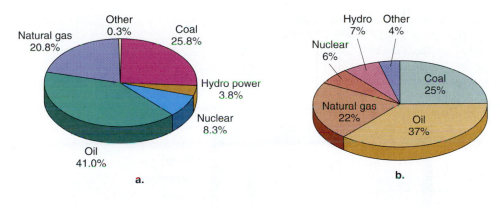

Figure 4.5

(a) United States energy consumption from various sources (1998)

(Source: Data from Department of Energy/EIA.)

(b) World energy consumption from various sources (1998).

(Data from various sources [1998] and U.S. EPA.)

Geothermal, wind, and solar sources are combined into a sliver marked "Other" in Figure 4.5 a. This pie chart indicates the percentage of the U.S. 1998 energy consumption derived from various sources. Figure 4.5b is a similar representation of world energy consumption for 1998. These global data indicate an 11% reliance on traditional energy sources and relatively less dependence on oil than is characteristic of the United States. Some of the currently underutilized alternate energy sources will be considered in Chapters 7 and 8.

> ### 4.8 Consider This: Future Talk Show
>
> Imagine that you are in charge of energy policy in the United States. You are put into a time machine and transported 200 years into the future. This makes you an instant celebrity and you are interviewed by the talk show host of the day. The first question is "How could the people of your century feel justified in using up so much of the world's store of non-renewable resources such as oil and coal?" What would your answer be?

4.3 Energy: Where From and How Much?

At a time when the nation is seeking new sources of energy, it is reasonable to ask what it is that makes some substances such as coal, gas, oil, or wood usable as fuels, while many others are not. To find an answer, we must consider the properties of fuels and the means by which energy is released from them. The most common energy generating chemical reaction is burning or combustion. **Combustion** is the combination of the fuel with oxygen to form product compounds. In such a chemical transformation, the potential energy (on a molecular scale) of the reactants is greater than that of the products. Because energy is conserved, the difference in energy is given off, primarily as heat.

We illustrate the process with the combustion of methane, CH_4, the principal component of natural gas, a major home heating fuel. The products are carbon dioxide and water. In Chapter 1 you encountered the reaction represented by this equation:

$$CH_4(g) + 2\ O_2(g) \longrightarrow CO_2(g) + 2\ H_2O(g) + \text{Energy} \qquad (4.1)$$
$$\text{methane}$$

The above reaction is said to be **exothermic**—a term applied to any chemical or physical change that is accompanied by the release of heat. The quantity of heat energy released in a combustion reaction such as this can be experimentally determined with a device called a calorimeter (Figure 4.6). Not surprisingly, the amount of heat generated depends on the amount of fuel burned. Therefore, a known mass of fuel and an excess

Oil fires burning out of control during the 1991 war with Iraq.

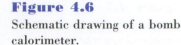

Figure 4.6

Schematic drawing of a bomb
calorimeter.

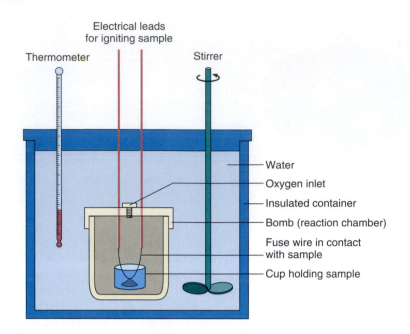

of oxygen are introduced into a heavy-walled stainless steel "bomb." The bomb is then
sealed and submerged in a bucket of water. The reaction is initiated with an electrical
current that burns through a fuse wire. The heat evolved by the exothermic reaction
flows from the bomb to the water and the rest of the apparatus. As a consequence, the
temperature of the entire calorimeter system increases. The quantity of heat given off by
the reaction can be calculated from this temperature rise and the known heat absorbing
properties of the calorimeter and the water it contains. The greater the temperature increase, the greater the quantity of energy evolved.

Experimental measurements of this sort are the source of most tabulated values of
heats of combustion. As the name suggests, the heat of combustion is the quantity of
heat energy given off when a specified amount of a substance burns in oxygen. Heats
of combustion are typically reported as positive values in kilojoules or kilocalories per
mole or per gram (kJ/mole, kJ/g, kcal/mole, or kcal/g). The energy equivalents of various foods are also usually determined by calorimetry. In the case of methane, experiment shows its heat of combustion to be 802.3 kJ. This means that 802.3 kJ of heat are
given off when one mole of $CH_4(g)$ reacts with two moles of $O_2(g)$ to form one mole
of $CO_2(g)$ and two moles of $H_2O(g)$ (see Equation 4.1). We can also calculate the number of kilojoules released when one gram of methane is burned. The molar mass of CH_4,
calculated from the atomic masses of carbon and hydrogen, is 16.0 g/mole. The heat of
combustion per gram of methane gas (kJ/g) is obtained as follows:

> Heats of combustion, by convention,
> are tabulated as positive values even
> though all combustion reactions
> *release* heat.

$$\frac{802.3 \text{ kJ}}{1 \text{ mole } CH_4} \times \frac{1 \text{ mole } CH_4}{16.0 \text{ g } CH_4} = \frac{50.1 \text{ kJ}}{\text{g } CH_4} = 50.1 \text{ kJ/g } CH_4$$

The fact that heat is evolved signals that there is a decrease in the energy of the chemical system during the reaction. In other words, the reactants (methane and oxygen) are
at higher energy than the products (carbon dioxide and water). The burning of methane
is thus somewhat like a waterfall or a falling object. In all these processes, potential energy decreases and is manifested in some other form of energy (heat, sound, etc.). This
decrease is signified by the negative sign that is traditionally attached to the energy
change for *all* exothermic reactions. For the combustion of methane, the energy change

> Energy$_{products}$ − Energy$_{reactants}$ = < 0
> for an exothermic reaction.

is listed as −802.3 kJ/mole. Figure 4.7 is a schematic representation of this process. The
downward arrow indicates the fact that the energy associated with 1 mole of $CO_2(g)$ and
2 moles of $H_2O(g)$ is less than the energy associated with 1 mole of $CH_4(g)$ and 2 moles

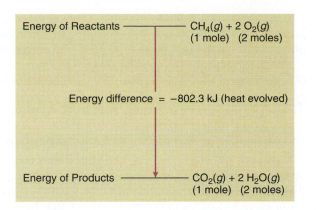

Figure 4.7
Energy difference in an exothermic reaction.

of $O_2(g)$. The energy difference between the products and the reactants is thus a negative quantity, as is the case for all exothermic reactions. In the combustion of methane, the energy difference is -802.3 kJ.

4.9 Your Turn

According to information in this section, the heat of combustion of methane is 802.3 kJ/mol. Methane is usually sold by the standard cubic foot, SCF. One SCF contains 1.25 moles of methane. Calculate the energy (in kJ) that is released by burning 1.00 SCF of methane.

Ans. 1000 kJ released

We still need to adequately explain the origin of the energy released in an exothermic reaction. To do that, we investigate the structure of the molecules involved. We have already encountered all of the reactants and products in the combustion of methane. Consequently we can write Lewis structures for all of the molecular species.

$$\text{(4.2)}$$

The reaction represented by this and any other chemical equation is a rearrangement of atoms. It involves the breaking and making of chemical bonds. Energy is required to break bonds, just as energy is required to break wood or tear paper. Bond breaking is thus an **endothermic** process, a term applied to any chemical or physical change that absorbs energy. On the other hand, the formation of chemical bonds is an exothermic process in which energy is released. The overall energy change associated with a chemical reaction depends on the net effect of the bond breaking and bond making. If the energy required to break the bonds in the reactants (endothermic) is greater than the energy released (exothermic) when the products form, the reaction is endothermic; energy is absorbed. If, on the other hand, the situation is reversed, the exothermic bond-making energy of the products is greater than the endothermic bond breaking in the reactants, then the net energy change is exothermic; energy will be released by the reaction.

Endothermic reaction	Exothermic reaction
$Energy_{products} > Energy_{reactants}$	$Energy_{products} < Energy_{reactants}$
Energy change is positive; gain of energy	Energy change is negative; loss of energy

The potential energy associated with any specific chemical species, for example, a CH_4 molecule, is in part a consequence of the interaction of the atoms via chemical bonds. When methane or any other fuel burns in oxygen, the energy released in bond formation exceeds the energy absorbed in bond breaking. The net result is the evolution of energy, mostly in the form of heat. Another way to look at such exothermic reactions is as a conversion of reactants involving weaker bonds (for example, CH_4 and O_2) to products involving stronger bonds (CO_2 and H_2O). In general, the products are more stable and less reactive than the starting substances.

Although the chemical reactions used to generate energy are all exothermic, there are also many naturally occurring endothermic reactions that absorb energy as they occur, such as photosynthesis. You have already encountered two that are very important in atmospheric chemistry. One is the decomposition of O_3 to yield O_2 and O and the other is the combination of N_2 and O_2 to yield two molecules of NO. Both reactions require energy, which can be in the form of electrical discharge, high-energy photons, or high temperatures. It is possible to experimentally determine the energy changes associated with many reactions—exothermic or endothermic. But sometimes it is easier to calculate values. We illustrate the process in the next section.

4.4 Calculating Energy Changes in Chemical Reactions

There is much interest in hydrogen as a fuel, because of the large amount of energy per gram produced when it burns. Hydrogen would become a plentiful fuel if an economical way could be developed to decompose water into its elements, hydrogen and oxygen (Figure 4.8). Methods currently are available to carry out the decomposition, but energy requirements and costs are too high to make hydrogen competitive with fossil fuels.

We can calculate the total energy change associated with the combustion of hydrogen to form water, as represented by equation 4.3.

$$2 \, H_2(g) + O_2(g) \longrightarrow 2 \, H_2O(g) + Energy \qquad (4.3)$$

To simplify matters for such calculations, we assume that all the bonds in the reactant molecules are broken and then the individual atoms are reassembled into the product molecules. In fact, the reaction does not occur that way. But we are only interested in the overall or net change, not the details. Therefore, we will proceed with our convenient fiction and see how well our calculated result agrees with the experimental value.

Figure 4.8

The Damiler-Benz NeCar (New Electric Car), a hydrogen-fueled vehicle. The car moves by the energy released when hydrogen (a "clean fuel") and oxygen react to form water. The hydrogen gas is stored in a tank in the roof of the car.

Table 4.1 Bond Energies (in kJ/mole)

Single Bonds								
	H	**C**	**N**	**O**	**S**	**F**	**Cl**	**Br**	**I**
H	432								
C	411	346							
N	386	305	167						
O	459	358	201	142					
S	363	272	—	—	226				
F	565	485	283	190	284	155			
Cl	428	327	313	218	255	249	240		
Br	362	285	—	201	217	249	216	190	
I	295	213	—	201	—	278	208	175	149

Multiple Bonds					
C=C	602	C=N	615	C=O	799
C≡C	835	C≡N	887	C≡O	1072
N=N	418	N=O	607		
N≡N	942	O=O	494		

From Ebbing, Darrell D., *General Chemistry*, Fourth Edition. Copyright © 1993 by Houghton Mifflin Company. DATA from *Inorganic Chemistry: Principles of Structure and Reactivity,* Third Edition by James E. Huheey. Copyright © 1983 by James E. Huheey. Reprinted by permission of Addison Wesley Longman.

The numbers we will need in the computation are given in Table 4.1, a listing of the **bond energies** associated with a large variety of covalent bonds. Bond energy is the amount of energy that must be absorbed to break a specific chemical bond. Thus, because energy must be absorbed, breaking bonds is an endothermic process, and all the bond energies in Table 4.1 are positive. Obviously, the amount of energy that is required will depend on the number of bonds broken—more bonds take more energy. Typically, bond energies are expressed in kilojoules per mole of bonds. Note that element symbols appear across the top of Table 4.1 and down the left side. The number at the intersection of any row and column is the energy (in kilojoules) needed to *break a mole of bonds* linking the atoms of the two elements thus identified. For example, the bond energy of an H—H bond, as in the H_2 molecule, is 432 kJ/mole. Similarly, the energy required to break one mole of O=O is 494 kJ, as noted from the bottom part of the table. Bond energies for other double bonds as well as triple bonds are also given in the table.

Because we are doing energetic bookkeeping, we need to keep track of the energy change involved in each step and whether the energy is taken up or given off. To do this, we assume that energy that is absorbed carries a positive sign, like a deposit to your checkbook. On the other hand, energy given off is like money spent; it bears a negative sign. Bond energies are positive because they represent energy absorbed when bonds are broken. But the formation of bonds releases energy, and hence the associated energy change is negative. For example, the bond energy for the O=O double bond is 494 kJ/mole. This means that when one mole of O=O bonds are broken, the energy change is +494 kJ; correspondingly, when one mole of O=O bonds are formed, the energy change is −494 kJ.

Now we are finally ready to apply these concepts and conventions to the burning of hydrogen gas, H_2. First we need to determine how many moles of bonds are broken and how many moles of bonds are formed. We can do so using Lewis structures relating them to the balanced equation (4.4) for this combustion reaction

$$2\ \text{H–H} + \ddot{\text{O}}{=}\ddot{\text{O}} \longrightarrow 2\ \underset{H}{\overset{\ddot{\text{O}}}{\diagup}}\diagdown_{H} \tag{4.4}$$

Remember that chemical equations are written in terms of moles. In this case, equation 4.4 indicates "2 moles of $H_2(g)$ plus 1 mole of $O_2(g)$ yields 2 moles of gaseous water (water vapor)." But, to use bond energies, we need to count the number of moles of *bonds* involved. Because each H_2 molecule contains one H—H bond, 1 mole of H_2 must contain 1 mole of H—H bonds. Similarly, equation 4.4 indicates that 1 mole of O_2 contains 1 mole of O=O bonds. Each mole of water contains 2 moles of H—O; thus, 2 moles of water contain 4 moles of H—O bonds. Therefore, we now have the total number of moles of bonds to be broken (2 moles of H—H and 1 mole of O=O) and those to be formed (4 moles of H—O). These number of bonds are then *multiplied* by the representative bond energy, using the appropriate sign convention (+ for bonds broken; − for bonds made).

Bonds Broken in Reactants			
Number of moles	Moles of bonds per mole of molecules	Energy	Energy Change (energy absorbed for bond breaking)
2 mole H—H	1 mole H—H per mole H_2	+432 kJ per mole H—H	2 × (+432 kJ) = +864 kJ
1 mole O=O	1 mole O=O per mole O_2	+494 kJ per mole O=O	1 × (+494 kJ) = +494 kJ

Total energy, bonds broken (energy *absorbed*) = (+864 kJ) + (+494 kJ) = +1358 kJ

Bonds Made in Products			
Number of moles	Moles of bonds per mole of molecules	Energy	Energy change (energy released by bond making)
2 moles H_2O	2 moles H—O per mole of H_2O	−459 kJ per mole H—O	4 × (−459 kJ) = −1836 kJ

Total energy, bonds made (energy *released*) = −1836 kJ

Consequently, the *overall* energy change in breaking bonds and forming new ones is:

$$(+1358 \text{ kJ}) + (-1836 \text{ kJ}) = -478 \text{ kJ}$$

A schematic representation of this calculation is presented in Figure 4.9. The energy of the reactants, 2 H_2 and O_2, is set at zero, chosen as an arbitrary, but convenient, value. The green arrows pointing upward signify energy absorbed to break bonds and convert the reactant molecules into individual atoms: 4 H, and 2 O. The red arrow pointing downward represent energy released as these atoms are reconnected with new bonds to form the product molecules: 2 H_2O. The heavy black arrow corresponds to the net energy change of −478 kJ signifying that the overall combustion reaction is strongly exothermic. The *release* of heat corresponds to a *decrease* in the energy of the chemical system, which explains why the energy change is *negative*.

We can also use bond energies from Table 4.1 to calculate the energy change for the combustion of methane.

$$CH_4(g) + 2\ O_2(g) \longrightarrow CO_2(g) + 2\ H_2O(g) + \text{Energy}$$

One mole of methane contains four moles of C—H bonds, each with a bond energy value of 411 kJ. Breaking two moles of O=O bonds requires 988 kJ (2 × 494 kJ). Bonds formed in the products are two moles of C=O bonds in one mole of CO_2 (2 ×

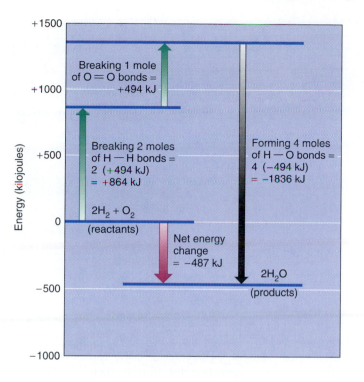

Figure 4.9
The energetics of the combustion of hydrogen to form water.

−799 kJ), and four moles of H—O bonds in two moles of water (4 × −459 kJ). Note again that bond formation is exothermic and the associated bond energies have minus signs. These energy changes can be summarized as follows.

Bonds Broken			
Number of moles	Moles of bonds per mole of molecules	Energy	Energy Change (energy absorbed for bond breaking)
1 mole CH_4	4 moles of C—H bonds per 1 mole of CH_4	+411 kJ per mole C—H bonds	4 × (+411 kJ) = +1644 kJ
2 moles O_2	1 mole O=O bonds per 1 mole O_2	+494 kJ per mole O=O bonds	2 × (+494 kJ) = +988 kJ

Bonds Formed			
Number of moles	Moles of bonds per mole of molecules	Energy	Energy Change (energy released with bond making)
1 mole of CO_2	2 moles of C=O bonds per 1 mole CO_2	−799 kJ	2 × (−799 kJ) = −1598 kJ
2 moles of H_2O	2 moles of H—O bonds per 1 mole H_2O (4 moles of H—O bonds per two moles of H_2O)	−459 kJ	4 × (−459 kJ) = −1836 kJ

Total energy change in breaking bonds = (+1644 kJ) + (+988 kJ) = +2632 kJ
Total energy change in making bonds = (−1598 kJ) + (−1836 kJ) = −3434 kJ
Net energy change = (+2632 kJ) + (−3434 kJ) = −802 kJ

Heats of combustion, by convention, are positive. Thus, the heat of combustion of methane calculated using bond energies is +802 kJ, a value close to that given in Section 4.3.

The Hindenburg, a German airship inflated with hydrogen, was destroyed in a spectacular fire as it prepared to dock at Lakehurst, NJ in 1937.

The energy changes we have just calculated from bond energies: -478 kJ for the burning of hydrogen; and -802 kJ for methane combustion, compare favorably with the experimentally determined values of -483 kJ and -802.3 kJ, respectively. This agreement justifies our rather unrealistic assumption that all of the bonds in the reactant molecules are first broken and then all of the bonds in the product molecules are formed. This is not at all what actually happens. But the energy change that accompanies a chemical reaction depends on the energy *difference* between the products and the reactants, not on the particular process, mechanism, or individual steps that connect the two. This is an extremely powerful idea for understanding chemical energetics and doing related calculations. Even so, not all calculations come out as well as this one did. For one thing, the bond energies of Table 4.1 only apply to gases, so calculations using these values only agree with experiment if all the reactants and products are in the gaseous state. Moreover, tabulated bond energies are really average values. The strength of a bond depends on the overall structure of the molecule in which it is found, in other words, on what else the atoms are bonded to. Thus, the strength of an O—H bond is slightly different in HOH (H_2O), HOOH (H_2O_2), and CH_3OH. Nevertheless, the procedure we have illustrated here is a useful way of estimating energy changes in a wide range of reactions. The approach also helps illustrate the relationship between bond strength and chemical energy.

This analysis also helps clarify why the H_2O or CO_2 formed in combustion reactions cannot be used as fuels. There are no substances into which these compounds can be converted that have stronger bonds and are lower in energy; we cannot run a car on its exhaust.

4.10 Your Turn

Use the bond energies in Table 4.1 to calculate the heat of combustion of propane, C_3H_8 (LP or "bottled gas"). Report your answer both in kJ/mole of C_3H_8 and kJ/g C_3H_8. This is the equation for the reaction, written with structural formulas.

$$H-\overset{\overset{\displaystyle H}{|}}{\underset{\underset{\displaystyle H}{|}}{C}}-\overset{\overset{\displaystyle H}{|}}{\underset{\underset{\displaystyle H}{|}}{C}}-\overset{\overset{\displaystyle H}{|}}{\underset{\underset{\displaystyle H}{|}}{C}}-H + 5\,\overset{..}{\underset{..}{O}}{=}\overset{..}{\underset{..}{O}} \longrightarrow 3\,\overset{..}{O}{=}C{=}\overset{..}{\underset{..}{O}} + 4\,H-\overset{..}{\underset{..}{O}}-H$$

Hint: Note that there are 8 moles of C—H bonds, 2 moles of C—C bonds, 5 moles of O=O bonds, 6 moles of C=O bonds and 8 moles of O—H bonds involved in the reaction.

Ans. Energy change = -2016 kJ/mole C_3H_8 or -45.8 kJ/g C_3H_8
Heat of combustion = 2016 kJ/mole C_3H_8 or 45.8 kJ/g C_3H_8.

4.11 Your Turn

Use the bond energies in Table 4.1 to calculate the heat of combustion of ethanol, C_2H_5OH, one of the components of "gasohol" fuel. The molecular structure of ethanol is

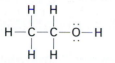

Ethanol, a renewable energy resource, was introduced in Section 4.1.

Ans.   Energy change = −1250 kJ/mole C_2H_5OH or −27.2 kJ/g C_2H_5OH
Heat of combustion = 1250 kJ/mole C_2H_5OH or 27.2 kJ/g C_2H_5OH.

4.12 Your Turn

Use the bond energies in Table 4.1 plus information from Chapter 2 to explain why the ultraviolet radiation absorbed by O_2 has a shorter wavelength than the UV radiation absorbed by O_3.

4.5 Getting Started: Activation Energy

Just because two substances can react in an exothermic process does not mean that they will do so, even if they are in intimate contact. For example, if you turn on the gas jet at a Bunsen burner, methane and oxygen will be present in a potentially combustible mixture. But they will not react unless a spark, a flame, or some other source of energy is supplied. This turns out to be a fortunate feature of matter. Wood, paper, and many other common materials are energetically unstable and potentially capable of exothermic conversion to water, carbon dioxide, and other simple molecules, but they do not suddenly burst into flame.

The energy necessary to initiate a reaction is called its **activation energy.** Figure 4.10 is a schematic of the energy changes that might occur in a typical exothermic reaction. It looks a little like the cross section of a hill. The activation energy corresponds to a peak over which a boulder must be pushed before it will roll downhill. Although energy must be expended to get the reaction (or the boulder) started, a good deal more energy is given off as the process proceeds to a lower potential energy state. Generally, reactions that occur rapidly have low activation energies; slower reactions have higher activation energies.

Activation energy is also involved in another aspect of chemical reactions that determines whether a given substance can be used as a fuel. Useful fuels react at rates that are neither too fast nor too slow. Slow reactions are of little use in producing energy

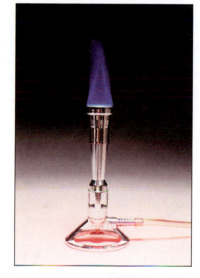

The methane from a laboratory burner ignites to produce a flame.

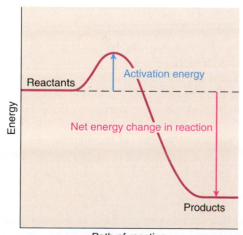

Figure 4.10
Energy–reaction pathway diagram.

because the energy is released over too long a time. For example, it would not be very practical to try to warm your hands over a piece of rotting wood, even though the overall reaction is similar to burning and forms CO_2 and H_2O. On the other hand, fast reactions can release energy too rapidly to be put to convenient use. In fact, such reactions often lead to explosions because the gases produced expand rapidly.

One way to speed up the rate of a reaction is to divide the fuel into small particles. This principle is used in fluidized-bed power plants in which pulverized coal is burned in a blast of air. The fine coal dust is quickly heated to the kindling point and the large surface area means that oxygen reacts rapidly and completely with the fuel. The combustion actually occurs at a lower temperature than that required to ignite large pieces of coal. As a result, the generation of nitrogen oxides is minimized. If finely divided limestone (calcium carbonate) is mixed with the powdered coal, sulfur dioxide is also removed from the effluent gas. Thus the amount of pollution is reduced while the efficiency of coal combustion is enhanced.

Increasing temperature also increases the rates at which reactions occur. The added heat energy helps the reactants over the activation energy barrier. Catalysts, including those used in automobile catalytic converters (Section 1.12) and in petroleum refining (Section 4.8), increase reaction rates by providing alternate reaction pathways with lower activation energies.

> ### 4.13 Consider This: The Striking Case of a Single Match
>
> It is not unusual for the electrical power to go out in a residential area during a summer thunderstorm. Imagine that the power went out in your house just as you were about to cook dinner. You scrambled around in the camping gear looking for an alternative to your electric stove. You found a portable gas stove, one match, a small grill, and a bag of charcoal briquettes. Because you have only one match and no other sources of fire, you must choose between using the portable gas stove or the charcoal grill. Which will you choose and why? Be sure to justify your decision based upon what you have learned about combustion.

4.6 There's No Fuel Like An Old Fuel

Coal, oil, and natural gas possess many of the properties needed in a fuel. Therefore, most of the energy that drives the engines of our economy comes from these remnants of the past. In a very real sense, these fossil fuels are sunshine in the solid, liquid, and gaseous state. The sunlight was captured millions of years ago by green plants that flourished on the prehistoric planet. The reaction is the same one that is carried out by plants today.

$$2800 \text{ kJ} + 6\ CO_2(g) + 6\ H_2O(l) \xrightarrow{\text{chlorophyll}} \underset{\text{glucose}}{C_6H_{12}O_6(s)} + 6\ O_2(g) \qquad (4.5)$$

This conversion of carbon dioxide and water to glucose and oxygen is endothermic. It requires the absorption of 2800 kJ of sunlight per mole of $C_6H_{12}O_6$ or 15.5 kJ/g glucose formed. The reaction could not occur without the absorption of energy and the participation of a green pigment molecule called chlorophyll. The chlorophyll interacts with photons of visible sunlight and uses their energy to drive the photosynthetic process.

$$\frac{2800 \text{ kJ}}{1 \text{ mole glucose}} \times \frac{1 \text{ mole glucose}}{180 \text{ g glucose}}$$
$$= 15.5 \text{ kJ/g glucose formed}$$

You are already aware of the essential role of photosynthesis in the initial generation of the oxygen in the Earth's atmosphere, in maintaining the planetary carbon dioxide balance, and in providing food and fuel for creatures like us. In our bodies, we run the above reaction backwards, like living internal combustion engines.

$$C_6H_{12}O_6(s) + 6\ O_2(g) \longrightarrow 6\ CO_2(g) + 6\ H_2O(l) + 2800 \text{ kJ} \qquad (4.6)$$

We extract the 2800 kJ that are released per mole of glucose "burned" and use that energy to power our muscles and nerves, though we do not do it with perfect efficiency

Explosions occur in grain elevators (storage tanks) when an inadvertent spark or flame causes finely powdered grain to burn explosively.

(see 4.6 Sceptical Chymist). The same overall reaction occurs when we burn wood, which is primarily cellulose, a polymer composed of repeating glucose units.

When plants die and decay, they are also largely transformed into CO_2 and H_2O. However, under certain conditions, the glucose and other organic compounds that make up the plant only partially decompose and the residue still contains substantial amounts of carbon and hydrogen. Such conditions arose at various times in the prehistoric past of our planet, when vast quantities of plant life were buried beneath layers of sediment in swamps or on the ocean bottom. There these remnants of vegetable matter were protected from atmospheric oxygen, and the decomposition process was halted. However, other chemical transformations did occur in Earth's high-temperature and high-pressure reactor. Over millions of years, the plants that captured the rays of a young Sun were transmuted into the fossils we call coal and petroleum.

In Equation 4.6, an *exo*thermic reaction, energy is shown as a product (2800 kJ) because energy is *given off* by the reaction. In an *endo*thermic reaction, energy is noted as a reactant because the reaction *absorbs* energy. Many chemical equations are written without including energy as a reactant or product. Including the energy in an equation is a way of emphasizing the energy change associated with the reaction.

4.7 Coal: Black Gold

The great exploitation of fossil fuels began with the Industrial Revolution, about two centuries ago. The newly built steam engines consumed large quantities of fuel, but in England, where the revolution began, wood was no longer readily available. Most of the forests had already been cut down. Coal turned out to be an even better energy source than wood because it yields more heat per gram. Burning one gram of coal releases approximately 30 kJ, compared to 10–14 kJ per gram of wood. This difference in heat of combustion is a consequence of differences in chemical composition. When wood or coal burn, a major energy source is the conversion of carbon to carbon dioxide. Coal is a better fuel than wood because it contains a higher percentage of carbon and a lower percentage of oxygen and water.

Coal is a complex mixture of compounds that naturally occurs in varying grades. Although coal is not a single compound, it can be approximated by the chemical formula $C_{135}H_{96}O_9NS$. This formula corresponds to a carbon content of 85% by mass. The carbon, hydrogen, oxygen, and nitrogen atoms come from the original plant material. In addition, samples of coal typically contain small amounts of silicon, sodium, calcium, aluminum, nickel, copper, zinc, arsenic, lead, and mercury. Soft lignite or brown coal is the lowest grade. The vegetable matter that makes it up has undergone the least amount of change, and its chemical composition is similar to that of wood or peat. Consequently, the heat of combustion of lignite is only slightly greater than that of wood (Table 4.2). The higher grades of coal, bituminous and anthracite, have been exposed to higher

"... You will die but the carbon will not; its career does not end with you... it will return to the soil, and there a plant may take it up again in time, sending it once more on a cycle of plant and animal life." Jacob Bronowski *Biography of an Atom— And the Universe*

Table 4.2 **Classification, Composition, and Fuel Value of Various U.S. Coals***

Fuel	State of Origin	Analysis, Weight % Before Drying				Heat Content (kJ/g)
		Moisture	Volatile Matter	Carbon	Ash	
Anthracite	PA	4.4	4.8	81.8	9.0	30.5
Bituminous						
Low volatile	MD	2.3	19.6	65.8	12.3	30.7
Medium volatile	AL	3.1	23.4	63.6	9.9	31.4
High volatile	OH	5.9	43.8	46.5	3.8	30.6
Sub-bituminous	WA	13.9	34.2	41.0	10.9	24.0
	CO	25.8	31.1	34.8	4.7	19.9
Lignite (brown coal)	ND	36.8	27.8	30.2	5.2	16.2
Peat	MS	—	—	—	—	13.
Wood**	—	—	—	—	—	10.4–14.1

*Most data from *Energy and the Future,* Table 1, A.L. Hammond, W.D. Metz, and T.H. Maugh II. ©1973, American Association for the Advancement of Science.

**Includes waste.

From J.W. Moore and E.A. Moore, *Environmental Chemistry,* p. 94, Academic Press, 1976. Used with permission of the authors.

pressures in the Earth. In the process they have lost more oxygen and moisture and have become a good deal harder—more mineral than vegetable. The percentage of carbon has increased, and with it, the heat of combustion. Anthracite has a particularly high carbon content and low concentrations of sulfur, both of which make it the most desirable grade of coal. Unfortunately, the deposits of anthracite are relatively small and the United States supply is almost exhausted. We now rely most heavily on bituminous and sub-bituminous coal.

Generally speaking, the less oxygen a compound contains, the more energy per gram it will release on combustion because it is higher up on the potential energy scale. This explains why burning one mole of carbon to form carbon dioxide yields about 40% more energy than that obtained from burning one mole of carbon monoxide. To be sure, coal is a mixture, not a compound, but the same principles apply. Anthracite and bituminous coals consist primarily of carbon. Their heat of combustion is, gram for gram, about twice that of lignite, which contains a much lower percentage of carbon.

The global supply of coal is large and it remains a widely used fuel, but it is not without some serious drawbacks. Coal is difficult to obtain, and underground mining is dangerous and expensive. In *Invention and Technology,* Summer 1992, Mary Blye Howe reports that since 1900 more than 100,000 workers have been killed in American coal mines by accidents, cave-ins, fires, explosions, and poisonous gases. Many thousands more have been injured or incapacitated by respiratory diseases. If the coal deposits lie sufficiently close to the surface, surface or strip mining can be used. In this method, the overlying soil and rock are stripped away to reveal the coal seam, which is then removed by heavy machinery.

Much care is necessary in mining to prevent serious environmental deterioration. Current regulations require the replacement of earth and topsoil and the planting of trees and vegetation at mine sites. But, in the past these regulations were not in place to prevent the great holes in the Earth and heaps of eroding soil that still dot regions of abandoned strip mines. Once the coal is out of the ground, its transportation is complicated by the fact that it is a solid. Unlike gas and oil, coal cannot be pumped unless it is finely divided and suspended in a water slurry.

Perhaps the most widely discussed drawback of coal is the fact that it is a dirty fuel. It is, of course, physically dirty, but its dirty combustion products may be more serious. The unburned soot from countless coal fires in the nineteenth and early twentieth centuries blackened buildings and lungs in many cities. Less visible but equally damaging

Underground mechanized coal mining.

are the oxides of sulfur and nitrogen that are formed when certain coals burn. If these compounds are not trapped, they can contribute to the acid precipitation that forms the subject for Chapter 6. In addition, coal suffers from the same drawback of all fossil fuels: the greenhouse gas, carbon dioxide, is an inescapable product of its combustion.

4.14 Your Turn

a. Assuming the composition of coal can be approximated by the formula $C_{135}H_{96}O_9NS$, calculate the mass of carbon (in tons) contained in 1.5 million tons of coal. This is approximately the quantity of coal that might be burned by a power plant in one year.

b. What mass of CO_2 will be produced by the complete combustion of 1.5 million tons of this coal?

c. Compute the amount of energy (in kJ) released by burning this mass of coal. Assume the process releases 30 kJ per gram of coal. Recall that 1 ton = 2000 lb and that 1 lb = 454 g.

Ans:

a. Start by calculating the approximate molar mass of coal from the given formula.

$$135 \text{ mole C} \times \frac{12.0 \text{ g C}}{1 \text{ mole C}} = 1620 \text{ g C}$$

$$96 \text{ mole H} \times \frac{1.0 \text{ g H}}{1 \text{ mole H}} = 96 \text{ g H}$$

$$9 \text{ mole O} \times \frac{16.0 \text{ g O}}{1 \text{ mole O}} = 144 \text{ g O}$$

$$1 \text{ mole N} \times \frac{14.0 \text{ g N}}{1 \text{ mole N}} = 14 \text{ g N}$$

$$1 \text{ mole S} \times \frac{32.1 \text{ g S}}{1 \text{ mole S}} = 32 \text{ g S}$$

The sum of these values gives the molar mass of $C_{135}H_{96}O_9NS$ = 1906 g/mole. Note that there are 1620 g C in every 1906 g of coal. The mass-to-mass relationship stays the same as long as the same mass unit is used for both; the ratio is just as useful expressed in tons.

$$\text{Mass C} = 1.5 \times 10^6 \text{ tons } C_{135}H_{96}O_9NS \times \frac{1620 \text{ tons C}}{1906 \text{ tons } C_{135}H_{96}O_9NS}$$

$$= 1.3 \times 10^6 \text{ tons C} = 1.3 \text{ million tons C}$$

b. 4.8 million tons

c. 4.1×10^{13} kJ

In spite of these less-than-desirable properties, the world's energy dependence on coal will likely increase rather than decrease. The recoverable world supply of coal is estimated as 20 to 40 times greater than world petroleum reserves. As the latter become depleted, reliance on coal will increase, unless alternative energy sources are developed. It is, however, possible that coal will not be burned in its familiar form, but rather converted to cleaner and more convenient liquid and gaseous fuels. That is a subject for a subsequent section, after we consider the properties of petroleum.

4.8 Petroleum: Black Liquid Gold

Children in the average American city or town would be hard-pressed to find lumps of coal. Indeed, the children may never have seen coal, but they have undoubtedly seen gasoline. Somewhere around 1950, petroleum surpassed coal as the major energy source in the United States. The reasons are relatively easy to understand. Petroleum, like coal, is partially decomposed organic matter, but it has the distinct advantage of being liquid. It is easily pumped to the surface from its natural, underground reservoirs, transported

Figure 4.11
An oil refinery, symbol of the
petroleum industry.

via pipelines, and fed automatically to its point of use. Moreover, petroleum is a more concentrated energy source than coal, yielding approximately 40–60% more energy per gram. Typical figures are 48 kJ/g for petroleum and 30 kJ/g for coal.

There is, however, one property of petroleum that impeded its initial acceptance. Unlike coal, crude oil is not ready for immediate use when it is extracted from the ground. Crude oil must first be refined—a process that has given gainful employment to many chemists and chemical engineers (and quite a few others). It has also provided an amazing array of products. Petroleum is a complex mixture of thousands of different compounds. The great majority are **hydrocarbons**, molecules consisting only of hydrogen and carbon atoms. The hydrocarbon molecules can contain from one to as many as sixty carbon atoms per molecule. Concentrations of sulfur and other contaminating elements are generally quite low, minimizing polluting combustion products.

The oil refinery has become a symbol of the petroleum industry (Figure 4.11). In the refining process, the crude oil is separated into individual compounds or, more often, into **fractions** that consist of compounds with similar properties. This fractionation is accomplished by **distillation.** Distillation is a purification or separation process in which a solution is heated to its boiling point and the vapors are condensed and collected. The petroleum is pumped into an industrial-sized retort or still, and the mixture is heated. As the temperature increases, the components with the lowest boiling points are the first to vaporize. The gaseous molecules escape from the liquid and move up a tall distillation column or tower. There the cooled vapors recondense into the liquid state, only this time in a much purer condition. By varying the temperature of the still and the fractionating column, a petroleum engineer can regulate the boiling point range of the fractions distilled and condensed. Higher temperatures mean higher boiling compounds. All fractions are not produced in equal proportions; market demand dictates which are maximized and which are decreased.

Figure 4.12 is a schematic drawing of a distillation tower and a listing of some of the fractions obtained. They include gases such as methane, liquids such as gasoline and kerosene, waxy solids, and a tarry asphalt residue. Note that the boiling point goes up with increasing number of carbon atoms in the molecule, and hence with increasing molecular mass and number of electrons. Heavier, larger molecules with an abundance of electrons are attracted to each other with stronger intermolecular forces than are lighter, smaller molecules having fewer electrons. Higher temperatures are required to vaporize the compounds of higher molecular mass to overcome their stronger intermolecular forces.

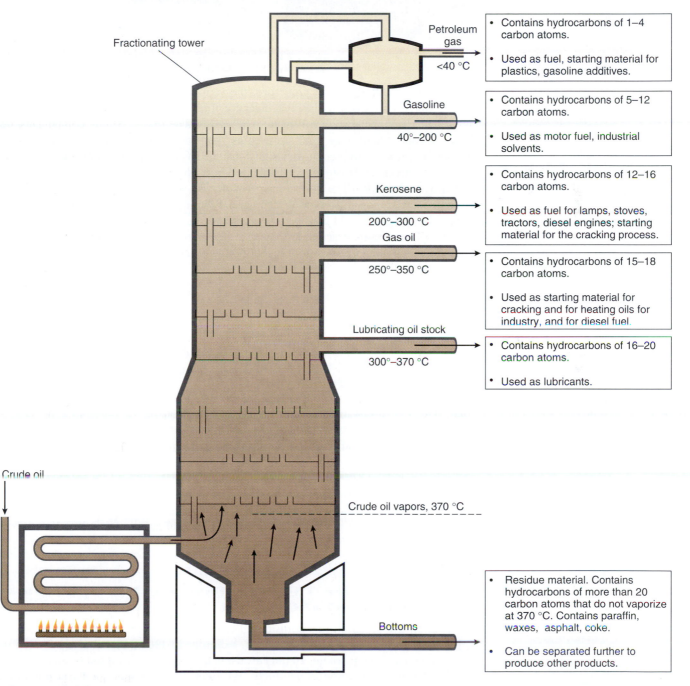

Figure 4.12

Diagram of a fractional distillation tower and various fractions.

Figure 4.13

End-uses for products from the refining of one barrel (42 gallons) of crude oil.

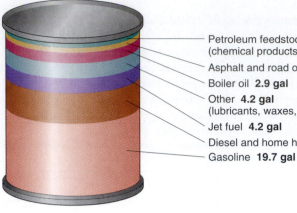

Barrel of Crude Oil

Petroleum feedstocks **1.25 gal** (chemical products, plastics)

Asphalt and road oil **1.3 gal**

Boiler oil **2.9 gal**

Other **4.2 gal** (lubricants, waxes, solvents)

Jet fuel **4.2 gal**

Diesel and home heating oil **8.4 gal**

Gasoline **19.7 gal**

Because of differences in properties, the various fractions distilled from crude oil have different uses. Indeed, the great diversity of products obtained has made petroleum a particularly valuable source of matter and energy. The lowest boiling components are gases at room temperature, and are used as "bottled gas" and other fuels. Sometimes flames at the tops of refinery towers signal that the gas is being burned off in what seems to be an unnecessary waste of a valuable resource. The gasoline fraction, containing hydrocarbons with 5 to 12 carbon atoms per molecule, is particularly important to our automotive civilization. Efforts at designing and mass producing self-propelled vehicles were largely unsuccessful until petroleum provided a convenient and relatively safe liquid fuel. The kerosene fraction is somewhat higher boiling, and it finds use as a fuel in diesel engines and jet planes. Still higher boiling fractions are used to fire furnaces and as lubricating oils.

The refining of a barrel of crude oil (42 gallons) provides an impressive array of products, the vast majority of which (20 gal) is gasoline (Figure 4.13). A staggering 35 gal of the 42 gal in a barrel of crude oil is simply burned for heating and transportation. The remaining 7 gal are for non-fuel uses, including only 5 quarts (1.25 gal) set aside to serve as non-renewable starting materials (reactants, commercially called feedstocks) to make the myriad of plastics, pharmaceuticals, fabrics, and other carbon-based industrial products so common in our society.

It is appropriate that a discussion of petroleum should also include natural gas. This fuel, which is mostly methane, currently provides heat for two-thirds of the single-family homes and apartment buildings in the United States. Recently there has also been increased interest in using natural gas as an energy source for generating electricity and for powering cars and trucks. A distinct advantage of natural gas is that it burns much more completely and cleanly than do other fossil fuels. Because of its purity it releases essentially no sulfur dioxide, it emits only very low levels of unburned volatile hydrocarbons, carbon monoxide, and nitrogen oxides, and it leaves no residue of ash or heavy metals. Moreover, on a per joule of energy basis, natural gas produces 30% less carbon dioxide than oil and 43% less carbon dioxide than coal.

4.15 Consider This: The Changing Mix of Products from a Barrel of Crude Oil

Chemical research has contributed to increasing the amount of gasoline derived from a barrel of crude oil. For example, in 1904 a barrel of crude oil produced 4.3 gal gasoline, 20 gal kerosene, 5.5 gal fuel oil, 4.9 gal lubricants, and 7.1 gal miscellaneous products. By 1954, the mix was 18.4 gal gasoline, 2.0 gal kerosene, 16.6 gal fuel oil, 0.9 gal lubricants, and 4.1 gal miscellaneous products. Compare these values with those shown in Figure 4.13 and offer some reasons why the mix of products has changed over time.

4.9 Manipulating Molecules

Research has shown that many of the compounds distilled from crude oil are not ideally suited for the desired applications. Nor does the normal distribution of molecular
masses correspond to the prevailing use pattern. The demand for gasoline is considerably greater than that for higher boiling fractions. Gasoline that comes directly from the
fractionating tower represents less than 50% of the original crude oil. Heavier and
lighter crude oil fractions can be manipulated to form still more gasoline. Therefore,
chemistry is used to rearrange large molecules by breaking them into smaller ones suitable to be used in gasoline, a process called **cracking**. For example, a hydocarbon with
16 carbons can be cracked into two almost equal fragments:

$$C_{16}H_{34} \longrightarrow C_8H_{18} + C_8H_{16} \tag{4.7}$$

or into different sized ones:

$$C_{16}H_{34} \longrightarrow C_5H_{12} + C_{11}H_{22} \tag{4.8}$$

Note that the numbers of carbon and hydrogen atoms are unchanged from reactants to
products; the larger reactant molecules simply have been rearranged (split) to form
smaller, more economically important molecules. When cracking is achieved by heating
the starting materials, the process is called thermal cracking. However, valuable energy
can be saved if catalysts are used to speed up the molecular breakdown at lower temperatures in an operation called catalytic or cat cracking. The catalysts employed in this
process are chemically similar to the ion exchangers used in water softening (see Chapter 5). Important cracking catalysts were developed by research chemists at Mobil Oil
and at Union Carbide.

If there is an excess of small molecules and a need for intermediate sized ones, such
as those in gasoline, the former can be catalytically combined to form the latter.

$$4\ C_2H_4 \longrightarrow C_8H_{16} \tag{4.9}$$

The refining process can also rearrange the atoms within a molecule. It turns out that
not all the molecules with a single chemical formula are necessarily identical. For example, octane, an important component of gasoline, has the formula C_8H_{18}. Careful
analysis discloses that there are 18 different compounds with this formula. Although the
chemical and physical properties of these forms are similar, they are not identical. For
example, the substance called octane has a boiling point of 125°C, whereas that of the
compound commonly known as isooctane is 99°C. Different compounds with the same
formula are called **isomers.** Isomers differ in molecular structure—the way in which the
constituent atoms are arranged. This is like the letters a, e, and t; they can be organized
to form three different recognizable words—ate, eat, and tea. The structures of octane
and isooctane are illustrated below.

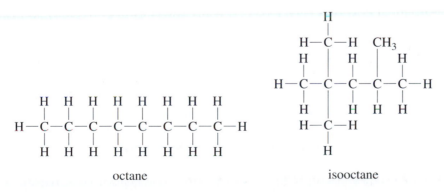

octane   isooctane

The molecules of both isomers consist of 8 carbon atoms and 18 hydrogen atoms, but
these atoms are arranged differently in the two compounds. In octane all the carbon
atoms are in an unbranched ("straight") line; in isooctane the carbon chain is branched.
Chapter 10 includes more information about isomers and how to interpret these
structures.

Figure 4.14

Smooth ignition and preignition. Preignition causes engine "knocking."

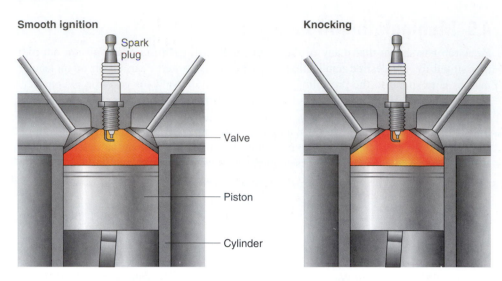

Smooth ignition

Knocking

Spark plug

Valve

Piston

Cylinder

Gasoline is available in 87, 89, and 92 octane.

Both octane and isooctane have essentially the same heat of combustion, but the former compound ignites much more easily. In a well-tuned car engine, gasoline vapor and air are drawn into a cylinder, compressed by a piston, and ignited by a spark. But compression alone is often enough to explode pure octane. This premature firing or preignition gives rise to a knocking sound (Figure 4.14). Knocking reduces the efficiency of the engine because the energy of the exploding and expanding gas is not applied to the pistons at the optimum time. Extensive knocking can cause engine damage.

On the other hand, isooctane is very resistant to preignition. It is the standard of excellence for rating the tendency of fuels to knock, though some compounds are even better. The performance of isooctane in an automobile engine has been measured and arbitrarily assigned an octane rating of 100. On this scale, octane has an octane rating of −20 (Table 4.3). When you go to the gasoline pump and fill up with 87 octane, the gasoline has the same knocking characteristics as a mixture of 87% isooctane (octane number 100) and 13% heptane (octane number zero). Higher grade gasolines are also available: 89 octane (regular plus) and 92 octane (premium); these contain a greater percentage of higher octane compounds.

It is possible to rearrange or "reform" octane to isooctane, thus greatly improving its performance. This is accomplished by passing octane over a catalyst consisting of rare and expensive elements such as platinum (Pt), palladium (Pd), rhodium (Rh), or iridium (Ir). Reforming isomers to improve octane rating has become particularly important because of the nationwide efforts to ban lead from gasoline. Lead does not occur naturally in petroleum. But in the 1920s, the compound tetraethyl lead, $Pb(C_2H_5)_4$, was first added to gasoline to reduce knocking and increase octane rating in more powerful engines. The strategy proved successful, but it was not without environmental consequences. In a very short time, automobile internal combustion engines became a major source of lead, emitted into the environment as volatile lead oxide through cars' tailpipes. Lead is a heavy metal poison, with cumulative neurological effects that are particularly damaging to young children. Therefore, major efforts have been launched in the United States to dis-

Table 4.3 **Octane Ratings of Several Substances**

Compound	Octane Rating
Octane	−20
Heptane	zero
Isooctane	100
Methanol	107
Ethanol	108
MTBE	116

continue the practice of adding lead compounds to motor fuel. Since 1976, all new cars and trucks sold in the United States have been designed to run on unleaded gasoline. The results have been dramatic. In 1970, approximately 200 thousand metric tons of lead were released into the atmosphere. Over the next thirty years, lead emissions dropped to less than 5% of that value. On balance, this decrease has been environmentally beneficial, but there have been associated costs, as the next activity suggests. About 50 countries worldwide still allow leaded gasoline containing 0.8 g of lead per liter, about 75% of the amount previously permitted in the United States.

4.16 Consider This: Leaded vs. Unleaded

Modern car engines designed to burn unleaded gasoline are somewhat less efficient than older engines designed to burn leaded fuel. It can therefore be argued that the switch to unleaded gasoline has contributed to greater fuel consumption and to air pollution by unburned exhaust residues. Moreover, some of the hydrocarbons introduced into gasoline when lead was phased out have been identified as possibly causing cancer. Some countries are still facing the switch from leaded to unleaded gasoline. To help with their decision making, draw up a list of risks and benefits associated with leaded and unleaded gasoline, and indicate the additional information you would need to appropriately weigh these risks and benefits.

Dr. Thomas Midgley (seated), receiving the Willard Gibbs Medal from American Chemical Society President Dr. Harry N. Holmes. The medal was awarded to Midgley for his research that developed, among other things, tetraethyl lead and the first CFCs.

4.10 Newer Fuels: Oxygenated and Reformulated Gasolines

This chapter began with a discussion about reformulated gasolines, focusing particularly on ethanol as an oxygenated fuel. Why is ethanol or other oxygenates needed in automobile fuels?

Eliminating the use of tetraethyl lead as an octane enhancer required finding substitutes for it to continue producing gasoline with sufficiently high octane ratings to meet the requirements of modern automobile engines. Several substitutes have come to the fore, including ethanol and MTBE (methyl-*tertiary*-butyl ether), each with an octane rating greater than 100.

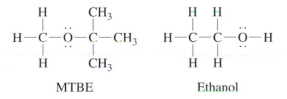

MTBE   Ethanol

These gasoline additives also are used in **oxygenated gasolines,** which are blends of gasolines with oxygen-containing compounds such as MTBE, ethanol, and methanol (CH_3OH). When oxygenated gasolines burn, they produce less carbon monoxide than their non-oxygenated counterparts, thereby reducing CO emissions. Cities with excessive wintertime carbon monoxide emissions are required by the Clean Act of 1990 to use oxygenated gasolines that contain 2.7% oxygen by weight. Oxygenated gasoline is also required in the 40 U.S. cities with the highest air pollution. Questions about the efficacy of MTBE and oxygenated gasolines in cold weather have been raised in Alaska and Wisconsin. Some data indicate an actual decrease in air quality and an increase in respiratory health problems since oxygenated gasolines have been introduced in these states. Whether there is a causal relationship among these factors is still being studied.

4.17 Your Turn

The molecular formula of MTBE is $C_5H_{12}O$. Calculate the percent (by mass) of oxygen in MTBE.

Hint: See Section 3.9 for a review of such calculations.

The molecular structure of benzene is discussed in Section 10.3.

Reformulated gasolines (RFG) are oxygenated gasolines that also contain a lower percentage of certain more volatile hydrocarbons such as benzene found in non-oxygenated conventional gasoline. RFGs cannot have greater than 1% benzene (C_6H_6) and must be at least 2% oxygenates. Because of their composition, reformulated gasolines evaporate less easily than conventional gasolines, and produce less carbon monoxide emissions. The more volatile hydrocarbons (benzene, etc.) in conventional gasoline also are involved in tropospheric ozone formation, especially in high-traffic metropolitan areas. To diminish ozone levels in these areas, RFGs have been required since 1995 in the nine cities in the United States with the highest ozone levels, along with voluntary compliance in nearly 90 other metropolitan areas with ozone levels above the standards.

4.18 Consider This: Your Contribution to Air Quality

According to the EPA, driving a car is "a typical citizen's most 'polluting' daily activity." **a.** Do you agree? Why? **b.** What pollutants do cars emit? Information on automobile emissions provided by the EPA (together with the information in your text) can help you fully answer this question. Check the *Chemistry in Context* web site for a direct link.

Reformulated gasolines play a role in reducing emissions. **c.** Where in the country are RFG's required? Check the current list published on the web by the EPA. When this book went to press, there were 10 cities listed, but more may have been added. **d.** Explain which emissions reformulated gasolines are supposed to lower.

In March 1999, California's governor Gray Davis decided to phase out the use of MTBE in gasoline sold in the state by the end of 2002. This action came as a result of evidence that MTBE was seeping into groundwater, not as a matter of MTBE's use to improve air quality.

The use of RFGs and oxygenated gasolines exemplifies a risk-benefit situation. The potential benefits are considerable; for example, using RFGs in the San Francisco Bay area could prevent 40 tons of hydrocarbon emissions and 160 tons of CO daily, the equivalent of removing over half a million vehicles from the road. If proportional reductions are realized in other cities across the country, the impact on air quality improvement truly will be significant. Yet, RFGs and oxygenated gasolines are not risk free; they have not been used long enough for possible long-term adverse effects, if any, to arise. Currently, risk levels are presumed to be low, with benefits outweighing risks. But the usefulness of RFGs and oxygenates to abate CO during the winter in cold climates has been called into question in several areas. Commenting on MTBE use, Dr. Sandra Mohr, a National Research Council committee member and occupational physician at the New Jersey Environmental and Occupational Health Sciences Institute, reported in the *Trenton Times:* "From a health effects point of view, we can't say there are any long-term health consequences [to MTBE use] but we recommend a cost-benefit analysis of the wintertime oxygenated fuel program."

4.19 Consider This: Sloppy Science

Consult the web to find two articles on gasoline additives. Find an example of an objective article and one that you believe to be biased or to contain other flaws such as sweeping generalizations that lead to misconceptions or misinterpretations of scientific data. Compare the two articles on the quality of their science and objectivity of their presentation. Provide their titles, authors, and URLs. On what basis do you rate the quality of an article? Describe your criteria.

4.11 Seeking Substitutes

Because the world's coal supply far exceeds the available oil reserves, there is interest in converting coal into gaseous and liquid fuels that are identical with or similar to petroleum products. As a matter of fact, some of the appropriate technology is quite old. Before large supplies of natural gas were discovered and exploited, cities were lighted with water gas. This is a mixture of carbon monoxide and hydrogen, formed by blow-

Energy, Chemistry, and Society **165**

ing steam over hot coke (the impure carbon that remains after volatile components have been distilled from coal).

$$C(s) + H_2O(g) \longrightarrow CO(g) + H_2(g) \qquad (4.10)$$

This same reaction is the starting point for the Fischer–Tropsch process for producing synthetic gasoline. The carbon monoxide and hydrogen are passed over an iron or cobalt catalyst, which promotes the formation of hydrocarbons. These can range from the small molecules of gases like methane, CH_4, to the medium-sized molecules (containing five to eight carbon atoms) typically found in gasoline. This process, which was developed during the 1930s in Germany, is only economically feasible where coal is plentiful and cheap and oil is scarce and expensive. This is the case in South Africa, where 40% of gasoline is obtained from coal. In the future, such technology may also become competitive in other parts of the world.

Concerns about the dwindling supply of petroleum have also led to the use of renewable energy sources. This generally means biomass—materials produced by biological processes. One such source, wood, was much touted during the 1970s energy crisis. But the energy demands of our modern society cannot possibly be met by burning wood. Burning trees would also destroy effective absorbers of carbon dioxide while adding that greenhouse gas and other pollutants to the atmosphere. In some parts of the country, the use of wood-burning stoves has been severely curtailed because the smoke and soot particles produced have a negative effect on air quality.

Ethanol is another alternative fuel produced from renewable biomass. It is formed by the fermentation of carbohydrates such as starches and sugars. Enzymes released by yeast cells catalyze the reaction that is typified by this equation.

$$\underset{\text{glucose}}{C_6H_{12}O_6} \longrightarrow 2\ C_2H_5OH + 2\ CO_2 \qquad (4.11)$$

The burning of ethanol as in the following equation releases 1367 kJ per mole of C_2H_5OH.

$$C_2H_5OH(l) + 3\ O_2(g) \longrightarrow 2\ CO_2(g) + 3\ H_2O(l) + 1367\ kJ \qquad (4.12)$$

The energy output corresponds to 29.7 kJ/g. This value is somewhat lower than the 47.8 kJ/g produced by C_8H_{18}, because the ethanol is already partially oxidized. Nevertheless, ethanol is already being mixed with gasoline to form "gasohol." At the usual concentration of 10% ethanol, gasohol can be used without modifying standard automobile engines. Higher concentrations of alcohol require changes in design, but these have already been made for racing engines that run on pure methanol (methyl alcohol), CH_3OH. Of the 13 million vehicles in Brazil, more than 4 million use pure ethanol, made from

An advertisement for gasohol made using ethanol.

fermenting sugar cane juice. Most of the rest of Brazilian cars operate on a mixture of ethanol and gasoline.

Whether ethanol will make a significant contribution to energy production depends on other factors, especially agriculture and politics. The great variety of alcoholic beverages indicates that C_2H_5OH can be prepared from almost any plant product—corn, wheat, barley, rice, sugar beets, sugar cane, grapes, apples, potatoes, dandelions, and so on. But these sources also serve as food for humans or other animals. Therefore, the use of agricultural products for the production of fuel must depend on supply and demand, surpluses and shortages. Currently, the United States produces a significant surplus of corn and other grains that could be converted to ethanol. But in a recent paper, Bernard Gilland, a civil engineer, estimates that meeting only 10% of the world's current primary energy demand with alcohol would require that one-quarter of the world's cropland be removed from food and feed production. Clearly, there are limits to the amount of energy we can obtain from biomass.

The use of ethanol as a petroleum substitute or supplement has also become a political hot potato in the United States. The issues involve not only ethanol, but other oxygen-containing compounds. Since January 1995, all of the gasoline sold in specified metropolitan areas with high ozone pollution has contained 2% oxygen by mass. Burning this reformulated gasoline is expected to reduce air pollution, principally carbon monoxide and the ozone generated in secondary reactions. Ethanol is a prime source of oxygen. So is methanol, CH_3OH, which is typically manufactured from natural gas. In addition, there are other oxygenated additives, some made from ethanol and some from methanol, such as MTBE. So the question comes down to the source of the mandated additives—agricultural products or fossil fuels. To further escalate the arguments, the EPA ruled in 1994 that 30% of the oxygenated compounds used in the reformulated gasoline must come from renewable (agricultural) sources. The ruling was immediately challenged.

The battle lines are clearly drawn, largely on the basis of self-interest. Supporting the greater use of ethanol are the EPA and over 20 farm groups, including the National Corn Growers Association. In opposition are the American Petroleum Institute, the petroleum refiners and gasoline companies, and the Sierra Club. U.S. Senators stake out positions on the issue depending on whether they are from agricultural states or ones tied closely to oil. There is a good deal at stake—among other things, 100 million to 200 million bushels of corn per year. The pro-petroleum faction responds that even with extensive farm support programs, ethanol is significantly more expensive per gallon and per joule than is conventional gasoline. Amid all the lobbying, the claims and counterclaims, the charges and the rebuttals, it is frustratingly difficult to find the facts, even the scientific ones. For example, some researchers have argued that when all aspects of production and processing are included, burning ethanol actually contributes more CO_2 to the environment than burning petroleum-based fuel. A related issue is the amount of energy that is required to produce a gallon of ethanol. The Sun is not the only source involved. Energy is required to plant, cultivate, and harvest the corn, to produce and apply the fertilizers, to distill the alcohol from the fermented mash, and to manufacture tractors. Hard data are difficult to get, but some sources claim that more energy goes into producing a gallon of ethanol than can be obtained by burning it.

4.20 Consider This: Gasohol

Unlike fuels obtained from petroleum, a nonrenewable source, ethanol can be produced from renewable resources. One method of producing ethanol is through the fermentation of crops such as sugar cane, potatoes, and corn. **a.** Draw up a list of reasons that argue for using crops for food and also a list of reasons for converting crops to ethanol for use in fuels. **b.** What factors will determine how you decide which is the optimal use of these crops?

Yet another potential energy source is a commodity that is cheap, always present in abundant supply, and always being renewed—garbage. No one is likely to design a car that will run on orange peels and coffee grounds, but there are approximately 140 power

Figure 4.15
Hennepin County Resource
Recovery Facility—a garbage-
burning power plant.

plants in the United States that do just that. One of these, pictured in Figure 4.15, is the Hennepin County Resource Recovery Facility in Minneapolis, Minnesota. Hennepin County produces about one million tons of solid waste each year. About one-third of that is burned in the waste-to-energy facility. The heat evolved is used to generate electricity via a process described in the next section. At full capacity, the Minneapolis plant produces 37,000 kJ per second, enough energy to meet the needs of 40,000 homes. One truckload of garbage (about 27,000 lb) will generate the same quantity of energy as 21 barrels of oil.

This resource recovery approach, as it is sometimes called, simultaneously addresses two major problems—the growing need for energy and the growing mountain of waste. The great majority of the trash is converted to carbon dioxide and water and no supplementary fuel is needed. The unburned residue is disposed of in landfills, but it represents only about 10% of the volume of the original refuse. Although some citizens have expressed concern about gaseous emissions from garbage incinerators, the stack effluent is carefully monitored and must be maintained within established composition limits. Both Japan and Germany are making considerably greater use of waste-to-energy technology than is the United States.

Perhaps the ultimate example of using waste as an energy source is provided by methane generators. In rural China and India there are over one million reactors in which animal and vegetable wastes are fermented to form biogas. This gas, which is about 60% CH_4, can be used for cooking, heating, lighting, refrigeration, and generating electricity. The technology lends itself very well to small-scale applications. The daily manure from one or two cows can generate enough methane to meet most of the cooking and lighting needs of a farm family. Two-thirds of China's rural families use biogas as their primary fuel.

4.21 Consider This: Hennepin County Waste Burning Plant

The Hennepin County Resource Recovery Facility in Minneapolis has been the subject of a great deal of controversy for the county residents. The idea of generating usable energy from trash sounds wonderful, until the facility is built in your neighborhood. This is the problem faced by homeowners and residents in the area surrounding the plant. To address residents' concerns, an open meeting between the residents and representatives of the plant is scheduled. Managers from the plant, engineers, and representatives of the state pollution control agency will be present. Prepare a list of questions that you, as a resident in this area, would like to see addressed at this meeting.

4.12 Transforming Energy

Essentially all of the fuels we have been considering in this chapter—coal, oil, alcohol, or garbage—give up their energy through combustion. They are burned to generate heat. For the most part, however, heat is not the form in which the energy is ultimately used. Heat is nice to have around on a cold winter day, but it is a cumbersome form of energy. It is dangerous if uncontrolled, difficult to transport, and hard to harness for other purposes. The industrialization of the world's economy began only with the invention of devices to convert heat to work. Chief among these was the steam engine, developed in the latter half of the eighteenth century. The heat from burning wood or coal was used to vaporize water, which in turn was used to drive pistons and turbines. The resulting mechanical energy was used to power pumps, mills, looms, boats, and trains. The smoke-belching mechanical monsters of the English midlands soon replaced humans and horses as the primary source of motive power in the Western world.

A second energy revolution occurred early in the 1900s with the commercialization of electrical power. Today, one-third of the energy produced in the United States is electrical. Most of it is generated by the descendants of those early steam engines. Figure 4.16 is a schematic of a modern power plant. Heat from the burning fuel is used to boil water, usually under high pressure. The elevated pressure serves two purposes: it raises the boiling point of the water and it compresses the water vapor. The hot, high-pressure vapor is directed at the fins of a turbine. As the gas expands and cools, it gives up some of its energy to the turbine, causing it to spin like a pinwheel in the wind. The shaft of the turbine is connected to a large coil of wire that rotates within a magnetic field. The turning of this dynamo generates an electric current—a stream of electrons that represents energy in a new and particularly convenient form. Meanwhile, the water vapor leaves the turbine and continues in its closed cycle. It passes through a heat exchanger where a stream of cooling water carries away the remainder of the heat energy originally acquired from the fuel. The water condenses into its liquid state and reenters the boiler, ready to resume the energy transfer cycle.

This process of energy transformation can be summarized in the three steps diagrammed in Figure 4.17. **Potential energy** in the chemical bonds of fossil fuels is first converted to **heat energy**. The heat released in combustion is absorbed by the water vaporizing it to steam. This heat is then transformed into **mechanical energy** in the spin-

Figure 4.16

Diagram of a power plant for the conversion of heat to work to electricity.

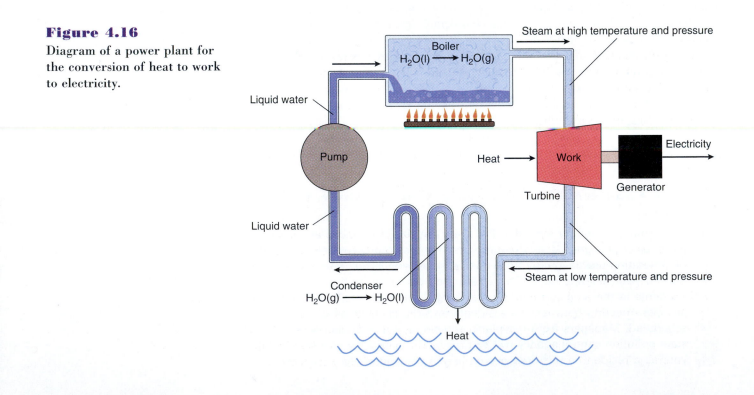

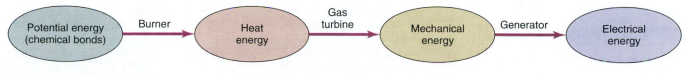

Figure 4.17
Energy transformation in an electric power plant.

ning turbine that turns the generator that changes the mechanical energy into **electrical energy.** In compliance with the first law of thermodynamics, energy is conserved throughout these transformations. To be sure, no new energy is created, but none is lost, either. We may not be able to win, but we can at least break even . . . or can we?

4.13 Energy and Efficiency

The last question is not as facetious as it might sound. In fact, we cannot break even. No power plant, no matter how well designed, can completely convert heat into work. Inefficiency is inevitable, in spite of the best engineers and the most sincere environmentalists. There are, of course, energy losses due to friction and heat leakage that can be corrected, but these are not the major problems. The chief difficulty is nature; more specifically, the nature of heat and work.

The **maximum theoretical efficiency** with which a power plant can convert heat to work depends on the difference between the highest temperature to which the water vapor is heated and the lowest temperature to which the condensed water is cooled. The mathematical expression for the efficiency is given below.

$$\text{Efficiency} = \frac{\text{Highest temp} - \text{Lowest temp}}{\text{Highest temp}}$$

It is important to note that the temperatures in this equation are expressed in the **absolute** or **Kelvin scale.** Zero on the Kelvin scale is absolute zero or $-273°C$, the lowest temperature possible. In fact, it cannot quite be attained. To convert a temperature from the Celsius scale to the Kelvin scale, you simply add 273.

$$\text{Temperature (in Kelvins)} = \text{Temperature (in degrees Celsius)} + 273$$

A well-designed modern fossil fuel-burning power plant operates between a high temperature of about 550°C and a low temperature of 30°C. This means that, on the Kelvin scale, highest temp = 550° + 273 = 823 K, and lowest temp = 30° + 273 = 303 K. It follows that

$$\text{Efficiency} = \frac{823 \text{ K} - 303 \text{ K}}{823 \text{ K}} = \frac{520 \text{ K}}{823 \text{ K}} = 0.63; \ 0.63 \times 100 = 63\%$$

This value of 0.63 is the maximum theoretical efficiency of a power plant operating between these two temperatures. It means that at best only 63% of the heat that is obtained from the burning fuel is actually converted to work. The remainder is discharged to the cooling water, which consequently warms up. There are, in addition, other inefficiencies associated with friction, loss over long-distance power transmission lines, and so forth. Table 4.4 lists the efficiencies of a number of steps in energy production. The overall efficiency is the product of the efficiencies of the individual steps: the individual efficiencies are multiplied. The net result is that today's most advanced power plants operate at an overall efficiency of only about 42%.

Consider, for example, the case of electrical home heating, sometimes touted as being clean and efficient. We will assume that the electricity is produced by a methane-burning power plant with a maximum theoretical efficiency of 60%. The efficiencies of the boiler, turbine, electrical generator, and power transmission lines are given in Table 4.4; converting the electrical energy back into heat in the home is 98% efficient.

Table 4.4 Some Typical Efficiencies in
Power Production

Maximum Theoretical Efficiency:	55–65%
Efficiency of Boiler:	90%
Mechanical Efficiency of Turbine:	75%
Efficiency of Electrical Generator:	95%
Efficiency of Power Transmission:	90%

Because efficiencies are multiplicative, to find the overall efficiency of the electricity generation–to-home heating sequence, we multiply the efficiencies of the individual steps, expressed as their decimal equivalents:

Overall efficiency = efficiency of (power plant) ×

(boiler) × (turbine) × (electrical generator) × (power transmission) × (home heat)

$$= 0.60 \times 0.90 \times 0.75 \times 0.95 \times 0.90 \times 0.98 = 0.34$$

The overall efficiency of 0.34 indicates that only 34% of the total heat energy derived from the burning of methane at the power plant will be available to heat the house. If the electrically heated house requires 3.5×10^7 kJ (a typical value for a northern city in January), how much methane (grams) has to be burned at the power plant to furnish the heat needed? The combustion of 1 gram of methane releases 50.1 kJ. Remember that only 34% of the energy from the burned methane is available to heat the house. So, because of inefficiencies, far more methane will have to be burned than the amount to release 3.5×10^7 kJ. The total quantity of heat that must be used can be calculated:

Heat needed = (Heat used) × efficiency

3.5×10^7 kJ = (Heat used) × 0.34

$$(\text{Heat used}) = \frac{3.5 \times 10^7 \text{ kJ}}{0.34} = 1.0 \times 10^8 \text{ kJ}$$

Because each gram of burning methane yields 50.1 kJ, the mass of methane that must be burned to furnish 1.0×10^8 kJ is:

$$1.0 \times 10^8 \text{ kJ} \times \frac{1 \text{ g CH}_4}{50.1 \text{ kJ}} = 2.0 \times 10^6 \text{ g CH}_4$$

You can compare the efficiencies of heating a home with electricity or a natural gas furnace by completing the following Sceptical Chymist activity.

4.22 The Sceptical Chymist: Clean Electric Heat

Is electric heat clean and efficient? The electricity must first be generated, usually by a fossil-fuel power plant.

a. The house could also have been heated directly with a gas furnace burning methane at 85% efficiency. Calculate the number of grams of methane required in January using this method of heating.

Ans. Because the only inefficiency is that of the furnace, we can do a calculation similar to that done above, but using 0.85 as the efficiency. The energy required is 4.1×10^7 kJ, which corresponds to 8.2×10^5 g of methane. This is only 41% of the methane needed to provide the heat electrically.

b. On the basis of your answer to part **a** and the discussion just before this exercise, comment on the claim that electric heat is "clean" and efficient.

4.23 Your Turn

A coal-burning power plant generates electrical power at a rate of 500 megawatts, that is, 500×10^6 or 5.00×10^8 joules per second. The plant has an overall efficiency of 0.375 for the conversion of heat to electricity.

a. Calculate the total quantity of electrical energy (in joules) generated in one year of operation and the total quantity of heat energy used for that purpose.

Ans. 1.58×10^{16} J generated; 4.20×10^{16} J used

b. Assuming the power plant burns coal that releases 30 kJ/g, calculate the mass of coal (in grams and metric tons) that will be burned in one year of operation. (1 metric ton = 1×10^3 kg = 1×10^6 g.)

Ans. 1.40×10^{12} g; 1.40×10^6 metric tons

4.14 Improbable Changes and Unnatural Acts

Although we have asserted that heat cannot be completely converted into work, we have offered little evidence and no explanation for this fact of nature. To illustrate the problem, push this book off the desk, and wait for it to come back up by itself. Be prepared to wait quite a while! The idea of a book picking itself off the ground and rising against the force of gravity is so bizarre, it is unbelievable. In fact, it is just unlikely—extremely unlikely. To understand why, let us examine the process in a little more detail.

You probably recall that a book resting on a table has potential energy by virtue of its position above the floor. When the book is dropped, the potential energy is converted to kinetic energy, the energy of motion. When the book strikes the floor, the kinetic energy is released with a bang. Some of it goes into the shock wave of moving air molecules that transmit the sound. Most of the energy goes to increase the motion of the atoms and molecules in the book and in the floor beneath it. This microscopic motion is the origin of what we call heat. A careful measurement would show that the book and the floor, and the air immediately around them are all very slightly warmer than they were before the impact. Energy has been conserved, but it has also been dissipated. Heat or **thermal energy** is characterized by the random motion of molecules. They chaotically move in all directions.

Now consider what would be necessary for the book to rise by itself, and thus to do work against the force of gravity. The molecules in the book would all have to move upward at the same time. At that instant, all of the molecules in the floor under the book would also have to move in an upward direction, giving the book a little shove. Needless to say, such agreement among the 10^{25} or so molecules involved is very unlikely. Yet, such a change is necessary to convert heat into work—to transform random thermal motion into uniform motion. The first law of thermodynamics may confidently assert that all forms of energy are equal, but the fact remains that some forms of energy are more equal than others! The chaotic, random motion that is heat is definitely low-grade energy.

4.15 Order versus Entropy

The inability of a power plant to convert heat into work with 100% efficiency or for dropped books to spontaneously rise are both manifestations of the same law of nature, the **second law of thermodynamics.** There are many versions of the second law, but all describe the directionality of the universe. One version states that it is impossible to completely convert heat into work without making some other changes in the universe. Another observes that heat will not of itself flow from a colder to a hotter body. The falling book provides another version of the second law. That, too, is a transformation of ordered kinetic energy into random heat energy. Like all naturally occurring changes, it involves an increase in disorder or randomness of the universe. This randomness in position or energy level is called **entropy.** The most general statement of the second law of thermodynamics is **the entropy of the universe is increasing.** This means that organized energy, the most useful kind for doing work, is always being transformed into chaotic motion or heat energy.

A helpful way to look at the increase in universal entropy that characterizes all changes is in terms of probability. Disordered states are more probable than ordered ones, and natural change always proceeds from the less probable to the more probable. Let's suppose you define perfect order as a beautifully organized sock drawer, all the socks matched, folded, and placed in rows. This would represent a condition of zero entropy (no randomness). If you are like most people, this is probably a rather unlikely arrangement. It certainly did not occur by itself; it took work to organize the socks. Without the continuing work of organization, it is quite possible that, over the course of a week or a month or a semester, the entropy and disorder of that sock drawer will increase. The point is that there are lots of ways in which the socks can be mixed up, and therefore disorder is more probable than order. Conversely, it is not very likely that you will open your drawer some morning and find that the previously jumbled socks are in perfect order and the entropy in that particular part of the universe has suddenly and spontaneously decreased without any external intervention. That sort of change from disorder to order is essentially what is involved in the conversion of heat to work. Professor Henry Bent has estimated that the probability of the complete conversion of one calorie of heat to work is about the same as the likelihood of a bunch of monkeys typing Shakespeare's complete works 15 quadrillion times in succession without a mistake.

Perhaps by now some Sceptical Chymist in the class has objected that there are many earthly instances in which order increases. Sock drawers do get organized; power plants convert heat to work; dropped objects get picked up; water can be decomposed into hydrogen and oxygen; refrigerators transfer heat from a colder to a hotter body; and students learn chemistry. All of these are "unnatural" events—**nonspontaneous** in the vocabulary of thermodynamics. They will not occur by themselves; they require that work be done by someone or something. An input of energy is necessary to reduce the entropy and increase the order. And in every case, the work that is done generates more entropy somewhere in the universe than it reduces in one small part of the universe. Even when entropy appears to decrease in a spontaneous change, for example the freezing of water at temperatures below 0°C, there are balancing increases in entropy. In this particular case, the heat given off by the freezing water adds to the disorder of its surroundings. In short, when the entire universe is considered, entropy always increases. THERE IS NO FREE LUNCH!

One word of caution: you need to be a little careful about the scientific meaning of "spontaneous" and "nonspontaneous." Spontaneous is often taken to mean a process that occurs all by itself, without any apparent initiation, as in the tabloid headline "SLEEPING MAN BURSTS INTO FLAME." In scientific usage, a spontaneous change is one that *could* occur, in other words it is thermodynamically possible. But it might not take place all by itself because a large activation energy barrier must be overcome to start the reaction. Let's return to our sleeping man. Human beings are thermodynamically unstable with respect to combustion products such as carbon dioxide and water. So the burning of a human is a spontaneous change in the scientific sense of that term. But fortunately for us, the activation energy barrier for that process is so high that we do not have to worry about bursting into flames without any provocation. In the case of the reported sleeping man, it would be a good idea to look for an outside agent, perhaps a disgruntled friend, who might have helped him over that barrier with a can of gasoline and a match.

4.24 Consider This: Humpty Dumpty

All around us there are examples of the natural tendency for things to get messed up. Some have even been enshrined in literature: "All the king's horses and all the king's men, couldn't put Humpty together again." Cite some examples of your own.

4.25 Consider This: Entropy Decrease–Entropy Increase

During midterm time, many students become very serious about their studying, and for hours on end will concentrate on the plays of Shakespeare or the causes of World War II. This decrease in intellectual entropy is often associated with an increase in the entropy of the student's room. Identify another process in which entropy appears to decrease but is actually coupled with an increase in entropy elsewhere in the universe.

4.16 The Case for Conservation

A fundamental feature of the universe is that energy and matter are conserved. However, the process of combustion converts both energy and matter to less useful forms. For example, the energy stored in hydrocarbon molecules is eventually dissipated as heat when those molecules are converted to carbon dioxide and water—essential compounds, to be sure, but unusable as fuels. As residents of the universe, we have no choice; we must obey its inexorable laws. Nevertheless, there are many options within those constraints. One of the most important is to make a human contribution to the conservation of energy and matter.

The planet's store of fossil fuels is, of course, finite, although our appetites for them seem infinite. Worldwide demand for oil is increasing at more than 2% per year. Over the past 15 years, energy use is up 30% in Latin America, 40% in Africa, and 50% in Asia. Oil forecasts by the U.S. Energy Information Administration project a 60% increase in global demand for oil by 2020 to a whopping 40×10^9 barrels per year. To meet this growth, the market share of oil from middle eastern countries will likely rise beyond 30%, approaching the levels that produced the oil crises of 1974 and again in 1979. The conventional oil reserves as reported recently in the *Oil and Gas Journal* (1.020×10^{12} barrels) are calculated to last 43 years at the current rate of consumption. In addition, there is enough petroleum locked up in heavy crude oil, bitumen, and oil shale to provide for another 170 years, though this reserve will be more difficult and more expensive to extract. Global coal reserves appear to be considerably greater, but they too are limited. And in any event, we can be certain that the world's energy consumption will not remain at its current level.

The Organization for Economic Co-operation and Development, an organization of western European nations, has estimated that between 1990 and 2010 world energy consumption will increase by 50%, oil consumption will increase by 40%, coal consumption will increase by 45%, and natural gas consumption will increase by 66%. One consequence of this increased use of energy will be a 50% rise in global CO_2 emissions. Not surprisingly, the greatest increases are expected to occur in the developing countries that, by 2010, will account for over half of the energy consumed. Growing populations, migration to the cities, and industrialization will drive up the energy demand of the Third World to unprecedented highs. The pattern has already been established. In China energy utilization in 1993 was 22 times what it was in 1952. If the Chinese demand for electricity were to grow at 7% per year, by the year 2000 the country would need to open one medium-sized power plant every week. The construction of the immense Three Gorges dam and hydroelectric power station in China is one attempt by that country to meets its exploding electrical energy demands.

The demands of conventional power plants for coal, oil, and gas are voracious. But fossil fuels are so important as feedstocks for chemical synthesis that it is a great waste to burn them. Late in the nineteenth century, Dmitri Mendeleev, the great Russian chemist who proposed the periodic table of the elements, visited the oil fields of Pennsylvania and Azerbaijan. He is said to have remarked that burning petroleum as a fuel "would be akin to firing up a kitchen stove with bank notes." Mendeleev recognized that oil could be a valuable starting material for a wide variety of chemicals and the products made from them. But he would, no doubt, be amazed at the fibers, plastics, rubber, dyes, medicines, and pharmaceuticals that are currently produced from petroleum. Yet, we continue to ignore Mendeleev's warning and burn nearly 85% of the oil pumped from the ground.

The Three Gorges Dam will supply China with significant hydroelectric energy.

In short, the arguments for conserving energy and the fuels that supply it are compelling. Fortunately, some promising strategies are available, and considerable savings have already been realized. Although energy production and fuel consumption have increased since the 1974 oil crisis, there has also been a significant increase in the efficiency with which the fuels are used. The production of electricity by power plants is the major use of energy in the United States, making up 38% of the total. The conversion of heat to work is, of course, limited by the second law of thermodynamics, but power plants currently operate well below the thermodynamic maximum efficiency. Better design will bring power plants closer to that upper limit, perhaps to overall efficiencies of 50 or 60%. A particularly appealing approach is an integrated system that uses "waste" heat from a power plant to warm buildings.

Once the electricity is generated, great savings can be realized in its use. Estimates of the technically feasible savings in electricity range from 10–75%. The wide range in these predictions is worrisome, but specific data are encouraging. For example, in an article in *Scientific American* in September 1990, Arnold P. Fickett, Clark W. Gellings, and Amory B. Lovins make the following statement: "If a consumer replaces a single 75-watt bulb with an 18-watt compact fluorescent lamp that lasts 10,000 hours, the consumer can save the electricity that a typical United States power plant would make from 770 pounds of coal. As a result, about 1600 pounds of carbon dioxide and 18 pounds of sulfur dioxide would not be released into the atmosphere." In the process, about $100 would be saved in the cost of generating electricity. Improvements in the design of electric motors and refrigeration units also hold considerable potential for increased efficiency.

It is noteworthy that some utility companies now promote consumer education and provide financial inducements for conserving electricity. Sophisticated economic planning, new financing arrangements, and pricing policy are all part of efforts to save energy. One particularly important concept is "payback time," the period necessary before a private consumer, an industry, or a power company recaptures in savings the initial cost of a more efficient refrigerator, manufacturing process, or power plant.

4.26 The Sceptical Chymist: Light Bulbs Revisited

The quotation from Fickett, Gellings, and Lovins provides a marvelous opportunity for the Sceptical Chymist to apply her or his knowledge of chemistry. For example, let us check their assertion that replacement of a 75-watt (W) bulb with an 18-W fluorescent lamp will save the electricity made from

770 lb of coal. That is, assume that 770 pounds of coal, containing 65% carbon and 2% sulfur, will not be burned.

First, we note that the difference in the rate of energy consumption of the two bulbs is 75 W − 18 W = 57 W or 57 J/s. The total projected energy savings over the life of the bulb (10,000 hr) is obtained in this way.

$$\text{Energy Savings} = \frac{57 \text{ J}}{s} \times 10,000 \text{ hr} \times \frac{60 \text{ min}}{hr} \times \frac{60 \text{ s}}{min} = 2.05 \times 10^9 \text{ J}$$

The energy comes from coal, and coal typically yields 30 kJ/g or 30 × 10^3 J/g. To determine the mass of coal that must be burned to obtain 2.05 × 10^9 J, the following operation is performed.

$$\text{mass coal} = 2.05 \times 10^9 \text{ J} \times \frac{1 \text{ g coal}}{30 \times 10^3 \text{ J}} \times \frac{1 \text{ lb coal}}{454 \text{ g coal}} = 150 \text{ lb coal}$$

This is a significant discrepancy from the quoted value of 770 lb. Possibly Fickett, Gellings, and Lovins made an error, or perhaps they made an assumption that we neglected. Explore the latter possibility, and suggest what the assumption might have been.

4.27 Your Turn

Now it is your turn to exercise your skepticism and your computational skills by checking the other two claims. Calculate the mass of CO_2 and SO_2 that would *not* be released into the atmosphere if a 75-W bulb were replaced by an 18-W compact fluorescent lamp. Assume that 770 pounds of coal, containing 65% carbon and 2% sulfur, would not be burned.

4.28 The Sceptical Chymist: Keep Checking These Assumptions

Fickett, Gellings, and Lovins also claim that the bulb replacement we have been discussing would save about $100 in the cost of generating electricity. What value are they assuming for the cost of electricity?

Electricity is generally priced per kilowatt-hour (kWh), so we need to know the number of kWh saved over the 10,000 hr lifetime of the bulb. First we multiply the 57 watts saved by 10,000 hr.

$$57 \text{ watts} \times 10,000 \text{ hr} = 570,000 \text{ watt hr (W} \cdot \text{h)}$$

Then we convert the answer to kilowatt-hours, recognizing that 1 kWh = 1000 W · h.

$$570,000 \text{ W} \cdot \text{h} \times \frac{1 \text{ kW} \cdot \text{h}}{1000 \text{ W} \cdot \text{h}} = 570 \text{ kWh}$$

If 570 kWh of electricity cost $100, as the writers imply, what is the cost per kilowatt-hour? Once you have calculated the answer, find the cost of electricity to consumers in your city. Compare the results.

Recent advances in information technology and data processing have also made possible sizeable energy savings. "Smart" office buildings or homes feature a complicated system of sensors, computers, and controls that maintain temperature, airflow, and illumination at optimum levels for comfort and energy conservation. Similarly, the computerized optimization of energy flow and the automation of manufacturing processes have brought about major transformations in industry. Over the past 20 years, industrial production in the United States has increased substantially, but the associated energy consumption has actually gone down. A case in point is the low-pressure, gas-phase process developed by Union Carbide chemical engineers for making polyethylene, which is the world's most common plastic. This new process uses only one-quarter of the energy required by previous high-pressure methods. Although the capital investment associated with such conversions is often substantial, consumers and manufacturers may ultimately enjoy financial savings and increased profits. Mention should also be made of the energy conservation that results from recycling materials, especially aluminum. Because of the high energy cost required to extract aluminum metal from its ore, recycling the metal yields an energy saving of about 70%. To put things in perspective, you could watch television for three hours on the energy saved by recycling just one aluminum can.

One final area where energy conservation has a direct impact on lifestyle is transportation. About 20% of the total energy used and one-half of the world's oil production goes to power motor vehicles. But even here, we are making some progress in conservation. From the mid-1970s to the early 1990s, gasoline consumption in the United States dropped by one-half. Much of this saving was attributable to lighter-weight vehicles, thanks to the use of new materials, and to new engine designs. Yet this environmentally friendly trend is under assault as the twentieth century comes to an end. Sport utility vehicles (SUVs) and heavier "light" trucks now command a substantial part of the traditional car market. Two decades ago, the average light truck and car weighed about the same and had similar engines. That is no longer the case; light trucks are now more than 10% heavier and have nearly 80% more horsepower than those of 20 years ago. Although car manufacturers must meet a 27.5 miles per gallon (mpg) average fuel consumption for cars, the requirement is only 20.7 mpg for light trucks; SUVs are classified as light trucks, not as cars, one way to get around the higher mpg requirements. The lower mpg standards for trucks and SUVs have allowed their weight and horsepower to increase. Comparison data for several 1997 vehicles are given in Table 4.5. Note that a full-size SUV emits nearly 80% more CO_2 in a year and a full-size pickup 67% more than a mid-sized car. Both the full-size SUV and pickup truck emit much more CO_2 than even a large automobile.

On the plus side, research to develop cars with higher gas mileage continues. The Volkswagen Eco-Polo test car has achieved combined city/highway fuel consumption of 62 mpg and the Volvo LCO 2000 has reached 81 mpg on the highway. Methane and propane are being used to power a growing number of cars and trucks, and Chapter 8 will explore such alternative energy sources as hydrogen, electric batteries, and photovoltaic cells.

Such potential improvements in fuel economy are impressive, but the fact remains that the automobile is an energy-intensive means of transportation. A mass transit system is far more economical, provided it is heavily used. In Japan, 47% of travel is by public transportation, compared to only 6% in the United States. Of course, Japan is a compact country with a high population density. The great expanse of North America is not ideally suited to mass transit, although some regions in the United States, such as the population-dense Northeast and some other metropolitan areas, are. And one must also reckon with the long love affair between Americans and their automobiles.

Table 4.5 Gas Mileage and Vehicle Type

Vehicle type*	Weight, lb	Mpg	Pounds of CO_2 per yr (15,00 miles/yr)
Small-size car	2000	32	9200
Mid-size car	3100	25	11,800
Large car	3400	22	13,400
Mid-size SUV	4700	14	21,100
Mid-size SUV	3800	15	19,700
Full-size SUV	5500	15	19,700
Full-size SUV	5300	14	21,100
Full-size pickup	3900	16	18,500
Full-size pickup	4300	15	19,700

*Source: Data from *The New York Times*, November 30, 1997.

4.29 Consider This: Gasoline Rationing

Imagine that you were transported back to 1974, the time of a deepening oil crisis that caused stringent nationwide gas rationing to be put into place in the United States. Each household in your area has been allotted 60 gallons of gasoline a month. Assume that your vehicle averages 19 miles per gallon. Describe the number of people in your family, how many of them go to school or work each day, and how far each must travel. Use this information to prepare a detailed gasoline budget of how you and your family would use your allotment of 60 gallons for the month.

4.30 Consider This: The Price of Gasoline

Oil is a valuable resource, even beyond its use for home heating and gasoline. If we continue to use our petroleum supply to extract gasoline from it, we may lose our starting materials to make other petroleum-based products such as many pharmaceuticals and plastics. Up to now, voluntary conservation of gasoline has not been effective. The government could force more conservation by rationing gasoline or by heavily taxing it. If our government increased the price of gasoline to $4.00 per gallon (a price typical of that in Western Europe and Japan), sales of gasoline likely would drop. This price would have serious consequences for the American work force because the price of a gallon of gasoline would be raised compared to the current minimum hourly wage.

Suppose that a bill has been introduced in Congress to raise the price of gasoline to $4.00 per gallon. Draft a letter to a friend at another college, either supporting or protesting the bill. Include the reasons for your position and your opinion on what should be done with this new revenue if the bill is passed.

Conclusion

To a considerable extent, choice ultimately influences what technology can do to conserve energy. As individuals and as a society, we must decide what sacrifices we are willing to make in speed, comfort, and convenience for the sake of our dwindling fuel supplies and the good of the planet. The costs might include higher taxes, more expensive gasoline and electricity, fewer and slower cars, warmer buildings in summer and cooler ones in winter, perhaps even drastically redesigned homes and cities. During the 1970s a series of energy crises occurred because of a dramatic rise in the cost of imported crude oil, principally from the Middle East. Our dependency on it remains high, one reason to ask whether supply and demand factors for crude oil could precipitate another energy crisis. One thing seems to be clear: the best time to examine our options, our priorities, and our will is before we face another full-blown energy crisis.

Chapter Summary

Having studied this chapter, you should be able to:
- Describe the factors related to the United States' dependency on fossil fuels for energy (4.2)
- Relate energy use to atmospheric pollution and global warming (4.7, 4.8)
- Apply the terms exothermic, endothermic, and activation energy to chemical systems (4.3–4.5)
- Interpret chemical equations and basic thermodynamic relations to calculate heats of reaction, particularly heats of combustion (4.4)
- Use bond energies to describe the energy content of materials (4.4)
- Evaluate the risks and benefits associated with petroleum, coal, and natural gas as fossil fuel energy sources (4.7–4.8)

- Understand the physical and chemical principles associated with petroleum refining (4.8–4.9)
- Describe "octane rating" and how refining, leaded gasoline, ethanol, and MTBE relate to it (4.9)
- Discuss approaches to alternative (supplemental) automobile fuels (4.9–4.11)
- Describe why reformulated and oxygenated gasolines are used (4.10)
- Relate the energy potentially available from a process with the efficiency of that process (4.13)

- Use entropy as a concept to explain the second law of thermodynamics (4.15)
- Take an informed stand on what energy conservation measures are likely to produce the greatest energy savings (4.16)
- With confidence, examine news articles on energy crises and energy conservation measures to interpret the accuracy of such reports (4.16)

Questions

Emphasizing Essentials

1. **a.** What is the origin of fossil fuels?

   b. Name some examples of fossil fuels.

   c. Are fossil fuels a renewable resource?

2. Consider Figure 4.1a showing U.S. oil production and oil imports.

   a. Calculate the percentage of total oil production and imports that was supplied by domestic oil production in 1970 and in 1990. Also calculate the predicted value for 2010.

   b. How are these values changing with time?

   c. How have the sources of oil changed from 1970 to 1990 to what is predicted for 2010?

3. The reaction of water vapor with ethylene is one way to produce ethanol for use as a gasoline additive.

   $$CH_2CH_2 + H_2O \longrightarrow CH_3CH_2OH$$

   a. Rewrite this equation using Lewis structures.

   b. Was it necessary to break *all* of the chemical bonds in the reactants to form the product ethanol? Give a reason for your answer.

4. Consider the water in each of these two containers.

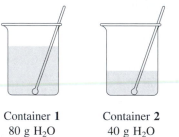

Container **1**   Container **2**
80 g H$_2$O   40 g H$_2$O
70°C   70°C

The temperature of the water is the same in each of the containers. Is the heat content of the water the same in each of these containers? How do you know?

5. The unit Calorie, used to express food heat values, is the same as a kilocalorie of heat energy. If you eat a chocolate bar from the United States with 600 Calories of food energy, how does the energy compare with eating a Swiss chocolate bar that has 3000 kJ of food energy? (*Note:* 1 kcal = 4.184 kJ)

6. A single serving bag of Granny Goose Hawaiian Style Potato Chips® has 70 Calories. Assuming that all of the energy from eating these chips goes towards keeping your heart beating, how long can the energy from these chips sustain a heartbeat of 80 beats per minute? (*Note:* 1 kcal = 4.184 kJ and each human heart beat requires 1 J of energy.)

7. The text states that an energy consumption of 650,000 kcal per person per day is equivalent to an annual personal consumption of 65 barrels of oil or 16 tons of coal. Use this information to calculate the amount of energy available in each of these quantities.

   a. one barrel of oil

   b. one gallon of oil (42 gallons per barrel)

   c. one ton of coal

   d. one pound of coal (2000 pounds per ton)

8. Use the information in question 7 to find the ratio of the quantity of energy available in one pound of coal to that in one pound of oil. *Hint:* one pound of oil has a volume of 0.56 quarts.

9. Use Figure 4.5 to compare the sources of U.S. energy consumption with the sources of world energy consumption. Arrange each in order of decreasing percentage and comment on the relative rankings.

10. Equation 4.1 represents the complete combustion of methane.

    a. Write a similar chemical equation for the complete combustion of ethane, C$_2$H$_6$.

    b. Represent this equation with Lewis structures.

    c. Represent this reaction with a sphere equation.

11. The heat of combustion for ethane, C$_2$H$_6$, is 52.0 kJ/g. How much heat will be released if one mole of ethane undergoes combustion?

12. **a.** Write the chemical equation for the complete combustion of heptane, C$_7$H$_{16}$.

    b. Given that the heat of combustion for heptane is 4817 kJ/mole, how much heat will be released if 250 kg of heptane undergo complete combustion?

13. Figure 4.7 shows energy differences for the combustion of methane, which is an exothermic chemical reaction. The combination of nitrogen gas and oxygen gas to form nitric oxide, NO, is an example of an endothermic reaction.

    $$180 \text{ kJ} + N_2(g) + O_2(g) \longrightarrow 2 \text{ NO}(g)$$

    Sketch an energy difference diagram for this reaction.

14. From your personal experience, predict whether each of these processes is endothermic or exothermic. Give a reason for each prediction.

a. a charcoal briquette burns

b. water evaporates from your skin

c. ice melts

d. wood burns

15. Use the bond energies in Table 4.1 to estimate the energy change associated with this reaction.

$$2 \; :C{\equiv}O: \; + \; :\ddot{O}{=}\ddot{O}: \; \longrightarrow \; 2 \; :\ddot{O}{=}C{=}\ddot{O}:$$

16. Draw a diagram like Figure 4.9 for the reaction in question 15.

17. Use the bond energies in Table 4.1 to explain why:

a. chlorofluorocarbons, CFCs, are so stable

b. it takes less energy to release Cl atoms than F atoms from CFCs

18. Use the bond energies in Table 4.1 to calculate the energy changes associated with each of these reactions. Remember to consider the Lewis structures for each of the reactants and products. Indicate which reactions are endothermic and which are exothermic.

a. $2 \; C_5H_{12}(g) + 11 \; O_2(g) \longrightarrow 10 \; CO(g) + 12 \; H_2O(l)$

b. $H_2(g) + Cl_2(g) \longrightarrow 2 \; HCl(g)$

c. $N_2(g) + 3 \; H_2(g) \longrightarrow 2 \; NH_3(g)$

19. Use the bond energies in Table 4.1 to calculate the energy changes associated with each of these reactions. Indicate which reactions are endothermic and which are exothermic.

a. $H_2(g) + O_2(g) \longrightarrow H_2O_2(g)$

b. $2 \; H_2(g) + O_2(g) \longrightarrow 2 \; H_2O(g)$

c. $2 \; H_2(g) + CO(g) \longrightarrow CH_3OH(g)$

20. Pentane, C_5H_{12}, has a boiling point of 36.1°C. Octane, C_8H_{18}, has a boiling point of 125.6°C. Will pentane and octane be gases at room temperature (25°C)?

21. Consider this reaction representing the process of cracking.

$$C_{16}H_{34} \longrightarrow C_5H_{12} + C_{11}H_{22}$$

a. What bonds are broken and what bonds are formed in this reaction? Use Lewis structures to help answer this question.

b. Use that information and Table 4.1 to calculate the energy change during this cracking reaction.

22. How many isomers are there for butane, C_4H_{10}? Write the Lewis structure for each isomer. *Hint:* Be careful not to repeat isomers. Remember that Lewis structures show how atoms are linked, but do not show spatial arrangement.

23. A premium gasoline available at most stations has an octane rating of 92.

a. What does the octane rating tell you about the knocking characteristics of this gasoline?

b. What does this tell you about whether the fuel contains oxygenates?

24. Assume that three power plants have been proposed, operating at the temperatures given here.

Plant **I**	T_{high} = 1200°C	T_{low} = 0°C
Plant **II**	T_{high} = 600°C	T_{low} = 20°C
Plant **III**	T_{high} = 450°C	T_{low} = −150°C

a. Calculate the maximum efficiency of each plant.

b. Identify the factors that affect the efficiency.

c. Discuss the practical limits that govern such efficiencies. Which plant is most likely to be built?

25. Which is the better analogy for a state of high entropy—an unopened deck of playing cards or a plate of cooked spaghetti? Explain your reasoning.

Concentrating on Concepts

26. How can you explain the difference between *temperature* and *heat* to a friend? Use some practical, everyday examples to help your friend understand. Assume your friend has not taken a chemistry course.

27. One risk of depending on oil imports is the shortage of gasoline in the event of unfavorable international events. Does a gasoline shortage only affect individual motorists? What are some of the ways that a gasoline shortage could affect your life?

28. How is the statement that "energy is neither created nor destroyed in a chemical reaction" related to the law of conservation of energy and the first law of thermodynamics?

29. A friend tells you that hydrocarbons containing larger molecules are better fuels than those containing smaller molecules.

a. Use these data, together with appropriate calculations, to discuss the merits of this statement.

Hydrocarbon	Heat of Combustion
Octane, C_8H_{18}	5450 kJ/mol
Butane, C_4H_{10}	2859 kJ/mol

b. Considering your answer to part **a**, do you expect the heat of combustion per gram of candle wax, $C_{25}H_{52}$, to be more or less than the heat of combustion per gram of octane? Do you expect the molar heat of combustion of candle wax to be more or less than the molar heat of combustion of octane? Justify your predictions.

30. Halons are synthetic chemicals similar to CFCs, but they also include bromine. Although halons are excellent materials for fire fighting, they are more effective at ozone depletion than are CFCs. The structural formula for halon-1211 is

$$:\ddot{Cl} - \overset{\displaystyle :\ddot{F}:}{\underset{\displaystyle :\ddot{F}:}{C}} - \ddot{Br}:$$

a. Which bond in this compound is most easily broken? How is that related to the ability of this compound to interact with ozone?

b. C_2F_4HCl is a compound being considered as a replacement for halons as a fire extinguisher. Draw the Lewis structure for this compound and identify the bond most easily broken. How is that related to the ability of this compound to interact with ozone?

31. During the distillation of petroleum, kerosene and hydrocarbons with 12–18 carbons used for diesel fuel will be located at position **C** marked on this diagram.

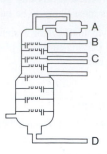

a. Separating hydrocarbons by distillation depends on the hydrocarbons having differences in a specific physical property. Which property is that?

b. How will the number of carbon atoms in the hydrocarbon molecules separated at **A**, **B**, and **D** compare with those separated at position **C**? Explain the basis for your prediction.

c. How will the uses of the hydrocarbons separated at **A**, **B**, and **D** differ from those separated at position **C**? Explain your reasoning.

32. Imagine you are at the molecular level, looking at what happens when liquid ethene, C_2H_4, boils. Consider a collection of four ethene molecules, representing each molecule with this sphere formula.

$$C_2H_4 =$$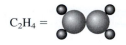

a. Draw a representation of ethene in the liquid state and then in the gaseous state. How will the collection of molecules change?

b. Estimate the temperature at which this transition from liquid to gas is taking place. What is the basis for your estimation?

33. It seems inefficient that petroleum first undergoes separation by distillation and then some fractions undergo further change through cracking.

a. Explain why cracking is necessary.

b. Hydrocarbons undergo physical changes during distillation in a fractionating tower. Does cracking also involve physical changes? Explain your reasoning.

34. Catalysts are used to speed up cracking reactions in oil refining and allow them to be carried out at lower temperatures.

a. Draw a sketch similar to Figure 4.10 in which you illustrate the energy changes for such a reaction in the absence and in the presence of the catalyst. Explain how your sketch illustrates the effect of the catalyst.

b. What examples of catalysts were given in the first two chapters of this text?

35. Consider these three structural formulas representing octane, C_8H_{18}; all hydrogen atoms and all C—H bonds have been omitted for simplicity.

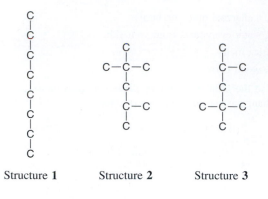

Structure **1**   Structure **2**   Structure **3**

a. Fill in the missing hydrogen atoms and demonstrate that these structures all represent C_8H_{18}.

b. Are there any duplicate representations here? How do you know?

c. Obtain a model kit and make one of the structures. What are the carbon-to-carbon-to-carbon bond angles in the structure?

d. If you make a different one of the structures, will the carbon-to-carbon-to-carbon bond angles change? Why or why not?

e. Write the structural formula of at least one more isomer of octane.

36. How is the growth in oxygenated gasolines related to:

a. restrictions on the use of lead in gasoline?

b. federal and state air quality regulations?

37. In each of the following pairs, select the substance that has the greater entropy. Explain the reasons behind your choice.

a. $H_2O(g)$ at 100°C and $H_2O(l)$ at 100°C

b. a solid piece of iron and an equal mass of iron powder

c. peanuts or an equal mass of peanut butter

38. State whether the entropy increases or decreases in each of these reactions or processes.

a. liquid water is converted to ice

b. solid sodium chloride is dissolved in water

c. hydrocarbons with 16 carbons are cracked into smaller hydrocarbons

39. A mole of diamond has an entropy value of 2.4 J/K at 25°C; one mole of methanol, $CH_3OH(l)$, has an entropy value of 127 J/K at 25°C. What generalization can be drawn from these two values?

40. a. Do oxygenate fuels have a higher energy content than non-oxygenate fuels? Explain your reasoning.

b. Do oxygenate fuels have a higher entropy than comparable non-oxygenate fuels? Explain your reasoning.

Exploring the Extensions

41. The text states that reformulated gasolines, RFGs, burn more cleanly by producing less carbon monoxide than non-oxygenated fuels. What is the evidence that supports this statement?

42. Consider this diagram.

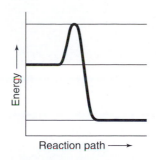

Reaction path ⟶

a. Does this representation show an exothermic or an endothermic reaction? Explain your reasoning by commenting on the shape of the curve.

b. Sketch this type of energy diagram for the type of reaction, exothermic or endothermic, *not* shown in the diagram above.

43. Bond energies such as those in Table 4.1 are sometimes found by "working backwards" from heats of reaction. A reaction is carried out and the heat absorbed or evolved is measured. From this value and known bond energies, other bond energies can be calculated. For example, the energy change associated with this reaction is +81 kJ.

$$NBr_3(g) + 3\ H_2O(g) \longrightarrow 3\ HOBr(g) + NH_3(g)$$

Use this information and the values found in Table 4.1 to calculate the energy of the N—Br bond. Assume all atoms other than hydrogen obey the octet rule.

44. Explain why it is possible to use a distillation tower to separate a mixture of hydrocarbons into different fractions but it is not possible to separate seawater, also a complex mixture, into all of its different fractions.

45. The text states that both octane and isooctane have essentially the same heat of combustion. How is that possible if they have different structures? Explain your thinking.

46. Why do you think that countries are willing to go to war over energy issues, but not over other environmental issues? Write a brief op-ed piece for your school newspaper discussing this issue.

47. What are the relative advantages and disadvantages associated with using coal and with using oil as energy sources? Which do you see as the better fuel for the twenty-first century? Give reasons for your choice.

48. What are the advantages and disadvantages of replacing gasoline with renewable fuels such as ethanol? Indicate your personal position on the issue and state your reasoning.

49. In December 1998 the EPA announced it would create a panel of experts from the public health and scientific communities, the automotive fuels industry, water utilities, and state and local governments to review the use of MTBE and other oxygenates added to gasoline. Find out more about why representatives of these groups have been included, and what the panel was asked to accomplish. Has the panel issued a final report at this time?

50. C.P. Snow, a noted scientist and author, wrote an influential book called *The Two Cultures,* in which he stated the following: "The question, 'Do you know the second law of thermodynamics?' is the cultural equivalent of 'Have you read a work of Shakespeare's?'" How do you react to this comparison? Discuss these questions in light of your own educational experiences.

Take a Drink: The Wonder of Safe Drinking Water

*Many types of bottled water are
available to consumers.*

Consider a day three or four years from now when you
receive a letter from your local water utility stating that
the maximum allowed levels of a regulated chemical
have been exceeded in the drinking water supplied to
your house on one or more occasions during the past
year. As a consumer, you will be expected to evaluate
this information and presumably demand better
performance from your utility. As a chemist, you may
be asked by your friends, neighbors, and family to help
them understand the hazards alluded to in the letter.
Everyone will have to decide whether the problem is
serious enough to warrant a change to a different
source of water, most likely bottled.

James Ryan
Today's Chemist at Work, 2/97

So begins an article by chemist James Ryan entitled "The Safe Drinking Water Act of 1996—And You." It was addressed to other chemists, but you might face a similar challenge in the near future. As a person who has studied chemistry and who will learn something about water quality and risk assessment, you might be asked by friends and family to help them understand the issues involved if you or they were to receive a letter like the one described above.

The letter raises a number of questions that we will attempt to answer in this chapter. What is a regulated chemical and what are maximum allowed levels? How do these chemicals get into water? Should we be concerned about any of them? Who establishes these rules? What is the Safe Drinking Water Act of 1996? What options are there for alternative sources of water? In particular, why is bottled water cited as the most likely alternative?

In spite of its commonness, this colorless liquid called water is amazing stuff. The noted anthropologist and essayist Loren Eiseley speaks poetically of the wonders of water: "If there is magic on this planet, it is contained in water... Its substance reaches everywhere; it touches the past and prepares the future; it moves under the poles and wanders thinly in the heights of the air." Arguably, water is the most important chemical compound on the face of the Earth. In fact, it covers about 70% of that face, giving the planet the lovely blue color in the famous "blue marble" photos taken from outer space by astronauts (p. 2). Water is essential to all living species; without it humans would die within a week. Our bodies are approximately 60% water, blood is at least 50% water, and a human brain is an astonishing 77% water! Water is so important to life as we know it that speculation about life elsewhere in the universe hinges first and foremost on the availability of water. Indeed, news stories in 1998 about the discovery of water on Mars fueled renewed speculation about life on our planetary neighbor. Water refreshes us, dominates weather systems, provides for many types of recreation on and in it, and even gives us aesthetic and restive pleasures (waterfalls and babbling brooks.)

Chapter Overview

Although we generally take water for granted, it is a remarkable chemical compound with unique properties that account for its essential life-supporting role. In this chapter, we will consider water from the perspective of those who drink it. First a question of aesthetics: What makes a glass of water pleasing to the eye and to the palate? There is more to water, however, than can be seen or tasted. Unseen impurities in water, depending on their identities and amounts, can impart a crisp, fresh taste or produce an unpleasant illness. And so, we next look at water as a solvent and at some of the things that may be dissolved in drinking water. How much of a substance dissolves in water makes concentration an important part of the story. The concentrations of substances dissolved in water can be expressed in several ways, including descriptions of the extremely low concentrations typical of toxic pollutants. To better comprehend aqueous solutions and why some substances dissolve in water while others do not, we will relate the properties of water molecules through such concepts as electronegativity, polarity, and hydrogen bonding.

Water dissolves various kinds of compounds, including those containing positive and negative ions, as well as some molecular compounds. We will consider the solubility of both types of substances and how it affects their presence in drinking water. Because the quality of drinking water is regulated by federal and state legislation, we will examine how drinking water is tested against standards, and how water is treated to make it potable (safe for drinking). Three case studies will give us the chance to look more closely at water quality: the effect of calcium on water hardness; lead in water; and a particular category of compounds called tri-halomethanes that are formed during water purification by chlorination. The questions posed early in the chapter about choices between tap water and bottled water are re-visited. Finally, because most of the world's people do not have good access or choices for safe drinking water, we look briefly at ways to purify water, including deionization, distillation, and reverse osmosis. And so, with water glasses or bottles raised, we extend the invitation to "take a drink."

5.1 Take a Drink of Water

Chapter 1 began with an invitation to "Take a breath of air." Without thinking, we do it automatically about 12–16 times a minute. Without air, we die in less than 15 minutes. Furthermore, we generally don't have any choice about *what* air we breathe; we must rely on that which surrounds us. On the other hand, there generally are choices with water: about *how frequently* we drink; *how much* we drink, and about the *source* of the water. Tap water, water from a lake or stream, bottled water, or water in bottled beverages are among the choices. If potable water is not available, we are in danger; without it, we would die in a few days.

Nothing could be more familiar than this clear, colorless, and (usually) tasteless liquid. And yet, we generally take water and water quality for granted. Unless there is a water emergency brought on by drought or contamination of our municipal water supply, we seldom think about where the water comes from, what it contains, how pure it is, or how long the supply will last. We turn on a faucet for a drink or a shower and simply expect a sufficient quantity of water to come flowing out of the tap. The 5.1 Consider This activity asks you to think a bit more about the water you drink.

We take safe drinking water for granted.

> ### 5.1 Consider This: Take a Drink of Water
>
> Obtain a glass of tap water and answer the following questions.
>
> **a.** Carefully describe the water—its taste, odor, appearance. What do you like or dislike about the water?
> **b.** Make a list of five qualities that you most want to have in your drinking water.
> **c.** What concerns, if any, do you have about the safety of the tap water or possible long-term health effects?
> **d.** Where do you think the water came from before it got to your tap or bottle?

5.2 Bottled Water

Most Americans obtain their drinking water from a water faucet or drinking fountain. This marvelous resource is remarkably inexpensive, costing only about 1/10 of a penny per quart. But not everyone drinks tap water. An increasing number of Americans are drinking bottled water instead. Indeed, the possible need for turning to bottled water was mentioned in the quotation opening this chapter. Because of its growing popularity, bottled water will be a sub-theme of this chapter.

Bottled water is big business, with annual revenues in the United States of over four billion dollars. Currently over three billion gallons of bottled water are sold each year in the United States, more than ten times that sold twenty years ago. In Europe, bottled water is now the biggest selling "soft drink." It is a common sight on campuses to see students carrying bottled water, and 18- to 24-year-olds consume most of it. Advertisements for bottled water tout its purity, often using words or images that conjure up nature, purity, and pristine beauty. For example, the label for Avalon bottled water urges:

In King Arthur's time, Avalon was a magical island of earthly paradise. *Imported Avalon Natural Spring Water* weaves its own magic. Drink it and experience *Refreshment. Pure & Simple*

Although popular, bottled water is also very expensive, relative to tap water. Typically, bottled water costs from $0.50 to $1.00 per quart, which is approximately 1000 times more expensive than the same volume of tap water. Are there important reasons why consumers are willing to ante up so much more for bottled water? One of the goals of this chapter is to help you learn enough about drinking water quality to make intelligent choices about whether there are compelling reasons to drink bottled water.

Bottled water is a very popular beverage.

5.2 Consider This: Bottled Water and You

Why do people buy bottled water, despite its high cost relative to tap water?

a. List what you perceive to be the advantages and the disadvantages of drinking bottled water.
b. Prioritize your lists in decreasing order of importance (most to least) for your personal decision about whether to drink bottled water.

5.3 Consider This: Bottled Water

If you search for "bottled water" on the web, you will get over a million "hits". Select two sites to explore, one provided by a supplier and the other provided as a source of consumer information. The former may flood you with statistics about the benefits of bottled water; the latter may raise questions such as, "Is bottled water safer?" or "Is it worth the cost?" For each site, list the title, author, URL, and two things from the site that you learned about water.

5.3 Where Does Drinking Water Come From?

What journey does water take to get from its natural source to your tap or bottle? Water is widely distributed on planet Earth (Figure 5.1). On the surface, it is found in oceans, lakes, rivers, snow, and glaciers (Figure 5.2). In the atmosphere it exists as water vapor and as tiny droplets in clouds that serve to replenish surface water by means of rain and snow.

5.4 Consider This: Drinking Water Sources

Consider the categories of water on Earth shown in Figure 5.2. From which category does your drinking water come? Trace your drinking water from the tap back through the distribution and purification system to its primary source. You may need to consult resources on the web or your local water district to help you.

Figure 5.1

Lakes and reservoirs provide much of our drinking water.

Water is also found underground in **aquifers,** great pools of water trapped in sand and gravel 50 to 500 feet below the surface. Some aquifers are enormous, such as the Ogallala Aquifer in the center of the United States that underlies parts of eight states from South Dakota to Texas (p. 187).

Water that can be made suitable for drinking normally comes from either surface water or groundwater. **Surface water**—lakes, rivers, and reservoirs—frequently contains substances that must be removed before it can be used as drinking water. By contrast, **ground water** is that pumped from wells that have been drilled into underground aquifers, and is usually free of harmful contaminants. Large-scale water supply systems for cities tend to rely on surface water resources; smaller cities, towns, and private wells

Figure 5.2
Distribution of water on the Earth.

Oceans
(97.4%)

Ice caps, glaciers,
groundwater
(2.59%)

Lakes, rivers,
atmosphere,
soil moisture
(0.014%)

Fresh water

tend to rely on ground water, the source of drinking water for a little over half of the U.S. population. If an aquifer becomes contaminated, it may take decades to become clean again. Prudence suggests that steps be taken to protect aquifers from contamination.

5.5 Consider This: Water Quality in Your State

Each state has different concerns about its surface and ground water.

a. Draw up a list of the issues in your state. For example, you might mention agricultural run off, leaking storage tanks, or pollution from other human activities.
b. Read the state fact sheet provided by the Office of Water at the EPA through the direct link at the *Chemistry in Context* web site. What does the sheet say about the surface and ground water quality in your state?
c. EPA notes that some states have lakes, rivers, and streams that support no aquatic life. What percent of the surface water falls in this category in your state?

The Ogallala Aquifer (dark blue area)

The Earth was not always as wet as it is now. Scientists believe that much of the water now on this planet originally was spewed as vapor from thousands of volcanoes that pocked the Earth. The vapor condensed as rain and over the ages the process was repeated. Water molecules cycled from sea to sky and back again. About three billion years ago, primitive plants, and then animals, extracted water from and contributed water to the cycle, a cycle that continues today. During an average year, enough precipitation falls on the continents to cover all the land area to a depth of more than 2.5 ft. As we are well aware, this precipitation does not fall all at once, nor is it distributed uniformly. The wettest place on record is Mount Waialeale on Kauia, Hawaii, which receives an annual average rainfall of 460 inches. In contrast is the 0.03 inches of rain per year average in Arica, Chile over 59 years! For 14 of those years, it did not rain at all. Closer to home, an average of 1.5×10^{13} liters of water falls daily on the continental United States—enough to fill about 400 million swimming pools.

This sounds like a great deal of water, but on a global scale, the amount of *fresh* water is relatively quite small. The great majority of the Earth's water, 97.4% of the total, is in the oceans, water that is undrinkable without expensive purification (Figure 5.2). The remaining 2.6% is all the fresh water we have, but 86% of the world's fresh water is permanently frozen in glaciers and polar ice caps. Only about 0.01% of the Earth's total water is conveniently located in lakes, rivers, and streams as fresh water. To make matters even worse, 80% of the fresh water in the United States is used to irrigate crops and cool electrical power plants. Thus, the bald fact is that the world's drinking water supplies are very limited, varying widely depending on locale.

5.4 Water as a Solvent

A major reason why we must consume water is because it is an excellent **solvent** for many of the chemicals that make up our bodies, as well as for a wide variety of other substances. A solvent is a substance that dissolves other substances. The substances that dissolve in a solvent are referred to as **solutes.** The resulting mixture is a **solution,** which is a homogeneous mixture of uniform composition. Solutions in which water is the solvent are called **aqueous solutions.** Later in this chapter we will examine why certain kinds of substances dissolve in water and others do not. For now we simply note that a remarkable variety of substances can dissolve in water and that this has important consequences for living organisms as well as for the environment. Table 5.1 summarizes some examples of water acting as a solvent.

Table 5.1 **Importance of Water as a Solvent in Living Organisms and in the Environment**

- Blood plasma is an aqueous solution containing a variety of life-supporting substances.
- Inhaled oxygen dissolves in blood plasma in the lungs where O_2 combines with hemoglobin.
- Blood carries dissolved CO_2 to the lungs to be exhaled from the body.
- Water transports nutrients into all the cells and organs of our body.
- Water can also transport *toxic* substances into, within, and out of living organisms.
- Water helps to maintain a chemical balance in the body by flushing waste products from it.
- Water-soluble toxic substances, such as pesticides, lead, and mercury, can be spread through the environment.
- By dilution, water reduces the concentrations of pollutants to safe levels and/or by carrying them away.
- Rainwater carries substances from the atmosphere down to Earth, including acid rain.

5.6 Your Turn

Use your experience to evaluate which of the following substances dissolve in water. In describing the solubility, use terms such as very soluble, partially soluble, or insoluble.

a. salt   **c.** sidewalk chalk   **e.** cooking oil
b. sugar   **d.** rubbing alcohol   **f.** aspirin

5.5 How Pure Is Drinking Water?

Regardless of its source, drinking water is rarely, if ever, just pure H_2O. You can be assured that almost certainly it contains other substances. For example, a label on Evian bottled water includes the following information:

Mineral Composition, mg/liter			
Calcium	78	Bicarbonates	357
Magnesium	24	Sulfates	10
Silica	14	Chlorides	4
		Nitrates (N)	1

Of these seven items for Evian water, all but one of them will be discussed in this chapter. We will also find out that the items listed on the label are themselves not chemical compounds, but are just parts of compounds. The number given with each item indicates how much of that substance (in milligrams) is present in one liter of Evian water. This raises a reasonable question: Should we be concerned about any of these substances and its amount? Calcium, for example, has a definite health benefit in producing stronger bones. Milk, not Evian water, is the preferred source for calcium; you would have to drink 4 liters of Evian water to get the same amount of calcium as that in one 8-oz glass of milk. In contrast, nitrate can be a health problem, especially for infants, depending on its concentration. The other listed substances in Evian bottled water are not likely to be a health problem. Elsewhere on the label it is noted that sodium, a health issue for some people, is present at less than 5 mg per bottle.

1 liter = 1.06 quart.

Such composition information can also be obtained for tap water supplied by municipal water companies. A typical analysis of tap water in the Midwest home of one of the authors of this text revealed the following:

Mineral Composition in mg/liter			
Calcium	66	Sulfate	42
Magnesium	24	Chloride	48
Sodium	18	Nitrate	0
		Fluoride	1

5.7 Your Turn

One 500-mL bottle of Evian water provides 4% of the recommended daily requirement of calcium.

a. Use the label information to find the number of milligrams of calcium that is recommended per day.
b. How many 500-mL bottles of Evian water would you have to drink to obtain your total daily supply of calcium?

Perhaps you have never considered drinking a glass of water as a risk/benefit act, yet it is. We pretty much consider water that has been chemically analyzed to have high potential benefit with very low risk. Fortunately, that is most often the case. But, however

The absence of evidence is not necessarily the evidence of absence. If a water pollutant is present below detectable limits, it does not mean that the pollutant is absent; it just means that it can't be detected at current levels of analytical chemical technology.

useful the information on either the bottled water label or in the tap water analysis may be, it is incomplete. The information indicates nothing about whether other substances, if any, are present in the water, nor how much of each is present (if present), or whether such substances pose health concerns. For example, even though a tiny amount of lead is found in almost all water samples, it is usually in such low amount as not to be a health problem. If the water has been chlorinated, it almost certainly has trace amounts of some chlorination by-products. (Chlorination and its by-products will be examined in detail later in this chapter.) Indeed, we rarely stop to think about what trace amounts of substances may be in the water because we tend to assume that the water is safe to drink. Usually we are correct in making this assumption. In part, this is because extensive federal and state regulations and standards govern water quality to protect the public. Some bottled water is regulated as well.

In assessing the healthfulness or risks of drinking water, it is not sufficient to know only *what substances* are present in the water and *how toxic* they are. We also need to know *how much* of each substance is present in a particular amount of the water. In other words, we need to be able to understand what is meant by the concentration of a solute and the usual ways of expressing it. And so, we turn to that topic next.

5.6 Solute Concentration in Aqueous Solutions

First introduced in Chapter 1 in relation to air quality, the concept of concentration warrants review and elaboration here in terms of substances dissolved in water. The composition of the major components of air was expressed in terms of the *percentages,* whereas the much lower concentrations of toxic air pollutants (ozone, sulfur dioxide, etc.) were given in *parts per million.* Although concentrations of air components might be a bit hard to visualize, solute concentrations in water solutions are more familiar and more easily imagined. For example, when a cooking recipe asks you to dissolve one teaspoon of an ingredient in one cup of water, a solution of a specific concentration results: 1 tsp per cup (tsp/cup). Note you would have the same 1 tsp/cup concentration if you also dissolved 2 tsps of the ingredient in 2 cups of water, 4 tsps in 4 cups, or 1/2 tsp in 1/2 cup. Even though you used larger or smaller quantities of the ingredient, the number of cups of water increased or decreased proportionally. Therefore, the **concentration**—the ratio of amount of ingredient to amount of water—would be the same in each case: 1 tsp *per* 1 cup. Solute concentrations in water solutions follow the same pattern as in the recipe, but are expressed in different units. We will use four ways of expressing concentration: percent; parts per million; parts per billion; and molarity. Each has particular application in various circumstances.

Percent. The simplest and most familiar way of expressing concentration is percent, defined as parts per hundred, that is, parts of solute per 100 parts of solution. A solution containing 5 g of sodium chloride in 100 g of solution would be a five percent (5%) solution. Hydrogen peroxide solutions, often found as an antiseptic in medicine cabinets, are usually 3% hydrogen peroxide, which indicates that they contain 3 grams of hydrogen peroxide in 100 grams of solution (or 6 grams in 200 grams of solution, etc.).

Ppm and ppb. Concentrations of substances in drinking water are normally far lower than 1% (1 part per hundred). Correspondingly, more appropriate concentration expressions are used to express such low concentrations. **Parts per million (ppm)** is the most common way of expressing the concentration of a solute in drinking water. A 1 ppm solution of calcium in water contains 1 gram of calcium in 1 million (1,000,000) grams of water. The same concentration, 1 ppm, could be applied to a solution with 2 grams of calcium in 2 million grams of water, 5 g in 5 million g of water, or 5 mg (0.005 g) in 5000 g of water. Although parts per million is a very useful concentration expression, measuring one million grams of water is not very convenient. Therefore, we look to find an easier, but equivalent way to establish ppm. We find it using the **liter,** the volume occupied by 1000 g of water at 4 °C, as the volume unit, rather than one million g of water. Proportionally, we use milligrams of solute, not grams, to express ppm in an alternative, but equivalent way: *1 ppm of any substance in water equals 1 mg of that substance per liter of water.*

1 ppm = 1 mg/L

The quart, not the liter, is a more familiar measure of volume for Americans, although not for the rest of the world. A liter is just a bit larger than a quart. (1 L = 1.06 qt.)

$$1 \text{ ppm} = \frac{1 \text{ g solute}}{1,000,000 \text{ g water}} = \frac{1 \text{ mg solute}}{1000 \text{ g water}} = \frac{1 \text{ mg solute}}{1 \text{ L water}}$$

Drinking water often contains various substances naturally present at concentrations in the ppm range, as illustrated on the Evian bottled water label. Certain toxic water pollutants may also be present in the ppm concentration range. For example, the acceptable limit for nitrate (often found in well water in agricultural areas) is 10 ppm and the limit for fluoride is 4 ppm.

Because some pollutants are of concern at concentrations much lower than even parts per million, they may be reported as **parts per billion (ppb)**. One part per billion of mercury in water means 1 gram of mercury in 1 billion grams of water. In more convenient terms, this means 1 *microgram* (1×10^{-6} g, abbreviated as 1 μg) of mercury in 1 liter (10^3 g) of water. For example, the acceptable limit for mercury in drinking water is 2 ppb.

1 ppb = 1 μg/L

$$2 \text{ ppb mercury} = \frac{2 \text{ g mercury}}{1 \text{ billion g water}} = \frac{2 \text{ g mercury}}{1 \times 10^9 \text{ g water}} = \frac{2 \times 10^{-6} \text{ g mercury}}{1 \times 10^3 \text{ g water}}$$

$$= \frac{2 \ \mu g}{1 \text{ L water}}$$

One part per million is a very small quantity. Several analogies to this concentration were given in Section 1.2, including the statement that 1 ppm corresponds to 1 second in nearly 12 days. Similar analogies can be offered for parts per billion: 1 ppb = 1 second in 33 years, or approximately 1 inch on the circumference of the earth.

5.8 Your Turn

a. If 80 μg of lead were detected in 5 L of water, what is the concentration of lead expressed in ppb and ppm?

b. If the maximum lead concentration in drinking water allowed by the federal government is 15 ppb, is the sample in **a.** in compliance with federal limits? Explain.

Molarity. Another concentration unit that is useful in chemistry is molarity which is based on the mole, chemists' favorite way of measuring matter. **Molarity,** abbreviated M, is defined as the number of moles of solute present in one liter of solution. Written more compactly, molarity is (moles solute)/L solution or simply moles/L. The great advantage of molarity is that a 1 molar (1 *M*) solution of *any* solute contains exactly the same number of chemical units (atoms or molecules) as any other 1 molar solution. The *mass* of solute may vary depending on the molar mass, but the *number* of chemical units will be the same for all 1 *M* solutions. Methods of chemical analysis of water (Section 5.16) frequently use molarity. It is also particularly useful for chemical reactions involving solutions, a topic that will be explored extensively in the next chapter. For now, we simply want to develop some familiarity with molarity itself.

Moles were introduced in Section 3.9.

As an example, consider a solution of sodium chloride in water. The molar mass of NaCl is 58.5 g; therefore, one mole of NaCl weighs 58.5 grams. If we dissolve 58.5 grams of NaCl in some water and then add enough water to make exactly 1.00 liter of solution, we will have a 1.00 molar NaCl solution (Figure 5.3). But, there are many ways to make a 1 molar NaCl solution. For example, we could use 1/10 of a mole of NaCl (5.85 g) in 1/10 of a liter (100 mL). Another possibility, among many others, would be to use 2 moles NaCl (117.0 g) in 2 liters of solution.

Molar mass of NaCl: (1 mole of sodium = 23.0 g) + (1 mole of chlorine = 35.5 g) = 58.5 g. Calculating molar masses was covered in Section 3.9.

$$1 \ M \text{ NaCl} = \frac{2 \text{ mole NaCl}}{2 \text{ L solution}} = \frac{1 \text{ mole NaCl}}{1 \text{ L solution}} = \frac{0.5 \text{ mole NaCl}}{0.5 \text{ L solution}} = \frac{0.1 \text{ mole NaCl}}{0.1 \text{ L solution}}$$

5.9 Your Turn

a. Consider a 1.5 *M* NaCl solution and a 0.15 *M* NaCl solution. How many moles of solute are present in 500 mL of each solution?

b. A solution is prepared by adding enough water to 0.5 mole of NaCl to form 250 mL of solution. A second solution is prepared by adding enough water to 0.6 mole of NaCl to form 200 mL of solution. Which solution is more concentrated? Explain your reasoning.

Figure 5.3
Preparing a 1.00 molar NaCl
solution.

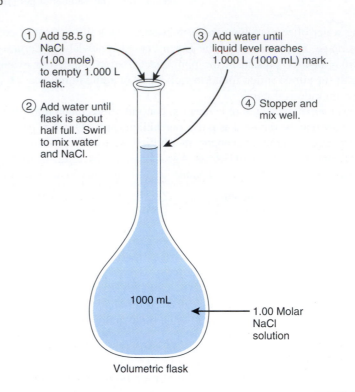

① Add 58.5 g
NaCl
(1.00 mole)
to empty 1.000 L
flask.

② Add water until
flask is about
half full. Swirl
to mix water
and NaCl.

③ Add water until
liquid level reaches
1.000 L (1000 mL) mark.

④ Stopper and
mix well.

1000 mL

1.00 Molar
NaCl
solution

Volumetric flask

Now that we have some idea about the importance of drinking water and about some
of the substances that may be present in it, we shift to a more detailed examination of
this vital elixir at the molecular level. Our aim is to understand its unique properties, including its behavior as a solvent.

5.7 H₂O: Surprising Stuff

The many and varied uses of water are a consequence of its properties. Not only is water an effective solvent for a wide range of materials, but it also has a number of unusual properties. Water is so ubiquitous that it has become the standard for many of the
units of modern science, including the Celsius temperature scale, the kilogram, and the
calorie. Yet, for all of that, the physical properties of water are quite peculiar, and we
are very fortunate that they are. If water were a more conventional compound, we would
be very different creatures.

This most common of liquids is full of surprises. First, it is noteworthy that water is
a liquid and not a gas at room temperature (about 25°C) and normal atmospheric pressure. The molar mass of water is 18.0 g/mole, whereas almost all other compounds with
molar masses that small are gases under these conditions. Consider three common atmospheric gases: N_2, O_2, and CO_2, whose molar masses are 28, 32, and 44 g/mole, respectively. All have molar masses greater than that of water, yet they are gases to breathe
rather than liquids to drink.

Moreover, not only is water a liquid under these conditions, it has an anomalously
high boiling point of 100°C. This temperature is one of the reference points for the Celsius temperature scale. The other is the freezing point of water, 0°C. And when water
freezes, it exhibits another bizarre property: it expands. Most liquids contract when they
solidify. These and other unusual properties derive from water's chemical composition
and its molecular structure (Table 5.2).

*Generally speaking, in a series of
similar compounds the higher the
molar mass, the higher the boiling
point. See Section 3.9 for more about
molar masses.*

5.8 Molecular Structure and Physical Properties

To understand the chemical and physical properties of water, we need information about
its chemical composition and molecular structure. The chemical composition is known
to practically everyone. Indeed, the formula for water, H_2O, is very likely the world's

Table 5.2   **Unusual Properties of Water**

Property	Comparison with Other Substances	Importance in Physical and Biological Environments
Specific heat (4.184 J/g°C)	Highest of all common liquids and solids except NH_3	Affects climate; moderates temperatures in organisms and the environment
Heat of fusion (freezing) (331 J/g)	Highest of all molecular solids except NH_3	Releases enormous amounts of heat upon freezing; saves crops from freezing by spraying them with liquid water.
Heat of vaporization (2250 J/g)	Highest of all molecular substances	Condensing water vapor releases very large quantities of thermal energy, creating storms

most widely known bit of chemical information. Recall from Chapter 2 that water is a covalent compound (Section 2.3). The oxygen atom is at the center of the bent molecule, attached by covalent bonds to two hydrogen atoms. Each of the two O—H bonds consists of one pair of electrons shared between the hydrogen and oxygen atoms. It turns out that the bonding electrons are not shared equally between these two atoms. Experimental evidence indicates that the oxygen atom attracts the electron pair more strongly than does the hydrogen. To use the appropriate technical term, oxygen is said to have a higher electronegativity than hydrogen. **Electronegativity** is a measure of an atom's attraction for the electrons it shares in a covalent bond. The greater the electronegativity, the more an atom attracts bonding electrons to itself. Figure 5.4 shows a periodic table of electronegativities for the first 18 elements.

An examination of Figure 5.4 reveals some useful generalizations about electronegativities. The highest electronegativities are associated with nonmetallic elements such as fluorine and chlorine. These halogens, members of Group 7A, have atoms with seven outer electrons. Recall (Section 2.3) that each of these atoms has a strong tendency to bond with another atom in such a way as to acquire a share in an additional electron, thus completing a stable octet of electrons. A similar argument explains the high electronegativities of other nonmetals. For example, oxygen, with six outer electrons per atom, also exhibits a powerful attraction for electrons. Conversely, the lowest electronegativities are associated with the metals found in Groups 1A and 2A. Atoms of these metallic elements have much weaker attractions for electrons than do nonmetals. In general, electronegativity values *increase* as you move across a row of the periodic table from *left to right* (from metals to nonmetals) and *decrease* as you move *down* a group of the table.

According to Figure 5.4, the electronegativity of oxygen is 3.5; that of hydrogen is 2.1. Because of this difference in electronegativities, the shared electrons are actually pulled by the more electronegative oxygen to itself and away from the less electronegative hydrogen. This unequal sharing gives the oxygen end of the bond a partial negative charge and the hydrogen end a partial positive charge. Because the bond has oppositely charged ends or poles, it is said to be a **polar covalent bond.** Polar bonds arise whenever atoms of differing electronegativities are covalently bonded to each other. The greater the electronegativity differences of the elements involved, the more polar the bond.

The periodic table was introduced in Section 1.6 and is found on the inside front cover of this text.

A polar covalent bond is an example of an *intra*molecular force, one that exists *within* a molecule.

Figure 5.4

Electronegativity values for the first 18 elements.

Group							
1A	2A	3A	4A	5A	6A	7A	8A
H							He
2.1							—
Li	Be	B	C	N	O	F	Ne
1.0	1.5	2.0	2.5	3.0	3.5	4.0	—
Na	Mg	Al	Si	P	S	Cl	Ar
0.9	1.2	1.5	1.8	2.1	2.5	3.0	—

Molecular structure of a water molecule.

Many of the unique properties of water are a consequence of its molecular shape and the polarity of its bonds. Both are shown in this drawing of an H_2O molecule. Arrows are used to indicate the direction in which the electron pairs are displaced. The δ^+ and δ^- symbols indicate partial positive and partial negative charges, respectively. Note that the hydrogen atoms are partially positive and that a partial negative charge appears to be concentrated in the two nonbonding pairs of electrons on the oxygen atom.

5.10 Your Turn

For each pair, identify the more polar bond. For that bond, identify the atom to which the electron pair of the bond will be more strongly attracted.

  a. H—F and H—Cl
  b. N—H and O—H
  c. N—O and S—O

Hint: Compare the electronegativities from Figure 5.4.

5.9 Hydrogen Bonding

Polar bonds can be used to understand the unusually high boiling point of water. Consider what happens when two water molecules approach each other. Because opposite charges attract, one of the partially positively charged hydrogen atoms of one water molecule will be attracted to one of the regions of negative charge associated with the two nonbonding electron pairs of the other water molecule. This is an *inter*molecular attraction, one that occurs *between* molecules. The fact that each H_2O molecule has two hydrogen atoms and two nonbonding pairs of electrons increases the opportunities for intermolecular attraction. Each of the two nonbonding electron pairs on oxygen can form a loose, long-distance kind of attraction, called a **hydrogen bond**, to a hydrogen atom of a neighboring water molecule. Similarly, each of the two hydrogen atoms in a water molecule can attract an electron pair from an adjacent water molecule to form hydrogen bonds. Thus, a single H_2O molecule can simultaneously hydrogen bond to as many as four other water molecules, as pictured in Figure 5.5.

Hydrogen bonds, such as those between water molecules, are only about one-tenth as strong as the covalent bonds that connect atoms together *within* molecules. Nevertheless, as intermolecular forces go, hydrogen bonds are quite strong. Their strength is manifested in the relatively high boiling point of water ($100°C$) and the large amount of energy required to convert liquid water into vapor (2250 J/g or $4.05*10^4$ J/mole). To boil water, the H_2O molecules must be separated from their relatively close contact in the liquid state and moved into the gaseous state where they are much farther apart. In other words, their intermolecular hydrogen bonds must be broken. If the hydrogen bonds in water were weaker, water would have a much lower boiling temperature and require less energy to boil. If water had no hydrogen bonding at all, it would boil at about $-75°C$, making life as we know it very uncomfortable, if not impossible. Although water is wet, boiling points may seem dry and impersonal until you reflect on the fact that our bodies are over 60% water. Because of hydrogen bonding, almost all of our body's water, whether in cells, blood, or other body fluids, is in the liquid state, well below the boiling point. Our very existence depends on hydrogen bonding; without it, we would be a gas!

| Hydrogen bonds are *inter*molecular forces, those occurring *between* molecules.

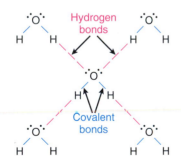

Figure 5.5
Hydrogen bonding in water. (distances not to scale)

5.11 Consider This: Bonds Within and Between Water Molecules

Use Figure 5.5 to help explain what bonds are broken when water boils. Draw a diagram to help illustrate your understanding.

Hint: Particulate diagrams were used extensively in Chapter 1 to help illustrate chemical reactions. Start with molecules of water in the liquid state and then show what happens to those molecules when water boils.

It should be noted that hydrogen bonds are not restricted to water. There is evidence for similar intermolecular attraction in many molecules that contain hydrogen atoms covalently bonded to oxygen, nitrogen, or fluorine atoms. Thus, ammonia, NH_3, and hydrogen fluoride, HF, also have unusually high boiling points, but lower than water. The net effect of hydrogen bonding in these compounds is weaker than in water. No common chemical compound other than water has the unique arrangement of two hydrogens bonded to another atom possessing two non-bonding electron pairs available to form hydrogen bonds. In water, each oxygen can be surrounded by a tetrahedral arrangement of hydrogen atoms involving two H—O covalent bonds and two hydrogen bonds. Hydrogen bonding is also important in stabilizing the shape of large biological molecules, such as proteins and nucleic acids. In proteins, which are major components of skin, hair, and muscle, hydrogen bonding occurs between hydrogen atoms and oxygen or nitrogen atoms. The coiled, double-helical structure of DNA (deoxyribonucleic acid) is stabilized by thousands of hydrogen bonds formed between particular segments of the linked DNA strands. So in this respect, too, hydrogen bonding plays an essential role in the life process.

The molecular structures of proteins and nucleic acids are discussed in Chapters 11 and 12.

Hydrogen bonding also explains why ice cubes and icebergs float in water. Ice is a regular array of water molecules in which every H_2O molecule is hydrogen bonded to four others. The pattern is pictured in Figure 5.6. Note that the pattern includes a good deal of empty space in the form of hexagonal channels. When ice melts, this regular array begins to break down and individual H_2O molecules can enter the open channels. As a result, the molecules in the liquid state are, on the average, more closely packed than in the solid state. Thus, a volume of one cubic centimeter (1 cm^3) of liquid H_2O contains more molecules than 1 cm^3 of ice. Consequently, liquid water has a greater mass per cubic centimeter than does ice. This is simply another way of saying that the *density* of liquid water is greater than that of ice.

One cubic centimeter (one milliliter) is about the same volume as one-fifth of a teaspoon.

The **density** of a substance is defined as its mass per unit of volume. For scientific purposes, mass is given in grams and the "unit of volume" is one cubic centimeter, which is identical to one milliliter (mL). One cubic centimeter of liquid water weighs 1.00 g. In other words, its density is 1.00 g/cm^3 or 1.00 g/mL. On the other hand, 1 cm^3 of ice weighs 0.92 g, so its density is 0.92 g/cm^3 or 0.92 g/mL. People often confuse mass with density. For example, you may hear someone say that iron is heavy or that

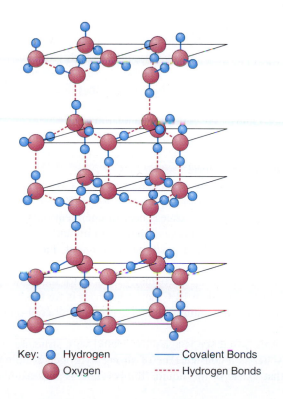

Figure 5.6
The hydrogen-bonded lattice structure of ice 1, the common form of solid H_2O. Note the open channels, which contribute to the fact that ice is less dense than liquid water.

Key: ● Hydrogen   —— Covalent Bonds
      ● Oxygen   ---- Hydrogen Bonds

metallic lead is very heavy. Pieces of iron and lead are indeed often quite heavy, but it is more accurate to say that iron has a high density (its density is 7.9 g/mL) and that lead has an even higher density (11.3 g/mL). On the other hand, popcorn has a low density and we are likely to say that a bag of popcorn feels very light.

For the great majority of substances, the solid state is more dense than the liquid. The fact that water shows the reverse behavior means that lakes freeze from the top down, not the bottom up. This topsy-turvy behavior is convenient for aquatic plants, fish, and ice skaters. On the other hand, the fact that water expands as it freezes is not so convenient for people whose water pipes and car radiators burst when the water inside them freezes.

5.12 Consider This: Oil and Water

Relative densities have practical consequences for water when it mixes with other substances in the environment. Crude oil has a density of approximately 0.8 g/cm^3. What implications does this have for cleaning up oil spills in the ocean?

Finally, we want to examine another of water's unusual properties, namely its uncommonly high capacity to absorb and release heat. On a global scale, this property helps determine worldwide climates. By absorbing vast quantities of heat, the oceans and the droplets of water in clouds help mediate global warming. The specific details of these processes are among the uncertainties that complicate efforts to model global warming accurately. We do know that heat is absorbed when water evaporates from seas, rivers, and lakes, and heat is released when water condenses as rain or snow. These changes between solid, liquid, and gaseous forms of water create the great thermal engine that helps drive weather patterns in the short term and regulates climates over longer periods of time. But when not changing its physical state, liquid water absorbs more energy than the ground if equal masses are used. This is because water has a higher capacity to store heat than do rocks and dirt. As a consequence, when the weather turns colder, the ground has less stored heat to lose than the water and therefore, cools more quickly. The water retains more heat and is able to provide more warmth for a longer time to the areas bordering it. Such properties should be familiar to anyone who has ever jumped into a warm lake on a cool fall day.

A quantitative measure of a substance's capacity to absorb heat is called its **specific heat.** Specific heat is defined as the quantity of heat energy that must be absorbed to increase the temperature of one gram of the substance by one degree Celsius. The specific heat of liquid water is 1.00 cal/g°C, which means that one calorie of energy will raise the temperature of one gram of $H_2O(l)$ by 1°C. In fact, the calorie was originally defined in this manner. Conversely, when the temperature of one gram of liquid water falls 1°C, one calorie of heat is given off. Because of the relationship between calories and joules (see Chapter 4), the specific heat of water can also be expressed as 4.18 J/g°C, which is equivalent to 1.00 calorie.

This 4.18 J/g°C (1.00 cal) may not sound like a very large value, but liquid water has one of the highest specific heats of any known liquid. Because of this, it is an exceptional coolant, used to carry away excess heat in chemical industry, power plants, and the human body. Most other compounds have significantly lower specific heats. For example, liquid benzene, C_6H_6, a gasoline component, has a specific heat of only 0.406 cal/g°C, less than half that of water. Most solid substances have lower specific heats than that of water. Metals are particularly notable for their low specific heats, most being below 0.1 cal/g°C, less than 10% that of water. It is for this reason that metals can be heated easily to high temperatures and, conversely, will cool quickly when the heat source is removed.

The reason for this difference is again associated with structure. The unusually high heat capacity of water is a consequence of strong hydrogen bonding and the resultant degree of order that exists in the liquid. Temperature is a measure of molecular mo-

Global warming is discussed in Chapter 3.

tion—the higher the temperature, the greater the average motion. When molecules are strongly attracted to each other, a good deal of energy is required to overcome these intermolecular forces and enable the molecules to move more freely. Such is the case with water. On the other hand, intermolecular forces are much weaker in non-hydrogen–bonded liquids, such as benzene, and they are much easier to overcome. Consequently, the specific heats for such compounds are lower.

5.13 Consider This: Showering Heat on Yourself

The high specific heat of water has important consequences for energy consumption or conservation in residences, where large amounts of energy are required to heat water for bathing and washing clothes and dishes. Suppose that the water enters the water heater in your residence at 20°C (68°F), and the heater is set to heat the water to 50°C (122°F).

a. How many calories of heat energy are needed to heat the 100 L of water that are used in a typical 5-minute shower?

b. To conserve energy, the heater is reset to 40°C (104°F). How many calories of heat energy would be saved during that 5-minute shower?

Ans.

a. $100 \text{ L} \times \left(\dfrac{10^3 \text{ g}}{1 \text{ L}} \right) \times \left(\dfrac{1 \text{ cal}}{\text{g°C}} \right) \times (50°C - 20°C)$

$= 3.0 \times 10^6 \text{ cal}$

5.10 Water As a Solvent: A Closer Look

One of the most important properties of water has already been discussed in Section 5.4, namely that water is an excellent solvent for a wide variety of substances. A great deal of chemistry occurs in water solutions. Because aqueous solutions are so important, we need some understanding of *why* substances dissolve in water.

"Sugar water" and "salt water" are examples of two main classes of aqueous solutions. A significant difference between the two can be demonstrated with a conductivity meter such as pictured in Figure 5.7. A source of electricity, a battery or house current, is attached by two wires to a light bulb. As long as the two separated ends of the wires do not touch, the bulb will not light; there is not a completed electrical circuit. If the separated ends are placed into pure "distilled" water or a solution of sugar in distilled water, the bulb remains dark. However, if the bare wires are placed into an aqueous solution of salt, the bulb illuminates. Perhaps the light has also gone on in the mind of the experimenter! Pure water or a solution of sucrose in water do not conduct

a. Distilled water (nonconducting)   **b.** Sucrose dissolved in water (nonconducting)   **c.** NaCl dissolved in water (conducting)

Figure 5.7
Conductivity in a water solution.

electricity and therefore do not complete the electrical circuit; the light does not glow. Sugar and other non-conducting solutes are called **non-electrolytes.** On the other hand, a water solution of salt (sodium chloride) is an electrical conductor and the light bulb lights. Sodium chloride and other conducting solutes are classified as **electrolytes.** But what accounts for this difference in properties between electrolytes and non-electrolytes? That is the next topic we consider.

5.11 Ions and Ionic Compounds

The flow of electric current involves the transport of electrical charge. Therefore, the fact that solutions of sodium chloride conduct electricity suggests they contain electrically charged species. These species are called **ions,** from the Greek for "wanderer." When solid sodium chloride dissolves in water, its positively charged ions (Na^+) called

| Cations are positively charged; anions are negatively charged.

cations, and negatively charged ions (Cl^-) called **anions,** separate. As these ions wander about in solution, they transport electrical charge.

It may be a little surprising to learn that Na^+ and Cl^- ions exist in the solid crystals of salt such as those in a salt shaker, as well as in a water solution of salt. Solid sodium chloride is a three-dimensional cubic arrangement of sodium and chloride ions occupying alternating positions (Figure 5.8). The attractions between cations and anions in the crystal are called **ionic bonds** that hold the crystal together. In an **ionic compound,** such as NaCl, there are no true covalently bonded molecules, only positively charged cations (Na^+) and negatively charged anions (Cl^-). Each Na^+ ion is surrounded by six oppositely charged Cl^- ions and, likewise, each Cl^- ion is surrounded by six positively charged Na^+ ions. A single, tiny crystal of sodium chloride consists of many billions of Na^+ and Cl^- ions held together in the arrangement shown in Figure 5.8.

Thus far we have described the structure and some of the properties of ionic compounds, but we have not explained why certain atoms lose or gain electrons to form ions. Not surprisingly, the answer involves the distribution of electrons within atoms. Recall that a sodium atom, with an atomic number of 11, has 11 electrons and 11 protons. Sodium is in Group 1A and each sodium atom has only one electron in its *outer* energy level. This electron is rather loosely attracted to the nucleus and can be easily removed from the atom by absorbing a small amount of energy. When this happens, the Na atom is transformed into an Na^+ ion by losing an electron (e^-), a process represented by the following equation.

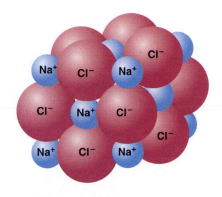

Figure 5.8

The arrangement of Na^+ and Cl^- ions in a crystal of sodium chloride.

$$Na \longrightarrow Na^+ + e^- \qquad (5.1)$$

A Na^+ ion bears a +1 charge because it contains the 11 protons originally present in the Na atom, but only 10 electrons. These 10 electrons are in a configuration that is essentially the same as the 10 electrons in an atom of the inert element neon (Ne).

Sodium atom	Sodium ion
Na	Na^+
11 protons	11 protons
11 electrons	10 electrons
Net charge: zero	**Net** charge: +1

An Na^+ ion, like the Ne atom, has eight *outer* electrons. This is a particularly stable arrangement. We may generalize by saying that *metallic elements tend to lose electrons to form cations.*

By contrast, a chlorine atom has a tendency to gain an electron. The electrically neutral Cl atom includes 17 electrons and 17 protons. It has seven *outer* electrons. Because of the stability associated with eight outer electrons, it is energetically favorable for a Cl atom to acquire an extra electron, such as one from a sodium atom, to become a Cl^- ion. In this ion there are 18 electrons and 17 protons; thus the net charge is −1.

$$Cl + e^- \longrightarrow Cl^- \qquad (5.2)$$

Chlorine atom	Chloride ion
Cl	Cl$^-$
17 protons	17 protons
17 electrons	18 electrons
Zero *net* charge	*Net* charge: −1

Because elemental chlorine consists of diatomic Cl_2 molecules, we can also write this gain in electrons in the following fashion.

$$Cl_2 + 2\ e^- \longrightarrow 2\ Cl^- \tag{5.3}$$

In general, *non-metallic elements—those on the right-hand side of the periodic chart—gain electrons to form anions.*

When sodium metal and Cl_2 gas are mixed together, electrons are transferred from sodium atoms to chlorine atoms with the release of a considerable amount of energy. The result is the aggregate of Na^+ ions and Cl^- ions known as sodium chloride. *In the formation of an ionic compound, such as sodium chloride, the electrons are actually transferred from one atom to another, not simply shared as they would be in a covalent compound.*

Is there *evidence* for electrically charged ions in pure sodium chloride? Experimental tests show that crystals of sodium chloride do not conduct electricity, but when these crystals are melted, the resulting liquid conducts electricity. This provides evidence that Na^+ and Cl^- ions from the solid NaCl also exist in the liquid state without the presence of water. Crystals of NaCl and other ionic compounds are hard and brittle. When hit sharply, they shatter rather than being flattened as would be true for a substance consisting of molecules with weak forces between them. This suggests the existence of strong forces that extend throughout the ionic crystal. Literally speaking, there is no such thing as a specific, localized "ionic bond" analogous to covalent bonds in molecules. Rather, there is a generalized ionic bonding that holds together a large assembly of ions.

Having established that sodium chloride is an ionic compound, we next ask what other elements form ions and ionic compounds. Electron transfer to form cations and anions, respectively, is likely to occur between metallic elements (those elements in the left and middle blocks of the periodic table) and non-metallic elements (those elements on the right side of the periodic table, except the noble gases, Group 8A). Sodium, lithium, magnesium, and other metallic elements have a strong tendency to give up electrons and form positive ions. On the other hand, chlorine, fluorine, oxygen, and other nonmetals have a strong attraction for electrons and readily gain electrons to form negative ions. Therefore, *ionic compounds are formed when elements from opposite sides of the periodic table react.* Potassium chloride (KCl) and sodium iodide (NaI) are two of many such compounds. Because ordinary table salt (NaCl) is such an important example of an ionic compound, chemists frequently refer to other ionic compounds simply as "salts," which are crystalline solids.

The idea of ion formation when atoms gain or lose electrons can help explain and even predict the formulas of many compounds. Back in Chapter 1 you were told that the formula of the compound formed from calcium and chlorine is $CaCl_2$, but no explanation was given. The reason should now be apparent. An atom of calcium, a member of Group 2A, readily loses its two outer electrons to form a Ca^{2+} ion.

Calcium atom   Calcium ion

· Ca · Ca^{2+}

Chlorine, as we have already seen, forms Cl^- ions. For the electrical charges to be balanced, two Cl^- ions are required for each Ca^{2+} ion. Hence the formula of calcium chloride is $CaCl_2$. In an ionic compound, the *sum* of the positive charges equals the *sum* of the negative charges. Your Turns 5.14 and 5.15 give you opportunities to apply similar logic to other elements and compounds.

5.14 Your Turn

Predict the charge on the ion that will form from each of these atoms. Draw the Lewis structure for each atom and for its ion, clearly labeling the charge on the ion.

a. Br   **b.** Mg   **c.** O   **d.** Al

Hint: The periodic table will help you determine the number of outer electrons in each atom. That in turn will help you determine how many electrons must be lost or gained for the atom to achieve stability by an octet of electrons.

Ans.   a. Bromine, being in the seventh group (column) of the periodic table, will most easily gain one more electron for an octet to form a stable ion. This will give it a charge of negative one, just as was the case for chlorine in the same chemical family. The Lewis structures are:

$$: \overset{..}{\underset{..}{Br}} \cdot \quad \text{and} \quad \left[: \overset{..}{\underset{..}{Br}} : \right]^{-}$$

5.15 Your Turn

Predict the formulas of the ionic compounds that would be formed by the combination of each pair of elements.

a. Ca and Br   **b.** K and F   **c.** Li and O   **d.** Sr and Br

Hint: The sum of the positive charges must match the sum of the negative charges for the compound to be neutral.

Ans.   a. We have already concluded that Ca forms Ca^{2+} ions and Br forms Br^{-} ions. To have an electrically neutral compound, there must be two Br^{-} ions for each Ca^{2+} ion. This means that the formula for calcium bromide is $CaBr_2$.

Some ionic compounds include **polyatomic ions,** ones that are themselves made up of more than one atom or element. A case in point is sodium sulfate, Na_2SO_4. This compound consists of Na^+ and SO_4^{2-} ions. In the sulfate ion, the four oxygen atoms are covalently bonded to a central sulfur atom in a symmetric tetrahedral arrangement. Counting the electrons in the Lewis structure (Figure 5.9) reveals that there are 32 electrons, two more than the 30 valence electrons in one neutral sulfur and four neutral oxygen atoms. Hence, the sulfate ion has a charge of -2.

Table 5.3 is an alphabetical list of some of the more common polyatomic ions. Most of them are anions, but polyatomic cations are also possible, as in the case of the ammonium ion, NH_4^+. Note that some elements (carbon, sulfur, and nitrogen) form more than one polyatomic anion.

The rules for predicting formulas of compounds containing polyatomic ions and naming them are similar to those that apply to simple compounds of two elements. Consider, for example, aluminum sulfate, a compound used in water purification; it is formed from

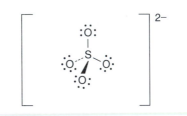

Figure 5.9

Structure of the sulfate (SO_4^{2-}) ion.

Table 5.3   Some Common Polyatomic Ions

Name	Formula	Name	Formula
acetate	$C_2H_3O_2^-$	nitrite	NO_2^-
bicarbonate	HCO_3^-	phosphate	PO_4^{3-}
carbonate	CO_3^{2-}	sulfate	SO_4^{2-}
hydroxide	OH^-	sulfite	SO_3^{2-}
hypochlorite	OCl^-	ammonium	NH_4^+
nitrate	NO_3^-		

Al^{3+} and SO_4^{2-} ions. Like all ionic compounds, this compound must be electrically neutral; thus the sum of the positive charges must equal the sum of the negative charges. This requires that two Al^{3+} ions combine with three SO_4^{2-} ions [$2 \times (+3) = 3 \times (-2)$]; the formula of aluminum sulfate must be $Al_2(SO_4)_3$. Note that the subscript 3 applies to the entire SO_4^{2-} ion that is enclosed in parentheses. The formula of the compound thus represents two Al ions along with three sulfate ions containing a total of three S atoms and twelve O atoms. As is always the case with ionic compounds, the positive ion is named first.

5.16 Your Turn

Give the correct name for each of these compounds.

a. KNO_3 **b.** $(NH_4)_2SO_4$ **c.** $NaHCO_3$ **d.** $CaCO_3$ **e.** $Mg_3(PO_4)_2$

Ans.

a. potassium nitrate This name is just the combination of the names of the two ions, potassium and nitrate.

b. ammonium sulfate *Note:* When naming ionic compounds, it is not necessary to use the prefixes such as *di-* or *tri-* as we did when naming covalently bonded compounds such as carbon dioxide or dinitrogen pentoxide.

5.17 Your Turn

Write the formula for these compounds.

a. calcium hypochlorite (used in bleaches)
b. lithium carbonate (treatment of manic disorders)
c. potassium nitrate (matches and fireworks)
d. barium sulfate (medical X-rays)

Ans.   a. $Ca(OCl)_2$ As in Your Turn 5.15, the formula should be written so the sums of the negative and positive charges are equal. Here the ions are Ca^{2+} and OCl^-; therefore, two hypochlorite ions are needed to equal the charge on a calcium ion, balancing the charges at 2^+ and 2^-.

5.12 Water Solutions of Ionic Compounds

Now we are in a position to understand one of the most important properties of ionic compounds, namely why many are quite soluble in water. Recall (Section 5.8) that water molecules are **polar;** they have both a partial positive side (hydrogen) and a partial negative side (oxygen). When a solid sample of an ionic compound is placed in water, the polar H_2O molecules are attracted to the individual ions. The partial negatively charged oxygen atom of a water molecule is attracted to the positively charged cations in the crystal. At the same time, hydrogen atoms in H_2O, with their partial positive charges, are attracted to the negatively charged anions of the solute. Thus the ions are surrounded by water molecules, which diminishes the anion–cation attraction in the solid. The substantial attraction between the ions and H_2O molecules results in the surrounding water molecules literally plucking the ions out of the solid and into solution. In dissolving, the ionic compound separates into its component cations and anions. Equation 5.4 and Figure 5.10 represent this process for sodium chloride and water.

$$NaCl(s) + H_2O(l) \longrightarrow Na^+(aq) + Cl^-(aq) \qquad (5.4)$$

The (*aq*) in the equation indicates that the ions are present in an *aqueous* solution.

When compounds containing polyatomic ions dissolve in water, the polyatomic ions remain intact. For example, when sodium sulfate dissolves in water, the sodium ions and sulfate ions separate, but the SO_4^{2-} ions remain intact.

$$Na_2SO_4(s) + H_2O(l) \longrightarrow 2\,Na^+(aq) + SO_4^{2-}(aq) \qquad (5.5)$$

Figure 5.10
The dissolving of sodium
chloride in water.

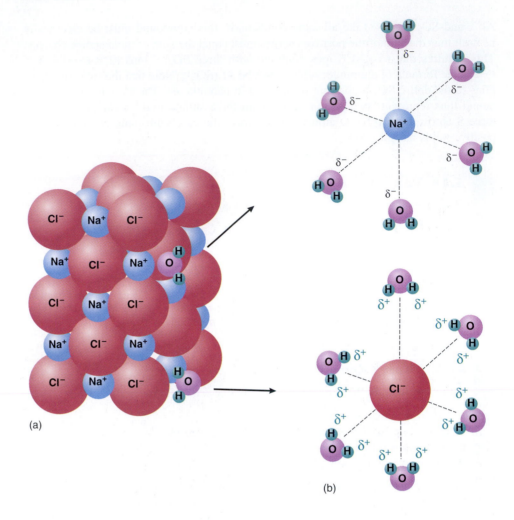

(a)

(b)

What has been described above for sodium chloride and sodium sulfate dissolving in water is true for many other ionic compounds. Indeed, this behavior is so common that the chemistry of ionic compounds is largely that of their behavior in water solutions. Conversely, almost all naturally occurring water samples contain various amounts of ions. Even our body fluids contain significant concentrations of ions.

5.18 Consider This: Electricity and Water Don't Mix

Small electrical appliances, such as a hair dryer or curling iron, carry a warning label prominently advising the consumer of the hazard associated with using the appliance near water. Why is this a problem if water does not conduct electricity? What is the best course of action if a plugged-in hair dryer does accidentally fall into a sink full of water?

We conclude the discussion of ionic compounds and their solutions with a brief excursion into the relative solubilities of various ionic compounds. In principle, the dissolving process in water, as described above, ought to be true for any ionic compound. Indeed, many ionic compounds are highly soluble in water. But some are insoluble or, at best only slightly soluble, and some have extremely low solubility in water. The reasons for this range of behavior would take us too far afield. Basically, they involve the sizes and charges of the ions, how strongly the ions attract each other, and how strongly the ions are attracted to water molecules. Nevertheless a few generalizations (Table 5.4) have proven to be quite useful for predicting the solubilities of common ionic compounds.

Table 5.4 **Some Generalizations about the Solubilities of Ionic Compounds in Water**

Note: "Insoluble" means that the compounds have extremely low solubility in water (less than 0.01 M). However, all ionic compounds have at least a very small solubility in water.

All **sodium, potassium,** and **ammonium (NH$_4^+$)** compounds are soluble.

All **nitrates** are soluble.

Most **chlorides** are soluble (exceptions: silver, mercury, and lead chlorides).

Most **sulfates** are soluble (exceptions: strontium, barium, and lead sulfates).

Most **carbonates** are insoluble (exceptions: those with Group 1A or NH$_4^+$ cations).

Most **hydroxides** and **oxides** are insoluble (exceptions: those with Group 1A or NH$_4^+$ cations).

Most **sulfides** are insoluble (exceptions: those with Group 1A or NH$_4^+$ cations).

It is possible to use Table 5.4 and a periodic table of the elements to predict the solubilities (or insolubility) of many compounds. For example, calcium nitrate, $Ca(NO_3)_2$, is predicted to be soluble in water because all nitrates are soluble. On the other hand, calcium carbonate, $CaCO_3$, should be insoluble because most carbonates are insoluble, and calcium is not one of the exceptions for carbonates. By similar reasoning, copper hydroxide, $Cu(OH)_2$, is insoluble, but copper sulfate, $CuSO_4$, is soluble.

5.19 Your Turn

Use the solubility generalizations in Table 5.4 to predict the solubility of these compounds.

a. ammonium nitrate, NH_4NO_3 (used in fertilizers)
b. sodium sulfate, Na_2SO_4 (used as an additive in detergents)
c. mercury sulfide, HgS (the mineral cinnabar)
d. aluminum hydroxide, $AlOH_3$ (used in antacid tablets)

Ans: **a.** All nitrates and all ammonium compounds are soluble, so ammonium nitrate is soluble.

The land masses on the earth are made up largely of minerals consisting of ionic compounds that have extremely low solubilities in water. If that were not the case, most of the land masses would have dissolved long ago! Table 5.5 summarizes some environmental consequences of the differing solubilities of minerals and other substances in water.

Table 5.5 **Environmental Consequences of Solubility Generalizations**

- Sodium chloride dissolves from the land and washes into the sea. Thus, oceans are salty and cannot be used for drinking water without expensive purification.
- Agricultural fertilizers contain nitrates. Because all nitrates are soluble in water, run off water from fertilized fields carries nitrates into surface and ground water. At high enough concentration, they may create a health hazard because of the toxicity of nitrate ion.
- Most metals (iron, copper, zinc, chromium, etc.) exist on Earth as insoluble sulfide (S^{2-}) or oxide (O^{2-}) minerals; iron ore (iron oxide, Fe_2O_3) is a common example. If these minerals were soluble in water, they would have been largely dissolved out of the ground and washed to sea long ago.
- Waste piles left from mining operations remain a visual as well as chemical blight on the landscape. They often contain small amounts of toxic ions, such as lead and mercury, which have low solubility. However, these ions may be leached very slowly from the waste into rivers and lakes, where they may contaminate water supplies.

5.13 Covalent Compounds and Their Solutions

From the previous discussion, you might get the impression that only ionic compounds dissolve in water. But, other kinds of compounds dissolve as well. Common experience tells us that sucrose (ordinary table sugar) dissolves readily in water. But sucrose contains no ions; it is a **covalent** or **molecular compound.** Like water, carbon dioxide, chlorofluorocarbons, and many of the other compounds you have been reading about, sucrose molecules consist of covalently bonded atoms. The formula for sucrose is $C_{12}H_{22}O_{11}$ and it exists as individual molecules consisting of 45 atoms, as shown in Figure 5.13.

When sugar dissolves in water, its molecules become uniformly dispersed among the H_2O molecules. As in all true solutions, the mixing is at the most fundamental level of the solute and solvent—the molecular or ionic level. The $C_{12}H_{22}O_{11}$ molecules remain intact and do not dissociate. Evidence for this is the fact that aqueous sucrose solutions do not conduct electricity. However, the sugar molecules do interact with the water molecules. In fact, solubility is always promoted when there is a net attraction between the solvent molecules and the solute molecules or ions. This suggests a general solubility rule: *"Like dissolves like."* Compounds with similar chemical composition and molecular structure tend to form solutions with each other. The intermolecular attractive forces between similar molecules are high, thus promoting solubility. Dissimilar compounds do not dissolve in each other.

Consider, for example, the following three familiar covalent compounds, all of which have high water solubilities: sucrose; ethylene glycol (the main ingredient in antifreeze); and ethanol (ethyl alcohol, the "grain alcohol" found in alcoholic beverages). Like all alcohols, they contain an —OH group (Figures 5.11 and 5.13.)

| *"Like dissolves like."*

Figure 5.11

Structures of ethanol and ethylene glycol.

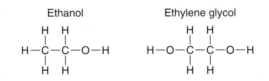

We start with the simplest, ethanol, C_2H_5OH. Its oxygen atom is covalently bonded to a hydrogen atom and to a carbon atom. The —OH group of a C_2H_5OH molecule can form hydrogen bonds with H_2O molecules (Figure 5.12). This hydrogen bonding causes these two polar compounds to have a great affinity for each other, a conclusion consistent with the fact that ethanol and water form solutions in all proportions. Ethylene glycol is also an alcohol, and it has two —OH groups available for hydrogen bonding with H_2O. Thus, ethylene glycol is highly water soluble, a necessary property for an anti-freeze ingredient.

5.20 Your Turn

Sketch a diagram to show hydrogen bonding between ethylene glycol and water.

Finally, we consider sucrose, the compound that introduced this section. Examination of its structure (Figure 5.13) discloses that the sucrose molecule contains eight —OH groups and three additional oxygen atoms that might also be able to participate in hydrogen bonding. This accounts for the high solubility of sugar in water.

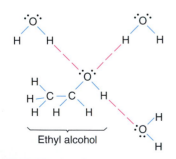

Figure 5.12

Hydrogen bonding of ethanol with water.

5.21 Consider This: Three-Dimensional Representations of Molecules

Three-dimensional representations of molecules can be viewed on the web with the aid of CHIME, a plug-in that you can download and install for free. Three-dimensional representations of ethanol, ethylene glycol, and sucrose are there. Can you identify the places in each compound where hydrogen bonding occurs? Has your mental picture of these molecules changed after seeing these 3-D representations? Explain.

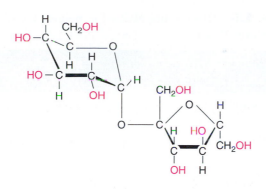

Figure 5.13

Molecular structure of sucrose.

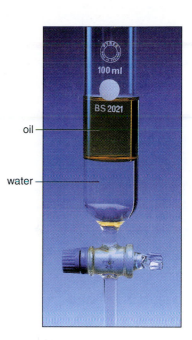

Figure 5.14

Oil and water do not dissolve in each other.

On the other hand, molecular compounds that differ in composition and molecular structure do not attract each other strongly. It has often been observed that "oil and water don't mix." They don't mix because they are very different; unlike compounds don't like each other, which is just the reverse of "like dissolves like." Water is a highly polar compound, whereas oil consists of nonpolar hydrocarbon compounds. When placed in contact, they remain aloof and segregated (Figure 5.14). But oily, nonpolar compounds generally dissolve readily in hydrocarbons or chlorinated hydrocarbons. For this reason, the latter compounds have often been used in dry cleaning solvents.

The tendency of nonpolar compounds to mix with other nonpolar substances affects the way the bodies of fish and animals store certain highly toxic substances such as PCBs (polychlorinatedbiphenyls) or the pesticide DDT. PCB and DDT molecules are nonpolar and so when fish absorb them from water, the molecules are stored in body fat (which is also nonpolar) rather than in the blood (which is a highly polar water solution).

Solvents used to dry clean clothes are usually chlorinated compounds such as tetrachloroethylene, C_2Cl_4, also known as "perc" (perchlorinated ethylene), which is a potential human carcinogen. These materials also have serious environmental consequences. Dr. Joe DeSimone of the University of North Carolina–Chapel Hill has discovered a substitute for chlorinated compounds by synthesizing cleaning detergents that work in liquid carbon dioxide. The key to the process are the detergents, whose molecules are designed so that one end of the molecule is soluble in nonpolar substances like grease and oil stains, while the other end dissolves in the liquid CO_2. The new process recycles carbon dioxide produced as a waste product from industrial processes. Replacing large volumes of "perc" by using recycled CO_2 reduces the negative impact of "perc" on the workplace and the environment. The breakthrough process is paving the way for designing replacements for conventional halogenated solvents currently used in manufacturing and in industries making coatings. For his work, Professor DeSimone received the 1997 Presidential Green Chemistry Challenge Award.

5.14 Protecting Our Drinking Water: Federal Legislation

We can now apply to drinking water what we know about the structure and properties of pure water and aqueous ionic solutions. What dissolves in water determines its quality and the potential for adverse health effects. Keeping public water supplies safe has long been recognized as an important public health issue. In 1974, the U.S. Congress passed the **Safe Drinking Water Act (SDWA)** in response to public concern about findings of harmful substances in drinking water supplies. The aim of the SDWA, as amended in 1996, is to provide public health protection to all Americans who get their water from public water supplies (over 200 million people). Contaminants that may be health risks are regulated by the EPA, as required by the SDWA. The EPA sets legal limits for such contaminants that reflect knowledge about health effects and risk

Table 5.6 MCLGs and MCLs for Selected Pollutants in Drinking Water

Pollutant	MCLG (ppm)	MCL (ppm)
Cadmium (Cd^{2+})	0.005	0.005
Chromium (Cr^{3+}, CrO_4^{2-})	0.1	0.1
Lead (Pb^{2+})	0	0.015
Mercury (Hg^{2+})	0.002	0.002
Nitrate (NO_3^-)	10	10
Benzene (C_6H_6)	0	0.005
Trihalomethanes ($CHCl_3$, etc.)	0	0.080

calculations (Table 5.6). These limits also take into account the practical realities of the concentration of contaminants likely to be present in drinking water sources, and the ability of water supply utilities to remove the offending contaminants by using available technology.

For each contaminant, the EPA has established a health goal, or **Maximum Contaminant Level Goal (MCLG).** This is the level, expressed in ppm or ppb, at which a person weighing 70 kg (154 lb) could drink two liters (about two quarts) of water containing the contaminant every day for 70 years without suffering any ill effects. Each MCLG includes built-in safety factors for uncertainties in the standardizing data and for individual differences in sensitivity to the contaminant. An MCLG is not a legal limit with which water systems must comply; it is based solely on considerations of human health. For known cancer-causing substances **(carcinogens),** the EPA has set the health goal at zero, under the assumption that *any* exposure to the substance could present a cancer risk.

It is important to recognize that the mere presence of a contaminant does not necessarily mean a serious health problem. For a problem to exist, the concentration of the contaminant must exceed the legal limit, expressed in ppm or ppb, referred to as a **Maximum Contaminant Level (MCL).** The EPA sets legal limits for each contaminant as close to the MCLG as possible, keeping in mind the practical realities of technical and financial barriers that may make it difficult to achieve the goals. Except for contaminants regulated as carcinogens, for which the MCLG is zero, most legal limits and health goals are the same. Even when they are less strict than the MCLGs, the MCLs provide substantial public health protection.

5.22 Consider This: Understanding MCLGs and MCLs

Most people are unfamiliar with these terms from the Safe Drinking Water Act. Assume you are making a presentation in another class to explain what these acronyms mean and how the information helps to safeguard our drinking water. Prepare a short outline of what you will say. Be prepared to answer questions from the audience, particularly dealing with why MCLs were not set to zero for all carcinogens.

Because of improved detection and quantitative analytical methods, the number of regulated contaminants in drinking water increases each time Congress updates the legislation. Lower limits for MCL values have been established as more accurate risk information becomes available. Currently, there are more than 80 regulated contaminants, which fall into several major categories: metals (such as cadmium, chromium, copper, mercury, and lead), a few nonmetallic elements (e.g., fluoride and arsenic), pesticides, industrial solvents, compounds associated with plastics manufacturing, and radioactive materials. Depending on the particular contaminant, MCLs vary from around 10 ppm to less than 1 ppb. Some contaminants interfere with liver or kidney functions. Others can affect the nervous system if ingested over a long period at levels consistently above the legal limit (MCL). Pregnant women and infants are at particular risk for some contaminants because of their effects on a developing fetus or the digestive system of an infant. In Section 5.16 we will look at case studies for two contaminants, lead and tri-halomethanes (THMs).

In addition to contaminants that can pose chronic health problems, there are two types of substances in drinking water that present acute health risks. The first are nitrate (NO_3^-) and nitrite (NO_2^-) ions that limit the blood's ability to carry oxygen. Even when consumed in tiny doses, these ions cause immediate health effects for infants. Thus, the EPA limit for nitrate and nitrite ions in drinking water specifically protects infants. The other acute health risk is biological, not chemical—from bacteria, viruses, and other microorganisms such as *Giardia*. News media warnings announcing a "boil water emergency" are typically the result of a "total coliform" violation. Coliforms are a broad class of bacteria, most of which are harmless, that live in the digestive tracts of humans and other animals. The presence of high coliform concentration in water usually indicates that the water treatment or distribution system is not working properly. Diarrhea, cramps, nausea, and vomiting, the symptoms of coliform-related illness, are not serious for a healthy adult, but can be life-threatening for the very young, the elderly, or those with weakened immune systems.

5.23 Consider This: A Drink of Water; What Is In It?

Table 5.6 is merely a starting point for the wealth of information available about six possible pollutants in drinking water. At the EPA Office of Ground Water and Drinking Water, you can access a consumer fact sheet on each of these pollutants, as well as dozens more. Look up one listed in Table 5.6 and find out how the pollutant gets into the water supply, and how you would know if it were in your drinking water. Is your state listed as one of the top states that releases the contaminant?

Note: Cadmium, lead, chromium, mercury, and nitrate/nitrite ions are found under the section on "Inorganic Chemicals". Benzene is listed under "Volatile Organic Chemicals". There are no trihalomethane such as $CHCl_3$ currently listed, but you can find a variety of other chlorinated compounds found in water such as CCl_4 and CH_2Cl_2. A Consumer Version and a technical version are available, and the latter is recommended.

5.24 Consider This: MTBE In Ground Water

The gasoline additive MTBE was discussed in Chapter 4 as one of the mandated oxygenates in certain metropolitan areas to reduce air pollution. There is some concern that MTBE from this use may be accumulating in surface and ground water, with the potential to compromise the quality of drinking water supplies. Consult the *Chemistry in Context* web site or other resources suggested by your instructor to find the latest information on MTBE in water.

In addition to the Safe Drinking Water Act, there is also Federal legislation to control pollution of surface waters, including lakes, rivers, and coastal areas. The **Clean Water Act (CWA),** passed by Congress in 1972 and amended several times, has provided the foundation for dramatic progress in reducing surface water pollution over the past three decades. The CWA establishes limits on the amounts of pollutants that industries can discharge into surface waters, resulting in actions that have removed over one billion pounds of toxic pollution from U.S. waters every year.

Improvements in surface water quality have at least two major beneficial effects: they reduce the amount of clean-up needed for public drinking water supplies and they result in a healthier natural environment for aquatic organisms. In turn, a healthier aquatic ecosystem has many indirect benefits for humans. In keeping with the new trend toward "green chemistry," industries are finding ways to convert these waste materials into useful products, as well as to design processes "up front" so that they neither use nor produce substances that degrade water quality.

5.15 How Is Drinking Water Treated to Make It Safe?

A large supply of water is no guarantee of its drinkability. Coleridge's shipwrecked Ancient Mariner knew this all too well, surrounded as he was by "Water, water every where, nor any drop to drink." So how is water treated to make it potable, that is, fit for human consumption?

The Safe Drinking Water Act requires that standards of purity and safety for public water supplies be established and enforced by the U.S. EPA. The first step in a typical municipal treatment plant for treating water to be used for drinking is to pass the water through a screen that excludes larger objects both natural (fish and sticks) and artificial (tires and beverage cans). Aluminum sulfate (alum, $Al_2(SO_4)_3$) and calcium hydroxide (slaked lime, $Ca(OH)_2$) are then added. These compounds react to form a sticky gel of aluminum hydroxide, $Al(OH)_3$, which collects suspended clay and dirt particles on its surface.

$$Al_2(SO_4)_3(aq) + 3\ Ca(OH)_2(aq) \longrightarrow 2\ Al(OH)_3(s) + 3\ CaSO_4(aq) \qquad (5.6)$$

The $Al(OH)_3$ gel settles slowly carrying with it the suspended particles down into a settling tank. Any remaining particles are removed as the water is filtered through gravel and then sand.

The next step, disinfection to kill disease-causing organisms, is the most crucial one for making drinking water safe. In the United States, this is most commonly done by **chlorination.** Chlorine is usually administered in one of three forms: chlorine gas, Cl_2; sodium hypochlorite, NaOCl (used in Clorox™ and other brands of laundry bleach); or calcium hypochlorite, $Ca(OCl)_2$ (which is also used to disinfect swimming pools). The antibacterial agent generated in solution by all three substances is hypochlorous acid, HOCl. The degree of chlorination is adjusted so that a very low concentration of HOCl, between 0.075 and 0.600 ppm, remains in solution to protect the water against further bacterial contamination as it passes through the pipes to the user.

Before chlorination was used, thousands died in epidemics spread via polluted water. In a classic study, John Snow was able to trace a mid-1800s cholera epidemic in London to water contaminated with the excretions of victims of the disease. A more contemporary example occurred in Peru in 1991. This cholera epidemic was traced to bacteria in shellfish growing in estuaries polluted with untreated fecal matter. The bacteria found their way into the water supply where they continued to multiply because of the absence of chlorination.

Chlorination, however, is not without some drawbacks. The taste and odor of residual chlorine may be objectionable to some and is a reason commonly cited as why people drink bottled water or use home water filters to remove chlorination at the tap. A possibly more serious drawback is the reaction of residual chlorine with other substances in the water to form by-products at potentially toxic levels. The most widely publicized of these are the trihalomethanes (THMs) such as chloroform, $CHCl_3$, which are the subject of a case study (Section 5.16.3).

Many European cities use gaseous ozone (O_3) to disinfect their water supplies. Chapter 1 discussed tropospheric ozone as a serious air pollutant. Chapter 2 described the beneficial effects of the stratospheric ozone layer. In water treatment, the toxic property of ozone is used for a beneficial purpose! The degree of antibacterial action necessary can be achieved with a smaller concentration of ozone than chlorine, and ozone is more ef-

A water purification plant in Oakland, CA.

fective than chlorine against water-borne viruses. But ozonation is more expensive than chlorination and becomes economical only for large water treatment plants. An additional major drawback of ozone is the fact that it decomposes quickly and hence does not protect the water from contamination after the water leaves the treatment plant. Consequently, a low dose of chlorine is added to ozonated water as it leaves the treatment plant.

Another disinfection method gaining in popularity is the use of ultraviolet (UV) radiation. In Chapter 2 it was pointed out that UV radiation is dangerous for living species, including bacteria. Ultraviolet disinfection is very fast; there are no residual by-products and it is economical for small installations (including rural homes with unsafe well water). Like ozonation, UV disinfection does not protect the water from contamination after it leaves the treatment site unless a low dose of chlorine is added.

5.25 Consider This: Risk/Benefit—To Chlorinate or Not?

Chlorination of water supplies has risks as well as benefits associated with it. Prepare to debate the issues involved. In preparing for the debate, list the major advantages and disadvantages of chlorination. Take a position either in favor or opposed to chlorination. If opposed, be ready to discuss alternative methods of disinfection. If in favor, be prepared to offer reasons why alternatives are not as effective as chlorination.

Depending on local conditions, one or more additional purification steps may be carried out at the water treatment facility after disinfection. Sometimes the water is sprayed into the air to remove volatile chemicals that create objectionable odors and taste. If the water is sufficiently acidic to cause problems such as corrosion of pipes or the leaching of heavy metals from pipes, calcium oxide (lime) is added to partially neutralize the acid. Many municipalities also add about 1 ppm of fluoride (as NaF) to protect against tooth decay.

In water, sodium fluoride, NaF, dissociates into $Na^+(aq)$ and $F^-(aq)$ ions. In teeth, fluoride ions are incorporated into a calcium compound called fluorapatite, which is more resistant to dental decay than is apatite, the usual tooth material.

5.16 Case Studies of Water Pollution

It should be clear by now that water is an excellent solvent for many different substances. Some solutes in water, such as oxygen, can be beneficial. Some solutes can change the properties of water. Some are highly toxic and are cause for concern. In this section, we will consider three case studies, each of which deals with a different aqueous impurity. Our aim is to understand the nature of the impurity, how it gets into water, how its concentration in water can be measured, how serious a health threat it poses, and what can or should be done to reduce the impurity.

5.16.1 Case Study: Hard Water

Our first case study is of hard water, which is not a toxic substance. Rather, the case study is an examination of some aqueous ions—principally calcium, Ca^{2+}, and magnesium, Mg^{2+}—frequently present at high concentrations in hard water. Although they pose no health threats, these ions can be a considerable nuisance, as well as expense unless removed from solution.

Water hardness is a property commonly reported for drinking water supplies. **Water that is "hard"** contains calcium, magnesium, and occasionally iron ions, in the form of their chloride, bicarbonate, and sulfate compounds. In contrast, **"soft"** water contains few of these ions. Because calcium ions, Ca^{2+}, are generally the largest contributors to hard water, hardness is usually expressed in parts per million of calcium carbonate by mass. This method of reporting does not mean that the water sample actually contains $CaCO_3$ at the indicated concentration. Rather, it specifies the mass of solid $CaCO_3$ that could be formed from the Ca^{2+} in solution, provided sufficient CO_3^{2-} ions were also present.

$$Ca^{2+}(aq) + CO_3^{2-}(aq) \longrightarrow CaCO_3(s) \qquad (5.7)$$

Thus, a hardness of 10 ppm indicates that 10 mg of $CaCO_3$ could be formed from the Ca^{2+} ions present in 1 L of water.

| Table 5.4 indicates that calcium carbonate is not very water soluble.

The source of hard water is limestone rock, which is composed of calcium carbonate or a mixture of calcium carbonate and magnesium carbonate. Limestone was formed from ancient inland seas in which calcium carbonate slowly deposited over millions of years. When surface or ground water flows over or through limestone rock, a small amount of calcium and magnesium carbonate dissolves in the water. Hard water results when sufficient magnesium and calcium ions dissolve. Water hardness in the United States varies from nearly 0 ppm in mountainous regions with mostly granite rock to over 400 ppm in parts of the Midwest, where limestone is prevalent.

5.26 Your Turn

Write the formulas for:

a. calcium bicarbonate
b. magnesium sulfate
c. magnesium chloride

The usual analytical method for measuring water hardness utilizes a **titration** in which a measured volume of hard water is reacted with an ethylenediaminetetraacetate ($EDTA^{2-}$) solution whose molarity is known accurately (Figure 5.15). In the analysis, calcium ions in the hard water react with a chemical known by the shorthand name "EDTA."

| $EDTA^{2-}$ is actually in a solution of sodium EDTA, Na_2EDTA.

$$Ca^{2+}(aq) + EDTA^{2-}(aq) \longrightarrow CaEDTA(aq) \qquad (5.8)$$

To carry out the analysis, a solution containing a known concentration of $EDTA^{2-}$ is added slowly to a water sample until an indicator in the solution changes color when

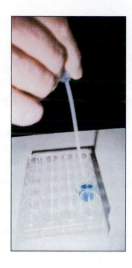

Figure 5.15

A titration of hard water with EDTA: (a) a large-scale titration using a buret; (b) a smaller-scale titration using a Beral-type pipet and a well plate.

the number of moles of $EDTA^{2-}$ added exactly equals the number of moles of calcium in the water sample. Note from equation 5.8 that 1 mole of $EDTA^{2-}$ is required for each mole of Ca^{2+} ions in the solution. The volume of water, the volume of $EDTA^{2-}$ solution needed, and the molarity of the EDTA are used to calculate the molarity of calcium ions in the sample.

For example, suppose that 27.3 mL of 0.0100 M $EDTA^{2-}$ were added to a 100-mL water sample before the indicator changed color. The number of moles of Ca^{2+} in the water sample can be determined from the moles of $EDTA^{2-}$ used for the reaction: 1 liter of 0.0100 M $EDTA^{2-}$ contains 0.0100 mole $EDTA^{2-}$; 27.3 mL (0.0273 L) of it were used:

$$\text{moles of } EDTA^{2-} \text{ used} = 0.0273 \text{ L} \times \frac{0.0100 \text{ mole } EDTA^{2-}}{1 \text{ L}}$$

$$= 2.73 \times 10^{-4} \text{ mole } EDTA^{2-}$$

Equation 5.8 points out that 1 mole of $EDTA^{2-}$ reacts with one mole of Ca^{2+}. Therefore, because 2.73×10^{-4} mole $EDTA^{2-}$ were used, the water sample contained 2.73×10^{-4} mole of Ca^{2+}. This 2.73×10^{-4} mole of Ca^{2+} would form 2.73×10^{-4} mole $CaCO_3$, because one mole of calcium reacts with one mole of CO_3^{2-} (equation 5.7).

To express the water hardness of this sample as ppm (mg/L) of $CaCO_3$, we must first convert the 2.73×10^{-4} moles of Ca^{2+} to an equivalent number of moles of $CaCO_3$, and then to milligrams of $CaCO_3$.

$$\text{moles } CaCO_3 = 2.73 \times 10^{-4} \text{ moles of } Ca^{2+} \times \frac{1 \text{ mole } CaCO_3}{1 \text{ mole } Ca^{2+}}$$

$$= 2.73 \times 10^{-4} \text{ moles of } CaCO_3$$

The molar mass of $CaCO_3$ is 100 g/mole (by adding up the molar masses of Ca, C, and three O).

$$\text{mg } CaCO_3 = 2.73 \times 10^{-4} \text{ moles of } CaCO_3 \times \frac{100 \text{ g } CaCO_3}{1 \text{ mole } CaCO_3} \times \frac{1000 \text{ mg } CaCO_3}{1 \text{ g } CaCO_3}$$

$$= 27.3 \text{ mg } CaCO_3.$$

The final step is expressing the concentration in ppm (mg/L) in the 100-mL (0.100 L) sample:

$$\frac{27.3 \text{ mg CaCO}_3}{0.100 \text{ L}} = 273 \text{ mg CaCO}_3/\text{L} = 273 \text{ ppm hardness}$$

Water containing Ca^{2+} can form a hard, insoluble deposit of $CaCO_3$ in water heaters, cooking ware, pipes, and industrial equipment. If you live in a hard water region, you have probably noticed a white deposit inside a tea kettle or other utensil that is used to heat water. In hot water pipes, the build-up can cause serious interference with water flow (Figure 5.16). In water heaters, the build-up interferes with heat transfer, resulting in wasted energy and greater cost for water heating.

Probably the most common manifestation of water hardness is the way in which calcium ions and magnesium ions interfere with the effectiveness of soaps. Magnesium and calcium are members of the same chemical family, Group 2A, of the periodic table of elements. Their ions, Mg^{2+} and Ca^{2+}, share the tendency to react with soap to form an insoluble compound that separates from solution. This insoluble compound is the stuff of bathtub rings and the scum deposited on clothing washed in hard water. Because much of the soap is tied up in the precipitate, more soap is required to form suds and cleanse things in hard water than is needed in soft water.

As shown in Figure 5.17, a soap "molecule" contains two parts: a sodium ion (Na^+), and a long hydrocarbon chain with a negatively charged ionic end (the "soap ion"). In aqueous solution, soap releases Na^+ ions and the negative soap ions. Soaps are used for cleaning because their long, nonpolar hydrocarbon tails dissolve readily in materials that are predominantly nonpolar, such as grease, chocolate, or gravy. The negatively charged ionic ends stick out of the surface of a grease globule because ionic substances are not soluble in nonpolar media. Thus, the grease becomes covered with negative charges. These negatively charged grease particles interact favorably with the partially positive regions of water molecules, are solubilized, and get carried away with the rinse water.

The insoluble precipitate (bathtub ring and scum on laundered clothes) arises when soap ions interact with Ca^{2+} or Mg^{2+} ions. Two of the negatively charged "soap" ions react with each of the Mg^{2+} or Ca^{2+} ions to form a precipitate (scum). One obvious way to avoid the formation of the precipitate is to remove the calcium and magnesium ions, in other words, to "soften" the water. This can be done by adding sodium carbonate (washing soda), Na_2CO_3, along with the soap. The carbonate ions (CO_3^{2-}) react with the Ca^{2+} to form insoluble calcium carbonate that is rinsed away (equation 5.7). Other water-softening compounds, such as sodium tetraborate or borax ($Na_2B_4O_7$) and trisodium phosphate (Na_3PO_4), work in a similar fashion. Calgon™ water softener contains hexametaphosphate ions, $P_6O_{18}^{6-}$, which tie up calcium and magnesium ions as large, soluble ions.

Another way to soften hard water is to remove Ca^{2+} and Mg^{2+} ions before they get to the washing machine or shower. This is often accomplished by a process called **ion exchange** that forms "soft" water. A water softener typically contains a **zeolite,** a clay-

The action of soap is an example of the generalization that "like dissolves like" (Section 5.13).

Figure 5.16

A pipe with hard-water scale build-up.

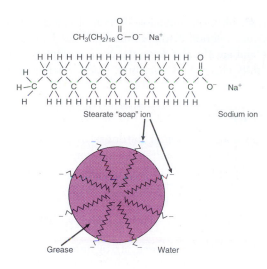

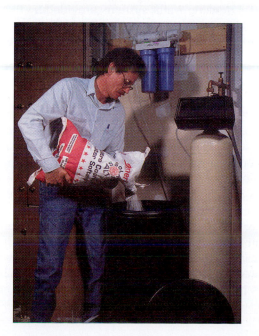

Grease   Water

Figure 5.17

Soap and its interaction with
grease.

like mineral made up of aluminum, silicon, and oxygen. These atoms are bonded into a
rigid, three-dimensional structure bearing many negative charges. All of these charges
must be balanced by positive charges, usually supplied by Na^+ ions associated with the
zeolite. However, when water containing Ca^{2+}, Mg^{2+}, or Fe^{3+} is passed through the zeolite, these ions displace the Na^+ ions because the multiply charged ions are more
strongly attracted to the negatively charged zeolite than are the singly charged Na^+ ions.
In other words, "hard water" ions (calcium, magnesium, iron) are exchanged for "soft
water" sodium ions. If we represent the zeolite as Z, we can write an equation representative of the process.

$$Na_2Z(s) \; + \; Ca^{2+}(aq) \; \longrightarrow \; CaZ(s) \; + \; 2\,Na^+(aq) \qquad (5.9)$$

Zeolite   (in hard water)   Zeolite   (in soft water)

(Na form)   (Ca form)

The Na^+ ions rather than Ca^{2+} ions flow through the tank and into the pipes of the
residence. Sodium ions do not interfere with the function of soap or result in the build-
up of scale, and so the problem of hardness has been left behind on the zeolite ion exchanger. When the exchanger becomes saturated with Mg^{2+}, Ca^{2+}, and other undesirable

*Adding salt to recharge an ion
exchange water softener.*

ions, it is back-flushed with a concentrated solution of sodium chloride. The high Na^+ ion concentration of this solution displaces the Mg^{2+} and Ca^{2+} from the zeolite, reversing equation 5.9, and are flushed down the drain as $MgCl_2$ and $CaCl_2$. The ion exchanger is left in its fully charged sodium form, ready to soften more hard water.

5.27 Consider This: Water Hardness in Your Area

Find information about the water hardness in your part of the country. Find out what the options are for softening the water, if needed. Possible follow-up search topic: Do sodium ions from softeners pose any health risk?

5.16.2 Case Study: Lead in Drinking Water

Standards for acceptable levels of pollutants are of little value unless accurate methods are available to measure the concentrations of pollutants. Newer analytical methods, in turn, have made possible more careful studies of pollutant concentrations in water and their toxicological effects. Chemical analysis of water involves a variety of analytical methods, depending on the impurity in question. Our second case study deals with lead, one of the most serious pollutants commonly found in drinking water. The source of the problem frequently arises within individual homes. Unless proper precautions are taken, lead from drinking water can have serious long-term health effects, especially tragic for young children.

All of the heavier metallic elements near the lower right side of the metals in the periodic table are very toxic. Several of them, including lead (Pb), mercury (Hg), and cadmium (Cd) are frequently encountered in drinking water. The metals themselves are not water soluble; rather, it is only their positive ions– Pb^{2+}, Hg^{2+}, and Cd^{2+}—in ionic compounds that are soluble and of concern. Because Pb^{2+} is the most common of these three and poses the most serious health risk due to its widespread occurrence, we will examine the lead story in some detail.

5.28 Consider This: Lead, Mercury, or Cadmium in Your Drinking Water

Find out whether lead, mercury, or cadmium ions are a significant problem in drinking water where you live or on your campus. You might begin with the map of local drinking water systems provided by EPA's Office of Ground Water and Drinking Water. Your local water utility company or state drinking water program should be able to provide information as well.

a. If these ions are present, what are some likely sources of them?
b. Are the concentrations of these ions above the MCLG or MCL?

The symbol Pb comes from the Latin name for lead, "plumbum," the origin of our word plumbing.

In its metallic form, lead is very dense (50% more dense than iron or steel). Because lead is an abundant, soft, and easily worked metal that does not rust, it has been used since ancient times for water pipes and roofs. Romans were the first to use lead for water pipes and as a lining for wine casks. Some historians attribute lead poisoning from such extensive use as a major factor contributing to the fall of the Roman Empire.

In more modern times, most U.S. homes built before 1900 had lead water pipes, now replaced by copper or plastic ones. Up until 1930, lead pipes were commonly used to connect homes to public water mains. There is no accurate way of knowing how many people suffered permanent health damage from living in residences with lead pipes. But, there are a few recorded cases of fatalities caused by lead poisoning in which the victim over many years habitually prepared a morning beverage using the "first draw" of water that had been standing in lead pipes overnight.

An interesting chemical connection exists between water hardness and how much lead dissolves from a lead pipe. Hard water forms a protective coating of calcium carbonate inside lead pipes that prevents water from coming into direct contact with the lead. On the other hand, with naturally soft water no such protective coating forms, and a small amount of lead (as Pb^{2+} ions) can dissolve in the water. Soft water also tends to be more acidic, which causes additional lead to dissolve. The effect is most severe in hot water pipes.

Some Pb^{2+} can get into drinking water even where there are no lead pipes. Solder used to join copper pipe contains 50–75% lead. Drinking fountains generally have a holding tank to store chilled water, and the seams in the tank and connections from it to the fountain are made with lead-based solder. Water for drinking fountains may stand in the tank for many hours, thus providing more contact time for lead from the solder to dissolve into the water.

When ingested, lead causes severe and permanent neurological problems in humans. This is particularly tragic for children who may suffer mental retardation and hyperactivity as a result of lead exposure, even at relatively low concentrations. Severe exposure in adults causes irritability, sleeplessness, and irrational behavior, including loss of appetite and eventual starvation. Unlike many other toxic substances, lead is a cumulative poison and is not transformed into a nontoxic substance. Once it enters the body it accumulates in bones and the brain.

Lead toxicity is a particular problem for children because lead ion, Pb^{2+}, is normally incorporated rapidly into bone along with calcium ion, Ca^{2+}. In children, who have less bone mass than adults, the lead remains in the blood longer where it can damage cells, especially in the brain. Besides lead in drinking water, young children are exposed to large amounts of lead from chewing on lead paint. This is especially the case in older houses where the paint is chipped and flaking. A national program monitoring blood lead levels in children is aimed at identifying children at risk. Health officials are required to investigate cases in which children are known to exceed the currently acceptable blood level of 15 $\mu g/dL$ (micrograms per deciliter). The EPA estimates that one of six U.S. children under six years of age has a blood lead level above this limit.

5.29 Your Turn

Two samples of drinking water were compared for their lead content. One had a concentration of 20 ppb and the other had a concentration of 0.003 mg/L.

a. Explain which one contains the higher concentration of lead.
b. Compare each sample to the current acceptable limit.

Since the 1970s, the federal government has had regulations for acceptable levels of lead in water and foods. These limits have gradually become more restrictive with the development of better analytical methods for measuring extremely low concentrations and as more has been learned about the health effects of lead. Lead is so widespread in the environment that older measurements suffered from unintentional contamination of both the equipment and the reference standards. Until recently the maximum contaminant level (MCL) for lead ion in drinking water was 15 ppb. In 1992 the U.S. EPA converted this to an "action level" meaning that the EPA will take legal action if 10% of tap water samples exceed 15 ppb. The hazard from lead is so great that the EPA has established a maximum contaminant level goal (MCLG) of zero, even though lead is not a carcinogen.

MCLs and MCLGs are discussed in Section 5.14.

The good news is that there is very little lead in most public water supplies. In 1987, the U.S. EPA estimated that approximately 600 public water systems using ground water and 215 public water systems using surface water may have water leaving the treatment plant with lead levels above 5 ppb. These represent less than 1% of public water supply systems and they serve less than 3% of the U.S. population. Most lead in drinking water comes from plumbing systems, not from the source water.

The limits for lead in drinking water are meaningless unless there are reliable analytical methods for measuring lead accurately in the low-ppb range. The almost universal method for lead analysis utilizes a variation of the spectrophotometric technique of analysis, which relies on absorption of light by a colored species in solution. A substance that forms a colored compound with Pb^{2+} is added to the water sample. The deeper the color, the greater the concentration of the species, say lead (or iron), in the sample. The intensity of the color is then measured with a **spectrophotometer** (Figure 5.18). Light of the desired wavelength passes through the sample and strikes a special detector where the light intensity is converted into an electrical voltage. The voltage is displayed on a meter or sent to a computer or other recording device. The amount of light absorbed by the solution, and which therefore does not reach the detector, is proportional to the concentration of the species of interest in the sample. The higher the species concentration, the more light that is absorbed by the sample.

To relate the concentration of the species (lead, iron, etc.) analyzed in a water sample to absorbance data taken by the spectrophotometer, an analyst first prepares a **calibration graph.** The graph is made by measuring the absorbances of a set of reference water samples containing *known* concentrations of the species to be analyzed. An example of a calibration graph for iron analysis is shown in Figure 5.19. Iron concentration is shown on the horizontal axis and absorbance at 505 nm is shown on the vertical axis. If a water sample gives an absorbance measurement of 0.45, an analyst can use that value to read directly from the graph that the concentration of iron is 2.7 ppm (see dashed lines, Figure 5.19). Looking at the horizontal scale on the graph, it is apparent that measurements of iron in water can be made down to low-ppm concentrations.

> ## 5.30 Your Turn
>
> Use Figure 5.19 to estimate the concentration of iron in these water samples. If the calibration graph cannot be used, state why not and suggest a possible way around the problem.
>
> **a.** Absorbance = 0.3
> **b.** Absorbance = 0.9
> **c.** Absorbance = 1.5
>
> **Ans:** **a.** The concentration of iron is approximately 1.8 ppm.

Figure 5.19 illustrates a caution about water analysis: the accuracy of the analysis is only as good as the accuracy of the calibration graph. Notice that the reference samples do not all lie on an absolutely straight line. Rather, there is some bit of uncertainty in each of the measurements, which contributes to a small uncertainty in the analysis of any water sample.

The specific method used for lead analysis is called **atomic absorption spectrophotometry.** A small water sample is vaporized at very high temperature into a beam of UV light coming from a lead-containing lamp. Radiation unique to lead atoms is emitted from the hot lead atoms in the lamp and is absorbed by lead atoms in the vaporized water sample. The sample's absorbance is then compared to a calibration graph

If the species is colorless, a chemical may be added to the sample to produce a colored species in the solution.

The nature of light and wavelength was discussed in Section 2.4. The use of a spectrophotometer to measure infrared radiation was described in Section 3.5.

Figure 5.18
Simplified diagram of a spectrophotometer.

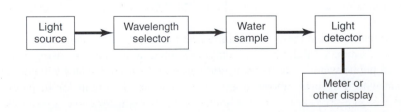

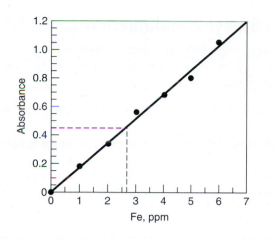

Figure 5.19
Calibration graph for spectrophotometric analysis of
iron in water.

to determine the Pb^{2+} concentration in the water. Conventional versions of the equipment can measure lead ion concentrations down to about 50 ppb, but that is not good
enough to meet the present MCL of 15 ppb. A more sophisticated and expensive version
can measure lead at well below 1 ppb.

5.31 Consider This: Shifting Limits for Lead

Before 1962, the recommended limit for lead in drinking water was 100 ppb.
In 1962, the limit was lowered to 50 ppb. In 1988, the MCL was lowered to
15 ppb.

a. Suggest some reasons for this trend. Give other examples where legal
limits have changed over time.
b. In addition to the current action level of 15 ppb of lead in tap water in
residences, the EPA recommends that source water from water utilities
should contain no more than 5 ppb lead, and water in school drinking
fountains should contain no more than 20 ppb. Suggest possible reasons
for these differences.

5.32 Consider This: Policy Trade-offs for Lead in Drinking Water

Establishing more stringent regulations on lead in drinking water may be desirable to protect public health—especially the health of young children and
the elderly—but it is costly.

a. Identify some of the costs associated with stricter limits.
b. Which costs do you think are justified?
c. Who should pay these costs? Give the reasons behind your opinions.

Based on what has been learned, there are several ways to minimize exposure to lead
from drinking water. The most obvious first step is to replace lead pipes where they are
known to exist. Even for residences without lead pipes, there are likely to be some sources
of lead, including brass plumbing fixtures. For this reason, the U.S. EPA recommends the
following two precautions: (1) Use only cold water for drinking, cooking, and especially
for making baby formula. (2) Any time water has stood in the pipes for six hours or
longer, flush the pipes by letting the water run until it gets as cold as it will get. It is especially important to not use the "first-draw" water in the morning, either hot or cold. And
if lead pipes are suspected or known to be present, extreme caution should be exercised,
allowing the water to run for several minutes before using it for cooking or drinking.

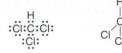

Structure of chloroform, CHCl₃.

Halogens are Group 7A elements in the periodic table. The common ones are fluorine, chlorine, bromine, and iodine.

5.16.3 Case Study: Trihalomethanes in Drinking Water

The third case study deals with substances known collectively as trihalomethanes (THMs), the most important of which is chloroform, $CHCl_3$. Trihalomethanes are derivatives of methane, CH_4.

Other trihalomethanes contain, as the name implies, any three "halogen" atoms, which are elements in Group 7A of the periodic table. The most significant trihalomethanes, in addition to $CHCl_3$, are $CHCl_2Br$, $CHClBr_2$, and $CHBr_3$.

Chloroform and other THMs are by-products of drinking water chlorination. Recall from Section 5.15 that chlorination leaves some residual HOCl in the water. This HOCl can lead to formation of THMs by its reaction with humic acids, which are breakdown products of plant materials. Humic acids are almost always present in surface waters and so, formation of THMs in chlorinated surface water is unavoidable. However, the concentration of THMs in drinking water is normally far less than 1 ppm. But even in very low concentrations, they contribute an unpleasant taste to chlorinated water. The presence of THMs can sometimes be detected by their hospital-like odor, especially in a hot shower. Recent EPA standards require municipal water treatment facilities to reduce the concentration of humic acids in water prior to its chlorination, which will decrease the likelihood of THMs formation.

The primary health concern about THMs is that chloroform is suspected of causing liver cancer. There is some epidemiological evidence of slightly greater cancer rates (including bladder and rectal cancer) in people living in communities with chlorinated drinking water compared to those that do not. The current MCL for total THMs established in 1998 by the EPA is 80 ppb (0.080 ppm), down from the previous level of 100 ppb. Most drinking water samples meet that standard. The national average for THMs is 51 ppb for municipal drinking water that comes from surface water. Well water has a lower THMs concentration because it contains little or no plant material. Although the EPA limit (MCL) is 100 ppb, the goal (MCLG) is 0 ppb.

5.33 Consider This: Health Risk From Chloroform

Chloroform is a toxic substance that may cause cancer. But, chloroform is also an effective cough suppressant that has been used in many over-the-counter medications. Until recently, many children's cough syrups contained several percent of chloroform. To what extent does this provide convincing evidence that chloroform in drinking water is not a health hazard, and therefore that chlorination of drinking water is safe?

It should be apparent that a reliable analytical method is needed that can accurately measure THMs in the ppb range. The analytical method of choice for THMs is called **gas chromatography (GC)**, a powerful technique for measuring trace amounts of various molecular substances in water. In GC, molecular solutes in water are first extracted from a large sample of water into a small volume of a non-ionic liquid such as octane. This extraction concentrates the solutes to be analyzed. A very small portion of the extract is injected into a flowing gas stream that passes down a long, heated tube coated with an absorbing material; a detector is at the far end (Figure 5.20a). Components in the sample move down the tube at various rates, thus reaching the detector at different times. The signal from the detector is displayed on a recorder, as shown in Figure 5.20b; each peak corresponds to a different substance. For each plotted peak, the time required for the substance to reach the detector identifies the substance, while peak area measures its concentration. Gas chromatography is the normal analytical method for trace amounts of a wide variety of toxic substances in drinking water. These include pesticides, PCBs, dioxins, industrial solvents, and gasoline or other petroleum products.

Trihalomethanes in chlorinated drinking water create a classic risk/benefit situation. On the one hand, chlorination of drinking water is an efficient, inexpensive, effective

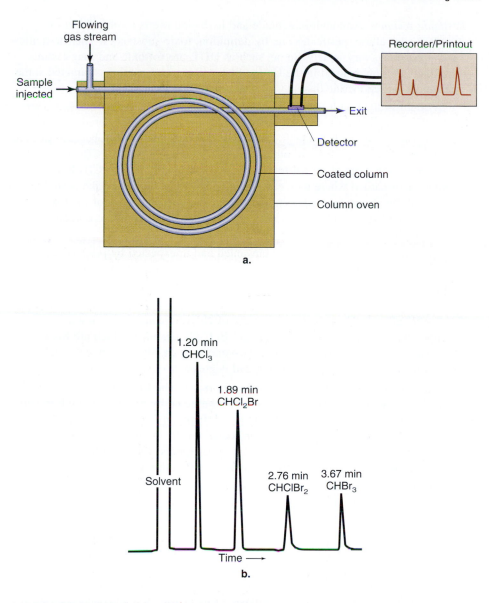

Figure 5.20
(a) A simplified gas chromatography apparatus.
(b) Gas chromatography analysis of a mixture of THMs.

method that greatly reduces the health risk from bacteria and other disease-causing organisms in the water. On the other hand, chlorination results in the formation of relatively low concentrations of THMs, which may cause cancer.

5.16.4 Some Other Important Water Pollutants

The previous three case studies illustrate some of the problems associated with substances dissolved in drinking water. At present, there are at least 83 substances in drinking water regulated by the U.S. EPA. Listed below are some of these pollutants, with very brief descriptions of their nature, their sources, and concerns about them.

> *Mercury.* Like its sister heavy metal ion, Pb^{2+} (Section 5.16.2), mercury ion (Hg^{2+}) is extremely toxic, causing severe neurological effects to humans and other animals. Once in the human body, it tends to remain there. Environmental mercury sources vary from fossil fuel combustion to a variety of consumer products. Mercury in drinking water is rarely a direct health threat, but mercury in lakes can result in harmful levels of mercury in fish. It is a cumulative poison for those who eat the fish. Pregnant women should avoid eating fish containing even low levels of mercury.

Pesticide residues. Although insecticide and herbicide use is a central part of modern agriculture, pesticides are, by definition, toxic substances. Their toxicities and persistence vary widely. Some, such as DDT, are so toxic and long-lasting that they are now banned in the United States. Recent research has targeted some pesticides as "endocrine disrupters," which may lead to reproductive abnormalities in certain species of animals.

PCBs. The letters stand for *polychlorinatedb*iphenyls. Once considered to be very useful compounds with desirable properties, they are now banned because of their long-term effects on birds and other animals. The largest use of PCBs was to transfer heat in large electrical transformers. Huge quantities of PCBs have been spilled or discarded where they can contaminate water supplies. Like pesticides, PCBs exhibit strong *biomagnification,* that is, the concentrations of the toxic material increase dramatically moving up the food chain from microorganisms to fish to higher animals.

Dioxins. These toxic substances are unwanted and unexpected by-products of certain industrial processes (such as pesticide and paper manufacturing). But, they are also formed in tiny amounts by natural processes and whenever carbon-based fuels are burned.

Industrial solvents. These include a variety of chlorine-containing carbon compounds, such as *t*richloroethylene (TCE, C_2HCl_3), all of which are toxic, very stable, and long-lasting in the environment. Such solvents are extremely useful in our society for dry cleaning and degreasing machined metal.

Phosphate build-up in lakes and other water bodies. The Chesapeake Bay and Lake Erie, among other locations, have been affected by phosphate build-up and are the subjects of extensive study and policy debates. Phosphate is not inherently toxic, but excessive phosphate in water bodies results in uncontrolled algae growth that destroys the natural ecosystem. Detergents have been targeted as the main culprit in such cases, but agricultural runoff from fertilized fields may be a much larger source.

Fluoride. Fluoride (F^-) is not a pollutant, but it may be naturally present in water. It is often deliberately added to public water supplies at 1 ppm to promote stronger teeth in children.

Nitrate. In rural agricultural areas, both well water and surface water are often contaminated with nitrate ion, NO_3^-. The sources include seepage water from manure and fertilizer. The economics of today's farming and its need for maximum productivity often lead to excessive fertilizer use. Nitrate concentrations are of particular concern for infants because nitrate is easily converted to nitrite ion, NO_2^-, which ties up hemoglobin, depriving tissues of oxygen.

5.34 Consider This: Copycat Chemicals

Environmental endocrine disrupters or endocrine mimics are a topic of much current debate. Use the web to search for additional information about these substances. In particular, your search should look for evidence whether endocrine disrupters are a problem, and why there is controversy about them. Write a brief summary of both positions, take a stand in favor of one of them, and outline the rationale for your choice.

5.17 Consumer Choices: Tap Water, Bottled Water, and Filtered Water

We have developed a chemical background sufficient to answer some questions about drinking water sources posed early in the chapter and to make informed choices about the water we drink.

Tap Water

1. Is safe drinking water generally available in the United States? The answer, a resounding "yes," is due to high standards mandated by federal regulations for drinking water provided by public water supply utilities. As importantly, the treatment technology is available to achieve these high standards; without such technology, standards would be merely hollow gestures. Very few people in our country suffer acute illness from drinking contaminated water unless they are using water from a private well that has not been properly tested. It should be noted that the Safe Drinking Water Act Amendments of 1996 enhanced protection including increased requirements for notifying consumers promptly of any problems with water safety. This chapter began with an example of a possible letter of this type from a local water utility.

2. Is tap water "pure" water? Certainly not; it almost surely contains small amounts of sodium, calcium, magnesium, chloride, sulfate, and bicarbonate ions, as well as trace amounts of other ions. Tap water also contains dissolved air, which is a mixture that includes N_2, O_2, CO_2, and air-borne particles.

3. What *problems* are likely to exist? In some parts of the country, hard water containing higher than normal concentrations of calcium and magnesium ions is a problem. Water softening options are readily available, if desired. Some tap water may contain dangerous Pb^{2+} concentrations, although lead is normally a problem only in buildings with lead pipes, and then only if the water is naturally soft. Other heavy metal elements, such as mercury and cadmium, may be present at dangerous concentrations, although this is extremely unlikely. Chlorinated tap water from surface water sources will contain a small amount of residual chlorine. It may also contain small amounts of trihalomethanes, by-products of chlorination. Depending on its source, the water may contain low concentrations of mercury, nitrate, pesticide residues, PCBs, and industrial solvents. By now you should understand that the presence of such substances in drinking water is not likely a cause for alarm. Rather, the crucial question is "How much?" If pollutant concentrations are below the maximum contaminant levels (MCLs), the EPA regards the water as safe, with an adequate margin of safety.

Bottled Water

This chapter began with a look at bottled water, drunk by people for a variety of reasons: taste, convenience, and/or the belief that it is healthier than tap water. Whatever its source, bottled water is expensive. We can raise the same questions about bottled water as those asked about tap water.

1. Is it safe? Bottled water is not regulated by the same laws that apply to public water supplies. Considered as a food, bottled water is instead regulated by the Food and Drug Administration (FDA). A provision of the SDWA amendments of 1996 requires the FDA to develop bottled water standards that are equal to EPA drinking water standards. In past years, critics have questioned the safety of bottled water. However, 85% of bottled water currently sold in the United States is produced by member companies of the International Bottled Water Association (IBWA), whose member companies must meet higher water quality standards than those imposed by the FDA (Figure 5.21). Springs and underground aquifers that do not require disinfection are the principal sources of bottled water. If disinfection is required, it is done with ozone or UV radiation, rather than with chlorine, thus leaving no objectionable taste and no THMs. In addition, most bottled water is subjected to either filtration, reverse osmosis, or distillation (see Section 5.18). The absence of chlorine, THMs, and the probable absence of various trace pollutants found in surface water provides much of the argument for bottled water as a healthier alternative to tap water.

2. Is bottled water pure? Because bottled water normally comes from springs or wells, we can be sure that it contains ions, dissolved as the water percolates through the surrounding rocks (Figure 5.21). In fact, bottled water from some well-known spas (Bath in England, Baden-Baden in Germany, and White Sulfur Springs in West

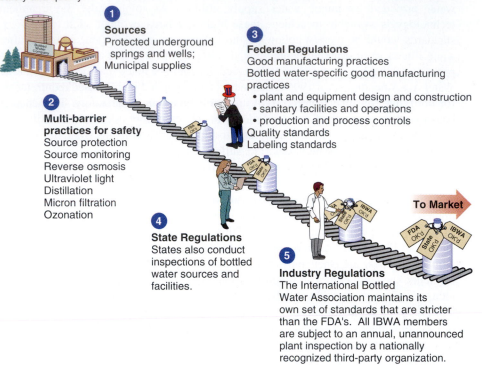

Figure 5.21
Bottled water's path to market.

Bottled Water's Path to Market

The International Bottled Water Association illustrates the process its members' products follow from the source to the consumer's satisfaction. Federal, state, and industry regulations guarantee safety and quality.

1 Sources
Protected underground springs and wells; Municipal supplies

2 Multi-barrier practices for safety
Source protection
Source monitoring
Reverse osmosis
Ultraviolet light
Distillation
Micron filtration
Ozonation

3 Federal Regulations
Good manufacturing practices
Bottled water-specific good manufacturing practices
• plant and equipment design and construction
• sanitary facilities and operations
• production and process controls
Quality standards
Labeling standards

4 State Regulations
States also conduct inspections of bottled water sources and facilities.

5 Industry Regulations
The International Bottled Water Association maintains its own set of standards that are stricter than the FDA's. All IBWA members are subject to an annual, unannounced plant inspection by a nationally recognized third-party organization.

To Market

Virginia) contains relatively large amounts of calcium and other ions, as well as dissolved carbon dioxide. In a few cases, dissolved hydrogen sulfide gas provides a characteristic "sulfur" odor.

Filtered Water

Water filter units are a readily available alternative to bottled water. These units generally attach to a kitchen or bathroom faucet and remove most trace pollutants. The units simultaneously employ two purification techniques. The first is "activated carbon," a special form of charcoal with a very high surface area that absorbs most of the non-ionic solutes including residual chlorine, THMs, pesticide residues, solvents, and other similar substances. The second component is an ion exchange resin (Section 5.16.1) that removes heavy metal ions (chiefly lead and copper) and hard water calcium and magnesium ions. Water from such a unit is free of objectionable taste and odor, and should be free of most hazardous substances. More correctly, the use of filters will reduce the concentrations of such substances to extremely low values, well below the concentrations of concern for human health. Filtered water typically costs only 20% as much as bottled water.

5.35 Consider This: Evaluating Bottled Water

In 5.2 Consider This near the beginning of this chapter you were asked to list and then prioritize the advantages and the disadvantages associated with drinking bottled water. Having now studied this chapter, check your list. Would the order of importance be the same? Explain how your reasoning may have changed based on the information and understanding gained in the study of this chapter.

5.18 International Needs for Safe Drinking Water

Those who live in the United States are privileged to have drinking water choices available. We can select from tap, bottled, or filtered water, all generally of high quality. Such is not the case for people in most of the rest of the world. The reality there is that more than a billion people (one in five), principally in developing nations, lack access to safe drinking water. Whereas bottled water is a discretionary option for many in the United States, the majority of the world's population does not have that option.

For those living in arid regions, such as the Middle East, fresh water is scarce. Seawater is readily available in many such areas, but its high salt concentration makes it unfit for human consumption. Coleridge's Ancient Mariner is more than just a poetic fantasy; it is a physiological reality. Ocean water contains 3.5% salt compared to only about 0.9% salt in body cells. Consequently, seawater can be drunk only after most of the salt is removed. Fortunately, there are ways to do this, but they require large amounts of energy. Collectively the methods are known as **desalination,** a broad general term describing any process that removes ions from salty water.

One desalination method is distillation, an old and rather common way of purifying water for laboratory and other uses. Distilled water is used in steam irons, car batteries, and other devices whose operation can be impaired by dissolved ions. **Distillation** is remarkably simple—a liquid is evaporated and then condensed, just as in the natural hydrologic cycle. An apparatus such as that shown in Figure 5.22 is used. Impure water is put into a flask, pot, or other container and heated to its boiling point, 100°C. As the water vaporizes, it leaves behind most of its dissolved impurities. The water vapor passes through a condenser where it cools and reverts back into a liquid, now free of contaminants. Not surprisingly, the product is usually called "distilled water." If distillation is done very carefully, extremely pure water, with no detectable amounts of contaminants, is produced.

Notice that to distill water, it must be heated to its boiling temperature (100°C) and then heated further to convert the liquid water to steam. Each step requires a large energy input. This would be true for the distillation of any liquid, but recall from Section 5.9 that water has an unusually high specific heat and an unusually large amount of heat required for evaporation. Both result from the uniquely extensive hydrogen bonding in water. The high energy cost for water purification by distillation suggests that it is only economically practical for countries or regions with abundant and cheap energy. **Ion exchange,** described in Section 5.16.1, is another method for desalination, although it is not very practical for large scale desalination.

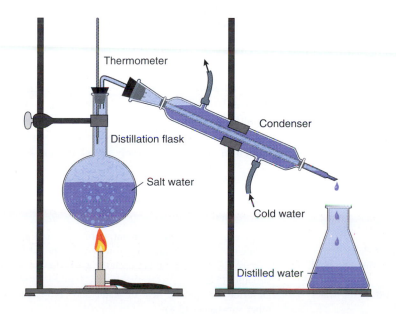

Figure 5.22
Water purification by distillation.

Figure 5.23
Reverse osmosis.

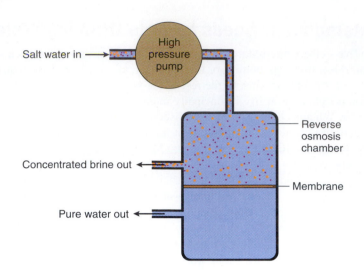

Another desalination technique gaining in popularity is **reverse osmosis.** To understand this method, we need to know that **osmosis** is the natural tendency for a solvent (water in this case) to move through a membrane from a region of higher solvent concentration to a region of lower solvent concentration. This tendency to equalize concentrations is involved in many biological processes. In this particular instance, the net effect is that cells lose water rather than gain it. However, osmosis can be reversed. If sufficient pressure is applied to the salt water side, water molecules can be forced through the membrane leaving ions behind. Figure 5.23 is a schematic representation of this process.

The world's largest desalination plant, located at Jubail, Saudi Arabia, provides 50% of that country's drinking water using reverse osmosis to desalinate Persian Gulf water. Although most such installations are in the Middle East, the number of reverse osmosis plants is increasing in the United States. Florida has 109 reverse osmosis desalination facilities, including the one that furnishes the city of Cape Coral with 15 million gallons of fresh water every day from brackish underground supplies. Small reverse osmosis installations are used in spot-free car washes and individual units are available for boaters. It must be pointed out, however, that reverse osmosis desalination is too expensive for use in most developing nations.

A small reverse osmosis apparatus for converting sea water to potable water.

Conclusion

This chemical compound H₂O is a very unusual substance, with many unique properties that contribute to its life-supporting role. Like the air we breathe, water is central to life, and we humans require large quantities of it. We take for granted that our drinking water, whether from the tap or bottle, typically is free of harmful contaminants. This chapter has focused almost exclusively on the quality of drinking water—its sources, substances dissolved in it, and potential contaminants and their concentrations. Federal and state regulations help make our drinking water safe. In three case studies, we considered how particular substances in water can be treated and analyzed. In the next chapter, we turn to an examination of rain water and the ways in which substances dissolved in rain can adversely affect the environment.

Chapter Summary

Having studied this chapter you should be able to:
- Recognize the sources and distribution of water (5.3)
- Discuss why water is such an excellent solvent for ionic and some covalent compounds (5.4, 5.12, 5.13)
- Describe the factors involved in providing pure drinking water (5.5, 5.15)

- Use concentration units: percent, ppm, ppb, and molarity (5.6)
- Relate the molecular structure of water to its properties (5.7, 5.8)
- Discuss the relationship between the properties of water and its structure (5.8)
- Describe the specific heat of water and compare it to that of other substances (5.8)

- Understand how electronegativity and bond polarity are related to the structure of water (5.8)
- Describe hydrogen bonding and its importance to the properties of water (5.9)
- Describe how the density of water is related to its molecular structure (5.9)
- Relate chlorination with water purification (5.15)
- Predict ion formation and the formulas for ionic compounds, including those with common polyatomic ions (5.11)
- Discuss the Maximum Contaminant Level Goal (MCLG) and the Maximum Contaminant Level (MCL) established by the EPA to ensure water quality (5.14)

- Discuss how drinking water is made safe to drink (5.15)
- Know the causes and effects of water hardness (5.16.1)
- Cite water softening methods (5.16.1)
- Describe atomic absorption spectrophotometry and gas chromatography as methods for analyzing contaminants in water (5.16.2, 5.16.3)
- Compare and contrast tap water, bottled water, and filtered water in terms of water quality (5.17)
- Understand distillation and reverse osmosis (5.18)

Questions

Emphasizing Essentials

1. **a.** The text states that currently over three billion gallons of bottled water are sold each year in the United States, more than 10 times that sold 20 years ago. Calculate the volume of bottled water sold 20 years ago.

   b. What percentage increase has there been in the sale of bottled water over the last 20 years?

2. **a.** What is an *aquifer?*

   b. Why is it important to prevent unwanted substances from reaching a clean aquifer?

3. If a 500-L drum of water represented the world's total water supply, how many liters would be water actually available for drinking in the United States? *Hint:* See Figure 5.2.

4. Based on your experience, predict the solubility of each of the following substances in water. Use terms such as very soluble, partially soluble, or not soluble. Cite some supporting evidence for your prediction.

   a. orange juice concentrate

   b. liquid clothes washing detergent

   c. household cleanser

   d. chicken broth

   e. chicken fat

5. A certain bottled water lists a calcium concentration of 55 mg/L. What is its calcium concentration expressed in ppm?

6. A certain vitamin tablet contains 162 mg of calcium and supplies 16% of the recommended daily amount of calcium required by a person on a typical 2000-Calorie diet. How many 500-mL bottles of Evian bottled water would you have to drink each day to obtain the same mass of calcium? *Hint:* See 5.7 Your Turn.

7. The acceptable limit for nitrate, often found in well water in agricultural areas, is 10 ppm. If a water sample is found to contain 350 mg/L, does it meet the acceptable limit? Show a calculation to support your answer.

8. One reagent bottle on the shelf in a laboratory is labeled 12 *M* H_2SO_4 and another is labeled 12 *M* HCl.

   a. How does the number of moles of H_2SO_4 in 100 mL of 12 *M* H_2SO_4 solution compare with the number of moles of HCl in 100 mL of 12 *M* HCl solution?

   b. How does the number of grams of H_2SO_4 in 100 mL of 12 *M* H_2SO_4 solution compare with the number of grams of HCl in 100 mL of 12 *M* HCl solution?

9. Methane, CH_4, and water are both compounds of hydrogen bonded with a nonmetallic element. Yet, methane is a gas at room temperature and pressure and water is a liquid. Offer a molecular explanation for the difference in properties.

10. Explain why the term "universal solvent" is applied to water.

11. Consult Figure 5.4 to help answer this question.

    a. Calculate the electronegativity difference between each pair of atoms.

    N and C

    S and O

    N and H

    S and F

    b. A single covalent bond forms between the atoms in each pair. Identify the atom that will attract the electron pair in the bond more strongly.

    c. Arrange the bonds in order of increasing polarity.

12. NaCl is an ionic compound, but chlorine and silicon are joined by covalent bonds in $SiCl_4$.

    a. Use Figure 5.4 to determine the electronegativity difference between chlorine and sodium, and between chlorine and silicon.

    b. What correlations can be drawn about the difference in electronegativity between bonded elements and their tendency to form ionic or covalent bonds?

    c. How can you explain on the molecular level the conclusion reached in part **b**?

13. Consider a molecule of ammonia, NH_3.

    a. Write the Lewis structure of NH_3.

    b. Are there polar bonds in NH_3?

    c. Is the NH_3 molecule polar? *Hint:* Consider the geometry of the ammonia molecule.

    d. Predict the solubility of NH_3 in water.

14. This diagram represents two water molecules in a liquid state. What kind of bonding force does the arrow indicate? Is this an *inter*molecular or *intra*molecular force?

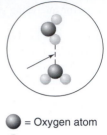

    ⬤ = Oxygen atom

    ◯ = Hydrogen atom

15. The density of liquid water at 0°C is 0.9987 g/cm³; the density of ice at this same temperature is 0.917 g/cm³.

  a. Calculate the volume occupied at 0°C by 100. g of liquid water and by 100. g of ice.

  b. Calculate the percentage increase in volume when 100. g of water freezes at 0°C.

16. Consider these liquids.

Liquid	Density, g/cm³
dishwashing detergent	1.03
maple syrup	1.37
vegetable oil	0.91

  a. If you pour equal volumes of these three liquids into a 250-mL graduated cylinder, in what order will you add the liquids to create three separate layers? Explain your reasoning.

  b. If an unknown liquid was poured into the cylinder and it formed a layer that was on the bottom of the other three layers, what can you tell about one of the properties of the unknown liquid?

  c. What will happen if a volume of water equal to the other liquids is poured into the cylinder in part **a** and the contents of the cylinder mixed vigorously? Explain.

17. Why is there the possibility of a water pipe breaking if the pipe is left full of water during extended frigid weather?

18. Calculate the quantity of heat absorbed (+) or released (−) during each of these changes.

  a. 250 g of water (about 1 cup) is heated from 15°C to 100°C

  b. 500 g of water is cooled from 95°C to 55°C

  c. 5 mL of water at 4°C is warmed to 44°C

19. Solutions can be tested for conductivity using this type of apparatus.

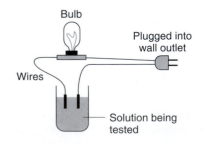

Predict what will happen when each of these dilute solutions is tested for conductivity.

  a. $CaCl_2(aq)$

  b. $C_2H_5OH(aq)$

  c. $H_2SO_4(aq)$

Explain your predictions briefly.

20. Predict the ion most likely to be formed by each of these atoms. Use the Lewis structure of each atom and its corresponding ion to show how the ion obeys the octet rule.

  a. Cl

  b. Ba

  c. S

  d. Li

  e. Ne

21. Predict the formula and give the name of the ionic compound formed by the reaction of each pair of elements.

  a. Na and S

  b. Al and O

  c. Ga and F

  d. Rb and I

  e. Ba and Se

22. Name each compound.

  a. $KC_2H_3O_2$

  b. $Ca(OCl)_2$

  c. LiOH

  d. Na_2SO_4

23. Based on the generalizations in Table 5.4, which compounds are likely to be water soluble?

  a. $KC_2H_3O_2$

  b. $Ca(NO_3)_2$

  c. LiOH

  d. Na_2SO_4

24. This represents the structural formula of a soap.

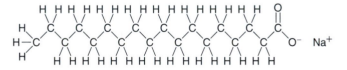

How is the ability of this soap to remove grease from clothes related to the structure of this soap?

25. Explain why desalination, despite its proven technological effectiveness, is not more widely used to produce drinking water.

Concentrating on Concepts

26. We take water for granted. How can you explain to a friend that we should value water as a unique substance, one essential for life?

27. Why is the concentration of calcium often given on the label for bottled water?

28. The label on Evian bottled water lists a magnesium concentration of 24 mg/L. The label of a popular brand of multivitamins lists the magnesium content as 100 mg per tablet. Which do you think is a better source of magnesium? Explain your reasoning.

29. There is a new sign posted at the edge of a favorite fishing hole that says "Caution: Fish from this lake may contain over 1.5 ppm Hg." Explain to a fishing buddy what this unit of concentration means, and why the caution sign should be heeded.

30. Four elements are identified by numbers in this periodic table.

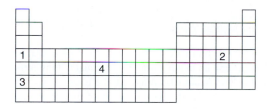

    a. Based on trends within the periodic table, which of the four elements do you expect to have the highest electronegativity value? Explain.

    b. Based on trends within the periodic table, rank the other three elements in order of decreasing electronegativity values. Explain your ranking.

31. A diatomic molecule **XY** that contains a polar bond *must* be a polar molecule. However, a triatomic molecule **XY₂** that contains a polar bond *does not necessarily* form a polar molecule. Use some examples of real molecules to help explain this difference.

32. Imagine you are at the molecular level, looking at what happens when gaseous water condenses.

    a. Consider a collection of four water molecules, using

         to represent each molecule.

Draw a representation of water in the gaseous state and then in the liquid state. How does the collection of molecules change?

    b. Discuss what happens to the bonding at the molecular level when water condenses to a liquid.

    c. What happens at the molecular level when water changes from a liquid to a solid?

33. Hydrogen bonding has been offered as a reason why ice cubes and icebergs float in water. Consider ethanol, C_2H_5OH.

    a. Draw its Lewis structure and use it to predict if pure ethanol will exhibit hydrogen bonding.

    b. A cube of solid ethanol will sink rather than float in liquid ethanol. Explain this behavior in view of your answer in part **a.**

34. The unusually high heat capacity of water is very important to regulating our body temperature and keeping it within a normal range despite time, age, activity, and environmental factors. Consider some of the ways that the body produces heat, and some of the ways that it loses heat. How would these functions differ if water had a much lower heat capacity?

35. How do solubility generalizations influence water quality? Suppose that you are in charge of regulating an industry in your area that manufactures agricultural pesticides. How will you decide if this plant is obeying necessary environmental controls? What criteria affect the success of this plant?

36. Health goals for contaminants in drinking water are expressed as MCLG or Maximum Contaminant Level Goals. Legal limits are given as MCL or Maximum Contaminant Levels. How are MCLG and MCL related for a given contaminant?

37. Explain how you would use Figure 5.19 to determine the absorbance of a solution containing 0.002% dissolved iron.

38. Borax, $Na_2B_4O_7$, can be used to soften hard water. Write an equation to represent the reaction of borax reacting with calcium ions in hard water.

39. An ion exchange resin totally charged with Na^+ ions holds 0.35 g of Na^+ per gram of resin. Use this information to determine the mass of water with a hardness of 40 ppm that could be deionized by 1 g of this resin.

40. Water quality in the chemistry building on the campus of one of the authors of this book is being continuously monitored because testing indicated water from drinking fountains in the building had dissolved lead levels above those established by the Safe Drinking Water Act.

    a. What is the likely major source of the lead in the drinking water?

    b. Does the chemical research carried out in this chemistry building account for the elevated lead levels found in the drinking water? Why or why not?

Exploring Extensions

41. Most people turn on the water tap with little thought about where the water comes from. In 5.4 Consider This, you investigated the source of your drinking water. Now take a more global view. Where does drinking water come from in other areas of the world? Investigate the source of drinking water in a desert country, in a developed European country, and in an Asian country. How do these sources differ?

42. Is there any such thing as "pure" drinking water? Discuss what is implied by this term, and how its meaning might change in different parts of the world.

43. One of the large aquifers in the United States is under the pine barrens of New Jersey.

    a. Where are the pine barrens in New Jersey?

    b. Why is this aquifer under increased political pressures?

44. In the mid-1990s, researchers in Canada and Australia reported that consumption of drinking water with more than 100 ppb aluminum can lead to neurological damage such as memory loss and perhaps to a small increase in the incidence of Alzheimer's disease. Has further research substantiated these findings? Find out more about this topic, and write a brief summary of your findings. Be sure to cite the sources of your information.

45. The text states that hydrogen bonds are only about one-tenth as strong as the covalent bonds that connect atoms within molecules. Check out that statement with this information. Hydrogen bonds vary in strength from about 4 to 40 J/mole. Given that the hydrogen bonds between water molecules are at the high end of this range, how does the strength of a hydrogen bond between water molecules compare to the strength of an hydrogen-to-oxygen covalent bond within a water molecule? *Hint:* Consult Table 4.1 for covalent bond energies.

46. The text states that mass and density are often confused.

    a. What do you think the term "heavy metal" implies when talking about elements on the periodic table?

    b. Compare the scientific definitions of this term that you may find in different sources, and discuss whether each definition is related to relative density or to relative mass.

47. We all have the amino acid glycine in our bodies. Its structural formula is

$$\underset{\underset{H}{|}}{\overset{\overset{H}{|}}{H-\ddot{N}-}}\underset{\underset{H}{|}}{C}\overset{\overset{:O:}{\|}}{-C}-\ddot{O}-H$$

   a. Is glycine a polar or nonpolar molecule? Use electronegativity differences to help with the answer to this question.

   b. Can glycine exhibit hydrogen bonding? Explain your answer.

   c. Predict the solubility of glycine in water. Explain the reasons for your prediction.

48. How hard is the water in your local area? One way to answer this question is to determine the number of water softening companies in your area. Use the resources of the web, as well as ads in your local newspapers and yellow pages, to find out if your area is targeted for marketing water softening devices.

49. Some areas have a higher than normal amount of trihalomethanes, THMs, in drinking water. Suppose that you are considering moving to such an area. Write a letter to the local water district asking relevant questions to be answered before deciding to move to that area.

50. PCBs (polychlorinatedbiphenyls) are very useful chemicals that may end up in the wrong place, causing long-term damage to birds and mammals. What are the uses of PCBs that made them desirable, and what are some of the negative effects of these materials?

Neutralizing the Threat of Acid Rain

Acid rain has caused significant
damage to this statue of George
Washington in New York City
during the past fifty-five years.

"ACID RAIN: The Problem That Won't Go Away... The
whole issue of acid rain has literally fallen from public
view. The public thinks the problem has been solved, and
I would suggest to you that it has not been solved."

Gene Lickens
Acid rain researcher
The Boston Globe December 23, 1996

"...The acid rain story is far from over. As emissions into
the atmosphere have decreased, omissions in the story
have become more evident. The 1990s have added a new
chapter of legislation and control, but a conclusion to this
hazardous environmental concern has yet to be written."

Don Muton
Environment July/August 1998

"...It is still too early to confirm any environmental benefits [of acid rain controls]... It's important to do this kind of monitoring. Just because a law has been passed doesn't mean a problem has been solved... It's going to be hard to see the improvements. Ecosystems don't respond immediately. They take 10 or 20 years..."

Donald Uhart, Program Director
National Acid Precipitation Program
August 1998

It is certainly true that the wave of publicity and papers produced on the topic of acid precipitation has slowed considerably over the past decade or so. In the 1980s, acid rain had become the poster child for environmental ills created by allegedly unbridled industrial production causing widespread destruction of lakes and forests. When the first edition of this textbook was written (1990), our files were bulging with articles about acid rain. Now, the topic does not command high media visibility. There are several possible explanations for this phenomenon. It could be that initially the problem was overstated. But the fact remains that many of the effects attributed to acid precipitation were and are genuine. You will see in this chapter that although much progress has been made in reducing the causes of acid rain, the problem has hardly been solved. Perhaps the neutralization of acid rain has simply become an unfashionable cause. But even if this latter explanation were true, the topic does warrant our attention, as evidenced by the words of Gene Lickens, Don Muton, and Donald Uhart; hence this chapter.

Comments about air being "acidic" were made as early as the beginning of the 1700s.

The fact that rain is often acidic apparently was first studied in detail in 1852 by a British chemist named Angus Smith. Twenty years later, he published a book entitled *Air and Rain,* but the book and Smith's ideas soon fell into obscurity. Then, in the 1950s, the effects of acid rain were rediscovered by scientists working in the northeastern United States, Scandinavia, and the English Lake District. Reports of damage attributed to acidic precipitation grew dramatically over the next three decades. Dozens of books, scientific papers, and popular articles were written describing the effects already observed and making dire predictions of more devastating damage yet to come.

Similar reports came from every part of the world. Many lakes in Norway and Sweden were reported as being effectively "dead," without fish or any other living things. Trees in northern Germany had been stripped of many of their leaves. The beautiful sculptures adorning the exteriors of cathedrals, other historic buildings, and monuments throughout Europe and elsewhere are eroding away. In all of these instances, acid rain has been charged with being one of the major causes of the damage. Yet, like a number of other air quality–related issues such as ozone depletion, global warming, some responsible authorities have counseled caution about taking action against acid rain. They argue that the scientific evidence linking acid precipitation and its attributed effects is tenuous or that the effects are relatively mild.

Chapter 1 dealt with ambient air quality, which is largely a *localized* problem and most serious in cities. Chapters 2 and 3 focused on two atmospheric phenomena that are *global* in nature: the greenhouse effect and destruction of ozone. On the other hand, acid rain, another air quality issue, tends to be *regional* in character—what goes up comes down in the same general region of the globe. But acid rain does not respect state or national boundaries and this causes acid rain to be mixed with intense political controversies and accusations. Acidic gases that originate in the Midwest, especially in the Ohio River valley, are carried to the northeast by prevailing winds and fall as acidic rain or snow on New York, New England, and eastern Canada. They are also carried to the southeast, producing acidic precipitation in Tennessee and North Carolina. In Europe, acids generated in Germany, Poland, and the United Kingdom are carried northward into Norway and Sweden.

The problem is compounded by what may well be an environmental paradigm of our times. By trying to use a technological fix for one problem we inadvertently create another. Taller smokestacks were built to eject pollutants high into the atmosphere where they could spread and thus improve local air quality; in effect, an "out-of-sight-out-of-mind" policy. But whatever goes up must come down somewhere, and it does, sometimes hundreds of miles away. Thus, potential local pollution problems are converted into regional ones. Once again, we are all caught in the same web.

Chapter Overview

The quotations that begin this chapter probably raise a number of questions: What is acid rain? Is it the same as acid deposition? What exactly is an acid? What's the source of acid rain? How does acid rain damage trees, fish, and statues? And perhaps most important: How do emissions into the atmosphere contribute to acid rain? These are primarily scientific questions, and in the pages that follow we will provide information and some answers.

We begin by considering acids and bases and their properties, and then look at acidic and basic solutions to find the ions responsible for these properties. The concept of pH is first introduced as an indication of acidity levels and then the focus shifts to measuring the pH of rain and the results of such measurements. Precipitation with low pH values (high acidity) generally falls in those regions of the country where atmospheric concentrations of sulfur oxides and nitrogen oxides are high. These gases are traced to coal-burning power plants and gasoline-burning automobiles, and the processes by which the compounds are generated and exert their acidic properties are considered. Then, we turn to the effects of acid precipitation on materials, visibility, human health, lakes and streams, and trees and forests.

Interwoven with these largely scientific issues are the social and political factors that have made the public debate over acid rain "as sour as the rain itself." Controversy surrounds the interpretation of the effects of acid rain and the best ways to deal with these effects. We consider several strategies, each with its own price tag. The financial impact of acid rain and the money required to curtail it introduce important economic dimensions. One nationwide pollution solution is that offered by the 1990 federal Clean Air Act amendments. This legislation and its amendments have already reduced the concentrations of the acidifying oxides. A discussion of this legislation and its current and potential impact brings the chapter to a close.

6.1 What Is an Acid?

The subject of acid rain brings together the atmospheric pollutant gases introduced in Chapter 1 with water, the principal topic of Chapter 5. Quite obviously, we need to define acids to understand this linkage. Like many other chemical concepts, acids can be

Citrus fruits contain acids.

defined either in terms of observable properties or conceptually, using theories of atomic and molecular structure. Chemists usually use both types of definitions, and we shall do so here.

Historically, chemists identified acids by their common properties—sour taste, color changes with indicators, or reactions with carbonate-containing materials. Acids have a characteristic sharp or sour taste. Although taste is not a safe way to test for acids, you undoubtedly know the sour taste of vinegar and lemon juice, two common consumer goods containing acids. Other tests for acids rely on their chemical reactivity. A familiar example is the litmus test. Litmus, a vegetable dye, changes color from blue to pink in the presence of acids. Indeed, the litmus test is so well known that it has become a figure of speech in our culture to describe other kinds of tests. For example, a newspaper article might read: "The litmus test for this particular legislation . . . " Another simple chemical test is to add a suspected acid to a carbonate-containing material such as marble or eggshell. Fizzing occurs due to the release of carbon dioxide gas produced by the reaction of the acid with the carbonate. This confirms the presence of an acid. We will return to these reactions later, after our conceptual definition of an acid.

| This definition of acids as H^+ donors applies to non-aqueous solutions as well.

From the standpoint of chemical structure, **acids** are substances that release hydrogen ions, H^+, in aqueous solution. You will recall from Chapter 5 that an ion is any atom or group of bonded atoms that has a net electric charge. Atoms and molecules normally have equal numbers of protons (positive charges) and electrons (negative charges), so each atom or molecule is electrically neutral. But if electrons are gained or lost, the atom or molecule acquires a charge. A hydrogen atom consists of one electron and one proton. If the electron is removed, all that remains of the formerly neutral hydrogen atom is a proton. The proton is, in effect, a hydrogen ion with a positive charge, designated as H^+.

There is a slight complication with the definition of acids as substances that release H^+ ions in aqueous solutions. H^+ ions (protons) are much too reactive to exist by themselves. They always attach to something else, such as water. As a simple example of this, pure hydrogen chloride is a gas, made up of HCl molecules. But when dissolved in water, each HCl molecule donates a proton to an H_2O molecule, forming an H_3O^+ **(hydronium)** ion and leaving a Cl^- (chloride) ion.

| The (aq) notation represents a species dissolved in water.

$$HCl(g) + H_2O(l) \longrightarrow (H^+) \cdot H_2O + Cl^-(aq) \longrightarrow H_3O^+(aq) + Cl^-(aq) \quad (6.1)$$

The resulting solution is called hydrochloric acid and has the characteristic properties of an acid because of the presence of H_3O^+ ions. Chemists often write simply H^+, but they understand this to mean H_3O^+ in aqueous solutions. Thus equation 6.1 is typically shortened to

$$HCl(g) \longrightarrow H^+(aq) + Cl^-(aq) \quad (6.2)$$

The HCl is said to ionize or dissociate completely in water to produce hydrochloric acid. Essentially no intact, undissociated HCl molecules are left in solution.

6.1 Consider This: Are All Acids Harmful?

The word "acid" conjures up all sorts of pictures in the minds of many people, but there many acids in the foods we eat and the beverages we drink. Check the labels of processed foods and beverages. List some of the acids found in common foods and beverages. What do you think is the function of the acid in each food or beverage on your list?

6.2 Your Turn

Write chemical equations showing the dissociation of one hydrogen ion from one molecule of each of these acids.

a. HI (hydroiodic acid)
b. HNO_3 (nitric acid)
c. H_2SO_4 (sulfuric acid)
d. H_3PO_4 (phosphoric acid)

6.2 Bases

No discussion of acids is complete without mentioning bases (or alkalies), *the chemical opposites of acids.* For our purposes, we will define a **base** as any compound that produces hydroxide ions, OH⁻, in aqueous solutions. Bases have their own characteristic properties that are attributable to the presence of OH^-. They generally taste bitter and have a slippery feel in water solution. Common examples of bases are solutions of ammonia, NH_3, or sodium hydroxide (lye), NaOH. The cautions on a can of household drain cleaner (mostly lye) give dramatic warning that some bases, like some acids, can cause severe damage to tissue and textiles.

Solid sodium hydroxide is an ionic compound, a crystalline arrangement of Na^+ and OH^- ions. When it dissolves in water, the sodium and hydroxide ions separate from the crystal.

$$NaOH(s) \longrightarrow Na^+(aq) + OH^-(aq) \qquad (6.3)$$

The source of the OH^- ions produced in an aqueous solution of ammonia, NH_3, is a little less obvious until we note the following reaction.

$$NH_3(g) + H_2O\ (l) \longrightarrow \quad NH_4^+(aq) \quad + OH^-(aq) \qquad (6.4)$$
$$\text{ammonium ion}$$

In this case, an H^+ ion is transferred from a water molecule to an ammonia molecule to form an ammonium ion (NH_4^+), as well as a hydroxide ion.

Ammonia
NH_3

Lye (sodium hydroxide) NaOH

Baking soda (sodium bicarbonate) $NaHCO_3$

Some common household bases.

The ammonium ion, NH_4^+, is analogous to the hydronium ion, H_3O^+. Reaction 6.4 occurs only to a limited extent.

6.3 Your Turn

Write chemical equations showing the release of ions as each of these solid bases dissolves in water.

a. KOH   **b.** LiOH   **c.** Ca(OH)₂

6.3 Neutralization: Bases are Antacids

The reaction of an acid and a base is called **neutralization.** We illustrate this familiar process with hydrochloric acid and sodium hydroxide.

$$HCl(aq) + NaOH(aq) \longrightarrow NaCl(aq) + H_2O(l) \qquad (6.5)$$

In this case, the products of neutralization are sodium chloride and water. The corrosive acid and the caustic base offset each other chemically and are thus transformed into salt and water, a neutral compound (neither an acid or a base).

The key to any neutralization reaction is the combination of hydrogen ions from an acid and hydroxide ions from a base to form water molecules. We can emphasize this by writing the above reaction in ionic form.

$$H^+(aq) + Cl^-(aq) + Na^+(aq) + OH^-(aq) \longrightarrow Na^+(aq) + Cl^-(aq) + H_2O(l) \quad (6.6)$$

This equation indicates that when HCl and NaOH dissolve in water, they dissociate into their positive and negative ions. The same is true for the NaCl formed by the reaction. Thus, the Na^+ and Cl^- ions in the reactants are unchanged as products in the acid-base reaction; they simply remain in solution. Because they appear on both sides of the equation, they can be canceled from equation 6.6. The overall chemical result of reacting an acid and a base—a neutralization reaction—can be represented as:

$$H^+(aq) \quad + \quad OH^-(aq) \quad \longrightarrow H_2O(l) \qquad (6.7)$$
$$\text{(from an acid)} \quad \text{(from a base)}$$

Although other ions must be present in solution, this generic equation applies equally well to reactions involving a wide variety of acids and bases.

Recall from Chapter 5 (Section 5.11) that NaCl is an ionic compound.

In cooking we use an acid-base neutralization reaction when we put lemon juice, which contains an acid, on fish. The reaction neutralizes the ammonia-like compounds that give fish their characteristic odor.

6.4 Your Turn

Write chemical equations showing the reaction of each acid and base pair. First write the complete balanced equation, and then rewrite it in ionic form.

a. HBr(aq) and Ba(OH)$_2$(aq)
b. H$_2$SO$_4$(aq) and NaOH(aq)
c. H$_3$PO$_4$(aq) and Mg(OH)$_2$(aq)

Ans. a. $2\ \text{HBr}(aq) + \text{Ba(OH)}_2(aq) \longrightarrow \text{BaBr}_2(aq) + 2\ \text{H}_2\text{O}(l)$

$2\ \text{H}^+(aq) + 2\ \text{Br}^-(aq) + \text{Ba}^{2+}(aq) + 2\ \text{OH}^-(aq) \longrightarrow \text{Ba}^{2+}(aq) + 2\ \text{Br}^-(aq) + 2\ \text{H}_2\text{O}(l)$

$2\ \text{H}^+(aq) + 2\ \text{OH}^-(aq) \longrightarrow 2\ \text{H}_2\text{O}(l)$

or by cancelling the coefficient 2:

$\text{H}^+(aq) + \text{OH}^-(aq) \longrightarrow \text{H}_2\text{O}(l)$

Complete neutralization requires that the concentrations of H$^+$ and OH$^-$ ions must be equal in the solution. This is the case for pure water or for any neutral solution. Acidic or basic solutions also contain H$^+$ as well as OH$^-$ ions, but in unequal concentrations. A simple, useful, and very important relationship exists between the molarity of hydrogen ions (M_{H^+}) and the molarity of hydroxide ions (M_{OH^-}) in any water solution. When these two molarities are multiplied, they give a constant value of 1×10^{-14} at 25°C. In equation form, the relationship is

$$\text{molarity of hydrogen ions} \times \text{molarity of hydroxide ions} = (M_{H^+}) \times (M_{OH^-})$$
$$= 1 \times 10^{-14} \qquad (6.8)$$

Thus, when the H$^+$ molarity is high, the OH$^-$ molarity must be low, and vice versa. The acid or base character of any aqueous solution can be summarized simply as:

Acidic solution: H$^+$ concentration > OH$^-$ concentration;

Neutral solution: H$^+$ concentration = OH$^-$ concentration;

Basic solution: H$^+$ concentration < OH$^-$ concentration

In **acidic solutions** the concentration of H$^+$ ions is greater than that of OH$^-$ ions; $(M_{H^+}) > (M_{OH^-})$;

In **basic solutions** the concentration of OH$^-$ ions is greater than that of H$^+$ ions, $(M_{OH^-}) > (M_{H^+})$;

In **neutral solutions,** the concentration of H$^+$ ions equals that of OH$^-$ ions, $(M_{H^+}) = (M_{OH^-})$.

Knowing one of these molarities, we can use equation 6.8 to calculate the other one. For example, a rain sample with a $1 \times 10^{-5}\ M$ H$^+$ concentration would have an OH$^-$ concentration of $1 \times 10^{-9}\ $M.

$$(1 \times 10^{-5}\ M_{H^+}) \times (M_{OH^-}) = 1 \times 10^{-14}$$

$$\text{Solving for } M_{OH^-}: \quad M_{OH^-} = \frac{1 \times 10^{-14}}{1 \times 10^{-5}} = 1 \times 10^{-9}$$

Because $1 \times 10^{-9}\ M$ is smaller than $1 \times 10^{-5}\ M$, the solution is acidic.

The value for $(M_{H^+}) \times (M_{OH^-})$ is somewhat temperature dependent. It is 1×10^{-14} at 25°C.

In pure water or a neutral solution, a particular situation applies: the molarities of hydrogen and hydroxide ions are equal to each other, each at $1 \times 10^{-7}\ M$. Applying equation 6.8 we can see that $(M_{H^+}) \times (M_{OH^-}) = (1 \times 10^{-7})(1 \times 10^{-7}) = 1 \times 10^{-14}$.

6.5 Your Turn

Classify the following solutions as acidic, neutral, or basic at 25°C.

a. $M_{H^+} = 1 \times 10^{-4}$
b. $M_{OH^-} = 1 \times 10^{-6}$
c. $M_{H^+} = 1 \times 10^{-10}$

6.6 Consider This: Acidic, Basic, or Neutral?

Consider each of these aqueous solutions. List all of the ions present in order of decreasing concentration, starting with the most abundant. Classify each solution as acidic, basic, or neutral.

a. $Ca(OH)_2(aq)$
b. $HNO_2(aq)$
c. $H_2SO_3(aq)$
d. $(NH_4)C_2H_3O_2(aq)$

Ans. **a.** $OH^-(aq)$, $Ca^{2+}(aq)$, $H^+(aq)$ Note there are two hydroxide ions for every calcium ion. $Ca(OH)_2(aq)$ is classified as a base because the concentration of $OH^-(aq)$ is greater (by far) than the concentration of $H^+(aq)$.

What we need now is some way of being able to say how acidic or basic a solution is. As you will soon see, the pH scale is just such a tool. It relates the acidity of a solution to its H^+ concentration.

6.4 Introducing pH

The letters pH show up almost every day in articles about acid rain and in advertisements for shampoo, skin care products, and other consumer goods. To understand such articles and advertisements, it is necessary to understand the significance of pH. The notation, always a lower case "p" and an upper case "H" written together, stands for "power of hydrogen." In simplest terms, **pH is a number between 0 and 14** (occasionally lower or higher than these limits) **that indicates the acidity of a solution.** Instruments called pH meters can readily measure the pH of a sample.

The pH scale has a particular quirk. As the acidity *increases,* the pH number *decreases. The higher the H^+ concentration, the lower the pH.* When the pH decreases by one unit, the H^+ concentration *increases* by a factor of ten. For example, a rain sample with a pH of 4.0 is ten times *more* acidic than rain with a pH of 5.0. A solution of pH of 6.0 is only one-tenth as acidic as a solution with a pH of 5.0.

The broad sweep of pH values includes three types of solutions. Solutions with a pH of less than 7.0 are **acidic;** those with a pH of 7.0 are **neutral;** and those with a pH greater than 7.0 are described as **alkaline or basic.** The pH values of various common substances are given in Figure 6.1.

Notice from Figure 6.1 that normal rain is slightly acidic (pH between 5 and 6); *acid rain has a pH even lower than normal rain.* Correspondingly, this means that acid rain is more acidic than normal rain, hence the term acid rain. We will consider this relationship more closely in Section 6.5.

Proper pH is important in consumer products.

As H^+ conc ↑, pH↓;
as acidity↑, pH↓.

pH < 7.0, acidic;
pH = 7.0, neutral;
pH > 7.0, basic (alkaline)

Figure 6.1

The pH scale and the pH of common substances. The acidity decreases with increasing pH.

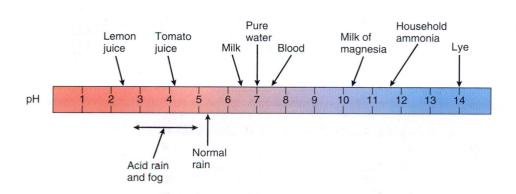

Ordinary drinking water is slightly acidic, with a pH of about 6. On the other hand, pure water has a pH of 7.0; it is neutral—neither acidic nor basic. So the obvious inference is that ordinary drinking water and normal rain are not pure H_2O. You will soon see what makes them acidic.

To many people, the term "acid" has a bad connotation, implying something dangerous and highly corrosive. To be sure, some acids (such as "battery acid" or sulfuric acid) do have such properties. But these acids have very high H^+ concentrations and very low pH values. You may be surprised to learn how many acids we eat, drink, and produce through metabolism. The naturally occurring acids in foods and metabolic by-products are much poorer H^+ donors than sulfuric acid, and often they contribute distinctive tastes. For example, vinegar contains acetic acid and has a pH of about 2.5. Apples (pH about 3.0) contain malic acid, and lemons (pH about 2.3) contain citric acid. Tomatoes are well known for their acidity, but in fact they are usually less acidic (pH about 4.2) than most fruits. Club soda has a pH of 4.8 because of the acids it contains, and the various cola soft drinks have a pH of about 3.1 because they contain phosphoric acid.

The pH of some other foods and household products: vinegar (2.5); sour pickles (3.0–3.4); oranges (3.0–4.0); black coffee (5.0); fresh egg white (7.6–8.0); detergents (10.5–10.8)

6.7 Your Turn

List vinegar, tomatoes, lemons, apples, Coca-Cola, pure water, and club soda in order of increasing acidity.

6.8 Consider This: On The Record

A legislator from a Midwestern state is on record as making an impassioned speech in which he argued that the environmental policy of the state should be to bring the pH of rain all the way down to zero. Assume that you are a legislative aide to this legislator and draft a brief memo to your boss on why his goal is incorrect.

The relationship between pH and hydrogen ion concentration (M_{H^+}) is an exponential one. For example, a solution in which $M_{H^+} = 0.0001 = 1 \times 10^{-4}$ has a pH of 4. The exponent (-4) equals the negative of the pH (4). This same pattern is repeated throughout the pH scale, as noted in Figure 6.2. A more detailed description of the relation between pH and M_{H^+} is given in Appendix 3.

Having established the pH scale as a measure of acidity, we now apply that concept to the nature of acid rain and its causes.

6.5 Measuring the pH of Rain

Before we describe the process for determining the pH of rain, it is important to realize that rain is only one of several ways that acids can be delivered to the surface and waters of the Earth. Snow obviously needs to be included, so **acid precipitation** is a more accurate description than acid rain. But even the term acid precipitation is

Figure 6.2

Relationship between pH and H^+ molarity. As the H^+ concentration decreases, the pH increases.

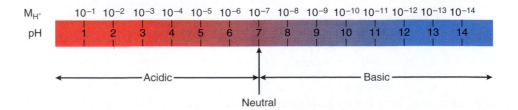

limited. A more inclusive term is acid deposition. **Acid deposition** includes fog, a cloud-like suspension of microscopic water droplets that is often more acidic and more damaging than acid rain. In addition, acidic deposition takes into account the action of acidic gases and the acidic solid particles that sometimes settle out on surfaces during dry weather. This "dry deposition" has been shown to be almost as important as the wet deposition of acids in rain, snow, and fog. It is also more difficult to measure.

The pH of a rain sample or any other solution is usually determined with a pH meter. This device includes a special probe capped with an H^+-sensitive membrane that is immersed in the sample. H^+ ions in a sample create a voltage across the membrane. The meter measures this voltage and converts it directly to pH, which is indicated on a dial or digital display (see photo, p. 241).

It is quite easy to measure the pH of rain samples, although certain precautions are necessary to obtain reliable results. The use of scrupulously clean containers is crucial, and the containers must be placed high enough to prevent "soil splash" that would contaminate rain samples collected at ground level. The pH meter must be calibrated carefully with solutions of known pH so that it will determine the correct pH.

Rain pH data have been collected at selected sites in the United States and Canada since about 1970. A more systematic study has been under way since 1978, with over 200 sites at which weekly samples are collected. The pH is measured immediately and then all samples are sent to a central laboratory for further analysis. Figure 6.3 was

pH measurements made at field laboratories, 1997

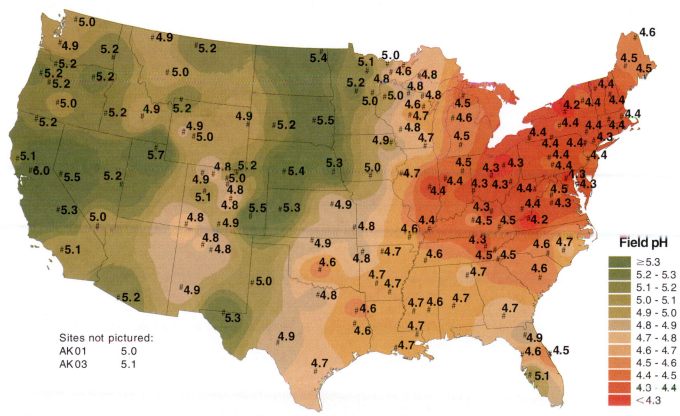

National Atmospheric Deposition Program (NRSR-3)/National Trends Network (1998)
http://nadp.sws.uiuc.edu/isopleths/oldpdf/phfield.pdf
NADP Program Office, Illinois State Water Survey, 2204 Griffith Dr., Champaign, IL 61820

Figure 6.3

Average annual pH of precipitation in the United States in 1997.

prepared from such data. It is a map showing the pH of precipitation during 1997. The different colored regions represent areas of the country with average pH values within a given range. Because this map contains a great deal of useful information, we will return to it several times in this chapter.

From the data of Figures 6.1 and 6.3, it appears that all rain is at least slightly acidic. At first thought, this seems surprising. If rain is pure water (as we tend to assume), we would expect it to have a pH of 7.0. But pure unpolluted rain always contains dissolved carbon dioxide, CO_2. Recall that CO_2 is a natural component of the Earth's atmosphere and its presence is essential for trapping solar energy close to the planet's surface. Carbon dioxide dissolves to a slight extent in water and reacts with it to form H^+ and HCO_3^- ions.

$$CO_2(g) + H_2O(l) \longrightarrow H^+(aq) + HCO_3^-(aq) \qquad (6.9)$$

The acid formed by the reaction of carbon dioxide with water is a poor H^+ donor; in water only about 1 out of 100 of the acid's molecules dissociate into ions, as depicted in equation 6.9. The remaining acid molecules are intact (undissociated).

But there are enough H^+ ions present to give a characteristic "tingle" to carbonated water ("soda" water), a saturated solution of CO_2 in water. Moreover, the hydrogen ions also contribute to the slightly acidic pH of even "pure" rainwater. At 25°C, a sample of water exposed to the normal atmospheric concentration of carbon dioxide has a pH of 5.6.

Figure 6.1 indicates that normal rain has a pH of about 5.3. It follows that CO_2 cannot be the sole source of H^+ in rainwater. Small amounts of other natural acids, including formic acid and acetic acid, are almost always present in rain and contribute to its acidity. However, even these acids cannot account for the fact that in many areas rain frequently has a pH significantly below 5.3 (Figure 6.3). We are now ready to try to find the source of this extra acidity.

6.9 Consider This: Acid Rain Across the Globe

Acid rain is of worldwide concern; the issues vary around the globe. A convenient way to check out the regional concerns is to search for documents on the worldwide web. **a.** Norway, Sweden, Germany, Mexico, and Canada are likely to have web sites dealing with acid rain. Use the web to search about acid rain in one of these countries or in another one of your choice. List the title and the URL for the site(s) you find. By what organization and for what purpose was the site posted? **b.** What acid rain damage is cited in the country you selected?

6.6 In Search of the Extra Acidity

According to Figure 6.3, the most acidic rain falls in the eastern third of the United States, with the region of lowest pH being roughly the states along the Ohio River Valley. The extra acidity must be originating somewhere in this heavily industrialized part of the country. Analysis of rain for specific compounds confirms that the chief culprits are the oxides of sulfur and nitrogen, which are sulfur dioxide (SO_2), sulfur trioxide (SO_3), nitric oxide (NO), and nitrogen dioxide (NO_2). These compounds are sometimes collectively designated as SO_x and NO_x and called "sox" and "nox."

If this interpretation of the origins of acid precipitation is correct, the geographical regions with the most acidic rain should also be heavy emitters of sulfur and nitrogen oxides. That relationship is generally confirmed by an examination of the maps in Figure 6.4. Emissions of sulfur dioxide are highest in regions where there are many coal-fired electric power plants, steel mills, and other heavy industries that rely on coal. Allegheny County, Pennsylvania is just such an area, and in 1990 it had the dubious distinction of leading the United States in atmospheric SO_2 concentration. Although power plants also generate nitrogen oxides, the highest NO_x emissions are generally found in states with large urban areas, heavy population density, and much automobile traffic. Therefore, it is not surprising that in 1990 the highest levels of atmospheric NO_2 were measured over Los Angeles County, the car capital of the country.

The circumstantial evidence linking acid precipitation with the oxides of sulfur and nitrogen appears compelling, but at this stage the Sceptical Chymist should be raising an important question. Given the definition of an acid as a substance that contains and releases H^+ ions in water, how can SO_2, SO_3, NO, and NO_2 qualify? None of these compounds even contains hydrogen! The objection is a sensible one. The explanation is that sox and nox react with water to release H^+ ions. Although they are not acids themselves,

Figure 6.4

(a) Variations of NO_2 emissions across the United States (1996). (b) Variations in SO_2 emissions across the United States (1996).

(a&b: Source: Data from EPA, *National Air Quality and Emission Trends Report, 1996.*)

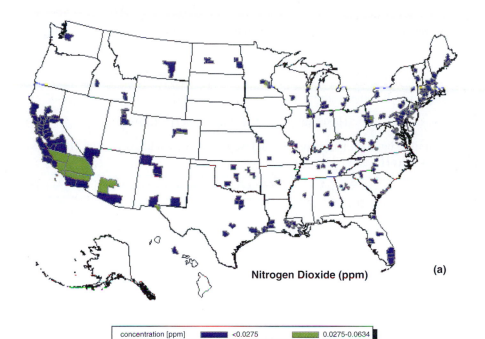

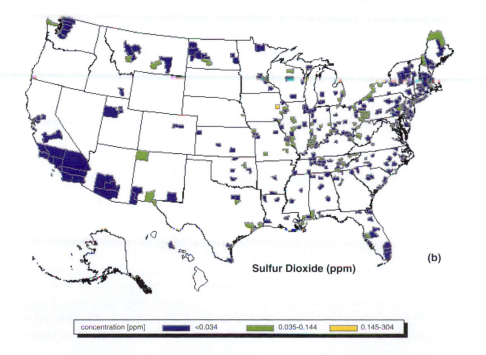

the oxides of sulfur and nitrogen are acid anhydrides, literally "acids without water." *When an acid anhydride is added to water, an acid is generated in solution.* For example, sulfur dioxide dissolves in water and reacts with the water to form sulfurous acid.

$$SO_2(g) \quad + H_2O(l) \longrightarrow H_2SO_3(aq) \tag{6.10}$$

sulfur dioxide   sulfurous acid

Reactions 6.10 and 6.11 are
analogous to the reaction of CO_2 with
water.

Similarly, sulfur trioxide reacts with water to yield sulfuric acid.

$$SO_3(g) \quad + H_2O(l) \longrightarrow H_2SO_4(aq) \tag{6.11}$$

sulfur trioxide   sulfuric acid

The sulfuric acid then dissociates to yield an H^+ ion and a hydrogen sulfate ion, HSO_4^-.

$$H_2SO_4(aq) \longrightarrow H^+(aq) + \quad HSO_4^-(aq) \tag{6.12a}$$

hydrogen sulfate ion

The hydrogen sulfate ion also dissociates into an H^+ ion plus a sulfate ion, SO_4^{2-}.

Not all of the HSO_4^- ions are
converted to SO_4^{2-} ions.

$$HSO_4^-(aq) \quad \longrightarrow H^+(aq) + SO_4^{2-}(aq) \tag{6.12b}$$

hydrogen sulfate ion   sulfate ion

The overall result is that sulfuric acid dissociates to yield two H^+ ions and a sulfate ion:

$$H_2SO_4(aq) \longrightarrow 2\ H^+(aq) + SO_4^{2-}(aq) \tag{6.12c}$$

In a similar, but a bit more complicated way, NO_2 can yield nitric acid, HNO_3, which dissociates into H^+ and NO_3^- ions.

$$4\ NO_2(g) \quad + 2\ H_2O(l) + O_2(g) \longrightarrow 4\ HNO_3(aq) \tag{6.13}$$

nitrogen dioxide   nitric acid

$$HNO_3(aq) \longrightarrow H^+(aq) + NO_3^-(aq) \tag{6.14}$$

nitrate ion

Now that we see how oxides of sulfur and nitrogen contribute to acid rain formation, we need to get a closer look at how these oxides are formed and released into the atmosphere.

6.7 Sulfur Dioxide and the Combustion of Coal

Thus far, this chapter has clearly established a relationship involving coal burning, atmospheric sulfur dioxide, and acid rain formation. Moreover, the fact that SO_2 and SO_3 react with water to yield acidic solutions is indisputable. What is not yet clear is why the combustion of coal should yield SO_2, the choking gas formed from burning sulfur (Figure 6.5). To answer this we need to know something about the chemical nature of coal. At first glance, coal appears to be just a black solid, not very different from charcoal or black soot, both of which are essentially pure carbon. When carbon is burned, it forms carbon dioxide and liberates large amounts of heat (which of course is the reason for burning it).

In ancient times sulfur was known as
brimstone, thus the biblical
admonition about "...fire and
brimstone...."

$$C\ (in\ coal)\ +\ O_2(g) \quad \longrightarrow \quad CO_2(g) \tag{6.15}$$

As you learned in Chapter 4, coal is quite complicated. No two samples have exactly the same composition. Although coal is not a pure chemical compound, we can approximate its composition with the formula $C_{135}H_{96}O_9NS$. In addition to these five elements, coal also contains small amounts of silicon and various metal ions such as sodium, calcium, aluminum, nickel, copper, zinc, arsenic, lead, and mercury. When coal is burned, oxygen reacts with *all* of the elements present to form oxides of those elements. Because carbon and hydrogen are the most plentiful, large quantities of gaseous CO_2 and H_2O are produced by the combustion. In addition, there is an unburned solid residue (ash) consisting of oxides of silicon, sodium, calcium, and the other trace ele-

(a)

(b)

Figure 6.5
(a) Pure water has a pH of 7.0.
(b) Burning sulfur in the same
flask as in (a) produces SO_2
gas, which dissolves in water
to produce sulfurous acid.
The resulting aqueous solution is acidic.

ments mentioned above. But the sulfur is our primary interest right now. The combustion reaction of sulfur from coal with oxygen yields sulfur dioxide, a poisonous gas with an unmistakeable acrid odor.

$$S \text{ (in coal)} + O_2(g) \longrightarrow SO_2(g) \tag{6.16}$$

Sulfur is present in coal because sulfur is present in all living things. Coal formed 100–400 million years ago from decaying vegetation. When plants decayed, the sulfur was left behind in the material that eventually became converted into coal. Coals from various parts of the world differ considerably in their sulfur content, but the combustion of almost all coals produces some sulfur oxides. This fact is central to the acid rain story. In large coal-burning electrical power generating stations and industrial plants, the sulfur dioxide goes up the smokestack (unless control measures are used) along with the carbon dioxide, water vapor, and various metal oxides. Once in the atmosphere, SO_2 can react with more oxygen to form sulfur trioxide, SO_3.

$$2 SO_2(g) + O_2(g) \longrightarrow 2 SO_3(g) \tag{6.17}$$

This reaction is fairly slow, but it is catalyzed (speeded up) by the presence of finely divided solid particles, such as the ash that goes up the stack along with the SO_2. Once SO_3 is formed, it reacts rapidly with water vapor or water droplets in the atmosphere to form sulfuric acid (equation 6.11). There are also a variety of other agents and pathways for the conversion of sulfur dioxide into sulfuric acid. Of particular importance are tropospheric ozone, hydrogen peroxide, H_2O_2, and the hydroxyl radical $\cdot OH$, which is formed from ozone and water in the presence of sunlight. The reaction of SO_2 with $\cdot OH$ accounts for approximately 20–25% of the sulfuric acid in the atmosphere. The reaction goes faster in intense sunlight, and thus is more important in summer and at midday.

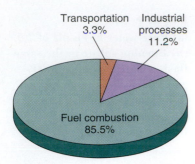

1995. SO_2 total emissions by source category

Figure 6.6

U.S. emission sources of sulfur oxides.

(Source: Data from EPA National Air Quality and Emission Trends Report, 1995 [issued in 1997]).

| Moles and molar masses were covered in Section 3.8.

We can use chemical calculations to better appreciate the vast quantities of SO_2 produced by a coal-burning power plant. Such a plant typically burns one million metric tons of coal a year:

1 metric ton = 1000 kg = 2200 lb; 1×10^6 metric tons = 1×10^9 kg = 1×10^{12} g. We will assume that the coal is low-sulfur coal containing 2.0% sulfur (2.0 g sulfur per 100 g coal). First we can calculate the grams of sulfur in one million metric tons of coal (1×10^{12} g):

$$1 \times 10^{12} \text{ g coal} \times \frac{2.0 \text{ g S}}{100 \text{ g coal}} = 2.0 \times 10^{10} \text{ g S}$$

Next, we use the fact that one mole of sulfur reacts with oxygen to form one mole of SO_2 (equation 6.16). Using molar masses, we determine that 1 mole of sulfur weighs 32.0 g, and 1 mole of SO_2 weighs 64.0 g, that is, 32.0 g + 2 (16.0 g). Therefore, 32.0 g of sulfur is converted into 64.0 g of SO_2.

$$2.0 \times 10^{10} \text{ g S} \times \frac{64.0 \text{ g } SO_2}{32.0 \text{ g S}} = 4.0 \times 10^{10} \text{ g } SO_2$$

This mass is equivalent to 40,000 metric tons of SO_2 (88 million pounds per year)! Plants burning high-sulfur coal emit twice this amount of SO_2 annually.

Although the largest source of sulfur dioxide in the United States is coal combustion for electrical power generation (70%), significant quantities of coal are also used in iron and steel production and other industrial processes (Figure 6.6). The large-scale production of nickel, copper, and certain other metals generates huge quantities of SO_2. The most common ores of these metals are sulfides, which are compounds of the metal plus sulfur. When heated to high temperatures in a smelter, the sulfides are decomposed and sulfur dioxide is released. The world's largest smelter, in Sudbury, Ontario, is used to convert nickel sulfide to nickel. The bleak, lifeless lunar landscape that was present in the immediate vicinity of the plant is mute testimony to earlier uncontrolled release of SO_2. Today, even with government controls, the Sudbury smelter emits about 2000 tons of sulfur dioxide per day from its 1250-foot smokestack. The fact that this is the world's tallest smokestack—equal in height to the Empire State Building—simply means that the emissions are carried farther away from Sudbury by the prevailing winds.

6.10 Your Turn

a. Assuming the composition of coal is represented by $C_{135}H_{96}O_9NS$, calculate the fraction and percent (by mass) of sulfur in the coal.

b. A power plant burns one million (1×10^6) tons of coal per year. Assuming the sulfur content calculated in **a**, calculate the number of tons of sulfur released per year.

c. Calculate the number of tons of SO_2 formed from this mass of sulfur.

d. Where does this SO_2 go, once it is released into the atmosphere?

Ans.   **a.** 0.0168 or 1.68%   **b.** 1.68×10^4 tons S (16,800 tons)

6.8 Nitrogen Oxides and the Acidification of LA

The combustion of coal has been indicted as a major environmental offender, contributing sulfur dioxide to the atmosphere and to acid deposition. But we know that SO_2 is not the only cause of acid precipitation and coal is not the only source. Another guilty party has been identified in California and other areas. The concentration of SO_2 in the smoggy air above the Los Angeles metropolitan area is relatively low, but so is the pH. For example, in January 1982, fog near the Rose Bowl in Pasadena was found to have a pH of 2.5. Breathing it must have been like breathing a fine mist of vinegar. This level is at least 500 times more acidic than normal, unpolluted precipitation and 10 times more acidic than required to kill all fish in lakes. And, in December 1982, fog at Corona del Mar, on the coast south of Los Angeles, was 10 times more acidic than that; it registered a pH of 1.5. In both cases, something other than sulfur dioxide was involved.

That acidic "other" is emitted by the millions of cars that jam the Los Angeles freeways day and night. The city literally runs on automobiles, and, as anyone who has ever visited there knows, the quality of the air can be poor. But it is not at all obvious why cars should contribute to acid precipitation. The compounds that make up gasoline blends are essentially all hydrocarbons (Section 4.8), and they are mostly converted to CO_2 and H_2O when burned. You recall from Chapter 1 that this conversion is not always complete, and that some CO and unburned hydrocarbon fragments escape in the exhaust. Nevertheless, gasoline contains almost no sulfur, and hence its combustion yields practically no SO_2. Consequently, we must look for another source of acidity.

Nitrogen oxides have already been identified as contributors to acid rain, but gasoline does not contain nitrogen, either. Therefore, logic (and chemistry) assert that nitrogen oxides cannot be formed from burning gasoline. Literally, that is correct—after all, you cannot make something out of nothing. Remember, however, that nitrogen is present in air. In fact, 78% of air consists of N_2 molecules. These molecules are remarkably stable and do not readily undergo chemical reactions under ordinary conditions. Thankfully, that is why nitrogen remains unchanged as we breathe it in and out of our lungs. Nevertheless, nitrogen can and does react directly with a few elements at high temperatures and pressures. One of these elements is oxygen. All that is needed is sufficient energy in the form of high temperatures or an electric spark. Under these conditions, the two elements combine to form nitric oxide, NO.

$$\text{Energy} + N_2(g) + O_2(g) \longrightarrow 2\,NO(g) \qquad (6.18)$$

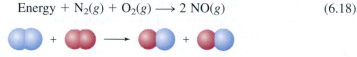

Heavy automobile traffic annually pumps millions of tons of nitrogen oxides into the atmsophere.

Because it is a mixture of nitrogen and oxygen, air is always a potential source for the production of nitric oxide. The energy necessary for the reaction can come from lightning bolts or the "lightning" that occurs in an internal combustion engine. In such an engine, gasoline and air are drawn into the engine cylinders and compressed to a high pressure. Moreover, the high pressure means that the nitrogen and oxygen molecules are closer together and thus even more likely to react. Then a spark from the spark plug ignites the gasoline, which burns rapidly. The energy released in this process is what moves the vehicle. But the unfortunate truth is that the energy also triggers reaction 6.18.

The reaction of N_2 with O_2 to form NO is not limited to lightning and the automobile engine. Chapter 2 mentioned the concern over NO production in jet aircraft engines. The same reaction occurs when air is heated to a very high temperature in the furnace of a coal-burning electrical power plant. Hence, such plants contribute vast amounts of both sulfur and nitrogen oxides to acidify precipitation. On a national basis, stationary sources (power plants) release more nitrogen oxides (49%) than mobile sources (motor vehicles—46%). But in urban environments, automobiles and trucks account for most of the atmospheric NO.

Air is 78% N_2. Therefore, NO is formed whenever air is used as an oxygen source in a combustion reaction.

A green chemistry solution to reducing NO emissions and energy consumption has been introduced into U.S. glass manufacturing by Praxair Inc. of Tarrytown, NY. The award-winning innovative technology substitutes pure oxygen for air in the large furnaces used to make glass. Switching from air (78% nitrogen) to pure oxygen eliminates NO production and also saves fuel by more efficient burning. Glass manufacturers using the Praxair Oxy-Fuel technology save enough energy annually to meet the daily needs of one million Americans.

Unlike nitrogen, nitric oxide is very reactive. In Chapter 2 you read that NO can react with ozone in the upper atmosphere, thus destroying the O_3. Close to the surface of the Earth, it reacts primarily with O_2 to form nitrogen dioxide, NO_2.

$$2\,NO(g) + O_2(g) \longrightarrow 2\,NO_2(g) \qquad (6.19)$$

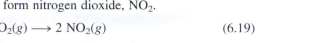

Several other oxides of nitrogen are formed from NO, with the most important being NO_2. It is a highly reactive, poisonous, red-brown gas with a nasty odor. For our purposes, its most significant reaction is the one that converts it to nitric acid, HNO_3. You saw one representation of that conversion in equation 6.13. Actually, a series of steps is involved in the chemistry that takes place in the urban atmosphere above Los Angeles and other sunny metropolitan areas. Unraveling this complex web of reactions has proved to be a fascinating scientific detective story. Sunlight is required, and volatile organic compounds, some released in the incomplete combustion of gasoline, are involved. An important intermediate is the hydroxyl radical, $\cdot OH$, which is formed in a reaction involving ozone, another common tropospheric pollutant. The hydroxyl radicals rapidly react with nitrogen dioxide to yield nitric acid.

$$NO_2(g) + \cdot OH(g) \longrightarrow HNO_3(l) \qquad (6.20)$$

As you have already read (equation 6.14), HNO_3 dissociates in water into H^+ and NO_3^- ions. The result is the alarmingly low pH values occasionally reported for Los Angeles rain and fog.

6.9 SO$_x$ and NO$_x$: Which Is Worse?

Now that we have identified the two major contributors to acid precipitation, it is reasonable to ask whether the oxides of sulfur or the oxides of nitrogen pose the greater problem. The annual U.S. anthropogenic emissions of SO_2 and NO_x are of roughly equal magnitude. Most (86%) of the sulfur dioxide emissions can be traced to coal-burning electric utilities (Figure 6.7). That same source accounts for 46% of the nitrogen oxides released, but transportation, powered by internal combustion engines, generates about 49% of the NO_x that enters the atmosphere from human sources.

Table 6.1 presents a global view of SO_2 and NO_x emissions from both natural and human sources. On this worldwide scale, human activities release almost twice as much SO_2 as NO_x. Furthermore, if only fossil fuel combustion is considered, the mass of nitrogen oxides emitted per year is less than 40% of the mass of sulfur dioxide. Unfortunately, reliable emissions information is difficult to obtain, and the data of Table 6.1 are somewhat out of date. According to *Vital Signs 94,* between 1980 and 1990, global SO_2 emissions from the burning of fossil fuels increased approximately 10% and NO_x emissions increased by 20%.

It is probably a significant indicator of things to come that during this period there was a major decrease in the mass of SO_2 generated by the industrialized nations. But this decrease has been more than offset by a massive increase in SO_2 emissions by the rapidly developing countries. For example, in 1970 the United States emitted about 30 million tons of sulfur dioxide and China emitted 10 million tons. In 1990, both coun-

Figure 6.7

U.S. emission sources of (a) sulfur dioxide and (b) nitrogen oxides. Data from EPA National Air Quality Trends report, 1995 (issued in 1997).

(Source: Data from EPA, *National Air Quality and Emission Trends Report, 1995,* [issued in 1997]).

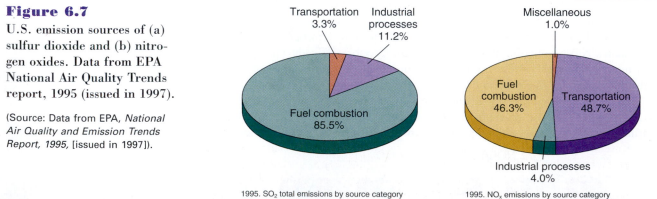

Table 6.1 Estimated Global Emissions of Sulfur and Nitrogen Oxides (in millions of metric tons per year)

Source	SO_2*	NO_x**
Natural:		
Oceans	22	1
Soil and plants	2	43
Volcanoes	19	
Lightning		15
Subtotals	43	59
Anthropogenic:		
Fossil fuels combustion	142	55
Industry (mainly ore smelting)	13	
Biomass burning	5	30
Subtotals	160	85
Totals	**203**	**144**

Sources: *Data from Spiro, *et al.*, "Global inventory of sulfur emissions with 1° × 1° resolution" in *Journal of Geophysical Research,* 97, No. D5, 6023, 1992.

**United States Environmental Protection Agency, *Air Quality Criteria for Oxides of Nitrogen,* EPA/600/8 – 91/049aA.

tries released about 22 million tons of SO_2. Thus far, the developing nations have been unable to afford the pollution-reduction technologies or low sulfur fuels that have been adopted by their more affluent neighbors. Nitrogen oxide emissions may pose an even more serious long-range problem. They are more difficult to control and appear to be increasing in most countries. Clearly, industrial development has had, and will continue to have, a major impact on both global economy and global environment.

It is also important to be aware that humans are not the only generators of sulfur and nitrogen oxides. As Table 6.1 indicates, oceans and volcanoes release large quantities of SO_2; soil, plants, and lightning are major sources of NO_x. In a typical year, natural emissions account for about 21% of the sulfur dioxide and 41% of the nitrogen oxides released into the atmosphere.

Occasionally, major geological events alter this pattern. The June 1991 eruption of Mount Pinatubo in the Philippines is a case in point. This eruption, the largest in a century, injected between 15 and 30 million tons of sulfur dioxide into the stratosphere. There the SO_2 reacted to form small droplets of sulfuric acid. For more than two years, much of this H_2SO_4 aerosol remained suspended in the atmosphere, reflecting and absorbing sunlight. The temporary drop in average global temperature that was observed in late 1991 and continued through 1992 has been attributed to the effects of the Mount Pinatubo eruption. Indeed, when the cooling effects of the Mount Pinatubo eruption are included in the computer programs used to model global temperature changes, the predictions agree well with the observations, thus validating the models. There is also evidence that droplets and frozen crystals of H_2SO_4 formed as a result of the eruption provided many new microsites for chemical reactions leading to the destruction of stratospheric ozone. Quite obviously, the topics of this text are tightly interwoven.

See Section 3.12 about global warming modeling.

6.10 The Effects of Acid Precipitation on Materials, Visibility, and Human Health

The evidence seems persuasive that much of the rain and snow in the United States is more acidic than normal, unpolluted precipitation. Fog, dew, and the bottom layers of clouds frequently have a pH of 3.0 or lower. And there are clear indications that, on a regional basis, the acidity of precipitation has increased significantly since the Industrial Revolution. But does it really matter? To answer that fundamental question, we need to know something about the effects of acid deposition and how serious they really are.

Figure 6.8

Acid rain can damage limestone statuary. This statue of George Washington was first put outside in New York City in 1944. During the next fifty-five years, acid rain caused significant damage to the statue.

In 1944   At present

Clearly, these issues are central to the acid rain debate. Scientific opinion about them is not unanimous, although a consensus is gradually emerging.

In an effort to gain the information necessary to make informed decisions, the U.S. Congress, during the 1980s, funded a major national research effort called the National Acid Precipitation Assessment Program (NAPAP). Over 2000 scientists were involved, with a total expenditure of $500 million. The project was completed in 1990 and the participating scientists prepared a 28-volume set of technical reports (NAPAP, *State of the Science & Technology,* 1991). Some of the material in the remainder of this chapter is drawn from the NAPAP reports plus other documents prepared for Congress. We first will consider possible damaging effects of acid rain on materials, visibility, and human health.

Limestone statues and monuments, such as those in the Gettysburg National Battlefield and New York City parks, are victims of irreparable pollution damage (Figure 6.8). Because they are made of limestone, which is calcium carbonate, $CaCO_3$, they slowly dissolve as the calcium carbonate reacts with H^+ ions in acid deposition.

$$CaCO_3(s) + 2\ H^+(aq) \longrightarrow Ca^{2+}(aq) + CO_2(g) + H_2O(l) \qquad (6.21)$$

Many other monuments and structures in the eastern United States are suffering similar fates. Some limestone tombstones are no longer legible. Even more serious is the fact that many priceless and irreplaceable marble and limestone statues and buildings are also being attacked by air-borne acids (Figure 6.9). The Parthenon in Greece, the Taj Mahal in India, the Mayan ruins at Chichén Itzá, and even the U.S. Capitol show signs of acid erosion. Visitors to the Lincoln Memorial in Washington learn that huge stalactites growing in chambers beneath the Memorial are the result of acid rain eroding the

Figure 6.9

Acid rain knows no geographical or political boundaries. Acid rain has eroded Mayan ruins at Chichén Itzá, Mexico.

marble. Much of the acid rain at these sites is due to NO_x produced by vehicular traffic, including tour buses.

Another damaging effect of acidic rain is the corrosion of metals, particularly iron, undoubtedly the most important structural metal. Buildings, bridges, railroads, and vehicles of all kinds depend on iron and steel. Unfortunately, iron will readily corrode or rust by undergoing a reaction with oxygen and water. The reaction requires hydrogen ions, but even in pure water there is sufficient H^+ concentration to promote slow rusting. In the presence of acid rain the corrosion is greatly accelerated. The role of H^+ is evident in equation 6.22, which represents the first of a two-step process. Iron (Fe) reacts with oxygen and hydrogen ions to yield Fe^{2+} ions.

$$4\ Fe(s) + 2\ O_2(g) + 8\ H^+(aq) \longrightarrow 4\ Fe^{2+}(aq) + 4\ H_2O(l) \qquad (6.22)$$

Then Fe^{2+} reacts with more oxygen to produce iron oxide, the familiar reddish brown material we call rust.

$$4\ Fe^{2+}(aq) + O_2(g) + 4\ H_2O(l) \longrightarrow 2\ Fe_2O_3(s) + 8\ H^+(aq) \qquad (6.23)$$

The net result, rust formation, is simply the sum of equations 6.22 and 6.23:

$$4\ Fe(s) + 3\ O_2(g) \longrightarrow 2\ Fe_2O_3(s) \qquad (6.24)$$

Because iron is inherently unstable when exposed to the natural environment, enormous sums of money are spent annually to protect exposed structural iron and steel in bridges, cars, ships, and other applications. Paint is the most common means of protection, although even paint degrades more rapidly when exposed to acid rain and acid gases. Another means of protection is to coat the iron with a thin layer of a second metal such as chromium (Cr) or zinc (Zn). Iron coated with zinc is called *galvanized iron*. There is widespread evidence that galvanized iron corrodes more rapidly in the presence of acid rain. As a consequence, galvanized structures must be replaced more frequently.

6.11 Your Turn

Show that the sum of equations 6.22 and 6.23 is the same as equation 6.24.

Automobile bumpers used to be coated with chromium metal (chrome plated) to protect against rusting. Reinforced plastic bumpers have largely replaced chrome-plated iron ones.

Car and truck paint can be affected by acid deposition, whether generated by SO_2 or NO_2, leaving etched spots or pits in the paint. To prevent this, automobile manufacturers have begun to use acid-resistant paints on new vehicles, adding approximately $5 cost per car or truck (about $61 million per year for all new vehicles). It is a bitter irony that trucks and automobiles—the very icon of vehicle-worshipping citizens of Los Angeles and elsewhere—create an air pollutant (NO_x) whose acidic by-products can attack the gleaming finish that many vehicle owners work so hard to maintain.

Less costly than rusting iron, but perhaps more obvious effects of acid deposition can often be observed by simply looking out of the window. Anyone living in the eastern half of the United States is familiar with the summer haze that usually clouds the landscape. (Ironically, you become more aware of it on the occasional really clear day when it does seem that you can see forever.) Travelers crossing the country by airplane can easily see the haze covering the east. And visitors to the Great Smoky Mountains National Park can view a prominent display of photographs showing reduced visibility in the mountains (p. 248). Power plants in the Ohio Valley are identified as the primary cause. This haze in the eastern part of the country has become steadily worse for several decades. It consists primarily of microscopic aerosol particles containing a mixture of sulfuric acid, ammonium sulfate [$(NH_4)_2SO_4$], and ammonium hydrogen sulfate (NH_4HSO_4). The haze is most pronounced in summer when there is more sunlight to accelerate the photochemical reactions leading to sulfuric acid, and it is particularly evident when the air is stagnant. As a consequence of this haze, average visibility in the east is now 25 miles or less and occasionally as low as 1 mile. By contrast, visibility in the western mountain states is typically 50–70 miles. It should be noted, however, that even the Grand Canyon is experiencing reduced visibility, probably as a result of SO_2 emissions from the huge Four Corners power plant in the northeast corner of Arizona.

Volatile *natural* hydrocarbons emitted by trees also contribute to the haziness in the Great Smoky Mountains.

A clear day and a hazy day in Great Smoky Mountains National Park.

20 DEAD IN SMOG;
RAIN CLEARING AIR
AS MANY QUIT AREA

Officials Study Cause of Plague
Apparently Borne by Heavy
Atmosphere in Donora, Pa.

A news headline about the deaths in Donora, PA in 1948.

Inhaling those same sulfate and sulfuric acid aerosols can cause illness and even death. The acidic droplets are deposited directly in the lungs where they attack sensitive tissue. The elderly, the ill, and those with pre-existing respiratory problems such as asthma, bronchitis, and emphysema are especially susceptible. One of the worst recorded instances of pollution-related respiratory illness occurred in London in December 1952. At that time, the English capital was still burning large quantities of sulfur-rich coal, much of it in home fireplaces. The deadly fog lasted five days and claimed approximately 4000 lives. In a similar incident in 1948, high tropospheric concentrations of sulfuric acid caused illness in 40% of the population of Donora, Pennsylvania, a coal-mining community in the western part of the state, and resulted in 20 deaths. To be sure, these were extreme and unusual situations. But the U.S. Environmental Program and the World Health Organization estimate that 625 million people are exposed to unhealthy levels of SO_2 released by burning fossil fuels.

Studies by the EPA have estimated that the reductions in SO_2 and associated acid aerosols pollution called for by the Clean Air Act of 1990 and its amendments could result in saving 12 to 40 billion dollars in health care costs. The savings would come principally from reduced costs to treat pulmonary diseases such as asthma and bronchitis, and the decrease in premature deaths caused by them.

Although acidic fogs are more immediately hazardous to health than is acid rain, there is growing concern over the indirect effects of acid precipitation. The solubilities of certain toxic heavy metals, including lead, cadmium, and mercury, are significantly increased in the presence of acids. These elements are naturally present in the environment, but normally they are tightly bound in minerals that make up soil and rock. Dissolved in acidified water and conveyed to the public water supply, these metals can pose serious health threats. Elevated concentrations of heavy metals have already been discovered in major reservoirs in western Europe.

6.11 Damage to Lakes and Streams

Healthy lakes have a pH of 6.5 or slightly above. As the pH is lowered below 6.0, various species are affected (Figure 6.10). Only a few species survive below pH 5.0, and at pH 4.0 a lake is essentially dead. When acidic precipitation falls on surface waters, it seems reasonable to predict that the waters will become more acidic. Numerous studies have reported the progressive acidification of lakes and rivers in certain geographic regions, along with reductions in fish populations. In southern Norway and Sweden, where the problem was first observed, one-fifth of the lakes no longer contain any fish, and half of the rivers have no brown trout. In southeastern Ontario, the average pH of lakes is now 5.0, well below the pH of 6.5 required for a healthy lake. In New England and the Adirondack Mountain region of northern New York, a puzzle has developed; the lakes remain as acidic as they had been, even though the rain and snow falling has become less acidic. Because of the significant reduction of SO_2 emissions from power plants along the Ohio River valley since 1990, the acidity of precipitation measured along that area and the eastern United States, particularly in the Middle Atlantic region, has decreased over the past decade.

Going westward, we find that many areas of the Midwest have no problem with acidification of lakes or streams, even though the Midwest is supposed to be the major source of the acids in acid precipitation. This apparent paradox can be explained quite simply. When acidic precipitation falls on a lake, the pH of the lake will drop (become more acidic) unless the lake or its surrounding soils contain bases that can neutralize the acid. The capacity of a lake to resist change in pH when acids are added is called its **acid neutralizing capacity (ANC).** The surface geology of much of the Midwest is mostly limestone (calcium carbonate, $CaCO_3$), which has a high acid neutralizing capacity, thus neutralizing the acid rain (equation 6.25).

$$CaCO_3(s) + 2\ H^+(aq) \longrightarrow Ca^{2+}(aq) + CO_2(g) + H_2O(l) \qquad (6.25)$$

But even more importantly, the lakes and streams have a relatively high concentration of calcium bicarbonate as a result of reaction of the limestone with carbon dioxide and water.

$$CaCO_3(s) + CO_2(g) + H_2O(l) \longrightarrow Ca^{2+}(aq) + 2\ HCO_3{}^-(aq) \qquad (6.26)$$
$$\text{calcium} \qquad \text{bicarbonate}$$

Calcium carbonate (limestone) is not soluble in water; calcium bicarbonate is soluble.

The bicarbonate ion accepts an H^+ ion, thus neutralizing acids:

$$HCO_3{}^-(aq) + H^+(aq) \longrightarrow CO_2(g) + H_2O(l) \qquad (6.27)$$
$$\text{bicarbonate}$$

Bicarbonate is also called hydrogen carbonate.

Because the added acid is consumed by this reaction, the pH of the lake will remain more or less constant.

Reaction 6.27 occurs when sodium bicarbonate (baking soda) is taken to neutralize excess stomach acidity. The CO_2 released produces a burp!

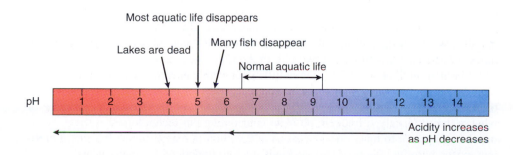

pH

Acidity increases as pH decreases

Figure 6.10
Aquatic life and pH. Recall that a change in one pH unit is a ten-fold change in H^+ concentration.

In contrast to the situation found in the Midwest, many lakes in New England and northern New York (as well as in Norway and Sweden) are surrounded by granite, which is a hard, impervious, much less reactive rock. Unless other local processes are at work, there is very little acid neutralizing capacity in these lakes. Consequently, they will most likely show a gradual acidification.

6.12 Consider This: The Rain in Maine (...or Texas or Alabama or...)

How acidic is the rain in your state? The answer depends both on the amounts of pollutants emitted into the air and their concentrations. EPA's Center for Environmental Information and Statistics (CEIS) can give you an estimate of the former. To obtain an environmental profile for a state, go to the CEIS web site (search or use the link at the *Chemistry in Context* web site). The CEIS provides a map of the United States on which you can select a state and follow directions to request emissions data for all counties in that state. When asked what data you would like to review, select "air quality." This will bring up a graph entitled "Tons of Criteria Air Pollution Emissions" for the state you selected.

a. For your state, which pollutant is emitted in the largest number of tons? Does this pollutant contribute to acid rain?

b. For NO_x and SO_2, record the values in 1986 and for the most current year given. What trend(s) do you observe? Suggest reasons for any increases or decreases.

c. Does the air in the state you selected blow in from neighboring states? If so, request the data for these states and see how their emissions compare.

There is experimental evidence that the adverse effect of increased acidity on fish populations is likely an indirect one involving aluminum ions (Al^{3+}). Aluminum is the most abundant metal and third most abundant element in the Earth's crust (after oxygen and silicon). Granite contains aluminum ions, and soil includes complicated aluminum-silicate structures. Natural aluminum compounds have very low solubility in water, but in the presence of acids, their solubilities increase dramatically. Thus, when the pH of a lake drops from say 6.0 to 5.0, the aluminum ion concentration in the lake may increase 1000-fold. When fish are exposed to a high concentration of aluminum ions, a thick mucus forms on their gills and the fish literally suffocate. Additionally, aluminum ions (Al^{3+}) react with water molecules to generate H^+ ions, increasing the acidity, which brings more aluminum ions into solution to further exacerbate the problem.

$$Al^{3+}(aq) + H_2O(l) \longrightarrow H^+(aq) + [Al(OH)]^{2+}(aq) \qquad (6.28)$$

> A shift from pH 6.0 to 5.0 is a ten-fold *increase* in acidity.

It turns out that analyzing the acidification of lakes is a good deal more complicated than simply measuring pH and acid neutralizing capacities. The source of the acid and the pH changes over time are often difficult to establish. Thus, although many reports claim that lakes are becoming more acidic, the evidence is frequently tenuous. We do not know with certainty how many acidic lakes have always been that way and how many have become acidic since the huge increase in SO_2 and NOx emissions beginning in the 1940s. The difficulties are twofold. In the first place, no one was aware of a possible problem until about 1970, so very little data were gathered prior to that time. Second, the equipment and procedures now routinely used for measuring pH and acid neutralizing capacity were not generally available until the 1970s. This is an example of what is, unfortunately, a frequent problem confronting our technological society. Without the equipment to make measurements and without a reason to suspect that a problem exists, harmful effects of various kinds may go unnoticed for many years.

> The effect of CFCs on stratospheric ozone is a good example of such a dilemma.

Liming lakes in southwestern Sweden Environment *40 (3), April 1998, p. 18.*

In spite of these complications, there is general agreement on some points. Approximately 10% of lakes and streams in the eastern United States appear to have been adversely affected by acid precipitation, with a corresponding reduction of aquatic life. Other lakes have low acid neutralizing capacities and are at risk of being damaged if present levels of acidic deposition continue unabated. But the words of Donald Uhart that started the chapter remind us that "...Ecosystems don't respond immediately. They take 10 or 20 years..." The already acidified and highly susceptible areas are mostly in the southern Adirondacks, New England, the forested mid-Atlantic highlands, and in the eastern Upper Midwest. A high percentage of lakes in northern Florida are acidic but the NAPAP studies have shown these to be naturally acidic and not a result of atmospheric deposition.

There is also some good news, in addition to the fact that acid deposition has decreased by 25% from Maryland to Maine since 1990. Evidence suggests that with the proper treatment many acidified lakes can be reclaimed. Recent experiments with adding calcium hydroxide, $Ca(OH)_2$, to acidified lakes are encouraging. Calcium hydroxide (lime) is a base, and it reacts to neutralize acid. If large-scale liming were carried out and further input of acids halted, many lakes could be restored to health.

Similar neutralization reactions also take place in the atmosphere. Calcium hydroxide, calcium carbonate, and other basic compounds of calcium, magnesium, sodium, and potassium are common in the soil. They are also abundant in the fly ash released from industrial smokestacks. Particles of these alkalis are swept into the air where they can react with droplets of acid. Thus, efforts to reduce air pollution by reducing particulate emission and controlling dust can lead to a decrease in the acid neutralizing capacity of the atmosphere and, ironically, a drop in the pH of precipitation. In fact, this may explain why the recent decrease in SO_2 emissions in this country has not resulted in a decrease in the acidity (increase in the pH) of rain.

6.13 Your Turn

a. Write an equation for the reaction of slaked lime, $Ca(OH)_2$, with hydrochloric acid, HCl.

b. Write an equation for the reaction that takes place when "quick lime," CaO, is added to water to produce $Ca(OH)_2$ ("slaked lime").

6.12 The Mystery of the Unhealthy Forests

Of all the ravages attributed to acid precipitation, the most widely discussed has been damage to trees and forests. Pictures of dead and dying trees provoked strong emotional responses and the ire of many environmentalists. But the fact is that the effects of acid deposition on trees are less well understood than any other aspect of the acid rain story. Consequently, the topic is fraught with controversy.

At least in certain cases, the reality of forest decline seems indisputable. The phenomenon was first observed in what was then the German Federal Republic (West Germany) in the 1960s, and extensive studies by German scientists have chronicled a steady decline. Fir and spruce trees at high elevations have been especially hard hit, first showing limp branches, then yellowing needles, and then loss of needles. This gradually spreads until the weakened trees finally are killed by drought, cold, insects, and winds. The decline was especially rapid from 1982 to 1985, when the number of damaged trees (defined as 10% needle loss) throughout West Germany increased from 8% to 52%.

Tree loss appeared to spread rapidly across western Europe during the 1980s, with damaged forests reported in Italy, France, the Netherlands, Sweden, Norway, and Britain. According to a 1993 study by the United Nations Economic Commission for Europe, almost one in four European trees has lost more than 25% of its leaves or needles. It is uncertain, however, when the reported damage actually began and how much is simply the result of better observing and reporting. For eastern Europe, where air pollution controls have been virtually nonexistent, information has only recently become available. Some reports claim that over 80% of the forests in eastern Germany, Poland, and the Czech Republic are damaged. In some areas of these countries, forests are said to have been totally destroyed by air pollution.

In North America, forest damage has been most dramatic in portions of the Appalachian Mountains. On Mount Mitchell and in the Great Smoky Mountains in North Carolina and Tennessee, you can now see the stark landscape of dead trees. In the Green Mountains of Vermont half of the red spruce trees on Camel's Hump died between 1965 and 1981. The sugar maple trees in New England and southeastern Canada, famed for autumn color and maple syrup, are claimed to be unhealthy, with some farmers reporting alarming decreases in maple syrup production.

It is tempting to blame acidic precipitation for all these effects. After all, the acids rain down directly on the trees, and the effects are generally most pronounced in regions with highly acidic precipitation. But the story is not that simple. It is difficult to prove cause-and-effect relationships, especially when there are many effects and many possible causes. In some cases (such as the New England and Canadian maple trees) weather-related stresses are considered responsible for most of the damage. In other regions, the cause may be insect infestations (as in the loss of the fir trees in the southern Appalachians) or tropospheric ozone and other air pollutants (as in the damage to pines in

Acid-rain damaged trees in the Smoky Mountains.

the San Bernardino Mountains). The dead trees on Mount Mitchell and other parts of the southern Appalachians are primarily Fraser firs that were killed by an infestation of the balsam woolly adelgid.

In mountainous regions losses are greatest at certain elevations, and only a few species of trees have shown complete die-back. Furthermore, careful field studies have shown that many of the effects are not as serious as had been believed. For example, although the famed Black Forest in Germany experienced a 10% needle loss, trees did not suffer the severe die-back that was widely reported in the press. The North American Sugar Maple Decline Project, initiated by the United States and Canada in 1987, has found that only 10% of sugar maples have experienced a 15% crown die-back and that this percentage actually decreased from 1988 to 1989.

Nevertheless, there is strong circumstantial evidence that acidic precipitation is at least a contributing factor to the declining health of trees. A careful study of tree growth rings in southern New Jersey pines showed a dramatic reduction in growth since 1965. This change, the greatest growth reduction in the 125-year record, correlates well with the acidity of nearby surface waters. Approximately 90% of the streams in the New Jersey Pine Barrens area are acidic, the highest percentage in the country. This may be an example of the synergistic effects of acids, ozone, SO_2, and NO_x.

According to this theory, which is gaining acceptance, the ozone and nitrogen oxides attack the waxy coating on leaves, permitting hydrogen ions to deplete nutrients. Acidification of soil beneath the trees mobilizes metals (especially aluminum) that attack the tree roots, thus preventing absorption of nutrients and water. Simultaneously, potassium, calcium, magnesium, and other minerals essential for plant growth are leached out of the soil. These effects leave the trees susceptible to destruction by natural factors such as disease, insects, drought, or high winds. A further factor in high elevation forests is that the mountains are often shrouded in fog, so-called cap clouds. As reported earlier in the chapter, cloud water often has a lower pH than even the most acidic rain, and the trees are perpetually exposed to an acidic mist. There is clear experimental evidence that acid deposition has contributed to the decline of red spruce at high elevations in the northern Appalachians by reducing the species' cold tolerance.

Whatever the complex causes and whatever the extent of involvement of acid rain, the damage to forests is unfortunate and expensive. However, it is important to realize that even in the case of severe decline and total die-back, most of the damaged regions have shown surprisingly good growth of new trees. The new growth appears to be thick and healthy. Nature does indeed regenerate itself.

New trees growing in a formerly acid rain–affected area.

6.13 Costs and Control Strategies

The acid rain debate has turned increasingly to economic considerations. Policymakers want to know the costs of the damage already incurred, the costs of cleanup, and the projected costs of future abatement. They also want to know who might benefit and, of course, who will pay. Unfortunately, it is extremely difficult to assess accurately the costs of the damage now occurring as a result of acid precipitation. Not only are the cause-and-effect relationships unclear, but many assumptions must be made. One such attempt was published in 1985 by Thomas Crocker, an economist, and James Regens, a natural resources scientist. They acknowledge that the only *conclusive* evidence for damage is to aquatic ecosystems. In spite of this caveat, Crocker and Regens estimate that for the eastern third of the United States, the maximum annual economic losses attributable to acid deposition exceed $5 billion. Others have concluded that all forms of air pollution in aggregate cost the United States over $40 billion each year in health care and lost productivity. And another study estimates $30.4 billion are lost annually because of sulfur dioxide damage to European forests. These huge numbers take on a more personal perspective when you consider another study, which reports that acid deposition does nearly $300 million worth of damage to buildings in Chicago each year, a cost that corresponds to $45 per resident.

During the 1980s, up to 10 pounds of SO_2 were released for every million BTUs of heat produced (1 BTU is equivalent to about 1 kJ). Phase I requires plants to meet a 2.5 lb SO_2 per million BTUs standard; Phase II lowers this to 1.2 lb SO_2 per million BTUs.

While efforts continue to gauge more accurately the current economic burden of acid rain, other studies are underway to evaluate strategies to control acid emissions and to estimate the costs of these measures. Most attention is being focused on ways to limit and eventually reduce the release of sulfur dioxide from coal-fired furnaces and nitrogen oxides from gasoline-fueled vehicles. The Clean Air Act Amendments of 1990 have made the reduction of SO_x and NO_x emissions a major national priority. That Act involves a two-phase restriction on SO_2 emissions from over 2000 fossil fuel–fired electrical power plants. The goal is to reduce SO_2 emissions by 10 million tons annually below 1980 levels. Phase I went into effect in 1995 affecting the dirtiest 110 plants in eastern and midwestern states. Phase II begins in the year 2000 with tightened SO_2 emission limits on these plants, and restrictions on remaining plants. The Act also calls for a 2 million ton reduction of NO_x emissions by the year 2000. A variety of techniques have been proposed, bearing a range of price tags.

See p. 29 for a photo of an automobile catalytic converter.

Reduction of nitrogen oxides is particularly challenging because most of the sources (motor vehicles) are small, individually owned, and, by design, mobile. And there are more than 200 million motor vehicles in the United States (about 1 billion worldwide). But it is chemically possible to reduce the NO emitted by these cars and trucks by fitting them with catalytic converters. We have already mentioned one of the functions of these catalysts: converting CO and unburned hydrocarbon fragments to CO_2. Other catalysts, typically in other parts of the catalytic converter, promote the reversal of the combination of nitrogen and oxygen that occurs in the engine at high temperatures. As the exhaust gases cool, there is a tendency for the NO to decompose into its constituent elements.

$$2\ NO(g) \longrightarrow N_2(g) + O_2(g) \tag{6.29}$$

Normally this reaction proceeds slowly, but the appropriate catalyst can significantly increase its rate and thus decrease the amount of NO emitted. An alternative, of course, is to eliminate the internal combustion engine altogether. One possibility would be to convert to electric-powered vehicles, so-called zero emission vehicles.

Electric-powered vehicles are discussed in Chapter 8.

6.14 Consider This: Electricity or Cars?

Your city has been warned by the EPA to cut sulfur dioxide and nitrogen oxide emissions or it will lose substantial federal funds. One citizen's group has advocated drastically reducing electricity consumption. A second group advocates reduced use of automobiles within the city limits. Choose one or the other position to support. After two groups are established and the research is done, a debate can be scheduled between the two groups. The goal is to come to a consensus of what action should be taken by the city.

The fact that most anthropogenic SO_2 comes from a limited number of point sources (coal-burning power plants and factories) makes the SO_2 problem easier to attack. Great strides have already occurred in reducing SO_2 emissions. Consider that in 1990, 10 million tons of SO_2 were emitted from coal-fired electrical plants. In 1995, the first year the 1990 Clean Air Act Amendments regulations took effect, a significant reduction of SO_2 from 445 coal-burning electrical utility plants occurred. The legislation set a 1995 target for allowed SO_2 emissions of 8.7 million tons. Only 5.3 million tons were released into the atmosphere from these sources, which represents a 47% reduction in SO_2 emissions from 1990 levels (Figure 6.11).

Three major strategies have been used to decrease SO_2 emissions: (1) switch to "clean coal" with lower sulfur content, (2) clean up the coal to remove the sulfur before use, and (3) use chemical means to neutralize the acidic sulfur dioxide in the power plant. We will briefly consider the effectiveness and the cost of each of these.

Coal switching is an option because coals vary widely in their sulfur content and their heat content. Anthracite or "hard" coal, found mainly in Pennsylvania, yields the greatest amount of energy and the smallest amount of sulfur dioxide per ton of fuel. But the anthracite supply is practically exhausted and more expensive. Bituminous or "soft" coal, which is abundant in the Midwest, has nearly the same heat content as anthracite but usually contains 3–5% sulfur. Western states have enormous deposits of low sulfur sub-bituminous coal and lignite (brown coal).

Coal cleaning is relatively easy to do and the technology is available. The coal is crushed to a fine powder and washed with water so that the heavier sulfur-containing minerals sink to the bottom. But the process removes only about half of the sulfur, and it is expensive—$500–$1000 per ton of SO_2 eliminated.

An alternative to using coal switching or coal cleaning is to chemically remove the SO_2 during or after combustion in the power plant. The chief method for doing this is called *scrubbing*. The stack gases are passed through a wet slurry of powdered limestone, $CaCO_3$. Calcium carbonate neutralizes the acidic SO_2 to form calcium sulfate, $CaSO_4$.

$$2\ SO_2(g) + O_2(g) + 2\ CaCO_3(s) \longrightarrow 2\ CaSO_4(s) + 2\ CO_2(g) \qquad (6.30)$$

Limestone is cheap and readily available. Although the process is highly efficient, installing scrubbers is expensive, so that the cost of this method has been estimated at $400–$600 per ton of SO_2 removed. Part of the expense is associated with the disposal of the $CaSO_4$ formed. We simply cannot avoid the law of conservation of matter. The sulfur must end up somewhere; either it goes up the stack as SO_2, or gets trapped as $CaSO_4$.

The carbon dioxide produced by reaction 6.30 adds to the concentration of atmospheric CO_2.

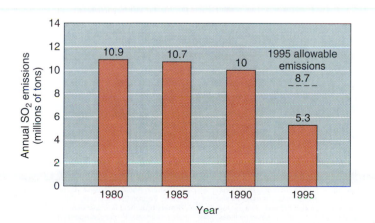

Figure 6.11

Reductions in SO_2 emissions.

(Source: Data from Environmental Protection Agency.)

6.15 Consider This: Excessive SO₂ Levels—An Action Plan

Imagine you live in a medium-sized city faced with a problem of excessive sulfur dioxide in the local air, in violation of state and federal regulations. The source of the problem is the local city-owned electrical power utility that burns high-sulfur coal obtained within the state. The three choices being considered by the city to rectify the situation are these.

1. Purchase low-sulfur coal from outside the state.
2. Install expensive scrubbers on the electrical power plant's smokestack.
3. Build a taller smokestack. The present smokestack is 150 feet tall. A smokestack of at least 300 feet would distribute the pollution from the power plant at a high enough altitude that prevailing winds would move it to another geographical location.

Take a position on this problem and prepare a presentation for the City Council outlining your position. Include in your presentation information on both the possible environmental and economic impacts of your position for the community.

The principal reason that compliance with the 1990 Clean Air Act Amendments regulations was achieved and even bettered was coal switching in which high-sulfur coal was replaced by low-sulfur coal. By the early 1990s, a new rail carrier and favorable railway tariffs made vast deposits of cheaper low-sulfur coal (even less than 1% S) in Montana and Wyoming available at costs lower than that for Midwestern or eastern low-sulfur coal. In 1991, western low-sulfur coal averaged just $1.30 per million BTUs; eastern low-sulfur coal was $1.60 to $1.70 per million BTUs. High-sulfur eastern coal cost $1.35–$1.55 per million BTUs. Given this price advantage, it is not surprising that nearly 60% of SO₂ reduction came from switching to low-sulfur western coal rather than using more expensive alternatives, such as scrubbing (Figure 6.12).

But there are hidden costs in this conversion to low-sulfur coal. It ignores the social and economic impact on the states that produce high-sulfur coal. It has been estimated that since 1990, coal switching has caused a 30% decline in employment in the high-sulfur coal–mining regions of Pennsylvania, Kentucky, Illinois, Indiana, and Ohio, although half of the drop has been because of automation and other market factors. Western states now produce nearly 33% of the coal mined in the United States, up from only 6% in 1970.

Figure 6.12

SO₂ compliance methods (percents).

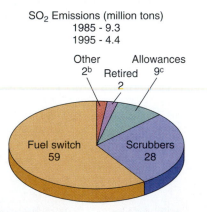

SO₂ Emissions (million tons)
1985 - 9.3
1995 - 4.4

Other 2[b]   Allowances 9[c]
Retired 2

Fuel switch 59

Scrubbers 28

[a] Does not include 174 substitution and compensating units.
[b] Includes switching to natural gas or petroleum and repowering.
[c] Nine percent of the 1995 SO₂ emissions reductions were at units that used allowances as their compliance method. The average sulfur content of coal consumed by these units was reduced by 16 percent from 1985 to 1995.

SO₂ = sulfur dioxide.
Note: Percent reductions of SO₂ emissions were computed using 1985 as the base year. (Source: **1985 Emissions:** U.S. Environment Protection Agency, *National Allowance Data Base*, Version 2.11 (January 1993). **1995 Emissions:** Acid Rain Division, U.S. Environmental Protection Agency.)

The shift to low-sulfur western coal has another side to it. Because the coal produces less heat per gram than eastern coals, power plants must burn more of it to generate the same amount of electricity, thus producing more CO_2 per unit of electricity generated. The increased carbon dioxide adds to the atmospheric burden and potential global warming. Mercury and other trace metals are more prevalent in western coal than in other coals, thereby increasing the atmospheric concentration of these metals unless steps are taken to remove them before they go up the smoke stacks (a costly proposition).

As the legislation that became the 1990 Clean Air Act Amendments moved through Congress, estimates were made about the costs that would be necessary for electrical utilities and other affected companies to comply with the proposed Act. The estimates varied widely (and wildly). The U.S. Congressional Office of Technology Assessment proposed a cost of between $3–$4 billion per year. Estimates of $4–$23 billion annually were made by the Electric Power Research Institute, an arm of the electrical utilities companies. In practice, the cost of acid rain controls has been much lower than estimated. Phase I costs to achieve compliance were actually less than 30% of the lowest estimated cost. In 1995, the first year for compliance, the annual cost for SO_2 reduction was $836 million, which works out to about $3 for each person in the United States.

Well in advance of the United States, Japan decided that SO_2 reduction was worth the cost. In 1968, the Japanese government issued strict SO_2 controls, and encouraged the use of low-sulfur fuels and desulfurization techniques. As a consequence, Japan's power plants cut their output of sulfur dioxide from almost 7 g/kWh in 1970 to less than 1 g/kWh in 1980. Quite obviously, the procedures work. Such dramatic decreases are related to the widespread use of scrubbers on Japanese power plants. In 1982, nearly 1200 scrubbers were in place, compared to about 200 in the United States. To be sure, Japan and the United States are different in many respects, including the fact that the U.S. electrical utilities chose coal switching rather than installing more expensive scrubbers. But the Japanese and United States experiences clearly indicate that the problems of reducing SO_2 emissions are political as well as technological.

6.14 The Politics of Acid Rain

It is not surprising that legislation to control acid rain has been the subject of intense political maneuvering. The neutralization of acid rain requires more than chemistry. For example, a few years ago the Indiana state legislature passed a bill requiring that a certain percentage of the coal burned in the state also had to be mined in Indiana. But the law was struck down after Wyoming coal producers filed suit against Indiana.

Nor is it surprising that the electrical power industry and the producers of high-sulfur coal, both strong and effective lobbies, have often resisted acid rain legislation and controls. Indeed, some electrical utilities representatives claimed that the effects of acid rain are grossly exaggerated. A typical response of the industry had been to call for more study rather than more regulations, arguing that it would be a grave mistake to spend billions of dollars before the causes and effects of acid deposition were fully understood. The costs and the risks of premature response were held to be too great for society to pay. On the other hand, environmentalists have argued that there is already sufficient evidence concerning acid deposition to pass the necessary legislation. They maintain that to delay would be to court disaster and much higher ultimate costs.

In 1990, the Congress of the United States entered into the controversy by passing major new environmental legislation. The Clean Air Act Amendments were signed into law by President Bush in November of that year. Although electrical utility companies initially opposed the new legislation, they soon realized the need to comply, and did so mainly through coal switching.

The 1990 legislation also sets up a unique system of "emissions trading" under which each company operates with a *permit* that specifies the maximum level of emissions that a company can legally release. Exceeding this maximum carries fines of up to $25,000 per day. Permits thus provide an environmental safety shield. In addition, companies are also assigned emission allowances that are set below the permit levels. The goal is for each company to achieve its individual allowance level. Some in fact perform better. A

Figure 6.13

Typical of Phase I states, Ohio retired mostly its own allowances. The band represents out-of-state allowances retired in 1996.

(Source: Data from EPA's Acid Rain web site: http://www.epa.gov/acidrain/effects/tradefx.htm)

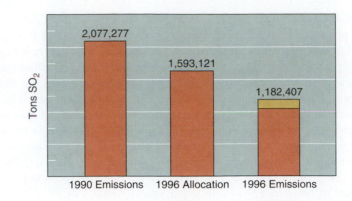

power plant with emissions below its allowance is assigned pollution credits, one credit per ton of SO_2. These credits can be sold to other power plants that have been unable to meet their emission allowances. There is thus a financial incentive for power producers to achieve significant reductions of acidic oxide emissions. On the other hand, the purchase of credits by those who cannot yet meet the more stringent standards allows them to continue operation, at or below the permit level, while the plant works to reduce emissions.

The first official trade of emissions credits under the provisions of the new law occurred in 1993. Since then, credits have been bought and sold in private transactions and at public auctions. There is even a commodity trading market in emission allowances on the Chicago Board of Trade. Prices have ranged between $400 (1993) to $68 (1996) per credit, nowhere near the $1000 per credit predicted by utility officials. Prices in 1998 were in the $90 to $100 range, which suggests that credits are not in as high demand as originally foreseen. This in turn indicates that progress has been made in overall reduction of SO_2 emissions, which it has, largely because of coal switching. Overall, companies have been able to achieve compliance without having to buy credits (Figure 6.13). Although some local areas of acid rain may persist, the net national air quality has improved. For example, Ohio emitted nearly 411,000 fewer tons of SO_2, a decrease of 46% from 1990 levels. Sulfur dioxide emissions in Indiana were down 37% from a decade ago.

Fears that home electrical utility bills would rise significantly because of the Clean Air Act Amendments have not been realized, even in eastern and midwestern states. Through a marriage of technology with economic forces, the average price charged by many large utilities for electricity has remained essentially constant over the past decade.

6.16 Consider This: Up for Auction

The year 2000 marks the eighth annual auction for sulfur dioxide allowances conducted for the EPA by the Chicago Board of Trade (CBOT). How have the sales been going? You can learn more about emissions credits at the web sites of both the EPA and the CBOT. For example, you can find recent information about allowance auctions and price trends at the EPA's site entitled, "Acid Rain Program." Do some detective work on the web and see if you can find out:

a. Are the allowances more costly or less costly this year than last?
b. How many allowances were auctioned last year?
c. Are most companies still achieving compliance without having to buy credits?

A more difficult web research question that you might want to consider is which emissions credits, if any, are now auctioned for pollutants other than SO_2. Some starting points for your search are provided at the *Chemistry in Context* web site.

Trading on the floor of the Chicago Board of Trade.

6.17 Consider This: Voting Record of Senators

Without consulting voting records or party affiliations, predict how senators from Kentucky, Wyoming, Texas, and Michigan might have voted on the Clean Air Act Amendments of 1990, and explain your reasoning. Then consult the Congressional Record to see how they actually did vote. If there is a difference between your expectation and the actual votes cast by these senators, offer some possible reasons for the difference.

6.18 Consider This: NAPAP Report

In January 1990, two of the nation's leading newspapers ran editorials about acid rain and the NAPAP study: the *Wall Street Journal* on January 26 and the *New York Times* on January 29. Obtain copies of these editorials (they are included in the Instructors Resource Guide that accompanies this text) and read them carefully. Compare and contrast the positions articulated. From this evidence, what can you infer about the general editorial policy of these two influential papers? Then either draft a letter to the editor of one of these newspapers, responding to the editorial, or write your own editorial on this topic.

Conclusion

If this chapter has taught anything, we hope it has been skepticism, prudence, and the recognition that complex problems cannot be solved by simple and simplistic strategies. "Acid rain" is not the dire plague described by many environmentalists and some journalists. Nor is it a matter to be ignored. It is sufficiently serious that federal legislation, the Clean Air Act Amendments of 1990, has been enacted to reduce SO_2 and NO_x emissions, precursors to acid deposition. The Act already has helped to clean the air.

Careful research has indicated that the damage caused by acid deposition, especially to trees and forests, is less severe than had been supposed. On the other hand, to fail to acknowledge the relationships involving the combustion of coal and gasoline, the production of sulfur and nitrogen oxides, and the reduced pH of fog and precipitation, is to deny some fundamental facts of chemistry. As is often the case with sharp and bitter controversy, the truth lies somewhere between the extremes. It is our task to discover it and to act accordingly.

One response that we as individuals and as a society might make to the problems of acid precipitation has hardly been mentioned in this chapter, yet it is potentially one of the most powerful. It is to conserve energy. Sulfur dioxide and nitrogen oxides are by-products of our voracious demand for energy—energy for electricity and energy for transportation. And, of course, carbon dioxide is an even more plentiful product. If our personal, national, and global appetite for fossil fuels continues to grow unchecked, our environment may well become a good deal warmer and a good deal more acidic, especially if developing countries fail to put into place environmental policies that restrict emissions from fossil fuel combustion. Moreover, the problem may be intensified as petroleum and low-sulfur coals are consumed and we become even more reliant on high-sulfur coal.

There are other sources of energy—nuclear fission, water and wind, renewable biomass, and the Sun itself. All of them are already being used, and their use will no doubt increase. We will explore nuclear fission and fusion in subsequent chapters. But we conclude this chapter with the modest suggestion that, for a multitude of reasons, the conservation of energy by industry and collectively by individuals could have profoundly beneficial effects on our environment.

Chapter Summary

Having studied this chapter, you should be able to:
- Define and apply the definitions of acids, bases, and neutralization (6.1–6.3)
- Use chemical equations to represent the dissociation of acids and bases (6.1–6.2)
- Describe solutions as acidic, basic, or neutral based on their pH or concentrations of H^+ and OH^- (6.3–6.4)
- Interpret pH values as being acidic, basic, or neutral (6.4)
- Describe acid rain (acid deposition) and factors causing it (6.5–6.8)
- Express the roles played by sulfur oxides and nitrogen oxides in causing acid rain, and describe regional variations (6.7–6.8)
- Discuss the contributions of man-made emissions of pollutants to the atmosphere linked to acid rain; compare these with natural emissions (6.7–6.9)

- Summarize the uncertainties associated with implicating acid rain as the cause for certain environmental degradation, that is, destruction of forests, and death of lakes (6.10–6.12)
- Express the effects and economic impact of acid rain on materials and the environment in general (6.13)
- Discuss the nature of the 1990 Clean Air Act and its amendments, and the impact they have had on SO_2 emissions (6.13–6.14)
- Outline the various alternatives proposed to control acid rain, noting the cost-benefit considerations to be made for each (6.13–6.14)
- Explain why coal switching was the method of choice used by electrical power companies (6.13)
- Explain why acid rain control is such a politically sensitive issue (6.14)

Questions

Emphasizing Essentials

1. **a.** What are the properties associated with acids?
   b. What is the structural feature that characterizes an acid?

2. **a.** Rewrite equation 6.1 using Lewis structures.
   b. Rewrite equation 6.1 using sphere representations.

3. Write a chemical equation showing the release of one hydrogen ion from a molecule of each of these acids.
   a. HBr (hydrobromic acid)
   b. $H_3C_6H_5O_7$ (citric acid)
   c. $HC_2H_3O_2$ (acetic acid)

4. **a.** What are the properties associated with bases?
   b. What is the structural feature that characterizes a base?

5. **a.** Rewrite equation 6.4 using Lewis structures.
   b. Rewrite equation 6.4 using sphere representations.

6. Write a chemical equation showing the release of ions as each solid base dissolves in water.
   a. RbOH
   b. $Ba(OH)_2$

7. Write a balanced chemical equation for each neutralization reaction.

   a. potassium hydroxide, KOH, with nitric acid, HNO_3

   b. barium hydroxide, $Ba(OH)_2$, with hydrochloric acid, HCl

8. Classify each of these aqueous solutions as acidic, neutral, or basic.

   a. $HI(aq)$

   b. $NaCl(aq)$

   c. $RbOH(aq)$

9. Classify each of these solutions as acidic, neutral, or basic. Give the criterion for your choice.

   a. $H^+ = 1 \times 10^{-8}\ M$

   b. $H^+ = 1 \times 10^{-2}\ M$

   c. $H^+ = 5 \times 10^{-7}\ M$

10. Formic acid, HCO_2H, is a natural component of rainwater. Write an equation showing how formic acid can contribute to the acidity of rainwater.

11. You just purchased a new car and are worried about whether its paint will be damaged by acid rain. Consult Figure 6.3 to find the data necessary to answer these questions.

    a. The atmosphere of which of these cities would be kindest to your car's paint in terms of acid rain—Chicago, Atlanta, Seattle, or San Francisco? Why?

    b. How does the average pH of rain for each of those cities compare with the average pH of rain in your location?

12. Write a balanced chemical equation for the reaction shown in Figure 6.5b.

13. The text states that the reaction of SO_2 with an · OH free radical accounts for about 20–25% of the sulfuric acid in the atmosphere.

    a. Write a balanced chemical equation for this reaction.

    b. Write the chemical equation in part **a** using sphere equations.

    c. What is the source of the · OH free radical in the atmosphere?

14. Give the formulas for the acid anhydride of each of these acids.

    a. carbonic acid, H_2CO_3

    b. sulfurous acid, H_2SO_3

    c. nitrous acid, HNO_2

15. Assume that coal can be represented by the formula $C_{135}H_{96}O_9N3$.

    a. What is the percent of nitrogen by mass in coal?

    b. If three tons of coal are burned completely, what mass of nitrogen in NO will be produced? Assume that all of the nitrogen in the coal is converted to NO.

    c. Will the mass of nitrogen in NO calculated in part **b** be the *maximum* amount of nitrogen in NO produced in this combustion reaction? Why or why not?

16. Acid rain can damage marble statues and limestone building materials. Write a balanced chemical equation to represent this destruction.

17. a. What does the phrase "pH balanced" imply on the label of a shampoo bottle?

    b. Will the presence of the phrase "pH balanced" influence your decision to buy a particular shampoo? Why or why not?

18. Consider the information in this graph.

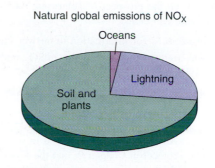

Natural global emissions of NO_X

    a. Approximately what percent of natural global emissions of NO_x are contributed by each source?

    b. To check the approximations made from the graph, calculate the percent contribution by using the data in Table 6.1. Compare these results with your estimations.

    c. The graph and Table 6.1 both convey the same information about the natural sources of NO_x emissions. What do you see as the strengths of each method as a means of communicating that information? Give a reason to support your opinions.

19. Figure 6.7a offers information about the percentage of SO_2 emissions from fuel combustion, mainly for electrical power production, and the percentage of SO_2 produced by transportation. Figure 6.7b offers information about the percentage of NO_x from fuel combustion and the percentage of NO_x produced by transportation. Is the relative importance of fuel combustion and transportation the same for emissions of both SO_2 and NO_x? Why might they differ?

20. Given that almost equal *masses* of SO_2 and NO_x are produced by human activities in the United States, how does their production compare based on a *mole* basis? Assume that all the NO_x is produced as NO_2.

21. About 13% of all global emissions of SO_2 and 21% of all global emissions of NO_x come from the United States.

    a. Suggest reasons why the U.S. percentage of global emissions is greater for NO_x than it is for SO_2.

    b. How do the percentages of U.S. SO_2 and NO_x emissions compare with the percentage of the world population that lives in the United States?

22. Calculate the mass of $CaCO_3$ (in tons) necessary to react completely with 1.00 ton of SO_2 according to the reaction shown in equation 6.30.

23. Cost estimates of reducing SO_2 emissions in the United States have varied considerably. This is one such estimate, stated in terms of percent reduction of emissions.

Emission Reduction, %	Cost (Billions of U.S. Dollars)
40	1–2
50	2–4
70	5–6

a. Prepare a graph to represent the relationship between the percent reduction in emissions and the cost.

b. Comment on the prospect of achieving 100% reduction, that is, zero emissions.

24. Consult Figure 6.12 for information to answer this question.

a. What has been the percent decrease in SO_2 emissions from 1985–1995?

b. Which strategy to reduce SO_2 emissions has been the most successful during this period?

c. What is meant by the term "allowances" in Figure 6.12?

25. **a.** The Clean Air Act has been discussed in this chapter, the Montreal Protocol in Chapter 2, and the Kyoto Accord in Chapter 3. What principal issue is each of these important pieces of legislation or international agreements trying to address?

b. Place each of these important legislative or international agreements, together with any significant amendments, on an appropriate time line. You may choose any format for the time line, but it should communicate the maximum amount of information.

Concentrating on Concepts

26. Judging by the taste, do you think there are more hydrogen ions in an equal volume of orange juice or in milk? Explain your reasoning.

27. The formula for acetic acid, the acid present in vinegar, is commonly written as $HC_2H_3O_2$. Many chemists write the formula as CH_3COOH.

a. Show that the two formulas both represent acetic acid.

b. What are the advantages and disadvantages of using each type of formula?

c. Draw the Lewis structure for acetic acid.

d. How many hydrogen atoms per acetic acid molecule are released as hydrogen ions? Explain.

28. Television and magazine advertisements remind us about the need for antacid tablets. A friend suggests that a good way to get rich quickly will be to market "antibase" tablets. Explain to your friend the purpose of an antacid tablet and offer some advice about the potential success of the antibase tablets.

29. In 6.6 Consider This you listed ions present in aqueous solutions of acids, bases, and common salts. Consider how the list changes if molecular substances are added to it.

a. List all molecular and ionic species in order of decreasing concentration in a 1.0 M aqueous solution of NaOH.

b. List all molecular and ionic species in order of decreasing concentration in a 1.0 M aqueous solution of HCl.

30. Which of these has the *smallest* concentration of hydrogen ions:

0.1 M HCl, 0.1 M NaOH, 0.1 M H_2SO_4, pure water? Explain your answer.

31. Explain why rain is naturally acidic, but not all rain is classified as "acid rain."

32. Do Figures 6.3 and 6.4 establish merely a *correlation* or a *causal relationship* between NO_x emissions and the average pH of rain? Explain your answer.

33. Ozone in the troposphere is an undesirable pollutant, while stratospheric ozone is beneficial. Does nitric oxide, NO, have a similar dual personality in these two atmospheric regions? Explain your thinking. *Hint:* Consult Chapter 2.

34. The mass of CO_2 emitted during combustion reactions is much greater than the mass of NO_x or SO_x, but there is less concern about the contributions of CO_2 to acid rain than from the other two oxides. Suggest two reasons for this apparent inconsistency.

35. The average pH of precipitation in New Hampshire or Vermont is relatively low, even though these states have low levels of vehicular traffic and virtually no industry that emits large quantities of air pollutants. How do you account for this low pH?

36. Consider these statements made at a public meeting discussing the issue of acid rain.

- Acid rain is a simple problem that can be solved by individuals.
- Acid rain is a somewhat straightforward problem that can be solved by industry alone.
- Acid rain is a somewhat complex problem but can be solved simply by banning all cars.
- Acid rain is a complex problem that cannot be solved by simplistic methods.

Explain which of these positions you support and which you do not, based on the information in this chapter.

37. **a.** Efforts to control air pollution by limiting the emission of particulates and dust can sometimes contribute to an increase in the acidity of rain. Offer a possible explanation for this observation.

b. In Chapter 2, stratospheric ice crystals in the Antarctic were involved in the cycle leading to the destruction of ozone. Is this effect related to the observations in part **a.**? Why or why not?

38. **a.** Several strategies to reduce SO_2 emissions are described in the text. The most effective ones in the last 10 years have been coal switching and stack gas scrubbing. Prepare a list of the advantages and disadvantages associated with each of these methods.

b. Explain why coal cleaning has not been an effective strategy.

39. Discuss the validity of the statement, "Photochemical smog is a local problem, acid rain is a regional one, and the enhanced greenhouse effect is a global one." Describe the chemistry behind each of these air quality problems and explain why the problems affect different geographical areas.

Exploring Extensions

40. The text makes this statement. "By trying to use a technological fix for one problem, we inadvertently create another."

a. Explain what problem associated with acid rain is the basis for this statement.

b. Pick another example connected with any of the issues explored in Chapters 1–5 that fits this statement. Briefly explain how your choice fits the statement and why you chose that example.

41. Each of these substances contains an OH group. Explain why you cannot write an equation similar to equation 6.3 for each substance.

   a. $Al(OH)_3$ *Hint:* Consult a solubility table.

   b. C_2H_5OH *Hint:* Consider bond energy and the nature of the bonds.

42. In 6.6 Consider This, you listed ions present in aqueous solutions of acids, bases, and common salts. In question 29 you added molecular substances to the list. To quantify this list:

   a. Calculate the molar concentration of all molecular and ionic species in a 1.0 *M* solution of NaOH.

   b. Calculate the molar concentration of all molecular and ionic species in a 1.0 *M* solution of HCl.

43. A representative of the Electric Power Research Institute, making a presentation in a workshop to establish research priorities and criteria on factors that govern precipitation chemistry, made this statement.

   "If whatever control strategy is hit upon is successful in cutting the acidity in half, an evil conspiracy of chemists will only allow the pH of precipitation to increase by 0.3."

   As a Sceptical Chymist in attendance at this workshop, how would you respond to hearing this statement? Explain the reasons for your response. *Hint:* See Appendix 3.

44. Equation 6.18 states that energy must be added to get N_2 and O_2 to react to form NO. A Sceptical Chymist wants to check this assertion and determine how much energy is required. Show the Sceptical Chymist how this can be done. *Hint:* Draw the Lewis structures for the reactants and products and then check Table 4.1 for bond energies.

45. 🖾 The text describes a green chemistry solution to reducing NO emissions for glass manufacturers.

   a. Identify the strategy.

   b. Use the web to research what other industries might use this green chemistry strategy. Write a report to summarize your findings.

46. The formation of rust is shown in equation 6.24.

   a. Find a different set of equations that represents the formation of rust. *Hint:* Try a general chemistry textbook.

   b. Which equations can be used more easily to explain why iron rusts more rapidly in moist, rather than dry, environments? Explain.

47. 🖾 How do researchers determine whether the negative effects of acid deposition on aquatic life are a direct consequence of low pH or the result of Al^{3+} released from rocks and soil? Find at least one web article that gives the details of such a study. In your own words write a summary of the experimental plan and its results.

48. One way to compare the acid neutralizing capacity (ANC) of different substances is to calculate the mass of the substance required to neutralize one mole of hydrogen ion, H^+.

   a. Write a balanced equation for the reaction of $NaHCO_3$ with H^+, and use it to calculate the ANC for $NaHCO_3$.

   b. Determine the cost to neutralize one mole of H^+ with $NaHCO_3$ if $NaHCO_3$ costs $9.50 per kilogram.

49. 🖾 Why is it that developing countries are more likely to emit an increasingly higher percentage of global SO_2? Pick one country and research the country's current emissions of SO_2 and calculate its percentage of global emissions. Speculate on whether this situation is likely to change in the future and offer an explanation for your prediction.

50. a. The 1990 Clean Air Act Amendments give electrical power generators emission reduction targets of 10 million tons of SO_2 by the year 2000. Do you predict that this target will likely be met? Explain the reasons for your prediction. *Hint:* Consult Figures 6.7 and 6.12 and question 24 to help with this prediction.

   b. How does the decrease in SO_2 emissions compare with that achieved in Japan? Suggest reasons for policy differences in these two nations.

The Fires of Nuclear Fission

An interior view of a nuclear power plant at St. Albans, France. Nuclear power plants produce nearly 80% of the electricity in France.

"Today, we Americans get more electricity from nuclear energy than from any other source, except coal. In many states, nuclear energy is the leading source of electricity. And because all the "lower 48" states are interconnected by power lines, the vast majority of Americans get some electricity from nuclear power plants. That means nuclear energy helps you enjoy the convenience of your microwave, the comfort of your electric blanket, and the fun of your CD player—even if there isn't a plant near your home."

The Nuclear Energy Institute, a Washington, D.C. nuclear energy trade group (1999)

A Greenpeace ship on a mission.

"The nuclear power industry is being squeezed out of the global energy marketplace for environmental and financial reasons. The promise of "Atoms for Peace" myth was never realized. Nuclear power is dirty, dangerous, uneconomic and has not provided any energy security or independence. Greenpeace believes that nuclear power should be phased out globally, starting with the immediate closure of the most dangerous reactors."

1998 position statement of Greenpeace, a worldwide pro-environment organization.

"It is the responsibility of our generation to take care of the waste generated by nuclear power. We have the technique needed to store the spent fuel, but we cannot get any farther if our method is not accepted by the public and the politicians. Our work is obstructed by those claiming to express major concern about the waste, whilst at the same time actively seeking to prevent us from gaining further knowledge."

Statement of scientists at the European Nuclear Society's 1998 conference

These quotes, reflecting very different outlooks on nuclear energy, raise a number of interrelated questions about nuclear energy (power) and nuclear waste: What is nuclear power? What is nuclear waste? Where does it come from? What is spent fuel? What is meant by long-term storage of nuclear waste? What is the connection between nuclear energy and nuclear waste?

The eight hemispheric-domed structures at this nuclear power plant in Ontario Canada are storage containers made of stainless steel and concrete for on-site storage of high-level nuclear waste from the nuclear reactor. The nuclear reactor is housed in the taller hemispheric-domed building.

7.1 Your Turn

Reread the quotes that open this chapter, paying special attention to terms and phrases related to nuclear energy and nuclear waste. Now make three lists of these terms or phrases, dividing them according to **(1)** those you feel you have a working knowledge of; **(2)** those you have some familiarity with, but are not quite as confident about, and **(3)** those you do not know. At the end of the chapter, you will revisit these lists in 7.21 Consider This.

Nuclear phenomena—probably no subject in all of physical science is more likely to provoke an emotional response. The word "nuclear" carries a tremendous baggage of disturbing associations, including the bombing of Hiroshima and Nagasaki, radioactive fallout from bomb tests, radiation-induced cancer and birth defects, the risks of accidents and meltdowns (Three Mile Island and Chernobyl), the difficulties of disposing of radioactive wastes, and the ultimate threat of nuclear annihilation. And yet, many benefits also spring from the nucleus, the very heart of matter—the production of electricity by nuclear power plants, the uses of radioactivity and other nuclear phenomena in medicine for the diagnosis and treatment of a wide variety of ailments, and the technological exploitation of nuclear materials in industry. The applications of nuclear phenomena, harmful at one extreme and beneficial at the other, present us with a dilemma of risks and benefits. It is a double-edged sword of Damocles that hangs precariously over our heads, described succinctly by Dr. Hans Blix, Director General of the International Atomic Energy Agency: "The dual challenge of the atom—to exploit in peace, not explode in war."

Certainly the largest, and arguably the most controversial, nonmilitary application of nuclear energy is the generation of electricity by nuclear power plants. When it was first demonstrated that electricity could be obtained from the splitting of atoms, a new age appeared to be dawning. The first commercial nuclear power generating station in this country was completed in 1957 at Shippingport, Pennsylvania, along the Ohio River near Pittsburgh. This new source held the promise of unlimited, cheap electricity. During the early 1960s, proponents of nuclear power were plentiful, including the former head of the U.S. Atomic Energy Commission who suggested naively that electricity produced by this method would be so inexpensive that it would be inconsequential to even meter consumers' use of it. There would be plenty of electricity for everyone! In spite of the optimism, the prediction of costless electricity has not come true. Although the percentage of electricity generated by nuclear power has increased since that time, its critics are numerous and outspoken.

In Greek legend, Damocles was courtier to a king, who made the courtier aware of the constant dangers associated with being king. At a banquet, he seated Damocles under a sword suspended by a single hair.

Chapter Overview

For nearly fifty years, many critical issues have arisen about using nuclear power to generate electricity, issues that have raised serious doubts. Citizens have asked and continue to ask important questions related to it: How does nuclear fission produce energy and how does a nuclear reactor produce electricity? What are the safeguards against a "meltdown"? Can a nuclear power plant explode like an atomic bomb? Is there a danger that nuclear fuel can be diverted to make nuclear weapons? What is radioactivity and what are the hazards associated with it? How long will nuclear waste products remain radioactive and how will such wastes be disposed? What is the current status of nuclear energy, nationally and internationally? What are the risks and benefits associated with nuclear power? And finally, how crucial is nuclear power to our future energy requirements? In this chapter, we address all of these questions. In every instance, we combine scientific fundamentals, application technology, and societal implications. Moreover, we try to temper emotionalism with understanding to help readers rationally weigh the risks and benefits of nuclear energy and radioactivity. We begin by considering a case study that will reappear throughout the chapter—a nuclear power plant in Seabrook, New Hampshire. But before we start, we ask you to consider your own position regarding nuclear power by completing 7.2 Consider This.

7.2 Consider This: Personal Opinion of Nuclear Power

Record your answers to the following questions about nuclear power. Save your answers because you will be asked to revisit them at the end of the chapter in 7.22 Consider This.

1. How does the electricity produced by a power plant differ from that produced by a coal-burning plant?
2. What is the greatest danger associated with nuclear power plants?
3. Given a choice between electricity generated by a nuclear power plant and a traditional coal-burning plant, which would you choose and why?
4. Would you be more willing to live near a nuclear plant or near a coal-burning plant? Give reasons to support your answer.
5. Would you support the burial of radioactive waste from a nuclear power plant in an appropriate site in your home state? Give reasons for your answer.
6. If you already have a position about the use of nuclear power in the United States, under what circumstances, if any, would you be willing to change your position?

7.1 The Seabrook Saga

In 1972, the Public Service Company of New Hampshire proposed building a nuclear power plant on the New Hampshire coast at Seabrook. Plans called for twin reactors, the first to become operational in November 1979 and the second to start operating two years later. Total costs for the project were estimated at $973 million. The site was selected for several reasons. Its location on the Atlantic Ocean offers easy barge access for transporting heavy equipment. Moreover, the ocean provides a supply of cooling water, which is necessary for the operation of any power plant that converts heat into work and electrical power. The underlying rock at Seabrook is a solid and stable foundation for vibration-sensitive machinery. Most importantly, Seabrook is within 40 miles of Boston and its more than 4 million people addicted to refrigerators, television sets, kitchen ranges, lights, and hundreds of other conveniences that require electricity.

But proximity to a population center also created problems. The proposed facility was met with prompt and vigorous opposition, and groups such as the Clamshell Alliance formed to combat construction of the plant. In January 1974, the Commonwealth of Massachusetts began legal action to block the project, and two years later the citizens of Seabrook voted to oppose the plant. People on both sides of the issue spoke with strongly held conviction.

At its core, the nuclear issue is a confrontation between corporate, technocratic domination and decentralized, community independence. The choice is closely linked to a broad spectrum of issues—to unemployment and high electric rates, the exploitation of Third World people and resources, to the plagues of nuclear armaments, environmental chaos, and our soaring cancer rates.

Harvey Wasserman, Organizer
Clamshell Alliance

Figure 7.1
Protesting the construction of the Seabrook, New Hampshire nuclear plant.

> There's no question that as time goes on even the people who opposed Seabrook will recognize its benefits to their region and their way of life. It will light the homes and run the factories of New England while emitting no greenhouse gases and while displacing 11 million barrels of oil every year.
>
> **Harold B. Finger, President**
> *United States Council for Energy Awareness*

Despite objections, construction of the nuclear facility began in July 1976. The first of several protests followed almost immediately. The largest protests occurred in April and May of 1977 when 2000 demonstrators occupied the site and 1400 were arrested (Figure 7.1).

The Seabrook project was plagued with other problems as well, and in 1984 the owners canceled plans to build the second unit. The initial and only reactor at the site was finally completed in July 1986. However, because of changing federal regulations, legal gambits, and the bankruptcy of the Public Service Company of New Hampshire, the reactor was not tested until June 13, 1989. This was 17 years from when the reactor was first proposed, and almost 10 years past its initial projected operational date. A major factor in the bankruptcy was the $6.45 billion price tag for the power plant—12 times the initial estimate for a single-reactor system.

The last hurdle for the Seabrook power station was cleared in March 1990, when the Nuclear Regulatory Commission voted to give the plant an operating license. At full capacity the plant generates 1160 megawatts of power, which equals 1160 million joules every second. A few pounds of uranium daily produces the same amount of energy that would consume 1,850,000 gallons of oil or 10,000 tons of coal. No carbon dioxide will be added to the atmosphere to contribute to the greenhouse effect, and no sulfur dioxide or nitrogen oxides will be released to create acid rain.

Although Seabrook has sophisticated safeguards to protect the environment and the nearby populace, feelings about the plant still run high. Whether the people of New Hampshire, Massachusetts, and the other New England states are winners or losers in this drama remains to be seen. In the pages that follow, we will attempt to assemble some of the evidence.

7.3 Consider This: *Boston Globe* Reporter

Imagine you are a reporter for the *Boston Globe* assigned to cover the anti-Seabrook demonstration of 1977. You have appointments to interview the protest leaders and then the officials of the Public Service Company of New Hampshire. List the questions that you will ask to obtain the information you need to write a balanced article. Make your questions as focused as possible so you can gain the maximum amount of information in the time available.

7.2 How Does Fission Produce Energy?

The key to answering this question is probably the most famous equation in all of the natural sciences, $E = mc^2$. This equation dates from the early years of the twentieth century and is one of the many contributions of Albert Einstein (1879–1955). The equation summarizes the equivalence of energy E and matter or mass m. The symbol c represents the speed of light, 3.0×10^8 m/s, so c^2 is equal to 9.0×10^{16} m²/s². The fact that this number is very large means that it should be possible to obtain a tremendous amount of energy, which we will soon consider, from a very small amount of matter, whether in a power plant or in a weapon.

For over 30 years, Einstein's equation was a curiosity. Scientists believed that it described the source of the Sun's energy, but as far as anyone knew, no one had ever observed on Earth a conversion of a substantial fraction of matter into energy. Then, in 1938, two German scientists, Otto Hahn (1879–1968) and Fritz Strassmann (b. 1902), discovered what appeared to be the element barium (Ba) among the products formed when uranium (U) was bombarded with neutrons. The observation was unexpected because barium has an atomic number of 56 and an atomic mass of about 137. Comparable values for uranium are 92 and 238, respectively. At first, the scientists were tempted to conclude that the element was radium (Ra, atomic number 88), which is a member of the same periodic family as barium. But Hahn and Strassmann were fine chemists, and the chemical evidence for barium was too compelling.

The German scientists were unsure of how barium could have been formed from uranium, so they sent a copy of their results to their colleague, Lise Meitner (1878–1968), for her opinion. Dr. Meitner had collaborated with Hahn and Strassmann on related research, but was forced to flee Germany in March 1938, because of the anti-Semitic policies of the Nazi government. When she received their letter she was living in Sweden. She discussed the strange results with her physicist nephew, Otto Frisch (b. 1904), as the two of them took a walk in the snow. In a flash of insight, the explanation became clear: under the influence of the bombarding neutrons, the uranium atoms were splitting into smaller atoms of lighter elements, such as barium. The nuclei of the heavy atoms were dividing, like biological cells undergoing **fission.**

That word from biology is applied to a physical phenomenon in the letter that Meitner and Frisch published on February 11, 1939, in the British journal *Nature.* In the letter, entitled "Disintegration of Uranium by Neutrons: a New Type of Nuclear Reaction," the authors state the following:

Lise Meitner.

> Hahn and Strassmann were forced to conclude that isotopes of barium are formed as a consequence of the bombardment of uranium with neutrons. At first sight, this result seems very hard to understand . . . On the basis, however, of present ideas about the behavior of heavy nuclei, an entirely different . . . picture of these new disintegration processes suggests itself . . . It seems therefore possible that the uranium nucleus . . . may, after neutron capture, divide itself into two nuclei of roughly equal size . . . The whole "fission" process can thus be described in an essentially classical way.

The letter is just over a page long, but it would be difficult to think of a more important scientific communication. Its significance was recognized immediately. Niels Bohr (1885–1962), an eminent Danish physicist, learned of the news directly from Frisch and brought it to the United States on an ocean liner several days before its publication. Within a few weeks of Meitner's and Frisch's letter in *Nature,* scientists in a dozen laboratories in various countries confirmed that the energy released by the fission of uranium atoms was that predicted by Einstein's equation. Lise Meitner's contributions to the discovery of nuclear fission have been honored by naming element 109 meitnerium. The other woman for whom an element is named is Marie Curie, an earlier nuclear pioneer (Section 7.7).

Energy is given off when an atom splits because the total mass of the products is slightly less than the total mass of the reactants. In spite of what you may have been taught, neither matter nor energy is *individually* conserved. Matter disappears and an equivalent quantity of energy appears as the former is converted to the latter. Alternately, one can view matter as a very concentrated form of energy, and nowhere is it more concentrated than in the atomic nucleus. Remember that an atom is mostly empty space. If the farthest distance of an electron from its nucleus formed a sphere half a mile in diameter, the nucleus would be the size of a baseball. Because almost the entire mass of an atom is associated with the nucleus, the density of the nucleus is incredibly high. Indeed, a pocket-sized matchbox full of atomic nuclei would weigh over 2.5 billion tons! Given the energy-mass equivalence of Einstein's equation, this means that the energy content of all nuclei is, relatively speaking, immense.

It is important to realize, however, that the nuclei of only certain elements undergo fission. Furthermore, not every atom of a fissionable element such as uranium is capable of splitting when struck by a neutron. Whether fission occurs depends on the relative number of protons and neutrons in the nucleus. Approximately 99.3% of uranium atoms consist of 92 electrons, 92 protons, and 146 neutrons. **The mass number of** each of these atoms is the sum of the number of protons and neutrons, 92 + 146, or 238, which is used to identify the isotope as uranium-238 (U-238). Naturally occurring uranium also contains about 0.7% uranium-235 (U-235), which is fissionable. These atoms contain 92 protons, 92 electrons, but only 143 neutrons (mass number = 92 + 143 = 235).

Physicists and chemists conventionally write the mass number as a superscript preceding the elementary symbol. The atomic number (the number of protons in the nucleus and hence its positive charge) is written as a subscript. Hence, uranium-238 is represented as follows:

Isotopes were discussed in Section 2.2.

$$\text{Mass number = number of protons + number of neutrons} \longrightarrow {}^{238}_{92}U$$
$$\text{Atomic number = number of protons} \longrightarrow 92$$

A wide variety of possible products or fission fragments can be formed when the nucleus of an atom of U-235 is struck with a neutron. One typical reaction is given by the equation below.

$$\tfrac{1}{0}n + {}^{235}_{92}U \longrightarrow \left[{}^{236}_{92}U\right] \longrightarrow {}^{141}_{56}Ba + {}^{92}_{36}Kr + 3\,{}^{1}_{0}n \qquad (7.1)$$

This nuclear equation makes use of the notation just introduced. Note that the subscript for a neutron (designated n) is 0, indicating zero charge. The superscript is 1 because the mass number of a neutron is one. In a balanced nuclear equation such as equation 7.1, *the sum of the subscripts on the left side equals that of the subscripts on the right side of the equation. Likewise, the sum of superscripts on each side of the equation must be equal.* Coefficients in a nuclear equation, such as the 3 preceding the neutron symbol in the products, are treated the same way as in chemical equations: The coefficient multiplies the term following it. In the particular case above, the coefficient indicates three neutrons. Where no coefficient is given explicitly, a 1 is understood. We can check the correctness of equation 7.1 by applying these rules.

Nuclear equations are similar but not identical to conventional chemical equations.

	Left	Right
Superscripts:	1 + 235 = 236	141 + 92 + (3 × 1) = 236
Subscripts:	0 + 92 = 92	56 + 36 + (3 × 0) = 92

7.4 Your Turn

Find the relevant atomic numbers from the periodic table to help write nuclear equations for the following fission reactions that occur when an atom of uranium-235 is struck by a neutron.

a. The conversion of U-235 to Ba-138, Kr-95, and neutrons.
b. The conversion of U-235 to an element with an atomic number of 52 and a mass number of 137, another element with an atomic number of 40 and a mass number of 97, and neutrons.

Ans: a. ${}^{1}_{0}n + {}^{235}_{92}U \longrightarrow {}^{138}_{56}Ba + {}^{95}_{36}Kr + 3\,{}^{1}_{0}n$

7.5 Your Turn

Strontium-90 (Sr-90) is a radioactive fission product that contaminated milk for some time after atmospheric nuclear bomb tests. It can be formed from the neutron-induced fission of U-235 in a reaction that also produces three neutrons and another element. Identify the other product element and write a nuclear equation for this reaction that occurs when U-235 is bombarded by a neutron.

Although the sum of the mass numbers of the particles on the reactant side of a balanced fission equation equals the sum of the mass numbers of the particles on the product side, the actual mass does in fact decrease slightly. As a consequence, the total potential energy of the product nuclei is less than the potential energy of the reactants, and the difference in the amount of energy is released. When atoms of U-235 split under neutron bombardment, about 0.1% or 1/1000th of the mass disappears and reappears as energy. We can use this information and $E = mc^2$ to calculate just how much energy would be produced by the fissioning of 1.0 kilogram (2.2 pounds) of U-235. Only 1/1000th of this mass is converted to energy. Therefore, the value for m that goes into the Einstein equation is $1.0 \text{ kg} \times 1/1000 = 1.0 \times 10^{-3}$ kg (or 1.0 g). As you already know, $c = 3.0 \times 10^8$ m/s. Substituting these values yields the following expression.

$$E = mc^2 = 1.0 \times 10^{-3} \text{ kg} \times (3.0 \times 10^8 \text{ m/s})^2$$
$$E = 1.0 \times 10^{-3} \text{ kg} \times 9.0 \times 10^{16} \text{ m}^2/\text{s}^2$$

Completing the calculation gives an energy answer in what appears to be unusual units.

$$E = 9.0 \times 10^{13} \text{ kg m}^2/\text{s}^2$$

$1 \text{ J} = 1 \text{ kg m}^2/\text{s}^2$

Although the unit kg m^2/s^2 may not look familiar, it is identical to a joule. Therefore, the energy released by the fissioning of 1.0 kg of uranium-235 is 9.0×10^{13} joules or 9.0×10^{10} kilojoules.

To put things into perspective, 9×10^{13} J is the amount of energy released by 33,000 tons (33 kilotons) of exploding TNT or 3300 tons (7.3×10^6 lb) of burning coal. It is enough energy to raise approximately 700,000 cars six miles into the sky, or turn 8.7 million gallons of water into steam. Yet, all of this comes from one kilogram of U-235, only one gram (0.1%) of which is actually transformed into energy.

All this energy is accessible because the fission of a uranium atom releases two or three neutrons; three neutrons are indicated in equation 7.1. Thus, there is a net production of neutrons. Each of these neutrons can strike another U-235 nucleus and cause it to split. The result is a rapidly branching and spreading chain reaction (Figure 7.2) that can, under certain circumstances, sweep through a mass of fissionable uranium in a fraction of a second. Such a chain reaction will occur spontaneously if a critical mass, about 15 kg (33 pounds), of pure U-235 is brought together in one place. But as you will soon see, the uranium in a nuclear power plant is far from pure U-235.

7.6 Your Turn

Earlier we reported that at full capacity, the Seabrook plant generates 1160 million joules of electrical energy per second. Calculate the total amount of electrical energy produced per day and the mass of U-235 actually converted to energy per day.

Hint: Start by calculating the quantity of energy generated not per second, but per day. Then use the equation $E = mc^2$, and solve for mass, m. Report your answer in grams.

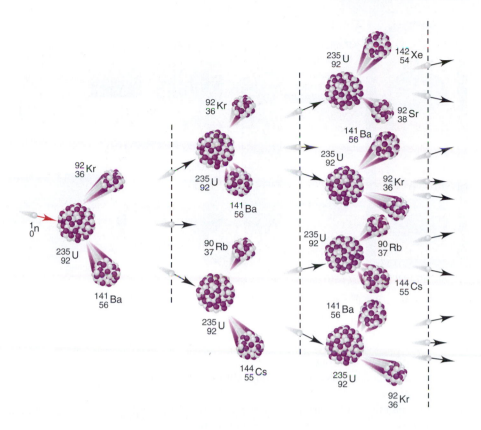

Figure 7.2
A chain reaction of U-235.

7.3 How Does a Nuclear Reactor Produce Electricity?

Chapter 4 includes a description of a conventional power generating station in which fuel such as coal or oil is burned to produce heat. The heat is used to boil water, converting it into a hot, high-pressure vapor that turns the blades of a turbine. The shaft of the spinning turbine is connected to large wire coils that rotate within a magnetic field, thus generating electrical energy. A nuclear power plant operates in much the same way, except that the water is heated not by fossil fuel, but by the energy released from the fission of nuclear "fuel" such as U-235. Like any power plant, a nuclear power plant is subject to the efficiency constraints imposed by the second law of thermodynamics. The theoretical efficiency for converting heat to work depends on the maximum and minimum temperatures between which the plant operates. This thermodynamic efficiency, typically 55–65%, is significantly reduced by other mechanical, thermal, and electrical inefficiencies.

A nuclear power station consists of two segments: a nuclear reactor and a nonnuclear portion. The latter contains the turbine and the electrical generator (Figure 7.3). The nuclear reactor is the hot heart of the power plant. It is housed in a special steel vessel within a separate reinforced concrete dome-shaped containment building.

The uranium fuel in the reactor core is in the form of uranium dioxide (UO_2) pellets, each about the size of a pencil eraser. These pellets are placed end to end in tubes made of a special metal alloy; about 200 pellets are in each tube. The tubes, in turn, are grouped into stainless-steel clad bundles (Figure 7.4). The rate of fission and the amount of heat generated by the fission are controlled using a principle employed in the first controlled nuclear fission reaction, which took place at the University of Chicago in 1942. Rods composed primarily of the element cadmium, an excellent neutron absorber, are interspersed among the fuel elements. Modern control rods also contain silver and indium. As long as these rods are in place, the reaction cannot become self-sustaining

See Figure 4.16 for the design of a power plant, and Section 4.13 for information on operating efficiencies.

Figure 7.3
Diagram of a nuclear power plant.

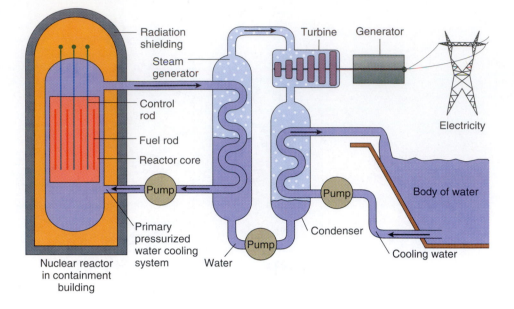

because insufficient neutrons are available. When the rods are withdrawn, the reactor "goes critical," but the rods can be rapidly adjusted to halt the chain reaction in the event of an emergency.

The fuel bundles and control rods are bathed in what is called the primary coolant. In the Seabrook reactor and in many others, the primary coolant is a water solution of boric acid, H_3BO_3. The boron atoms absorb neutrons and thus serve to control the rate of fission and the temperature. The water solution also serves as a "moderator" for the reactor, slowing the speed of the neutrons and making them more effective in causing fission. Of course, a major function of the primary coolant is to absorb the heat generated by the nuclear reaction. Because the solution is at a pressure of more than 150 times normal atmospheric pressure, it does not boil, but it is heated far above its normal boiling point. It circulates in a closed loop from the reaction vessel to the steam generators, and back again. This closed loop thus forms the link between the nuclear reactor and the rest of the power plant (Figure 7.3).

The heat from the primary coolant is transferred to the water in the steam generators, sometimes called the secondary coolant. At Seabrook, 33,000 gallons of water are converted to vapor each minute. The energy of this hot, pressurized gas is transferred to the blades of a turning turbine and to the attached electrical generator. The water vapor must then be cooled and condensed back into the liquid state before it is returned to the steam generator to continue its heat transfer cycle. In many nuclear facilities the cooling is done using large cooling towers, which are commonly mistaken for the reactors. The reactor is actually housed in a relatively small dome-shaped building (Figure 7.5). Moreover, cooling towers are not necessarily an indication of a nuclear power plant. Many fossil–fuel burning plants also use them.

The Seabrook facility does not have cooling towers because ocean water is used to cool the condenser. Each minute, 398,000 gallons of ocean water flow through a tunnel 19 feet in diameter and 3 miles long, bored through rock 100 feet beneath the floor of the ocean. A similar tunnel from the plant carries the water, now 22°C warmer, back to the ocean. Special nozzles distribute the hot water so that the observed temperature increase in the immediate area of the discharge is about 2°C. Because the primary coolant comes in contact with the steel-clad fuel elements, there is a possibility that the coolant may become radioactive. However, this boric acid solution is kept isolated in a closed circulating system, which makes the transfer of radioactivity to the secondary coolant water in the steam generator highly unlikely. Similarly, the tertiary cooling system does not come in direct contact with the

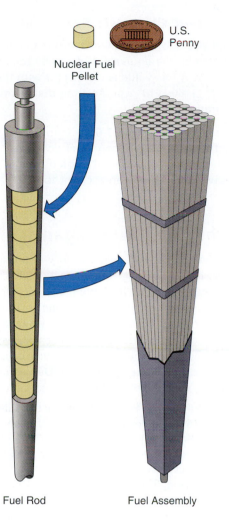

Nuclear Fuel
Pellet

U.S.
Penny

Fuel Rod

Fuel Assembly

Figure 7.4
Fuel pellet, fuel rod, and fuel
assembly making up the core
of a nuclear reactor.

secondary system, so the ocean water is well protected from radioactive contamination. It should be obvious that the electricity generated by a nuclear power plant is identical to the electricity generated by a fossil-fuel plant; the electricity is not radioactive, nor can it be.

Figure 7.5
Cooling tower and containment building at the Union Electric Callaway nuclear power plant. The nuclear reactor is in the domed containment building.

7.4 What Are the Safeguards against "Meltdown"?

A 1979 film called *The China Syndrome* told the story of a near-disaster in a fictitious nuclear power plant. The heat-generating fission reaction almost got out of control. If such a thing were to happen, the intense heat might cause a "meltdown" of the uranium fuel and the reactor housing. Fancifully, the underlying rock might even melt "all the way to China." But in spite of various human and instrumental errors, the safety features of the system worked in the film and fictional disaster was averted. Seven years later, the engineers of the very real Chernobyl power plant in Ukraine, which was then part of the Soviet Union, were less fortunate (Figure 7.6). During an electrical power safety test at Chernobyl reactor 4, the flow of cooling water to the core was deliberately interrupted as part of the test. The temperature of the reactor rose rapidly. The plant operators re-inserted the control rods according to procedure, but could not regain control of the runaway nuclear reaction due to faulty reactor design. Graphite, used to slow neutrons in the reactor to fissionable speeds, caught fire at these high temperatures. Although no nuclear explosion occurred, the fire was followed by a chemical explosion that figuratively shook the world. The blast blew off the 4000-ton steel plate covering the reactor and spewed radioactive products over a wide area (Figure 7.7). Although no nuclear explosion occurred, the effects included the generation of extremely high temperatures and a chemical fire and explosion that blew vast quantities of radioactivity from the reactor core. A major part of the plant was destroyed. The head of the crew on duty at the time of the accident has written: "It seemed as if the world was coming to an end . . . I could not believe my eyes; I saw the reactor ruined by the explosion. I was the first man in the world to see this. As a nuclear engineer I realized the consequences of what had happened. It was a nuclear hell. I was gripped with fear." (*Scientific American,* April 1996, p. 44.)

Chernobyl is the Russian spelling; Chornobyl is the more accurate Ukrainian usage.

Figure 7.6
Chernobyl, in Ukraine of the former Soviet Union.

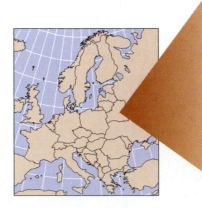

In the aftermath, approximately 190 people fell ill as a result of acute radiation sickness, and 31 of them died as a direct result of the accident. Nearly 150,000 people living within 60 kilometers of the power plant were permanently evacuated after the meltdown. An estimated 260,000 people have been exposed to levels of radiation that could ultimately shorten their lives. Included in this figure are 30,000 "liquidators," people who buried the most hazardous wastes and constructed a 10-story concrete structure ("the sarcophagus") to surround the failed reactor. It will be extremely difficult to accurately determine just how many have died as an indirect result of the Chernobyl disaster. Yuri Shcherbak, an epidemiologist and Ukraine ambassador to the United States, suggests that about 30,000 have died as an indirect result of the accident.

The Chernobyl zone continues to be among the most highly radioactive places on Earth. The exploded reactor and its nuclear fuel are entombed in concrete, but three other reactors of similar design continue to operate at the Chernobyl site. In response to international pressure, Ukraine has agreed to shut down the entire complex by 2000, but only if an alternative source of electrical energy is in place. The remaining three Chernobyl reactors produce about 5% of Ukraine's electricity, 40% of which is generated by nuclear power. The Ukrainian government estimates that a replacement gas-burning power station and associated expenses will require $4 billion. Thus far, Western countries have pledged $2.3 billion towards the project, and the United States is providing financial and technical assistance to establish an international nuclear safety and environmental research center near Chernobyl. But there are many hidden costs. A study by a Russian economist estimates the total cost of the Chernobyl meltdown at $358 billion, a figure that includes the expense of the clean-up and the loss of farm production.

Figure 7.7

The Chernobyl reactor 4 after the explosion.

7.8 Consider This: Chernobyl's Legacy

The Why? Files, a web site of the National Institute for Science Education, specializes in giving the science behind the news. If you have questions about Chernobyl, you can find answers as well as information that will raise many more questions. Check out The Why? Files on the web, either by doing a search or by using the direct link at the *Chemistry in Context* web site. Find the news story about Chernobyl, entitled "Radiation Reassessed." What happened back in 1986? Review the photos on the web and read through the scenario. News about the survivors' exposure to low-level radiation is still coming in today. What are the latest reports?

This recounting of the solemn facts concerning the Chernobyl tragedy leads to an inevitable question: "Could it happen here?" America's closest brush with nuclear disaster occurred in March 1979, when the Three Mile Island power plant near Harrisburg, Pennsylvania, lost coolant and only a partial meltdown occurred. There were no fatalities and no serious release of radiation. In spite of the initial failure, the system held and the damage contained. Since then, refinements in design and safety have been made to existing reactors and those under construction. Nuclear engineers agree that no commercial nuclear reactors in the United States have the design defects that led to the Chernobyl catastrophe.

The Seabrook nuclear power plant has been hailed as an example of state-of-the-art engineering, with multiple safety features. The energetic heart of the station is the 400-ton reaction vessel with 44-foot high walls made of eight-inch carbon steel. It is surrounded by a reinforced concrete, dome-shaped containment building, a feature the Chernobyl plant did not have, but all reactors operating in the United States must have. As the name suggests, this structure is built to withstand accidents of natural or human origin and prevent the release of radioactive material. It is clearly visible in the photograph (Figure 7.8). The inner walls of the building are 4.5 feet thick and made of steel-reinforced concrete; the outer wall is 15 inches thick. Information supplied by North Atlantic Energy Service Corporation (NAESCO), the company that manages the

Unlike at Chernobyl, an individual at the gates of the Three Mile Island plant at the time of the accident would have been exposed to less radiation in two weeks than that received from a single chest X-ray.

Figure 7.8

Seabrook nuclear power plant. The dome is part of the containment building, which houses the reactor.

Seabrook station, states that the containment building is constructed to withstand hurricanes, earthquakes, 360-mph winds, and the direct crash of a United States Air Force FB-111 bomber.

7.5 Can a Nuclear Power Plant Undergo a Nuclear Explosion?

This is a question many people ask. The devastation and destruction that atomic bombs brought to Hiroshima and Nagasaki are painfully etched in the memory of anyone who has even seen the pictures of those cities and their survivors. Therefore, it is reassuring that the answer to the question is "No."

Obviously, the purposes of a nuclear power plant and a nuclear weapon are not the same. Correspondingly, the desired rates of their reactions are very different. A nuclear power plant requires a slow, controlled energy release; in a nuclear weapon, the release is rapid and uncontrolled. In both cases, the fuel is U-235 and the reaction is essentially the same, with one important difference. Commercial nuclear power plants typically operate using uranium that is about 3–5% of the fissionable U-235 isotope and 95–97% U-238. Most of the neutrons given off by fissioning U-235 nuclei are absorbed by atoms of U-238 and elements such as cadmium and boron. As a consequence, the neutron stream cannot build up enough to establish a spontaneously explosive chain reaction, such as that in a nuclear fission bomb.

As we have noted, a spontaneously explosive chain reaction will occur only if about 33 pounds of highly purified U-235 are quickly brought together in one place. Fortunately for our troubled world, it is not easy to prepare pure U-235. The separation of this fissionable isotope from the nonfissionable U-238 that makes up 99.3% of naturally occurring uranium requires extensive and expensive processing. Chemical reactions cannot help much in this process because chemical properties are determined by the number and arrangement of the electrons in an atom. Because all the isotopes of any given element, such as uranium, are identical in this respect, they all have essentially identical chemical reactivity.

For more than four decades, uranium isotopes were separated by gaseous diffusion at the Oak Ridge National Laboratory in Tennessee. The method, developed during World War II, takes advantage of the fact that lighter molecules, on average, move faster than heavier molecules. A uranium sample is first reacted with fluorine to form uranium hexafluoride, UF_6, a yellow liquid that boils at 56°C. About 99.3% of the UF_6 molecules contain U-238 atoms and have a molecular mass of 352 [238 + 6(19) = 352]; 0.7% of the molecules contain U-235 and have a molecular mass of 349 [235 + 6(19) = 349]. The process is carried out at a temperature above 56°C, so that all of the UF_6 is in a gaseous state. The average molecule containing a U-235 atom moves only about 0.4% faster than the average molecule containing U-238. But if the diffusion is allowed to occur over and over through a long series of permeable barriers, significant separation of the fissionable and the nonfissionable isotopes can be achieved. Other separation methods, including centrifugation of UF_6 molecules, have also been developed. Inspectors attempting to determine a nation's nuclear capabilities will often look for the apparatus necessary to concentrate U-235.

7.6 Could Nuclear Fuel Be Diverted to Make Weapons?

Given the amount of processing that would be required to extract highly purified U-235 from reactor grade fuel, such a diversion from peaceful to military uses would be difficult and costly. A more likely fissionable material for clandestine weapons manufacturing is plutonium-239 (Pu-239). This isotope is formed in a conventional reactor when a nucleus of the plentiful U-238 absorbs a neutron and subsequently emits two electrons as beta particles. These particles, designated as $_{-1}^{0}e$ in equation 7.2, will be discussed later when we consider radioactivity.

$$_{0}^{1}n + _{92}^{238}U \longrightarrow _{94}^{239}Pu + 2\,_{-1}^{0}e \qquad (7.2)$$

This transformation was discovered early in 1940. The chemical and physical properties of plutonium were determined with an almost invisible sample of the element on the stage of a microscope. The chemical processes devised on such minute samples were scaled up a billionfold and used to extract plutonium from the spent fuel slugs from a reactor built on the Columbia River at Hanford, Washington. **Spent fuel** is the material remaining in fuel rods after they have been removed from a reactor. The reactor was called a **breeder reactor** because it was designed primarily to convert U-238 to fissionable Pu-239 by means of the reaction given in equation 7.2. The plutonium was chemically separated from the uranium and used in the first fission test explosion on July 16, 1945 and in the bomb dropped on Nagasaki a little less than a month later.

Plutonium-239 can also be used to power nuclear reactors. Thus, a breeder reactor is one that creates both energy and new fuel (Pu-239) as it fissions the old fuel (U-235). This seems like a dream come true to an energy-hungry planet. France, the United Kingdom, Russia, Japan, and the United States have all conducted research on breeder reactors that permit recovery of plutonium from spent fuel. However, such reactors represent another example of the very mixed blessings of modern technology. The problems are largely associated with the product, Pu-239. The radiation emitted by Pu-239 cannot penetrate the skin. Moreover, solid metallic plutonium is not easily absorbed into the body. But when plutonium is exposed to air, it reacts with oxygen to form plutonium oxide, PuO_2, a powdery compound. The PuO_2 dust can be easily dispersed and inhaled. Once it enters the body, plutonium is one of the most toxic elements known. A few micrograms (10^{-6} g) of PuO_2, lodged in the lungs, can induce lung cancer. Plutonium oxide can also slowly dissolve in the blood and be transported to other parts of the body, especially bone and liver, where its long-lived radioactivity can do serious damage.

Plutonium-239 poses an international problem because the plutonium produced in power reactors could possibly wind up in bombs. It has been widely speculated that the 1981 bombing of a nuclear facility in Iraq by war planes from Israel was done to prevent Iraq from being able to produce plutonium-containing nuclear weapons. If Iraq had succeeded in developing a nuclear arsenal, the outcome of Operation Desert Storm in 1991 might have been quite different. A more recent international crisis involved efforts to dissuade North Korea from building a reactor to produce plutonium. And in 1998,

United Nations weapons inspectors were denied access to sites in Iraq suspected of being nuclear weapons production facilities. Given the potential risks associated with Pu-239 and U-235, it is essential that the supplies and distribution of these isotopes be carefully monitored nationally and around the world. The United States for many years banned the reprocessing of commercial fuel elements. That ban was lifted in 1981, but no plutonium is currently being recovered from commercial reactors in this country. The price of uranium is currently so low that plutonium recovery is not competitive.

One not-so-beneficial side effect of the halt of the nuclear arms race between the United States and the former Soviet Union is the problem posed by the plutonium and uranium removed from warheads. Given current projections, 40,000 American and Russian weapons will be dismantled by 2003, resulting in the recovery of up to 150 tons of plutonium. Plutonium recovered from dismantled weapons is now stored as pits—hollow, metal-clad spheres of plutonium—as well as in other forms. According to a report issued by the National Academy of Sciences (NAS) in January 1994, this surplus fissionable material poses a "clear and present danger to national and international security."

The recovered plutonium and uranium are already in a highly enriched form, unlike reactor grade plutonium that would have to be processed before fabrication into weapons. Moreover, the NAS study expresses concern over the laxity with which the reclaimed weapons-grade plutonium and uranium are being monitored and guarded in the former Soviet Union. Low salaries, corrupt officials, and organized crime can be catalysts for the transfer of fissionable isotopes to terrorists or countries eager to gain nuclear arms. There have already been a number of documented cases of theft or disappearance of plutonium. No doubt the threat will persist until the nations of the world devise safe methods to store and dispose of plutonium and radioactive products of nuclear fission. The challenge posed by dismantled nuclear weapons is only part of a larger problem that also involves the disposal of spent fuel from nonmilitary reactors. We will consider some of the proposed solutions in a later section, but first we need to know more about radioactivity.

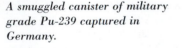

7.9 Consider This: The Reality of Reprocessing

The supply of uranium in the United States (and the rest of the world) is large, but not limitless. By failing to reprocess spent nuclear fuel, we are discarding a potential source of energy that some European countries are presently tapping. Is the current American practice justified? Why is it different from the European practice? List arguments on both sides of this issue and then take a stand.

A smuggled canister of military grade Pu-239 captured in Germany.

7.7 What Is Radioactivity?

Radioactivity was discovered accidentally in 1896 by Antoine Henri Becquerel (1852–1908). The French physicist found that when a uranium-containing mineral sample was placed on a photographic plate that had been wrapped in black paper, the plate's light-sensitive emulsion darkened. It was as though the plate had been exposed to light. Becquerel immediately recognized that the mineral itself was emitting a powerful form of radiation that penetrated the light-proof paper. Further investigation by Marie Curie revealed that the rays were coming from the element uranium, a constituent of the mineral. In 1899 Marie Curie applied the term **radioactivity** to this spontaneous emission of radiation by certain elements. Subsequent research by Ernest Rutherford (1871–1937) in Canada and England led to the identification of two major forms of radiation. Rutherford named them after the first two letters of the Greek alphabet, alpha (α) and beta (β). Their properties are summarized in Table 7.1.

Beta radiation consists of negatively charged (-1) particles; a **beta** (β) **particle** is an electron with a mass of about $1/2000$ on the standard atomic mass scale. Alpha (α) radiation is made up of particles with a mass of 4 units on the same scale and a charge twice as large as that of an electron, but with a positive sign ($2+$). An **alpha particle**

Marie Sklowdowska Curie won two Nobel Prizes—one in chemistry, the other in physics—for her basic research on radioactive elements.

Table 7.1   Radioactive Emissions

Radiation	Composition	Charge	Change in Atomic Number of Parent Nucleus	Change in Mass Number of Parent Nucleus	Change in Number of Neutrons of Parent Nucleus
Alpha, $^4_2$He	2 protons + 2 neutrons	+2	−2	−4	−2
Beta, $^0_{-1}$e	electron	−1	+1	0	−1
Gamma, $^0_0\gamma$	photon of energy	0	0	0	0

is the nucleus of a helium atom—two protons and two neutrons. It was subsequently discovered that a third form of radiation, gamma (γ) radiation, is frequently associated with the emission of alpha or beta radiation. Unlike alpha and beta radiation, **gamma rays** do not consist of ordinary particles. Rather, they are made up of high-energy, short wavelength photons of energy and are part of the electromagnetic spectrum, as are infrared (IR), visible, and ultraviolet (UV) light rays. Gamma rays have no charge and no mass.

> Alpha radiation is identical to the nucleus of a helium-4 atom; alpha radiation is also symbolized as $^4_2$He. Beta radiation, being an electron, is also symbolized as $^0_{-1}$e.

> See Section 2.4 for the electromagnetic spectrum.

Whenever an alpha or beta particle is given off during radioactive decay, a remarkable transformation occurs: the emitting atom changes to an atom of a different element! For example, an atom of uranium-238 becomes converted into an atom of thorium-234 (Th-234) when it loses an alpha particle. Such a change might be understood by the ancient alchemists, who sought to transmute lead and other common metals into gold. But according to modern chemistry, elements and atoms are supposed to be unchanging and unchangeable. Yet, there is ample experimental evidence that *whenever an atom emits an alpha particle, it is converted into an atom of the element with an atomic number two less than the original.* Such a transformation can be represented with a nuclear equation, as in the case of alpha emission by uranium-238 to form thorium-234.

$$^{238}_{92}\text{U} \longrightarrow {}^{234}_{90}\text{Th} + {}^4_2\text{He} \qquad (7.3)$$

The species undergoing radioactive decay (here U-238) is called the **parent,** and the product species is called the **daughter** (here Th-234). Note that the emission of an alpha particle means that the mass number of the daughter isotope is 4 less than the mass number of the parent isotope. The loss of 4 nuclear particles—2 protons and 2 neutrons—accounts for this change.

> The sum of the mass numbers on both sides of equation 7.3 are equal: 234 + 4 = 238; the total of atomic numbers are also equal: 90 + 2 = 92.

The Th-234 formed in equation 7.3 is also radioactive. It undergoes beta particle emission to yield protactinium-234 (Pa-234).

$$^{234}_{90}\text{Th} \longrightarrow {}^{234}_{91}\text{Pa} + {}^0_{-1}\text{e} \qquad (7.4)$$

Thus, emission of a beta particle also results in the parent nucleus being changed into a different element as the daughter product. In this case, Th-234 with atomic number 90 changes to the element Pa-234 with an atomic number of 91, one greater than that of the parent. *The increase in atomic number means that the number of protons increases by one when a beta particle is ejected from the nucleus.* This suggests that you can regard a neutron as consisting of a proton plus an electron. The loss of an electron (a beta particle) converts a neutron to a proton.

> In beta emission, the daughter product has an atomic number one higher than that of the parent nucleus.

> As with alpha emission, beta emission also converts the parent nucleus into a new element as the daughter product.

$$^1_0\text{n} \longrightarrow {}^1_1\text{p} + {}^0_{-1}\text{e} \qquad (7.5)$$

The total number of neutrons plus protons in the nucleus remains constant (at 234 in this instance) during beta emission.

Whether an isotope is an alpha emitter, a beta emitter, or nonradioactive depends on the stability of the nucleus. Nuclear stability, in turn, is related to the ratio of neutrons to protons. Radioactive nuclei adjust this ratio by emitting alpha, beta, or other radiation until a stable neutron/proton ratio is achieved. At that point, the nucleus is no longer radioactive. For example, the radioactive decay of U-238 and Th-234 (equations 7.3 and

Figure 7.9
U-238 decay series.

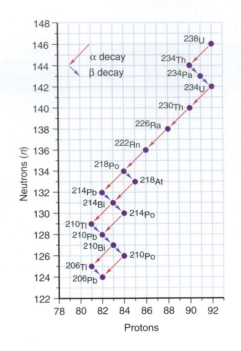

7.4) are the first two steps in a natural series of 14 steps, ending in the nonradioactive lead isotope, Pb-206 (Figure 7.9). For most elements, the most plentiful isotope is nonradioactive. However, all the isotopes of the elements with an atomic number of 90 or greater are radioactive. Their atoms, with many more neutrons than protons, are all unstable.

7.10 Your Turn

a. Rubidium-86, Rb-86, is produced in nuclear power plant reactors. Rb-86 is a beta emitter. Identify its daughter product by name, symbol, atomic number, and mass number. Then write a nuclear equation for the reaction.

b. Radon was discussed in Section 1.14 as an indoor air pollutant. Radon-222, Rn-222, is a natural alpha emitter. Identify its daughter product by name, symbol, atomic number, and mass number. Then write a nuclear equation for the reaction.

c. Iodine-131 is an isotope that is used in medicine for measurement of thyroid activity. It is a natural beta emitter. Identify the daughter product produced by beta emission from I-131 by name, symbol, atomic number, and mass number. Then write a nuclear equation for the reaction.

Ans: **a.** When rubidium 86 emits a beta particle, the mass number remains unchanged. The atomic number increases by one, forming element 38, strontium, Sr. This is the nuclear equation for the reaction.

$$^{86}_{37}\text{Rb} \longrightarrow ^{86}_{38}\text{Sr} + ^{0}_{-1}\text{e}$$

7.8 What Hazards Are Associated with Radioactivity?

It may come as a surprise that the radiation exposure an American citizen receives from living near a nuclear power plant is about one-tenth the radiation he or she would get during a coast-to-coast jet plane trip. Even under adverse circumstances, radiation exposure from nuclear power plants is low. If you had been at the gate of the Three Mile

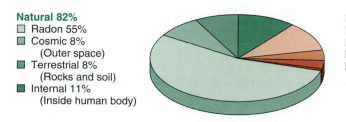

Natural 82%
- ☐ Radon 55%
- ☐ Cosmic 8%
  (Outer space)
- ☐ Terrestrial 8%
  (Rocks and soil)
- ☐ Internal 11%
  (Inside human body)

Man-made 18%
- ☐ Medical X-rays 11%
- ☐ Nuclear medicine 4%
- ☐ Consumer products 3%
- ☐ Other <1%:

Occupational	0.3%
Fallout	<0.3%
Nuclear fuel cycle	0.1%
Miscellaneous	0.1%

Figure 7.10

U.S. sources of background radiation.

(Source: Data from National Council on Radiation Protection and Measurements)

Island plant for the last week of March 1979 (the time of the accident), you would have been exposed to less radiation than from a single chest X-ray. But the evidence of the past makes it clear that it would be a serious mistake to dismiss radioactivity as harmless.

Unfortunately, some of the first scientists to study radioactivity, including Marie Curie (1867–1934), were not fully aware of the potential dangers inherent in the phenomenon. Madame Curie died of a form of leukemia that was very likely induced by her overexposure to radiation. Often the energy of alpha or beta particles or gamma rays is sufficient to produce ionization in atoms and molecules struck by the radiation. As in the case of bombardment by ultraviolet rays, the resulting changes in molecular structure can have profound effects on living things. Rapidly growing cells are particularly susceptible to damage, a fact that has led to the use of radiation as a treatment for some kinds of cancer. But bone marrow and white blood cells are also easily damaged, and anemia and susceptibility to infection are among the early symptoms of radiation sickness. Radiation-induced transformations of DNA can give rise to cancer or genetic mutation.

Today, considerable care is taken to shield medical, laboratory, and other workers from nuclear radiation. Protective shielding made of lead and other dense metals is used to absorb radioactive emissions. However, it is important to recognize that it is impossible to be fully protected from exposure to radioactivity. The Earth itself, the building materials quarried or manufactured from it, our food, and even you and your best friends contain naturally occurring radioactive atoms. Because of this, all of these sources emit what is called background radiation (Figure 7.10). Note that the vast majority of background radiation (82%) is natural radiation. Most of that arises from radon released during natural radioactive decay of rocks and minerals. Radon was discussed as an indoor air pollutant in Section 1.14.

The amount of background radiation you receive depends on where you live, the type of residence you live in, the number of people you live with, and how close you get to them. The late Isaac Asimov, a prolific science writer, pointed out in one of his many books that a human contains approximately 3×10^{26} carbon atoms, of which 3.5×10^{14} are radioactive carbon-14 atoms. With each breath you inhale three and a half million C-14 atoms.

7.11 The Sceptical Chymist: Radioactive Carbon in Your Body

Assume that Isaac Asimov's figures are correct, and that 3.5×10^{14} of the 3.0×10^{26} carbon atoms in your body are radioactive. Calculate the fraction of carbon atoms that are radioactive carbon–14.

The extent of biological damage that can be caused by radiation depends, in part, on the total amount of energy absorbed. This energy is measured in a unit called the rad, short for radiation absorbed dose. **One rad** is defined as the absorption of 0.01 joule of radiant energy per kilogram of tissue. But not all radiation is equally harmful to living organisms; some types are more damaging than others. Therefore, to estimate the potential physiological damage, the number of rads is multiplied by a factor characteristic of the particular radiation involved. This factor is symbolized with an *n*. Very damaging radiation, such as alpha particles and high energy neutrons, has an *n* value of 10. Less harmful forms, including beta, gamma, and x-radiation are assigned an *n* of 1. When *n* is multiplied by the number of rads, the product is called **rem** for roentgen equivalent applied to mammals.

$$\text{number of rems} = n \times (\text{number of rads})$$

Thus, a dose of 10 rad of alpha radiation ($n = 10$) equals 100 rem. This is equivalent to 100 rad of beta radiation ($n = 1$), which also equals 100 rem. The number of rems in a dose of radiation exposure is thus a measure of the power of the radiation to cause damage to human tissue. The likely effects of a single dose of radiation at various levels are given in Table 7.2.

Because most doses of radiation are significantly less than one rem, a smaller unit, the millirem, is commonly employed. One **millirem** (mrem) is 1/1000th of a rem ($1 \text{ mrem} = 10^{-3} \text{ rem}$). Table 7.3 uses millirems to report the radiation exposure associated with various activities and lifestyle factors. This information is the basis for estimating your personal annual radiation exposure in 7.12 Your Turn. Once you have completed this exercise, you can check Table 7.3 to see how your exposure compares with that of the average American.

7.12 Your Turn

Use Table 7.3 to estimate the approximate radiation dosage you receive each year.

Note that nearly 295 mrems (82%) of the approximately 360 millirems absorbed in a year by a typical resident of the United States comes from natural background sources—mostly cosmic rays, soil, and rock—not from human-related sources. Of the 67 mrem of artificial radiation absorbed annually, about 60 are attributable to medical procedures such as diagnostic X-rays. As is evident from Table 7.3, the radiation emitted by properly operating nuclear power plants is negligible compared to normal background radiation, including the natural radiation of your own body. For example, about 0.01% of all the potassium ions (K^+) that are essential to your internal biochemistry are K-40, a radioactive isotope. These K-40 ions give off about 20 mrem per year, approximately 1000 times more radioactivity than that received as a result of living within 20 miles of a nuclear power plant. In fact, because bananas are rich in potassium, a steady diet of them can contribute to your personal radioactivity.

To put things further in perspective, it is useful to note that the immediate physiological effects of radiation exposure are generally not observable below a single dose of

Table 7.2   **Physiological Effects of a Single Dose of Radiation**

Dose (rem)	Likely effect
0–25	No observable effect
26–50	White blood cell count decreases slightly
51–100	Significant drop in white blood cell count; lesions
101–200	Nausea, vomiting, loss of hair
201–500	Hemorrhaging, ulcers, possible death
>500	Death

Table 7.3   **Your Annual Radiation Dose**

Source of Radiation	Mrem^a/yr
1. Location of your town or city	
a. Cosmic radiation at sea level (U.S. average 26 mrem^b) (Cosmic radiation is radiation emitted by stars across the universe. Much of this is deflected by the Earth's atmosphere and ionosphere.)	_____ mrem
b. Add an additional millirem value based on your town or city's elevation above sea level: 1000 m (3300 ft) above sea level = 10 mrem 2000 m (6600 ft) above sea level = 30 mrem 3000 m (9900 ft) above sea level = 90 mrem (Estimate the millirem value for an intermediate elevation.)	_____ mrem
2. House construction Choose the material from which your house is made; enter the correct value. (Building materials contain a very small percentage of radioisotopes.) Brick, 70 mrem; wood, 30 mrem; concrete, 7 mrem	_____ mrem
3. Ground Radiation from rocks and soil (U.S. average)	<u>26</u> mrem
4. Food, water, and air (U.S. average)	<u>40</u> mrem
5. Fallout from nuclear weapons testing (U.S. average)	<u>4</u> mrem
6. Medical and dental X-rays	
a. Chest X-ray (number of visits times 10 mrem per visit)	_____ mrem
b. Gastrointestinal tract X-ray (number of visits times 500 mrem per visit)	_____ mrem
c. Dental X-rays (number of visits times 10 mrem per visit)	_____ mrem
7. Jet travel (Jet travel increases exposure to cosmic radiation.) Number of flights (five-hour flights at 30,000 ft or 9000 m) times 3 mrem per flight.	_____ mrem
8. Nuclear power plants If your home is adjacent to a plant site add 1 mrem.	_____ mrem
Your Total Annual Dose of Radiation	**_____ mrem**

Compare your annual dose to the U.S. annual average of 360 mrem.

^{a}1 millirem = 10^{-3} rem.

^{b}Based on the "BEIR Report III"—National Academy of Sciences, Committee on Biological Effects of Ionizing Radiation 1987. *The Effects on Populations of Exposure to Low Levels of Ionizing Radiation.* Washington, DC: National Academy of Sciences.

Reprinted with permission from *Chemistry in the Community.* (ChemCom), 1988. Copyright © 1988 American Chemical Society.

25 rems (25,000 mrems; Table 7.2). This is nearly seventy times the average annual exposure. The more a given number of rems is spread out over time, the less harmful it appears to be. However, there is still uncertainty about the long-term effects of low doses of radiation. The assumption is usually made that there is no threshold below which no damage occurs. However, the effects of low doses are so small and the time span so great that scientists have not been able to make reliable measurements. Moreover, tests with animals are not always reliable because there is considerable species-to-species variation in the effect of radiation.

The issue then is how to extrapolate the known high-dose data to low doses. Two dose-response models are illustrated in Figure 7.11. The assumption of a linear relationship between the incidence of cancer and radiation dose is represented by curve A. In this model, doubling the radiation dose doubles the incidence of cancer, tripling it causes three times the number of cancers, and so on. This is the relationship currently used by federal agencies in setting exposure standards. Many scientists believe that the biological effect is relatively less at low levels of radiation because of the self-repairing mechanism of cells. This model is represented by curve B, which drops below the straight line of curve A.

Figure 7.11

Dose-response curves for radiation. (The curves level off and then decrease as radiation doses get very high because more cells die than become cancerous.)

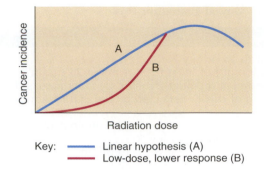

Key: ——— Linear hypothesis (A)
     ——— Low-dose, lower response (B)

7.13 Consider This: Radiation Dose Response

You have just read about two models for the dose-response curve for low level radiation (Figure 7.11). The two response curves represent the more stringent linear hypothesis used by federal agencies in setting exposure limits and the low-dose, lower response curve, believed by many scientists to be closer to our true biological susceptibility.

The ramifications of adopting a specific model are both biological and economical. The more stringent linear dose model requires stricter limits on workers' acceptable radiation dose limit than the lower dose model. By using the linear model, are we being "better safe than sorry" or are we wasting a lot of money protecting ourselves from an emotional issue without looking at the science behind it?

As an X-ray technician in a hospital who must operate under the stricter federal limits for radiation safety, write a letter to an interested friend giving your position on this issue and the reasons for it.

7.9 How Long Will Nuclear Waste Products Remain Radioactive?

Here is one area in which popular magazines, in spite of their frequent hyperbole, seldom exaggerate the problem; some of the products formed in nuclear reactors have dangerously high levels of radioactivity for thousands of years. There is no way to speed up the radioactive decay process. The fact that most of us experience considerably more radioactivity from natural sources than from artificial ones should not lull us into a false sense of security. Nuclear waste disposal presents formidable problems because one radioactive isotope often generates others. As noted in the previous section, daughter products from decays act as parents for other radioactive emissions, for example, U-238 decays to Th-234, which in turn yields Pa-234 (Figure 7.9).

See Equations 7.3 and 7.4

A particularly significant consideration in the disposal of radioactive waste is the rate at which the level of radiation declines. Depending on the particular isotope, the decline can occur very rapidly over a short time, or very slowly over a long period. The rate of decay is typically reported in terms of the **half-life,** the time required for the level of radioactivity to fall to one-half of its initial value. For example, plutonium-239, the alpha-emitting fissionable isotope formed in uranium reactors, has a half-life of 24,400 years. This means that it will take 24,400 years for the radiation intensity of a freshly generated sample of Pu-239 waste to drop to one-half its original value. At the end of a second half-life of another 24,400 years, the radiation will be one-fourth the original level. And in three half-lives (a total of 73,200 years), the level will be one-eighth of the original (Figure 7.12).

The half-life of a particular isotope is a constant, and is *independent* of the physical or chemical form in which the element is found. Moreover, the rate of radioactive decay is essentially unaltered by changes in temperature and pressure. But when various

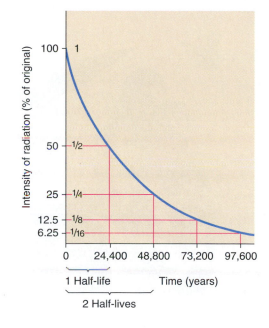

Figure 7.12
Radioactive decay of Pu-239.

radioisotopes are compared, their half-lives are found to range from millennia to milliseconds. For example, the half-life of uranium-238 (equation 7.3) is 4.5 billion years. (Coincidentally, this is approximately the age of the oldest rocks on Earth, a determination made by measuring their uranium content.) By contrast, the beta decay of thorium-234 (equation 7.4) occurs much more rapidly; its half-life is 24 days. The half-lives of different isotopes of the same element differ. Consider that although the half-life of plutonium-239 is 24,400 years, that of Pu-231 is only 8.5 minutes. Even faster decay is exhibited by polonium-214 (Po-214), an alpha emitter with a half-life of 0.00016 seconds.

Hydrogen-3 (H-3 or tritium), which is sometimes formed in the primary coolant water of a nuclear reactor, is a beta emitter with a half-life of 12.3 years. Iodine-131, in the form of iodide ions, with a half-life of 8 days, is used to treat hyperthyroidism in persons with Graves' disease. In this procedure, the orally administered radioactive iodide ions concentrate in the overactive thyroid gland, fully or partially destroying it (Figure 7.13). In most patients, thyroxin, the iodine-containing hormone normally secreted by the thyroid, must then be supplemented with a synthetic substitute.

Strontium-90 (Sr-90) proved to be a particularly dangerous isotope in the fallout from nuclear weapons testing. Strontium ions are chemically similar to calcium ions; both elements are in Group 2A of the periodic table. Thus, strontium (Sr^{2+}), like Ca^{2+}, concentrates in milk and bone. There, the radiation from the Sr-90 can pose a lifelong threat to an affected individual because of the 28.9-year half-life of Sr-90. Significantly, I-131 and Sr-90 are among the fission products produced in nuclear reactors, and there is concern that many persons living near Chernobyl were exposed to potentially harmful levels of both isotopes. The incidence of thyroid cancer among children in the Chernobyl radioactive fallout area seems significantly higher than normal, and iodine-131 has been implicated (Figure 7.14).

Carbon-14 (C-14), with a half-life of 5730 years, is a beta emitter that decays to nitrogen-14 (N-14). This isotope of carbon is well known because it is often used to determine the age of the remains of once-living things or objects made from them. Atmospheric carbon dioxide contains a constant steady-state ratio of one radioactive carbon-14 atom for every 10^{12} atoms of nonradioactive carbon-12. Living plants and animals incorporate the isotopes in that same ratio. However, when the organism dies, exchange of CO_2 with the environment ceases. Thus, no new carbon is introduced to replace the C-14 converted to nitrogen-14 by beta decay. As a consequence, the concentration of C-14 decreases with time, dropping by half every 5730 years. In the 1950s, W. F. Libby first recognized that by experimentally measuring the C-14/C-12 ratio in a sample, the time at which the organism died could be estimated. Human remains and

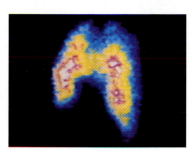

Figure 7.13
An I-131 thyroid scan.

Figure 7.14
A child awaiting a thyroid examination at the Kiev Institute of Endocrinology.

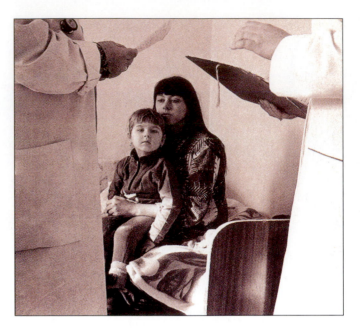

Carbon-14 was used to establish the age of the Shroud of Turin.

many human artifacts contain carbon, and fortunately, the rate of decay of C-14 is a convenient one for measuring human activities. Charcoal from prehistoric caves, ancient papyri, mummified human remains, and suspected art forgeries have all revealed their ages by this technique. The C-14 technique provides ages that agree to within 10% of those obtained from historical records, thus validating the legitimacy of the technique.

7.14 Your Turn

There has recently been considerable publicity about the pollution of indoor air, especially in basements, with radon-222. Rn-222 is a radioactive noble gas element formed by the decay of radium-226,(Ra-226), which is naturally present in hard rocks.

a. Write a nuclear equation for the formation of Rn-222 from Ra-226. (Note that the type of nuclear radiation emitted has not been specified, but you should be able to determine it from the information given.)
b. Suppose that the radioactivity from Rn-222 in your basement was measured as 32×10^{-5} rem. If no additional radon entered the basement, how much time would have to pass before the radiation level fell to 2×10^{-5} rem? The half-life of Rn-222 is 3.8 days.

Hint: **b.** Note that in dropping from 32×10^{-5} rem to 2×10^{-5} rem, the radiation level is cut in half four times: $32 \times 10^{-5} \longrightarrow 16 \times 10^{-5} \longrightarrow 8 \times 10^{-5} \longrightarrow 4 \times 10^{-5} \longrightarrow 2 \times 10^{-5}$. This corresponds to four half-lives of 3.8 days each.

7.10 How Will We Dispose of Waste from Nuclear Plants?

The experience of more than 50 years seems to suggest that the answer to this vitally important question is "slowly and with difficulty." Whether that is the correct answer is another matter. In a June 1997 *Physics Today* article, John Ahearne, past Chairman of the U.S. Nuclear Regulatory Commission, reminds us that "...Like death and taxes, radioactive waste is with us—it cannot be wished away..."

High-level nuclear waste (HLW) is used-up fuel from commercial nuclear reactors as well as waste from nuclear weapons production. In 1992, the U.S. Department of Energy reported that the accumulated high-energy radioactive wastes that had been generated by the Defense Department occupied a total volume of 400,000 cm^3. This volume corresponds to a cube 240 feet on each side, or an acre of land covered to a depth of 320 feet. Military waste is in the form of solutions, suspensions, slurries, and salt cake stored in barrels, bins, and tanks. In addition, about 30,000 tons of radioactive spent fuel have been removed from the nation's commercial reactor sites since the 1950s, a figure that is expected to grow to 52,000 tons by 2005. Most of this waste is currently under water in storage pools on site at the nuclear plant where it was used (Figure 7.15). It is estimated that by 2010, the country will face a total disposal problem of more than 100,000 tons of high-level military and civilian nuclear waste. The year is significant, because it is the earliest date at which a permanent underground repository can possibly open. Given past and recent experience, 2010 seems an unrealistic goal.

Radioactive spent fuel is an unavoidable by-product of nuclear reactors. After about three or four years of use, the U-235 concentration in the fuel rods of a power reactor, initially at 3–5%, drops to the point where it is no longer effective in sustaining the fission process. Approximately one-fourth to one-third of the fuel rods are replaced annually, on a rotating schedule. Nevertheless, the spent fuel rods are still "hot"—both in temperature and in radioactivity. They contain various isotopes of uranium plus plutonium-239 formed by the capture of neutrons by U-238, and a wide variety of fission products such as iodine-131, cesium-137, and strontium-90. Remotely controlled machinery, operated by workers protected by heavy shielding, removes the spent rods from the reactor and replaces them with new fuel rods. The spent rods are transferred to on-site deep pools where they are cooled by water containing a neutron absorber. For example, the storage facility at Seabrook is a 34-foot deep steel-lined concrete pool in a secure building. It has the capacity to hold up to 25 years' worth of nuclear waste.

The on-site storage of high-level radioactive waste for 25 years is hardly ideal. John Ahearne points out that "Almost all of the [spent fuel] waste is currently being stored at the sites where it was generated, in facilities that were not built for long-term storage." The initial plans, begun in the 1950s and early 1960s, had been to reprocess the spent fuel to extract plutonium and uranium from it and recycle these elements as nuclear fuel to produce additional energy. Storage capacity for spent fuel rods on site was designed with such reprocessing in mind. However, only one of several planned reprocessing plants actually went into operation, and then only briefly (1967–75). Thus, reprocessing never was capable of keeping up with the rate of spent fuel production, about 2000 tons every year. In 1977, then-President Jimmy Carter, a nuclear engineer, declared a moratorium on commercial nuclear fuel reprocessing; it would be postponed indefinitely. Long-term geological storage of high-level nuclear waste (HLW), first proposed in 1957 by the National Academy of Sciences, became the alternative option. Yet, no long-term high-level nuclear waste storage facility currently exists in the United States (or in any other country).

Each of the 109 U.S. commercial nuclear reactors produces about 20 tons of spent fuel annually.

A geological repository for HLW is projected to be in operation by 2008 in Sweden.

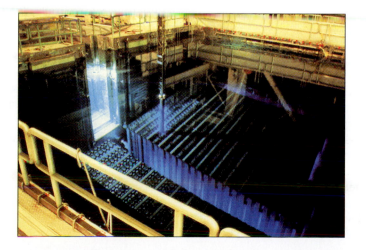

Figure 7.15
Spent fuel rods in a cooling chamber.

The absence of a long-term repository is becoming a significant impediment to the use of nuclear power. In fact, in the 1970s, some states passed laws prohibiting the construction of any new nuclear power plants until the federal government demonstrated that radioactive wastes could be disposed of safely and permanently. The Department of Energy contracted with electrical utility companies to begin accepting spent fuel elements for underground long-term storage in 1998. But progress in preparing a national underground disposal site has been painfully slow. Geologists have been studying the problem for over 30 years. The site selected for radioactive waste disposal must safely contain the radioactive material for an extended time and prevent it from entering the underground water supply. HLW would be stored in the chamber for at least 10,000 years, which is long enough for the high level of radioactivity to decrease significantly. It is estimated that the beta-emitting fission products need to be isolated for 300 to 500 years, which is about 10 half-lives for species such as Sr-90 and Cs-137. Uranium and heavier elements such as plutonium typically have much longer half-lives, but their radiation intensity is much lower. Most plans call for encasing the spent fuel elements in ceramic or glass, packing the product in metal canisters, and burying them deep in the Earth. A method called vitrification has been developed to contain reprocessed defense wastes, including Pu-239, for future geological burial. The wastes are mixed with finely ground glass and melted to about 1150°C. The molten glass and wastes are then poured into stainless steel canisters, cooled, and capped for on-site storage until development of a long-term underground repository (Figure 7.16). More than one million pounds of waste have been treated this way.

The concept behind long-term HLW storage is to carve out a chamber at least 1000 feet below ground, 1000 feet above the water table, in an appropriate rock formation. Salt, basalt, tuff, granite, and shale have all been considered. Salt domes, which are geological formations entirely of salt, are particularly attractive because they are very stable, extremely dry, and self-sealing if cracks should appear. Granite and basalt always contain cracks, but they have a great capacity to chemically absorb most wastes. A 1983 report by the National Academy of Sciences concluded that it is possible to identify rock formations from which a drop of water would require millions of years to travel to the point where it could become incorporated into living things. Supporting evidence is provided by a self-sustaining natural fission reactor in Gabon, Africa. Measurements taken there indicate that fission products have moved less than six feet in two million years.

The federal government must deal with state legislatures and Indian tribes whose land rights are affected. The "not in my backyard" (NIMBY) syndrome has even led a number of states to adopt legislation prohibiting the disposal of high-level nuclear waste within their boundaries. The Mescalero Apaches of southern New Mexico may be unique in bucking this trend. In 1995 the tribe voted to implement a plan for storing high-level commercial radioactive waste on reservation lands. Over 30 utility companies are willing to pay a high price for temporary storage of spent fuel rods. But the plan has been opposed by the New Mexico state legislature and Congressional delegation, and the issue was unresolved at press time. Southeastern New Mexico is also the location of a five-year waste isolation pilot project in which salt beds 2150 feet below ground are being tested as storage reservoirs. If the tests prove successful, the site may be used to store transuranic wastes, elements with lower levels of radioactivity and atomic numbers greater than that of uranium.

Meanwhile, tunnels are being dug 1400 feet beneath Yucca Mountain, Nevada (Figure 7.17) to determine its adequacy for deep, long-term storage of HLW. The Department of Energy has already spent an estimated $54 billion on the project. If completed, the site will be the largest radioactive storage facility in the world, with a capacity of 70,000 tons of spent fuel and 8000 tons of high-level military waste. It has been estimated that at least 20–25 years will be required just to transport the waste to the Nevada site, at a rate of 20 shipments per day. But it is not certain that the Yucca Mountain depository will ever become operational. Warner North, senior vice president of Decision Focus Inc., summarizes the situation: "The most formidable problems associated with using Yucca Mountain are political ones."

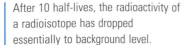

After 10 half-lives, the radioactivity of a radioisotope has dropped essentially to background level.

Figure 7.16
Encapsulating reprocessed HLW in glass canisters.

Figure 7.17
(*a*) Map of Yucca Mountain, Nevada and nearby area; (*b*) An aerial view of the proposed HLW repository at Yucca Mountain, Nevada.

(Source for (a): Data from the Department of Energy).

Geological formations suitable for deep burial of HLW were identified in six states— Nevada, Texas, Washington, Utah, Louisiana, and Mississippi. The selection of Yucca Mountain as the only site for study of underground burial of HLW required overcoming difficult political barriers as well as technical ones. The 1982 Nuclear Waste Policy Act projected that the initial geological repository would be constructed in time to receive high-level nuclear waste by 1998. In 1987, the Nuclear Waste Policy Amendments Act designated Yucca Mountain as the sole site to be studied as an underground long-term high-level nuclear waste repository. To fulfill the requirements of the 1982 Act, the Department of Energy (DOE) contracted with nuclear utilities to begin accepting spent fuel beginning in 1998. To date, utility companies have paid over $14 billion to fund the development of a repository to do so. According to a recent court decision, that responsibility remains with DOE, even though no such facility is available now.

Congress again got involved with HLW disposal by proposing legislation to create the Nuclear Waste Policy Act of 1997. Under this legislation, an *interim* above-ground storage site would be created for 40,000 tons of spent fuel at the Nevada Test Site, adjacent to the unfinished Yucca Mountain site. Spent fuel would be stored at the interim site, to be operational by January 2002, in metal canisters inside concrete bunkers until a permanent repository is ready. In discussing the bill, Nevada Representative Jim Gibbons complained that "The people in support of this bill are the ones who have nuclear waste in their districts and want to get it out—get it from wherever it is into the state of Nevada." Nevada has no commercial nuclear reactors, and Nevada officials are concerned that a permanent site might never be approved, thereby leaving Nevada stuck with the temporary facility indefinitely.

Congressional opponents of the bill have dubbed it the "mobile Chernobyl" bill. They are concerned that the spent fuel high-level waste would be transported by rail and highway through 43 states within half a mile of 50 million Americans before it reached the proposed Nevada Test Site interim repository. Representative Dennis Hastert of Illinois, a supporter, declared that the bill "assures that another 15 years will not pass before the federal government lives up to its responsibility of accepting spent fuel." His position is not surprising given that Illinois has more commercial nuclear reactors (12) than any other state. President Clinton promised to veto the legislation, if passed by Congress. The veto may not be necessary. In June 1998, unable to work out a satisfactory compromise between the House and Senate versions of the bill, Congress failed to take final action on the bill. Such controversial political maneuvering is sometimes described as the "NIMTO" phenomenon—*N*ot *I*n *M*y *T*erm in *O*ffice.

Some members of Congress have argued that a decision on a 10,000-year storage facility should be postponed and a 100-year storage facility should be developed. Congress (or its successor) may still be debating the issue 24,400 years from now, when the plutonium-239 completes its first half-life.

Other disposal methods seem even less promising. Disposal in deep-sea clay sediments, under 3000–5000 m of water was investigated. Proposals to bury the radioactive waste under the Antarctic ice sheet or to rocket it into space have largely been discredited. But one thing is sure: whatever disposal methods are ultimately adopted, they must be effective over the long term.

7.15 Consider This: Nuclear Waste Warning Markers

The Department of Energy recently asked 13 experts to design a system of markers to be installed near an underground nuclear waste repository in New Mexico, warning future generations of the existence of nuclear waste. The markers must last for at least 10,000 years, which is more than four times the age of the pyramids of Egypt. The message on the markers must be intelligible to Earthlings in the year 12,000 A.D. Try your hand at designing these warning markers, keeping in mind the changes that have occurred in *Homo sapiens* during the past 10,000 years and those that might occur in the next ten millennia.

7.11 What Is Low-level Nuclear Waste?

Not all nuclear waste is high-level nuclear waste, the kind produced by nuclear power plant reactors or obtained from nuclear weapons. In fact, nearly 90% of the volume of all nuclear waste is *low-level* waste, nuclear waste other than HLW. **Low-level nuclear waste (LLW)** is described as waste contaminated with relatively small quantities of radioactive materials. Such waste includes a wide range of materials such as laboratory clothing, gloves, and cleaning tools from medical procedures using radioisotopes, development and testing of new drugs, used air filters from nuclear power plants, and discarded smoke detectors. Approximately twice the volume of LLW comes from military sources as that from commercial and medical sources. It is estimated that in the United States there will be 4.5 million m^3 of low-level nuclear waste by the year 2030.

Most low-level nuclear waste does not contain radioisotopes with long half-lives, and is far less radioactive than high-level waste. Therefore, LLW can be disposed of in a much different way than high-level nuclear waste. LLW can be put into sealed canisters and buried in lined trenches 30 m deep (Figure 7.18). Low-level military nuclear waste is disposed at federally owned sites maintained by the Department of Energy. Nonmilitary low-level waste disposal is the responsibility of the state where it is generated.

Although low-level waste poses significantly less danger than does high-level waste, the NIMBY syndrome operates with low-level nuclear waste as well. The very idea of radioactive waste, even if low-level, is sufficient to generate considerable opposition to proposed sites for LLW. Congress expected that states, through the Low Level Radioactive Waste Policy Act of 1980, would form regional compacts by which compact members would send their low-level waste to a disposal site in one state. Such compacts have not been successful, including failed efforts in Illinois and in New York, each costing $55 million over eight years.

Figure 7.18
Burial of low-level nuclear waste.

A battle continues to establish a LLW site at Ward Valley (CA) in the Mojave Desert, 250 miles east of Los Angeles. The site is supported by then-Governor Pete Wilson and opposed by Senator Barbara Boxer. Concern focuses on the possibility of radioisotopes in the waste leaching from the disposal site into the groundwater and the Colorado River. Opposition continues in spite of studies indicating that the short half-life radioisotopes would decay to harmless levels during migration from the site before reaching the groundwater. The studies also point out that even if all of the longer-lived radioisotopes were to reach the Colorado River over a thirty year period (a highly unlikely event according to geologists), alpha radiation in the water would be 33% below that allowed by current California drinking water standards. Currently, LLW is stored at Richland, WA, Clive, UT, and Barnwell, SC in depositories that predate the 1980 legislation.

If sufficient places cannot be found in a country for LLW disposal, why not send the waste to another country? This is not a rhetorical question; such transfer is actually being considered. In 1997, the Taiwan Power Company (Taipower) began negotiations to ship low-level nuclear waste to the Democratic People's Republic of Korea (North Korea) for burial there in abandoned coal mine shafts. The North Korean government, badly in need of money, has agreed to accept an initial shipment of 60,000 barrels over two years, with the option of receiving a total of 200,000 barrels. The agreement is for a cash settlement of about $1150 per barrel for a total payment of approximately $227 million U.S. dollars. Taiwan has not yet approved the safety standards for the site. South Korea opposes the shipments because it fears that North Korea could fail to ensure adequate safeguards to prevent the waste from harming the environment of the Korean Peninsula. In contrast, a proposed LLW waste site in the United States would be examined by the Nuclear Regulatory Commission, typically taking about eight years at a cost of $1 million or more, to review the proposed site and facility before taking final action.

> The 7.16 Consider This exercise related to high-level nuclear waste disposal has a possible counterpart for low-level waste, the transferring of LLW from Taiwan to North Korea.

7.12 Nuclear Power in the United States and Worldwide

> The United States has about 25% of the world's nuclear power plants. Illinois has 12 reactors, more than any other state.

Cost overruns, changing federal regulations, years of litigation, and public opposition have slowed the construction of new fission power plants in the United States. In 1998, 109 power reactors were operating in this country in 32 states at 72 different sites to produce about 22% of our total electric power. Six states generate more than 50% of their electricity using nuclear power and 13 more states get 25–50% of their electricity that way. By contrast, coal is used to generate 57% of our electrical energy, along with acid rain–producing sulfur dioxide and the greenhouse gas carbon dioxide. Only eight additional nuclear power plants are currently being built in this country or are in various stages of testing and start-up. However, construction is no guarantee of operation. The Shoreham reactor on Long Island has been completed, but by agreement it sits forever unfueled and unused. The Haddam Neck (CT) nuclear plant is being closed and is to be decommissioned after 28 years of service. The Maine Yankee plant, New England's biggest electricity producer, also has been closed following 25 years of operation. A number of proposed nuclear power plants have recently been canceled.

> Eighty different designs have been used to build the U.S. reactors, each of which must meet strict design and operating codes.

Globally, about 17% of the electricity produced and consumed is generated in 440 nuclear power plants. If this amount of energy were to be replaced, it would require the entire annual coal production of the United States or the former Soviet Union. Thus, there is already a relatively small but significant international reliance on nuclear energy. Table 7.4 gives nuclear power statistics, as of 1998, for some of the countries that are among the largest users and a few others. The United States is by far the largest generator of electrical power from nuclear sources. This country also has the largest number of operating reactors. But numbers of reactors and total power output do not tell the complete story. A more interesting measure is the percentage of electrical power a country obtains from fission reactors. Such information is graphed in Figure 7.19. On a percentage basis, France leads the world in nuclear power. It has 57 operational nuclear power plants, compared to almost twice that number in the United States. These French reactors generate 78% of the electricity used in France. The Swiss also have a high nu-

Table 7.4 Nuclear Power Statistics for Selected Countries as of December 1998

	Reactors		Electrical Power from Nuclear Reactors (MW)
	Operating	Under Construction	
Brazil	1	1	626
Canada	21	0	14,902
China	3	2	2,167
France	57	4	59,948
Germany	20	0	22,282
Hungary	4	0	1,729
India	10	6	1,695
Japan	53	3	42,369
Mexico	2	1	1,308
Russia	29	2	19,843
Sweden	12	0	10,040
United Kingdom	35	1	12,928
United States	109	6	100,685

(Source: Data from International Nuclear Safety Center; Argonne National Laboratories/DOE, 1998.)

clear dependency, producing nearly 42% of their electricity with only five reactors, less than 5% of the total number of fission-fueled power plants in the United States. Most of the countries that generate over 40% of their electricity from nuclear power plants are in western Europe; Hungary and South Korea are exceptions.

Also noteworthy are the countries not included in Figure 7.19. Thus far, only industrialized nations have been able to afford major commercial development of nuclear fission. Although Cuba has reactors under construction, none are currently operative. As

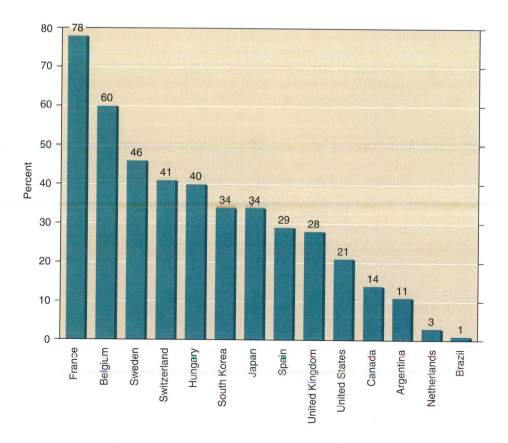

Figure 7.19

Percent of electrical power generated by nuclear reactors in selected countries, 1997.

(Source: Data from International Atomic Energy Agency PRIS database: http://www.iaea.org/programmes/a2/nucshare.html)

Pakistan's and India's nuclear capabilities extend to military applications as well, as evidenced by the nuclear tests each conducted near the other's border in 1998. These tests raised the specter of nuclear weapons being used again after a period of 53 years without nuclear warfare.

of 1998, Mexico had two operating reactors, Pakistan and Romania each had one operating reactor, and India obtained less than 2% of its electricity from its ten reactors. China had three operating nuclear power reactors in 1998, with two more to come on stream by 2001. Ironically, in spite of the fact that much of the world's uranium comes from Africa, the great majority of the nations on that continent use no nuclear energy whatsoever.

7.17 Consider This: Nuclear Production of Electrical Power in Selected Countries

Use the data in Table 7.4 to answer these questions:

a. Who are the top three international producers of electricity from nuclear power in terms of megawatts of energy?
b. Of the nations listed, which three produce the smallest amount of their electricity from nuclear power?
c. In general, how would you characterize the countries that produce large amounts of electricity from nuclear power?
d. How would you characterize the three lowest producers?
e. What reasons can you offer to explain the difference?
f. How does this list of producers compare with the "Top 20" in carbon dioxide emissions found in 3.24 Consider This? Explain this relationship.

7.18 Consider This: Electricity Produced by Nuclear Power

Use the data in Figure 7.19 to answer the following questions:

a. Which countries produce 50% or more of their electricity with nuclear reactors?
b. Which countries produce less than 20% of their electricity with nuclear reactors?
c. How does the United States compare to these two groups of countries?
d. What reasons can you suggest to explain the percentage of electric power generated by nuclear reactors in the United States?

7.19 Consider This: Nuclear Neighborhood

The number of nuclear reactors both in the United States and worldwide keeps changing: some reactors are operating, others are under construction, and still others are being decommissioned. The Nuclear Regulatory Commission's (NRC) web site provides a map of all reactors now licensed to operate in the United States. How many are there? Identify the three reactors closest to where you live. You can find a direct link to the NRC at the *Chemistry in Context* web site.

7.13 Living with Nuclear Power: What Are the Risks and Benefits?

As of 1998, there were 440 commercial nuclear reactors operating in more than 30 nations. It is obvious from Table 7.4, Figure 7.19, and the accompanying text that countries differ markedly in the extent to which they use nuclear power to generate electricity. What is not so obvious are the reasons for this variability. Some nations have fiscal problems so severe that it is difficult or impossible for them to fund the construction or expansion of nuclear power facilities. For others, an adequate supply of relatively cheap electricity is available from water power, fossil fuels, or other sources. Therefore, there is little need for them to use atomic energy. Nuclear power provides a means for some countries to gain a greater independence from needing to import fossil fuels. In still other countries, such as France, a conscious choice has been made to use nuclear energy to produce the bulk of electrical energy to reduce dependency on imported fossil fuel. And, just the opposite conclusion has been reached by other nations. In Sweden, which currently obtains 46% of its electricity from fission, a referendum has called for the halt of nuclear power generation by 2010.

Regardless of whether a country contains many nuclear powered electrical generators or only a few, associated risks and benefits must be weighed. Such risk-benefit analyses are never easy, though in a sense we do it every day. We commonly regard risk as the probability of being injured or losing something, but there are many types of risk. They can be voluntary, such as those associated with bungee jumping, or involuntary, such as inhaling someone else's cigarette smoke. When we drive a car we control the risks (at least to some extent), but we have no control over the increased risk of radiation exposure at high altitudes or of a commercial plane crash. Counterbalancing risks are benefits such as the improvement of health, increased personal comfort or satisfaction, saving money, or reducing fatalities. Everyday living inevitably involves risks and their related benefits: crossing a street, riding a motorcycle or in a car, cooking a meal or eating one, and even the simple act of getting up in the morning. Because there is some element of risk in everything we do, we almost automatically make judgments about what level of risk we consider acceptable. The instinct for survival is such that most people do not intentionally put themselves at high risk, even when the potential benefit is also high such as going into a burning building to save a child. On the other hand, there is an alarming increase in the number of people who expect "zero risk" in whatever they do or in whatever surrounds them, although it is impossible to achieve; there is no such thing as zero risk.

In the case of nuclear energy, we are dealing with social benefits in relation to technological risks, but we must not make the mistake of only considering the risks and benefits that relate directly to fission. We must also weigh the risks associated with the alternatives—the coal-fired power plants that nuclear reactors are designed to replace. Recall, for example, the estimate in Chapter 4 that over 100,000 workers have been killed in American coal mines since 1900. Table 7.5 summarizes the risk of fatalities from the annual operation of a 100-megawatt power plant using either coal or nuclear power. The conclusion is that, at least for the hazards identified here, the risk associated with energy produced by nuclear power is considerably less than that of coal-burning plants.

Paradoxically, coal-fired power plants release more radioactivity than do nuclear plants. According to the EPA, a ton of coal typically contains 1.3 ppm of uranium and 3.2 ppm thorium, both radioactive elements. W. Alex Gabbard, a physicist at the Oak Ridge National Laboratory, has estimated that in 1982, power plants in the United States burned a total of 616 million tons of coal and released 801 tons of uranium and 1971 tons of thorium into the environment. In fact, the quantity of uranium emitted by the coal-fired plants exceeded the mass of uranium consumed in nuclear plants. Globally, the coal burned in 1982 released about 12,600 tons of radioactive waste, widely dispersed in ash and the atmosphere.

However, as we have noted, nuclear energy carries tremendous emotional overtones, made of mystery, misunderstanding, and mushroom clouds. The risks of radiation, involuntary and uncontrolled, and the possibility of a major disaster, however remote, loom large in human consciousness. The accidents at Three Mile Island and Chernobyl, even though hardly equivalent, make the public wary. We have limited trust in technology, and perhaps even less in people. We are apprehensive about human error in the design, construction, and management of nuclear power plants. After all, human errors and technicians' responses to them were the weak points in the prescribed safety procedures that caused the accidents at Three Mile Island and Chernobyl.

The nineteenth century poet William Wordsworth spoke of technological risks and benefits as " . . . Weighing the mischief with the promised gain . . . " In this case, he was speaking about the railroad, a technology new in his time.

Table 7.5   **Risks from Coal and Nuclear-Powered Electricity Generation**

Hazard Type	Coal	Nuclear
Routine occupational hazard	Coal-mining accidents and black-lung disease constitute a uniquely high risk.	Risks from sources not involving radioactivity dominate.
Deaths	2.7	0.3 to 0.6
Routine population hazards	Air pollution produces relatively high, though uncertain, risk of respiratory injury. Significant transportation risks.	Low-level radioactive emissions are more benign than corresponding risks from coal. Significant transportation risks incompletely evaluated.
Deaths	1.2 to 50	0.03
Catastrophic hazards (excluding occupational).	Acute air pollution episodes with hundreds of deaths are not uncommon. Long-term climatic change, induced by CO_2, is conceivable.	Risks of reactor accidents are small compared to other quantified catastrophic risks. The problem lies in as yet unquantified risks for reactors and the remainder of the fuel cycle.
Deaths	0.5	0.04
General environmental degradation	Strip mining and acid runoff; acid rainfall with possible effect on nitrogen cycle, atmospheric ozone; eventual need for strip mining on a large scale.	Long-term contamination with radioactivity; eventual need for strip mining on a large scale.

(Deaths are the number expected per year for a 100-megawatt power plant. In all cases, 6000 man-days lost are assumed to equal one death.)

(Source: Modified from *Perilous Progress: Managing the Hazards of Technology,* by Robert W. Kates, Ed., 1985, Westview Press, Boulder, Colorado.)

7.14 What Is the Future for Nuclear Fission?

This final question is, in many ways, the most difficult one posed in this chapter. The answer is very uncertain. But, as the cover of *Time* magazine on April 29, 1991, asked, "Do we have a choice?" Then-President George Bush's proposed energy plan relied heavily on licensing a new generation of nuclear power plants. The National Academy of Sciences, in an April 1991 report, called for the rapid development of such new plants as a means of reducing the emission of carbon dioxide and other greenhouse gases. And yet, the people of this country are not completely at ease with such a forecast. But, there could be another reason for developing new nuclear power plants. Alan Waltar, past-president of the American Nuclear Society, predicts that at some point in the twenty-first century a significant energy shortage will occur because there will not be sufficient energy from conventional sources to meet the world's demands. Waltar and other nuclear power advocates predict that major blackouts will happen in many countries during these periods. Although nuclear power could help to meet future energy demands, "the future of nuclear energy depends on public acceptance" says Alvin Weinberg, former director of the Oak Ridge (TN) National Laboratory. "Because a large segment of the U.S. population does not like nuclear energy, it is very difficult for politicians to push it."

7.20 Consider This: Public Perception of Nuclear Energy

a. What do you need to know to be an informed citizen about nuclear power plants, that is, ones using a nuclear fission reactor? Make a list of questions (at least five) that would be important to ask about a specific reactor.

b. Then check out the specifications of a reactor using the data on the web site provided by the Nuclear Regulatory Commission. Choose any reactor in the country you wish. Does the information provided answer the questions you posed? Comment on what you would like to know that you were unable to find. The *Chemistry in Context* web site has a direct link to the reactor specifications. If others in your class selected different reactors, you may wish to compare notes. How are the reactors alike? How are they different?

Nuclear power has been touted as a way to reduce global warming and acid rain. Because heat from the fission of uranium produces the steam used to generate electricity, a nuclear power plant releases no carbon dioxide, a major greenhouse gas. Nor does it emit the acidic oxides of sulfur and nitrogen. Recall that the Seabrook plant generates electricity at a rate of 1160 megawatts (1160 million joules per second) or 1×10^{14} joules per day. Approximately 10,000 tons of coal would have to be burned in a conventional power plant to generate this daily energy output. Burning this quantity of coal could easily release 300 tons of SO_2 and perhaps 100 tons of NO_x. Whether the risks associated with nuclear power outweigh those of global warming and acid rain is a difficult question for which there are no clear-cut answers, in spite of extended study and debate. Proponents line up on each side of the argument.

In August 1988, during a summer of record heat, 15 U.S. senators co-sponsored a bill to fund research to combat the greenhouse effect by developing carbon dioxide–free energy sources, including safer and more cost-effective nuclear power plants of standardized design. Alan Crane, an energy policy specialist, appeared before a House subcommittee hearing and spoke of global warming and nuclear energy risks in these terms:

> There is significant, though not yet quantifiable, risk that the resulting climate changes will wreak devastating changes in agricultural production throughout the world, among other problems. Such changes could lead to the death of far more people and cause far greater environmental damage than any nuclear reactor accident, and appear to be considerably more likely.

Others conclude that nuclear power can reduce global warming only slightly and suggest that it would be much less expensive to invest in enhanced energy efficiency. Bill Keepin and Gregory Kats of the Rocky Mountain Institute have estimated that to reduce carbon dioxide emissions significantly through the use of nuclear power would require the completion of a new nuclear plant every 2 days for the next 38 years. Oak Ridge National Laboratory staff members reported that before any massive replacement of fossil fuel with nuclear power could occur, several new techniques would have to be developed. These include commercial-scale recycling of nuclear fuels, breeder reactors to extend existing fuels, and possibly uranium recovery from seawater.

Hans Blix, director general of the International Atomic Energy Agency, in describing the international dimensions of the problem, noted that developing nations will not likely build nuclear plants in the near future: "If nuclear power is to be relied on to alleviate our burdening of the atmosphere with carbon dioxide, it is therefore to the industrialized countries that we must first look. They are in the position to use these advanced technologies—and they are also the greatest emitters of carbon dioxide."

The original nuclear era, synonymous with the growth of nuclear power in the United States, began in the early 1960s and lasted until 1979. It fell victim to stabilized demand for electricity, which was brought on by enormous oil price hikes in 1975, and the Three Mile Island crisis in 1979. As a consequence of that accident, the required number of nuclear plant personnel and their training requirements grew significantly. In addition, the mandatory retrofitting of existing nuclear facilities to enhance their safety added significantly to the cost of an already capital-intensive industry. As a result, the electrical power industry became understandably reluctant to invest further in proposed and planned nuclear facilities.

If a second nuclear era occurs, it likely will be characterized by a period of cheap nuclear energy delivered by reactors that hold public confidence in terms of their safety

Note the linkage of this chapter with Chapters 3 and 6.

Two tons of U-235 (40–60 tons of enriched uranium fuel) can fuel a 1000 megawatt–producing commercial nuclear reactor for about one and a half years. To produce this amount of electricity, a coal-fired plant would use the coal carried by a train with 200 coal cars each carrying 15 tons of coal every day for one and a half years.

and economy. Some experts believe that such a rebirth is possible through the development of smaller, more efficiently designed reactors in the 500–600 megawatt range, rather than the current 1000–1200 megawatt facilities. Unlike the many different designs used to build the 109 current reactors, these new reactors would be of a standardized, easily replicated design, have a longer operational lifetime (60 years versus the current 30), and be demonstrably safer and more economical in operation. One proposal advocates a series of developmental stages over a 13–15 year span. The initial stage would involve experimental construction, a prototype operational plant seven years later, and commercial use after another six years. If the proposed timetable is adhered to, it might be possible to begin a second nuclear era by the second decade of the twenty-first century.

But in all of this, the unsolved problem of the safe disposal of radioactive waste remains perhaps the greatest impediment. Nuclear engineers William Kastenberg and Luca Gratton conclude their *Physics Today* article (June 1997) with the sobering thought: "For a high-level waste depository of the type proposed for Yucca Mountain, it is clear that natural processes will eventually redistribute the waste materials. Present design efforts are directed toward ensuring that, at worst, the degraded waste configurations will eventually resemble stable, natural ore deposits, preferably for periods exceeding the lifetimes of the more hazardous radionuclides. Perhaps that's the best we can hope for."

7.21 Consider This: Risks and Benefits of Nuclear Fission

Now that we have looked at nuclear fission in some detail, it is time to analyze the risks and benefits of this situation. To help in the analysis, start by referring to your three lists made in 7.1 Your Turn. Now make new lists of the risks and benefits associated with currently operating nuclear fission reactors. Using this analysis, take a stand on the question of future use of nuclear fission–powered plants and write an editorial for your local newspaper proposing your view of a viable twenty-year national policy on the issue.

7.22 Consider This: Second Opinion Survey

Now that you have studied the benefits and risks of nuclear power, return to the personal opinion survey of 7.2 Consider This and answer the questions one more time.

After completing the survey for a second time, compare your answers to the second survey with those from the first. Are there any striking differences in your opinions of nuclear power between the first and second surveys? If so, which of your opinions about nuclear power changed the most? What was responsible for this shift?

Conclusion

Over 40 years have passed since the first commercial nuclear power plant began producing electricity in the United States. The glittering promise of boundless, unmetered electricity, drawn from the nuclei of uranium atoms, has proved illusory. But the needs of our nation and our world for safe, abundant, and inexpensive energy are far greater today than they were in 1957. So scientists and engineers continue their atomic quest. Where the search will lead is uncertain, but it is clear that people and politics will have a major say in ultimately making the decision. Reason, not emotionalism, must govern our actions.

Chapter Summary

Having studied this chapter, you should be able to:

- Tell how nuclear fission occurs (7.2)
- Write balanced nuclear equations for alpha and beta decay, and for nuclear fission (7.2)
- Use mathematical relationships to calculate the amount of energy produced by a fission reaction (7.2)
- Compare and contrast how electricity is produced by a conventional power plant with how it is produced by a nuclear power plant (7.3)
- Summarize the reasons why a nuclear power reactor cannot undergo a nuclear explosion (7.5)
- Relate the issues surrounding the use of nuclear power in this country and abroad (7.12)
- Describe the issues associated with the production and storage of high-level nuclear waste, including spent fuel (7.10)
- Summarize the nature of low-level nuclear waste and it storage (7.11)
- Develop a personal radiation dose inventory and describe the biological effects of nuclear radiations (7.8)
- Understand and apply the concept of half-life to the use of radioisotopes, radio-carbon dating techniques, and nuclear waste storage (7.9)

- Report on the use of nuclear power for electricity generation globally and the reasons why several countries have very high percentages of electrical power production from nuclear reactors compared to the United States (7.12)
- Describe the risks and benefits of the use of nuclear power (7.13)
- Take an informed stand on the use of nuclear power for electricity production (7.14)
- Discuss why the 1950s – 60s promise of abundant and cheap nuclear energy was not realized in this country (7.14)
- Outline the factors that will allow or oppose the growth of nuclear energy in the next decade (7.14)
- Take an informed stand on the storage of high-level nuclear wastes (7.10)
- Read and hear news articles on nuclear power and nuclear waste issues with confidence in your ability to interpret the accuracy and efficacy of such reports (7.10 and 7.11)

Questions

Emphasizing Essentials

1. $E = mc^2$ is one of the most famous equations of the twentieth century. What do each of the letters in the equation represent?

2. What is the difference between the symbol N and the symbol ^{14}N?

3. a. How many protons are represented by $^{239}_{94}Pu$?

   b. How many neutrons are represented by $^{90}_{38}Sr$?

   c. How many protons and neutrons are represented by $^{238}_{92}U$?

4. For each of these nuclei, find the number of protons (from a periodic table) and the number of neutrons (from the atomic and mass numbers).

   a. Na-23

   b. Cl-37

   c. Cu-65

   d. Hg-200

5. C-12, Ca-40, Zn-64, and Sn-119 are stable isotopes that have an even number of protons. Based on this information, what generalization can you propose about the relative number of protons and neutrons in stable nuclei with an even number of protons as the atomic number increases?

6. a. Na 22 is a radioactive isotope, Na-23 is stable, and Na-24 is a radioactive isotope. Based on this information, what generalization can you propose about the relative number of protons and neutrons in stable nuclei with an odd number of protons?

   b. What additional information would be needed to draw a generalization about the relative number of protons and neutrons in stable nuclei with an odd number of protons as the atomic number increases?

7. In what ways does a nuclear equation differ from a chemical equation?

8. a. Boron, element number 5, exists in only two common stable isotopes. They are B-10 and B-11. Given that the periodic table lists the average atomic mass as 10.81, which isotope must be more abundant? Why do you think so?

   b. The average atomic mass of chlorine is 35.453. Does this mean that there are just two stable isotopes, one with atomic mass 35 and one with atomic mass 36? Explain your answer.

9. Show that for this nuclear equation, the sum of the subscripts on the left is equal to the sum of the subscripts on the right. Then show that the sum of the superscripts on the left is equal to the sum of the superscripts on the right.

   $$^{239}_{94}Pu + ^{4}_{2}He \longrightarrow [\,^{243}_{96}Cm\,] \longrightarrow ^{242}_{96}Cm + ^{1}_{0}n$$

10. Write equations to represent each of these nuclear reactions.

    a. Two H-2 nuclei join to form another nucleus and a neutron.

    b. U-238 is bombarded with N-14 to produce another nucleus and 5 neutrons.

    c. Pu-239 is bombarded with a neutron to form Ce-146, another nucleus, and 3 neutrons.

11. When 4.00 g of hydrogen nuclei fuse to form helium in the Sun, 0.0265 g of matter is converted into energy. Use Einstein's equation, $E = mc^2$, to calculate the energy equivalent of this change in mass.

12. Consider a reaction in which an isotope of hydrogen fuses with an isotope of helium to form a different isotope of helium and another isotope of hydrogen. The exact mass of a mole of each isotope is given below the isotope.

$$\underset{2.01345 \text{ g}}{^2_1\text{H}} + \underset{3.01493 \text{ g}}{^3_2\text{He}} \longrightarrow [^5_3\text{Li}] \longrightarrow \underset{4.00150 \text{ g}}{^4_2\text{He}} + \underset{1.00728 \text{ g}}{^1_1\text{H}}$$

   a. What is the difference between the mass of the reactant isotopes and the combined mass of the product isotopes?

   b. What is the energy released in this reaction?

13. Einstein's equation, $E = mc^2$, also applies to chemical change as well as to nuclear reactions. An important chemical change studied in Chapter 4 was the combustion of methane, which releases 50.1 kJ of energy for each g of methane burned.

   a. Assume that 50.1 kJ of energy were released by a nuclear transformation of methane. What mass of methane would have been converted to energy?

   b. To produce the same amount of energy, what is the ratio of the mass of methane burned in a chemical reaction to that converted in a nuclear reaction?

   c. Use your results in parts a and b to comment on why Einstein's equation, although correct for chemical change and nuclear change, is usually only applied to nuclear change.

14. This schematic diagram represents the reactor core of a nuclear power plant.

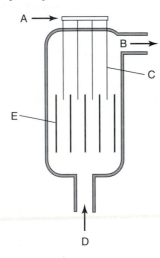

   Match each letter with one of these terms.

   fuel rods

   cooling water into the core

   cooling water out of the core

   control rod assembly

   control rods

15. Identify the segments of the nuclear power plant diagrammed in Figure 7.3 that are nuclear and those that are not nuclear. Briefly explain your choices.

16. Compare the photograph in Figure 7.5 with the diagram in Figure 7.3. Are they oriented in the same direction? Identify the segments of the nuclear power plant shown in Figure 7.5 that are nuclear and those that are not nuclear. Briefly explain your choices.

17. One important distinction between the Chernobyl reactors and those in the U.S. is that those in Chernobyl used graphite as a moderator to slow neutrons whereas U.S. reactors use water. Give two reasons for believing that U.S. reactors are safer than those at Chernobyl.

18. One of the biggest challenges in preparing fuel for nuclear power plants is to separate the isotopes of uranium.

   a. Why is it not possible to separate the isotopes of uranium by chemical means?

   b. How are the isotopes of uranium separated?

19. Write equations to represent each of these nuclear reactions.

   a. reaction of U-235 with a neutron to form Br-87, La-146, and neutrons

   b. U-238 is bombarded with a nucleus to produce Fm-249 and 5 neutrons

   c. neutron-induced fission of U-235 to form one nucleus with 56 protons, a second with a total of 94 neutrons and protons, and 2 extra neutrons

20. Pu-239 is most hazardous when it has been exposed to air and then inhaled. Explain the reasons behind this observation.

21. What is the difference between α, ^4_2He, and $^4_2\text{He}^{2+}$?

22. Consider emission of an alpha particle, a beta particle, and a neutron during a naturally occurring nuclear change. Identify the types of radioactive emission that typically accompany each of these nuclear changes.

   a. the mass number changes

   b. the atomic number changes

   c. both the atomic number and the mass number change

   d. neither the atomic number nor the mass number change

23. Write nuclear equations to represent each of these transformations.

   a. I-131 (used in nuclear medicine and tracer studies) releases a beta particle.

   b. U-238 decays with the loss of an alpha particle.

   c. Tl-206 emits a beta particle.

   d. Mo-98 is bombarded with neutrons to form Tc-99 and another particle.

24. Given that the average U.S. citizen receives 360 mrem of radiation exposure per year, use the data in Table 7.3 to calculate the percentage of radiation exposure the average U.S. citizen receives from each of these sources.

   a. food, water, and air

   b. a dental X-ray twice a year

   c. the nuclear power industry

25. Determine the fraction of a radioactive isotope that would remain after two half-lives, four half-lives, and six half-lives.

26. What is the half-life of the radioisotope X in this graph?

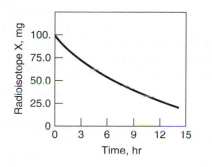

Concentrating on Concepts

27. In 7.2 Consider This, you were asked to answer several question about nuclear power. Extend this survey by asking the same questions of someone at least one generation older than you, and of someone still in middle school or high school. What similarities and differences did you find in their answers compared with your opinions?

28. **a.** Why were citizens' groups in Massachusetts concerned about construction of the Seabrook Nuclear Power Plant in New Hampshire? *Hint:* Consult a road atlas to find the exact location of Seabrook.

  b. What aspects of the site chosen for the Seabrook plant were advantages in the minds of the designers and builders of the Seabrook plant, but were disadvantages in the minds of some protesting citizens' groups?

29. The Seabrook power plant at full capacity uses only a few pounds of uranium to generate 1160 megawatts of power, which is equivalent to 1.16×10^9 joules every second. To produce the same amount of energy would take about two million gallons of oil or about 10,000 tons of coal in a conventional power plant. What is the fundamental difference in the way that energy is produced in the Seabrook plant, compared with conventional power plants?

30. Considering that Einstein had proposed the equation $E = mc^2$ over 30 years earlier, why were Otto Hahn and Fritz Strassmann puzzled when, in 1938, they discovered the element barium among the products formed when uranium was bombarded with neutrons?

31. In a chemical reaction, it is often said that matter is conserved. Why is it incorrect to say that mass is conserved in a nuclear reaction?

32. If you look at nuclear equations in sources other than this textbook, you may find that the subscripts have been omitted. For example, you may see an equation for a fission reaction written this way.

$$^{235}U + {}^{1}n \longrightarrow \left[{}^{236}U \right] \longrightarrow {}^{87}Br + {}^{146}La + 3\ {}^{1}n$$

  a. How do you know what the subscripts should be?

  b. Why are the superscripts not omitted?

33. More than two decades have passed since the incident at Three Mile Island. Use the web to find this information. **a.** Is Reactor 2, the site of the problem in 1979, back on line producing electricity? **b.** How has the accident's nuclear waste been treated?

34. Many people assume that the reactors at Chernobyl were completely shut down after the nuclear accident; this is not the case. Draft a position statement on this point, giving your viewpoint on what has happened there and what you think should happen by the year 2010.

35. What are the similarities between the incidents at Chernobyl and at Three-Mile Island? What are the differences?

36. What happens to the neutrons in a nuclear power plant that are not used for fission reactions?

37. Is spent nuclear fuel still radioactive? Give the reasons for your answer.

38. It is generally believed that terrorists would be more likely to construct a nuclear bomb using Pu-239 reclaimed from breeder reactors than using U-235. Use your knowledge of chemistry to offer reasons for this.

39. Consider these factors: temperature, external pressure, catalysts, concentration of reactants.

  a. Which can change the half-life of naturally occurring radioactive isotopes?

  b. Which can change the rate of a chemical reaction?

40. The 40-year licenses of several U.S. nuclear power plants will expire early in the next century. At least two plants have started the application procedures to extend their licenses for another 20 years, whereas others have decided to stop operations when their current licenses expire.

  a. Offer some possible circumstances under which a utility would apply for an extension of their licenses.

  b. Offer some possible circumstances under which a utility would decide to stop the operation of a nuclear power plant.

Exploring Extensions

41. This chapter starts with this quotation, "Today, we Americans get more electricity from nuclear energy than from any other source, except coal." Find statistics that either prove or disprove this assertion.

42. Lise Meitner and Marie Curie were both pioneers in developing an understanding of atomic nuclei. You likely have heard of Marie Curie and her work, but may not have heard of Lise Meitner. How are these two women related in time and in their scientific work?

43. Gallium consists of just two stable isotopes, Ga-69 and Ga-71.

  a. If the atomic mass of elemental gallium is 69.72, which isotope is present in a larger percentage?

  b. If Ga-69 has a mass of 68.9257 and Ga-71 has a mass of 70.9249, what percentage of each isotope is present?

44. Alchemists in the Middle Ages dreamed of converting base metals, such as lead, into precious metals—gold and silver. Why did they never succeed? Has the situation changed since then?

45. In question 12, the energy of a mole of H-2 joining with He-3 was calculated. What is the ratio of the energy released in this nuclear reaction to the energy released in the combustion of a mole of hydrogen gas? *Hint:* This value can be calculated using bond energies from Table 4.1.

46. a. Californium, element number 98, was first synthesized by bombarding an element with alpha particles. The products were californium-245 and a neutron. What was the target isotope used in this nuclear synthesis?

   b. Evidence for the existence of element 114 was recently gathered by Russian scientists. Research the experiment used to form element 114, and write a nuclear equation for the reaction leading to its discovery.

47. Consider this representation of a Geiger counter, a device commonly used to detect ionizing radiation. The probe contains a gas.

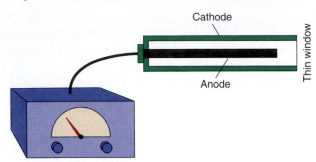

   a. How does radiation enter the Geiger counter?

   b. Why does this device only detect radiation that is capable of ionizing the gas contained in the probe?

   c. What are other methods for detecting the presence of ionizing radiation?

48. A selected group of utilities running nuclear power stations have recently been asked by the Department of Energy to consider if they are willing to participate in an experimental program to use Pu-239 as part of the fissionable material in their power plants.

   a. What are the advantages and disadvantages for a utility in participating in this experimental program?

   b. As of today, how many utilities have actually signed up to participate in this program?

49. There have been several advertisements for Swiss Army watches that have stressed the role of tritium. One says that the "...hands and numerals are illuminated by self-powered tritium gas, 10 times brighter than ordinary luminous dials...". Another advertisement boasts that the "...tritium hands and markers glow brightly making checking your time a breeze, even at night....". Evaluate these statements and after doing some web research, discuss what form the tritium takes in these watches and what its role is.

50. MRI, or magnetic resonance imaging, is a very important tool for some types of medical diagnoses.

   a. What is the scientific basis for this technique?

   b. What information can an MRI give a physician that cannot be obtained through direct examination of a patient?

   c. This MRI method used to be called NMR, nuclear magnetic resonance. Why do you think the name was changed?

New Energy Sources for the New Century

A bank of photovoltaic cells at a solar power plant in the Swiss Alps. These cells convert sunlight into electricity.

FOR LEASE: 1998 General Motors EV1; 2-door, 2-seat coupe; 137 horse power; accelerates from 0 to 60 miles per hour in 8.5 seconds, maximum speed 80 miles per hour, nearly noiseless operation, low maintenance; no emissions; no gasoline or other fuel required, but must be electrically charged periodically; cruises approximately 90 miles between recharging; energy cost only one-third that of gasoline-powered car; $33,995 ($399 – $549 per month lease).

The automobile industry is changing. Perhaps not since the gasoline-burning internal combustion engine was introduced has there been such a revolution in motor vehicle energy sources under development and in production. The advertisement above describes the first modern all-electric passenger car to be mass produced and marketed by a major American automobile manufacturer. EV1s are now in U.S. showrooms, but only in Phoenix and Tucson, Arizona and

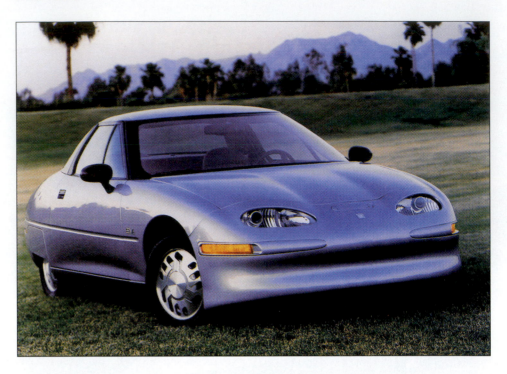

The General Motors EV1 electric car. This car runs exclusively on electricity produced by a bank of batteries.

in large metro areas in California. Although the leasing of these cars has been low, the EV1 and other so-called zero emission vehicles may symbolize the beginning of a major transformation in the way in which the world considers alternatives to fossil fuels as energy sources. This chapter is about that transformation as it applies to devices that manufacture, transfer, and store electrical energy.

8.1 Consider This: Electric Vehicles

Is General Motors really leasing and/or selling electric vehicles? Go to the GM web site, www.gm.com, and search for "electric vehicle"(or perhaps "advanced technology vehicle" or "EV1"). This will connect you to the latest information on electric cars being produced in Detroit. The *Chemistry in Context* web site also has direct links to web sites pertaining to electric vehicles. **a.** What do EVs now cost to lease or buy? **b.** What do their owners say about them? **c.** What does the manufacturer say about them?

In two earlier chapters we considered two sources of energy that are used extensively in the United States and in most other countries to produce electrical energy. Chapter 4 emphasized the energy obtained from the burning of fossil fuels, and Chapter 7 focused on the energy released by the splitting of certain fissionable atomic nuclei. Centralized power plants, whether fueled by coal or fission, distribute electricity regionally through vast power networks to offices, classrooms, and residences. But the supplies of both fossil and fission fuels are limited, and when they are gone, they are gone forever. Moreover, these fuels also bear a huge environmental cost. The combustion of coal and petroleum releases vast quantities of carbon dioxide, which contribute to global warming, and sulfur dioxide and nitrogen oxides, which give rise to acid precipitation. Nuclear fission is bedeviled by the as-yet unsolved problem of disposing of high-level nuclear wastes that will continue to emit dangerous radiation for thousands of years. The conclusion seems obvious: if our species is to continue to inhabit this planet, we must develop other sources of energy.

Chapter Overview

What do we turn to for electricity when we flip a light switch and nothing happens due to a power failure? We typically use battery-powered flashlights and lamps. Other portable devices, such as fuel cells, also produce electricity. Hydrogen is one of the fuels used in fuel cells. And so the chapter begins by considering hydrogen, a non-fossil fuel. Once obtained from a variety of possible sources, hydrogen can either be burned as a fuel or used to generate electricity in a fuel cell. The discussion of fuel cells leads to a related section on the electrochemical cells and batteries that power many modern devices, including the so-called zero emission electric vehicles (ZEVs), such as the EV1. Although electric cars are now commercially available, questions about them remain: Why have them? How do they operate? How do they compare to gasoline-powered vehicles? Are electric vehicles competitive in the marketplace? Why are they leased rather than sold, and are people leasing them? We will address all these questions in this chapter, as well as discuss other electric vehicle technologies.

In the long run, the most promising alternate technology for generating electicity could well be photovoltaic cells, devices that convert the Sun's radiation directly into electricity. The principles governing semiconductors in photovoltaic cells, their operation, and their current and potential applications are discussed. The chapter ends with a brief account of another solar process, nuclear fusion, the source of the energy from the Sun and other stars.

8.1 Water: "An Inexhaustible Source of Heat and Light?"

In Jules Verne's 1874 novel *Mysterious Island,* a shipwrecked engineer speculates about the energy resource that will be used when the world's coal supply has been used up. "Water," the engineer declares, "I believe that water will one day be employed as fuel, that hydrogen and oxygen which constitute it, used singly or together, will furnish an inexhaustible source of heat and light."

Is this simply science fiction, or is it energetically and economically feasible to break water into its component elements? Can hydrogen really serve as a useful fuel? And what does this have to do with solar energy? To answer these questions and to assess the credibility of the claim by Verne's engineer, a Sceptical Chymist needs to examine the energetics of the reaction represented by equation 8.1.

$$2\ H_2(g) + O_2(g) \longrightarrow 2\ H_2O(l) \tag{8.1}$$

Experiment shows that the reaction, as written above, gives off 572 kJ when two moles of liquid water are formed from the combination of two moles of hydrogen and one mole of oxygen. It follows that the burning of one mole of H_2 will yield one mole of H_2O and 1/2(572) or 286 kJ of energy. This indicates a highly exothermic reaction, represented by equation 8.2.

$$H_2(g) + \tfrac{1}{2} O_2(g) \longrightarrow H_2O(l) + 286\ kJ \tag{8.2}$$

Because energy is *evolved*, the energy change in this combustion reaction is -286 kJ for each mole of H_2 burned to form liquid water. This is equivalent to releasing 143 kJ per gram of H. In comparison, the heat of combustion of coal is 30 kJ/g, octane (a major component in gasoline) is 46 kJ/g, and methane (natural gas) is 54 kJ/g when the products of combustion are $CO_2(g)$ and $H_2O(l)$. Clearly, hydrogen has the potential of being a powerful energy source. In fact, on a per gram basis, hydrogen has the highest heat of combustion of any known substance. It is used (along with oxygen) to launch the space shuttle and other rockets. The extraordinary energy per gram of hydrogen when it burns raises the tantalizing prospect—using it as a fuel to power motor vehicles that would produce only water vapor, and no pollutants such as the carbon monoxide and nitrogen oxides created by burning fossil fuels. Simple as this might seem, there are difficulties to overcome before burning hydrogen as fuel in automobiles becomes common practice.

A fundamental question to be answered is: How can we obtain a sufficient supply of hydrogen if it were to be used for fueling motor vehicles? On the one hand, things look promising because hydrogen is the most plentiful element in the universe. Over 93% of all atoms are hydrogen atoms. Although hydrogen is not nearly that abundant on Earth, there is still an immense supply of the element. But essentially all of it is tied up in chemical compounds. Hydrogen is too reactive to exist for long in its diatomic form, H_2, in the presence of the other elements and compounds that make up the atmosphere and the Earth's crust. Therefore, to obtain hydrogen for use as a fuel, it is necessary to extract hydrogen from hydrogen-containing compounds, and this requires energy.

8.2 Your Turn

Calculate the number of moles and grams of H_2 that would have to be burned to yield an American's daily energy share of 260,000 kcal. (1 kcal = 4.18 kJ)

Ans.   3800 mole or 7600 g

8.3 Your Turn

You have just learned that the formation of one mole of liquid water from hydrogen and oxygen releases 286 kJ of energy. In contrast, the direct formation of one mole of gaseous water from hydrogen and oxygen releases 242 kJ of energy. Offer a possible explanation for this observed difference in energy.

8.2 Splitting Water

The traditional laboratory method of generating hydrogen involves the action of certain acids on certain metals. Perhaps the most common combination is sulfuric acid and zinc.

$$H_2SO_4(aq) + Zn(s) \longrightarrow ZnSO_4(aq) + H_2(g) \tag{8.3}$$

Although convenient as a small-scale source of hydrogen, the reaction of a metal with an acid is far too expensive to scale up for industrial applications. Therefore, we turn to the most abundant earthly source of hydrogen—water.

Because the formation of one mole of water from hydrogen and oxygen releases 286 kJ (equation 8.2), an identical quantity of energy must be absorbed to reverse the reaction to produce hydrogen. Figure 8.1 summarizes the processes.

Figure 8.1

Energy differences in the hydrogen-oxygen-water system.

$H_2(g) + \frac{1}{2} O_2(g)$

(Energy evolved)
−286 kJ

(Energy absorbed)
+286 kJ

$H_2O(l)$

Figure 8.2
Electrolysis of water.

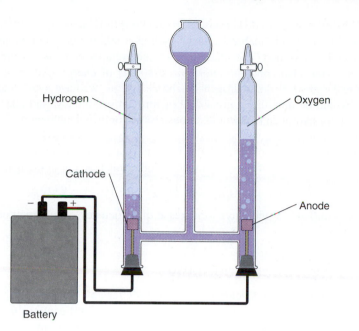

All that is needed to bring about this reaction is a source of 286 kJ. The most convenient method of decomposing water into hydrogen and oxygen is by **electrolysis,** which is the passage of a direct current of electricity of sufficient voltage to break the O—H bonds in water (Figure 8.2). When water is electrolyzed, the volume of hydrogen generated is twice that of oxygen. This suggests that a water molecule contains twice as many hydrogen atoms as oxygen atoms, testimony to the formula H_2O.

$$286 \text{ kJ} + H_2O(l) \longrightarrow H_2(g) + \tfrac{1}{2} O_2(g) \qquad (8.4)$$

Of course, the question remains: "How will the electricity for large-scale electrolysis be generated?" Currently, most of it is produced by the burning of fossil fuels in conventional power plants. If we only had to deal with the first law of thermodynamics, the best we could possibly do would be to burn an amount of fossil fuel equal in energy content to the hydrogen produced in electrolysis. But recall from Chapter 4 that we must also deal with the consequences of the second law of thermodynamics. Because of the inherent and inescapable inefficiency associated with transforming heat into work, the maximum possible efficiency of an electrical power plant is about 60%. When we add the additional energy losses caused by friction, incomplete heat transfer, and transmission over power lines, it would require at least twice as much energy to produce the hydrogen than we could obtain from its combustion. This is comparable to buying eggs for ten cents each and selling them for five; it's no way to do business. Furthermore, most methods of generating electricity have a variety of negative environmental effects. At one time it was thought that the "cheap" extra electricity from nuclear fission could be used to produce hydrogen to fuel the economy, but that energy utopia has hardly been realized. It should be apparent, therefore, that electricity generated from fossil fuels or nuclear fission does not offer a practical way to split water to produce hydrogen for use as a fuel.

A second possibility is to use heat energy to decompose water. One process used to produce commercial hydrogen does, in fact, use heat. You have already encountered it, in Chapter 4, in the discussion of substitutes for fossil fuels. Hot steam is passed over coke (essentially pure carbon) at 800° C.

$$131 \text{ kJ} + H_2O(g) + C(s) \longrightarrow H_2(g) + CO(g) \qquad (8.5)$$

The mixture of hydrogen and carbon monoxide that is produced can be burned directly. It can serve as the starting material for the synthesis of hydrocarbon fuels and other important compounds, or the hydrogen can be separated from the mixture and used as needed. Reaction 8.5 is being studied in an effort to find catalysts that will make it possible to carry it out at lower temperatures.

Simply heating water to decompose it thermally into H_2 and O_2 is not commercially promising. To obtain reasonable yields of hydrogen and oxygen, temperatures of over 5000°C would be required. The attainment of such temperatures is not only extremely difficult, it would also consume enormous amounts of energy—at least as much as would be released when the hydrogen was burned. Thus, we have again reached a point where we would be investing a great deal of time, effort, money, and energy to generate a quantity of hydrogen that would, at best, return only as much energy as we had invested, and in practice a good deal less.

Methane, CH_4, the major component of natural gas, is currently the chief source of hydrogen. The hydrogen is formed from the endothermic reaction of methane with steam.

$$165 \text{ kJ} + CH_4(g) + 2 \text{ } H_2O(g) \longrightarrow 4 \text{ } H_2(g) + CO_2(g) \tag{8.6}$$

Researchers continue to find ways to increase the efficiency of this method of producing hydrogen.

8.3 The Hydrogen Economy

If and when we succeed in developing economically feasible methods for producing hydrogen cheaply and in large quantities, we will still face significant problems in storing and transporting it. Although H_2 has a high energy content per gram, it occupies a very large volume—about 12 liters (a bit over 12 quarts) per gram at normal atmospheric pressure and room temperature. If H_2 is to be stored and transported in its gaseous state, large, heavy-walled metal cylinders will be required, thus eliminating much of the advantage of the favorable energy/mass ratio. To save space, gases are typically converted to the liquid state under high pressure, as in the case of "bottled gas" or "liquid propane." But hydrogen must be cooled to −253°C before it liquifies. This means that keeping it in liquid form requires low temperatures and high costs.

A number of attempts have been made to circumvent these problems by storing H_2 in other forms. One of these, proposed recently, involves absorbing gaseous H_2 on a solid, such as activated carbon, which can hold a great deal of H_2 at only moderate pressures. The carbon can be heated as needed to release the H_2 for combustion. A different approach involves reacting the H_2 with certain metals to produce compounds called hydrides, which are relatively stable solids that have a reasonably high storage capacity. For example, when 10 liters (slightly over 10 quarts) of H_2 gas at 25°C and 1 atmosphere react with lithium metal, the resulting LiH occupies a mere 4.3 mL, or slightly less than a teaspoon.

$$Li(s) + \tfrac{1}{2} H_2(g) \longrightarrow LiH(s) \tag{8.7}$$

When such hydrides are reacted with H_2O, they produce H_2, which can then be burned in the usual fashion.

$$LiH(s) + H_2O(l) \longrightarrow H_2(g) + LiOH(aq) \tag{8.8}$$

Prototype vehicles that operate on this principle have been built and used in a number of locations. Clearly, such approaches would greatly improve the safety and convenience of handling H_2, and perhaps be the decisive factor in determining the extent of its acceptance as a fuel. A hydrogen-fueled car would produce only water vapor, and none of the carbon monoxide or nitrogen oxides emitted from using gasoline-fueled internal combustion engines.

Even if we manage to solve all of the production, storage, and transport problems identified above, we must consider how best to extract energy from our hydrogen. The most obvious way would be to burn it in power plants, vehicles, and homes. A stream of pure hydrogen burns smoothly, quietly, and safely in air, delivering 143 kJ per gram and forming only non-polluting water as an end product. But when hydrogen is mixed with oxygen, a spark is often sufficient to detonate a devastating explosion. The vivid photographs of the exploding space shuttle Challenger and the burning hydrogen-filled dirigible Hindenburg are unforgettable reminders of the risks associated with the many benefits of a hydrogen economy (Figure 8.3).

Figure 8.3

The uncontrolled reaction of hydrogen with oxygen caused the explosion that destroyed the space shuttle Challenger, killing seven astronauts.

8.4 Fuel Cells: A Slow Burn

Suppose someone were to suggest that there is a way to combine H_2 and O_2 to form H_2O without the hazards of combustion. Suppose further that this individual claimed that the reaction could be carried out without allowing the hydrogen and oxygen to come in contact with each other. The Sceptical Chymist might well dismiss such assertions as sheer nonsense—an outright impossibility. And yet, sometimes what appears to be completely contrary to common sense can in fact happen in the natural world. The operation of a fuel cell is a case in point. *In a fuel cell, the chemical energy of a fuel is converted directly into electricity without burning the fuel.* Fuel cells were invented in 1839 by Sir William Grove, an English physicist, but they remained a mere curiosity until the advent of the U.S. space program in the 1960s. Such devices are routinely used as sources of electrical energy in the space program. The space shuttle, for example, carries three sets of 32 cells each that use hydrogen as a fuel, but never burn it.

A fuel cell functions somewhat like a conventional flashlight battery. But unlike batteries, fuel cells use an external constant supply of fuel such as hydrogen. Therefore, they do not need to be recharged, and they produce electricity as long as fuel is provided. In a fuel cell, the chemical reaction is physically separated into two parts, each of which occurs in a segregated region of the cell. Recently developed hydrogen fuel cells use a solid polymer to separate the reactants and to act as the electrolyte. The polymer is a proton exchange membrane (PEM) permeable to H^+ ions, and coated on both sides with a platinum-based catalyst (Figure 8.4).

In a fuel cell, one of the reactions is always **oxidation,** in which a reactant loses electrons. The other reaction is a **reduction,** involving the gain of electrons by some other reactant. In a hydrogen fuel cell using a PEM, hydrogen gas (H_2) is the fuel. The oxidation and reduction are represented by half reactions such as equations 8.9 and 8.10. As the hydrogen passes through the membrane, it loses electrons to form H^+ ions. This oxidation half reaction is:

$$H_2(g) \longrightarrow 2\ H^+(aq) + 2\ e^- \tag{8.9}$$

The hydrogen ions flow through the proton exchange membrane to the other side where they combine with oxygen (O_2) and electrons to form water in the reduction half reaction.

$$\tfrac{1}{2}\ O_2(g) + 2\ H^+(aq) + 2\ e^- \longrightarrow H_2O(l) \tag{8.10}$$

Thus, there is a transfer of electrons—from H_2 to O_2. An important thing to note is that oxidation cannot occur alone; that would be rather like one hand clapping. Oxidation

The Challenger explosion was a consequence of the violent reaction of hydrogen with oxygen in the propellant system, not from the fuel cells.

Polymers are the subject of Chapter 9.

Oxidation = Loss of electrons
Reduction = Gain of electrons

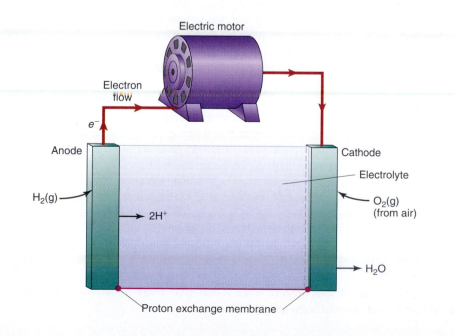

Figure 8.4
A PEM fuel cell. The anode reaction is:
$H_2 \longrightarrow 2\ H^+ + 2\ e^-$;
the cathode reaction is:
$\tfrac{1}{2}\ O_2 + 2\ H^+ + 2\ e^- \longrightarrow H_2O$;
the overall (net) reaction is:
$H_2 + \tfrac{1}{2}\ O_2 \longrightarrow H_2O$.

(electron loss) must always be paired with a reduction reaction (electron gain). The overall reaction (equation 8.11) is one that takes into account gain and loss of electrons in the two half reactions by combining the oxidation and the reduction half reactions.

$$H_2(g) + \tfrac{1}{2} O_2(g) + 2\ H^+(aq) + 2\ e^- \longrightarrow 2\ H^+(aq) + 2\ e^- + H_2O(l) \quad (8.11)$$

The two electrons lost in the oxidation half reaction provide the two electrons gained in the reduction process. The two electrons and 2 H$^+$ that appear on each side of the arrow in equation 8.11 can be canceled to give the net equation.

$$H_2(g) + \tfrac{1}{2} O_2(g) \longrightarrow H_2O(l) \quad\quad\quad\quad (8.12)$$

This is clearly the equation for the burning of hydrogen in oxygen. But in a fuel cell, it is "burning" without a flame and with relatively little heat. Water is the only product, besides a transfer of electrons, which is what we call electricity. Hence, fuel cells are a more environmentally friendly way to produce electricity than from burning fossil fuel or fissioning uranium or plutonium atoms.

To produce electricity, the two half reactions of the fuel cell must be connected in such a way that the electrons released during the oxidation reaction are transferred to the reduction reaction (Figure 8.4). This is accomplished by using **electrodes,** electrical conductors placed in the cell as sites for chemical reaction. The electrode at which oxidation takes place is called the **anode.** The electrons given up in the process flow from the anode through a wire to the **cathode.** At the cathode, the electrons are used in the reduction half-reaction. The electrical energy appears because of the spontaneous reaction that occurs in the fuel cell.

The electrons flowing from the anode to the cathode of a fuel cell can be channeled through an external circuit to do work, which is the whole point of the device. On the space shuttle these electrons are used to illuminate bulbs, power small motors, and operate computers. The tendency of the electrons to flow through the external circuit depends on the difference in electrical potential between the anode and the cathode. That in turn depends upon the chemistry that is taking place. The difference in electrode potential is the voltage of the cell, and it is measured and reported in **volts (V).** The greater the difference in potential, the higher the voltage. The rate at which the electrons flow is called the **current,** and it is measured in **amperes (amps).** You are probably already familiar with these terms because they are used to describe household electricity —110 volts and 15 amps for many purposes. Household electricity is alternating current or AC, but the electricity generated by a fuel cell or any other battery is direct current or DC.

The net reaction represented by equation 8.12 releases 286 kJ of energy per mole of water formed. But instead of liberating most of this energy in the form of heat, the fuel cell converts 70–75% of it into electrical energy. This direct production of electricity eliminates the inefficiencies associated with using heat to do work to produce electricity. Internal combustion engines are only 20–30% efficient in converting energy from fossil fuels. Moreover, a fuel cell does not "run down" or require recharging. It keeps functioning as long as the fuel (hydrogen) and oxygen are supplied. To generate useful amounts of electrical energy, fuel cells are stacked in layers.

Because PEM fuel cells are compact, light, and do not require caustic electrolytes such as potassium hydroxide used in previous fuel cells, they are prime candidates for use in electric vehicles. To get around the bulky storage of hydrogen problem that limits the vehicle's driving range (page 305), fuels such as methanol (CH$_3$OH) can be used in PEM fuel cells. A device, called an "on-board methanol reformer" converts the hydrogen from methanol into H$_2$O (equation 8.15). A new catalyst, an alloy made from four metals, replaces the platinum-based one used with hydrogen-only fuel cells, which cannot be used with methanol and water. At the anode of a methanol fuel cell, a 3% solution of methanol and water reacts to produce carbon dioxide, six hydrogen ions, and six electrons.

$$H_2O(l) + CH_3OH(aq) \longrightarrow CO_2(l) + 6\ H^+(aq) + 6\ e^- \quad (8.13)$$

In the cathode compartment, air or oxygen is blown in where it reacts with the electrons and hydrogen ions to produce water.

Anode: Oxidation electrode
Cathode: Reduction electrode

Potential energy difference is measured in volts, a unit honoring Alessandro Volta. Current is measured in amperes to honor the physicist André Ampère.

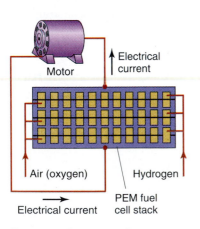

Proton exchange membrane (PEM) fuel cells are frequently "stacked" to achieve useful amounts of electricity.

$$\tfrac{3}{2}\,O_2(g) + 6\,H^+(aq) + 6\,e^- \longrightarrow 3\,H_2O(l) \qquad (8.14)$$

The net reaction is: $CH_3OH(aq) + \tfrac{3}{2}\,O_2(g) \longrightarrow CO_2(l) + 2\,H_2O(l)$ (8.15)

As with hydrogen fuel cells, the electricity produced by methanol fuel cells is used to power electric motors that provide the motor power for the vehicle. Note from equation 8.15 that electric vehicles powered by these fuel cells produce no nitrogen oxides, such as those produced from internal combustion engines. Also, the amount of CO_2 generated per unit of useful energy is lower than that emitted in the direct combustion of the fuel. In addition, methanol is a renewable fuel, unlike gasoline. Furthermore, fuel cells have no moving parts, so the vehicles should require little repair and last longer. According to Dr. Halpert, Program Manager for Batteries and Fuel Cells at the Jet Propulsion Laboratory, "This [methanol] fuel cell may well become the power source of choice for energy-efficient, non-polluting electric vehicles."

8.4 Consider This: Fuel Cells in Your Future?

Many different types of fuel cells are under development. Another promising type is the propane-oxygen fuel cell. Propane is $C_3H_8(g)$. The equation for the chemical reaction that takes place in a propane-oxygen fuel cell is:

$$C_3H_8(g) + 5\,O_2(g) \longrightarrow 3\,CO_2(g) + 4\,H_2O(l)$$

a. Identify the substance that undergoes oxidation and the substance that undergoes reduction in this reaction.
b. Unlike batteries, fuel cells do not store chemical energy. Explain the significance of this statement for the future of fuel cells.
c. What are some of the reasons that fuel cells have not been the energy source of choice in the past, but may become a viable choice in the future?

Electric vehicles powered by fuel cells is not a far-fetched idea. Buses carrying up to 60 passengers have been powered by PEM fuel cells in Chicago, IL and at Georgetown University in Washington, DC (Figure 8.5). In 1997, Mercedes-Benz unveiled NECAR 4 (New Electric Car), an electric vehicle that operates on methanol fuel cells (Figure 8.6). NECAR 4 averages about 25 miles per gallon of methanol and has a 250-mile range without refueling, close to that of more conventional vehicles. Dr. Ferdinand Panik with Daimler-Benz remarked, "In the end, I believe the fuel cell can

Figure 8.5
A methanol-based fuel cell bus at Georgetown University. The bus is funded by the Federal Transit Administration.

Figure 8.6
The NECAR 4 (New Electric Car 4) by Daimler-Benz (now Daimler-Chrysler) operates using methanol fuel cells.

be done for the same price as the piston engine, or lower. And I believe it can let the owner travel 50% farther for the fuel used, with an engine that will be truly maintenance-free."

Whether Panik's prediction will be the case remains to be seen. Reliable predictions of the true costs of automobiles powered by fuel cells are difficult to obtain, but the prices of the cell components are declining. Also true is that the "big three" domestic automobile manufacturers each have put significant resources behind developing fuel cell–powered electric cars under the Department of Energy's Partnership for a New Generation of Vehicles. General Motors plans to have a fuel cell electric car ready for production by 2004, Ford Motor Company expected to have a test car ready by 2000, and Chrysler hoped to have a prototype vehicle developed by that time as well. A fuel cell manufacturer, Ballard Power Systems, predicts that competitively priced, zero-emission vehicles will start to show up in automobile showrooms for the 2004–2005 model year. To make the prediction a reality, Ballard Power Systems has formed a worldwide alliance with Daimler-Chrysler and Ford Motor Company to become the largest commercial producer of fuel cell–powered electric drivetrains and components for cars, trucks, and buses. The alliance is gearing up to hit their target of mass producing 100,000 fuel cell–powered electric vehicles by 2004–2005. "The beauty of fuel-cell vehicles is that they are pollution-free and energy efficient, and we can make the fuel right here in America," said Paul Lehman, a fuel cell researcher at Humboldt State University (CA). "In electric cars, fuel cells offer important advantages over batteries: They have greater range, and they take minutes to refuel—not hours to recharge."

In 1998, Chrysler Corporation merged with Daimler-Benz to form Daimler Chrysler.

Arthur D. Little, an energy consulting firm, has announced the development of a prototype fuel cell that converts gasoline to hydrogen. In the prototype fuel cell, gasoline vapor is converted into hydrogen and carbon monoxide. The carbon monoxide, in contact with a special catalyst, is then reacted with steam to produce carbon dioxide and additional hydrogen. The hydrogen can then be used to make electricity, as in a conventional fuel cell. As a fuel to produce hydrogen for fuel cells, gasoline would have the advantage of using a pre-existing fuel distribution system, a feature not yet available for distributing hydrogen directly. Still in development, the gasoline fuel cell is not likely to be available commercially until at least 2003.

There is also a good deal of interest in using fuel cells to generate power at stationary locations. The largest fuel cell assembly in the United States supplies electrical power to 1000 homes in Santa Clara, CA. The health maintenance organization giant Kaiser Permanente has installed fuel cell units to furnish electricity in three of its California hospitals. Engineers and developers also project applications of methanol fuel cell technologies for laptop computers, cellular phones, and other consumer electronics, and for lawn mowers and portable electric generators.

8.5 Electrons, Cells, and Batteries

Although you may not have had personal experience with fuel cells, the chances are very good that you have a close relative of a fuel cell with you at the present time. If you are wearing a battery-powered watch or carrying a calculator, cellular phone, or laptop computer, you have a system for the direct conversion of chemical energy to electrical energy. Most people refer to these devices as batteries, and we will discuss them in this section. There are several reasons for our interest. In the first place, you will soon see that batteries are similar to fuel cells in some significant respects. Secondly, batteries are found everywhere in today's society because they are convenient, transportable sources of stored energy. And finally, battery uses will likely become even more important, including supplying power to run electric vehicles such as the General Motors EV1, which introduced this chapter.

The first thing to note is that the word "battery" refers to a collection of similar objects—as in a battery of cannons. A standard flashlight "battery" is more correctly called a cell, or an electrochemical cell. A collection of several electrochemical cells wired together constitutes a true battery. **An electrochemical cell** is a device that converts the energy released in a spontaneous chemical reaction into electrical energy. It is the opposite of an **electrolytic cell,** in which electrical energy is converted to chemical energy.

Electrochemical cells produce electricity in fundamentally the same way that fuel cells operate. Again, two half-reactions—one oxidation and the other reduction—are involved. Electrons are generated by the oxidation reaction, collected at the anode, and flow through an external wire to the cathode, where reduction occurs. The difference in voltage or potential between the two electrodes is proportional to the energy evolved.

The two half-reactions for the familiar alkaline cell, pictured in Figure 8.7, are given below.

$$\text{Anode (Oxidation): } Zn(s) + 2\ OH^-(aq) \longrightarrow Zn(OH)_2(s) + 2\ e^- \qquad (8.16)$$
$$\text{Cathode (Reduction): } 2\ MnO_2(s) + H_2O(l) + 2\ e^- \longrightarrow Mn_2O_3(s) + 2\ OH^-(aq) \qquad (8.17)$$

As in a fuel cell, the overall cell reaction is the sum of the two half-reactions.

$$Zn(s) + 2\ MnO_2(s) + H_2O(l) \longrightarrow Zn(OH)_2(s) + Mn_2O_3(s) \qquad (8.18)$$

The voltage produced by this cell is 1.54 volts. The voltage of a cell depends primarily on which elements and compounds are participating in the reaction. It does not depend on factors such as the overall size of the cell, the amount of material that it contains, or the size of the electrodes. This is apparent from the fact that all alkaline cells, from the tiny AAA size to the large D cells, give this same 1.54 volts. On the other hand, the current or electron flow does depend on the size of the cell. Larger cells generate larger currents. And, because power is obtained by multiplying the voltage by the current, larger cells are more powerful than smaller ones.

AAA to D cells all produce 1.54 volts, but the larger cells generate larger currents.

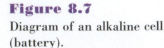

Figure 8.7

Diagram of an alkaline cell (battery).

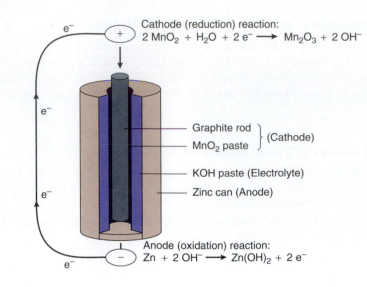

Cathode (reduction) reaction:
$2 MnO_2 + H_2O + 2 e^- \longrightarrow Mn_2O_3 + 2 OH^-$

Graphite rod } (Cathode)
MnO₂ paste

KOH paste (Electrolyte)

Zinc can (Anode)

Anode (oxidation) reaction:
$Zn + 2 OH^- \longrightarrow Zn(OH)_2 + 2 e^-$

Many different electrochemical cells have been developed for different specific purposes. Some of these cells and their uses are listed in Table 8.1, along with their voltages and an indication of whether they are rechargeable. Because mercury batteries can be made very small, they are used widely in watches, camera equipment, hearing aids, calculators, and other devices that use transistors and integrated circuits that do not require large currents. Unfortunately, the toxicity of mercury (Chapter 5) makes the disposal of these cells a potential hazard. The EPA estimated that in 1989, 88% of the 1.4 million pounds of mercury in urban trash came from non-rechargeable batteries. Burning trash may exacerbate and extend the problem by releasing mercury vapor into the atmosphere.

8.7 Consider This: How Many Batteries Do You Use and Throw Away?

An indirect cost of using batteries in small appliances is the cost of battery disposal in terms of putting them into landfills and discarding recyclable metal resources. Just how big a problem is this? To help grasp the magnitude of this problem, answer these questions.

a. Make a list of things that you own that run on disposable batteries.
b. Estimate the number of times each year that you replace these batteries.
c. Expand your estimate to include the other members of your immediate family.
d. Given that the population of the United States is about 260 million people, approximately how many batteries are disposed of each year in this country?
e. Will this number of batteries be representative of other countries? Why or why not?

Table 8.1 Some Common Electrochemical Cells and Their Associated Voltages

Type	Voltage	Rechargeable?
Dry cell	1.5	No
Alkaline	1.54	No
Mercury	1.3	No
Lithium-iodine	2.8	No
Lead storage	2.0	Yes
Nickel-cadmium	1.46	Yes

Lithium-iodine cells are so reliable and long-lived that they are used to power cardiac pacemakers. The lithium battery takes advantage of the low density of lithium to make a lightweight battery. A lithium-iodine pacemaker battery implanted in the chest can last as long as 10 years before it needs to be replaced. In fact, the widespread use of such pacemakers today has been due, in large part, to the improvements that have been made in the batteries used to power them, rather than in the pacemakers themselves.

Most batteries convert chemical energy into electrical energy with an efficiency of about 90%. This may be compared with the much lower efficiencies that typically characterize the conversion of heat to work (30–40%). However, it is important to remember that considerable energy is required to manufacture electrochemical cells. Metals and minerals must be mined and processed, and the various components manufactured and assembled. Moreover, a battery has a finite life. Sooner or later, the chemical reaction will reach completion, the voltage will drop to zero, and electrons will no longer flow. The battery will be "dead" and ready for disposal, and disposal is a formidable problem. In February 1993, *National Geographic* reported that some two and a half billion household batteries are purchased each year in the United States at a total price of $3.3 billion. Of these, over 90% are single-use batteries that find their way into landfills or incinerators.

8.8 Consider This: Do You Know Where Your Used Batteries Are?

Many mercury-containing batteries are disposed of in household trash that eventually ends up in local or regional incinerators. When such batteries are incinerated, the mercury vaporizes and can be released to the atmosphere.

a. What are some of potential hazards associated with the release of mercury vapor?

b. What are alternatives to incinerating mercury-containing batteries?

c. What are some of the practical limits to your proposed alternatives?

d. How common are the alternatives to used battery incineration in your area?

On the other hand, some batteries are rechargeable, a characteristic that greatly extends their lifetimes and increases their cost-effectiveness. The best known rechargeable battery is the lead storage battery used in almost all American cars. It is a true battery because it consists of six cells, each generating 2.0 V for a total of 12.0 V. The overall cell reaction is given by equation 8.19. Lead batteries are *storage* batteries because they store electrical energy.

$$Pb(s) + PbO_2(s) + 2\ H_2SO_4(aq) \longrightarrow 2\ PbSO_4(s) + 2\ H_2O(l) \qquad (8.19)$$

  lead   lead dioxide   sulfuric acid   lead sulfate   water

The anode is made of lead and the cathode of lead dioxide, PbO_2. The electrolyte is a concentrated sulfuric acid solution (Figure 8.8). Although the weight of the lead and the corrosive properties of the acid are disadvantages, the lead storage battery is dependable and long lasting. The key to its success is the fact that reaction 8.19 is readily reversible. As it spontaneously proceeds in the direction indicated by the arrow, the reaction produces the energy necessary to power a car's starter, headlights, and various devices. But as the reaction proceeds, the battery "discharges." The electrical demands of a modern car are so great that in a short time, most of the reactants would be converted to products, significantly reducing the voltage and the current. To counter this, the battery is attached to a generator, or alternator, turned by the engine. The alternator generates direct current electricity, which is run back through the battery. This input of energy reverses the reaction represented by equation 8.19 and recharges the battery. In a high-quality lead storage battery, this process of discharging and recharging can go on over a period of five years or more.

Figure 8.8

Cutaway view of a lead-acid storage battery.

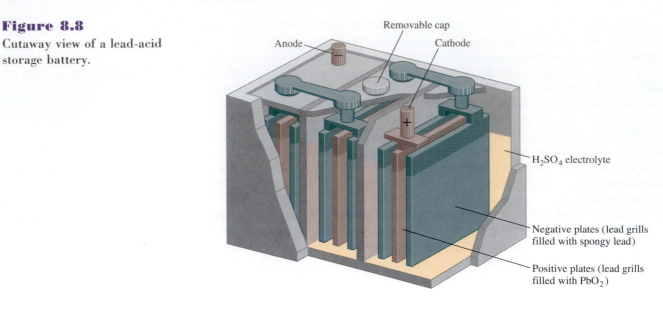

In environments where the fumes from internal combustion engines cannot be tolerated, batteries often provide the only source of energy for locomotion. Thus, forklifts in warehouses, passenger carts in airport terminals, golf carts, and wheelchairs are typically powered by lead storage batteries. Because of the dependability of lead batteries, they are sometimes used in conjunction with wind turbine electrical generators. The generator charges the batteries when the wind is blowing, and the batteries provide electricity when the wind stops.

We turn next to how batteries can be used to actually power motor vehicles and not simply as the energy source to start them as in conventional vehicles.

8.9 Consider This: Batteries—Functions and Uses

The match between technical capability and intended function determines the optimal type of battery used for a specific application. Consider each of these uses and discuss the criteria that will determine which battery is best for the intended use: (1) TV remote controls; (2) heart pacemakers; (3) deep-space probes; (4) cellular phones; (5) automobiles.

8.6 Batteries, The EV1, and Electric Cars

Twenty-six lead storage batteries, weighing a total of about 1100 pounds, are at the energetic heart of the 137-horsepower General Motors EV1 (Figure 8.9). This new car has been created in response to legislation enacted in California and subsequently in New York and Massachusetts. By 2003, 10% of the new cars sold by major auto manufacturers in these states must meet "zero emission" standards. This goal is to be achieved through a series of intermediate steps. Each automobile corporation has a quota of zero-emission cars that must be sold in a given year. If the number actually sold is below that minimum, the company will be fined $5,000 for each car short of that goal. The regulations proposed for southern California are even more stringent. In an effort to remedy what has been called the worst air pollution in the United States, a tentative plan has been adopted that would require by 2007 that all cars in the Los Angeles basin be converted to electric power or other clean fuel. Given current technology, the only way to achieve this goal is by relying on lead storage batteries. Fuel cell research and rapidly developing new automobile battery technologies may permit manufacturers to meet the 2007 deadline.

Figure 8.9
The EV1 electric car by General Motors at a charging station in San Francisco.

The good news is that the batteries and the cars they power release no carbon dioxide or carbon monoxide, no oxides of sulfur or nitrogen, and no unburned hydrocarbons or ozone into the atmosphere. The cars require no gasoline or other fuel. Unfortunately, zero emission on the road does not translate to a similar absence of pollution elsewhere. The batteries in an electric car must be recharged frequently, usually after only 90 miles of highway driving or 70 miles of city driving. And if the temperature drops to 20°F, battery efficiency drops and the driving range falls to about 25 miles. To re-energize the car, it must be plugged into a source of electricity; and the recharging requires at least three hours. Moreover, charging stations are currently few and far between, creating the electric car equivalent of "running out of gas" with no gas station nearby. However, 220-volt chargers to use at home are available at a cost of about $2000, which is built into the cost of the vehicle.

But recharging is only part of the problem. As you know, electrical power generation is notoriously inefficient. Less than half of the energy released from burning fuel is converted into electricity. Furthermore, power plants that draw their energy from fossil fuels release sulfur dioxide, nitrogen oxides, and carbon dioxide. In fact, calculations indicate that the SO_2 and NO_x emitted from power plants generating the electricity to keep a fleet of battery-powered cars operational will exceed the amount of these two gases that would have been released by the gasoline-powered cars that have been replaced. The overall CO_2 emission does decline when electric cars are substituted for internal combustion automobiles, but by less than 50%.

Another serious criticism of "pollution free" electric cars was published in *Science* in May 1995 by Lester Lave, Chris Hendrickson, and Francis McMichael. These three Carnegie Mellon University professors argue that a switch to cars powered exclusively by lead storage batteries would dramatically increase the amount of lead released into the environment. Their calculations include estimates of lead dispersed in mining, processing, and battery manufacture; and they conclude that with current technology, 1.34 grams of the toxic metal would be emitted per kilometer traveled by an electric car. This corresponds to 2.16 grams of lead per mile or 47.5 pounds in 10,000 miles, a typical annual mileage driven in the United States. Ironically, this amount of lead is 60 times that released over the same distance by a car burning leaded gasoline. To be sure, critics have questioned some of the assumptions made by Lave, Hendrickson, and McMichael and argue that their conclusions greatly exaggerate the problem. Moreover, power plants, lead mines, lead refineries, and battery factories are point, not mobile, sources of pollution. This makes it easier to control emissions from them than from the millions of cars that clog the California and other freeways and spew their emissions over thousands of square miles. But experience should alert us to the fact that it is dangerous to focus on one part of a problem without considering broad systemic environmental impact, lest the cure be worse than the ailment it sets out to remedy.

Power plant efficiency is discussed in Section 4.13.

Kilowatt hours were discussed in Section 4.16.

Batteries for the EV1 are included in the three-year overall warranty and are recyclable, as are the lead storage batteries from conventional automobiles. GM estimates energy costs for the EV1 to be 2.48 cents per mile at a recharging cost of 10 cents per kilowatt hour. Based on those figures, a 75 mile trip in an EV1 would cost $1.86. This compares to $5.11 for a gasoline-powered car making the same trip averaging 22 miles per gallon of gas costing $1.50 per gallon. Under these conditions, the energy cost for the EV1 is only one-third that of a conventional car. But the manufacturer's suggested retail price for the EV1, $33,995, is near the upper end of the cost spectrum, considering that the vehicle is based on an emerging technology. Because electric car technology is still being developed, GM only leases, not sells, the EV1 through 33 Saturn dealers to provide a better cost consistency as further technological developments occur. About 500 EV1s have been leased so far.

It is unlikely that electric cars powered by lead storage batteries alone will become a major, long-term answer to the problem of automotive air pollution. At best they appear to be a temporary solution. But the development of lighter, more efficient, more environmentally benign batteries would be a major step forward. Among the designs being used is a nickel-metal hydride cell that would double the range of an electric car and cut recharging time to 15 minutes. Unlike lead batteries that must be replaced after 25,000–50,000 miles, these batteries last the life of the vehicle. Other possible battery combinations include nickel-cadmium, sodium-nickel-chloride, sodium-sulfur, zinc-air, aluminum-air, and lithium polymer. The U.S. Advanced Battery Consortium (USABC), a joint venture of major U.S. automakers and the Department of Energy, continues to sponsor intensive research aimed at developing advanced battery systems for motor vehicles, including development of nickel-metal hydride batteries. USABC Chairman Robert L. Davis said that USABC-sponsored research "has identified lithium-based batteries as the long-range solution to competitive electric vehicles and has worked successfully to demonstrate technical design feasibility for these technologies."

8.10 Consider This: Nickel versus Lithium-based Batteries

Are lithium-based batteries better than nickel-metal hydride ones? Use the web to find details about these two types of batteries. Then, write a brief summary of your findings and give your conclusion about which battery would be better to use in an electric vehicle.

A compromise, perhaps an ideal one, that would get past the limitations of depending solely on batteries would be a car with the convenience and range of a gasoline-powered car combined with the environmental advantages of an electric vehicle. Such a vehicle, called a **hybrid car,** has been developed by Toyota, and is in development by other auto manufacturers as well. The Toyota Prius has been available in Japan since 1998 (Figure 8.10). With a 1.5-liter gasoline engine sitting side-by-side with nickel-metal hydride batteries, an electric motor, and an electric generator, the Prius does not need to be recharged. It consumes only about half the gasoline, emits 50% less carbon dioxide and far less nitrogen oxides than a conventional car, while delivering 66 miles per gallon of gasoline (650 miles per tankful). The electric motor draws power from the batteries to get the car moving or when it is traveling at low speeds. Using a process called regenerative braking, the kinetic energy of the car is transferred to the generator, which charges the batteries during deceleration and braking. The gasoline engine assists the electric motor during normal driving, with the batteries boosting power when extra acceleration is needed. The attractive $18,000 price tag has been set artificially low to stimulate sales in Japan, which have been brisk enough for demand to outpace supplies. The Prius is expected to be available in the United States by 2000.

Figure 8.10

(*a*) The Toyota Prius hybrid automobile. (*b*) The "engine" of the Prius.

8.11 Consider This: Hybrid Cars

Earlier in the chapter, you visited the General Motors web site to check out their EV1 electric vehicles and what owners thought of them. Now check out Toyota's hybrid car. Search for "hybrid car" or "future vehicles" to find the latest information. The *Chemistry in Context* web site has links to other sites describing hybrid vehicles.

a. What information can you find about test driving, leasing, or buying hybrid cars?
b. What does the manufacturer say about them?
c. Are the cars available in the U.S.?

8.12 Consider This: Batteries, Fuel Cells, and Hybrids

The text describes two potential sources of electrical power for cars and other motor vehicles; rechargeable batteries and fuel cells. Both use chemical reactions to generate electricity, but differ in the details. There also are hybrid systems under development. Demonstrate your knowledge of the chemical principles involved by explaining the similarities and differences in the operation of rechargeable batteries and fuel cells to a classmate. Then explain the advantages that a hybrid car, such as a Prius, could bring to the consumer.

In the long run, even the new generation of batteries may not be able to compete with fuel cells. But Figure 8.11 does present an intriguing possibility. It is a photograph of a battery-powered car charging up outside the offices of the Sacramento (CA) Municipal Utilities District. If you look closely, you will note that the car is plugged into the Sun via a bank of photovoltaic cells.

Automobile industry leaders agree that there will be no mass market for alternative energy vehicles, like hybrid cars or ones using fuel cells, unless the performance and price of such vehicles match those of conventional models. Jack Smith, CEO of General Motors, says: "People are too practical. There's a certain element who will fall in love with new technology, but technology won't survive unless it's cost effective."

Figure 8.11

A battery-powered electric car being charged from photovoltaic cells at the Sacramento Municipal Utilities District. The bank of photovoltaic (solar) cells is in the panels mounted on the columns.

> ### 8.13 Consider This: Electric Cars—Visit the Showroom!
>
> Many of the major manufacturers are already touting their versions of electric cars. Assume that electric cars are about to become available in your area. What features would convince consumers to buy these cars, even if the price were several thousand dollars more than the conventionally powered car? Design either a poster or a radio or television announcement that would help in the campaign to convince consumers to buy these cars.

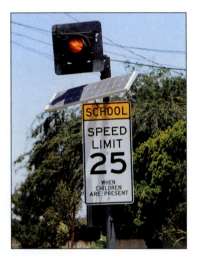

Photovoltaic (solar) cells are used to power this traffic signal.

See Sections 2.4 and 3.5.

8.7 Photovoltaics: Plugging in the Sun

Each day the Earth receives more energy from the Sun than is consumed by our planet's nearly six billion inhabitants in 27 years. But currently, less than 0.5% of the power generated in the United States comes directly from the Sun. Electricity supplies about 35% of U.S. energy needs, two-thirds of it used in residential and commercial buildings. Therefore, it would be to our advantage if solar radiation could be directly converted into electricity, without the intermediary of hydrogen or some other fuel. This is the function of **photovoltaic cells,** also called solar cells. Such devices have already demonstrated their practical utility for both large- and small-scale electrical generation such as for satellites, highway signs, street lights, navigational buoys, automobile recharging stations, and remote residences. They presently may represent the best hope for capturing and using solar energy.

Electricity involves a stream of electrons, flowing from a region of higher electrical potential (voltage) to one of lower voltage. For a photovoltaic cell to generate electricity, light must induce such a flow in the cell. This flow depends on the interaction of matter and photons of radiant energy—a topic already treated in considerable detail in Chapters 2 and 3. In those chapters, we pointed out that the portion of the Sun's radiation reaching the Earth's surface is mainly in the visible and infrared regions of the spectrum, with a maximum intensity near a wavelength of 500 nm. Light of this wavelength is green and has an energy of about 4×10^{-19} J per photon, corresponding to 240 kJ per mole of photons. A photovoltaic cell must be made of atoms or molecules that will release electrons when struck by radiation of approximately this same wavelength.

Among the substances that will do this is a class of materials known as **semiconductors**—materials that do not normally conduct electricity well, but do so under certain conditions. One of the first semiconductors identified was the element silicon, which you may recognize as being used in computers, radios, pocket calculators, and digital watches. A crystal of silicon consists of an array of silicon atoms, each bonded to four others by means of shared pairs of electrons, as represented in Figure 8.12a. These shared electrons are normally fixed in bonds and are unable to move about

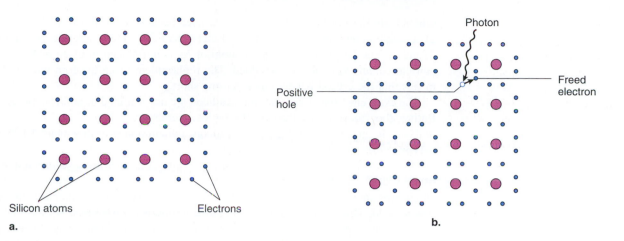

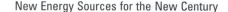

Figure 8.12

(*a*) Schematic of bonding in silicon. (*b*) Photon-induced release of a bonding electron in a silicon semiconductor.

through the crystal. Consequently, silicon is not a very good electrical conductor under ordinary circumstances. However, if an electron absorbs sufficient energy, it can be excited and released from its bonding position, as indicated in Figure 8.12b. Once freed, the electron can move throughout the crystal lattice, thus making the silicon an electrical conductor. The energy required for silicon to release an electron from a bond is 1.8×10^{-19} J per photon, which is equivalent to radiation with a wavelength of 1100 nm. Visible light has a wavelength range of 350 to 700 nm. Recall that the shorter the wavelength of radiation, the greater the energy per photon. Therefore, photons of visible sunlight have more than enough energy to excite electrons in silicon semiconductors. Indeed, hand-held calculators are now powered with solar cells, thus eliminating the added expense of buying batteries.

The fabrication of photovoltaic cells is not without some major problems. Although the starting material from which silicon is extracted (common sand—silicon dioxide) is cheap and abundant, purifying silicon to the appropriate level (99.999% purity) is fairly expensive. A second complication is that the direct conversion of sunlight into electricity is not very efficient. A photovoltaic cell could, in principle, transform into electricity up to 44% of the radiant energy to which it is sensitive. But more than a third of this (16% of the total) is lost to internal cell processes. This leaves a theoretical efficiency limit of 28%. In practice, efficiencies between 10% and 20% have been achieved.

In Chapter 4 we lamented the 30–40% efficiency of converting heat to work in a conventional power plant. It might seem that we should be even more distressed at the lower limits that can be achieved by photovoltaics. But, remember that the Sun is an essentially unlimited energy source, and there is a good deal of empty space on the planet that is well suited for large arrays of photovoltaic cells and for little else. Moreover, the fact that solar energy is free of many of the environmental problems associated with burning fossil fuels or with nuclear fission adds impetus to research and development of solar cells.

The first use of solar cells was to provide electricity in NASA spacecraft, where cost seems to not be a major concern and the intensity of radiation is so high that the low efficiency is not a serious limitation. But because most commercial applications must be more cost-conscious, a great deal of effort is being directed to increasing the efficiency of solar cells and lowering manufacturing costs. One promising innovation is replacing crystalline silicon with the non-crystalline form of the element. Photons are more efficiently absorbed by the less highly ordered Si atoms, a phenomenon that permits reducing the thickness of the silicon semiconductor to 1/60th of its former value. The cost of materials is thus significantly reduced. More common is the "doping" of silicon with other elements. This process consists of intentionally introducing about 1 ppm of elements such as gallium (Ga) or arsenic (As) into the silicon. These two elements and

An ingot of purified silicon.

Si is in Group 4A;
Ga is in Group 3A;
As is in Group 5A.

others from the same periodic families are used because their atoms differ from silicon by a single outer electron. Silicon has 4 electrons in its outer energy level, gallium has 3, and arsenic has 5. Thus, when an atom of As is introduced in place of Si in the silicon lattice, an extra electron is added. The replacement of an Si atom with a Ga atom means that the crystal is now one electron "short."

The extra electrons in arsenic-doped silicon are not confined to bonds between atoms. Rather, they move easily through the lattice, thereby increasing the electrical conductivity of the material over that of pure silicon. Silicon doped in this manner is called an **n-type** semiconductor because its electrical conductivity is due to negative carriers—electrons. On the other hand, for each silicon atom replaced with a gallium ion, an electronic vacancy or "hole" is introduced into what is normally a two-electron bond. When an electron moves into this hole, a new hole appears where the mobile electron formerly was located. Thus, holes and electrons move in opposite directions. Because holes can be regarded as positive carriers of electricity, gallium-doped silicon is called a **p-type** semiconductor. Figure 8.13 illustrates both n-and p-type semiconductors. Both types of doping increase the conductivity of the silicon because less energy is needed to get extra electrons or holes moving. This means that photons of low energy (light of longer wavelength) can induce electron release and transport in doped crystals.

"Sandwiches" of n- and p-type semiconductors are used in transistors and many of the other miniaturized electronic devices that have revolutionized communication and computing. Similar structures are central to the direct conversion of sunlight to electricity. A photovoltaic cell typically includes sheets of n- and p-type silicon in close contact (Figure 8.14). The n-type semiconductor is rich in electrons and the p-type is rich in positive holes. When they are placed in contact, there is a tendency for the electrons to diffuse from the n-region into the p-region, and for the positive holes to move from the p-region to the n-region. This generates a voltage or potential difference at the junction between the semiconductors. This voltage difference accelerates the electrons released when sunlight strikes the doped silicon. If the two layers are connected by a wire or other conductor, electrons will flow through the external circuit from the n-semiconductor where their concentration is higher to the p-semiconductor where it is lower. The result is a direct current of electricity that can be intercepted to do essentially all the things that electricity does. As long as the cell is exposed to light, the current will continue to flow, powered only by radiation.

In addition to making an effort to improve the performance of silicon semiconductors by doping, scientists have been searching for other substances that exhibit the same or better semiconductor properties. Among promising substitutes are germanium, an element found in the same family of the periodic table as silicon, and compounds that have the same number of outer electrons as these elements. Included in this latter list are gallium arsenide (GaAs), indium arsenide (InAs), cadmium selenide (CdSe), and cadmium telluride (CdTe). Some of these new semiconductors have enhanced the

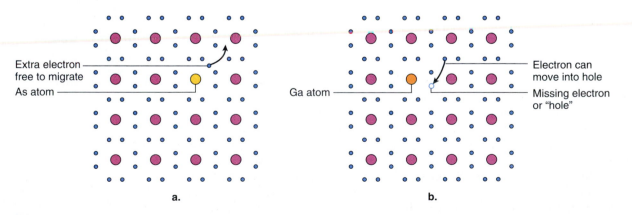

Figure 8.13

(a) An arsenic-doped n-type silicon semiconductor. (b) a gallium-doped p-type silicon semiconductor.

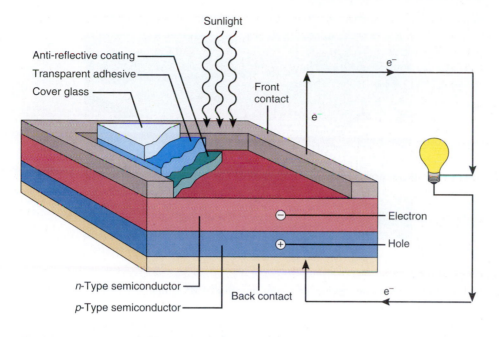

Figure 8.14
Schematic diagram of a photo-voltaic (solar) cell.

efficiency of radiation-to-electricity conversion and made possible photovoltaic cells that are responsive to certain regions of the spectrum. Siemens Solar Industries has recently developed a copper indium selenide semiconducting thin film that has significant advantages over amorphous silicon and cadmium telluride.

8.14 Your Turn

Using the periodic table as a guide, show why the electron arrangement and bonding in GaAs would be very similar to that in pure Ge.

As a result of these and other advances in photovoltaic technology, the cost of producing electricity in this manner has dropped dramatically from $3 per kilowatt-hour in 1974 to 28 cents per kilowatt-hour in 1998. Although electricity produced by this technology is still not competitive with that produced from fossil fuels, the long-range prospects for solar energy are encouraging. Its cost is decreasing while the cost of electricity generated from fossil fuels is increasing. In 1993, the average cost of electricity produced by conventional power plants was 6–8 cents per kilowatt-hour. But the increasing cost of clean fuel and the added expense of pollution controls will drive the cost higher. Given this situation and the expected improvements in the performance and decreases in the cost of solar cells, electricity from photovoltaic systems could become competitive with that from fossil fuels by early in the twenty-first century. Photovoltaic power has grown at an average rate of 16% per year since 1990 (Figure 8.15). "World solar markets are growing at ten times the rate of the oil industry, which has expanded only 1.4% per year since 1990" write Christopher Flavin and Molly O'Meara of the Worldwatch Institute. Major energy companies, such as Amoco and British Petroleum have invested heavily in the solar business, now valued at about $1 billion a year.

The largest solar installation in the United States, located in Carrisa Plains, CA, was built by ARCO Solar, Inc. and Pacific Gas and Electric Company. It generates about seven megawatts at peak power. Although this generating capacity is small compared to that of fossil fuel, nuclear, and hydroelectric plants, much larger solar installations are expected in the future. A 200-megawatt plant, which could provide household power for 300,000 people, could be erected on a square mile of land at relatively modest capital expense and minimal maintenance. At currently attainable levels of operating efficiency, all the electricity needs of the United States could be supplied by a

Using solar cells rather than batteries in navigational buoys saves the U.S. Coast Guard an estimated $6 million annually through reduced maintenance and repair.

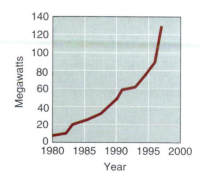

Figure 8.15
The growth of world photovoltaic shipments since 1980.

(Source: Data from G. Maycock, Worldwatch Institute)

Figure 8.16
This array of 1600 photo-voltaic cells in Sacramento, CA produces 2 megawatts (2 million watts) of electricity.

photovoltaic generating station covering an area of 85 miles $\times$ 85 miles, roughly the area of New Jersey.

Photovoltaic technology is already in use in a number of countries. From small, remote villages in developing countries to upscale suburbs in Japan and the United States, 500,000 homeowners worldwide use solar cells to generate their own electricity. In sunny Sacramento, CA, the Municipal Utility District has constructed an array of 1600 photovoltaic cells that produces two megawatts (2×10^6 watts) of electricity, sufficient to serve 600 homes (Figure 8.16). An additional 420 homes have rooftop photovoltaic systems. The homeowners sell back excess electricity to the utility company. The district has voted to close down its nuclear reactors in favor of using photovoltaics and other "cleaner" energy technologies. Aided by generous tax credits, over 23,000 homes in Japan have rooftop solar units, installed to overcome the high cost of electricity.

The European Union and the United States are partners in the "Million Roofs" program, an initiative to install a million rooftop photovoltaic systems by 2010 (Figure 8.17). The U.S. program proposes a 15% tax credit and funds to subsidize partnerships among builders, utilities, and local governments. Already supported by private companies, as well as local and state governments, it remains to be seen whether the program will receive adequate funds from Congress to achieve the goal. A study in western Germany pointed out that half the electricity needs of that region could be met using rooftop solar cells.

More that one-third of the Earth's population is not hooked into an electrical network because of the costs associated with connecting and maintaining equipment, and supplying the fuel to generate the electricity. Because photovoltaic installations are essen-

Figure 8.17
Rooftop photovoltaic cells as part of the "Million Roofs" program.

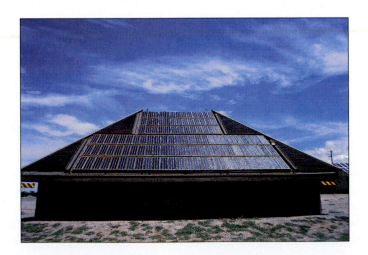

Figure 8.18
Solar (photovoltaic) cells bring electricity to remote areas.

tially maintenance free and can be used almost anywhere, they are particularly attractive for electrical generation in remote regions of the Earth. For example, the highway traffic lights in certain parts of Alaska, far from power lines, operate on solar energy. A similar, but more significant application of photovoltaic cells may be to bring electricity to isolated villages in developing countries. In recent years, more than 200,000 solar lighting units have been installed in residential units in Colombia, the Dominican Republic, Mexico, Sri Lanka, South Africa, and India. An exemplary application of the use of photovoltaic technology is in Indonesia, an archipelago of more than 13,000 islands. About 70% of all households there do not have access to electrical lines. Therefore, installing photovoltaic cells is a legitimate alternative. In the village of Leback, solar electric units have been installed in 500 homes, as well as public buildings, shops, 11 public television units, streetlights, and a satellite antenna (Figure 8.18). Before the photovoltaic systems, Leback villagers used kerosene for lighting and batteries for radios. Kerosene costs $6–12 per month, depending on availability. Under a loan/purchase agreement, villagers pay only $4.25 per month for their home solar electric system.

Solar cells even power cars. Sunrayce USA, a nine-day, 1250-mile race sponsored by the U.S. Department of Energy and private corporations, has become a popular activity for engineering students. Student teams design, build, test, and drive cars that are powered only by photovoltaic cells (Figure 8.19). In spite of their vanguard designs, the cars are not ready to be put on the roads for everyday use. The World Solar Challenge, an 1865-mile race across Australia, from Darwin in the north to Adelaide in the south, attracts commercial entrants. In the 1996 race, a car built by Honda using photovoltaic cells manufactured in Australia covered the course in 35.5 hours, for a record average speed of 53.1 miles per hour. There is also an intercollegiate solar boat competition called Solar Splash. In addition, an airplane powered only by battery-charging amorphous (non-crystalline) silicon solar cells has flown more than 2400 miles in less than 120 hours, and a solar-powered boat has made its appearance.

Earlier in the chapter, we mentioned the potential importance of hydrogen as a fuel from which electricity could be produced. Recently, researchers at the National Renewable Energy Laboratory in Golden, Colorado, developed a photovoltaic cell that in one step generates sufficient electricity to extract hydrogen by decomposing water (Figure 8.20). The cell uses paired gallium-based semiconductors to create the electricity to decompose the water. The efficiency of the process, 12.5%, is nearly double that of the previous record. In that case, the process took two steps in separate devices. The first step generated the electricity; the second step used the electricity to split water into hydrogen and oxygen.

Although prodigious amounts of sunshine hit the Earth daily, the rays do not strike any specific spot on our planet for 24 hours a day, 365 days a year. This means that the electricity generated by photovoltaic cells during the day must be stored for use at night. Given current technology, electrical storage requires batteries, electrochemical devices

Energy costs are high in remote areas, such as in Alaska, or in developing nations. Thus, electricity from photovoltaics is economically competitive in these areas.

Figure 8.19
The Solar Eagle III built by students at California State University-Los Angeles won the 1997 Sunrayce, a 1250 mile race from Indianapolis to Colorado Springs for solar-powered vehicles.

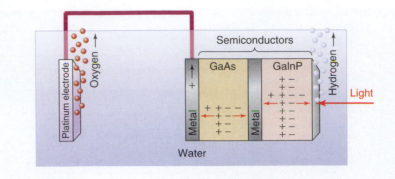

afflicted with the disadvantages discussed in Section 8.5. Nevertheless, the direct conversion of sunlight to electricity has many advantages. In addition to freeing us from our dependence on fossil fuels, an economy based on solar electricity would decrease the environmental damage that frequently occurs when these fuels are extracted or transported. Further, it would help to lower the levels of atmospheric pollutants such as sulfur oxides and nitrogen oxides, and it would also help avert the dangers of global warming by decreasing the amount of carbon dioxide released into the atmosphere. It is likely that fossil fuels will continue to be the preferred form of energy for certain applications. On balance, however, the future looks sunny for solar-based energy. Indeed, the only better source of energy might be if some modern Prometheus could steal a part of the Sun and bring it down to Earth.

| Prometheus was a Greek mythic figure who stole fire from the gods.

8.8 Stealing the Sun

Essentially all of the energy used by the inhabitants of the Earth is nuclear in origin. The energy stored in the fossil fuels we burn and the foods we eat originated in the furnace of the Sun, where it was born in a nuclear reaction. Every second, five million tons of the Sun's matter are converted into 3×10^{23} kJ (7×10^{22} kilocalories) of energy. Yet, despite the profligate rate at which our star is consuming its substance, astrophysicists estimate that it is only about halfway through its expected nine-billion-year life cycle.

The primary reaction that fuels the fires of the Sun is believed to be the combination or **fusion** of four hydrogen atoms to yield one helium atom and two positrons. **A positron** has a mass equal to that of an electron, but carries a positive charge instead of a negative one.

$$4 \, {}_{1}^{1}\text{H} \longrightarrow {}_{2}^{4}\text{He} + 2 \, {}_{+1}^{0}\text{e} \qquad (8.20)$$

As in the case of nuclear fission, the mass of the products of nuclear fusion is slightly less than that of the reactants, and the difference is manifested as energy. The nuclear fusion of one gram of hydrogen releases as much energy as the combustion of approximately 20 tons of coal. Gram for gram, this is eight times as energetic as the fission of uranium-235, *making nuclear fusion the most concentrated energy source known.* If we could only capture and control this same reaction on Earth, our energy problems would be solved forever!

8.15 The Sceptical Chymist

Check the claim that the fusion of one gram of hydrogen will release approximately as much energy as the combustion of 20 tons of coal. To do so, you will need to make use of the fact that when 1.00 g of hydrogen undergoes fusion (as in equation 8.20), 6.72×10^{-3} g of matter is converted into energy. Burning coal releases 30 kJ/gram.

Hint: Remember that $E = mc^2$; $c = 3.0 \times 10^8$ m/s; 1 J = 1 kg · m²/s²

For one thing, there would be no need to worry about running out of fuel. Hydrogen is the most abundant element in the universe. Even on Earth, the waters of the oceans contain a hydrogen supply that is, for all practical purposes, limitless. Furthermore, because the fusion reaction produces only a stable, nonradioactive, and nonreactive element (helium), concerns about air pollution or storing long-lived radioactive waste products would decrease. There are, however, some formidable technical problems that must be overcome before we can create and control a mini-Sun here on Earth. Temperatures greater than 100,000,000°C are required to overcome the repulsion of positively charged hydrogen nuclei and cause them to fuse. The vexing problem is thus twofold: how to create such extreme temperatures and how to contain the incredibly hot matter produced. Large international research teams are working to develop a controlled nuclear fusion reaction that produces more energy than it requires.

To be sure, humans have already carried out fusion reactions on Earth by testing hydrogen bombs, devices that use nuclear fusion. But such tests hardly can be considered controlled fusion. In such thermonuclear devices the heat necessary to trigger fusion is generated by the fission of uranium-235 or plutonium-239. The fusing species are two isotopes of hydrogen: deuterium, which has one proton, one electron, *and* one neutron per atom; and tritium, which has one proton, one electron, and *two* neutrons per atom. Deuterium is symbolized as $_1^2H$ or $_1^2D$, while tritium is represented as $_1^3H$ or $_1^3T$. When atoms of these two isotopes fuse, a helium atom and a neutron are formed (Figure 8.21).

Nuclear equations were discussed in Section 7.7.

$$_1^2H \ + \ _1^3H \ \longrightarrow \ _2^4He \ + \ _0^1n \qquad (8.21)$$

deuterium   tritium   helium   neutron

This reaction is being investigated because it is thought to be somewhat easier to initiate than that involving four atoms of the simplest hydrogen isotope, each with only a proton and an electron. Reaction 8.21 does suffer from the disadvantage that tritium, a radioactive isotope with a half-life of 12.3 years, does not occur in nature in any appreciable amount. But tritium can be produced by bombarding lithium-6 ($_3^6Li$) with neutrons.

$$_3^6Li + _0^1n \longrightarrow _1^3H + _2^4He \qquad (8.22)$$

Deuterium alone is being investigated as an alternative fusion fuel. It forms either tritium and normal hydrogen ($_1^1H$), or helium-3 ($_2^3He$) and a neutron.

$$_1^2H + _1^2H \longrightarrow _1^3H + _1^1H \qquad (8.23)$$

$$_1^2H + _1^2H \longrightarrow _2^3He + _0^1n \qquad (8.24)$$

8.16 Your Turn

A fusion reaction in the sun involves fusing two helium-3 nuclei to form a helium-4 nucleus and two protons. Write the nuclear equation for this reaction.

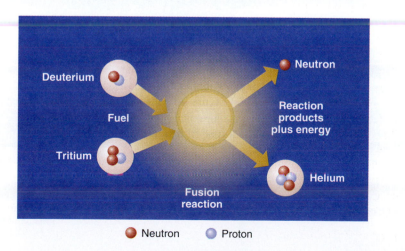

Neutron   Proton

Figure 8.21

Stylized diagram of a nuclear fusion reaction:

$$_1^3H + _1^2H \longrightarrow _2^4He + _0^1n$$

(Source: Data from Department of Energy, *National Ignition 3 Facility*: http://lasers.llnl.gov/lasers/nif/nif_ife.html#fusion)

Figure 8.22

Magnetic containment of a fusing plasma in a tokamak.

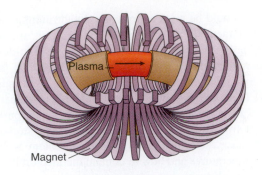

Controlling the reaction once fusion has begun is the other major problem associated with using nuclear fusion as a manageable energy source. For almost 50 years scientists have been trying to accomplish this task, but success has evaded them. The principal difficulty is one of containing the superhot gases, because most materials that might be used as containers vaporize at temperatures above 6000°C, far below temperatures needed for nuclear fusion. A special bottle must be created to enclose this genie. The enormous mass of the Sun provides sufficient gravity to contain its fusion reactions. The mass and gravity of Earth are far less, and so another confinement method must be found here. Two such schemes are currently under investigation: magnetic containment and inertial confinement.

The **magnetic containment** method involves injecting atoms of the appropriate hydrogen isotopes into a donut-shaped metallic chamber, called a **tokamak,** where they are subjected to strong magnetic and electrical fields (Figure 8.22). The electrons are stripped from the atoms to produce a new state of matter, a **gas-like plasma** that consists of electrons and positively charged ions. The electric field heats the plasma until the nuclei fuse. This incredibly hot matter is contained by a strong magnetic force that prevents it from touching and vaporizing the metal walls of the container. Huge amounts of energy are required to generate this plasma. The 40-foot tokamak at Princeton University has achieved temperatures high enough to initiate fusion, and in December 1993 it set a new world's record with a power output of six megawatts. Unfortunately, the power input was 20 megawatts. Since then, a British fusion reactor has broken even, generating as much energy as it consumed, but only for an instant. It may be a long time before a tokamak produces any excess energy. A current estimate is that commercial fusion power plants are another 40 years away.

In **inertial confinement,** diagrammed in Figure 8.23, tiny glass spheres are filled with deuterium or deuterium-tritium mixtures at a pressure of several hundred atmo-

Figure 8.23

Diagram of steps in inertial confinement to achieve nuclear fusion.

(Source: Data from Department of Energy, *National Ignition 3 Facility*: http://lasers.llnl.gov/lasers/nif/nif_ife.html#fusion)

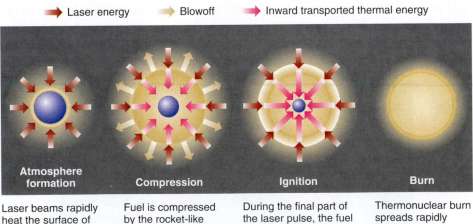

| Laser energy | Blowoff | Inward transported thermal energy |

Atmosphere formation	**Compression**	**Ignition**	**Burn**
Laser beams rapidly heat the surface of the fusion target forming a surrounding plasma envelope.	Fuel is compressed by the rocket-like blowoff of the hot surface material.	During the final part of the laser pulse, the fuel core reaches 20 times the density of lead and ignites at 100,000,000°C.	Thermonuclear burn spreads rapidly through the compressed fuel, yielding many times the input energy.

spheres. These fuel pellets are then subjected to ultraviolet radiation generated by high-energy lasers. The hope is that under this photon bombardment, the atoms are squeezed together tightly enough to cause them to fuse. Up to the present time, this goal has not been achieved and inertial confinement remains unfulfilled as a means of controlling nuclear fusion.

A Washington, DC pundit has said "Nuclear fusion is the energy source of tomorrow, and it always will be." Over nearly 50 years, more than $10 billion, and trillions of joules have been spent in a largely fruitless effort to force hydrogen nuclei to fuse and emit excess energy. Yet, the goal is tantalizing because the potential benefits, like the fuel supply, are almost boundless. Although the natural abundance of deuterium is only about 1.6 ppm, just one cubic kilometer of seawater would furnish enough deuterium to produce the energy equivalent of 2 trillion barrels of oil, roughly the planet's total oil reserves. Considering the vastness of the oceans, there is enough readily accessible deuterium to supply the Earth's energy needs until the dying Sun engulfs our blue-green marble.

8.17 Consider This: Starlight, Star Bright: Nuclear Fusion

"The future of energy on Earth is the energy of the stars," (U.S. Department of Energy). How close are we to this future? What are the latest developments in nuclear fusion? To answer these questions, visit the home page of the U.S. Fusion Energy Sciences Program. You can locate this either by searching the web or by using the link provided at the *Chemistry in Context* home page. Scroll down to see a list of the fusion projects in the United States and abroad.

a. Browse through this award-winning web site on nuclear fusion. List three facts or issues about fusion that catch your attention.

b. What other countries are working on nuclear fusion? Select a link to visit a country of your choice. Summarize the types of information that the site offers.

c. Finally, compare your notes from **a** and **b** with others in your class. Is "the energy of the stars" likely to meet our energy needs in the near future?

Conclusion

Fossil fuels, the Sun's ancient investments on Earth, are fast disappearing and we must seek alternatives for tomorrow. For that future, we are looking to our star for energy or for an example. It is fortunate that what seems to be one of the most promising methods for capturing and transforming solar energy is also the most direct. Photovoltaic cells convert sunlight directly into electricity. There is no need for inefficient intermediate steps in which fuels are synthesized and burned, and the resulting heat energy is transformed first into mechanical and then into electrical energy. Of course, for some purposes, conventional fuels are more convenient than electricity. Perhaps advances in research and changes in global economies will make it fiscally and energetically feasible to use solar radiation to extract hydrogen from water or some other hydrogen source. The hydrogen can either be burned directly as a clean fuel or combined with oxygen in a fuel cell that generates electricity rather than heat. Automobile engineers and designers are actively creating prototypes that use hydrogen, batteries, and fuel cells as energy sources for motor vehicles. Battery-powered cars, such as the EV1, have moved beyond the prototype stage to being mass produced.

But the laws of thermodynamics and human nature are such that these transformations will not occur spontaneously. Energy alternatives cannot be developed without hard work and the investment of intellect, time, and money. Yet, in the United States, the amount of effort and money devoted to research on new energy sources appears to be directly proportional to the cost of oil. When international crises drive up the price of petroleum, there is a sudden flurry of official interest in energy conservation and the development of alternate technologies. When oil supplies are plentiful and prices are low at the gasoline pump, few seem to care about preparing for the time when fossil fuels will be depleted or much too polluting or too expensive to burn. Someday, maybe teams of chemists, physicists, and engineers will succeed in simulating the Sun by controlling nuclear fusion as an affordable energy source, not a mere laboratory curiosity. But until then, we need to establish national and personal priorities, and act on them. We have been the beneficiaries of a bountiful nature, but in turn, we have an obligation to assure energy sources for unborn generations.

Chapter Summary

Having studied this chapter, you should be able to:
- Describe the advantages and disadvantages of hydrogen as a fuel (8.1–8.3)
- Relate the energetics of producing hydrogen and using it as a fuel (8.3)
- Discuss issues related to a hydrogen economy (8.3)
- Describe the design, operation, applications, and advantages of fuel cells (8.4)
- Discuss the principles governing the operation of electrochemical cells, including oxidation and reduction (8.5)
- Compare and contrast the costs and benefits of battery-powered cars (8.6)
- Describe the principles governing the operation of photovoltaic (solar) cells and their current and potential uses (8.7)
- Consider nuclear fusion as an energy source (8.8)

Questions

Emphasizing Essentials

1. **a.** How are equations 8.1 and 8.2 the same and how are they different?

   b. How will the energy released in the reaction shown in equation 8.1 compare to the energy released in the reaction represented by equation 8.2? Explain your reasoning.

2. If 286 kJ of energy are released per mole of H_2 that burns, how much energy will be released when 375 kg of H_2 are used?

3. The text states that 286 kJ of energy are released per mole of hydrogen undergoing combustion according to this equation.

$$H_2(g) + \tfrac{1}{2} O_2(g) \longrightarrow H_2O(l)$$

   a. Use bond energies in Table 4.1 to calculate the energy released in this reaction when one mole of hydrogen burns.

   b. Account for the difference between your results in part **a** and the stated value of 286 kJ of energy released.

4. Consider this table of heats of combustion.

Fuel	Heat of Complete Combustion, kJ/g	Molar Mass g/mol
Hydrogen	143	2.0
Methane (major component of natural gas)	54	16
Octane (major component of gasoline)	46	114

   Will the list be in the same order if the heats of complete combustion are listed in units of kJ/mol? Show calculations to support your answer.

5. How does equation 8.4 relate to equation 8.2? Explain, using appropriate equations.

6. Is the conversion of oxygen gas, $O_2(g)$, to $H_2O(g)$ in a fuel cell an example of oxidation or reduction? Use electron loss or gain to support your answer.

7. Two common units associated with electricity are volts and amps. What is each unit measuring?

8. Electrical energy can be measured in joules, calculated by finding the product of amps × volts × seconds. Calculate the number of joules of electrical energy that can be obtained from a 1.5-volt D cell drawing a current of 1.8 amps for 4.0 hours.

9. Electrical energy can be measured in kilowatt-hours.

   a. Use the fact that there are 1000 watts/kilowatt and 3600 seconds/hour to calculate the number of joules in one kilowatt-hour.

   b. Calculate the number of kilowatt-hours of electrical energy available from the D cell in question 8.

10. Calculate which device uses more kilowatt-hours of electrical energy:

   a 6.0-V lantern battery drawing 2.5 amps of current for 8.0 hours

   a 12.0-V lead storage battery drawing 6.0 amps of current for 2 hours.

11. Consider this diagram of a hydrogen/oxygen fuel cell used in earlier outer space missions.

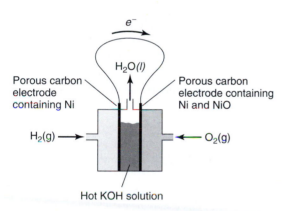

   a. How does the reaction between hydrogen and oxygen in a fuel cell differ from the combustion of hydrogen with oxygen?

   b. Write the half reaction that takes place at the anode in this fuel cell.

   c. Write the half reaction that takes place at the cathode in this fuel cell.

12. In addition to the studies on hydrogen fuel cells, experiments are being done on fuel cells using methane as a fuel. Balance the given oxidation and reduction half reactions and write the overall equation for a methane-based fuel cell.

   Oxidation:
   $$\underline{\quad} CH_4 + \underline{\quad} OH^- \longrightarrow \underline{\quad} CO_2 + \underline{\quad} H_2O + \underline{\quad} e^-$$
   Reduction: $\underline{\quad} O_2 + \underline{\quad} H_2O + \underline{\quad} e^- \longrightarrow \underline{\quad} OH^-$

13. Is there a difference between an electrochemical cell and an electrochemical battery? Explain, giving examples to support your answer.

14. In the lithium-iodine cell, Li is oxidized to Li^+; I_2 is reduced to $2 I^-$.

   a. Write equations for the two half reactions that take place in this cell, labeling the oxidation half reaction and the reduction half reaction.

   b. Write an equation for the overall reaction in this cell.

   c. Identify the half reaction that occurs at the anode and the half reaction that occurs at the cathode.

15. These are the *unbalanced* equations for the half reactions in a lead storage battery. These half reactions do not show the electrons either lost or gained.

   $$Pb(s) + SO_4^{2-}(aq) \longrightarrow PbSO_4(s)$$
   $$PbO_2(s) + 4 H^+(aq) + SO_4^{2-}(aq) \longrightarrow PbSO_4(s) + 2 H_2O(l)$$

   a. Balance both equations with respect to charge by adding electrons on either side of the equations, as needed.

   b. Which half reaction represents oxidation and which represents reduction?

   c. One of the electrodes is made of lead, the other of lead dioxide. Which is the anode and which is the cathode?

16. This *unbalanced* equation represents the last step in the production of pure silicon for use in solar cells.

   $$\underline{\quad} Mg(s) + \underline{\quad} SiCl_4(l) \longrightarrow \underline{\quad} MgCl_2(l) + \underline{\quad} Si(s)$$

   a. How many electrons are transferred per atom of pure silicon formed?

   b. Is the production of pure silicon an oxidation or a reduction reaction? Why do you think so?

17. The mercury battery is used extensively in medicine and electronics industries. The overall reaction can be represented by this equation.

   $$HgO(s) + Zn(s) \longrightarrow ZnO(s) + Hg(l)$$

   a. Write the half reaction of oxidation.

   b. Write the half reaction of reduction.

18. What is meant by the term "hybrid car"?

19. a. What is the "Million Roofs" program?

   b. Who are the partners in this program?

   c. What does this program hope to achieve?

20. The symbol ● represents an electron and the symbol ◯ represents a silicon atom. Does this diagram represent a gallium-doped *p* type silicon semiconductor, or does it represent an arsenic-doped *n*-type silicon semiconductor? Explain your answer.

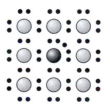

21. a. Where is photovoltaic technology already in widespread use?

   b. What conditions in those areas made photovoltaic use possible?

22. What is the difference between nuclear fission and nuclear fusion? Give an example of an equation representing each process to help explain the difference.

23. a. What is a positron and how is it related to an electron?

   b. What is the symbol for a positron?

Here is a table of nuclear masses for several isotopes. Use these data to answer questions 24 and 25.

Isotope	Nuclear Mass, g/mol
^1_1H	1.00728
^2_1H	2.01355
^3_1H	3.01550
^1_0n	1.00867
^3_2He	3.01493

24. Calculate the mass difference and the energy released in this fusion reaction.

$$^2_1H + {}^2_1H \longrightarrow {}^3_1H + {}^1_1H$$

25. Does this fusion reaction release more energy or less energy than the reaction in question 24? Show your calculation.

$$^2_1H + {}^2_1H \longrightarrow {}^3_2He + {}^1_0n$$

Concentrating on Concepts

26. Assuming that electric cars are available in your area, what questions would you ask the car dealer before deciding to purchase this type of car? Which questions do you consider most important? Offer reasons for your choices.

27. a. What is meant by the phrase "the hydrogen economy"?

   b. Even if methods for producing hydrogen cheaply and in large quantities become available, what problems still remain for the hydrogen economy?

28. Although hydrogen gas can be produced by the electrolysis of water, this reaction is usually not carried out on a large scale. Suggest a reason for this fact.

29. Consider this diagram of two water molecules in the liquid state.

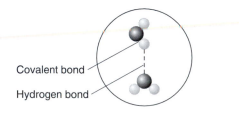

Covalent bond

Hydrogen bond

   a. Which of these bonds are broken when water boils? Are these intermolecular or intramolecular bonds?

   b. What bonds are broken when water is electrolyzed? Are these intermolecular or intramolecular bonds?

30. You have seen several examples of redox reactions in this chapter, and have identified the exchange of electrons taking place. Now examine these equations and decide which are re-

dox reactions and which are not. Give the reasons for your decisions.

Equation 1. $Zn(s) + 2 MnO_2(s) + H_2O(l) \longrightarrow Zn(OH)_2(s) + Mn_2O_3(s)$

Equation 2. $HCl(aq) + NaOH(aq) \longrightarrow NaCl(aq) + H_2O(l)$

Equation 3. $CH_4(g) + 2 O_2(g) \longrightarrow CO_2(g) + 2 H_2O(g)$

31. The text describes a prototype fuel cell that converts gasoline to hydrogen and carbon monoxide. The carbon monoxide, in contact with a special catalyst, then reacts with steam to produce carbon dioxide and additional hydrogen.

   a. Write a set of reactions that describe this prototype fuel cell, using octane (C_8H_{18}) to represent the hydrocarbons found in gasoline.

   b. When is this fuel cell expected to be commercially available?

   c. Speculate as to the future economic success of this prototype fuel cell.

32. Fuel cells were invented in 1839, but were never developed into practical devices for producing electrical energy until the advent of the U.S. space program in the 1960s. What advantages did fuel cells have over other power sources that led to widespread utilization of fuel cells in the space program?

33. Hydrogen, H_2, and methane, CH_4, can each be used with oxygen in a fuel cell. Hydrogen and methane also can be burned directly. Which has a higher heat content when burned, 1.00 g of H_2 or 1.00 g of CH_4? *Hint:* Write the balanced chemical equation for each reaction and use the bond energies in Table 4.1 to help answer this question.

34. Very small mercury batteries are useful in calculators, hearing aids, and camera equipment. Why is the EPA concerned about the use of such batteries?

35. Why are electric cars powered by lead storage batteries alone only a short-term solution to the problem of air pollution emissions from automobiles? Outline your reasons.

36. 5.6×10^{21} kJ of energy come to Earth from the Sun every year. Why can't this energy be used to meet all of our energy needs?

37. At the present time, the cost of electricity generated by solar thermal power plants is greater than that of electricity produced by burning fossil fuels. Given this economic fact, suggest some strategies that might be used to promote the use of environmentally cleaner electricity.

38. Prepare a list of the environmental costs and benefits associated with hybrid vehicles. Compare that list with the environmental costs and benefits of vehicles powered by gasoline. On balance, which energy source do you favor, and why?

39. William C. Ford, Jr., the chief executive officer of Ford Motor Company, is quoted as saying that going "totally green" with zero-emission vehicles will be a real challenge. Regular drivers won't buy high-tech clean cars, Ford admits, until the industry has a "no-tradeoff" vehicle widely available. What do you think he means by a "no-tradeoff" vehicle? Do you think he is justified in this opinion?

40. Are commercial nuclear power plants (Chapter 7) powered by fusion reactions? Why or why not?

Exploring Extensions

41. A laboratory method to prepare hydrogen gas is by reacting metallic sodium with water as shown in this equation.

$$2 \text{ Na}(s) + 2 \text{ H}_2\text{O}(l) \longrightarrow \text{H}_2(g) + 2 \text{ NaOH}(aq)$$

   a. Calculate the grams of sodium needed to produce 1.0 mole of hydrogen gas.

   b. Calculate the grams of sodium needed to produce sufficient hydrogen to meet an American's daily energy requirement of 1.1×10^6 kJ.

   c. The price of sodium is $94 per kilogram. Determine the cost of producing 1.0 mole of hydrogen by the reaction of sodium and water, assuming the water is free.

42. a. There are several advantages and disadvantages of using hydrogen as a fuel. Set up parallel lists that give the advantages and the disadvantages of using hydrogen as the fuel for transportation and for producing electricity.

   b. Do you advocate the use of hydrogen as a fuel for transportation and/or for the production of electricity? Explain your position by writing a short editorial to a classmate.

43. The aluminum-air battery is being explored for potential use in automobiles. In this battery, aluminum metal undergoes oxidation to Al^{3+} and forms $Al(OH)_3$. Oxygen, O_2, from the air undergoes reduction to OH^- ions.

   a. Write equations for the oxidation and reduction half-reactions. Use H_2O as needed to balance the number of hydrogens present, and add electrons as needed to balance the charge.

   b. Add the half reactions to obtain the equation for the overall reaction in this cell.

   c. Specify which half reaction occurs at the anode and which occurs at the cathode in the battery.

   d. What are the potential benefits of the widespread use of the aluminum-air battery? What are some of the limitations? Write a brief summary of your findings.

   e. What is the current state of development of this battery? Is it in use in any vehicles at the present time? What is its projected future use?

44. The text discusses projections made by Ballard Power Systems, in worldwide alliances with Daimler-Chrysler and Ford Motor Company, to become the largest commercial producer of fuel cell–powered electric drive trains and components for cars, trucks, and buses. The alliance plans to produce 100,000 fuel cell–powered electric vehicles by 2004–2005. What is the progress toward that goal? See if you can find whether there have been any revisions to their predictions. Write a brief report explaining their progress, gearing your report towards an investor who wants to know the potential future growth of the company.

45. There has not been much publicity about the "Million Roofs" program. Design a poster to explain to the general public what this program is all about and to enlist their support.

46. Offer an explanation for each of these statements.

   a. Nuclear fusion appears to be possible only at high temperatures, whereas nuclear fission occurs spontaneously at room temperature.

   b. Nuclear fusion releases more energy per gram than nuclear fission.

   c. Electricity produced by solar energy is expected to decrease in cost within twenty years. Electricity produced from fossil fuels is expected to increase in price.

47. Consider three artificial sources of light: A candle, a battery-powered flashlight, and an electric light bulb. For each source, provide this information.

   a. The origin of the light.

   b. The immediate source of the energy that appears as light.

   c. The original source of the energy that appears as light. (*Hint:* Trace this back stepwise as far as possible.)

   d. The end-products and by-products of using each of the light sources.

   e. The environmental costs associated with each light source.

   f. The advantages and disadvantages of each light source.

48. Figure 8.16 shows an array of photovoltaic cells installed by the Municipal Utility District in Sacramento, CA. Where else in the United States or in the world is there a comparable array? Use the web to learn of other large-scale photovoltaic cell installations. What factors help to influence this approach, one that uses a centralized array rather than using individual rooftop solar units?

49. The text refers to fossil fuels as the " . . . Sun's ancient investment on Earth." How would you interpret this statement to a friend who is not enrolled in this course?

50. If hybrid cars become the choice of consumers in the near future, the popular saying "fill 'er up" will have to be replaced by "plug 'er in." Design a one-minute TV public service announcement that conveys what the new phrase means and why the former phrase no longer applies. Present this announcement to your class either live or on film and gather feedback on its effectiveness.

The World of Plastics and Polymers

Synthetic polymers have many uses, depending on their properties.

"I'm on my way to go hiking and camping this weekend. Oh, to get away from classes and studying for a long weekend! I can just put all that stuff on the back burner until next week.

Although the forecast isn't too nice, I don't even care whether it will be cold or rainy. I'm really going to be well equipped for this trip. I've got these new polyester and Lycra pants and a warm acrylic sweater. And of course if it really gets very cold I'll be wearing my microfiber polypro thermal tights. What I'm really proud of is this jacket. It's Gore-Tex so that the rain won't wet it at

all, and the Thinsulate lining makes it
warm, but it weighs almost nothing. My
hiking boots are terrific—lightweight,
waterproof, and they fit like a glove. And
I'm borrowing a new tent made of nylon
with plastic poles. It all folds up so I can
stick the whole thing into my backpack.
Well, I'm out of here to commune with
Ma Nature. It's great to get away from all
that synthetic, artificial junk for a
change. Catch ya later."

An example of using polymers to play.

9.1 Consider This: Hiking Gear Analysis

Plastics, polymers, and other synthetic materials have revolutionized sports equipment in recent years. Such technology has created lighter, stronger, and more responsive materials to match recreational needs. This is especially true with hiking and camping. Look a little more closely at the campers shown in Figure 9.1. Identify the parts of their equipment and clothing that are created by chemists, and describe the properties of these materials that make them well-suited for their intended uses.

9.2 Consider This: Sporty Polymers

Choose a favorite outdoor activity—on the land, water, or even in the air. Do a web search to find a company that manufactures or sells equipment for this activity. Which polymers can you find mentioned in this web site? Make a table of their names, the equipment they are used in, and any desirable properties cited as advertising points. *Hint:* Many polymer names start with "poly," such as polyester or polypropylene. Other polymers have trade names, such as Gore-Tex, Orlon, or Styrofoam. Still other polymers are coatings and resins, and may be mentioned as epoxides or acrylics.

Chapter Overview

We begin the chapter by stating the obvious: plastics are all around us. After considering a few examples, we identify plastics with polymers—materials whose molecules consist of long chains of atoms. Polymers can be natural or synthetic, but here we are more concerned with the latter. A brief history of the development of synthetic polymers is followed by an examination of the great growth in the use of these materials and their impact on the American economy, lifestyle, and leisure. Observation and discussion of some of the properties of polymers lead to the identification of the six strategies employed by chemists to vary these properties by modifying molecular structure. These strategies are illustrated in our coverage of the six most common polymers. Because polyethylene is the simplest and most widely used plastic, we look at it in some detail. Its composition, structure, production, properties, modifications, and uses are all considered. This study leads to briefer treatments of the other members of the "Big Six." In this context, polymerization by addition and by condensation reactions is explained. A short but significant aside addresses proteins, an important class of natural polymers, and nylon, a related synthetic polymer.

The phenomenal success of plastics and their widespread distribution has not been without cost. Therefore, the final third of the chapter is devoted to the raw materials that go into the manufacture of plastics and the problems associated with

the disposal of used plastics. The issue of plastic versus paper supermarket bags provides a focus and a brief case study. We examine disposal options including incineration, biodegradation, reuse, recycling, and source reduction. Not surprisingly, we find that there are no easy solutions to the problems posed by these useful and ubiquitous forms of matter.

9.1 Polymers: The Long, Long Chain

It should be obvious that our camping student missed the point. You cannot get away from synthetic polymers—especially not when hiking or camping; or anywhere else, for that matter. At this moment you are probably wearing or carrying at least a half dozen different materials that did not exist 70 years ago, perhaps even 20 years ago. Your shoes alone may consist of six or more different kinds of plastic: in the sole, the trim, the foam padding, the sock liner, the upper, the laces, and even the lace tips. It may be printed right on the shoe: "All man-made materials." Or, the label might tell you what is not synthetic: "Leather uppers." Very likely at least some of your clothing contains synthetic fibers. The pen that you are taking notes with is made of plastic. Your calculator has a plastic case, and so does your cellular phone and computer. And the CDs that you play are made of plastic, as are their containers. Several kinds of plastics are also essential components of your car, which contains nearly 400 pounds of them. And that coffee cup you hold is likely made of plastic, whether it is the light, white kind or the environmentally friendly "travel" mug that you get refilled at the coffee shop.

We apply the term "plastic" to a wide range of materials with an equally broad range of properties and applications. According to a standard dictionary definition, *plastic* is an adjective meaning "capable of being molded" or a noun referring to something that is capable of being molded. More specifically, the *Merriam-Webster Collegiate Dictionary Tenth Edition* mentions "any of numerous organic synthetic or processed materials that are mostly thermoplastic or thermosetting polymers of high molecular weight and that can be molded, cast, extruded, drawn, or laminated into objects, films, or filaments." You are familiar with the common or brand names of many different plastics: rayon, nylon, Lycra, polyurethane, Teflon, Saran, Styrofoam, and Formica to list only a few. In this text we will reserve the term plastics for such synthetic substances—all are polymers and all are creations of chemists. What polymers have in common is evident at the molecular level.

9.3 Consider This: How Much Plastic Do You Throw Away?

Keep a journal of all the plastic and plastic-coated products you throw away (not recycle) in one week. Include plastic packaging from food and other products that you purchase. Keep the journal handy because you will be asked to review it in a later activity (9.29 Consider This).

All plastics are large molecules made up of long chains of atoms covalently bonded together. Like a linked strand of paper clips, the molecular chain in a plastic consists of subunits that are repeated many times. This basic subunit is called a **monomer** (from *mono* meaning "one" and *meros* meaning "unit"). Many monomers join together to form a long molecular strand called a **polymer** (*poly* means "many"). These polymer molecules can be very long indeed. Sometimes they involve thousands of atoms, and molecular masses can reach over a million. No wonder that polymers are sometimes referred to as **macromolecules.**

Although this chapter will focus primarily on synthetic polymers, it is important to note that chemists did not invent polymers. Polymeric materials were here long before chemists were, and chemists are also made up of many polymers. Natural polymers are found in plants as well as in animals: wood, wool, cotton, starch, rubber, skin, hair; even some minerals, such as asbestos and quartz, are polymers. Polymeric molecules give strength to an oak tree, delicacy to a spider's web, softness to goose down, and flexibility to a blade of grass. Much of the motivation for the synthesis of plastics has been a desire to reproduce such properties in artificial materials. Indeed, many synthetic polymers were originally created as substitutes for expensive or rare naturally occurring materials, or to improve on natural polymers.

Nature provides the pattern and prototype for polymers. Large molecules are possible only when certain atoms join to form long chains. Chief among these atoms are those of the element carbon. Carbon atoms combine with each other not only in chains, but also in rings, meshlike networks, and three-dimensional structures. Even pure carbon possesses these characteristics, which are evident in its two common allotropes— graphite and diamond. In diamond, the upscale allotrope, each of the carbon atoms is covalently bonded to four others as illustrated in Figure 9.2a. The shared electron pairs are held tightly between adjacent atoms. In the low-priced form, graphite, the atoms are arranged at the corners of six-membered rings (Figure 9.2b). These rings are interconnected so that a sheet of graphite looks like chicken wire. Although the bonding between adjacent atoms in graphite is strong, the forces between the sheets are weak. Hence, it is easy to rub off layers of carbon. You do so every time you use a "lead" pencil. The "lead" isn't lead at all, but graphite with a clay binder. Not too surprisingly, the carbon fibers in recreational items, such as skis and tennis rackets, are more like graphite than like diamond. The carbon fibers are long chains of hexagonally arrayed carbon atoms, well aligned in a polymeric resin. The carbon provides the strength and flexibility, and the resin holds the composite material together.

Until around a decade ago, this brief discussion of diamond and graphite would have told almost the entire story of carbon and its structures. But chemists have become very excited about a new form of carbon called "buckyballs," discovered in 1985. In spite of the name, these balls have nothing to do with sport and recreation, at least not yet. But they have a good deal to do with the arrangement of atoms. Chemists at Rice University succeeded in forming and isolating carbon molecules composed of 60 atoms. The shape of one of these C_{60} molecules is beautiful to behold (Figure 9.2c). It looks like a soccer ball, with the carbon atoms bonded in 20 six-membered rings and 12 five-membered rings. Each carbon is located at a corner where two six-membered rings and one five-membered ring come together. Dr. Richard Smalley, Dr. Harry Kroto, and Dr. Robert Curl, Jr., the chemists who pioneered work on this new form of carbon, won the 1996 Nobel Prize in chemistry for their groundbreaking creations. They also demonstrated a sense of humor in giving the new carbon allotrope the whimsical name "buckminsterfullerene," after the visionary designer and thinker, Buckminster Fuller. Fuller was a pioneer in designing geodesic domes, rigid structures with the same three-dimensional geometry as his namesake molecules.

Right now, fullerenes and compounds made from them are mostly chemical curiosities. But chemists are already thinking of dozens of potential uses, including possible application of fullerenes in superconductors. And close relatives to buckyballs called buckytubes, symmetrical tubes of carbon atoms, are being studied for applications in electronics and medicine.

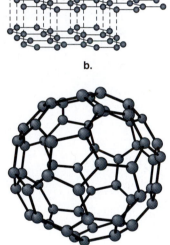

Figure 9.2a,b,c

Structures of diamond (*a*), graphite (*b*). (*c*) Structure of C_{60}, buckminsterfullerene. The spheres represent the carbon atoms and the solid lines indicate the covalent bonds between them.

(a) From left: Drs. Robert Curl, Jr., Harold Kroto, and Richard Smalley, 1996 Nobel Prize winners in chemistry. (b) The geodesic dome at Epcot Center.

9.2 The Plastic Economy

Celluloid, the first commercial plastic, was developed over a century ago by John Wyatt in response to a $10,000 prize offered for a synthetic substitute for ivory in billiard balls. In 1870, Hyatt obtained a patent for a mixture of cellulose nitrate, alcohol, and camphor that was heated, molded, and allowed to harden forming celluloid. Cellulose nitrate, made by treating cotton with nitric and sulfuric acids, is better known as "gun cotton." It is highly flammable and, under some conditions, sufficiently explosive to be used in smokeless gunpowder. Such properties are somewhat less than desirable in billiard balls, though the story of exploding pool balls may be a myth. Back in 1870, the primary motivation for seeking substitutes for ivory was probably economic. Today, the motivation has changed. Elephant herds have been drastically depleted by poachers collecting tusks for the ivory trade. In response, the U.S. government has banned the import of ivory. New plastics provide the starting materials, not only for billiard balls, but for many art objects as well.

The next major breakthrough in plastics, and the first totally synthetic one, was an invention by Leo Baekeland, a Belgian-born chemist who spent most of his career in the United States. In 1907 Baekeland combined two common carbon-containing compounds, phenol and formaldehyde, under high temperatures and pressures. The result was an opaque brittle black polymer that the inventor called Bakelite. Bakelite was not particularly attractive, but it soon found uses in electrical devices, containers, and, as with celluloid, billiard balls. Polyvinyl chloride was introduced in 1912. The 1930s and early 1940s saw a great growth in new polymers with the invention and commercialization of polystyrene, Plexiglas, Melamine, nylon, Teflon, polyethylene terephthalate, and Orlon (Figure 9.3). Many of the more specialized polymers were introduced after World War II.

To some people, the word "plastic" may carry the connotation "cheap" or "tacky." But the fact remains that synthetic polymers have revolutionized modern life. Few advocates of natural materials would be willing to give up sports shoes, synthetic rubber tires, camping gear, and the dozens of other plastic objects that have become an accepted part of today's lifestyle. As chemists have developed new polymers, the variety of properties and uses have expanded dramatically. For example, plastics have become increasingly important in automobile manufacturing. Some new plastics are stronger than steel and much more resistant to corrosion. Hence, they can be substituted for steel and other metals and materials in various parts of a car, including gasoline tanks, bumpers, body parts, and exterior and interior trim. Because the plastics are considerably less dense than structural metals, such substitution has led to significant reductions in vehicle weight. As a result, plastics help make cars more fuel efficient.

Figure 9.3

Items made from polystyrene, Plexiglas®, Melamine®, nylon, Teflon®, polyethylene terephthalate, and Orlon®, respectively.

Automobiles, of course, are one of many examples of the uses of polymers. Plastic packaging reduces weight, eliminates breakage, and helps to save fuel during shipping. Grocery bags made of plastic are another example of altered materials; they now use 30% less material than they did in 1994. Plastic construction materials have replaced wood in some applications, and plastic pipes substitute effectively for ones made from lead, iron, copper, and tile. As the introductory episode indicates, recreation has been revolutionized by the introduction of synthetic polymers (Figure 9.4). Football is played on artificial turf by players wearing plastic helmets, padding, and pants. Tennis balls,

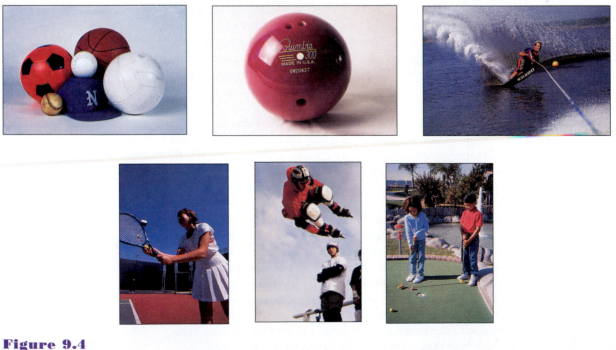

Figure 9.4

Plastics are used in many recreational activities.

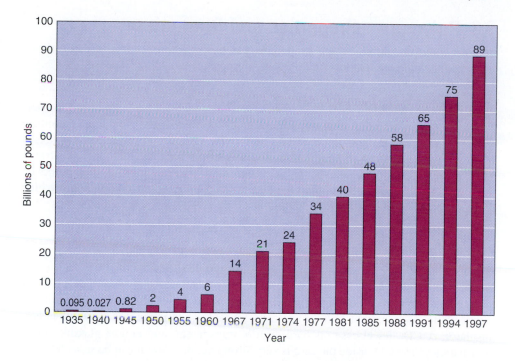

Figure 9.5

Annual U.S. production (in billions of pounds) of plastics from 1935 to 1997.

(Source: Data from Society of the Plastics Industry.)

tennis rackets, and strings are all made from synthetic polymers. In-line skates are made almost entirely of synthetic polymers. Carbon fibers embedded in plastic resins provide the strength and flexibility required in fishing rods, golf-club shafts, and sailboat hulls and sails. Ice skaters and hockey players skate without ice on rinks of Teflon or high-density polyethylene. Most modern canoes are made of synthetic polymers, not birch-bark, wood, or aluminum. Professional baseball, that bastion of conservatism and tradition, still clings to natural polymers in the form of wooden bats, leather gloves, and a ball made of cowhide covering woolen yarn and a cork center. But even here, cotton/polyester blend uniforms have replaced the hot, scratchy wool ones worn by Babe Ruth, Joe DiMaggio, and even Hank Aaron early in his career.

All of this adds up to a formidable economic opportunity. Figure 9.5 indicates that in the period between 1935 and 1997, production of plastics in the United States increased over 900-fold. In 1935, a mere 95 million pounds of plastics were produced. By 1997, that amount had increased to 88.8 billion pounds of plastics produced in this country. This represents an 18% increase over the 1994 production level, stimulated by growth in both the consumer and industrial markets. The plastic manufactured in 1997 had a value of over $40 billion, and was used in products valued at $91 billion. In fact, plastics production has eclipsed that of metals. Since 1976, the United States has manufactured a larger volume of synthetic polymers than the volume of steel, copper, and aluminum combined. Six polymeric materials account for about 66% of all the plastics used in the United States. Much of this chapter will deal with the "Big Six."

9.4 Consider This: Personal Polymers

A Teflon ear bone, fallopian tube, or heart valve? A Gore-Tex facial implant or hernia repair? Some polymers are biocompatible and are now used to replace or repair body parts. List four properties that would be desirable for polymers used *within* the human body.

Other polymers may be used outside your body — but in close contact with it. For example, no surgeon is needed for you to use your contact lenses—you insert, remove, clean, and store them yourself. From which plastics are contact lenses made? What properties are desirable in these plastics? Either a call to an eyewear store or a search on the worldwide web may provide some answers. For the latter, a search using "contact lenses" and "plastics" will provide a useful start.

9.3 Polymers—Properties and Strategies

A good way to begin your study of polymers is by collecting various types of plastic and making some observations.

> ## 9.5 Consider This: Plastics You Use
>
> Gather up a variety of plastic items from your residence or your dormitory room—plastic bags, soda bottles, whatever happens to be at hand. Make a list of the objects and note the properties of the plastics. Include color, transparency, flexibility, elasticity, hardness, and other properties that could be used to classify and identify the plastics. Try to draw conclusions about which objects are made from the same material.
>
> *Hint:* Table 9.1, on page 350, will be of some help.

You have no doubt discovered from the 9.5 Consider This activity (or from prior experience) that plastics exhibit a wide range of properties. We can illustrate this with a few objects that might well be found in your room or residence. The ubiquitous Styrofoam cup is white, opaque, light, soft, and easily deformed and torn, but it is an excellent heat insulator. A CD "jewel case" is hard, brittle, transparent, and almost glasslike in its transparency. The CD inside the case is very strong yet delicate; it can be scratched easily. Videotape is shiny, flexible, and durable. Your backpack is water resistant, lightweight, and sturdy, chosen from a myriad of styles and colors, possibly even your school's colors. The plastic that makes up most soft-drink bottles is transparent and has moderate hardness and flexibility. A plastic milk bottle is somewhat opaque or at least translucent. Although it can be deformed, it is not as soft and flexible as many plastics. Finally, in our brief survey, a typical grocery plastic fruit bag is light, transparent, flexible, and easily stretched.

Investigations of this sort yield useful information about the properties of plastics. Additional data can be obtained in the laboratory by quantitative determination of density, hardness, tensile strength, melting point, and so on. But it is sometimes difficult to correlate these properties unambiguously to the chemical composition of the plastics. If you were trying to sort the items described in the previous paragraph for recycling, you might find it hard to do so. In fact, it may be surprising that objects as different as the foam coffee cup and a CD case are made of the same plastic—polystyrene. The videotape and soda bottle are composed primarily of polyethylene terephthalate, and the plastic bag and the milk bottle are both polyethylene (Figure 9.6).

It follows that the properties of a plastic must be a consequence of more than just its chemical composition—the atomic ratio of the elements that make up the material. How the atoms of those elements are linked together is an important factor. Indeed, the great variety in the properties of plastics is a consequence of variations in the molecular structure of a polymer. Therefore, a major goal of polymer science (and this chapter) is to correlate the properties and uses of plastics with their molecular structures.

The widespread applicability of plastics is a consequence of the ability of chemists to customize the properties of plastics by altering their molecular structures. Such activities provide creative challenges and meaningful employment to many of our industrial colleagues. As a matter of fact, more chemists are employed in the polymer and plastics sector than in any other branch of the chemical industry. In a sense, these scientists "design" the desired properties of plastics into their constituent molecules. In doing so, they follow one or more of only six general strategies for modifying polymer chains—a remarkably small number when you consider the thousands of plastics known. Yet these six ways produce stunning results.

The strategies involve altering the following features of the polymer chain:

1. the length of the chain (the number of monomer units),
2. the three-dimensional arrangement of the chains in the solid,

Figure 9.6
Items made from polyethylene (milk jug, plastic bag) and polystyrene (CD case).

3. the branching of the chain,

4. the chemical composition of the monomer units,

5. the bonding between chains, and

6. the orientation of monomer units within the chain.

All of these options will be illustrated in the pages that follow. We begin with polyethylene, the most common plastic of all, which is made using ethylene as the monomer.

9.4 Polyethylene: The Most Common Plastic

We will use polyethylene to exemplify the use of strategies 1 and 2 for modifying polymer chains. You probably encounter **polyethylene** every day of your life. Over 20 million tons of it are produced in the United States each year. It is found in grocery bags for fruits and vegetables, dry-cleaner garment bags, squeeze bottles, TV cabinets, toys, and hundreds of other objects. This wide variety of uses suggests a similarly wide range of properties for this single polymer. Yet all polyethylene is made from the same starting material—ethylene, C_2H_4 (also represented as CH_2CH_2).

Ethylene is a compound extracted from petroleum. At ordinary temperatures and pressures, ethylene is a gas. However, in the 1930s it was discovered that by using a special catalyst to initiate the reaction, individual ethylene molecules could be made to bond to each other to form a polymer. The key to this behavior is the structure of the ethylene molecule. The two carbon atoms in ethylene are linked with a double bond that is capable of reacting with another ethylene molecule. The production of polyethylene begins with the joining of two ethylene molecules to form a four-carbon chain (equation 9.1.)

Curiously, ethylene acts as a natural plant hormone used to ripen fruit.

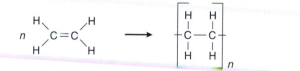

$$(9.1)$$

An additional ethylene molecule joins by adding to the growing chain (equation 9.2).

$$(9.2)$$

By continuing to add ethylene molecules, the chain grows rapidly to form a long polymer containing n monomeric units (equation 9.3).

$$(9.3)$$

The numerical value of n and hence the length of the chain varies with the reaction conditions (strategy one of modifying the properties of a polymer). Often n is in the hundreds or thousands. Moreover, it can vary because a typical synthetic polymer is a mixture of individual polymer molecules of varying length and mass. Molecular masses of polyethylene are generally between 10,000 and 100,000. In every case, however, the carbon atoms are attached to each other by single bonds, and the hydrogen atoms are bonded to the carbon atoms. A polyethylene molecule is thus a macromolecular version of a hydrocarbon molecule, such as those in petroleum.

Strategy 1: Changing the length of the chain.

9.6 Your Turn

Write the structural formula of a polyethylene chain containing eight ethylene units.

9.7 Your Turn

The average molar mass of a sample of polyethylene is 84,500 g.

a. How many monomer units, *n*, of ethylene are present in this polymer molecule?

b. How many atoms are present in this polymer?

Ans.   a. The monomeric unit in polyethylene is CH_2CH_2, with a molar mass of 28.0 g. Therefore, *n*, the number of monomer units, is obtained dividing the average molar mass of the polymer by the molar mass of the monomer and then multiplying by Avogadro's number.

$$n = \frac{84500\ \cancel{g}}{1\ \text{mole polymer}} \times \frac{1\ \text{mole monomer}}{28.0\ \cancel{g}}$$

$$n = \frac{3000\ \text{mole monomer}}{\text{mole polymer}};$$

$$3000\ \text{mole monomer units} \times \frac{6.02 \times 10^{23}\ \text{monomer units}}{1\ \text{mole monomer units}} = 1.82 \times 10^{27}\ \text{monomer units}$$

b. 1.09×10^{28} atoms

Free radicals are very reactive species, as you may recall from the role of · Cl atoms in stratospheric ozone destruction (Section 2.15).

How ethylene polymerizes involves what happens to its electrons. The reaction is initiated by a catalyst that is a free radical that has one unpaired electron. In Figure 9.7, a representation of the polymerization of polyethylene, the free radical is represented by "R·" (the dot indicates an unpaired electron). The radical reacts readily with a CH_2CH_2 molecule. One of the two bonds between the carbon atoms in ethylene breaks, and one of the electrons from that bond pairs with the unpaired electron of the radical to form a covalent bond (Figure 9.7). The new molecule that is formed, RCH_2CH_2·, is also a free radical because it carries an unpaired electron left over from the broken carbon-carbon bond. Therefore, RCH_2CH_2· can react with another ethylene molecule that bonds to the carbon atom with the unpaired electron at the reactive, growing end of the polymer. This process is repeated many times over in many chains at the same time. Occasionally, the free radical ends of two polymers interact to form a bond and stop the chain growth. The result of all this chemistry is that gaseous ethylene is converted to solid polyethylene.

A tetrahedral arrangement around each carbon atom makes the carbon atoms in polyethylene align in a zig-zag arrangement rather than a straight chain.

Many of the properties of polyethylene are related to the presence of these long chains of polymer molecules. Relatively speaking, they are very long indeed. If a polyethylene molecule were as wide as a piece of spaghetti, the molecular chain could be as much as half a mile long. To continue the analogy, in the polyethylene used to make plastic bags these chains are arranged somewhat like spaghetti on a plate. The strands are jumbled up and not very well aligned, though there are quasi-crystalline regions where the molecular chains are parallel. Moreover, the polyethylene chains, like spaghetti strands, are not bonded to each other. Evidence of this molecular arrangement can be obtained by doing a little experiment. Cut a strip from a heavy-duty transparent polyethylene bag, grab the two ends of the strip, and pull. A fairly strong pull is required to start the plastic stretching, but once it begins, less force is needed to keep it going. The length of the plastic strip increases dramatically as the width and thickness decrease

Figure 9.7

The polymerization of ethylene.

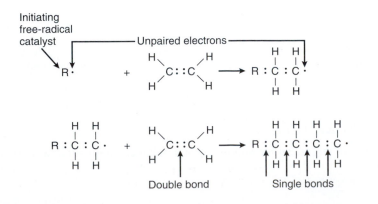

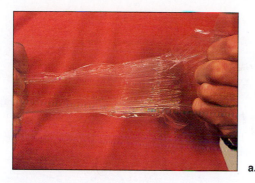

a.

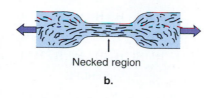

Necked region

b.

Figure 9.8
(*a*) A plastic bag stretched until it "necks."
(*b*) Molecular rearrangement as polyethylene is stretched.

(Figure 9.8). A little shoulder forms on the wider part of the strip and a narrow neck almost seems to flow from it in a process called "necking." Unlike the stretching of a rubber band, the necking effect is not reversible and eventually the plastic thins to the point where it tears.

Figure 9.8b is a representation of the necking of polyethylene from a molecular point of view. As the strip narrows and necks down, the previously mixed-up molecular chains move. They shift, slide, and align parallel to each other and the direction of the pulling force. In some plastics, such stretching or "cold drawing" is carried out as part of the manufacturing process to obtain ordered polymer chains. *This is an example of the second general strategy—altering the three-dimensional arrangement of the chains in the solid.* Of course, as the force and stretching continue, the polymer eventually reaches a point at which the strands can no longer realign, and the plastic breaks. Paper, another polymeric material, tears when pulled because the strands (fibers) in paper are rigidly held in place and are not free to slip like the long molecules in polyethylene.

Strategy 2: Changing the *three-dimensional arrangement* of the chains.

9.8 Consider This: Necking Polyethylene

Necking polyethylene changes the properties of polyethylene.

a. Does necking affect the number of monomer units, *n*, in the average polymer?

b. Does necking affect the bonding between the monomer units within the polymer? Explain your reasoning.

9.5 Low- and High-Density Polyethylene: Chain Branching

A third strategy to control the molecular structure and physical properties of polymers is to regulate the branching of the polymer chain. This approach is used to produce two general types of polyethylene. The version found in clear plastic bags, such as for supermarket fruits and vegetables, is **low-density polyethylene** or **LDPE.** It is soft, stretchy, transparent, and not very strong. This low-density form was the first type of polyethylene to be manufactured. Study of its structure reveals that the molecules consist of about 500 monomeric units and that the central polymer chain has many side branches, like the limbs radiating from a central tree trunk (Figure 9.9).

About 20 years after the discovery of LDPE, chemists were able to adjust reaction conditions to prevent branching and make another form of polyethylene called **high-density polyethylene (HDPE).** In their Nobel Prize–winning research, Karl Ziegler and Giulio Natta developed new catalysts that enabled them to make linear (unbranched) polyethylene chains consisting of about 10,000 monomer units. Having no side branches, these long chains can be arranged parallel to one another (Figure 9.9). The structure of HDPE is thus more like a regular crystal than the irregular tangle of the polymer chains

Strategy 3: Controlling the *branching* of the chain.

Figure 9.9

(*a*) Detail of bonding in high-density (linear) polyethylene and low-density (branched) polyethylene.
(*b*) Representation of high-density (linear) polyethylene and low-density (branched) polyethylene.

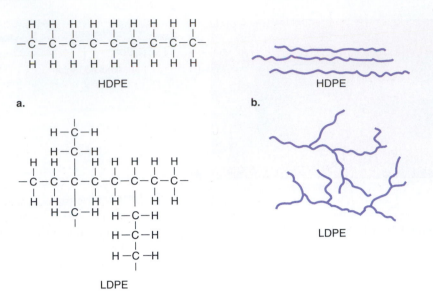

in LDPE. The highly ordered structure of HDPE gives it greater density, rigidity, strength, and a higher melting point than LDPE. Furthermore, the high-density form is opaque; the low-density form tends to be transparent.

9.9 Consider This: High- and Low-Density Polyethylene

Use the structures of HDPE and LDPE found in Figure 9.9 to explain why the density of HDPE is greater than that of LDPE.

The differences in properties of high- and low-density polyethylene give rise to different applications. HDPE is used to make toys, gasoline tanks, baby bottles, radio and television cabinets, and heavy-duty pipes. One new use of HDPE has been spurred by the AIDS epidemic. A surgeon who breaks her or his skin during an operation on an HIV-positive patient runs the risk of acquiring the human immunodeficiency virus through contact with the patient's blood. Allied-Signal Corporation has produced a linear polyethylene fiber called Spectra that can be fabricated into liners for surgical gloves. Spectra gloves are said to have 15 times the cut resistance of medium-weight leather work gloves, but they are so thin that a surgeon can retain a keen sense of touch. A sharp scalpel can be drawn across the glove with no damage to the fabric or the hand inside. Such strength is in marked contrast to the properties of the common plastic grocery bag, which is made of low-density polyethylene. LDPE is also used for squeeze bottles, cling wrap for foods, plastic flowers, and as a liner for disposable diapers.

It would be a mistake, though, to conclude that polyethylene is restricted to the extremes represented by highly branched or strictly linear forms. By modifying the extent and location of branching in LDPE, its properties can be varied from soft and wax-like (coatings on paper milk cartons), to stretchy (plastic food wrap), to fairly rigid (plastic milk bottles). Unfortunately, the consumer is sometimes unaware of the consequences of such structural tinkering. For example, the higher melting point of HDPE (130°C) permits plasticware made from it to be washed in automatic dishwashers. But objects made of LDPE, with a melting point of 120°C, melt in dishwashers.

It is the high temperature of the dishwasher's heating element, not hot water, that can melt the plastic.

Finally, we should note that one of the first and most important uses of polyethylene was a consequence of the fact that it is a good electrical insulator. During World War II, polyethylene was used by the Allies as insulation to coat electrical cables in aircraft radar installations. Sir Robert Watt, who discovered radar, described polyethylene's critical importance in these words.

The availability of polythene [polyethylene] transformed the design, production, installation, and maintenance problems of airborne radar from the almost insoluble to the comfortably manageable . . . A whole range of aerial and feeder designs otherwise unattainable was made possible, a whole crop of intolerable air maintenance problems was removed. And so polythene played an indispensable part in the long series of victories in the air, on the sea, and on land, which were made possible by radar.[1]

9.10 Consider This: The Importance of Polyethylene

The last quote conveys the importance of the electrical insulating properties of polyethylene in World War II. Use the molecular structures in Figure 9.9 to help explain why polyethylene is such a good electrical insulator. Speculate why natural polymers such as silk were not used as electrical insulators during this same period.

Hint: Recall from Chapter 8 that the passage of an electrical current means the movement of electrons.

9.6 The "Big Six"

In spite of polyethylene's range of properties and many uses, it cannot fill all the roles we assign to plastics. Polyethylene melts at a relatively low temperature, is permeable to gases, swells in the presence of oil or organic solvents, and is not very transparent. It is very expensive to make polyethylene sufficiently crystalline to be exceptionally rigid and strong. A serious limitation is the fact that polyethylene, the simplest of polymers, is made up of carbon chains with attached hydrogen atoms. Because the ethylene monomer is so simple, the only options chemists have for changing the structure and properties of the polymer are to alter the branching and, within limits, the length of the polymer chain. Such alterations can, in turn, result in changes in the three-dimensional arrangement of the chains in the solid. As we have just seen, this has been done brilliantly. *But to obtain greater variety in properties and greater control over them, chemists have made frequent use of one of the most important strategies of molecular manipulation—the use of different monomers to form different polymers (item four on our strategies list).*

Strategy 4: Change the *composition* of the monomer units.

Today, more than 60,000 plastics are known. Most have been developed for special purposes ranging from frying pan coatings to resins for restoring antiques. Yet, the two types of polyethylene (LDPE and HDPE) and four other polymers—**polypropylene (PP), polystyrene (PS), polyvinyl chloride (PVC),** and **polyethylene terephthalate (PET or PETE)**—make up the bulk of the plastics we regularly encounter. In addition to low- and high-density polyethylene, the other four plastics all are ultimately derived from petroleum. Over 26 million tons of these six polymers are made annually in the United States.

Table 9.1 is a summary of information about the "Big Six." Six different monomers are involved. Ethylene, vinyl chloride, styrene, and propylene molecules are similar in that they each contain two carbon atoms connected by a double bond. In ethylene, two hydrogen atoms are attached to each of the double-bonded carbon atoms (C=C). But in vinyl chloride, styrene, and propylene molecules, one of the hydrogen atoms has been replaced with something else. In the case of vinyl chloride, hydrogen is replaced by a chlorine atom. In styrene the replacement is a phenyl group, $-C_6H_5$. The phenyl group consists of six carbon atoms bonded in a ring with a hydrogen atom attached to five of them. The replacement for a hydrogen in propylene is a methyl group, $-CH_3$. These replacements create variety in the monomers and the polymers formed from them. Moreover, the substituents give the chemist greater latitude in designing plastics for

A phenyl group, $-C_6H_5$

---

[1] Quoted by J.C. Swallow in "The History of Polythene" from *Polythene—The Technology and Uses of Ethylene Polymers* (2d ed.) A. Renfrew (ed). London: Iliffe and Sons, 1960.

Table 9.1 The Big Six (Including Identifying Code of the Polymer)

Polymer	Monomer	Properties of Polymer	Uses of Polymer
Polyethylene (LDPE) ♳ 4 LDPE	Ethylene $H_2C=CH_2$	Opaque, white, soft, flexible, impermeable to water vapor, unreactive toward acids and bases, absorbs oils and softens, melts at 100°–125°C, does not become brittle until –100°C, oxidizes on exposure to sunlight, subject to cracking if stressed in presence of many polar compounds.	Plastic bags, toys, electrical insulation
Polyethylene (HDPE) ♴ 2 HDPE	Ethylene $H_2C=CH_2$	Similar to LDPE, more opaque, denser, mechanically tougher, more crystalline and rigid.	Milk and water jugs, gasoline tanks, cups
Polyvinyl chloride ♵ 3 V	Vinyl chloride $H_2C=CHCl$	Rigid, thermoplastic, impervious to oils and most organic materials, transparent, high impact strength.	Shampoo bottles, garden hoses, "bubble" package wrap, plumbing pipe
Polystyrene ♶ 6 PS	Styrene $H_2C=CH(C_6H_5)$	Glassy, sparkling clarity, rigid, brittle, easily fabricated, upper temperature use 90°C, soluble in many organic materials.	Styrofoam insulation, inexpensive furniture, drinking glasses, packing "peanuts"
Polypropylene ♷ 5 PP	Propylene $H_2C=CHCH_3$	Opaque, high melting point (160°–170°C), high tensile strength and rigidity, lowest density commercial plastic, impermeable to liquids and gases, smooth surface with high luster.	Battery cases, indoor-outdoor carpeting, bottle caps, auto trim
Polyethylene terephthalate ♸ 1 PETE	Ethylene glycol $HOCH_2CH_2OH$ Terephthalic acid $HOOC-C_6H_4-COOH$	Transparent, high impact strength, impervious to acid and atmospheric gases, not subject to stretching, most costly of the six.	Clothing, soft drink bottles, audio- and videotapes, film backing

particular uses. Polyethylene terephthalate is a special case among the "Big Six," and we will return to it, but only after spending more time with the other members of this sextet.

Table 9.1 also lists some of the more important properties of these six polymers. They are all **thermoplastic,** that is, they can be melted and shaped, and all tend to be flexible. Three of them, low- and high-density polyethylene and polypropylene, have both crystalline and amorphous regions. The crystalline regions, with their structural regularity, impart toughness and resistance to mechanical abrasion and make polypropylene and high-density polyethylene opaque. The amorphous regions promote flexibility. The other three polymers—polyethylene terephthalate, polystyrene, and polyvinyl chloride—are not crystalline. Their molecular chains are bonded together tightly but more or less randomly. *This process, called cross-linking, is number five in the list of strategies identified for modifying polymer structure and properties* (p. 345). Cross-linking is somewhat like the arrangement of strands in a net, but there is a wide variety of ran-

Strategy 5: *Cross-linking* between chains.

Figure 9.10
Items made by polymerization of vinyl chloride (garden hose), propylene (luggage), and styrene (Styrofoam insulation) monomers.

domly sized holes. Because of cross-linking the chains cannot move or slip, creating a plastic that is rigid and hard to stretch. Another property of these amorphous polymers is their transparency and clarity. The range of properties among different polymers means that they are differently suited for specific applications. These uses are also indicated by the pie charts of Figure 9.11. The next Consider This gives you an opportunity to match polymers with their properties and uses.

9.11 Consider This: Uses of the "Big Six" Polymers

For each of these uses, specify the desirable properties of a plastic and, using the information in Table 9.1, suggest the most suitable polymer or polymers.

a. a bread bag
b. a soft-drink bottle
c. "bubble" packaging around a toy
d. bottle caps
e. outdoor lawn furniture
f. a drinking water bottle

Whatever use is made of them, the six major plastics also generally have small amounts of other materials added to them. Because all six are colorless, coloring agents are often introduced. Plasticizers, substances that improve the flexibility of the polymer, are commonly added, as are a variety of other substances that enhance the performance and durability of the plastic. Indeed, the smell associated with certain plastics (and interiors of new cars) is sometimes due to vaporizing plasticizers.

9.7 Addition Polymerization: Adding up the Monomers

We have already noted that ethylene, vinyl chloride, styrene, and propylene molecules each contain a carbon-carbon double bond. All of these monomers polymerize by a process called **addition polymerization.** In every case, the reaction involves the unpairing and re-pairing of electrons described above for polyethylene (Figure 9.7). In **addition polymers,** the monomers simply add to the growing polymer chain in such a way that the product contains all the atoms of the starting material. No other products are formed, and no atoms are eliminated. Thus, vinyl chloride molecules become bonded together to form polyvinyl chloride (PVC). In the process, the C=C double bonds are converted to C—C bonds, and the polymer contains only C—C single bonds.

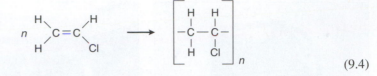

(9.4)

Ethylene and polyethylene are made up only of carbon and hydrogen atoms. But the fact that a vinyl chloride molecule contains a chlorine atom, CH_2CHCl, introduces an opportunity for variability in the structure of polyvinyl chloride (PVC). The presence of the chlorine atom creates an asymmetry in the molecule. Let us arbitrarily think of the carbon atom bearing two hydrogens (CH_2) as the "head" of a vinyl chloride molecule and the chlorinated carbon atom (CHCl) as its "tail." (We could just as easily have made the reverse assignments.) Because of the chlorine atom, when vinyl chloride molecules

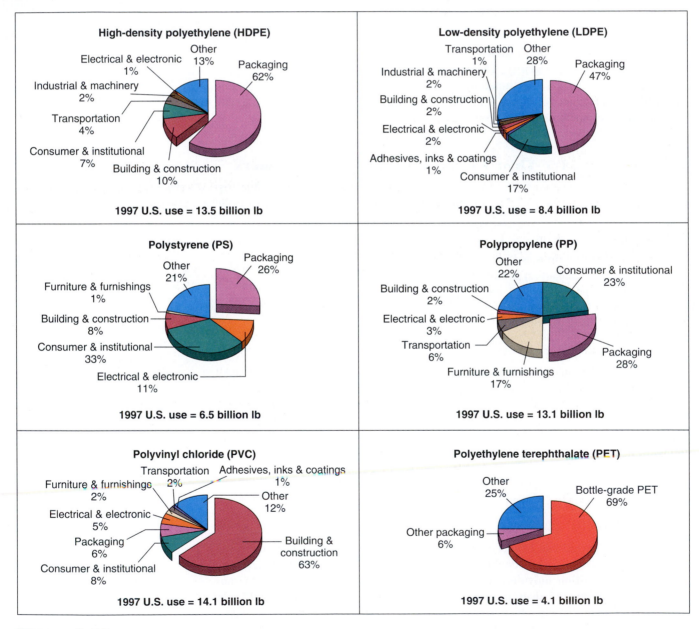

Figure 9.11

Uses of the "Big Six" polymers in the U.S. (1997).

(Source: Data from Society of the Plastics Industry.)

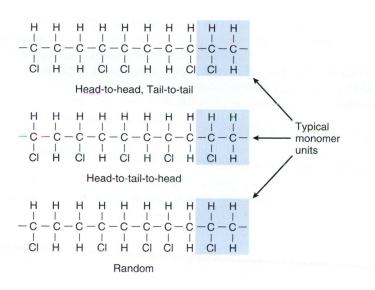

Head-to-head, Tail-to-tail

Head-to-tail-to-head

Random

Typical monomer units

Figure 9.12

Three possible arrangements of monomer units in PVC.

add to each other to form polyvinyl chloride, the molecules can be oriented in three possible arrangements: alternating head-to-head and tail-to-tail; repeating head-to-tail; and a random distribution of heads and tails. Figure 9.12 should help make this molecular interplay more evident.

In a head-to-head/tail-to-tail arrangement of PVC, chlorine atoms are on adjacent carbons. In the head-to-tail structure, chlorine atoms are on alternate carbons. And in the random polymer, an irregular mixture of the previous two types occurs. In each case, the properties are somewhat different. *Controlling monomer orientation within the chain was strategy number six in our list of methods to influence polymer properties.* The head-to-tail arrangement is the usual product for polyvinyl chloride. Depending on its formulation, PVC can be stiff or flexible. Stiff PVC finds use in drain and sewer pipes, credit cards, house siding, toys, furniture, and various automobile parts. The flexible version is familiar in wall coverings, upholstery, shower curtains, garden hoses, insulation for electrical wiring, and packaging films.

The familiar plastic foam hot beverage cup is the most common example of **polystyrene (PS)**. The styrene monomer has the C_6H_5 ring in place of a hydrogen atom on one of the double-bonded ($C\!=\!C$) carbons (equation 9.5). The $—C_6H_5$ **phenyl group** is one of the most common structures in organic chemistry. It consists of six carbon atoms, each at the corner of a hexagon. Five of the six carbon atoms are bonded to hydrogen atoms; the sixth carbon atom is usually linked to some other atom. Typically the symbols for the carbon and hydrogen atoms in the phenyl group are omitted, and the entire $—C_6H_5$ structure is represented by a hexagon. The structure on the right of equation 9.5 indicates that bonding in the C_6H_5 ring can be viewed as consisting of three carbon-carbon single bonds and three carbon-carbon double bonds, alternating around the hexagon. This arrangement of electrons conforms to the octet rule. Experiment shows that the electrons are in fact uniformly distributed around the ring, with all the carbon-carbon bonds of equal strength and length. This uniformity is implied by the circle within the hexagon.

Under appropriate catalytic conditions, styrene polymerizes to polystyrene, usually with the head-to-tail arrangement. The by-now-familiar type of addition equation applies; here *n* equals about 5000.

Strategy 6: Controlling the *orientation* of the monomer units.

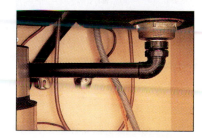

Flexible (shower curtain) and stiff (pipe) PVC items.

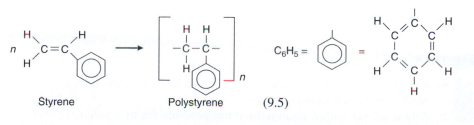

Styrene   Polystyrene   (9.5)

$$C_6H_5 =$$

A polystyrene (styrofoam) cup.

We noted earlier that the hard, brittle, transparent "jewel" cases for CDs are chemically almost identical to light, white, opaque foam coffee cups. Both are polystyrene. Styrofoam is made by expansion molding. Polystyrene beads containing 4–7% of a low-boiling liquid are placed in a mold and heated using steam or hot air. The heat causes the liquid to vaporize and the expansion of the gas also expands the polymer. The expanded particles are fused together into the shape determined by the mold. Because it contains so many bubbles, this plastic foam is not only light, but it is also an excellent thermal insulator. Until relatively recently, chlorofluorocarbons were used as foaming agents, but concern over the involvement of CFCs in the destruction of stratospheric ozone (Chapter 2) led to their replacement in 1990. Gaseous pentane (C_5H_{12}) and carbon dioxide are now frequently used for this purpose. The hard, transparent version of polystyrene is made by molding the melted polymer without the foaming agent. It is used to fabricate wall tile, window moldings, and radio and television cabinets, in addition to CD cases.

Dow Chemical Company has developed a new process that uses pure carbon dioxide as a blowing (foaming) agent to produce styrofoam for packaging material. Using the Dow 100% CO_2 technology eliminates the use of three and a half million pounds of CFC-12 or HCFC-22 (Section 2.14) as blowing agents. The CO_2 used in the process is a by-product from existing commercial and natural sources such as ammonia plants and natural gas wells, thus it does not contribute additional CO_2 to global warming. Because it is nonflammable, carbon dioxide is preferred as a blowing agent over pentane, which is flammable.

Polypropylene (PP), like PVC and polystyrene, is also formed by addition polymerization, in this case using propylene monomers (see Table 9.1). A particularly useful form of polypropylene has the monomeric units bonded in a head-to-tail fashion. This regularity imparts a high degree of crystallinity and makes the polymer strong, tough, and able to withstand high temperatures. Its uses reflect these properties. Polypropylene is found in indoor-outdoor carpeting, videocassette cases, and cold weather underwear. Strength and chemical resistance make polypropylene a good choice for applications in which structural ruggedness is required, such as in indoor-outdoor carpeting.

Carpets and videocassettes made of polypropylene.

9.12 Consider This: Polypropylene—A Tough Plastic

Polypropylene, one of the "Big Six," is used to construct a number of items in which toughness counts. It may not be as familiar to you as polyethylene and PET, because many polypropylene items are not marked with a recycling symbol (and are not collected at curbside). Figure 9.11 shows that polypropylene has numerous uses, but does not provide many specifics. Find these out for yourself. Search for "polypropylene" on the web to identify half a dozen specific items manufactured from polypropylene.

9.13 Your Turn

Write the structural formula of a polystyrene chain containing eight styrene units arranged in the head-to-tail arrangement. Why do you think this arrangement is favored rather than the head-to-head arrangement?

9.14 Your Turn

The monomer used to form Teflon is tetrafluoroethylene, CF_2CF_2.

a. Write the structural formula of a polytetrafluoroethylene chain containing eight tetrafluoroethylene units.
b. Why is a head-to-tail arrangement not possible for this polymer?

Roy Plunkett, a Dupont chemist, accidentally discovered Teflon while experimenting with gaseous tetrafluoroethylene. Plunkett was curious enough about the solid that formed accidentally to study it sufficiently to realize that he had made a previously unknown polymer. This exemplifies Louis Pasteur's maxim that "Chance favors only the prepared mind."

9.8 Condensation Polymers: Bonding by Elimination

Unlike the other polymers described above, **polyethylene terephthalate (PET or PETE)** is not formed by an addition reaction. Rather, it is produced via a condensation reaction. Many polymers are formed by condensation reactions: natural ones such as cellulose, starch, wool, silk, and proteins; and synthetics like nylon, Dacron, Formica, Kevlar, and Lexan.

In condensation polymerization, monomer units join by eliminating (splitting out) a small molecule, often water. Thus, a condensation polymerization has two products—the polymer itself plus the small molecules split out during the polymer's formation. Polyethylene terephthalate is compounded of two monomers, ethylene glycol and terephthalic acid, and hence it is called a copolymer. Ethylene glycol, the chief ingredient in automobile antifreeze, is a dialcohol. Its formula, $HOCH_2CH_2OH$, reveals that it has an —OH on each end of the molecule. The —OH is a **functional group** found in all alcohols, and this group is responsible for the chemical and physical properties that characterize alcohols. A molecule of terephthalic acid, $HOOCC_6H_4COOH$, has a —COOH group at each end. This functional group signifies an organic acid. Because terephthalic acid has two organic acid groups per molecule, it is a diacid. The six carbon atoms of the C_6H_4 group are arranged in the same hexagonal ring you just encountered in styrene. In equation 9.6 the ring is again represented by a hexagon.

A key point about this condensation polymerization is that each time a monomer reacts, the —OH of an acid group and the H from an alcohol group react to form a water molecule, which is given off by the polymerization reaction. The remaining portions of the alcohol and the acid join to form an **ester,** another class of compounds. The condensation polymerization reaction of ethylene glycol with terephthalic acid is represented by equation 9.6; the ester linkage is enclosed in a shaded box.

—OH is the alcohol functional group; —COOH is the functional group of an organic acid. Other functional groups are given in Table 10.2 and Section 10.4. See Section 4.10 about alcohols in fuels.

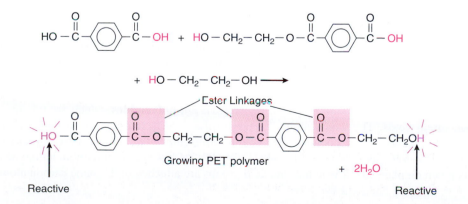

$$\text{(9.6)}$$

Note that the ester produced in equation 9.6 still has a —COOH group on one end and an —OH on the other. The acid group can react with an alcohol group of another ethylene glycol molecule; likewise, the alcohol group of the growing polymer can react with an acid group of another terephthalic acid molecule. This process, represented in Figure 9.13, occurs many times over to yield a long polymeric chain of polyethylene terephthalate.

Polyethylene terephthalate (PET) is classified as a **polyester** because it contains many ester linkages. Since their introduction, polyester fibers have found many uses in

Figure 9.13

A growing PET polymer chain.

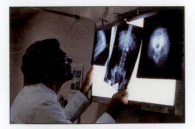

Polyester items.

fabrics and clothing. The polymer is perhaps most familiar under the trade name Dacron. This polyester is frequently mixed with cotton, wool, or other natural polymers, but it has many other uses. Indeed, about five million pounds of PET are produced annually in the United States. Narrow, thin-film ribbons of it (under the trade name Mylar) are coated with metal oxides and magnetized to make audiotapes and videotapes. Dacron tubing is used surgically to replace damaged blood vessels, and artificial hearts contain parts made of PET. Photographic and X-ray film are made from PET, and containers are made from it for medical supplies to be sterilized by irradiation. The most common use for this plastic is in soft-drink bottles because PET is semi-rigid, colorless, and gas-tight.

9.15 Consider This: Can All Acids and Alcohols Form Polyesters?

You have seen that terephthalic acid and ethylene glycol are capable of forming a polyester. Consider the following structures for another organic acid, acetic acid, and another alcohol, ethyl alcohol.

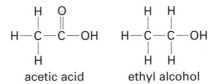

acetic acid   ethyl alcohol

a. Can these two substances form an ester?
b. Can they form a polyester? Explain your reasoning.

9.16 Your Turn

Polyethylene naphthalate (PEN) is a newly developed polymer used in bar code labels, which its manufacturers hope will convert the Big Six into the Big Seven. In both PET and PEN, the alcohol monomer is ethylene glycol, but the organic acid monomers differ slightly. The structural formula of the organic acid monomer used to produce PEN is:

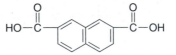

Naphthalic acid

Draw the structural formula to show the formation of the PEN polymer after two units of naphthalic acid and two units of ethylene glycol have polymerized.

9.9 Polyamides: Natural and Nylon

No discussion of condensation polymerization can be complete without including one of the most important classes of natural polymers and the synthetic substitute that brilliantly duplicates some of the properties of the natural material. The naturally occurring polymer is protein. A wide variety of these biological macromolecules make up our skin, hair, muscle, and enzymes. All proteins are **polyamides**, which are polymers of amino acids. The word "amino" refers to the —NH_2 functional group, which is found in a class of compounds called amines. *Molecules of amino acids contain amine groups (—NH_2) as well as acid groups (—COOH).* A general formula for an amino acid is given on page 357. The amine and acid groups are attached to the same carbon atom.

| An amine group: —NH_2

In addition, a hydrogen atom and another group (represented by an R) are bonded to that carbon as well.

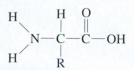

The 20 amino acids found in most proteins differ in the identity of their R groups. In some amino acids, R consists of carbon and hydrogen atoms, as in alanine, where R is a methyl group, —CH₃. In others, R also includes oxygen, nitrogen, or sulfur atoms. Some R groups are acidic and others are basic.

Chapters 11 and 12 include a good deal of additional information about amino acids and their polymers, proteins. At present we will focus on some fundamentals. The amino acids are the monomers for protein polymerization. The crucial point in protein formation is the fact that the —COOH group of one amino acid reacts with the —NH₂ group of another in a condensation reaction analogous to polyester formation. In the protein-forming reaction, an OH from the acid group of one amino acid reacts with an H from the —NH₂ group of another amino acid to form an H₂O molecule, which is eliminated. A **peptide bond** forms between the remaining portions of the two amino acids. The reaction is represented by equation 9.7, and the peptide bond is enclosed in a box. One amino acid contains the substituent R₁, the other R₂.

See Sections 11.6 and 12.4 for more about amino acids and proteins.

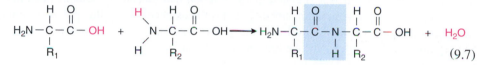

(9.7)

Reaction 9.7 is similar to the neutralization of an acid by a base (see Section 6.3).

In the sophisticated chemical factories called biological cells, this condensation reaction is repeated many times over to form long polymeric protein chains. Given the fact that there are 20 different naturally occurring amino acid building blocks, a great variety of proteins can be synthesized. Some proteins are made up of hundreds of amino acids whereas others, like the hormone oxytocin, contain only eight.

Chemists are often well advised to attempt to replicate the chemistry of nature. In the 1930s, a brilliant chemist working for the DuPont Company set out to do just that. Wallace Carothers (1896–1937) was studying a variety of polymerization reactions, including the formation of peptide bonds. Instead of using amino acids, Carothers tried combining adipic acid, HOOC(CH₂)₄COOH, and hexamethylene diamine, H₂N(CH₂)₆NH₂ (also known as 1,6-diaminohexane). Note that a molecule of adipic acid has an acid group on both ends and the hexamethylene diamine molecule has an amine group on each end. As in the case of protein synthesis, the acid and amine groups react to eliminate water and form peptide bonds. But in this instance, the polymer consisted of alternating adipic acid and hexamethylene diamine monomer units.

Wallace Carothers, the inventor of nylon.

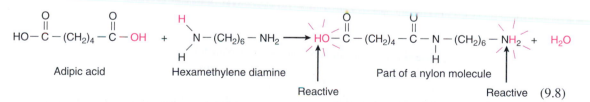

(9.8)

DuPont executives decided the new polymer had promise, especially after company scientists learned to draw it into thin filaments. These filaments were strong and smooth, and very much like the protein spun by silkworms. Therefore, it was as a substitute for silk that "Nylon" was first introduced to the world. The world greeted it with bare legs and open pocketbooks. Four million pairs of nylon stockings were sold in New York City on May 15, 1940, the first day that they became available (Figure 9.14). But in spite of consumer passion for "nylons," the civilian supply soon dried up, as the polymer was diverted from hosiery to parachutes, ropes, clothing, and hundreds of other wartime

Figure 9.14
Women eagerly lined up to
buy nylon stockings in 1940
when they were first available
commercially.

Figure 9.14
Women eagerly lined up to
buy nylon stockings in 1940
when they were first available
commercially.

Items made from nylon.

uses. By the time World War II ended in 1945, nylon had repeatedly demonstrated that
it was superior to silk in strength, stability, and rot resistance. Today this polymer, in its
many modifications, continues to find wide applications in clothing, sportswear, camping equipment, the work room, the kitchen, and the laboratory.

9.17 Your Turn

Kevlar is a condensation polymer used to make bulletproof vests. It is made
from terephthalic acid and phenylenediamine.

$$H_2N - \bigcirc - NH_2$$

Phenylenediamine

Use the structural formulas of terephthalic acid (see page 355) and phenylenediamine to draw a segment of a Kevlar molecule containing three units of
terephthalic acid and three phenylenediamine units.

9.10 Plastics: Where from and Where to?

Given the constraints of the law of conservation of matter, it is obvious that we should
pay close attention to the raw materials that are incorporated into plastics and the disposal or recycling of this matter after its use. You have already read that petroleum is
the source of the monomers used to make most synthetic polymers (Section 4.9). Crude
oil is a mixture of many compounds that is refined into various fractions on the basis of
boiling point and molar mass. Figure 9.15 graphically depicts those fractions and their
primary uses. Not surprisingly, given our discussion in Chapter 4, the great majority of
petroleum is burned as fuel. Only 3% of it is reserved as a chemical feedstock (reactants) for manufacturing polymers and other chemicals, including new medicines.

This 3% is essential to current methods of manufacturing the polymers that have re-
shaped modern life. But the planet's supply of petroleum is limited and non-renewable,
a fact that creates a serious dilemma. You already know from previous chapters (4 and
6) that petroleum is not an ideal energy source. Burning it releases carbon dioxide that
can contribute to global warming, and unburned carbon fragments and other compounds
that give rise to smog and air pollution. But compared to coal, petroleum is quite clean

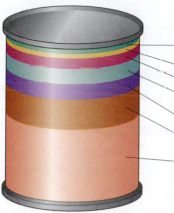

Barrel of Crude Oil

Petroleum feedstocks **1.25 gal**
(chemical products, plastics)

Asphalt and road oil **1.3 gal**

Boiler oil **2.9 gal**

Other **4.2 gal**
(lubricants, waxes, solvents)

Jet fuel **4.2 gal**

Diesel and home heating oil **8.4 gal**

Gasoline **19.7 gal**

Figure 9.15

Fractions and their uses from a barrel of crude oil.

and convenient. Therefore, we return to a question posed earlier: "To burn or not to burn?" Should petroleum be burned, or should more of it be diverted for use as a raw material for the synthesis of materials that cannot now be made from any other source? What are the risks and benefits—the economic and social trade-offs—in using oil as a source of energy or as a source of synthetic products? If a greater share of petroleum is not reserved for uses other than fuel, the age of plastics may be very short.

It is important to note that chemistry may rescue society from this dilemma. In principle, at least, polymers can be made from any carbon-containing starting material. Crude oil is simply the most convenient and the most economical. But it might also prove possible to convert renewable biological materials such as wood, cotton fibers, straw, starch, and sugar into new polymers. After all, chemists at the Arthur D. Little Company once actually made a silk purse out of a sow's ear. But to transform biomass into synthetic polymers, new methods and new technologies would have to be developed. The cost of the research and the manufacturing would likely be substantial. Moreover, it would be essential to estimate the supply of the starting materials and the demand for the finished products. There are implications for land use, crop productivity, the environment, and no doubt much more.

There seems to be a good deal more concern about where plastics go than where they come from. Much of the plastic we use eventually ends up in a landfill, along with lots of other types of municipal and domestic solid wastes, in the usual "out of sight, out of mind" approach. As a nation, we daily discard enough trash to fill two Superdomes. Of course that's just a ballpark figure, but it corresponds to about three tons of trash annually per family. The EPA estimates that about 75% of all municipal solid waste is put into landfills. Of the remainder, 10% is recycled and 15% is incinerated. Figure 9.16 provides information about the contents of a typical landfill given in terms of percent by volume. It is the volume of the buried materials, not the mass, that causes landfills to reach their capacity.

You will note from Figure 9.16 that only 9% of municipal solid waste is plastic, about 30% of which is used in packaging. But the largest percentage of municipal solid waste

Food waste
10%

Yard waste
13%

Glass
6%

Metal
8%

Other
16%

Plastics
9%

Paper
38%

Figure 9.16

What's in your garbage? Composition of municipal solid waste in volume percent.

(Source: Data from U.S. Environmental Protection Agency, 1997.)

(38%) is paper and paper products. This raises a question that has been in the news: Which constitutes the lesser environmental burden, paper or plastic? The section that follows is a real-life glimpse into this controversy.

> ### 9.18 Consider This: The Polymer Age: From Start to Finish
>
> It has been stated that "there seems to be a good deal more concern about where plastics go than where they come from."
>
> **a.** What are some of the reasons why this is true within the United States?
> **b.** How do you expect this perception to change within your lifetime? Explain your reasoning.

9.11 Paper or Plastic? The Battle Rages

The rather melodramatic title of this section appeared as a headline in the *Syracuse Herald-Journal*. It introduced a heated exchange of opinions, forcefully expressed by readers. At issue was whether grocery bags should be made of paper or plastic. The actual letters that follow are selected from many more, and reprinted here to illustrate this controversy in the context of a supermarket.

> In these environmentally conscious times I am often appalled at the number of people who choose plastic bags over paper ones at the supermarket. They must be aware that plastic bags are neither biodegradable nor as easily recycled as the paper variety.
>
> It's common knowledge that plastic bags are not biodegradable, and although some forms of plastic are recyclable, no recycling center in the local area takes plastic bags. What most consumers fail to realize is that the production and processing of plastic involves a great amount of highly toxic chemicals.
>
> Improper land disposal of hazardous wastes, emissions of toxic chemicals into the air, and discharges of toxic industrial effluents into waterways as a result of plastic production seriously threaten the public health and the environment.[2]

The local supermarket took the opposite position, and argued that by using plastic bags it was acting in an environmentally responsible manner. As evidence, it printed the following message on its grocery bags:

> Thank you for using plastic bags. If all of the Wegmans shoppers using plastic bags last year had insisted on paper, they would have increased the amount of solid waste by over eight million pounds and taken up nearly seven times more space in landfills.
>
> | 1000 plastic bags equal | 1000 paper bags equal |
> |---|---|
> | 17 lbs and 1219 cubic inches | 122 lbs and 8085 cubic inches[3] |

(See Figure 9.17).

A second letter to the editor, based on information such as that just cited, provides the perspective of a consumer and an employee.

> As an employee of Wegmans Food markets you may determine that my opinion is biased and it certainly is . . . Clearly the plastic bags take up less space than paper . . . In our backroom of the store, an entire pallet of paper bags takes up as much space as the plastic, however, there are only approximately 10,000 paper bags on a pallet compared to approximately 30,000 plastic sacks . . . (T)he cost certainly is a benefit. Plastic bags cost 1.5 cents whereas paper ones cost 3 cents. If all customers would insist on plastic bags, the savings would certainly be passed on to the consumer.
>
> Finally, I would like to address the landfill issue. As mentioned on our plastic bags, the use of paper sacks fills landfills much faster than the use of plastic bags. Plastic bags take up considerably less space than paper . . . While environmental groups claim that paper bags degrade at a rapid rate, they are simply misleading the public.[4]

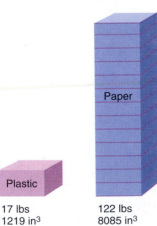

Figure 9.17

A scale representation of volumes occupied by 1000 plastic bags and 1000 paper bags.

Paper
122 lbs
8085 in³

Plastic
17 lbs
1219 in³

[2,3,4]*Syracuse Herald-Journal,* Syracuse, N.Y.

There is much in these letters to engage the Sceptical Chymist, but we may not yet have enough information to pass critical judgment on the many complex issues involved. Indeed, we may not achieve such knowledge and wisdom within the limitations of this chapter. Nevertheless, in the next section we press on in our efforts to become better informed.

9.19 The Sceptical Chymist: Analyzing Letters to the Editor

Analyze the two letters printed above with their opposing views. Pay particular attention to the initial assumptions, the evidence cited, the logic used, and the conclusions drawn. **a.** Which makes the more compelling case and why? **b.** On the basis of these two letters only, which position would you support?

9.20 Consider This: Plastic vs. Paper—Where Do You Stand?

When you are in a supermarket, which do you usually request, plastic or paper grocery bags? List the advantages and disadvantages of plastic vs. paper bags, and decide which is preferable. Then prepare a letter to the editor of your local newspaper clearly stating your position and explaining your reasons for it.

9.12 Dealing with Plastics

Every year, about 89 billion pounds of plastic are produced in the United States—nearly 350 pounds for every woman, man, and child. Most of this ultimately finds its way into landfills. Given this huge quantity, there is little consolation in the fact that landfills contain considerably more paper than plastic. The reduction of the amount of plastic going into landfills remains a high priority. Five strategies suggest themselves: **incineration, biodegradation, reuse, recycling,** and **source reduction.** In the paragraphs that follow, we will examine each of these approaches and attempt to weigh their relative merits.

Because the Big Six and most other polymers are composed primarily of carbon and hydrogen, **incineration** is an excellent way to dispose of used plastics. Indeed, a recent study in Germany led to the conclusion that burning waste plastic does less damage to the environment than any other method of disposal. The chief products of combustion are carbon dioxide, water, and a good deal of energy. In fact, pound for pound, plastics have a higher energy content than coal. Plastics account for about 7% of the weight of municipal solid waste, but approximately 30% of its energy content. The German study found that the greater the percentage of plastic in the refuse burned in a garbage incinerator, the more efficient the burning, the greater the quantity of energy released, and the lower the emission of airborne pollutants. By contrast, recycling polymers requires energy, and if the waste plastic is dirty or of low quality, more energy is needed to recycle it than to produce a comparable quantity of new virgin plastic. It has been estimated that incineration can decrease the volume of plastic headed for landfills by as much as 90%.

But incineration of plastics is not without some drawbacks. The repeated message of Chapters 1–4, that burning does not destroy matter, applies here as well. Effluent gases produced by combustion may be "out of sight," but they had best not be "out of mind." Burning plastics produces CO_2, which potentially contributes to global warming. Of special concern in incineration are chlorine-containing polymers such as polyvinyl chloride, which release hydrogen chloride during combustion. Because HCl dissolves in water to form hydrochloric acid, such smokestack exhaust could make a serious contribution to acid rain. Moreover, some plastic products have inks containing heavy metals

Plastics are an important source of the energy produced in garbage burning power plants (Section 4.11).

Polyacrylonitrile, a polymer made from acrylonitrile, CH_2CHCN, is in Orlon and Acrilan fibers used to make rugs and fabrics. When acrylonitrile burns, it produces the poisonous gas hydrogen cyanide, HCN.

such as lead and cadmium. These toxic elements concentrate in the ash left after incineration and thus contribute to a secondary disposal problem. Burning plastics also converts the carbon they contain into carbon dioxide, just as if petroleum, the raw material from which plastics are made, had been burned directly. Thus, burning plastics converts their carbon into a by-product that is not easily converted directly into other useful materials. On balance, however, if carefully monitored and controlled, incineration can lead to a large reduction in plastic waste, generate much-needed energy, and have little negative impact on the environment.

9.21 Your Turn

When polypropylene burns completely, it produces just carbon dioxide and water. Write a balanced chemical equation for the combustion of polypropylene. Assume an average chain length of 2500 monomer units.

Hint: Check the molecular formula of propylene in Table 9.1.

Another potential strategy for disposing of plastic wastes is to enlist bacteria to do the job—in other words, to employ **biodegradation.** The problem is that bacteria and fungi do not find most plastics very appetizing. These microorganisms do not have the enzymes necessary to break down plastics. However, because bacteria and fungi evolved in our natural environment, they possess enzymes to break down naturally occurring polymers into simpler molecules. Indeed, many strains of bacteria use cellulose from plants or proteins from plants and animals as their primary energy sources. You have already encountered several instances of such processes in this text. In Chapter 3 you read about the release of methane by belching cattle. Actually, the methane is produced when bacteria decompose cellulose in the cow's rumen. In the same chapter, we also mentioned that methane is generated by natural decomposition of organic material in landfills.

| See Section 3.11.

Scientists are now engineering biodegradability into some synthetic polymers. Certain bonds or groups are introduced into the molecules to make them susceptible to fungal or bacterial attack, or to decomposition by moisture. Recently, research scientists at DuPont have developed a biodegradable polymer called Biomax, which decomposes in about eight weeks in a landfill. This new polymer is a close chemical relative of PET. Biomax uses other monomers in conjunction with those conventionally used to prepare PET (ethylene glycol and terephthalic acid). When polymerized, these co-monomers create sites in the polymer chains that are susceptible to degradation by water. Once the moisture does its job of breaking the polymer into smaller chains, naturally occurring microorganisms feed off the smaller chains, converting them to CO_2 and water. Biomax could be used in a variety of applications such as lawn bags, bottles, liners of disposable diapers, disposable eating utensils, and cups.

9.22 Consider This: Will Biomax Work?

a. Identify some appropriate applications of Biomax, other than those given above.

b. Identify some applications of Biomax that would be inappropriate.

Achieving biodegradability in polymers raises some concerns, as an EPA report cautions:

> Before the application of these technologies can be promoted, the uncertainties surrounding degradable plastics must be addressed. First, the effect of different environmental settings on the performance (e.g., degradation rate) of degradables is not well understood. Second, the environmental products or residues of degrading plastics and the environmental impact of degradables on plastic recycling is unclear.[5]

---

[5]From EPA Report to Congress "Methods to Manage and Control Plastic Wastes," February 1990.

Figure 9.18
Some buried wastes can re-main intact for a long time.

Part of the difficulty is that even natural polymers do not decompose as completely in landfills as was suggested earlier in this chapter. Modern waste disposal facilities are covered and lined to prevent leaching of waste and waste by-products into the surrounding ground. Landfill linings and coverings create anaerobic (oxygen-free) conditions that impede bacterial and fungal action. As a result, many supposedly biodegradable substances decompose slowly or not at all when buried. Recent excavations of old landfills have found 37-year-old newspapers that are still readable and five-year-old hot dogs that, while hardly edible, are at least recognizable (Figure 9.18).

9.23 Consider This: Landfill Liners

A search on the worldwide web for "landfills" can bring up high quality information about garbage. If you browse through these sites, you will see that there is a fair amount of controversy about the plastic materials that can be used to construct the liner. Search for "landfill liners" and check out the plastics involved.

a. Which polymers are now used for liners?
b. What are their drawbacks?
c. Are there new polymers that offer more desirable properties? Cite the author and the URL of the web sites.

Reusing a plastic product is one way to divert it from a landfill. Although not all plastics are directly reusable, many are. Plastic bottles can be reused by cleaning them and filling them with the same substance (milk, water, shampoo, etc.), or used in other ways. A recent study reported that 80% of Americans reuse plastic products, such as food storage containers and refillable bottles.

In the United States, nearly 50% of certain plastic parts from damaged or discarded cars are repaired and reused. In a bold move to foster reuse on a large scale, new cars in Germany must be designed and built so that when the life of the car is over, its plastic parts can be removed easily and used to build other automobiles, thus creating automobiles that renew themselves (Table 9.2.) Through reusing rather discarding, far less petroleum is needed to make new plastics in the first place. The savings in this case can be substantial considering that modern automobiles each have nearly 200 kg (440 lb) of plastics in them.

9.24 Consider This: Plastic Cars

A considerable number of different polymers used in automobiles are noted in Table 9.2. Pick one of the polymers in the table that is not discussed in this chapter and find out what the polymer is made of and how its properties relate to its use.

Table 9.2   Some of the Different Plastics in an Automobile

Plastic	Items
Polyethylene	Fuel tanks, bottles for other liquids
Polyester	Grill panels, interior fibers
Polypropylene	Interior fabrics, bumpers
Polyvinyl chloride	Interior trim
Nylon	Fuel lines, manifolds, fibers
Polymethacrylate	Lenses
Polyurethane	Foam, bumpers
ABS (acrylonitrile/butadiene/styrene)	Interior and exterior trim, housings
Others	Hoses, tires

Given the problems associated with landfill disposal of natural and synthetic polymers, attention has logically turned to **recycling** both. Although recycling plastics does not literally dispose of them as does incineration or biodegradation, it helps to reduce the amount of new plastic entering the waste stream. In 1998, more than 7000 American cities and towns participated in programs of curbside collecting and recycling of plastics. Nearly 20,000 communities, representing 78% of the nation's population, provide either curbside or drop-off collection of plastics. Through these efforts, more than one billion pounds of plastic packaging, about 2% of the total produced in the United States, were recycled. Although this represents a 12% increase over 1992, this country still lags far behind Germany and other developed nations in the percentage of plastics recycled. Germany has a particularly aggressive program, that, in 1994, resulted in recycling 52% of the plastics used in packaging. That same year, the Germans collected 10.9 billion pounds of packaging materials, including glass, paper, and tin cans, and recycled 10.1 billion pounds of it.

In the United States, more than 200 million milk and water jugs have been recycled to convert their high-density polyethylene into a fiber. The fiber is then made directly into TYVEK, a material used in such broad applications as sports clothing, durable mailing envelopes, and insulating wrap for new buildings.

> 200 million milk jugs joined end to end would form a chain long enough to reach nearly twice around the Earth.

a.

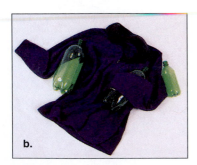

b.

(a) T-shirts can be made from recycled PET bottles.
(b) Activewear from 100% recycled PET.

9.25 Consider This: Plastic and Aluminum Recycling

Currently, only about 2% of all plastic in the United States is recycled, but nearly 30% of aluminum is recycled. Suggest some scientific, economic, and sociological reasons for this difference.

Increasing quantities of "post-consumer" plastics are being used in the United States. The demand for recycled "Big Six" polymers grew by 32% between 1988 and 1993, and it is expected to grow. In 1996, 25% of all plastic bottles made in the United States were recycled (1.3 billion pounds). Recycling rates for PET and HDPE are 29% and 24%, respectively. Polyethylene terephthalate soft drink bottles are particularly easy to melt and reuse. Nearly 600 million pounds of PET plastic containers were recycled in the United States in 1997, amounting to more than half a billion pounds of PET that were not buried in landfills (Figure 9.19). That amount represents 22% of all PET bottles manufactured.

Much of recycled PET was converted into polyester fabrics, including carpeting, T-shirts, the popular "fleece" used for jackets and pullovers, and for the fabric uppers in jogging shoes. Five recycled 2-liter bottles can be converted into a T-shirt or the insulation for a ski jacket; it takes just about 450 such bottles to make polyester carpeting for a 9 × 12 foot room.

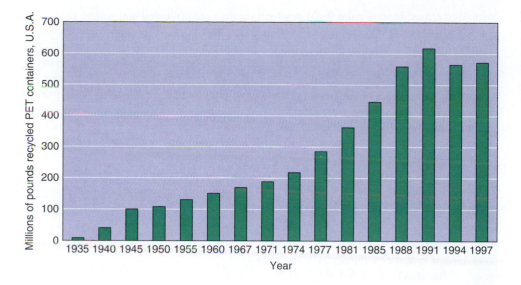

Figure 9.19
Recycling containers made of PET (millions of pounds annually).

The chemist Nathaniel Wyeth used his creativity to develop the plastic soda bottle in contrast with his artist brother Andrew Wyeth who expresses his creativity on canvas. Nathaniel Wyeth has an even bigger vision than T-shirts or rugs from recycled PET. In *ChemMatters* magazine (October 1994) he said: "One of my dreams is that we're going to be able to melt the returned bottles down, mix them with reinforcing fibers, and make car bodies out of them. Then, once the car has served its purpose, rather than put it in the junk pile, melt the car down and make bottles out of it."

Although such technology is possible, its use depends on economic factors as well. These factors have helped increase the demand for recycled polyethylene terephthalate and other polymers. Cleaned, recycled PET sells for 28 to 30 cents per pound (1998 prices). By contrast, virgin PET sells for about $1.50 per pound. Not surprisingly, the laws of supply and demand work here as they do throughout the economy. When virgin PET prices rise, recycled PET suppliers raise their prices in response to the market.

Another major recycling initiative involves national supermarket chains recycling their HDPE grocery bags. In fact, if you read the labels on plastic materials, you will increasingly find them made of a mixture of virgin and recycled (post-consumer) plastics. Some of the use of post-consumer plastics is now mandated by law. For example, since 1995, all HDPE packaging used in California has been required to contain 25% recycled material.

9.26 Consider This: Polymers in Your Computer

Twenty years ago, recycling personal computers was not a concern because there weren't enough of them around to matter. These days, given the number of monitors, keyboards, and "mice" in circulation, there is good reason to keep these out of the landfill.

a. What polymers are used in making your computer?
b. Is it possible to recycle the polymers in your computer? Use the web to find out.
c. What is being done currently to recycle plastics from computers? You might want to search for "computers" and "recycling" to get started. Which plastic-containing computer supplies and accessories should have recycling programs?

However, simply collecting plastics to be recycled is not enough. For recycling to be successful and self-sustaining, a number of factors must be coordinated. They involve not only science and technology, but economics and sometimes politics. True recycling

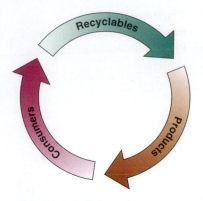

Figure 9.20
True recycling involves a
never-ending loop.

| Density is discussed in Section 5.8.

Automated sorting methods are
needed to accommodate the huge
volume of plastic waste, especially in
urban areas. An estimated 40 million
plastic bottles are discarded *daily* in
metropolitan New York City.

involves a closed loop (Figure 9.20) in which plastics are collected and sorted, then converted into products that consumers buy, use, and later recycle. First of all, there must be a dependable *supply* of used plastic, consistently available at designated locations. This creates a formidable task of *collecting* the discarded objects, along with the associated job of *sorting and separating* the various polymers. The codes that appear on plastic objects (Table 9.1) are provided to help facilitate this process. But, because of the large volume of material to be sorted, automated sorting methods are being developed. The plastics industry has spent more than $1 billion nationally on recycling research and developing environmentally responsible and sustainable plastic recycling programs.

9.27 Your Turn

Plastics vary in density. When placed in a liquid, a substance, such as a plastic, will float if its density is less than that of the liquid, and the substance will sink if it is more dense than the liquid.

The densities (grams/mL) of four common plastics are: High-density polyethylene 0.95–0.97; polyethylene terephthalate 1.38–1.39; polypropylene 0.90–0.91; and polyvinyl chloride 1.18–1.65. The densities of six liquids are:

Substance	Density (g/mL)
Methanol	0.79
An ethanol/water mixture	0.92
A second kind of ethanol/water mixture	0.94
Water	1.0
Saturated $MgCl_2$	1.34
Saturated $ZnCl_2$	2.01

Develop a procedure that uses the densities of the liquids and of the plastics given by which the individual plastics can be separated from a mixture containing pieces of all four plastics.

Once the entropy and disorder have been overcome by sorting, the *reprocessing* is relatively simple. Almost any polymer that is not extensively cross-linked can be melted. If the supply of waste is homogeneous, that is, if it contains only one type of plastic, the molten polymer can be used directly in the *manufacturing* of new products. Alternatively, it can be solidified, pelletized, and stored for future use. However, when mixtures of various polymers are melted, the product tends to be dark with varying properties, depending on the nature of the mixture. Although this reprocessed material does not have outstanding working properties, it is good enough for general lower grade uses such as parking lot bumpers, disposable plastic flower pots, and cheap plastic lumber. Such mixed material is obviously not as valuable as the pure, homogeneous recycled polymer. Hence, the importance of sorting plastics. For similar reasons, manufacturers prefer to use only a single polymer in a product to avoid having to separate the various polymers.

A variant on this method of reprocessing plastics is to decompose the polymers into simpler molecules, in some cases back to the actual original monomers. Fanciful as Nathaniel Wyeth's bottles-to-cars-to-bottles idea might seem, it is now a reality. Dupont chemists have won a Presidential Green Chemistry Challenge award for improving on Wyeth's dream. They developed a proprietary process for treating post-consumer PET which, like pulling beads apart from a necklace, unlocks (depolymerizes) the polymer back to its original monomers. These can then be reused to make new PET for other products. Mary Johnson, a Dupont employee, says "Because these monomers retain their original properties, they can be reused over and over again in any first-quality application. A popcorn bag can become an overhead transparency, then a polyester peanut butter jar, then a snack food wrapper, then a roll of polyester film, then a popcorn bag again."

The essential final step in recycling plastics is *marketing*. But a company or a city would be well advised to determine or create the demand for recycled polymers before completing all the other steps. Without a product and buyers, recycling programs are doomed to fail. In fact, recycling laws in a number of cities have not been implemented and enforced because one of the links in this polymeric chain of supply, collecting, sorting, processing, manufacturing, and marketing is missing. Without all of these, the system will not work, unless it is heavily subsidized. Most municipalities have been unwilling to provide the necessary funds. As this third edition of *Chemistry in Context* went to press, the market for post-consumer plastics was strong and prices were high enough to justify significant recycling activity.

The patented Dupont process is called Petretec Polyester Regeneration Technology.

9.28 Consider This: Recycling Mandates and Incentives

Laws have been passed in several states and the District of Columbia making recycling mandatory; the mandates are often coupled with incentives. Because of these laws, the volume of recycled material has increased substantially in several states. Yet, in spite of these laws, recycling has not always proved profitable. Comment on why the laws are not the same in every state and why, even if there are incentives, recycling has not always proven to be popular or profitable.

The remaining option for dealing with plastics, **source reduction**—less waste in the first place—appears to be simplest and most direct: simply decrease the quantity of plastics produced and used. The advantages are many—resources would be conserved, pollution would be reduced, and potentially toxic materials would be minimized. Examples of source reduction are the improvements made in plastic technologies that have reduced the amounts of plastic needed to make high-volume products; the 2-liter soda bottle now uses 25% less plastic than when it was introduced in 1975, and the 1-gallon milk jug weighs 30% less than a decade ago.

To reduce the amounts of materials, solvents, and labor, engineers at Chrysler Corporation have developed a car, the Plymouth Pronto Spyder, with a body made entirely from plastic. There are already three plastic-bodied cars on the market—Dodge Viper, Chevrolet Corvette, and General Motor's Saturn—but they use expensive plastic composites. What sets the Plymouth Pronto Spyder apart is that its body is made entirely of low-cost PET. This innovation could cut production expenses enough for the Spyder to cost only half the price of a regular car. Using just six molded PET body pieces that can be glued together eliminates using the usual eighty or so pieces, and does away with expensive surface painting, stamping, and forming operations. Paint is molded directly into the PET, coloring the plastic all the way through, thus eliminating conventional surface painting, a costly and time-consuming step in auto production.

Yet source reduction, a seemingly innocuous option, is far more complicated than it appears. The problem is that something else is generally used to replace the plastic, and this substitution can be fraught with hidden pitfalls. In making choices between alternative materials, the decisions must be informed by the source and nature of chemical feedstocks (reactants), the method of manufacturing, waste products produced during manufacturing and their disposal, and many other factors. Energy costs as well as economic costs must be taken into account. How much energy must be expended in the entire life cycle of a product from raw material to final disposal?

Obviously, the identification and proper weighing of all of the possible variables is a complex and difficult task. But when the job is done properly, one sometimes discovers that attempts to reduce the amount of plastic waste by substitution for the plastic may actually increase the overall amount of waste and the associated negative environmental

Table 9.3   Paper versus Styrofoam Cups

Item	Paper Cup	Plastic Cup
Per cup		
Raw materials		
Wood and bark, g	33	0
Petroleum, g	4.1	3.2
Finished mass, g	10	11.5
Cost	2.5 times that of plastic	1
Per million grams of material		
Utilities		
Steam, kg	9000–12,000	5000
Power, 10^9 J	3.5	0.4–0.6
Cooling water, m^3	50	154
Water effluent		
Volume, m^3	50–190	0.5–2
Suspended solids, kg	35–60	Trace
Air emissions, kg	7–22	35–50
Recycling potential		
Primary user	Possible	Easy
After use	Low	High
Ultimate disposal		
Heat recovery (million J/kg)	20	40
Mass in landfill, g	10.1	1.5
Biodegradable	Slowly, if at all	No

impact. For example, two pounds of plastic is enough to make containers able to hold about eight gallons of juice, soft drinks, or water. To hold that same amount of beverage would require three pounds of aluminum, eight pounds of steel, or 27 pounds of glass.

As another, more detailed example, consider the replacement of a plastic cup with one made of paper. Each occupies about the same volume in a landfill, where both will probably remain undecomposed for a long time. The "conventional wisdom" of public opinion has it that paper cups are more environmentally friendly than styrofoam ones. But a detailed analysis of the issue made by Martin Hocking of the University of British Columbia counters that position (*Science* Vol 251, 1 February, 1991, pp 504–50). Hocking did a cradle-to-grave type of life cycle analysis in which he considered all aspects of the production and disposal of the two types of cups (Table 9.3). Hocking's conclusion is that paper cups are not as environmentally friendly as commonly thought. The paper cups require more raw material and consume as much petroleum as that used to produce the styrofoam cups. The harsh nature of the chemicals used, the large volume of water required, and the nature of effluents generated into the air and water in paper making are far greater than those affiliated with producing polystyrene cups. Both types of cups are not very biodegradable in sealed landfills, so paper offers no significant advantage there. The styrofoam cups are easier to reuse and recycle, and about as easy to incinerate as paper ones.

9.29 Consider This: Reuse and Alternatives

To complete 9.3 Consider This, you kept a journal of all the plastic and plastic-coated products you discarded in a week. Review that list and indicate ways you could reuse those discarded products or suggest alternatives for the plastic in the product.

Of course, the best method of source reduction is not to replace plastics, but to do without them or their substitutes whenever possible. One correspondent in the great plastic versus paper battle said it well:

> There is a danger in this grocery bag controversy of losing sight of issues of greater importance. One of these is the matter of legitimate, responsible use of resources. Plastics are made from one of the most precious resources, one which cannot be renewed or replaced. In many respects we should regard it as more precious than gold or diamond. There are products essential to human health and well-being which can be made only from petroleum. There are also non-essential, wasteful uses of this priceless commodity. Where did we get the idea that it is our right to waste millions of barrels of oil each year exceeding the speed limit? Who said we're justified in manufacturing and using plastic items like shopping bags, burger boxes, and disposable diapers which are instant garbage? How did we get hooked on the consumer habits that are destroying not only a level of comfort we take for granted, but the very air and water we need to survive?[6]

There is no single best solution to the problems posed by plastic waste, and solid waste in general. Incineration, biodegradation, reuse, recycling, and source reduction all provide benefits and all have associated costs. Therefore, it is likely that the most effective response will be the development of an integrated waste management system that will employ all four of these strategies. The goal of such an integrated system would be to match the methods to the composition of the waste stream, thus optimizing efficiency, conserving energy and material, and minimizing cost and environmental damage.

Conclusion

The letter quoted in the last section goes well beyond a choice of plastic or paper shopping bags. Once more we come to an issue of lifestyle. Over the past 60 years, chemists have created an amazing array of polymers and plastics—new materials that have made our lives more comfortable and more convenient. Many of these plastics represent a significant improvement over the natural polymers they replace. Furthermore, many products we take for granted today would be impossible without synthetic polymers and plastics. There would be no audiotape and videotape, no compact discs, no kidney dialysis apparatus, and no heart/lung machines. We have become dependent on plastics, and it would be difficult if not impossible to abandon their use. Chemical industry has given consumers what they want. But there now appears to be rather more of it than we would like or perhaps can deal with responsibly—mountains of soft drink bottles and miles of plastic bags. We must learn to cope with this glut of stuff while saving matter and energy for tomorrow. To create a new world of plastics and polymers will require the intelligence and efforts of policy planners, legislators, economists, manufacturers, consumers, and, above all, chemists.

Chapter Summary

Having studied this chapter, you should be able to:
- Understand the nature of plastics and polymers, their typical properties and molecular structure (9.1)
- Identify the allotropic forms of carbon: diamond, graphite, and fullerenes (9.1)
- Describe a brief history of the synthetic polymer industry and its economic impact (9.2)
- Discuss trends in production of plastic since 1935 (9.2)

- Recognize the six strategies for altering the molecular structure of polymers and their resulting properties (9.3)
- Describe typical uses for the "Big Six" polymers (9.6)
- Understand the molecular mechanism of addition polymerization (9.7) and condensation polymerization (9.8)
- Recognize the chemical composition and molecular structure of the "Big Six" polymers:
    Low-density polyethylene (LDPE) and high-density polyethylene (HDPE) (9.4, 9.5)
    Polyvinyl chloride (PVC) (9.7)

---

[6]*Syracuse Herald-Journal*, Syracuse, N.Y.

Polystyrene (PS) (9.7)
Polypropylene (PP) (9.7)
Polyethylene terephthalate (PET) (9.8)
- Tell how amino acids and proteins are related chemically (9.9)
- Use structural formulas to write the chemical equation for the synthesis of nylon (9.9)

- Identify sources of materials for manufacturing plastics (9.10)
- Compare the relative costs and benefits of plastic and paper grocery bags (9.11)
- Relate the technical, economic, and political issues in methods for disposing of waste plastic: incineration, biodegradation, reuse, recycling, and source reduction (9.12)

Questions

Emphasizing Essentials

1. This chapter deals with the world of plastics and polymers. Do these terms mean the same thing?

2. Why are polymers sometimes referred to as macromolecules?

3. Figure 9.2 shows the structures of diamond and graphite, two allotropes of carbon. Examine the structures closely to see the number of bonds around any carbon in the central part of the structure. Does the number of bonds match the number predicted from the Lewis structure of carbon? Explain.

4. Survey ten friends and ask what descriptive adjectives they associate with the noun "plastic." Share your responses with others in the class to see if others have found the same associations. Does it seem that the word "plastic" still has the connotation of "cheap" or "tacky" among your friends?

5. What is the major reason to substitute plastics for metal in automobiles?

6. Use Figure 9.5 in answering these questions.
   a. How many pounds of plastic were produced per person in the United States in 1997, when the population was approximately 269 million persons?
   b. How many pounds of plastic were produced per person in the United States in 1977, when the population was approximately 220 million persons?
   c. What is the percent change in the total number of pounds of plastic produced per year between 1977 and 1997?
   d. What is the percent change in the number of pounds of plastic produced per person per year between 1977 and 1997?

7. Do you expect the heat of combustion of polyethylene, as reported in kJ/g, to be most similar to that of hydrogen, coal, or octane, C_8H_{18}? Explain your prediction.

8. Equations 9.1 and 9.2 show the polymerization of two ethylene monomers to form a small segment of polyethylene. Use the bond energies of Table 4.1 to calculate the energy change during the reaction in equation 1. Is the reaction endothermic or exothermic?

9. How will your result from question 8 differ if, rather than using ethylene as the monomer, tetrafluoroethylene is used as the monomer, forming a small segment of the Teflon polymer, polytetrafluoroethylene. This is the structure of the monomer.

10. a. Determine the number of CH_2CH_2 monomeric units, n, in one molecule of polyethylene with a molar mass of 40,000 g.
    b. What is the total number of carbon atoms in this polyethylene?

11. Explain the role of a free radical in the polymerization of ethylene.

12. This is a representation of a small segment of polyethylene. The hydrogen atoms are omitted for the sake of clarity.

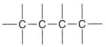

Does this representation predict that the carbon-to-carbon-to-carbon bond angles in polyethylene are all 180°? Explain your reasoning.

13. This is the usual two-dimensional representation for the formation of polyvinyl chloride polymer from vinyl chloride. As this process takes place at the molecular level, how does the approximate Cl—C—H bond angle change?

14. Describe how each of these strategies would be expected to affect the properties of polyethylene and give an atomic/molecular level explanation for each effect.
    a. increasing the length of the polymer chain
    b. aligning the polymer chains with one another
    c. increasing the degree of branching in the polymer chain

15. Both bottles are made of polyethylene. How do the two bottles differ at the molecular level?

16. The manufacturers of some plastic household containers say that it is fine to place the item in the dishwasher, particularly on the top shelf. Others do not claim that their products are dishwasher safe.

    a. Why is it recommended that dishwasher-safe plastics be put only on the top shelf?

    b. Assume that you have lost the information sheet that originally accompanied the dishwasher in your kitchen. Also, assume that you want to use the dishwasher as much as possible, rather than washing plastic containers by hand. What general properties of the plastic containers should help to guide you in avoiding problems in the dishwasher?

17. Do all of the "Big Six" have a common structural feature (see Table 9.1)? If so, identify that feature. If not, identify which of the six do share a common structural feature.

18. Table 9.1 lists features of the "Big Six" polymers.

    a. Which of the "Big Six" are the most flexible?

    b. Which of the "Big Six" do not have crystalline regions?

    c. Which of the "Big Six" are soluble in organic materials?

    d. Which is the most costly of the "Big Six"?

19. **a.** The structure for the styrene monomer is given in Table 9.1. Rewrite this shorthand structure showing all of the atoms.

    b. What is the molecular formula for styrene?

    c. What is the molar mass of a polystyrene molecule consisting of 5000 monomer units?

20. Vinyl chloride monomers can join in several different orientations to form polyvinyl chloride. Several different arrangements are shown in Figure 9.12. Which of those arrangements is shown here?

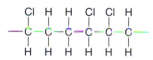

21. Butadiene, $H_2C{=}CH{-}CH{=}CH_2$, is polymerized to make buna rubber. Write an equation representing this process. Is this an example of addition or condensation polymerization?

22. Dow Chemical Company has developed a new process that uses CO_2 as the blowing or foaming agent to produce Styrofoam packaging material. What compound does CO_2 likely replace in the process, and why is this substitution environmentally beneficial?

23. Kevlar is a type of nylon called an *aramid* that contains aromatic rings. Because of its great mechanical strength, Kevlar is used in radial tires. The two monomers for producing Kevlar are these.

Monomer A   **Monomer B**

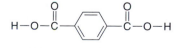   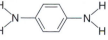

    When these two monomers join, what linkage between monomers forms the resulting polymer?

24. Polyacrylonitrile is a polymer made from the monomer acrylonitrile, CH_2CHCN.

    a. Draw a Lewis structure of this monomer.

    b. Polyacrylonitrile is used in making Acrilan fibers used widely in rugs and upholstery fabric. What danger do rugs or upholstery made of this polymer create in the case of house fires?

25. Section 9.10 has this heading: "Plastics: Where From and Where To?" Answer both questions posed in this heading, concentrating on the major source and major means of disposal.

Concentrating on Concepts

26. What are the relationships among these terms?

    Natural, synthetic, polymers, nylon, protein

    Show these relationships by constructing a concept web if you are familiar with that technique. Alternatively, write an outline with enough information to show how the terms are related.

27. You were asked in 9.3 Consider This to keep a journal of all the plastic and plastic-coated products you *throw away* in one week. Now consider all of the plastic items that you *recycle* in one week. Are there any from your first list of items thrown away that could be on your second list? Why or why not?

28. Celluloid was the first commercial plastic, developed in response to the need to replace ivory for billiard balls and piano keys.

    a. Speculate on what specific properties were required for celluloid that allowed it to be a suitable substitute for ivory in these products.

    b. Another early plastic was Bakelite. Was it also developed in response to a specific need? Explain your reasoning.

29. Reconsider Figure 9.5 showing the U.S. production of plastics from 1935–1997.

    a. Is there a uniformly spaced scale representing billions of pounds of plastics along the y-axis? If so, what does each division represent?

    b. Is there a uniformly spaced scale of years along the x-axis? If so, what does it represent?

    c. Redraw the representation in Figure 9.5 as a line graph showing the relationship of the year to the number of pounds of plastics produced.

    d. Starting in 1960, how many years have been required for plastics production to double? How has this doubling time changed from 1960 onward?

    e. Discuss whether Figure 9.5 or the graph you drew was easier to use to establish these doubling times.

30. The properties of a plastic are a consequence of more than just its chemical composition. What are some of the other features of polymer chains that have an influence on the properties of the polymer formed?

31. Consider the polymerization of a thousand ethylene monomers to form a large segment of polyethylene.

$$1000\ H_2C{=}CH_2 \longrightarrow {-}(CH_2CH_2){-}_{1000}$$

    a. Calculate the energy change during this reaction. (Use Table 4.1 of bond energies.)

    b. Should heat be supplied or must heat be removed from the polymerization vessel to promote this reaction? Explain.

32. Catalysts are used to help control the average molar mass of polyethylene, an important strategy to control polymer chain length. During World War II, low-pressure polyethylene production used varying mixtures of triethyl aluminum, $Al(C_2H_5)_3$, and titanium tetrachloride, $TiCl_4$, as a catalyst. Here are some data showing how the molar ratio of the two components of the catalyst affects the average molar mass of the polymer produced.

Moles $Al(C_2H_5)_3$	Moles $TiCl_4$	Average Molar Mass of Polymer, g
12	1	272,000
6	1	292,000
3	1	298,000
1	1	284,000
0.63	1	160,000
0.53	1	40,000
0.50	1	21,000
0.20	1	31,000

   a. Prepare a graph to show how the molar mass of the polymer varies with the mole ratio of $Al(C_2H_5)_3/TiCl_4$.

   b. What conclusion can be drawn about the relationship between the molar mass of the polymer and the mole ratio of $Al(C_2H_5)_3/TiCl_4$?

   c. Use the graph to predict the molar mass of the polymer if an 8:1 ratio of $Al(C_2H_5)_3$ to $TiCl_4$ were used.

   d. What ratio of $Al(C_2H_5)_3$ to $TiCl_4$ would be used to produce a polymer with a molar mass of 200,000?

   e. Can this graph be used to predict the molar mass of a polymer if either pure $Al(C_2H_5)_3$ or pure $TiCl_4$ were used as the catalyst? Explain.

33. When you try to stretch a piece of plastic bag, the length of the piece of plastic being pulled increases dramatically and the thickness decreases. Does the same thing happen when you pull on a piece of paper? Why or why not? Explain on a molecular level.

34. Consider the new polymer, Spectra, Allied-Signal Corporation's new HDPE fiber used as liners for surgical gloves. Although the Spectra liner has a very high resistance to being cut, the polymer allows a surgeon to maintain a delicate sense of touch. The interesting thing is that Spectra is *linear* HDPE, which is usually associated with being rigid and not very flexible.

   a. Suggest a reason why branched LDPE cannot be used in this application.

   b. Offer a molecular-level reason for why linear HDPE is successful in this application.

35. One limitation of the "Big Six" is the relatively low temperatures at which they melt, 90–170°C (see Table 9.1). Suggest ways to raise the upper temperature limits while maintaining the other desirable properties of these substances.

36. All of the "Big Six" polymers are insoluble in water, but some of them dissolve or at least soften in hydrocarbons or in chlorinated hydrocarbons (Table 9.1). Use your knowledge of molecular structure and solubility concepts to explain this behavior.

37. Vinyl chloride monomers can join in several different orientations to form polyvinyl chloride (see Figure 9.12). Do these two structures represent the same possible arrangement?

Explain your answer by identifying the orientation in each arrangement.

38. What structural features must a monomer possess to undergo addition polymerization? Explain, giving an example.

39. What structural features must a monomer possess to undergo condensation polymerization? Explain, giving an example.

40. The uses of the "Big Six" polymers are shown in Figure 9.11. Overall, how significant is the contribution of plastics used for packaging to the total numbers of pounds of plastics used in the United States?

Exploring Extensions

41. The text described "buckminsterfullerene," a carbon allotrope, as being shaped somewhat like a soccer ball. Each carbon is located at a corner where two six-membered rings and one five-membered ring come together. Locate a structural drawing of buckminsterfullerene on the web and find a corner that fits this description.

42. Did Dr. Richard Smalley, Dr. Harry Kroto, and Dr. Robert Curl, Jr., the pioneering chemists who won the 1996 Nobel Prize in chemistry, all work together in the same laboratory? Research the connections among their work and write a short report to describe why they won the Nobel Prize.

43. What is the difference in the material used in "hard" contact lenses and in "soft" contact lenses? How do the differences in properties affect the ease of wearing of contact lenses?

44. Two terms that have been added to the vocabulary of plastics are "virgin plastic" and "post-consumer waste" plastics. What do these terms imply about the plastics being used?

45. Free radical peroxides promote the polymerization of ethylene into polyethylene. They also play a key role in tropospheric smog formation. Use the web to learn more about how the peroxides promote ethylene polymerization and how peroxides are involved with photochemical smog formation in the troposphere. Write a brief report comparing the types of peroxides important to each of these cases. Give references for the web information.

46. Synthetic rubber is usually formed through addition polymerization. An important exception is silicone rubber, which is made by the condensation polymerization of dimethylsilanediol. This is a representation of the reaction.

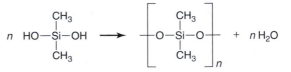

   a. Predict some of the properties of this polymer. Explain the basis for your predictions.

b. "Silly Putty" is a popular form of silicone rubber. What are some of the properties of "Silly Putty"?

47. 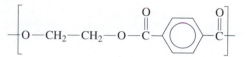 Who first synthesized Kevlar? What was the background and academic training of these scientists? Was the potential for using this polymer in radial tires immediately understood? What are other applications of Kevlar? Write a short report on the results of your findings, giving references either to books or web information.

48. This is the structural formula for Dacron, a condensation polyester.

$$\left[O-CH_2-CH_2-O-\overset{\overset{\displaystyle O}{\|}}{C}-\underset{}{\bigcirc}-\overset{\overset{\displaystyle O}{\|}}{C} \right]$$

Dacron is formed by the reaction of an alcohol with an —OH group at each end of its molecule and a dicarboxylic acid, one with —COOH groups at opposite ends of the acid molecule. Write the structural formulas for the alcohol and the acid monomers used to produce dacron.

49. Cotton, rubber, silk, and wool are natural polymers. Consult other sources to identify the monomer unit in each of these polymers; specify which are addition and which are condensation polymers.

50. How does your college's or university's community address the problem of disposing of plastics? Of all the strategies for disposing plastics described in this chapter, which are used? How are the alternatives presented to the people in the community? Find out the current practices, and then offer some suggestions for improving the current practices in your community.

Manipulating Molecules and Designing Drugs

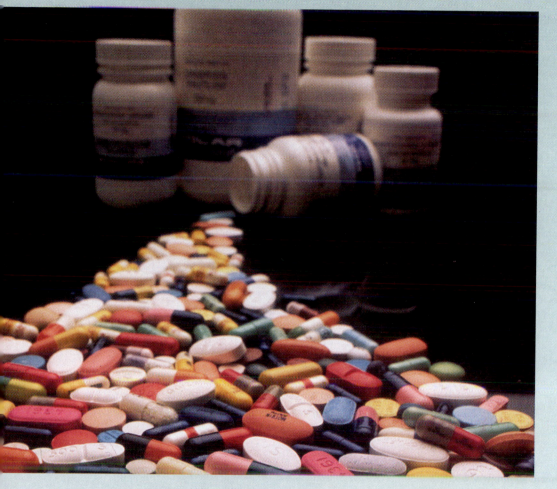

Chemists have designed and developed a vast array of prescription and "over-the-counter" drugs.

In the fourth century B.C., Hippocrates, perhaps the most famous physician of all time, described a "tea" made by boiling willow bark in water. The concoction was said to be effective against fevers. Over the centuries, that folk remedy, common to many different cultures, ultimately led to the synthesis of a true "wonder drug"—one that has aided millions of people.

Figure 10.1

The willow tree *Salix alba,* source of a miracle drug.

10.1 The Origin of a Miracle Drug

One of the first systematic investigators of willow bark (Figure 10.1) was Edmund Stone, an English clergyman. His report to the Royal Society set the stage for a series of further chemical and medical investigations. Chemists were subsequently able to isolate small amounts of yellow, needle-shaped crystals of a pure compound from the willow bark extract. Because the tree species was *Salix alba,* this new compound was named *sali*cin. Experiments showed that salicin could be chemically separated into two compounds. Clinical tests provided evidence that only one of these components reduced fevers and inflammation. It was also demonstrated that the active component was converted to an acid in the body. Unfortunately, the clinical testing revealed some troubling side effects. The active component not only had a very unpleasant taste, but its acidity also led to acute stomach irritation.

The active acid was used as a treatment for pain, fever, and inflammation. But because of its serious side effects, chemists set out to modify the structure of the active acid to form a related compound that still would be effective, but without the undesirable taste or stomach distress. The first modification attempt took a very simple approach. The acid was neutralized with a base, either sodium hydroxide or calcium hydroxide, to form a salt of the acid. It turned out that the resulting salts had fewer side effects than the parent compound. Based on this finding, chemists correctly concluded that the acidic part of the molecule was responsible for the undesirable properties. Consequently, the next step was to seek a structural modification that would lessen the acidity of the compound without destroying its medicinal effectiveness.

One of the chemists working on the problem was Felix Hoffmann, an employee of a major German chemical firm. Hoffmann's motivation was more than just scientific curiosity or assigned work. His father regularly took the acidic compound as treatment for arthritis. It worked, but he suffered nausea. The younger Hoffmann succeeded in converting the original compound into a different substance, a solid that reverted back to the active acid once it was in the body. This molecular modification greatly reduced nausea and other adverse reactions; a new drug had been discovered.

Extensive hospital testing of Hoffmann's compound began along with simultaneous preparation for its large-scale manufacture by a well-known pharmaceutical company. The new drug itself could not be patented because it was already in the chemical literature. However, the company hoped to recoup its investment by patenting the manufacturing process. Clinical trials showed the drug to be nonaddicting and relatively nontoxic. Its toxicity by ingestion is classed as low, but 20–30 g ingested at one time may be lethal. At the suggested dose of 325–650 mg (0.325–0.650 g) every four hours, it is a remarkably effective antipyretic (fever reducing), analgesic (anti-pain), and anti-inflammatory agent. Data from clinical tests uncovered the side effects noted in Table 10.1. The drug was also found to increase blood clotting time and to cause at least some small, almost always medically insignificant, amounts of stomach bleeding in about 70% of users.

10.1 Consider This: Miracle Drug

In the United States, the final step for approval of a drug is the submission of all its clinical test results to the Food and Drug Administration (FDA) for a license to market the product.

a. If you were an FDA panel member presented with the information in Table 10.1, Side Effects of the "Wonder Drug," would you vote to approve this drug that treats pain, fever, and inflammation?

b. If approved, should this drug be released as an over-the-counter drug or would you restrict its availability by making it a prescription drug? Write a one-page report stating your position and defending it.

Perhaps you have already guessed the identity of the miracle drug related to willow bark tea. Its chemical name 2-(acetyloxy)-benzoic acid or (more commonly) acetylsalicylic acid, may not help much. But the power of advertising is such that, had we revealed that the firm that originally marketed the drug was the Bayer division of I.G. Farben, we

Table 10.1   Side Effects of the "Wonder Drug"

The severity scale ranges from 1 — life threatening, seek emergency treatment immediately to 5 — continue the medication and tell physician at next visit.

Symptoms	Frequency	Severity
Drowsiness	rare	4
Rash, hives, itch	rare	3
Diminished vision	rare	3
Ringing in the ears	common	5
Nausea, vomiting, abdominal pain	common	2
Heartburn	common	4
Black or bloody vomit	rare	1
Black stool	rare	2
Blood in the urine	rare	1
Jaundice	rare	3
Anaphylaxis (severe allergic reaction)	rare	1
Unexplained fever	rare	2
Shortness of breath	rare	3

Source: Data from H. W. Griffith, *The Complete Guide to Prescription and Non-Prescription Drugs,* 1983, HP Books, Tucson, Arizona.

would have let the tablet out of the bottle. The compound in question is the world's most widely used drug. A century after its discovery, Americans annually consume nearly 80 billion tablets of this miracle medicine. You know it as aspirin.

Admittedly, we have compressed the time somewhat. Most of the development, testing, and design of aspirin occurred in the eighteenth and nineteenth centuries. Stone's letter to the Royal Society was written in 1763, and Felix Hoffmann's modification of salicylic acid to yield aspirin was done in 1898. Furthermore, the clinical testing of aspirin was somewhat less systematic than our account implies. But the basic facts and the steps that led to aspirin's full development are essentially correct. We must also add one more very important fact: aspirin did not have to receive drug approval before being put on the market; no such certifying process was in place at that time. Had such approval based on clinical test results been necessary, it is quite likely that aspirin might only be available on a prescription basis.

10.2 Consider This: What Should a Drug Be Like?

Make a list of the properties you would like a drug to have. Then compare your list with those of your classmates. Note similarities and differences in the lists.

a. Are there items that are listed that should not be?

b. After further consideration, are there items that were missing from the original lists that should be there?

Chapter Overview

Drugs or pharmaceuticals are substances that prevent, moderate, or cure illnesses. We began the chapter with a drug that has probably been used by all readers of this book. We chose aspirin not only because of its familiarity, but because its discovery and development demonstrate how a new drug often comes into existence. This process can be generalized and extended to other drugs such as penicillin. All of the drugs mentioned (indeed, most of the drugs used today) contain the element carbon. Therefore, we embark on a brief excursion into organic chemistry, the realm of carbon compounds. The principles governing the structure of organic molecules are those we have already applied to other molecules in previous chapters. The molecular structures of organic compounds offer the opportunity to explore new features such as isomers and functional groups. Both are of great importance in linking molecular structure to drug function. An example of characteristic drug activity is again provided by aspirin, but a more general treatment of the topic introduces the concept of the fit between the drug and the biochemical site at which it acts. Sometimes, activity depends on the subtle property of chirality, also called optical isomerism.

Discussion then shifts to steroids, one of the most interesting and important families of biologically active compounds. Members of the steroid family that are treated in some depth are cholesterol, sex hormones, contraceptives, aborting agents, and anabolic steroids. These compounds, or at least their uses, are familiar to almost everyone because they have often been steeped in controversy. By setting the steroids in their social context we explore a number of these issues. The case of thalidomide introduces drug testing and approval, the lengthy and demanding process that is necessary to obtain approval to sell, distribute, and use a new drug. Here again, we will find that risks and benefits are present, whether using non-generic or generic drugs.

10.2 Drug Discovery: By Accident and By Design

The curative powers of certain chemicals have been discovered by various means: from concocting folk remedies to targeted research, and from lucky accidents to systematic investigation. Aspirin is obviously an example of a drug derived from a folk remedy.

Throughout recorded history and in all societies, the identification of substances effective against illness and disease has been an important activity. Tribal shamans and other traditional healers have been keen observers of the effects of plant extracts on their patients. Some of these extracts have proven useful in modern medicine. In addition to aspirin, the best known are probably digitalis, quinine, morphine, and many of the mind-altering drugs. One argument for the preservation of tropical rain forests is that they may contain plant species with still undiscovered medicinal properties.

On the other hand, curative powers have been attributed to some folk remedies that have no beneficial effects. The spectacular medical and rejuvenative properties attributed to powdered rhinoceros horn or dried bear spleen are without medicinal validity and pose serious threats to endangered species. Moreover, natural materials are not necessarily benign; folk remedies can do considerably more harm than good. For example, tea made from sassafras contains safrole, a known carcinogen. Toxic compounds of mercury and arsenic also have been part of the primitive pharmacopoeia.

Some drugs have been discovered accidentally, through the correct interpretation of a lucky observation or curiosity about an unusual occurrence. Included in this group of serendipitous discoveries are LSD and some tranquilizers. But the most famous and perhaps the luckiest discovery of this type was that of penicillin by the British bacteriologist Alexander Fleming in 1928. Fleming's curiosity was aroused by the chance observation that in a container of bacterial colonies, the area contaminated by the mold *Penicillium notatum* was free of bacteria (Figure 10.2). He correctly concluded that the mold gave off a substance that inhibited bacterial growth, and he named this biologically active material penicillin.

A careful reconstruction has indicated that a series of critical, but fortuitous events, had to occur for the discovery to be made. A colleague in a laboratory one floor below Fleming's London laboratory happened to be working on *Penicillium notatum,* a rare strain of mold, and Fleming happened to be working on *Staphylococcus,* a bacterial strain subsequently found to be particularly sensitive to penicillin. Spores from the mold drifted into Fleming's laboratory and accidentally contaminated some Petri dishes containing *Staphylococcus* growing on a nutrient medium. Then came a series of chance incidents involving poor laboratory housekeeping, a vacation, and a spell of cool weather. The Petri dishes were left unwashed in Fleming's laboratory while he was on vacation. During this time, the uncommonly cool weather slowed the growth of the bacteria, but not the mold growth. Then the weather turned warmer, permitting bacterial growth, except in the area around the mold. Back from his vacation, Fleming fortunately noticed the dishes in which the *Staphylococcus* had been killed. Using former experience, he correctly interpreted the phenomenon, recognizing that the unknown substance being given off by the *Penicillium* was a potential antibacterial agent for the treatment of infection.

Fleming, who shared a Nobel Prize for his work on penicillin, later said that the chance that put the right mold in the right spot at the right time was about like the slim

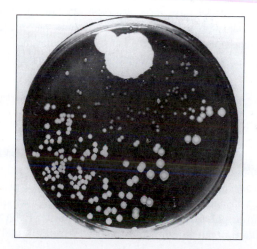

Figure 10.2
A Petri dish showing the antibacterial properties of penicillin. The large white area at the 12 o'clock position is the mold *Penicillium notatum*; the smaller white spots are areas of bacterial growth. The relative absence of bacterial colonies in the area immediately surrounding the mold is evidence that penicillin, a compound produced by the mold, inhibits bacterial growth.

chance of winning the Irish Sweepstakes. "The story of penicillin," he went on, "has a certain romance in it and helps to illustrate the amount of chance, or fortune, of fate, or destiny, call it what you will, in anybody's career." But of course, the discovery would not have happened without Fleming's powers of observation and insight in response to the unexpected. The episode admirably illustrates the often misquoted maxim of the great French scientist, Louis Pasteur: "In the fields of observation, chance favors only the prepared mind." Most versions of this famous aphorism neglect the "only." It was *only* because Fleming's mind was prepared that he was able to capitalize on this chain of unlikely events.

Not all drug discoveries are the results of lucky breaks. Even when an accident provides the initial clue, much careful planning and thorough investigation are necessary to develop the full potential of the drug. Taking penicillin from the Petri dish to the pharmacy was not easy. The first step was a systematic effort to isolate the active agent produced by *Penicillium notatum.* Once identified, the substance had to be separated, purified, and concentrated by new, sophisticated techniques. Also, the efficacy of penicillin in treating humans had to be demonstrated. The first person to be treated with penicillin was an Oxford policeman suffering from a very serious infection. Over four days of treatment, his condition improved dramatically. Then the penicillin supply was exhausted, in spite of the fact that excreted unused penicillin was recycled from the patient's urine. Without the drug, the policeman died. World War II gave increased impetus to this research and to the development of new methods for preparing large quantities of penicillin. Because the scientists were successful, thousands of lives were saved during the war, and millions since then.

The discovery of one successful drug often leads to the creation of many more. The initial drug serves as a prototype for others, and researchers typically investigate the effectiveness of related compounds, called analogs. Thus, once it became clear that molds produce antibiotic compounds, several companies launched programs to collect mold samples from all over the world and screen them for antibacterial activity. Though costly and time-consuming, this approach yielded a number of different molecular members of the penicillin family. Cyclosporin, a major anti–tissue rejection drug that has revolutionized organ transplant surgery, was discovered in this way.

Even more sophisticated research is needed to determine the molecular structure of a biologically active compound and to duplicate it in the laboratory and the factory. The first structural determination of a naturally occurring penicillin compound was done in 1941 by the British crystallographer Dorothy Crowfoot Hodgkin. She measured the pattern in which X-rays were reflected by the atoms in a crystal of the compound. Then she translated this information into a three-dimensional molecular model showing the molecular structure in exquisite detail. With this knowledge, chemists set out to prepare synthetic versions of the compound. The first total synthesis of a penicillin was completed in 1964. Many of the penicillins currently in use are chemically created without the necessity of molds and fermentation. In fact, some antibiotics do not even exist in nature. They are elaborations and improvements on naturally occurring compounds. Sometimes, subtle changes in complex molecular structures can alter or enhance biological activity or change the specificity with which the antibiotic acts on various bacterial strains. Chemists who customize molecules for use against a particular medical problem are skillful molecular architects, masters of the fascinating subdiscipline called organic chemistry.

10.3 A Brief Excursion into Organic Chemistry

The great majority of earthly chemical compounds contain the element carbon. This element is so widely distributed throughout nature that the largest subdiscipline of chemistry, **organic chemistry,** is devoted to the study of carbon compounds. The name "organic" suggests a biological origin for the substances under investigation, but this is not necessarily true in every instance. In practice, most organic chemists confine themselves to compounds of carbon and a relatively small number of other elements—hydrogen, oxygen, nitrogen, sulfur, chlorine, phosphorus, and bromine. Even with this restriction, there are over 11 million organic compounds out of 13 million total known compounds.

Dr. Dorothy Crowfoot Hodgkin, an Oxford University professor, won the 1964 Nobel Prize in Chemistry for her x-ray crystallography to determine the structure of complex molecules, including that of penicillin.

In this chapter we concentrate on only a few organic compounds and stress their important role in interactions with living things. Molecular shape will prove to be of special significance.

Organic compounds must be named, and chemists use a formal set of nomenclature rules established by an international committee. However, many of these compounds have been known for a long time by common names such as alcohol, sugar, or morphine. When a headache strikes, even chemists do not call out for 2-(acetyloxy)-benzoic acid; they simply say "give me some aspirin!" Likewise, prescriptions specify penicillin-N rather than 6[(5-amino-5-carboxy-1-oxopentyl)amino]-3,3-dimethyl-7-oxopentyl-4-thia-1-azabicyclo[3,2,0]-heptane-2-carboxylic acid. Mouthfuls like this are the cause of great merriment to those who like to satirize chemists, but they are important and unambiguous to those who know the system. You can rest easy because in this chapter we will use common names in almost all cases.

The incredible variety of organic compounds exists because of the remarkable ability of carbon atoms to bond in multiple ways. They can bond with other carbon atoms or with atoms of other elements. To better understand such possibilities, we need a few basic rules for bonding in organic molecules. The most fundamental generalization is one you used as early as Chapter 2—the octet rule. *Each bonded carbon atom shares in eight electrons.* These electrons are paired to form covalent bonds and can be grouped in four different common bonding patterns: (a) four single bonds, (b) two single bonds and one double bond, (c) one single bond and one triple bond, or (d) two double bonds. These arrangements are illustrated in Figure 10.3. Other elements exhibit different bonding behavior. A hydrogen atom is always attached to a molecule by a single covalent bond. An oxygen atom in a molecule typically has two pairs of bonding electrons, either in the form of two single bonds or one double bond. A nitrogen atom shares in three pairs of bonding electrons and hence can form three single bonds, one triple bond, or one single and one double bond.

Molecular formulas, such as C_4H_{10}, indicate the kinds and numbers of atoms present in a molecule, but do not show how the atoms are arranged. To get that higher level of detail, structural formulas are used. These representations show the atoms and their arrangement with respect to each other in a molecule. In the case of C_4H_{10} (butane, a hydrocarbon used in cigarette lighters and camp stoves) a structural formula can be written as follows.

$$\begin{array}{cccc} H & H & H & H \\ | & | & | & | \\ H-C-C-C-C-H \\ | & | & | & | \\ H & H & H & H \end{array}$$

Notice that in this representation, the bonding and position of each atom relative to all others is specified. But a drawback to writing structural formulas, at least in a textbook, is that they take up considerable space. Instead, modified or condensed structural formulas can be used to convey the same information. In these, carbon-to-hydrogen bonds are not drawn out explicitly, but simply understood to be single bonds. Condensed structural formulas for C_4H_{10} are given below.

$$CH_3-CH_2-CH_2-CH_3 \quad \text{or} \quad CH_3CH_2CH_2CH_3$$

This representation implies that carbon atoms are bonded directly to other carbon atoms in a straight chain. The hydrogen atoms do not intervene in the chain. Rather, two or three are attached to each carbon atom, depending upon its position in the molecule.

The octet rule is discussed in Section 2.3.

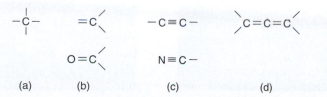

(a)   (b)   (c)   (d)

Figure 10.3

Representation of carbon with some of its bonding possibilities.

Structural isomers are described in Section 4.8.

One reason why there are so many different organic molecules is because the same number and kinds of atoms can be arranged in unique ways called isomers. **Isomers** are compounds with the same chemical formula (same number and kinds of atoms) but different molecular structures and properties. You have already encountered isomers in the discussion of octane, C_8H_{18} in Chapter 4. Here we will illustrate the idea with C_4H_{10}. One way the atoms can be arranged is given on page 381, the linear isomer called butane. However, another arrangement is possible in which the four carbon atoms are not in a "straight" line. This other isomer is represented by the following structural formula.

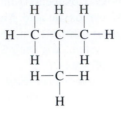

This isomer is known as iso-butane, and its formula can also be written in condensed form.

$$CH_3-\underset{\underset{CH_3}{|}}{CH}-CH_3 \quad \text{or} \quad CH_3CH(CH_3)CH_3$$

The parentheses around the CH_3 indicate that its carbon is attached to the carbon to its left. Note from the structural formula that the CH_3 introduces a "branch" into the molecule. Molecular models of both isomers of C_4H_{10} are pictured in Figure 10.4.

These two compounds are the only isomers of C_4H_{10}. It might be tempting to draw another structural formula that looks something like the following.

$$\underset{CH_3-\underset{\underset{}{|}}{CH}-CH_3}{\overset{\overset{CH_3}{|}}{}}$$

However, a bit of inspection should reveal that this is simply the previous structure for iso-butane written upside down, and not a different compound. As the number of atoms in a hydrocarbon increases, so do the number of possible isomers. Thus, there are 18 isomers of C_8H_{18} and 75 isomers of $C_{10}H_{22}$.

10.3 Your Turn

The formula C_5H_{12} represents three different compounds, all structural isomers of pentane. Draw structural formulas for these isomers.

Figure 10.4

Photographs of molecular models of the two isomers of butane, C_4H_{10}: butane (left) and iso-butane (right).

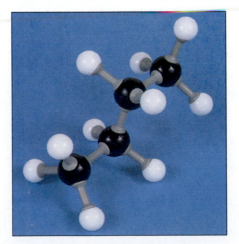
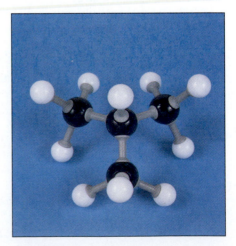

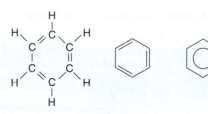

Figure 10.5
Three representations of benzene, C_6H_6.

Carbon atoms are not limited to being bonded in a linear or branched fashion. In many molecules, including aspirin, carbon atoms are also arranged in a ring. Such rings most commonly contain five or six carbon atoms. In aspirin, the carbon atoms are joined in a six-member hexagonal ring called a benzene ring after the compound with the formula C_6H_6. The Lewis octet rule applied to C_6H_6 predicts alternating single and double bonds between adjacent carbon atoms. The complete representation of this structure appears on the left in Figure 10.5. But the benzene ring is so widely distributed in organic molecules that symbols for individual carbon and hydrogen atoms are generally not written. Instead, the C_6H_6 molecule is written as a hexagon with a ring of electrons. One carbon atom with one attached hydrogen atom is assumed to occupy each vertex (corner) of the hexagon. The representation in the center of Figure 10.5 indicates single and double bonds connecting the carbon atoms, but this picture is somewhat misleading. Experiment shows all carbon-carbon bonds in benzene to be identical. In strength and in length, these bonds are somewhere between single bonds and double bonds. This means that the electrons connecting the carbon atoms must be uniformly distributed around the ring. The circle within the hexagon in the drawing on the right is an effort to convey this idea. This same hexagonal structure is found in the $-C_6H_5$ phenyl group that is part of many molecules, including styrene and polystyrene, which you encountered in Chapter 9.

The uniform distribution of electrons in the benzene ring is an example of resonance (Section 2.3).

10.4 Functional Groups

Fortunately, the plethora of organic compounds is somewhat simplified by the existence of a relatively small number of functional groups that appear with considerable frequency. **Functional groups** are distinctive arrangements of groups of atoms that impart characteristic chemical properties to the molecules that contain them. Indeed, these groups are so important that we often focus our formulas on them and symbolize the remainder of the molecule with an R. The R is generally assumed to include at least one carbon atom that is connected to the functional group, but it can be practically anything. You already encountered some functional groups in Chapter 9. The generic formula for an alcohol is ROH, as in methanol (wood alcohol), CH_3OH, and ethanol (grain alcohol), CH_3CH_2OH. The presence of the $-OH$ group makes the compound an alcohol.

Section 9.8 discussed alcohols in polymerization reactions.

Similarly, acidic properties are conveyed by a carboxylic acid group

commonly written as $-COOH$. In aqueous solution, an H^+ ion (a proton) is transferred from the $-COOH$ group to an H_2O molecule to form a hydronium ion, H_3O^+. We represent an organic acid with the general formula RCOOH. In acetic acid, the acid in vinegar, R is $-CH_3$, a methyl group. Table 10.2 lists eight of the most important functional groups found in drugs and other organic compounds. Each functional group is characteristic of an important class of compounds. The table presents the general, generic formula of the class. In every case, the functional group is highlighted in color. In addition, Table 10.2 includes the formula and molecular structure of an example of each compound class.

The presence and properties of functional groups are responsible for the action of all drugs. Aspirin has three such subunits, boxed and numbered in Figure 10.6. You will recognize that box 1 encloses a benzene ring. Its presence makes aspirin soluble in fatty

Table 10.2   **Some Important Classes of Organic Compounds and Their Characteristic Functional Groups**

Class of Compound	Generic Formula	Example		
		Structural Formula	Name	Condensed Structural Formula
alcohol	R—OH	H—C—C—OH (with H's)	ethyl alcohol (ethanol)	CH_3CH_2—OH
ether	R—O—R′	H—C—C—O—C—C—H (with H's)	diethyl ether	CH_3CH_2—O—CH_2CH_3
aldehyde	R—C(=O)—H	(benzene ring)—C(=O)—H	benzaldehyde	C_6H_5—C(=O)—H
ketone	R—C(=O)—R′	H—C—C(=O)—C—H (with H's)	acetone	CH_3—C(=O)—CH_3
carboxylic acid	R—C(=O)—OH	H—C—C(=O)—OH (with H's)	acetic acid	CH_3—C(=O)—OH
ester	R—C(=O)—OR′	H—C—C(=O)—O—C—C—H (with H's)	ethyl acetate	CH_3—C(=O)—OCH_2CH_3
amine	R—NH_2	H—C—N(H)(H) (with H's)	methyl amine	CH_3—NH_2
amide	R—C(=O)—NH_2	H—C—C—C(=O)—N(H)(H) (with H's)	ethyl amide	CH_3CH_2—C(=O)—NH_2

Note: The characteristic functional groups are printed in color. R or R′ usually signifies a group of atoms including at least one carbon atom bonded to the functional group. R and R′ can be the same or different, depending on the compound.

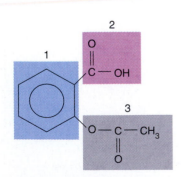

Figure 10.6
Structural formula of aspirin.

compounds that are important cell membrane components. The other two portions, boxes 2 and 3, are responsible for the drug activity. You have just been reminded that the —COOH group indicates an organic acid. The other functional group (box 3) is an ester. You will recall from our discussion of polyethylene terephthalate in Chapter 9 that an ester is formed by the reaction of an acid and an alcohol. Water is eliminated in the process.

Felix Hoffmann prepared aspirin by modifying the structure of salicylic acid. But note that he did not modify the carboxylic acid group on the molecule. Salicylic acid also contains an alcoholic —OH group, and it was this part of the molecule that Hoffmann reacted with acetic acid via equation 10.1. The product was an ester of acetic acid and salicylic acid, which accounts for one of aspirin's names—acetylsalicylic acid.

Because aspirin retains the —COOH group of the original salicylic acid, it still has some of the undesirable acidic properties of the parent compound. However, the presence of the ester group reduces the strength of the acid group and makes the compound more palatable and less irritating to the stomach lining. Once aspirin is ingested and reaches the site of its action, reaction 10.1 is reversed. The ester splits into acetic acid and salicylic acid, and the latter compound exerts its antipyretic (fever-reducing) and analgesic (pain-reducing) properties.

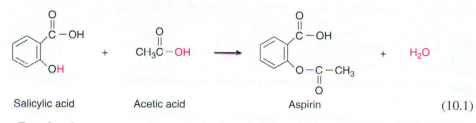

Salicylic acid   Acetic acid   Aspirin   (10.1)

Functional groups can play a role in the solubility of a compound, an important consideration in the uptake, rate of reaction, and residence time of drugs in the body. The general solubility rule, "like likes like," applies in the body as well as in the test tube. When that rule was introduced in Chapter 5, a distinction was made between polar and nonpolar molecules. A polar molecule has a nonsymmetrical distribution of electric charge. This means that a negative charge builds up on some part (or parts) of the molecule, while other regions of the molecule bear a positive charge. Water is an excellent example of a polar molecule. Relatively speaking, the oxygen atom is slightly negatively charged and the hydrogen atoms are slightly positive. Because the molecule is bent, it has a nonsymmetrical charge distribution. Functional groups containing oxygen and nitrogen atoms (for example, —OH, —COOH, and —NH_2) usually increase the polarity of a molecule. This in turn enhances its solubility in a polar substance such as water.

By contrast, compounds whose molecules do not contain such atoms, but consist primarily or exclusively of carbon and hydrogen atoms, are typically nonpolar. A hydrocarbon such as octane, C_8H_{18}, is a good example; this compound is insoluble in water. However, it does dissolve in nonpolar solvents that are structurally similar to it. For the same reason, drugs with significant nonpolar character tend to accumulate in cell membranes and fatty tissues, which are themselves largely hydrocarbon and nonpolar.

Drugs with similar physiological properties often have similar molecular structures, including some of the same functional groups. Of the approximately 40 alternatives to aspirin that have been produced, ibuprofen and acetaminophen (Tylenol), are the most familiar. Figure 10.7 gives the structural formulas of the three leading analgesics. All are built on a benzene ring with two substituents, but the substituents differ in detail. In 10.4 Your Turn you have an opportunity to identify the structural similarities and differences of these compounds.

The current commercial method for producing ibuprofen is a stunning application of green chemistry. Conventional methods of ibuprofen production required six steps, used large amounts of solvents, and generated significant quantities of wastes. By using a catalyst that also serves as a solvent, BHC Company, a 1997 Presidential Green Chemistry Challenge Award winner, makes ibuprofen in just three steps with a minimum of solvents and waste. In the BHC process, virtually all of the reactants are converted to ibuprofen or another usable by-product; any unreacted starting

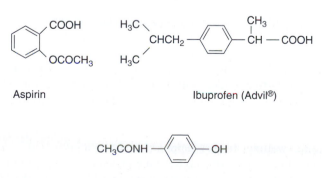

Aspirin   Ibuprofen (Advil®)

Acetaminophen
(Tylenol®)

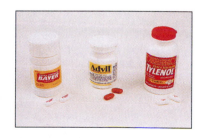

Figure 10.7
Structural formulas and samples of some common analgesics.

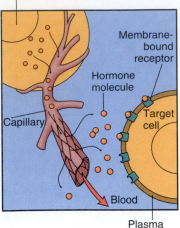

Figure 10.8

Chemical communication in the body. Hormone molecules travel from the cell where they are made, through the bloodstream, to the target cell.

Section 12.6 decribes the use of genetic engineering to obtain human insulin from bacteria.

You have encountered catalysts in several other contexts, including automobile emissions control (Section 1.11), petroleum refining (Section 4.5), and addition polymerization (Section 9.7).

materials are recovered and recycled. Nearly eight million pounds of ibuprofen, enough to make 18 billion 200-mg pills, are produced annually in Bishop, Texas at the BHC facility, built specifically for the commercial production of the drug.

10.4 Your Turn

Look at the structural formulas given in Figure 10.7. Identify the structural features and functional groups that aspirin, ibuprofen, and acetaminophen have in common.

10.5 How Aspirin Works

To understand the action of aspirin it is necessary to know something about the body's chemical communication system. We normally think of internal communication as consisting of electrical impulses traveling along a network of nerves. This is certainly true for the system that triggers movement, breathing, heartbeats, and reflex actions. Most of the body's messages, however, are conveyed not by electrical impulses, but through chemical processes. In fact, your very first communication with your mother was a chemical signal saying "I'm here; better get your body ready for me." It is much more efficient to release chemical messengers into the bloodstream, which then circulates them to appropriate body cells, than to "hardwire" each individual cell with nerve endings.

These chemical messengers are called **hormones** and they are produced by the body's endocrine glands. Figure 10.8 is a representation of such chemical communication. Hormones encompass a wide range of functions and a similarly wide range of chemical composition and structure. Thyroxine, an iodine-containing amino acid, is one of the simpler ones, but is essential for regulating metabolism. The chemical breakdown of "blood sugar" (glucose) requires insulin. This hormone, a small protein of only 51 amino acids, is secreted by the pancreas. Persons who suffer from diabetes are often required to take daily injections of insulin. Yet another well-known hormone is adrenaline (epinephrine), a small molecule that prepares the body to "fight or flee" in the face of danger. And the hormonal messages that are so compelling in adolescents are carried by steroids, a sexy set of molecules that we will visit in a few pages.

Aspirin and other drugs that are physiologically active, but not anti-infectious agents, are almost always involved in altering the chemical communication system of the body. A significant problem is that this system is very complex, allowing many compounds to be used to send more than one message simultaneously. The wide range of aspirin's therapeutic properties, as well as its side effects, are clear evidence that the drug is involved in several chemical communication systems. It works in the brain to reduce fever, it relieves inflammation in muscles and joints, and it apparently decreases the chances of stroke and heart attack. It may even lessen the likelihood of colon, stomach, and rectal cancer.

In large measure, the versatility of aspirin and similar *non*steroidal *a*nti-*i*nflammatory *d*rugs (NSAIDs) is related to their remarkable ability to block the actions of other molecules. Research on the activity of aspirin indicates that one of its modes of action involves blocking cyclooxygenase (COX) enzymes. These enzymes, like all others, are biochemical **catalysts.** They are proteins that influence the rate of a chemical reaction. Most enzymes speed up reactions and channel them so that only one product (or a set of related products) is formed. In the case of cyclooxygenases, the reaction is the synthesis of a series of hormone-like compounds called prostaglandins. Prostaglandins cause a variety of effects. They produce fever and swelling, increase sensitivity of pain receptors, inhibit blood vessel dilation, regulate the production of acid and mucus in the stomach, and assist kidney functions. By preventing prostaglandin production, aspirin reduces fever and swelling. It also suppresses pain receptors and so functions as a painkiller. Because the benzene ring conveys high fat solubility, aspirin is also taken up into cell membranes. In certain specialized cells, the drug blocks the transmission of chemical signals that trigger inflammation. This process also appears to be related to aspirin's effectiveness as a pain reliever.

The aspirin substitutes exhibit these same properties in varying degrees. For example, because acetaminophen blocks COX enzymes, but does not affect the specialized cells, it reduces fever but has little anti-inflammatory action. On the other hand, ibuprofen is a better enzyme blocker and specialized cell inhibitor. Consequently, ibuprofen is both a better pain reliever and fever reducer than aspirin. Ibuprofen has fewer functional groups than aspirin, which may be the reason why ibuprofen has fewer side effects. With fewer functional groups, ibuprofen is less polar and more lipid soluble than aspirin. Its anti-inflammatory activity is five to fifty times that of aspirin.

All of these structurally related anti-inflammatory drugs appear to affect the way cell membranes respond to stimuli. Research has shown that this is yet another possible mode of action for aspirin and its chemical relatives. On the other hand, aspirin is unique among these three compounds in its ability to inhibit blood clotting. This property has led to the suggestion that low regular doses of aspirin can help prevent strokes or heart attacks. Of course, these anticoagulation characteristics also mean that aspirin is not the painkiller of choice for surgical patients or those suffering from ulcers. That is why "more hospitals use Tylenol." You have already read that some people experience stomach irritation when they take aspirin. Another drawback of the drug is that in rare cases it can trigger a sometimes fatal response known as Reye's syndrome.

Recently, scientists have been able to better understand how aspirin, ibuprofen, and acetaminophen affect the two cyclooxygenase (COX) enzymes. These drugs block one of the enzymes, COX-2, which makes prostaglandins associated with inflammation, pain, and fever, thereby reducing these symptoms. But the drugs also inhibit the other enzyme in the pair, COX-1, which primarily makes hormones that maintain proper kidney function and keep the stomach lining intact. Thus, the drugs are not sufficiently selective to affect COX-2 without shutting down COX-1 as well. By determining the crystal structure of COX-2 in 1992 , researchers were then guided in making nearly a dozen new candidate drugs that block COX-2 alone, thereby creating new medicines that have been termed "superaspirins." Regarding the future of superaspirins, medicinal chemist Phil Portoghese predicts that "Eventually all those [NSAID] molecules on the market will become dinosaurs." A compound that inhibits both enzymes, but preferentially acts on COX-2, is now available in Europe. One of the other new COX-2 inhibitors involves some clever molecular modification. While retaining the acetyl group, as in aspirin, a long molecular tail is added as a side chain that includes a sulfur atom and seven carbon atoms, two of which have a triple bond between them (Figure 10.9.)

A few final comments about aspirin seem appropriate. Because it is a specific chemical compound, aspirin is aspirin—acetylsalicylic acid. But although all aspirin molecules are identical, not all aspirin tablets are the same. The commercial products are mixtures of various components, including inert fillers and bonding agents that hold the tablet together. Buffered aspirin tablets also include weak bases that counteract the natural acidity of the aspirin. These differences in formulation can influence the rate of uptake of the drug and hence how fast it acts, and perhaps the extent of stomach irritation. Furthermore, although standards for quality control are high, it is conceivable that individual lots of aspirin may vary slightly in purity. Aspirin also decomposes with time, and the smell of vinegar can signify that such a process has begun. Fortunately, none of this poses a significant threat to health, and the benefits of aspirin far outweigh the risks for the great majority of people.

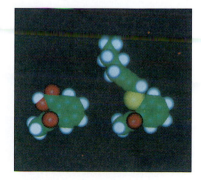

Figure 10.9
A COX-2 inhibitor (right) is more selective than aspirin (left) at blocking the action of COX-2.

10.5 Consider This: Aspirin in the Large Bottle

A friend who suffers from heart disease has been told by the doctor to take two aspirin tablets a day to "thin the blood." To save money, your friend often buys the large 500-tablet bottle of aspirin. You, on the other hand, rarely take aspirin, but cannot pass up a good bargain. You also buy the large 500-tablet bottle of aspirin.

a. Why is the "giant economy size" bottle of aspirin not as good a deal for you as it is for your friend?

b. What chemical evidence supports your opinion?

10.6 Drug Function and Drug Design

The modern approach to chemotherapy and drug design probably began early in this century with Paul Ehrlich's search for an arsenic compound that would cure syphilis without doing serious damage to the patient. His quest was for a "magic bullet" that would affect only the diseased site and nothing else. He systematically varied the structure of many arsenic compounds, simultaneously testing each new compound for activity and toxicity using experimental animals. He finally achieved success with Salvarsan 606, so named because it was the six hundred sixth compound investigated. Since then, medicinal chemists have adopted Ehrlich's strategy of carefully relating chemical structure and drug activity. The goal remains to produce a drug that meets the desired therapeutic need while exhibiting minimum side effects.

Drugs can be broadly classified into two groups: those that produce a physiological response in the body and those that kill or inhibit the growth of substances that cause infections. You have already learned that aspirin falls in the first group. So do synthetic hormones and psychologically active drugs. These compounds typically initiate or block a chemical action that generates a cellular response, such as a nerve impulse or the synthesis of a protein. Antibiotics exemplify drugs that kill foreign invaders. They do so by inhibiting an essential chemical process in the infecting organism. Thus, they are particularly effective against bacteria.

Although drugs vary in their versatility, many of them act only against particular diseases or infections. This specificity is consistent with the relationship that exists between the chemical structure of a drug and its therapeutic properties. Both the general shape of the molecule and the identity and location of its functional groups are important factors in determining its physiological efficacy. This correlation between form and function can be explained in terms of the interaction between biologically important molecules. Although many of these molecules are very large, consisting of hundreds of atoms, each molecule often contains a relatively small **active site** or **receptor site** that is of crucial importance in the biochemical function of the molecule. A drug is often designed to either initiate or inhibit this function by reacting with the receptor site.

An example is provided by a receptor site that controls whether a cell membrane is permeable to certain chemicals. In effect, such a site acts as a lock on a cellular door. The key to this lock may be a hormone or drug molecule. The drug or hormone bonds to the receptor site, opening or closing a channel through the cell membrane. Whether the channel is open or closed can significantly influence the chemistry that occurs in the cell. In fact, under some circumstances, the cell may be killed, which may or may not be beneficial to the organism.

This lock-and-key analogy is often used to describe the interaction of drugs and receptor sites. Just as specific keys fit only specific locks, a molecular match between a drug and its receptor site is required for physiological function. The process is illustrated in Figure 10.10. If a perfect lock-and-key match were required in the body, it would mean that each of the millions of physiological functions would have a unique receptor site and a specific molecular segment to fit it. Simple logic suggests that such rigid demands would not promote cellular efficiency. Consequently, the lock-and-key model, although a good starting point that works in a limited number of cases, must be modified.

Using another analogy, a receptor site is like a size 9 right footprint in the sand. Only one foot will fit it exactly, and many feet (all left feet and all right feet much larger or smaller than size 9) will not fit it at all. But many other right feet can fit into the print reasonably well. So it is with receptor sites and the molecules or functional groups that bind to them. Some active sites can accommodate a variety of molecules including drugs. Indeed, the way most drugs function is by replacing a normal protein, hormone, or other substrate in the invading organism. The presence of the drug molecule thus prevents the enzyme, cell membrane, or other biological unit from carrying out its required chemistry. As a result, the growth of an invading bacterium is inhibited, or the synthesis of a particular molecule is turned off (Figure 10.11).

Generally speaking, the drug that best fits the receptor site has the highest therapeutic activity. In some cases, however, a drug molecule does not need to fit the receptor

The lock-and-key analogy was first proposed in 1894 by Emil Fischer, a famous biochemist.

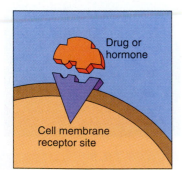

Drug or hormone

Cell membrane receptor site

Figure 10.10
Lock-and-key model of biological interaction.

site particularly well. The bonding of functional groups of the drug to the receptor site may even alter the shape of the drug, the site, or both. Often what counts is for the drug to have functional groups of the proper polarity in the right places. Thus, one important strategy in designing drugs is to determine the specific part of the molecule that gives the compound its activity. Medicinal chemists then synthesize a molecule having that specific active portion, but with a much simpler nonactive remainder. These researchers custom design the molecule to meet the requirements of the receptor site. In effect, they design feet to fit footprints.

An outstanding example of this approach is provided by opiate drugs such as morphine. Morphine, a very complex molecule, is difficult to synthesize. However, the particular portion of the molecule responsible for opiate activity has been identified and is highlighted in Figure 10.12. The flat benzene ring fits into a corresponding flat area of the receptor, and the nitrogen atom binds the drug molecule to the site. Incorporating this particular portion into other less complex molecules, such as demerol, creates opiate activity.

The discovery that only certain functional groups are responsible for the therapeutic properties of pharmaceutical molecules has been an important breakthrough. Sophisticated computer graphics are now used to model potential drugs and receptor sites. Thanks to these representations, with their three-dimensional character, medicinal chemists can "see" how drugs interact with a receptor site. Computers can then be used to search for compounds that have structures similar to that of an active drug. Chemists can also modify structure in the computer models and visualize how the new compounds will function.

Normal + Enzyme
substrate

↓

Normal + Enzyme
product

Drug + Enzyme

⚹

↓

Normal product

Figure 10.11
Reaction of a normal substrate with an enzyme to give a normal product; a drug blocks the action of an enzyme so that the normal product does not form.

10.6 Consider This: 3-D Drugs

See for yourself how drug molecules appear in three dimensions by visiting the Three-Dimensional Drug Structure Data Bank at the National Institute of Health (NIH). A direct link is provided at the *Chemistry in Context* web site, or you can locate the NIH site by searching for "Center for Molecular Modeling" and/or "NIH." Note: To view the molecules, you will need to install a "plug-in" called Chime. Directions for using Chime are provided at the *Chemistry in Context* web site.

a. Select several drugs and examine their three-dimensional structure. How do these computer representations differ from the structural formulas of drugs shown in this chapter?

b. What are the advantages of the computer representations over two-dimensional drawings? What are their limitations compared to "real" molecules?

Such techniques help to minimize the time it takes to prepare a so-called lead compound, one that shows high promise for becoming an approved drug. Combinatorial chemistry is a recent development that accelerates the creation of lead compounds. Combinatorial chemistry uses the fact that organic molecules contain functional groups,

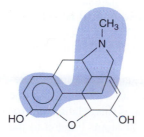

Morphine

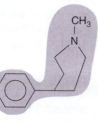

Active area

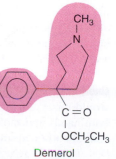

Demerol

Figure 10.12
Molecular structures of morphine and demerol, with active areas outlined.

that is, "pieces" of a molecule responsible for the chemical property of the molecule, and non-functional groups. Small-scale reactions among 96 combinations of reactants in an 8 × 12 array of small wells can be carried out quickly to produce potential drugs. Each of the 96 wells is examined to see whether any products have been formed, and what they are. If no product has been formed, or if products form without having any functional groups of value, the materials are discarded. In other wells, products of note are saved for future reference and use.

The process can be repeated several times, each time seeking to determine when reactions have formed potential drugs containing diverse functional groups. Unpromising reactions can be screened out quickly. From the continuing candidates, the company can develop a so-called library of molecular diversity for a huge array of synthesized compounds, any of which might become a lead compound in the search for a drug (Figure 10.13). Used in conjunction with computers, combinatorial chemistry can minimize the trial-and-error aspects and expense, thus speeding up drug design and development. Using traditional methods, a medicinal chemist could prepare perhaps four lead compounds per month at an estimated cost of $7000 each. With combinatorial chemical methods, the chemist can prepare nearly 3300 compounds in that same time for only about $12 each. Figure 10.14 is a computer model. It shows how using combinatorial chemistry and knowledge of the shape of a protein receptor site and the structure of a known drug, methotrexate for arthritis, can lead to other candidate compounds. As shown in Figure 10.14, the methotrexate molecule fits a receptor site. Another drug (shown in yellow in Figure 10.14) that mimics the shape of methotrexate, but with slight alterations in its chemical structure, could perhaps be even more effective than methotrexate.

Figure 10.13
Research on prospective new drugs uses combinatorial libraries developed from reactions in well plates.

10.7 Consider This: Orphan Drugs

Aspirin and other drugs have a large and profitable market around the world. However, development and marketing of essential drugs needed by only a small number of people suffering from rare diseases can be an enormous economic drain on a pharmaceutical company. If the pharmaceutical companies decide not to make and market these "orphan drugs" because of their low economic return, how will people who need these drugs obtain them? Should the government step in and require successful drug companies to contribute a percentage of their profits to a fund for research, development, and production of these "orphan drugs"? Take a position on this issue and outline the reasons that support your position.

10.7 Left- and Right-handed Molecules

Drug design is further complicated when drug-receptor interaction involves a common but subtle phenomenon called optical isomerism or chirality. **Chiral or optical isomers have the same chemical formula, but they differ in their molecular structure and their interaction with light, hence the name.** Chirality most frequently arises when four *different* atoms or groups of atoms are attached to a carbon atom. A compound having such a carbon atom can exist in two different molecular forms that are non-superimposable mirror images of each other. These are **chiral isomers,** also called **optical isomers.**

Non-superimposable mirror images should be familiar to you. You carry two of them around with you all the time—your hands. If you hold them with the palms up you can recognize them as being mirror images. For example, the thumb is on the left side of the left hand and on the right side of the right hand. Your left hand looks like the reflection of your right hand in a mirror. But your two hands are not identical. No one would mistake a left hand for a right hand. Figure 10.15 illustrates this relationship for both hands and molecules. Note that the four atoms or groups of atoms bonded to the central carbon atom are in a tetrahedral arrangement. *The positions of these four atoms*

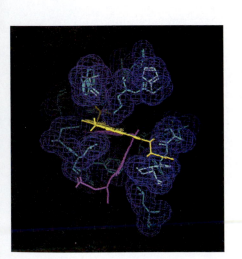

Figure 10.14
The arthritis drug methotrexate (pink) binds at a protein receptor site. A search of a combinatorial library found 225 additional compounds, including the molecule shown (yellow) that would be expected to bind at the same receptor site.

Figure 10.15
Mirror images of molecules
and hands. In the molecule
CHClFBr, each bond connects
a different kind of atom to the
central carbon atom.

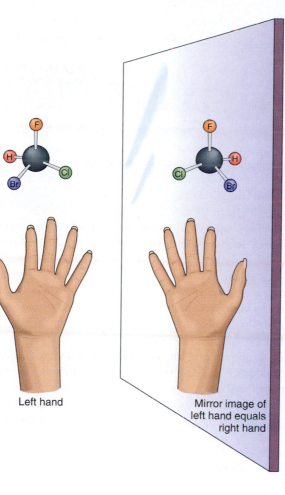

Left hand

Mirror image of
left hand equals
right hand

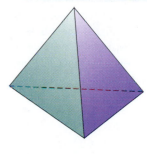

*A tetrahedron has four
equilateral triangular faces.*

correspond to the corners of a three-dimensional figure with equal triangular faces. The
"handedness" of these molecules gives rise to the term chirality, from the Greek word
for hand.

It turns out that many biologically important molecules, including sugars and amino
acids, exhibit chirality. This is significant because, although most chemical and physical properties of a pair of optical isomers are very nearly identical, their biological behavior can be profoundly different. (Maybe Lewis Carroll's Alice had some inkling of
this when, in *Through the Looking Glass,* she remarked to her cat, "Perhaps looking-
glass milk isn't good to drink.") Generally, the explanation for this difference is related
to the necessity of a good molecular fit between a molecule and its receptor site.

You can illustrate this relationship between chirality and biological activity by taking things in your own hands. Your right hand will only fit a right-handed glove, not a
left-handed one. Similarly, a right-handed drug molecule will only fit a receptor site that
complements and accommodates it. Any drug containing a carbon atom with four different atoms or groups attached to it will exist in chiral isomers, only one of which will
usually fit into a particular asymmetrical receptor site (Figure 10.16).

The extreme molecular specificity created by chirality makes the medicinal chemist's
job more complex. A drug molecule must include the appropriate functional groups, and
these groups must be arranged in the biologically active configuration. Often the "right"
and "left" optical isomers are produced simultaneously. Such a situation results in a
racemic mixture consisting of equal amounts of each optical isomer. But frequently
only one optical isomer is pharmaceutically active. For example, many opiate drugs exist as optical isomers, only one of which may have opiate activity. Levomethorphan, the
left-handed (levo) isomer of methorphan, is an addictive opiate. On the other hand, its
right-handed (dextro) mirror image is a non-addictive cough suppressant. This permits

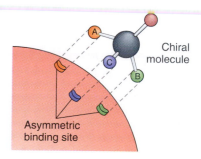

Chiral
molecule

Asymmetric
binding site

Figure 10.16
Representation of a chiral
molecule binding to an asymmetrical site.

Figure 10.17
D- and L-ibuprofen are optical
(chiral) isomers.

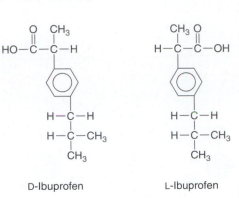

D-Ibuprofen   L-Ibuprofen

Figure 10.17
D- and L-ibuprofen are optical
(chiral) isomers.

Two isomeric forms of chiral
molecules: right-handed, dextro (D);
left-handed, levo (L).

The vitamin E sold in stores is
generally a racemic mixture of D- and
L-isomers. The D-isomer is the
physiologically active one, which can
be purchased in pure form at more
than twice the cost of the racemic
mixture.

Sales of chirally-pure antibiotics
alone were over $20 billion
worldwide in 1997.

the use of dextromethorphan in many over-the-counter cough remedies, but the right-handed (dextro) isomer must either be synthesized in pure dextro form or separated from a mixture with its levo isomer.

Many other drugs exhibit chirality and are active only in one of the isomeric forms. This is true for some antibiotics and hormones, and for certain drugs used to treat a wide range of conditions including inflammation, cardiovascular disease, central nervous system disorders, cancer, high cholesterol levels, and attention deficit disorder. Among the widely used chiral drugs are ibuprofen, the antirejection drug cyclosporin used in organ transplants, and the antidepressant Prozac. Ibuprofen is sold as a racemic mixture of D- and L-isomers. L-ibuprofen is a pain reliever whereas D-ibuprofen is not (Figure 10.17). However, in the body the D-form is converted to the L-isomer. Therefore, it is likely that someone taking ibuprofen is just as well off taking the racemic mixture rather than the more expensive L-ibuprofen. On the other hand, Naproxen, a common pain reliever, is one example of many in which one isomer is preferred, even required. One form of Naproxen relieves pain; the other causes liver damage.

Consequently, drug companies have active research programs designed to create chirally "pure" drugs, those having only the beneficial isomer of a drug in pure form. In fact, half of the top 100 drugs sold worldwide are distributed as chirally pure, rather than as a racemic mixture. Although making the proper, single isomer might seem like an exercise of interest only to chemists, it is big business. Worldwide sales of chirally-pure drugs are near $90 billion annually, not a bad piece of change for knowing how to produce molecules of the correct chirality.

10.8 Your Turn

Carefully examine the structural formula for ibuprofen given in Figure 10.17. Which is the chiral carbon atom?

10.8 Steroids: Cholesterol, Sex Hormones, and More

Certain cellular components, contraceptives, muscle-mass enhancers, abortive agents; what do they have chemically in common? They are all **steroids,** a family of compounds that arguably best illustrates the relationship of form and function. Certainly no other group of chemicals is more controversial than steroids because of their uses ranging from contraception to vanity promoters. The naturally occurring members of this ubiquitous group of substances include structural cell components, metabolic regulators, and the hormones responsible for secondary sexual characteristics and reproduction. Among the synthetic steroids are drugs for birth control, abortion, and bodybuilding.

In spite of their tremendous range of physiological functions represented in Table 10.3, all steroids are built on the same molecular skeleton. Thus, these compounds also provide a marvelous example of the economy with which living systems use and reuse certain fundamental structural units for many different purposes. The body synthesizes the many large molecules that are necessary for life by combining smaller molecular

Table 10.3   **Steroid Functions**

Function	Example
Regulation of secondary sexual characteristics	Estradiol and testosterone (an estrogen and an androgen)
Reproduction and control of the reproductive cycle	Progesterone and other gestagens
Regulation of metabolism	Cortisol and cortisone derivatives
Digestion of fat	Cholic acid and bile salts
Cell membrane component	Cholesterol

fragments. Once such a process is established, the same fundamental biochemical reactions are used to incorporate these molecular fragments into a variety of complex compounds. This wonderfully efficient process is rather like having a standardized house plan that can be reproduced readily—a unit that gains individuality by changes in the types of windows and doors or by the interior decorations.

The common characteristic of steroids is a molecular framework (nucleus) consisting of 17 carbon atoms arranged in four rings—"three rooms and a garage" if you like. This steroid nucleus is illustrated below. Recall that in such a representation, carbon atoms are assumed to occupy the vertices of the rings, but they are not explicitly drawn. The three six-membered carbon rings of the steroid nucleus are designated A, B, and C, and the five-membered ring is designated D.

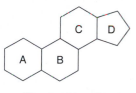

The steroid nucleus

The dozens of natural and synthetic steroids are all variations on this theme. They differ only slightly in structural detail, but can differ profoundly in physiological function. Extra carbon atoms and/or functional groups at critical positions on the rings are responsible for this variation.

A shorthand system is used to represent the molecular structures of the steroids. This system concentrates on the backbone of connected carbon atoms. One carbon atom is assumed to occupy each vertex of the structure. A line protruding from a molecular structure also signifies a carbon atom (actually a —CH_3 group), unless the symbol for another element is attached to it. Hydrogen atoms, which also are not indicated, are bonded to the carbon atoms as is necessary to satisfy the octet rule. This notation is illustrated below with estradiol, a female sex hormone. The figure on the left includes all the atoms in the molecule; the one on the right gives the skeletal representation.

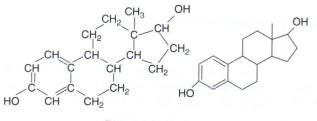

Estradiol $C_{18}H_{24}O_2$

This same system is used in Figure 10.18 to represent the molecular structures of six vitally important steroids. The boxes enclose the regions in which structural variations occur. Careful examination of the figure indicates that some very subtle molecular differences can result in profoundly altered properties. For example, the only differences between a molecule of estradiol and one of testosterone are associated with ring A. The

Figure 10.18
Molecular structures of some
important steroids.

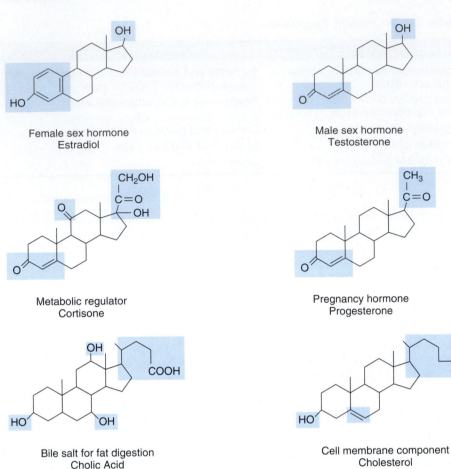

female sex hormone has three double bonds in the ring and an attached —OH group; the male hormone has only one double bond in the A ring, an ═O in place of the —OH, and a —CH₃ group (represented by the vertical line where the A and B rings come together). It is, of course, naive to suggest that the differences between men and women are all due to a carbon atom and a few hydrogen atoms, but it is tempting.

10.9 Your Turn

Carefully examine the structural formulas given in Figure 10.18. Identify the similarities in the structures of each of these pairs.

a. Estradiol and testosterone
b. Estradiol and progesterone
c. Cholic acid and cholesterol

In this chapter we concentrate on only a small number of the many steroid compounds. We begin with cholesterol, the most abundant steroid in the body and probably the best known. The average-sized adult has about half a pound of cholesterol in his or her body. Cholesterol is a starting point for the production of steroid-related hormones and a major component of cell membranes. Because their shape is relatively long, flat, and rigid, cholesterol molecules help to enhance the firmness of cell membranes. Although cholesterol is essential for human life, there are concerns that too much of the compound in the blood can lead to the build-up of plaque, fatty deposits in the blood vessels. This plaque restricts blood flow and can lead to a stroke or heart attack. Therefore, people are advised to regulate their dietary intake of cholesterol, which is found in milk, butter, cheese, egg yolks, and other foods rich in animal fats. But one must keep in mind that some "cholesterol-free" foods can nevertheless contribute to the build-up of cholesterol in the body, where it is synthesized from fatty acids of animal or vegetable origin. A diet rich in "saturated fats" (those without double bonds between carbon atoms) is particularly likely to lead to elevated serum cholesterol.

More information about dietary cholesterol appears in Section 11.5.

The role of cholesterol in the body is relatively passive, but steroid hormones are involved in a tremendous range of physiologically vital processes, including such popular pastimes as digestion and reproduction. Because of the importance of these functions, medicinal chemists have, over the past 50 years, synthesized many derivatives of naturally occurring steroid hormones. These drugs, developed to mimic or inhibit the activities of the hormones in the body, have been variously described as "miracle drugs," "killer compounds," or "sleazy therapeutic agents." Perhaps more than any other type of pharmaceutical, steroid-related drugs are involved with social and ethical issues. These issues include birth control, abortion, diet, bodybuilding, drug abuse, and drug testing. We begin by looking at drugs related to sex hormones.

10.9 "The Pill"

Sex hormones are the chemical agents that determine the secondary sex characteristics of individuals. Female sex hormones are classified as **estrogens;** male sex hormones as **androgens.** All males have a low concentration of female sex hormones, and there are low levels of male sex hormones in all females. However, androgens predominate in males and estrogens in females.

Because of their importance, androgens and estrogens were the first steroidal hormones studied in great detail. When this work was just beginning in the 1930s, techniques for determining molecular structure were in their infancy. A sample of several milligrams of the pure substance was required—much more than is needed today. Because sex hormones occur only in very small quantities, Herculean efforts were required to obtain sufficient amounts for the early chemical studies. For example, one ton of bull testicles was processed to yield just 5 mg of testosterone, and four tons of pig ovaries provided only 12 mg of estrone, a precursor of estradiol. Fortunately, improved technology and instrumentation allow modern chemists to determine molecular structures with samples weighing only a fraction of a milligram. After the molecular structures of the sex hormones were determined, work could proceed on the synthesis of drugs of similar structure. These efforts ultimately led to the creation of "the Pill"—the oral contraceptive that has had such a profound effect on modern society by launching the so-called sexual revolution.

A precursor is a molecule that can be converted directly to a different molecule.

As with aspirin, birth control drugs came about through molecular modifications, in this case, changing substituents selectively on the steroid nucleus. Interestingly, the initial motivation of the research that ultimately led to oral contraceptives was the enhancement of fertility in women who found it difficult to conceive. When fertilization occurs, the hormone progesterone is released, carrying a number of chemical messages. Some of these messages help prepare the uterus for the implantation of the embryo. Others block the release of pituitary hormones that stimulate ovulation. The reason for this is clear: ovulation during pregnancy could lead to very serious complications. Gregory Pincus and John Rock injected progesterone into patients to block ovulation and stimulate body changes related to pregnancy. Their hope was that when the therapy was discontinued, a kind of rebound would occur and ovulation would be stimulated. Such a response, now known as the "Rock rebound," does, in fact, take place and fertility increases.

Unfortunately, progesterone was expensive and not very effective when administered orally. It also caused some serious side effects in a small percentage of patients. Therefore, chemists working in a number of pharmaceutical firms set out to develop a synthetic analog for progesterone that could be taken orally, would reversibly suppress ovulation, and would have few side effects. The ultimate goal of these efforts soon became the inhibition of fertility, not its enhancement. In the mid-1950s, Frank Colton, a chemist at G.D. Searle, synthesized norethynodrel. The molecular structure of this compound

Figure 10.19

Molecular structures of progesterone and norethynodrel.

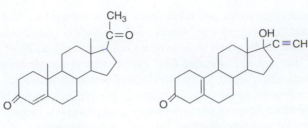

Progesterone   Norethynodrel

(Figure 10.19) shows some subtle but significant differences from progesterone, notably the replacement of —COCH$_3$ on the D ring with —OH and —C≡CH. As a consequence of these changes, the norethynodrel molecule is tightly held on a receptor site, which prevents its rapid breakdown by the liver and permits its oral administration. Norethynodrel became the active ingredient in Enovid, the first commercially available oral conceptive, which was approved for sale in 1960.

The drug's availability had an immediate and substantial impact. In 1962, 1.2 million women in the United States used the "Pill," then containing 150 micrograms (μg) of estrogen and approximately 10 mg of progestin, a close chemical relative of progesterone. Since the development of the original birth control pill, further molecular modifications (many of them minor) have led to decreased dosage and minimized side effects. The nearly 16 million users in 1997 ingested a Pill with just 20 μg of estrogen and approximately 1 mg of progestin. The current dosages are evidence of how just scant amounts of the hormones act as major chemical messengers with profound effects. Recent research into an alternative birth control delivery system culminated in a plastic implant that releases a progesterone analog so slowly that it can be effective for several years. A major contributor to the development of the Pill has been Carl Djerassi, professor of chemistry at Stanford University and president of Zoecon Corporation. Djerassi, the author of hundreds of scientific papers and the holder of many patents for modified steroids, is also a novelist.

The mechanism for the action of steroid-based contraceptives is diagrammed in the simple schematic of Figure 10.20. In effect, the drug "fools" the female reproductive system by mimicking the action of progesterone in true pregnancy. Birth control steroids, being progesterone-like molecules, send a chemical message that is similar to the message carried by progesterone. Because pregnancy is simulated, ovulation is inhibited. In effect, the message this time is not "Hey, Mom, I'm here!" but rather "Hey, you think I'm here, but I'm not!"

Figure 10.20

Action of a steroid contraceptive.

Drugs and the Human Body With Implications for Society by Liska, Ken. © 1994. Reprinted by permission of Prentice-Hall, Inc., Upper Saddle River, NJ.

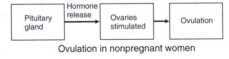

Ovulation in nonpregnant women

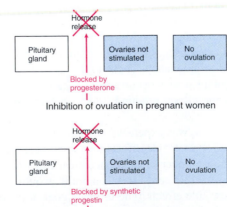

Inhibition of ovulation in pregnant women

Inhibition of ovulation in nonpregnant women
using a synthetic progestin

10.10 A "Morning After" Pill

Some controversy still surrounds the synthetic steroidal hormones that control fertility by inhibiting ovulation. But a far more controversial approach to birth control is a drug that can induce abortion with relative ease and safety—a "morning-after pill." One such drug is mifepristone, better known as RU-486. Since its discovery was announced in France in 1982, it has been used by over 400,000 women in France, Sweden, the United Kingdom, and China. RU-486 has been extensively tested and found to be 96% reliable. Moreover, it appears to have a relatively low incidence of serious side effects. Risk assessment studies suggest that the drug is probably the safest method to terminate a very early pregnancy of no more than nine weeks.

RU-486 and progesterone have very similar molecular structures. Thus, RU–486 is an **antagonist** for progesterone—it occupies the progesterone binding site, but shows no activity. Consequently, progesterone is not released and no pregnancy protein production signal is sent. Because progesterone activity is essential for implantation of the embryo in uterine cells, the developing embryo is spontaneously aborted.

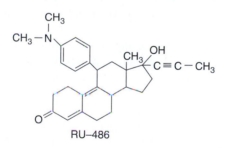

RU–486

Roussel-Uclaf, the company that manufactured RU-486, established five criteria that a country must meet before the company will seek approval of RU-486 in that country:

1. Abortion must be legal.
2. Abortion must be accepted by the public and the medical community.
3. A suitable synthetic prostaglandin must be available for use in the country. (RU-486 is given in conjunction with the prostaglandin for maximum efficacy and safety.)
4. Distribution must be strictly controlled.
5. A patient must agree in writing that if induced abortion fails, she will proceed with a surgical abortion.

Roussel-Uclaf has sold RU-486 in Sweden, France, Germany, and England. Although abortion is legal in the United States, it is certainly not universally accepted by the American public and the medical community. Because of threats of boycotts and protests, Roussel-Uclaf did not introduce RU-486 into this country until the company had arranged to donate the rights to the drug to the Population Council, a nonprofit research organization. The Council began clinical testing of mifepristone in October 1994, administering it at a dozen clinics and hospitals in various parts of the country. The women involved were all 18 or older and no more than nine weeks pregnant. The RU-486 protocol is carried out under the supervision of a physician and requires three outpatient visits to the clinic or hospital. After a physical examination and an ultrasound confirmation of the stage of pregnancy, three RU-486 tablets are administered. Two days later, the patient takes two tablets of the prostaglandin misoprostol. This latter drug induces uterine contractions, and in 70% of cases, abortion occurs within four hours. A follow-up visit is also required. In the 4–5% of cases studied in which the drugs fail to result in abortion, a surgical procedure must be employed.

In March 1996, after completion of its clinical trials, the Population Council requested approval of RU-486 from the FDA. Between 1996 and 1998, the FDA, with President Clinton's encouragement, seemed on the verge of approving the distribution of RU-486. But the FDA has since backed off, and placed contingencies that an RU-486 manufacturer must meet before approval is granted. Given these contingencies, manufacturers have been hesitant to deal with the politics associated with marketing the drug. In addition to liability and profitability questions, opposition to RU-486 by anti-choice organizations has also been a factor in reducing companies' interest in manufacturing the drug. Currently, few companies, not even Roussel-Uclaf, manufacture RU-486. Without a manufacturer, there is no guaranteed supply. Thus, clinics in Sweden, France, and England rely solely on remaining supplies of the drug, estimated to run out by the year 2000 or sooner.

China manufactures its own version of RU-486, a copy of the Rousell-Uclaf product.

An alternative method for drug-induced abortion has been developed in the United States that could make the potential availability of RU-486 in the United States a moot point. In August 1995, gynecologists announced that two widely available prescription drugs used in combination had been shown to have results very similar to those associated with mifepristone for non-surgical abortions no more than seven weeks after conception. The medical protocol is much like that followed with RU-486, but the initial drug administered is methotrexate. Unlike RU-486, methotrexate is not a progesterone mimic. Rather, it works as an abortifacient by blocking a B vitamin called folic acid required for normal cell growth and division. Hence, methotrexate inhibits the development of the embryo and placenta. As with RU-486, the expulsion of the fetus is induced by misoprostol.

What makes the use of methotrexate and misoprostol for this purpose particularly interesting is the fact that both were already approved by the FDA, though for different uses. Methotrexate has been used for some time, in large doses to treat some cancers and in small doses for rheumatoid arthritis and psoriasis. Misoprostol is prescribed to protect the stomach linings of people who require daily doses of powerful anti-inflammatory drugs or anti-ulcer medications. Once a drug has FDA approval, a licensed physician can use it for any purpose, including "off label" uses not originally specified or intended. Thus, no exhaustive approval process was required before these drugs could be employed in combination as abortive agents.

10.11 Consider This: "Off Label" and the FDA

Off-label uses are on the FDA web site (www.fda.gov.). This site contains a wealth of information that is organized in a user-friendly fashion; check out some of its features.

Drugs such as methotrexate and misoprostol can be prescribed for "off-label" uses (also called "unapproved," "unlabeled," or "extra-label" uses).

a. In general, does prescribing medications to be used this way strike you as a reasonable thing for physicians to do? Explain your reason. Compare your opinion with what the FDA thinks.

b. Use the search engine provided at the site and enter "off label." From the lengthy list of "hits," pick two that interest you, and summarize their contents.

c. Explain how what you learned from this exercise strengthened or modified your opinion of "off-label" uses.

10.12 Consider This: Losing Weight the Phen-Fen Way

More than 20 years ago, fenfluramine (fen), an appetite depressant, and phentermine (phen), an amphetamine-type drug, were each FDA approved for short-term use as diet aids, although they were not very effective. In the 1980s, the drugs were taken together in a single study of 121 obese patients, resulting in an average weight loss of 30 pounds. When the off-label study results appeared in the medical literature in 1992, physicians began prescribing the drugs in combination for weight loss. This was done even though the drugs had not been approved for long-term use in combination for weight management. The fen-phen craze was on. Eventually, an estimated six million Americans, most of them women, took the combination drug. In 1995, the FDA, after a one-year study, approved the use. However, in 1997 the FDA removed its approval because of serious heart valve abnormalities occurring among fen-phen users.

a. The individual who conducted the original combination study said: "I figured, gee whiz, these drugs have been on the market for 10, 12 years. Everything must be known about them." (*New York Times* 9/23/97) With the advantage of hindsight, what is your response to his statement?

b. Assume that you were among the many who considered taking fen-phen prior to 1997. What questions would you have asked your physician at that time regarding the medication?

c. Physicians are free to prescribe licensed drugs however they see fit. Make a list of questions you have about such a policy.

10.11 Anabolic Steroids: What Price Glory?

Androstenedione—is it or is it not a steroid? Attention was drawn to this controversy by the 1998 major league baseball home run race and the declared use of "andro" by the home run champion, Mark McGwire. To answer this question, we need a closer look at **anabolic** ("building up") **steroids.** Like birth control drugs, anabolic steroids are controversial. Moreover, like some of their contraceptive chemical cousins, anabolic steroids also were created for quite a different purpose than their ultimate use. These steroids were developed initially to help patients suffering from wasting illnesses to regain muscle tissue. Ironically, their use has now become perverted by the strong who seek to become even stronger.

It has long been known that testosterone promotes muscle growth as well as the development of male secondary sexual characteristics. Drug companies sought to pursue this avenue to produce a testosterone-like drug that would stimulate muscle growth in debilitated patients, such as those recovering from long-term illness. The intent was to modify the testosterone molecule in such a way that its analog would have the desired effects on muscle development without serious negative side effects. Anabolic steroids were the result.

This research, as any involving sex hormones, required the use of a suitable animal model to evaluate the effectiveness and safety of the drugs. Ethical considerations and public opinion preclude the use of human subjects for testing in such unpredictable circumstances. Therefore, castrated rats were used to test anabolic steroids. Various trials compared prostate gland weight with the weight of an isolated abdominal muscle. The idea was to develop a drug that increased muscle mass (an anabolic effect) without increasing prostate mass (an undesirable side effect). Side effects such as this are said to be androgenic. They result from changes in the level of sex hormones, and they often accentuate female characteristics in males (enlarged breasts) and male characteristics in females (deep voices, beards.)

Several drug companies eventually succeeded in greatly reducing the androgenic effects of synthetic steroids, while not affecting their desired anabolic effects. The struc-

Figure 10.21
Molecular structures of testosterone and two anabolic
steroids.

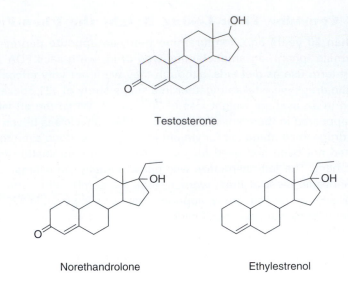

Testosterone

Norethandrolone   Ethylestrenol

tural formulas of the two most potent anabolic steroids, norethandrolone and ethylestrenol, are given in Figure 10.21, along with that of testosterone for comparison.
Note their very close similarities.

In other synthetic anabolic steroids, the molecular shape has been altered by adding
substituents to the A ring. These substituents interfere with the fit of the molecule on
the androgen activity receptor, but they do not impair anabolic activity. Two examples
are stanozol and oxymetholone.

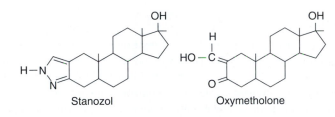

Stanozol   Oxymetholone

Research efforts to properly balance anabolic and androgenic effects in synthetic anabolic steroids have not been completely successful. Unfortunately, this serious drawback has not precluded the widespread use of these drugs. A vast new market has sprung
up on the world's playing fields and in athletic training facilities, even though the drugs
can be obtained legally only by prescription. Because anabolic steroids increase muscle
mass, they appeal to some athletes who compete in strength-related sports such as football, weight lifting, certain track and field events, even baseball. The desire to win has
led athletes to take anabolic steroids to gain a purported competitive edge. And the problem seems to be pervasive. It has been estimated that half of recent Olympic athletes, in
sports ranging from weight lifting to figure skating, have used anabolic steroids at some
time in their careers. Annual sales of illegal steroids to athletes are estimated to be in
excess of $200 million—in spite of the fact that legitimate experts in exercise physiol-
ogy and related fields hold opposing opinions about the merits of steroid use to enhance
athletic performance.

On the other hand, it is well established that using large doses of anabolic steroids
over time can cause a variety of undesirable side effects. In males, these include shrinking testes, difficulty in urination, impotence, fluid retention, baldness, high blood pressure, and heart attack (in short, the symptoms of aging). Women suffer from masculin-
ization in which female secondary characteristics are lost. Both sexes show increased
aggressiveness, unpredictable periods of violent mood changes, and other behavior disorders, the so-called roid rage. Heavy users often compound the problem of drug abuse
by using more than one anabolic steroid at a time—sometimes in untested combinations.
Many abusers seem to be convinced that "more is better." For example, steroids that are
prescribed in legitimate therapeutic doses of a few milligrams have been taken in doses

twenty times larger, in spite of the fact that such a large overdose may be lethal. Lyle Alzado, an NFL football star who died in May 1992, attributed his brain cancer to consuming $20,000–30,000 worth of steroids per year. Although Alzado championed a national campaign against steroid abuse, physicians have concluded that there is no evidence linking his use of these drugs and his cancer. Even without this connection, the physiological effects of anabolic steroid abuse are bad enough.

One of the most challenging competitions in sports is that between athletes who use performance-enhancing illegal drugs and the chemists who test for them. Very sophisticated chemical separation and analytical techniques have been developed to detect banned substances in blood and urine samples at the parts per billion level. Detection of synthetic anabolic steroids is difficult because they are generally used in relatively small amounts. But the real detection problem arises because they are chemically very similar to compounds that occur normally in the body. The success of synthetic chemists now becomes the analytical chemists' burden.

Athletes who use anabolic steroids illegally have resorted to many strategies to avoid detection. These range from simple substitution of a "clean" urine sample for their own, to the rather extreme practice of draining the bladder and then using a catheter to fill the bladder with a sample of "pure" urine just before the drug test. Methods for steroid testing must take into consideration the fact that the fat-soluble steroids take some time to completely clear the body. Because of this time lag, elaborate schemes have been developed for tapering off illegal drugs to get below allowable limits just before competitions. Not surprisingly, some coaches and athletes with very little previous curiosity of medicinal chemistry now make it a significant area of interest—at least the part that applies to steroids.

Returning to the question posed at the beginning of this section: Is androstenedione a steroid? It was developed in the 1970s in East Germany to enhance the Olympic performances of its athletes. "Andro," as it is called, is sold in the United States as a nutritional supplement, not a steroid. Therefore, it can be purchased "over the counter" without a prescription. Its purveyors tout its effectiveness, as one advertisement says: "Think of one compound that will let you put on 10 pounds of muscle in three weeks while adding 20 pounds to your bench press. Now try and come up with a compound that will do that legally . . . One customer gained 18 pounds of muscle in six weeks! $29.17 for 100 capsules (100 mg androstenedione). Typical doses are one capsule three to six times daily." Although gains in muscle mass are debatable and may vary with "andro" use, what is not in question is that androstenedione increases testosterone levels. Classified as a steroid, although technically not an anabolic steroid, androstenedione is a very close chemical relative of testosterone (Figure 10.22). Given the similarities, once ingested, "andro" is converted directly to testosterone, which is an anabolic steroid.

It is well established that testosterone stimulates the creation of more protein, which replaces and rebuilds muscle fibers broken down by exercise. This allows athletes to exercise harder, yet recover more quickly. Dr. Linn Goldberg, head of Health Promotion and Sports Medicine at the Oregon Health Science University stated in the September 8, 1998 *New York Times* that "When androstenedione is converted to testosterone, it is no different than taking anabolic steroids. It's an attempt to cheat to improve your performance, by unnaturally boosting your testosterone levels."

Although allowed by major league baseball, the National Hockey League, and the National Basketball Association, "andro" is banned by the National Football League, the National Collegiate Athletic Association (NCAA), and the International Olympic Committee. The Association of Professional Team Physicians has issued a position statement recommending that androstenedione be removed from over-the-counter status and be banned from all competitive sports. Yet some trainers and coaches insist that no compelling research evidence exists that the side effects from androstenedione are as adverse as those from excessive testosterone use. Patrick Arnold, the person who introduced androstenedione commercially into the United States, contends that using it is "a very safe way of performance enhancement, as safe as drinking a cup of coffee." Countering this is Dr. Gary Wadler, a steroid authority, who cautions: "It would be foolish to delude anyone that this is an innocuous substance."

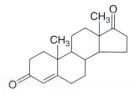

Figure 10.22

Structural formula of androstenedione

10.13 Your Turn

Carefully examine the structural formula for androstenedione given in Figure 10.22 and the structural formula for testosterone given in Figure 10.18. Identify the similarities in these two structures.

10.14 Consider This: Olympic Advantage

In a survey of world-class athletes, 50% said they would take a drug that would enable them to win an Olympic gold medal, even though it would probably cause their death within ten years. Assume you are the editor of the sports pages for your college newspaper. Write an editorial that will help your classmates understand the potential risks and benefits that are associated with taking performance-enhancing drugs.

10.15 Consider This: Andro: At What Price?

As this edition went to press, "andro" could be purchased via the web. Is this still possible? To find out, search for androstenedione and see what turns up.

a. Is "andro" still sold over-the-counter (over the web)?
b. What do web sites claim about "andro?"
c. What does this steroid cost you, both in dollars and in effects on your health? As always, cite both the sponsor and the URL of the web sites you visit.

10.16 Consider This: FDA in Handcuffs?

Writing about "andro," Pamela Zurer, a columnist for *Chemical & Engineering News* (September 28, 1998, page 37) comments: "I lay the blame squarely on Congress, which handcuffed FDA with the Dietary Supplement Health & Education Act (DSHEA) of 1994. Until some regulatory body is given the authority to require that the safety . . . of so-called nutritional supplements . . . we're going to see a lot more of these controversies. I hope not too many people ruin their health in the meantime." It may take a bit of detective work, but you can find the details of the DSHEA on the web.

a. See for yourself—did this really handcuff the FDA? Give the rationale for your choice.
b. Whose needs was the DSHEA intended to meet?

10.12 The Thalidomide Story

The type of drug testing mentioned earlier refers to determining the presence or absence of an illegal drug in a biological fluid and, if it is present, measuring its concentration. Another very different and far broader testing program is required of all new drugs. In accordance with the prevailing laws and regulations in this country, drugs are subjected to an intensive, extensive, and expensive screening process before they can be approved for sale and public use. Ultimately, the question to be answered is, "Is the drug safe to use?" Considering the variability within target populations and the need to minimize unwanted side effects, it is somewhat remarkable that any drug ever receives approval. However, if an error must be made, it is preferable to make it on the side of rejecting rather than approving a drug that does not fully meet rigorous testing requirements. Unfortunately, such was not the practice in Europe in the case of the drug thalidomide.

In 1956, thalidomide was put on the European market without sufficient screening because of an erroneous conclusion reached on the basis of incomplete testing data. The

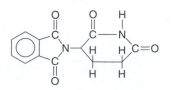

The structural formula of thalidomide.

results were tragic. Discovered and developed by a small German drug company, Chemie Grunenthal, thalidomide was a by-product of research aimed at developing new antibiotics. A medicinal chemist at Chemie Grunenthal recognized thalidomide as an analog of a drug that had recently been put into use as a sedative. Testing of thalidomide on four species—mice, rats, guinea pigs, and rabbits—led to the conclusion that the drug was a remarkably safe sedative. Unfortunately, testing was not done to determine if thalidomide was a **teratogen,** that is, whether it could damage a developing embryo sufficiently to cause birth defects. Assuming the new drug to be safe, Chemie Grunenthal approached several companies who were interested in gaining a greater share of the sedative market. At least one U.S. drug company rejected the compound as worthless and possibly unsafe after carrying out its own testing of the drug. However, several pharmaceutical firms accepted Chemie Grunenthal's offer and marketed thalidomide in different parts of the world.

Soon after thalidomide was introduced, several physicians reported cases of nerve damage in patients who had taken the drug, but these reports were largely ignored. Five years later, a German pediatrician reported a large increase in the number of infants suffering from phocomelia. In this condition, the development of the bones in the arms and legs is severely arrested, producing flipper-like limbs, badly deformed limbs, or no limbs at all. Because phocomelia is one of the rarest birth defects known, the sudden increase in its occurrence caused alarm. Subsequently, the outbreak of phocomelia was traced to thalidomide taken during the first three months of pregnancy by mothers who used the drug to control morning sickness and nausea. All together, about 10,000 deformed infants, know as "thalidomide babies," were born in 46 countries.

News of the thalidomide debacle brought expressions of grief and outrage from people around the world. Investigations indicated that greed plus inept, inadequate, and fraudulent testing were responsible for the disaster. Exceptional bad luck was also involved in that humans are more sensitive to the drug's teratogenic effects than are most test animals.

Thalidomide was not sold in the United States because Dr. Frances Kelsey, a pharmacologist and physician with the FDA had been unconvinced by the limited safety data supplied by the manufacturers. Ironically, thalidomide was the first drug application Kelsey was assigned when she became an FDA employee in 1960. She notes that "They gave it to me because they thought it would be an easy one to start on." Because of her stubborn skepticism, scientific integrity, and insistence on good testing procedures, relatively few thalidomide babies were born in the United States. Although the thalidomide story is probably the darkest chapter in the history of drug design, it was responsible for the establishment of greatly improved drug testing procedures throughout the world. Teratogenicity testing is now a standard part of new drug approval procedures in most Western countries. The FDA testing procedures, considered slow and overly methodical by some, have been widely adopted.

In spite of its tragic history, thalidomide is making a comeback. Recent research has disclosed that the drug has a number of potentially beneficial properties. Thalidomide appears to inhibit replication of the AIDS virus, stop drastic weight loss in AIDS and tuberculosis patients, and clear up canker sores in those with AIDS. The drug may even be effective against certain kinds of tumors, a useful supplement in bone-marrow transplants, and a means of preventing several common forms of blindness. Thalidomide may become the drug of choice in treating at least some of these conditions, as long as it is not taken by pregnant women. In 1998, the FDA approved thalidomide, following thirty years of research and testing, for treating the painful, disfiguring lesions of people with leprosy. The marketing of the drug for this purpose is tightly controlled; it is available only by prescription from physicians registered with an FDA thalidomide safety and education program. Patients receiving the drug must follow explicit guidelines for its use, including rigorous birth control methods, if necessary. In this case, the benefits of thalidomide are high for this application which, when used properly, carries low risk. The drug has also been approved in Mexico and Brazil for treating leprosy.

The thalidomide story is complicated by chirality. Studies carried out in 1979 at the University of Bonn in Germany seemed to indicate that only one of the two optical

Dr. Frances Kelsey, Chief, Investigation Drug Branch Division of New Drugs, FDA. Dr. Kelsey's work prevented the sale of thalidomide in the United States.

isomers of the drug caused birth defects. But other research has suggested that both isomers have that undesirable effect. It is likely that once in the human body, each isomer readily changes into the other form. Thus, even if only one isomer is teratogenic, ingesting either of them could potentially give rise to birth defects.

10.17 Consider This: Thalidomide—Curse or Cure?

In mid-1998, the FDA approved thalidomide for treating serious inflammatory conditions in patients with leprosy (Hansen's disease). This has rekindled the discussion of using thalidomide to treat other major medical problems, such as to prevent the growth of tumors. It is thought that thalidomide acts by the same mechanism to prevent tumor growth as that causing birth defects in the 1950s and early 1960s. The drug prevents the formation of extensive new blood vessels. Without blood vessels to limb tissue, a developing fetus cannot form arms or legs. Likewise, tumors cannot grow unless fed by thousands of new blood vessels. Even with the many restrictions placed on its use, there is the chance that a pregnant woman will take thalidomide and a deformed baby result. Take a position on the relative risks and benefits of approving thalidomide for additional applications in this country. Communicate your position in an e-mail message to a friend at another university.

10.13 Drug Testing and Approval

Manufacturing prescription drugs is done on a colossal scale to meet patients' demands. About two billion prescriptions are filled annually in the United States, and worldwide sales of drugs come to about $300 billion a year and growing. But the pathway for a new drug from a laboratory to a pharmacy shelf is long and complicated. All proposed new drugs, whether extracted from natural materials or synthesized in the laboratory, are subjected to exacting series of tests before they obtain FDA approval. Current law requires evidence that the drugs are safe as well as effective before such approval is granted. The steps for approval are summarized in Figure 10.23.

From discovery to approval, the development of a new drug takes, on average, nearly twelve years and more than $350 million—over twice the cost of a decade ago. The expenses are principally for the various stages of drug testing, probably the most complicated and thorough pre-marketing process ever developed for any product. Although the number of pills getting through the funnel of Figure 10.23 gets progressively smaller with time, the diagram does not begin to convey the high mortality rate of proposed drugs. Currently, the odds of getting a candidate drug from identification to approval are

Figure 10.23

Schematic of the drug approval process in the United States.

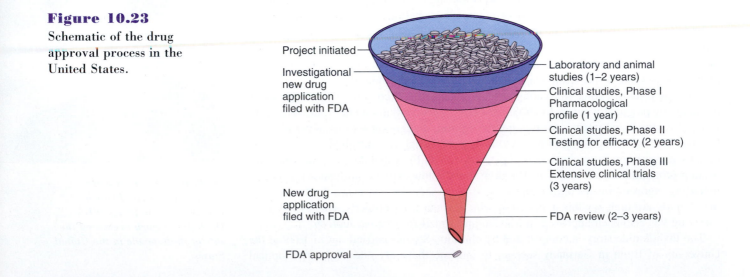

Project initiated

Investigational new drug application filed with FDA

Laboratory and animal studies (1–2 years)

Clinical studies, Phase I Pharmacological profile (1 year)

Clinical studies, Phase II Testing for efficacy (2 years)

Clinical studies, Phase III Extensive clinical trials (3 years)

New drug application filed with FDA

FDA review (2–3 years)

FDA approval

1 in 10,000. For every 10,000 trial compounds that begin the process, 20 make it to the level of animal studies, half that many get clearance for use in clinical testing with humans, and finally one gets FDA approval.

Examples already encountered in this chapter have suggested the long process of chemical hide-and-seek that often precedes the identification of a compound as possibly having therapeutic properties. Once the promising candidates have been identified, they are subject to *in vitro* studies, those carried out in laboratory flasks. Simultaneously, a wide range of activity is undertaken by the pharmaceutical company. Chemists and chemical engineers investigate whether the compound can be produced in large volume with consistent quality control. Pharmacists carry out studies of the most effective way to formulate the drug for administration—as capsules, pills, injection, syrup, or perhaps something more unusual such as a nasal spray, skin patch, or implant. Chemical stability and shelf life are evaluated. Economists, accountants, patent attorneys, and market analysts conduct research on the likelihood of deriving a profit from the product. A fair, responsible price must be established that allows the corporation to recapture the extensive development costs while keeping the drug affordable.

In vitro literally means "in glass."

Only a small fraction of compounds survive this scrutiny to move on to animal testing. Such *in vivo* ("in life") tests are designed to determine the drug's efficacy, safety, dosage, and side effects. It is typically at this stage that pharmacologists determine the drug's mode of action, how it is metabolized, and its rate of absorption and excretion. The tests are carefully controlled, requiring the collection of very specific kinds of data. For example, drugs are evaluated for their short-and long-term effects on particular organs (such as the liver or kidneys) and on more general systems (such as the nervous or reproductive system). Perhaps the most controversial toxicity testing involves the determination of the lethal dose-50 (LD_{50}), the minimum dose that kills 50% of the test animals.

10.18 Consider This: Animals and Drug Testing

Animal rights groups often target the LD_{50} standard as an example of callous indifference to animal welfare. Other groups argue that standards such as LD_{50} are necessary to ensure drug safety and effectiveness. Take a position on the issue and write a letter to a classmate defending your position.

Results of animal tests must be submitted to the FDA for evaluation before permission is granted to proceed to the next stage—clinical testing of the drug on humans. In addition, approval must be obtained from local agencies and authorities such as a hospital's ethics panel or medical board. The FDA must establish whether the drug is effective *and* safe before it can be sold to the public. What needs to go on the label regarding use, side effects, warnings, etc., must also be determined. Typically, clinical studies involve the three phases identified in Figure 10.23: Phase I, Developing a pharmacological profile, Phase II. Testing the efficacy of the drug, and Phase III. Carrying out the actual clinical tests. Most of the safety tests of Phase I are done with healthy male volunteers, who are given single and repeat doses of the drug in various amounts. It is also at this stage that researchers look for interactions with other drugs. Double-blind placebo tests are administered to small patient groups in Phase II to test the drug's effectiveness on patients having the condition that the drug is designed to affect. In this protocol, neither the patient nor the physician knows which patients are receiving the drug and which are receiving a placebo, an inactive imitation that looks like the "real thing." Such tests are designed to eliminate bias from the interpretation of the results. Long-term toxicity studies are also initiated during Phase II. The clinical trials are expanded in Phase III, while manufacturing processes are scaled up and tests are carried out on the stability of the drug. The entire process often requires six years or more. A decade ago an average of 40 clinical trials were done on drugs that were ultimately approved; that number has risen to 60, the trials have become more complex, and the number of patients treated per trial has more than doubled over the past twenty years.

Large-scale clinical trials are desirable because a large pool will more likely include a wide range of subjects. Variety is important because the drug in question may have markedly different effects on the young and the old; men and women; pregnant or lactating women; infants, nursing infants, and unborn infants; and persons suffering from diabetes, poor circulation, kidney problems, high blood pressure, heart conditions, and a host of other maladies.

10.19 Consider This: Double-Blind Testing

Double-blind protocols have other uses than testing for the effectiveness of a drug. For example, physicians may use a double-blind test for diagnosing food allergies. In such tests, the physician administers a series of foods and placebos in disguised form. The test substances are labeled in code known only to a third person. Why do you think double-blind tests for the effectiveness of a drug or for establishing a food allergy are necessary? Compared to single-blind tests in which only the patient is unaware of the drug or food being administered, how do double-blind tests affect the reliability of the information gained?

Once clinical trials have been completed successfully—typically by only 10 drugs out of an original pool of 10,000 compounds—the test data are submitted to the FDA as part of a new drug application. This document can easily exceed 3500 pages. Upon review, the Agency may require the repetition of experiments or the inclusion of new ones, thus adding years to the approval process. Of the drugs submitted to clinical testing, only about one in ten is finally approved.

Once it receives the FDA's imprimatur, a drug can be sold in the United States. Nevertheless, it still remains under scrutiny, monitored through reports from physicians. Drugs are removed from the market if serious problems occur. Some side effects show up only when large numbers of users are involved. Former FDA Commissioner Dr. David Kessler said "There is simply no way that we can anticipate all possible effects of a drug or device during clinical trials that precede approval. A new drug application, for example, typically includes safety data on several hundred to several thousand patients. If an adverse event occurs in 1 in 15,000 or even 1 in 1000 users, it could be missed in clinical trials. But it could pose a serious safety problem when the drug is used by many times that number of patients." Such an example is temafloxacin, an antibiotic that had been clinically tested before approval in more than 4000 patients. Less than four months after its February 1992 release, temafloxacin was withdrawn from the market after fifty serious adverse events occurred from its use, including three deaths. Only after the drug reached the market and many thousands of people used it did its serious, even lethal, side effects became apparent.

Another case is Seldane, the first antihistamine for treating seasonal allergies without causing drowsiness. A small, but statistically significant, number of patients who took Seldane along with certain antibiotics or antifungal medicines developed abnormal heart rhythms. Seldane is normally broken down in the liver to another antihistamine by-product. Particular antibiotic and antifungal medications prevent this breakdown and so high concentrations of Seldane remain in the blood causing cardiac arrhythmia in these patients. In 1996 the FDA removed Seldane from the approved list because other non-sedating antihistamines had become available. By substituting a carboxylic acid group for a methyl group on Seldane, chemists developed Allegra, an antihistamine that is identical to the liver breakdown product of Seldane.

The lengthy process for drug testing and approval is not without controversy. Influenced by events such as the thalidomide tragedy, most people probably favor thorough screening of any proposed drug. But the price of such protection is high. The most obvious costs are monetary. Bringing a new drug to market is incredibly expensive, and the number of new drugs being developed has risen even as research and development (R&D) costs have risen (Figure 10.24).

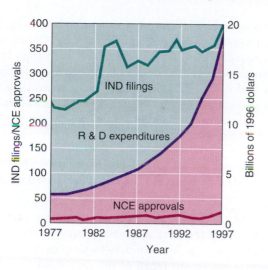

Figure 10.24

Drug development: investigational new drugs (IND); new certified approvals (NCE); and research and development (R&D) costs.

Reprinted with permission from *Science*, Vol. 280, May 22, 1998. Copyright © American Association for the Advancement of Science.

Of course much of the expense is passed on to the consumer. For example, a single dose of streptokinase, a medication that dramatically increases the likelihood of surviving a heart attack, costs $1000. Such prices have driven the costs of medical care and medical insurance to astronomical levels. One issue in the debate over health care reform is who will pay for the research and development that ultimately leads to new medication.

10.20 Consider This: Who Should Pay?

Streptokinase, a medication that dramatically increases the likelihood of surviving a heart attack, costs $1000 per dose. Who should pay for this life-prolonging drug—individuals, medical insurance, pharmaceutical companies, or the government? Take a stand and defend your position.

But more than money is at stake. In some cases, the costs of the protracted drug approval process may be human lives. When a patient is suffering from an almost certainly fatal disease such as some cancers, the risk/benefit equation changes. When there is nothing to lose, people are willing to take great risks, including imperfectly tested drugs. Some mortally ill patients have smuggled drugs from countries where the approval process is less demanding than it is in the United States. Some have grasped at the straw of largely unproven remedies. And within the system, some advocates have urged that the FDA approval process be short-circuited to permit the use of experimental drugs on patients who have no other options. In one case, described in 10.21 Consider This, a lottery was held to determine who would have access to a promising new drug for the treatment of AIDS.

10.21 Consider This: A Lottery for Life?

In 1996, a lottery was held to select 2000 patients with advanced AIDS to receive an experimental drug called ritonavir. In other tests, similar drugs had proved effective in almost completely suppressing the AIDS virus in most subjects. Therefore, thousands of people suffering from the disease signed up for what might be a chance at prolonged life. Responses to this unconventional means of distributing a limited supply of the drug were mixed. Ben Cheng, a San Francisco AIDS activist supported it: "I think everybody agrees the lottery is not the best way of doing this, (but) it's the only fair and equitable way of distributing what little of the drug is available." In contrast, Bob Chapman, another AIDS victim and volunteer, had this to say: "I'm opposed to these lottery things. They are playing with people with terminal illness and putting people in competition with each other. It's medicine hitting a new low." List the ethical arguments on both sides of the issue, and indicate which position you favor.

The FDA has responded appropriately within the limits of its legal responsibilities to balance benefits with risks. Ten years ago an FDA review for a new drug required nearly three years whereas today it is less than a year. A new "fast-track" system has been instituted for priority drugs—those that address life-threatening ailments or new drug therapies for conditions that had no such therapies. The fast-track policy promises to have priority drugs, if found to be acceptable, approved within six months of application. Action on non-priority drugs is to be taken within ten months, down from the initial target of twelve months.

10.22 Consider This: Safety and Standards on the Fast Track

The credibility of the FDA's fast-track drug testing program is called into doubt when five drugs released under the program have had to be withdrawn within a period of a year from mid-1997 to mid-1998. For example, some would argue that the recent recall of the pain killer Duract shows that adequate testing had not taken place on this drug. Others could counter that standards of testing have not changed, only the speed with which data are evaluated and that no set of tests can produce data that are 100% reliable in humans. Search the FDA and other web sites to gather information about this particular drug. Use what you learn to reach a decision as to whether you think fast-track drug testing is overall a benefit or a risk to the health of Americans.

People suffering from rare diseases may not be able to purchase appropriate medication at any price because it may not exist. There is a significant financial disincentive for a pharmaceutical company to invest heavily in developing "orphan drugs" that will be used by only a small fraction of the population.

To such considerations, one must add the objections some have to standard test protocols. Animal rights advocates are highly critical of the use of any animal subjects in drug screening. The sacrificing of test animals in establishing LD_{50} values is especially controversial. For others, the generally accepted methods of human testing are at issue. The argument is that because the drug may have some benefit and will probably do no significant harm, it is unethical to withhold it from a control population. Some terminally ill AIDS patients have refused to cooperate with double-blind clinical studies by mixing and sharing test drugs and placebos.

An intriguing new approach to drug design is the use of "virtual" patients, ones derived through computerized data analysis. Up to 50% of actual trial testing provides no meaningful contributions to the data regarding a tested drug. By using "virtual" patients, non-promising drugs can be weeded out earlier in the trials so as to increase the percentages of drugs that are successful during the clinical phases. In this method, preclinical data on toxicity, uptake rates, and other attributes of test drugs are used together with known behaviors of similar drugs to create a very large data base of computer-simulated "patients" and how they are expected to respond to the test drugs. Statistics and computer modeling are then applied to the huge number of "patients" to simulate real-world behavior and to predict the most promising lead candidates to submit for clinical trials.

10.14 Brand-Name or Generic Prescriptions?

A customer gives a prescription to a pharmacist and is asked "Brand name or generic?" This scenario is played out daily in thousands of cases across the country. How is the person to decide? For millions of Americans, the cheaper generic version can mean the difference between getting the necessary medication and not being able to afford it, although not all approved drugs are available in generic form.

The two forms can be differentiated rather simply. A **pioneer** drug is the first version of a drug that is marketed under a brand name, such as Valium for the anti-anxiety drug.

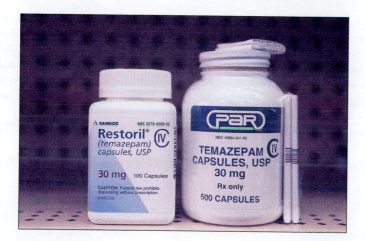

A brand name drug (Restoril) and its generic counterpart (Temazepam).

The **generic** version is a drug that is equivalent to the pioneer drug, but cannot be marketed until the patent protection on the pioneer drug has run out after twenty years. The lower-priced drug is commonly marketed under its generic name, in this case Diazepam instead of Valium. The 20-year patent protection on the pioneer drug begins when it is patented, not when first put on the market. In cases requiring a long pre-approval time, the actual marketing period can be relatively short, even less than six years. In such a situation, a drug company has very little time to recapture its research and development costs before a generic competitor can be manufactured. Almost 80% of generic drugs are produced by brand-name firms. Like pioneer drugs, generic drugs must also be approved by the FDA.

In 1984, Congress passed the Drug Price Competition and Patent Restoration Act, which greatly expanded the number of drugs eligible for generic status. This Act eliminated the need for generics to duplicate the efficacy and safety testing done on counterpart pioneer drugs. Doing so saves drug manufacturers considerable time and money. The FDA also issued specific guidelines for a generic drug's comparability to the pioneer drug. By FDA mandate, the generic and pioneer versions must be bioequivalent, and deliver the same amount of active ingredient into a patient's bloodstream at the same rate.

Health insurance companies and the FDA suggest that policyholders choose generic rather than brand-name drugs when possible, for obvious economic reasons. The concern for health care costs, along with the graying of baby boomers, will likely accelerate the use of generics, as will patents that expire on additional brand-name drugs, making their generic versions possible.

Conclusion

Molecular manipulations by chemists have created a vast new pharmacopeia of wonder drugs that have significantly increased the number and quality of our days. Thanks to penicillin, sulfa drugs, and more recent antibiotics, the great majority of bacterial infections are easily controlled. Once dreaded killers such as typhoid, cholera, and pneumonia have been largely eliminated—at least in wealthy, industrialized societies. Synthetic steroids, used from birth control to bodybuilding, have transformed society. And the humble aspirin tablet has been supplanted by newer NSAIDs, which likely will be replaced by a new generation of innovatively designed "superaspirins."

But no drug can be completely safe and almost any drug can be misused. Taking a medication is a conscious choice between the benefits derived from the drug and the risks associated with its side effects and limits of safety. Because most drugs have very wide, carefully established margins of safety, their benefits far outweigh their risks. For some drugs, however, the trade-off between effectiveness and safety involves a different balance. A drug with severe side effects may be the only treatment available for a life-threatening disease. Someone suffering from AIDS or advanced, inoperable cancer will understandably have a different perspective on drug risks and benefits than a person with a severe cold. And the impersonal anonymity of averages takes on new meaning at the bedside of a loved one. When chemistry is applied to medicine, science must be guided by morality and reason must be tempered with compassion.

Chapter Summary

Having studied this chapter, you should be able to:

- Describe the discovery, development, and physiological properties of aspirin (10.1)
- Discuss the general sources of drugs and strategies for drug development, and the example of penicillin (10.2)
- Understand bonding in carbon-containing (organic) compounds (10.3)
- Apply the concept of isomerism to organic compounds (10.3)
- Recognize functional groups and the classes of organic compounds that contain them (10.4)
- Relate the molecular structure of aspirin to other analgesics (10.4)
- Understand the mode of action of aspirin and other analgesics (10.5)
- Describe the lock-and-key mechanism of drug action (10.6)
- Understand differences in molecular structure between chiral (optical) isomers (10.7)

- Recognize the structure of the steroid nucleus (10.8)
- Consider the chief functions of steroids, and some specific examples: sex hormones (testosterone, progesterone); metabolism regulators (cortisone); cell-membrane components (cholesterol) (10.8)
- Understand how birth control pills inhibit ovulation (10.9)
- Understand the mechanisms by which RU-486 and methotrexate induce abortion (10.10)
- Compare uses and abuses of anabolic steroids (10.11)
- Discuss the ethical issues in the use of steroids for birth control, abortion, and muscle building (10.9, 10.10, 10.11)
- Describe the lessons learned from the thalidomide case (10.12)
- Discuss the procedure for drug testing and approval and the associated benefits and costs (10.13)
- Understand the similarities and differences between brand name and generic drugs (10.14)

Questions

Emphasizing Essentials

1. **a.** What is the intended effect of an antipyretic drug?
   b. What is the intended effect of an analgesic drug?
   c. What is the intended effect of an anti-inflammatory drug?
   d. Can any single drug exhibit all of these effects?

2. There are many subdisciplines within the field of chemistry.
   a. What do organic chemists study?
   b. How does this differ from what biochemists study?

3. Write condensed structural formulas for the three isomers of pentane assigned in 10.3 Your Turn.

4. Write the structural formula of each different isomer of hexane, C_6H_{14}. *Hint:* Be sure the bonding really is different, not just a different paper-and-pencil representation in two dimensions of the same structure.

5. Consider the isomers of butane shown in Figure 10.4. How many different isomers could be formed by replacing a single hydrogen atom with an —OH group, that is, how many different alcohols have the formula C_4H_9OH? Write the structural formula for each possible isomer.

6. For each compound, identify each functional group present and name the class of compounds to which it belongs.

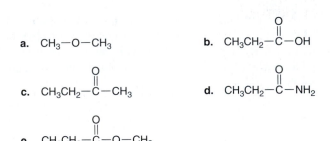

7. Which of these classes of compounds have a compound that contains only one carbon atom?
   a. alcohol   **d.** ester
   b. aldehyde   **e.** ether
   c. carboxylic acid   **f.** ketone

   Write a structural formula for each of these one-carbon compounds, and explain why the other classes of compounds are limited to compounds containing more than one carbon atom.

8. Some organic compounds exist in isomeric forms that are members of different classes of compounds. For example, in some cases the compound is an alcohol or an ether. For each

of these, identify the class of compound represented by the formula as written.

a. CH₃CH₂—OH

b. CH₃CH₂—C—H (with =O above C)

c. CH₃CH₂—C—O—CH₃ (with =O above C)

Then, write the formula of an isomer with the same composition that is a member of a *different* class of compound. Identify the new class.

9. Histamine causes runny noses, red eyes, and other symptoms in hay fever sufferers. Here is the structural formula of histamine.

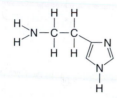

a. What is the molecular formula of this compound?

b. Identify the amine functional group in histamine.

c. Which part (or parts) of the molecule do you think make the compound water soluble?

10. Figure 10.7 shows a somewhat condensed structural formula for acetaminophen, the active ingredient in Tylenol.

a. Write the complete structural formula for acetaminophen, showing all atoms and all bonds.

b. What is the molecular formula for this compound?

c. Children's Tylenol is a flavored aqueous solution of acetaminophen. Predict what part or parts of the molecule make acetaminophen dissolve in water.

11. Identify the functional groups in each of these drugs or medicines.

a. PABA (an ingredient in sunscreens)

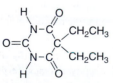

b. Barbital (a sedative)

c. Penicillin-G

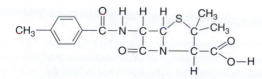

12. Ibuprofen is relatively insoluble in water but readily soluble in most organic solvents. Explain this solubility behavior based on its structural formula found in Figure 10.7.

13. This is the structural formula for methamphetamine, a stimulant.

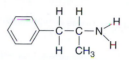

Judging from its structural formula, do you expect it to be more soluble in lipids or in aqueous solutions? Why?

14. Decode this sentence by giving the meaning of each abbreviation and explaining the effect being stated. "NSAIDs have an effect on COX enzymes."

15. Compare the physiological effects of aspirin with those of acetaminophen and ibuprofen. Relate differences to the nature of each compound at the molecular and cellular levels.

16. Would aspirin be more active if it were to interact with prostaglandins directly, rather than by blocking the activity of COX enzymes? Explain your reasoning.

17. What are "superaspirins"? How do they differ from regular aspirin and other NSAIDs?

18. Identify the functional groups in morphine and demerol, using the structural formulas found in Figure 10.12. Can these molecules be assigned to a particular class of compound? Why or why not?

19. Sulfanilamide is the simplest of antibiotics known as sulfa drugs. It appears to act against bacteria by replacing *para*-aminobenzoic acid, an essential nutrient for bacteria, with sulfanilamide. Use these structural formulas to explain why this substitution is likely to occur.

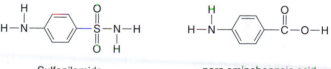

Sulfanilamide   *para*-aminobenzoic acid

20. Which of these compounds can exist in chiral forms?

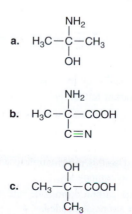

21. Which of these compounds can exist in chiral forms?

a. $C_2H_5-\underset{\underset{H}{|}}{\overset{\overset{OH}{|}}{C}}-CH_3$ c. $C_6H_5-\underset{\underset{NH_2}{|}}{\overset{\overset{H}{|}}{C}}-H$

b. $H_3C-\underset{\underset{SH}{|}}{\overset{\overset{NH_2}{|}}{C}}-COOH$

22. Use the structural formulas in Figure 10.18 to answer these questions.

   a. Identify the type and number of functional groups in cortisone.

   b. Suggest a reason why cholic acid is more soluble in water than cholesterol is.

23. Molecules as diverse as cholesterol, sex hormones, and cortisone all contain common structural elements. Write the structural formula that represents the common structural elements.

24. In 10.13 Your Turn, you compared the structural formulas of androstenedione and testosterone.

   a. What is the argument for banning the use of "andro" by athletes, which is the current policy of the International Olympic Committee?

   b. What is the argument for *not* banning the use of "andro" by athletes, which is the current policy of major league baseball?

25. This is the molecular structure of thalidomide. Identify the location of any chiral carbon atoms.

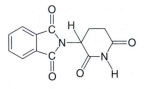

Concentrating on Concepts

26. The text states that some remedies based on the medications of earlier cultures contain chemicals that have been verified to be effective against disease, others are ineffective but harmless, and still others are potentially harmful. Describe how it might be determined into which of these three categories a recently discovered substance fits.

27. Draw structural formulas and determine the number and type of bonds (single, double, or triple) used by each carbon atom in these molecules.

   a. H_3CCN (acetonitrile; used to make a type of plastic)

   b. $H_2NC(O)NH_2$ (urea; an important fertilizer)

   c. C_6H_5COOH (benzoic acid; used as a food preservative)

28. Carbon usually forms four covalent bonds, nitrogen usually forms three bonds, oxygen usually forms two bonds, and hydrogen can only form one bond. Use this information to write structural formulas for all of these compounds.

   a. a compound that contains one carbon atom, one nitrogen atom, and as many hydrogen atoms as needed.

   b. a compound that contains one carbon atom, one oxygen atom, and as many hydrogen atoms as needed.

29. In 10.3 Your Turn, students are asked to draw the structural formulas for each of the three isomers of pentane, C_5H_{12}. One student submitted this set of isomers, with a note saying that six isomers had been found. Help this student to see why some of the answers are incorrect.

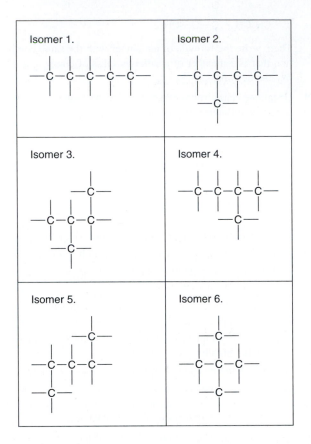

30. Styrene, $C_6H_5CHCH_2$, is another compound that contains a ring similar to benzene, C_6H_6. One hydrogen has been replaced by the side chain $-CHCH_2$. Write a set of structural formulas for styrene to show that this molecule, like benzene, also has resonance structures.

31. If aspirin is a specific compound, what justifies the claims for the superiority of one brand of aspirin tablets over another?

32. Consider Figure 10.8, which shows a schematic representation of chemical communication within the body. Write a paragraph explaining what this figure means to you in helping to explain chemical communication.

33. Consider this statement. "Drugs can be broadly classed into two groups: those that produce a physiological response in the body and those that kill or inhibit the growth of substances that cause infections." In which class does each of these drugs fall?

   a. aspirin   c. antibiotics

   b. superaspirin   d. hormones

34. Consider the structure of morphine in Figure 10.12. Codeine, another strong analgesic with narcotic action, has a very similar structure. The only difference is that the —OH group at-

tached to the 6-membered ring is replaced by an —OCH₃ group.

a. Draw the structural formula for codeine and label the functional groups present.

b. The analgesic action of codeine is only about 20% as effective as morphine. However, codeine is less addictive than morphine. Is this enough evidence to conclude that replacement of —OH groups with —OCH₃ groups in this class of drugs will always change the properties in this way? Why or why not?

35. Dopamine is found naturally in the brain. The drug L-dopa is found to be effective against the tremors and muscular rigidity associated with Parkinson's disease. Identify the chiral carbon in L-dopa, and comment on why L-dopa is effective whereas D-dopa is not.

36. Vitamin E is often sold as a racemic mixture of the D- and L-isomers.

a. Which is the more physiologically active isomer?

b. How does the cost of the racemic mixture compare with the price of the pure, physiologically active isomer?

37. Consider the fact that L-methorphan is an addictive opiate, but D-methorphan is safe enough to be sold in many over-the-counter cough remedies. Explain how this is possible from a molecular point of view.

38. Figure 10.20 diagrams the action of a steroid contraceptive. Study this diagram and then explain it in your own words to an interested friend.

39. Thalidomide, a sedative drug first originally introduced in Europe in 1956, was later discovered to be a potent teratogen.

a. Suggest a possible reason why the teratogenic effects of thalidomide might not have shown up in animal tests, even if those effects had been targeted for investigation.

b. If that is the case, why are doctors now interested in this drug?

40. Why is it that many projects to isolate or synthesize new drugs are started in this country, but few actually receive FDA approval for general use?

Exploring Extensions

41. One avenue for successful drug discovery has been to use the initial drug as a prototype for the development of other similar compounds, called analogs. The text states that cyclosporin, a major anti–tissue rejection drug used in organ transplant surgery, is an example of a drug discovered in this way. Research the discovery of this drug to verify this statement. Write a brief report describing your findings.

42. Dorothy Crowfoot Hodgkin first determined the structure of a naturally occurring penicillin compound. What was her background that prepared her to make this discovery? Write a short report on the results of your findings, giving your references.

43. Before the cyclic structure of benzene was determined (Figure 10.5), there was a great deal of controversy about how the atoms in this compound could be arranged.

a. Count the number of outer electrons that are available for C₆H₆, and then draw the structural formula for a possible linear isomer.

b. Give the condensed formula for the possible structure.

c. Compare your structure with those drawn by your classmates. Are they all the same? Why or why not?

44. Antihistamines are widely used drugs for treating allergies caused by reactions to histamine compounds. This class of drug competes with histamine, occupying receptor sites on cells normally occupied by histamine. Here is the structure for a particular antihistamine.

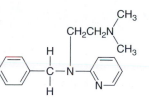

a. What is the molecular formula for this compound?

b. What similarities do you see between this structure and that of histamine (shown in question 9) that would allow the antihistamine to compete with histamine?

45. Over the next few years, the FDA may consider deregulating more than a dozen drugs, nearly as many as have already been approved for over-the-counter sales during the past decade. The products that have led this trend have been the widely advertised drugs for heartburn.

a. What questions need to be answered before a drug is deregulated?

b. Will these questions change if you are considering this need from the viewpoint of the FDA, a pharmaceutical company, or as a consumer?

46. Find out more about the new process for the manufacture of ibuprofen that won the 1997 Presidential Green Chemistry Challenge Award. How does this process differ from the earlier process for manufacturing ibuprofen? Write a brief report on your research.

47. Testosterone and estrone were first isolated from animal tissue. One ton of bulls' testicles were needed to obtain 5 mg of testosterone and four tons of pigs' ovaries were processed to yield 12 mg of estrone.

a. Assuming complete isolation of the hormones was achieved, calculate the mass percentage of each steroid in the original tissue.

b. Explain why the calculated result very likely is incorrect.

48. Herbal remedies are prominently displayed in supermarkets, drug stores, and discount stores.

a. What influences your decision whether to buy one of these remedies?

b. Choose one such remedy and carefully examine its label for as much information as possible about the active ingredients, any inert ingredients, any anticipated side effects, the suggested dosage, and the cost per dose.

c. How confident are you that the safety and efficacy of these remedies is assured? Explain your answer.

49. Habitrol was the most successful smoking-cessation prescription until the introduction of Zyban in 1997, which now has over 50% of the market. How are these two drug therapy approaches different? What is the current market share of each drug?

50. The drug approval laws of other nations are not the same as those of the United States. Choose a country and find out how their drug-approval process works compared to the process in the United States. Construct relative time lines that reveal what steps must be taken and approximately the length of time each step may require. Also, find out if there are specific drugs available in that country that are not available in the United States and comment on what factors may be influencing the policies of each country.

Nutrition: Food for Thought

Fruits and vegetables are an important part of a balanced diet.

"**Y**our choice of diet can influence your long-term health prospects more than any other action you might take." This statement was made by Dr. C. Everett Koop in 1988 when he was Surgeon General of the United States. If the Federal official charged with assuring the health of the American people feels that strongly, it might be a good idea for you to take a careful look at what you eat.

You can gain valuable information and insight about the food you eat by doing a bit of nutritional research at the breakfast table. All you need to do is inspect the labels on the cereal boxes that happen to be available. Processed and packaged foods are all required by law to list ingredients in order of decreasing concentration. In addition, a good deal of nutritional information is also reported. Because you may not have a box of breakfast food immediately at hand, we have copied the labels of five popular cereals in 11.1 Consider This. The cereals and the labels are real, but we have changed the names of the products. The activity includes some questions to help focus your study.

11.1 Consider This: Belly Up to the Breakfast Bar

Listed here is nutritional information from five popular cereals, disguised by pseudonyms. All values are listed for a 1-ounce serving. Remember that the Calories listed in this table are kilocalories: 1 Calorie = 1 kcal.

Cereal Brands: (1 oz. serving)	"Ko-ko Krunchies"	"Vita-Max"	"Health Nuts"	"Oaties"	"Best Choice"
Calories	110	100	107	110	100
Protein, g	1	3	3	4	2
Carbohydrates, g	25	22	22	20	20
Fats, total, g	1	1	2	2	2
Fats, unsaturated, g	—	—	2	—	2
Fats, saturated, g	—	—	0	—	—
Cholesterol, g	0	0	0	0	0
Sodium, mg	180	200	190	290	80
Potassium, mg	55	105	83	105	120
Percent of U.S. Recommended Daily Allowance (RDA)					
Vitamin A	< 2	100	10	25	2
Vitamin C	25	100	0	25	0
Thiamin	25	100	2	25	12
Riboflavin	25	100	2	25	12
Niacin	25	100	2	25	12
Calcium	4	20	0	4	1
Iron	25	100	7	45	12
Vitamin D	< 2	10	10	10	1
Vitamin B_6	25	100	20	25	12
Folic acid	25	100	20	25	12
Phosphorus	4	20	7	10	7
Magnesium	< 2	8	5	10	5
Zinc	< 2	100	20	6	12
Copper	< 2	6	3	6	2

By comparing this information from various brands, you can form opinions about the claims made by the cereal manufacturers. Consider each of these points.

a. Notice that both the nutritional information and the percent U.S. RDA are given for a 1-oz serving. Given that a single-serving box contains 0.75 oz of cereal, estimate how many ounces constitute a normal serving for you.

b. Based on your normal serving size, how many Calories-worth would you consume of an "adult" cereal such as "Vita-Max" or "Best Choice" compared to a "kids" cereal like "Ko-ko Krunchies"? Remember this does not include the Calories you get from milk or fruit used with the cereal.

c. Given that adults require an average of 60 g of protein per day in their diets, what percentage of the daily requirement is provided by each cereal?

d. Eating too much sodium can result in water retention and high blood pressure (hypertension). If you are avoiding excess sodium for health reasons, which of these cereals should you **not** eat?

e. "Health Nuts" and "Best Choice" are usually marketed as being the choices of a health-conscious population. No such claims are made for

"Ko-ko Krunchies." Compare the percent U.S. RDA for these cereals. Make a case either supporting or refuting the health claims for "Health Nuts" and "Best Choice."

f. Now that you have considered all the cereal data, which of these cereals would you choose to eat and why? Which would you avoid and why?

If you actually completed 11.1 Consider This at the breakfast table, your own cereal probably ended up as a rather soggy lump at the bottom of the bowl. But, in return, you learned quite a bit about cereals and reading nutrition labels. This chapter is devoted to several aspects associated with nutrition and the food we eat.

Chapter Overview

In this chapter we approach nutrition from both personal and molecular structure perspectives. We begin with an examination of what we eat and why we eat. This leads to a general consideration of the three main classes of food components: carbohydrates, fats, and proteins. Sources of these macronutrients are considered and we examine the current dietary recommendations for these three groups. What follows is a series of sections that investigate carbohydrates, fats, and proteins in considerably greater detail. In every case, we devote particular attention to molecular structure and its relationship to properties and functions. And we also include individual and public health issues. Sugars, starches, and other carbohydrates are discussed and lactose intolerance is treated briefly. Our study of fats includes information about saturated and unsaturated fats and oils, dietary cholesterol and heart disease, and fat substitutes. The two sections on proteins treat such topics as amino acids, phenylketonuria, and the importance of a complete diet with adequate essential amino acids.

Because food is the source of the energy that powers our bodies and our brains, we next consider the caloric content of various foods, the recommended food energy intake for men and women of various ages and weights, and the energy expenditures associated with a number of activities. But calories are not enough to ensure a balanced diet, which also requires the correct amounts of a wide array of vitamins and minerals. Therefore, we devote individual sections to the roles of a few of these micronutrients and the hazards of an insufficient or an excessive supply of them. The chapter concludes with a consideration of how food supplies are preserved, including the use of gamma radiation, a controversial technology in the eyes of some critics. The chapter contains a wide range of information, but our goals are straightforward: (1) to help you see the connection between chemistry and nutrition by applying chemical principles in the context of the composition and reactions of foodstuffs; (2) to provide information that you can use in making daily choices about personal nutrition and health; and (3) to analyze a number of nutrition-related controversies that have appeared in the popular press.

11.1 You Are What You Eat

The title of this section states the obvious. During your lifetime, you will eat foods that collectively weigh about 700 times your body weight. Some of this prodigious mass of stuff will consist of delicious culinary creations elegantly prepared to celebrate some memorable social occasion. Some, no doubt, will be junk food gobbled on the run. But whatever the circumstances or the cuisine, all of us eat because food provides the four fundamental types of materials required to keep our bodies functioning. *These materials are water, energy sources, raw materials, and metabolic regulators.*

11.2 The Sceptical Chymist: A Lifetime of Food

During a lifetime, you will eat a truly prodigious amount of food, estimated to be about 700 times your adult body weight. This statement is itself quite a prodigious assertion. Do calculations to check that the statement is in the ballpark. State all of your assumptions clearly.

Hint: Start by assuming a lifespan of approximately 78 years and that your present weight is your adult weight. Estimate the weight of food eaten daily at present, and use these data to project your lifetime consumption of food.

> Many of these important properties of water are discussed in Chapter 5.

Water serves as both a reactant and a product in metabolic reactions, as a coolant and thermal regulator, and as a solvent for the countless substances that are essential for life. Our bodies are over 60% water, but H_2O cannot be burned in the body or elsewhere. Therefore, we need food as a source of energy to power processes as diverse as muscle action, brain and nerve impulses, and the movement of molecules and ions in suitable ways at appropriate times and places. We also eat because raw materials are needed for the syntheses of new bone, blood, enzymes, muscles, hair, and for the replacement and repair of cellular materials. And finally, food supplies chemical regulators such as enzymes and hormones that control the biochemical reactions associated with metabolism and all other vital processes.

Eating properly is a matter of consuming the proper foods, not simply a case of eating sufficient amounts of food. It is possible to consume food regularly, even to the point of being overweight, and still be malnourished. The meaning of this commonly used term is important. **Malnutrition** is caused by a diet lacking in the proper mix of nutrients, even though the energy content of the food eaten may be adequate. Malnutrition contrasts with undernourishment. The daily caloric intake of people who are **undernourished** is insufficient to meet their metabolic needs. According to the Food and Agricultural Organization of the United Nations (FAO) 1998 report, 828 million people worldwide were undernourished during 1994–1996, up from 822 million in 1990–92 (Figure 11.1). This represents a very slight decrease in terms of the proportion of the world's people who are undernourished. The greatest increase in terms of sheer numbers is in South Asia; the largest change in percentage terms is the 7% increase in sub-Saharan Africa. Translating these sterile statistics into terms of human misery means that nearly one in five people in the developing countries is undernourished, evidence of the magnitude and tenacity of hunger in the world.

Contrast this grave accounting with the dietary circumstances of most people in the United States. Although there are malnourished and undernourished people in this country, hunger is typically not a consideration in a nation where nearly 30% of the popula-

Figure 11.1

Proportion (%) of undernourished people in the world by region.

(Source: Data from United Nations Food and Agriculture Organization, *State of Food and Agriculture, 1998* report.)

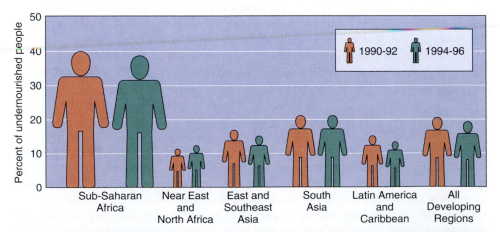

Overall, the percentage of undernourished people in the world decreased slightly between 1990–92 and 1994–96. But this total decrease masks regional and national increases. Regionally, the Near East and Sub-Saharan Africa and south Asia give most cause for concern.

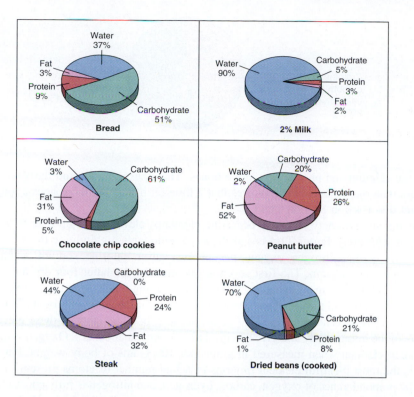

Figure 11.2

Percentage of water, carbohydrates, proteins, and fats in several foods.

(Source: Data from Y. Hui, *Principles and Issues in Nutrition*, 1985, Wadsworth Publishing Co., Belmont, CA.)

tion is up to 40% over their ideal weights, each consuming nearly 1000 excess dietary Calories (kilocalories) per day. One result of such excess is that about 50% of adult women and 25% of adult men annually attempt to practice girth control by dieting, generally with very limited long-term success. No wonder that Dr. Koop said "Americans are zany about food and diet. No other country gorges itself on junk food the way we do, and no other country has as many "experts" on health diets. We have become more concerned about what we should not eat than what we should eat."

For many foods, what we eat is conveyed by food labels, which prominently display the content of carbohydrates, fats, and proteins. These are the **macronutrients** that provide essentially all of the energy and most of the raw material for repair and synthesis. Sodium and potassium ions are present in much lower concentrations, but these ions of metallic elements are essential for the proper electrolytic balance in the body. A number of other minerals and an alphabet soup of vitamins are listed in terms of the percent of recommended daily requirements supplied by a single serving of the product. It should be self-evident that all of these substances, whether naturally occurring or added during processing, are chemicals. Unfortunately, this fundamental fact is apparently lost on those who pursue the impossible dream of a "chemical free" diet. All food is organic, whether claimed to be so or not; *all* food is inescapably and intrinsically chemical.

Vitamins and minerals are discussed in Sections 11.10 and 11.11.

Figure 11.2 indicates the percentages of water, carbohydrates, proteins, and fats in six familiar foods. The pie charts reveal that for this particular selection of foods, the variation in composition is considerable. But in every case, these four components account for almost all of the matter present. The percentage of water ranges from a high of 90% in 2% milk to a low of 2% in peanut butter. Peanut butter beats out steak in percent of protein and also leads these six foods in fat content. Chocolate chip cookies have the highest percentage of carbohydrate because of high sugar and starch concentrations.

Now compare Figure 11.2 with Figure 11.3, which presents similar data for the human body. It is not surprising that the composition of the human body is roughly similar to the composition of the stuff we stuff into it. We are wetter and fatter than bread, and contain more protein than milk; we are more like steak than like chocolate chip cookies. From the data of Figure 11.3, we can calculate that a 150-pound human being consists of 90 pounds of water (150 lb × 60 lb water/100 lb body) and 30 pounds of fat. The remaining 30 pounds are almost all comprised of various proteins and carbohydrates

Figure 11.3
Composition of the human body.

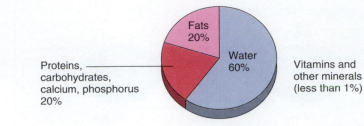

plus the calcium and phosphorus in the bones. The other minerals and the vitamins total less than one pound. This indicates that a little bit of each of them goes a long way, a point discussed in Section 11.11.

Of the nearly ninety naturally occurring elements, eleven make up over 99% of the mass of your body. Figure 11.4 is a skeleton periodic table highlighting these elements: hydrogen, carbon, nitrogen, oxygen, phosphorus, sulfur, chlorine, sodium, magnesium, potassium, and calcium. The first seven are nonmetals, the latter four are metals. Hydrogen, carbon, nitrogen, and oxygen are the "building-block" elements used to construct body cells and tissues; the other seven elements are the macronutrients. Table 11.1 lists the mass percentages (grams of element/100 g body weight) of these eleven elements in the human body and gives their relative atomic abundances. Oxygen is the most abundant element when measured in grams per 100 grams of body weight, but hydrogen is the most plentiful element in terms of actual number of atoms present.

The preponderance of oxygen, carbon, hydrogen, and nitrogen is fully consistent with the composition of the major chemical components of the human body. Hydrogen and oxygen are, of course, the elements of water. Moreover, along with carbon atoms, they constitute all carbohydrates and all fats. Finally, these three elements plus nitrogen are found in all proteins. Thus, nature uses very simple units—aggregates of oxygen, carbon, hydrogen, and nitrogen atoms—in a myriad of elegantly functional combinations to produce the major constituents of a healthy body and a healthy diet.

11.3 Your Turn

Table 11.1 gives both the mass of the major elements per 100 grams of body weight and the relative abundance in the number of atoms per million atoms in the body.

a. Why is oxygen the most abundant element when measured in grams per 100 g of body weight, but only the second most abundant when measured in terms of relative abundance per million atoms in the body?

b. How are these two values related?

Hint: Consider the atomic masses of each element.

Figure 11.4
Periodic table highlighting the eleven major elements of the human body.

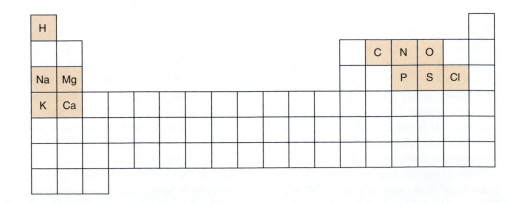

Table 11.1 **Major Elements of the Human Body**

Element	Symbol	Grams/100 g Body Weight	Relative Abundance in Atoms/Million Atoms in the Body
Oxygen	O	64.6	255,000
Carbon	C	18.0	94,500
Hydrogen	H	10.0	630,000
Nitrogen	N	3.1	13,500
Calcium	Ca	1.9	3,100
Phosphorus	P	1.1	2,200
Chlorine	Cl	0.40	570
Potassium	K	0.36	580
Sulfur	S	0.25	490
Sodium	Na	0.11	300
Magnesium	Mg	0.03	130

The current recommendations for a healthy diet, approved in 1991 by the U.S. Department of Agriculture, are represented by the food pyramid of Figure 11.5. The pyramid incorporates the traditional basic four food groups of the 1958 pie chart, but with different emphases. In particular, the pyramid calls for eating habits that increase the proportions of foods at the base of the pyramid and decrease those near or at the top. We are now urged to eat proportionally more bread, cereal, rice, and pasta, and more fruits and vegetables. Simultaneously, we are urged to reduce the percentage of fats, oils, and sweets in our daily diet.

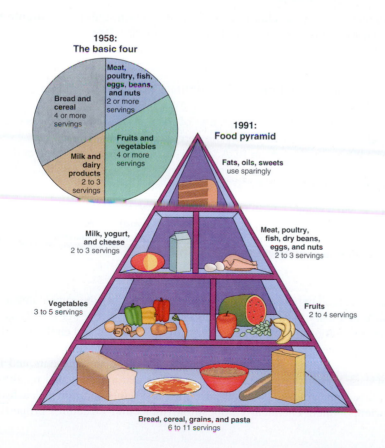

Figure 11.5

The basic four food groups and the food pyramid.

(Source: United States Department of Agriculture.)

11.4 Consider This: The Food Pyramid

Politics entered the picture when the U.S. Department of Agriculture (USDA) initially released the food pyramid (Figure 11.5). The meat and dairy industries pressured the USDA to delay public dissemination of the pyramid. Look at the 1958 basic-four pie chart and the new food pyramid. Although both charts list approximately the same number of servings for meat and dairy products, the number of some of the servings suggested on the pyramid are smaller than those on the pie chart. In addition to the smaller serving sizes, the pyramid includes a category not found on the pie chart, namely "fats, oils, and sweets." The category of "Fruits and Vegetables" found on the pie chart is separated into two categories, "Vegetables" and "Fruits," on the pyramid.

Some nutritionists claim that the USDA, which is obligated to both promote and regulate agricultural products, has a conflict of interest between this responsibility and its obligation to promote the health of American citizens. Imagine that you are an American cattle rancher.

a. How would you reconcile your concern for your livelihood, which depends on the quantity and price of the beef you sell, and your interest in preserving your health?

b. Would you support or oppose the USDA's food pyramid, which encourages eating more grains and less red meat? Give reasons for your answer.

c. Do you think the USDA should be the agency to make this determination? Draft a letter to a friend who is a cattle rancher stating and defending your position on this issue.

11.5 Your Turn

Suppose one of your classmates decided to eat nothing for a month but the "Ko-ko Krunchies" cereal listed in 11.1 Consider This. Assume that the student's daily food energy intake was 2500 Cal. Use the data from 11.1 Consider This to answer these questions.

a. How many one-ounce servings of the cereal did the student eat daily to meet the 2500 Cal?

b. How many grams (or milligrams) of the various food components listed did the student obtain from the cereal diet?

c. What percentage of the RDA values for the various vitamins and minerals did the "Ko-ko Krunchies" supply?

d. Critique this particular diet, identifying strengths and weaknesses.

Ans. **a.** 23 servings; **b.** 23 g protein; 580 g carbohydrates; 23 g fat; 0 g cholesterol; 4.1 g sodium (4100 mg); 1.3 g potassium (1300 mg); **c.** vitamin C, thiamin, riboflavin, niacin, iron, vitamin B_6, folic acid 25%; calcium, phosphorus 4%; vitamin A, vitamin D, magnesium, zinc, copper < 2%.

Although the food pyramid guidelines recommend roughly the same number of servings of meat and dairy products as did the 1958 scheme, the newer dietary guidelines call for smaller portions. Because of concerns over dietary fat and cholesterol, the consumption of red meat, butter, and whole milk has decreased nationally. There is still some controversy surrounding the ideal balance of the macronutrients and the best sources of these compounds, but there is no question that a healthy diet requires carbohydrates, fats, and proteins. We therefore turn to a consideration of each of these macronutrients—their chemical composition, molecular structure, properties, and sources.

11.2 Carbohydrates—Sweet and Starchy

The best known dietary carbohydrates are sugars and starch. **Carbohydrates** are compounds containing carbon, hydrogen, and oxygen, the latter two elements in the same 2:1 atomic ratio as found in water. Glucose, for example, has the formula $C_6H_{12}O_6$. This composition gives rise to the name, "carbohydrate," which implies "carbon plus water." But the hydrogen and oxygen atoms are not bonded together to form water molecules. Rather, carbohydrate molecules are built of rings containing carbon atoms and an oxygen atom. The hydrogen atoms and —OH groups are attached to the carbon atoms. This general arrangement provides many opportunities for differences in molecular structure. Thus, there are 32 distinct isomers (including chiral optical isomers) with the formula $C_6H_{12}O_6$. The isomers differ slightly in their properties, including intensity of sweetness.

The simplest sugars are *mono*saccharides or "single sugars" such as fructose and glucose, both $C_6H_{12}O_6$. Figure 11.6 indicates that molecules of these compounds include rings consisting of four or five carbon atoms and one oxygen atom. The best way to consider these two-dimensional representations of a three-dimensional structure is to imagine that the ring is perpendicular to the plane of the paper, with the bold-print edges facing you. The H atoms and —OH groups are thus either above or below the plane of the ring. This results in two forms of glucose– *alpha* (α) and *beta* (β). In *alpha* glucose, the OH group on carbon 1 is on the opposite side of the ring from the CH_2OH group attached to carbon 5. As shown in Figure 11.6, the *beta* glucose form has the two groups on the *same* side of the ring. This is also the case for *beta* fructose, with the OH on carbon 2 and the CH_2OH group at carbon 5 being on the same side of the ring.

Ordinary table sugar, sucrose, is an example of a *di*saccharide, a "double sugar" formed by joining two monosaccharide units. In a sucrose molecule, an *alpha* glucose and a *beta* fructose unit are connected by a C—O—C linkage created when an H and an OH are split out from the monosaccharides to form a water molecule. This reaction and the structure of the $C_{12}H_{22}O_{11}$ sucrose molecule are also shown in Figure 11.6.

The linking of monosaccharide molecules by this reaction is by no means restricted to the formation of disaccharides. Some of the most common and abundant carbohydrates are *poly*saccharides, polymers made up of thousands of glucose units. As the name implies, these macromolecules consist of "many sugar units." Polysaccharides form when monsaccharide monomer units join into a chain through splitting out a water molecule each time two monomer units combine. This is a condensation

These foods are a rich source of carbohydrates.

Chirality and optical isomerism are discussed in Section 10.7.

Disaccharides and polysaccharides are formed by the condensation polymerization of monosaccharides.

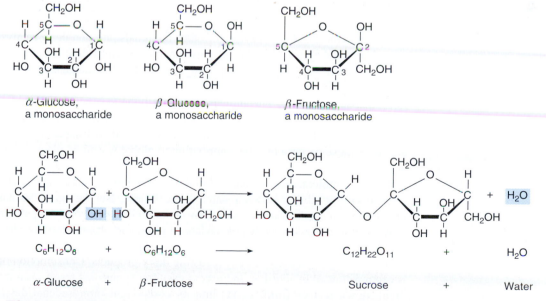

Figure 11.6
Molecular structures of some sugars.

polymerization reaction such as that also described in Section 9.8. Water produced in this way is used for other reactions in the body.

Starch, glycogen, and cellulose are three familiar examples of polysaccharides, sometimes called complex carbohydrates. A healthy diet derives more of its carbohydrates from starch than from simple sugars. Enzymes in our saliva initiate the process of breaking down the long polysaccharide chains into glucose molecules, an important first step in metabolism. But the body also synthesizes glucose from a variety of precursors, including other sugars. Some of the glucose molecules are polymerized to form another polysaccharide called glycogen, whose molecular structure is similar to that of starch. But the chains of glucose units in glycogen are longer and more branched than those in starch. Glycogen is vitally important because it is a storehouse of energy in molecular form. It accumulates in muscle and especially in the liver, where it is available as a quick source of internal energy.

11.6 Your Turn

Sustained chewing of an unsweetened, unsalted cracker results in a sweet taste. What is the molecular explanation for this phenomenon?

Like starch, cellulose is a polymer of glucose, but these two polysaccharides behave very differently in the body. Humans are able to digest starch by breaking it down into individual glucose units; on the other hand we cannot digest cellulose. Consequently, we depend on starchy foods such as potatoes or pasta as carbohydrate sources rather than literally devouring toothpicks or textbooks. The reason for this is a subtle difference in how the glucose units are joined in starch and in cellulose. In the *alpha* (α) linkage in starch, the bonds connecting the glucose units have an angular orientation, whereas the *beta* (β) linkage between glucose units in cellulose is linear.

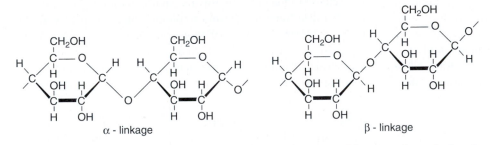

α - linkage β - linkage

Our enzymes, and the enzymes of other mammals, are unable to catalyze the breaking of beta linkages in cellulose. Consequently, we can't dine on grass, textbooks, or trees. However, cows, goats, sheep, and other ruminants manage to break down cellulose with a little help. Their digestive tracts contain bacteria that decompose cellulose into glucose monomers. The animals' own metabolic systems then take over. Similarly, the fact that termites contain cellulose-hungry bacteria means that wooden structures are sometimes at risk. And, as you have already read in Chapter 3, the methane produced by these bacteria may be contributing to global warming, placing the entire planet at risk.

See Section 3.11.

Lactose intolerance, a common metabolic anomaly, is somewhat related to the difference in digestibility between starch and cellulose. But "anomaly" is hardly the correct term for a condition that is shared by about 80% of the world's population. Although most Northern Europeans, Scandinavians, and people of similar ethnic background eat milk, cheese, and ice cream with no ill effects, they are the exception. The great lactose-intolerant majority have difficulty digesting dairy products. Consumption of these foods is often followed by diarrhea and excess gas. The symptoms result from the inability to break down lactose (milk sugar) into its component monosaccharides, glucose and galactose. The linkage between the two monosaccharides in lactose is a *beta* form, similar to that in cellulose. People who are lactose intolerant have a lack of or a low concentration of lactase, the enzyme that catalyzes the breaking of this bond. In such individuals, the intact lactose is instead fermented by intestinal bacteria. This process generates carbon dioxide and hydrogen gases, and lactic acid, the principal cause of the diarrhea. Given milk's importance for growing bones and teeth, it is significant that infants of all ethnic groups generally produce sufficient lactase to digest a milk-rich diet. But, as we age, this production decreases. By adulthood, most people in the world do not have enough of the enzyme to accommodate a diet heavy in dairy products.

11.7 Your Turn

Speculate why the slight difference in molecular structure between starch and cellulose is enough to make the latter polysaccharide indigestible to human beings.

Hint: Section 10.6 can be helpful.

11.8 Consider This: Lactose Intolerance — A Closer Look

Here are three questions on which to try your chemical detective skills using the web by searching for "lactose intolerance." Reading the labels in your cupboard or at a nearby store can also help you milk out some information.

a. Over-the-counter digestive aids allow you to increase your intake of dairy products. How do these work? What are their advantages and disadvantages?
b. Even with digestive aids, you may risk not getting enough calcium, an essential mineral that you will learn more about in Section 11.11. What other foods can you eat to obtain enough calcium in your diet?
c. Sometimes lactose turns up in foods in which you least expect it, such as bread. Although lactose may not be listed on the label, you will see ingredients such as whey, milk products, nonfat dry milk, or dry milk solids—all of which contain lactose. Find three other non-dairy foods that you may have to watch out for if you are lactose intolerant.

11.3 The Fat Family

Everyone knows from personal experience the properties of fats. They are greasy, slippery, soft, low-melting solids that are not water soluble. Butter, cheese, cream, whole milk, and certain meats are loaded with them. But margarine and some shortenings are evidence that fats can also be of vegetable origin. Oils, such as those obtained from olives or corn, exhibit many of the properties of animal-based fats, but in liquid form. The fact that these properties are also shared by petroleum-based oils and greases suggests a chemical and structural similarity. But there are some important differences. You are well aware that petroleum is made up almost exclusively of hydrocarbons. In the molecules of these compounds, carbon atoms are bonded to each other (often in chains) and to hydrogen atoms. Hydrocarbon molecules are nonpolar, and hence they do not mix well with water or other polar substances.

The inference that edible fats and oils must also be nonpolar is fully justified. The molecules of fats of animal and vegetable origin also include long hydrocarbon chains. But biological fats are a bit more structurally complex than their petroleum-based chemical cousins. Of particular significance is the fact that edible fats and oils always contain some oxygen. Most of these compounds are classified chemically as **triglycerides.** **Fats** are triglycerides that are solid at room temperature, whereas **oils** are obviously liquid under these same conditions. Whether solid or liquid, triglycerides are a major portion of a broader class of compounds called **lipids.** Cholesterol and other steroids (see Chapter 10) are also classified as lipids, and molecules of some complex compounds, such as lipoproteins, contain fatty segments.

Hydrocarbons in petroleum were discussed in Section 4.8. Reasons why nonpolar substances do not dissolve in water were examined in Section 5.13.

Hot dogs generally have a high fat content.

Esters are first discussed in Section 9.8, and functional groups, including esters, are summarized in Table 10.2.

To a chemist, the term 'triglyceride' reflects the composition and the formation of fats. A **triglyceride** is formally defined as an ester of three fatty acid molecules and one glycerol molecule. The formation of a triglyceride can be represented by a word equation:

$$3 \text{ Fatty acid} + \text{Glycerol} \longrightarrow \text{Triglyceride} + 3 \text{ Water} \qquad (11.1)$$

In this process, as in the formation of polysaccharides, smaller units join to form more complex molecules by splitting out water molecules. The presence of three fatty acid units in the final product molecule makes it a *tri*glyceride. If you have studied Chapter 10, you have already encountered esterification in the formation of polyesters from acids and alcohols. To see the connection, it is necessary to consider the molecular structures of fatty acids and glycerol.

Naturally occurring fatty acid molecules are characterized by two structural features: a long hydrocarbon chain containing an even number of carbon atoms (generally 12 to 24) including a carboxylic acid group (—COOH) at the end of the chain. This functional group is what puts the *acid* in fatty acid because —COOH can release a hydrogen ion (H^+). The long hydrocarbon chains, on the other hand, give fats most of their characteristic properties. Stearic acid, $C_{17}H_{35}COOH$, is a fatty acid found in animal fats; its structural formula and condensed molecular formula are given below.

$$CH_3CH_2CH_2CH_2CH_2CH_2CH_2CH_2CH_2CH_2CH_2CH_2CH_2CH_2CH_2CH_2CH_2 - \overset{\overset{\displaystyle O}{\|}}{C} - OH$$

A fatty acid, stearic acid

$$CH_3(CH_2)_{16}\overset{\overset{\displaystyle O}{\|}}{C} - OH$$

Condensed chemical formula of stearic acid

Glycerol, or glycerine as it is commonly called, is a sticky, viscous liquid that is sometimes added to soaps and hand lotions. The figure below indicates that a molecule of this compound includes three —OH groups, which classifies it as an alcohol.

$$\underset{\underset{\displaystyle H}{|}}{HO} - \overset{\overset{\displaystyle H}{|}}{\underset{\underset{\displaystyle H}{|}}{C}} - \overset{\overset{\displaystyle H}{|}}{\underset{\underset{\displaystyle OH}{|}}{C}} - \overset{\overset{\displaystyle H}{|}}{\underset{\underset{\displaystyle H}{|}}{C}} - OH$$

Glycerol

Organic acid + Alcohol ⟶

$$-\overset{\overset{\displaystyle O}{\|}}{C} - O - \overset{|}{\underset{|}{C}}-$$

Ester + Water; the group is characteristic of an ester.

In fatty acids and in glycerol, we have the acid and alcohol functional groups, respectively, required to form an ester. In fact, each of the three —OH groups in a glycerol molecule is able to react with a fatty acid molecule. Thus, equation 11.2 represents the combination of three stearic acid molecules with a glycerol molecule to form a triester or triglyceride. This is the process involved in the formation of most animal and vegetable fats and oils. Variety is built in by having up to three different fatty acids incorporated into the same triglyceride rather than just one, stearic acid, as in the example given.

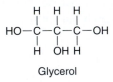

$$(11.2)$$

11.4 Saturated and Unsaturated Fats and Oils

Animal and vegetable fats and oils exhibit considerable variety. A chief reason for this diversity is that not all fatty acids are identical. As we have already noted, they vary in the number of carbon atoms and hence the length of the hydrocarbon chain. Additionally, fatty acids can also contain one or more C=C double bonds and can differ where these double bonds are located in the molecule. If the hydrocarbon chain contains only single C—C bonds between the carbon atoms, and no double bonds, the fatty acid is said to be **saturated**. This is the case with stearic acid. If, however, the molecule contains one or more C=C double bonds between carbon atoms, the fatty acid is **unsaturated**. Oleic acid, with one double bond per molecule, is classified as **monounsaturated**. Those fatty acids containing more than one C=C double bond per molecule are called **polyunsaturated**. Linoleic acid, which contains two double bonds per molecule, and linolenic acid with three double bonds per molecule, are polyunsaturated. Note that each of these three different unsaturated fatty acids contains the same number of carbon atoms, 18.

$$CH_3(CH_2)_{16}COOH$$

Stearic acid, a saturated fatty acid

$$CH_3-(CH_2)_7-CH=CH-(CH_2)_7-COOH$$

Oleic acid, a monounsaturated fatty acid

$$CH_3-(CH_2)_4-CH=CH-CH_2-CH=CH-(CH_2)_7-COOH$$

Linoleic acid, a polyunsaturated fatty acid

$$CH_3-CH_2-CH=CH-CH_2-CH=CH-CH_2-CH=CH-(CH_2)_7-COOH$$

Linolenic acid, a polyunsaturated fatty acid

The overwhelming majority of fatty acids in the body, almost 95%, are transported and stored in the form of triglycerides. The three fatty acids in a single triglyceride molecule can all be identical, two can be the same and the third can be different, or all three can be different. Moreover, the fatty acids in a triglyceride molecule can exhibit varying degrees of unsaturation. The fatty acids a given fat or oil contains govern its extent of unsaturation.

The physical properties of fats also depend upon their fatty acid content. Table 11.2 indicates that within a given family of fatty acids, for example, those that are saturated, melting points increase as the number of carbon atoms per molecule (and the molecular mass) increase. On the other hand, in a series of fatty acids with a similar number of carbon atoms, increasing the number of C=C double bonds decreases the melting point. Thus, when the melting points of the 18-carbon fatty acids are compared, saturated stearic acid (no C=C double bonds per molecule) is found to melt at 70°C, oleic acid (one C=C double bond per molecule) melts at 16° C, and linoleic acid (two C=C double bonds per molecule) melts at 5°C. These trends are carried over to the

Stearic acid is a solid at body temperature, whereas oleic and linoleic acids are liquids.

Table 11.2   Melting Points of Some Fatty Acids

Name	Carbon Atoms per Molecule	Melting Point, °C
Saturated fatty acids		
Capric	10	32
Lauric	12	44
Myristic	14	54
Palmitic	16	63
Stearic	18	70
Unsaturated fatty acids		
Oleic (1 double bond/molecule)	18	16
Linoleic (2 double bonds/molecule)	18	5
Linolenic (3 double bonds/molecule)	18	−11

A comparison of the fat contents of butter and of margarine.

Coconut oil has been used for making popcorn at movie concession stands and for cooking French fries at fast food establishments.

triglycerides containing the fatty acids. This explains why fats rich in saturated fatty acids are solids at room or body temperature, whereas highly unsaturated ones are liquids.

Figure 11.7 gives evidence of this generalization. The bar graphs present the composition of various dietary fats and oils in terms of saturated, polyunsaturated, and monounsaturated components. Typically, these naturally occurring lipids are mixtures of various triglycerides. In general, solid or semisolid animal fats, such as lard and beef tallow, are high in saturated fats. In contrast, olive, safflower, and other vegetable oils consist mostly of unsaturated triglycerides. However, the figure reveals that there are some surprising differences in the composition of oils. For example, palm and coconut oil contain much more saturated fat than do corn and canola oil. Ironically, the coconut oil used in some non-dairy creamers is 92% saturated fat, far more than the percentage found in the cream it replaces. In fact, coconut oil contains more saturated fats than does pure butterfat. Concern over the high degree of saturation in coconut and palm oil accounts for the statement sometimes printed on food labels: "Contains no tropical oils."

11.9 Your Turn

Using Figure 11.7 and information from this section, identify the predominant fatty acids likely to be present in: **a.** canola oil; **b.** olive oil; and **c.** lard.

Some food labels also reveal that not all the fats and oils in our diet are consumed in their naturally occuring molecular forms. Unless you eat "natural" peanut butter, the jar on your shelf probably is labeled something like: "oil modified by partial hydrogenation." Peanuts ground to make peanut butter always release a quantity of peanut oil. This oil separates on standing and must be stirred back into the solid before use. The oil extracted from peanuts is rich in mono- and polyunsaturated fats, and thus is a liquid. It can be treated chemically and converted into a semisolid that does not separate from peanut butter. This is done by reacting the oil with hydrogen gas over a metallic catalyst. The hydrogen adds to the double bonds in the oil, converting some, but not all, of the C=C bonds into C—C bonds, such as with the linoleic acid in peanut oil.

$$CH_3(CH_2)_4-CH=CH-CH_2-CH=CH-(CH_2)_7COOH + H_2 \longrightarrow$$
$$CH_3(CH_2)_4-CH_2-CH_2-CH_2-CH=CH-(CH_2)_7COOH$$

Figure 11.7
Saturated and unsaturated fats.

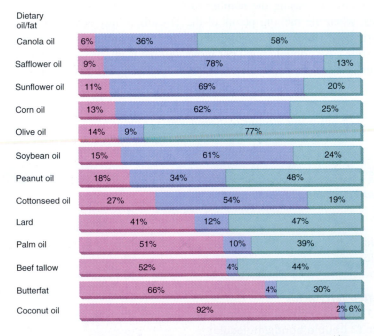

(*Source: Data from Food Technology, April 1989.*)

As a result of this partial hydrogenation, the number of double bonds in the lipid decreases, it becomes less unsaturated, and it is transformed from an oil into a semisolid fat. The extent of hydrogenation can be carefully controlled to yield products, in addition to peanut butter, of desired unsaturation and resultant melting point, softness, and spreadability. Such customized fats and oils are in many products, including margarines, cookies, and candy bars. But "natural food" advocates can take some comfort in the realization that these triglycerides are not really artificial. Their degree of saturation may be chemically altered, but the fats formed through partial hydrogenation certainly also exist somewhere in nature in some other food.

In our preoccupation with dietary fat, it is important to realize that fats often enhance our enjoyment of food. They improve "mouth feel" and intensify certain flavors. Of prime significance is the fact that fats are essential for life. They are the most concentrated source of energy in the body (Section 11.9), and they provide insulation that retains body heat and cushions internal organs. Moreover, triglycerides and other lipids, including cholesterol, are the primary components of cell membranes and nerve sheaths. Although "fathead" is hardly a compliment, in fact our brains are rich in lipids. These various functions require a variety of triglycerides incorporating a wide range of fatty acids—saturated, monounsaturated, and polyunsaturated. Fortunately, our bodies can synthesize almost all of the necessary fatty acids from the starting materials provided by normal diet. The exceptions are linoleic and linolenic acids. These two **essential fatty acids** must be obtained directly from the foods we eat; our body cannot produce them. Generally this does not create a problem because linoleic and linolenic acids are found in many foods including plant oils, fish, and leafy vegetables.

Cholesterol is discussed in Section 10.8.

11.10 Consider This: Spreadables and Fat Content

Here are the fat contents for Crisco (a partially hydrogenated shortening) and Brummel & Brown Spread (a soft, butter substitute.)

	Crisco	Brummel & Brown
Total fat	12 g	5 g
Polyunsaturated	3 g	2 g
Saturated	3 g	1 g

As advertised, Brummel & Brown has less fat than butter. Notice that the sum of the saturated and polyunsaturated fats does not add up to the total fat value of the two products. The difference is the amount of monounsaturated fats. Thus, Crisco contains 12 − (3 + 3) or 6 g monounsaturated fats, and Brummel & Brown contains 5 − (2 + 1) or 2 g monounsaturated fats per serving.

Conduct a mini-survey of butter and margarine products in a supermarket in your area. List the total fat content and the content of polyunsaturated, saturated, and monounsaturated fats for examples of the following five products: Butter stick; regular margarine stick; regular margarine tub; "light" margarine stick; and "light" margarine tub.

a. How does Crisco, listed above, compare to the butter and margarine samples in your survey in terms of total fat content and amounts of various types of fat?

b. What are the major differences in total fat content and amounts of various types of fat among the samples in your survey?

c. Which butter or margarine product would you choose based on its fat content and why?

11.5 Controversial Cholesterol

Although dietary fat is an essential part of a balanced diet, the fact remains that many Americans are consuming too much of it, and too much of the wrong kind. Fats provide about 40% of the calories in the average American diet. Health care specialists recommend that this value should be 30% or less. Much of the concern and controversy regarding cardiac health problems is focused on cholesterol, one of the steroids introduced in Chapter 10.

Cholesterol has drawn heavy media attention, but not always with total accuracy or objectivity. At issue is the connection between cholesterol in the blood (serum cholesterol) and cardiovascular disease. Over two decades, several national reports appeared and made conflicting recommendations. In 1980, the Food and Nutrition Board of the National Academy of Science issued "Toward Healthful Diets." This report cited the inconsistency of data collected in medical studies done up to that time. A direct causal connection had not been unambiguously established between dietary cholesterol and atherosclerosis, the thickening of arterial walls. Therefore, the report concluded that healthy adults did not need to reduce dietary cholesterol. However, by the end of the 1980s, a reversal of this position had occurred, based on the pooled data from a wide variety of new and continuing studies. These studies led to the conclusion that high serum cholesterol levels *appear* to predict the potential for a stroke or heart attack. Although an irrefutable direct linkage has not yet been demonstrated, the data seem compelling. Waxy deposits of excess cholesterol (plaque) cause arteries to narrow and harden, which may elevate blood pressure and increase the risk of heart disease (Figure 11.8).

Today there is general agreement that elevated blood cholesterol levels are associated with atherosclerosis, although there is not quite consensus on what constitutes dangerously high concentrations. Many medical researchers and the American Heart Association (AHA) consider values greater than 200 mg cholesterol per 100 mL of blood as the critical point for medical intervention. Other investigators are more generous, citing concentrations over 240 mg/100 mL as the threshold for such action.

One obvious response to elevated serum cholesterol is to restrict consumption of cholesterol. The American Heart Association recommends a maximum intake of 300 mg of cholesterol per day. This means cutting back on animal fats, which are rich in the compound. Included are fatty red meats, cream, butter, and cheese. Table 11.3 reveals that egg yolks are particularly rich in cholesterol, each yolk containing on average a whopping 213 mg. In contrast, egg whites contain no cholesterol, nor do fruits, vegetables, or vegetable oils.

But restricting dietary cholesterol is only one part of a two-part situation. It is possible that even a strict vegetarian with negligible cholesterol intake might have elevated serum cholesterol. This is because most of the body's cholesterol does not come from the diet directly, but is synthesized by the body. About one gram of cholesterol is synthesized daily in the liver to maintain the minimum concentration required for use in cell membranes and to produce estrogen, testosterone, and other steroid hormones. The

Vegetables, vegetable oils, and fruits contain no cholesterol.

Figure 11.8

Cross sections of a healthy artery (left) and an artery clogged with atherosclerotic plaque (right).

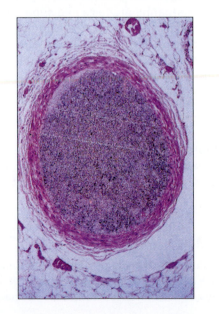
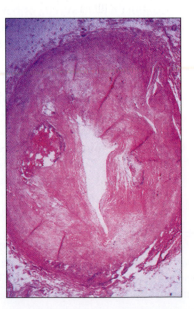

Table 11.3 Cholesterol Content of Various Foods

Food	Cholesterol (mg)	Food	Cholesterol (mg)
Fruits and vegetables	0	Milk, whole (8 oz)	33
Pork chop (3 oz)	83	Low-fat milk (8 oz)	22
Chicken, skinless (3 oz) white meat	71	Low-fat yogurt (8 oz)	14
Steak (3 oz)	70	Am. processed cheese (1 oz)	27
Shrimp (3 oz)	166	Cheddar cheese (1 oz)	30
Hot dog (3 oz)	43	Ice cream (4 oz)	29
Egg yolk	213	Butter (1 tbsp)	33

liver produces cholesterol principally from dietary saturated fats. Consequently, a high intake of saturated fats can result in a high concentration of cholesterol. Although cutting down on cholesterol consumption is an important step in lowering serum cholesterol, reducing the amount of saturated fats in the diet may be even more significant. The AHA recommends that only 8–10% of total calories should come from such fats, and to especially limit the intake of those with certain numbers of carbon atoms per molecule: 12 (lauric acid), 14 (myristic acid), and 16 (palmitic acid).

Diet, however, is only one of several factors influencing cholesterol synthesis. Perhaps the most important factor is genetic. This may explain why some people seem to eat fatty foods without suffering from heart disease, while others who carefully watch their diets are afflicted with it. Because we are not in complete control of our genes (at least not yet), physicians urge us to do what we can to lower our serum cholesterol. Reducing dietary cholesterol and fatty acids, exercising regularly, decreasing weight and stress, and eating certain types of dietary fiber appear to be important for maintaining good health.

It is possible that some of these habits also serve to lower the concentration of **low density lipoprotein** (LDL) and increase the concentration of **high density lipoprotein** (HDL). These compounds combine with cholesterol and triglycerides and carry them through the bloodstream, thus preventing the build-up of plaque in the arteries. The HDLs are the "good" lipoproteins, and are more effective in transporting cholesterol than are the LDLs. The AHA recommends a concentration of greater than 35 mg HDL/100 mL of blood, and an LDL value of less than 130 mg/100 mL. It appears that people with high values for the LDL/HDL concentration ratio are particularly susceptible to heart disease. No doubt other discoveries linking diet and heart disease will continue to be made. But it is important to remember that the most effective way to reduce the risk of heart disease has to do with inhaling as well as ingesting. Dr. Richard Peto, an Oxford University epidemiologist writing in the *Harvard Medical School Health Letter,* December 1989, minces no words: "You can't offer eternal life to old people. But what you can do is to avoid death in middle age. At the moment, about a third of all Americans die in middle age, and that isn't necessary. About half of those premature deaths could be avoided if people took smoking, blood cholesterol, and blood pressure more seriously."

11.11 Your Turn

Have the American Heart Association's recommended values for cholesterol, HDL, and LDL changed from those reported in this section? Check the current recommendations at their web site. If there are any differences, explain why.

Perhaps you are one of those people who is always looking for a way to have your cake and eat it too. That's easily done: if you eat the cake, you are likely to retain part of it in the form of increased fat deposits. Therefore, you might be interested in the development of Olestra, a nonfattening fat developed by the Procter & Gamble Company. In January 1996, Olestra was approved by the FDA for use in salty snack foods such as potato chips and tortilla chips. In many of its properties and in its molecular structure,

Figure 11.9
Schematic representation of an Olestra molecule.

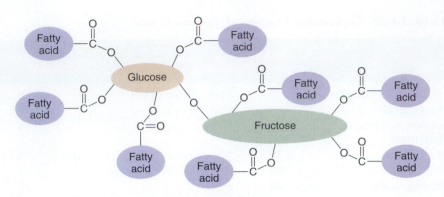

Recall that sucrose is made of glucose and fructose units.

This snack food contains Olestra, a fat substitute.

Olestra is similar to a triglyceride. It, too, is a fatty acid ester in which the alcohol groups are provided not by glycerol, but by the sugar sucrose. Each sucrose molecule has eight —OH groups. In a typical Olestra molecule, there are ester linkages formed by six, seven, or all eight of these OH groups with a fatty acid (Figure 11.9). The resulting sucrose polyester molecule is much larger and bulkier than a triglyceride molecule. In fact, it is so bulky that lipase, the enzyme that snips the fatty acids from a fat, cannot get close enough to do the job. As a result, Olestra molecules pass unabsorbed through the intestines. No absorption means no additional Calories, no new pounds, no new cholesterol. A serving of potato chips fried in Olestra provides 75 Calories, compared to 150 for conventional fat-fried chips.

The research that led to Olestra began in 1959. Olestra was originally developed to be a nutritious baby food additive, but a major problem arose—olestra was not digested; so much for its use as a nutritious additive! It remained a laboratory curiosity until applications research tested it as a fat substitute. In 1971, Procter & Gamble took out its first patents on sucrose polyesters and began conversations with the FDA. Six years later the company submitted a formal food use petition. The review of such a petition is supposed to take 180 days. It usually takes three to six years; in this case it took almost two decades. By the time of its FDA approval in 1996, Procter & Gamble had invested $200 million in Olestra and had generated 150,000 pages of studies. Dr. David Kesler, FDA Commissioner when Olestra was approved, stated that Olestra was "probably one of the most extensively studied food substances to date . . . research data on Olestra demonstrate reasonable certainty of no harm for use in certain snack foods." These studies showed that the compound is nontoxic, but not without a few less than desirable side effects. Olestra contributes to loose stools in some people and causes anal leakage in rare cases. It also inhibits the absorption of the fat-soluble vitamins A, D, E, and K. To overcome this shortcoming, Olestra contains supplemental amounts of these vitamins. In spite of these drawbacks, the majority of the FDA evaluators agreed that the nonnutritious fat is harmless, and its limited use was approved. The use of Olestra in any food products other than salty snack foods will require separate FDA approval.

Procter & Gamble sells the non-metabolizable fat under the name Olean to other manufacturers. Because Americans eat about 22 pounds of salty snack foods per capita every year, the annual market for Olestra-based snacks could exceed $1 billion. Of course if consumers simply increase the amount of Olestra-containing foods they eat, nothing will be gained, except perhaps pounds. Bryan McLeary, a Procter & Gamble representative states this clearly: "Olestra is a replacement for fat, not common sense."

11.12 The Sceptical Chymist: Leaning Towards or Away From Olean?

In spite of FDA approval, the use of Olestra (Olean) remains controversial, with its supporters and detractors. To learn more about Olean, check out the Olean web site sponsored by the company that markets Olean at the *Chemistry in Context* web site. Then search for "Olean" and "Olestra" together with terms such as "controversy", "opposition to", and "drawbacks" to view web sites that present other viewpoints. Use this information along with that in the text to answer these questions.

a. What are the main points of contention between the two sides?

b. What complaints have actual Olestra users registered?

c. Do you think that their complaints were justified, given the information provided? Explain.

d. Will consuming food containing fat substitutes always result in weight loss? Explain.

e. What is your personal decision about using products containing Olestra? Explain your reasoning.

A food with relatively high protein content.

11.6 Proteins: First Among Equals

The word "protein" derives from *protos,* Greek for "first." The name is misleading. Life depends on the interaction of thousands of chemicals, and to assign primary importance to any single compound or class of compounds is naive. Nevertheless, proteins are an essential part of every living cell. They are major components in hair, skin, and muscle; and they transport oxygen, nutrients, and minerals through the bloodstream. Many of the hormones that act as chemical messengers are proteins, as are all of the enzymes that catalyze the chemistry of life.

Proteins are polyamides or polypeptides, polymers made up of amino acid monomers. The great majority of proteins are made from various combinations among 20 different naturally occurring amino acids. Molecules of all of these amino acids share a common structural pattern. Four chemical species are attached to a carbon atom: 1) an acid group, —COOH; 2) an amine group, —NH_2; 3) a hydrogen atom, —H; and 4) a side chain designated as R in the following structure.

Amino acids and proteins were mentioned briefly in Section 9.9.

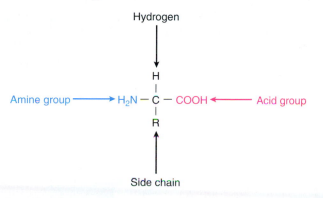

Variations in the R side-chain group differentiate the individual amino acids. In glycine, the simplest amino acid, the R is a hydrogen atom. In alanine, R is a —CH_3 group, in aspartic acid (found in asparagus) it is —CH_2COOH, and in phenylalanine it is a group with the formula —$CH_2(C_6H_5)$. Here C_6H_5 designates the hexagonal phenyl ring first introduced in Chapter 9. Note the structural relationship between alanine and phenylalanine.

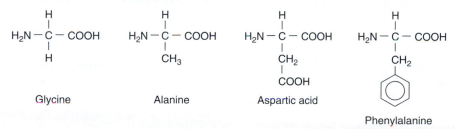

Two of the 20 naturally occurring amino acids have R groups that bear a second acidic —COOH functional group, three have R groups containing amine groups, and two others contain sulfur atoms. Because all amino acids except glycine involve four different units bonded to a central carbon atom, they all exhibit chirality or optical isomerism.

See Section 10.7 for a discussion of chirality.

The naturally occurring amino acids that are incorporated into proteins are all in the left-handed isomeric form.

The combination of amino acids to form proteins depends on the presence of the two characteristic functional groups that give this family of compounds its name—the amine group and the acid group. Equation 11.3 represents the reaction of glycine and alanine to form a **dipeptide,** a compound composed of the two amino acids. Here, the acidic —COOH group of a glycine molecule reacts with the —NH₂ group of alanine, and an H_2O molecule is eliminated. In the process, the two amino acids become linked by a **peptide bond** (indicated in the box that follows). Once they have been incorporated into the peptide chain, the amino acids are known as **amino acid residues.** The reaction in equation 11.3 is another example of condensation polymerization, already encountered in the formation of polysaccharides (Section 11.2) and some synthetic polymers (Section 9.8).

Equation 11.3 is equivalent to equation 9.7.

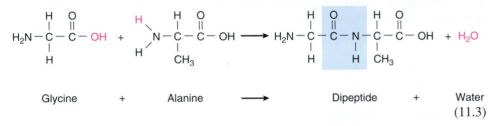

$$\text{Glycine} + \text{Alanine} \longrightarrow \text{Dipeptide} + \text{Water} \tag{11.3}$$

Because each amino acid bears an amine group and an acid group, there are two ways the amino acids can join. Hence, two dipeptides are possible. We will illustrate the options with simple block diagrams for the amino acids. The first case is that of equation 11.3: glycine acts as the acid and alanine as the amine.

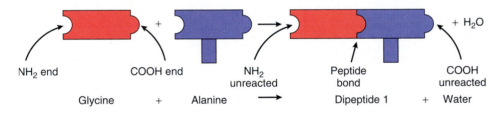

In the second case, the amino acids reverse roles; alanine provides the —COOH and glycine the —NH₂ for the reaction.

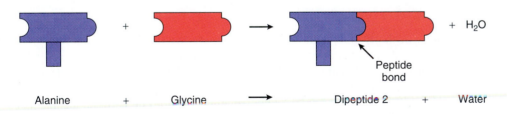

Examination of the molecular structures of the two dipeptides indicates that they are not identical. In dipeptide 1 the unreacted amine group is on the glycine residue and the unreacted acid group is on the alanine residue; in dipeptide 2 the —NH₂ is on the alanine and the —COOH on the glycine residue.

Putting the amino acids of a protein into their proper order is like assembling a train correctly by placing each car in the right sequence.

The point of all this is that the order of amino acid residues in a peptide makes a difference in its properties. The particular protein formed depends not only on what amino acids are present, but also on their sequence in the protein chain. Assembling the correct amino acid sequence to make a particular protein is like putting letters in a word; if they are in a different order, a completely new meaning results. Thus, a tripeptide consisting of three different amino acids is like a three-letter word containing the letters *a*, *e*, and *t*. There are six possible combinations of these letters. Three of them—*ate*, *eat*, and *tea*—form recognizable English words; the other three—*aet*, *eta*, and *tae*—do not. Similarly, some sequences of amino acids may be biological nonsense.

Still restricting ourselves to three-letter words and only the letters *a, e,* and *t,* but allowing the duplication of letters, we can make perfectly good words such as *tee* and *tat,* and lots of meaningless combinations such as *aaa* and *tte.* There are, in fact, a total of 27 possibilities, including the six identified earlier. Just as many words use letters more than once, most proteins contain specific amino acids incorporated more than once. More information about the structure and synthesis of proteins is included in Chapter 12.

11.13 Your Turn

You can see from the block diagrams in this section that one glycine (Gly) and one alanine (Ala) molecule can combine to form two dipeptides: GlyAla and AlaGly. If one permits multiple use of each of the two amino acids, two other dipeptides are possible: GlyGly and AlaAla. Thus, there are a total of four different dipeptides that can be made from two amino acids, if each amino acid can be used more than once. Eight different tripeptides can be made from supplies of two different amino acids, assuming that each amino acid can be used once, twice, three times, or not at all. Use the symbols Gly and Ala to write down representations of the amino acid sequence in all eight of these tripeptides.

Hint: Start with GlyGlyGly.

11.14 Your Turn

Structural features of amino acids are more readily apparent if you look at their three-dimensional representations. At the *Chemistry in Context* web site, with the help of Chime, you can view and rotate the molecular structures for several different amino acids.

a. How is the three-dimensional structure of glycine different from the two-dimensional structure shown in your text?
b. Glycine is the most simple amino acid. It contains only two functional groups (—NH$_2$ and —COOH) and only the elements C, H, O and N. Browse through the Chime collection of amino acids and then describe two ways in which their structures are more complex than glycine's.
c. In leucine, what four different groups are bonded to a central carbon atom? Is this molecule optically active? Explain your answer.

11.7 Enough Protein: The Complete Story

Dietary protein requirements are usually expressed in terms of grams of protein per kilogram of body weight per day, which varies with age, size, and energy demand. Infants require 1.8 g/kg/day, middle school children about 1.0 g/kg/day, and adults 0.8 g/kg/day. Therefore, a 20-lb (9 kg) child needs 16 g of protein daily to provide the raw materials for body growth and development. A 165-lb (75 kg) adult requires 60 g each day to maintain proper physiological function.

The body does not normally store a reserve supply of protein, so foods containing these nutrients must be eaten every day. As the principal source of nitrogen for the body, proteins are constantly being broken down and reconstructed. A healthy adult on a balanced diet will be in **nitrogen balance,** excreting as much nitrogen (primarily as urea in the urine) as she or he ingests. Growing children, pregnant women, and persons recovering from long-term debilitating illness have a positive nitrogen balance. This means that they consume more nitrogen than they excrete because they are using the element to synthesize additional protein. A negative nitrogen balance exists when more protein is being decomposed than is being made. This occurs in starvation, when the

Table 11.4 The Essential Amino Acids

Histidine	Lysine	Threonine
Isoleucine	Methionine	Tryptophan
Leucine	Phenylalanine	Valine

energy needs of the body are unmet from the diet, and muscle is metabolized to maintain physiological functions. In effect, the body feeds on itself.

Another cause of a negative nitrogen balance may be a diet that does not include enough of the **essential amino acids.** Of the 20 natural amino acids that make up our proteins, we can synthesize 11 from simpler molecules, but nine must be ingested directly. If any of these nine essential amino acids, identified in Table 11.4, are missing from the diet, many important proteins cannot be produced in the body in sufficient quantity. The result can be severe malnutrition.

Good nutrition thus requires protein in sufficient quantity and suitable quality. Beef, fish, poultry, and other meats contain all of the essential amino acids in approximately the same proportions found in the human body. Therefore, meat is termed a **complete protein.** However, most people of the world depend on grains and other vegetable crops rather than meat as their major sources of protein. If such a diet is not sufficiently diversified, some essential amino acids may be lacking. For example, many Mexicans and Latin Americans consume large quantities of corn, a protein source that is *incomplete* because it is low in tryptophan, an essential amino acid. A person may eat enough corn to meet the total protein requirement, but still be malnourished because of insufficient tryptophan.

Fortunately for millions of vegetarians, a reliance on vegetable protein does not necessarily doom one to malnutrition. The trick is to apply a principle nutritionists call *complementarity,* combining foods that complement each others' essential amino acid content so that the total diet provides a complete supply of amino acids. You do this, probably unknowingly, every time you eat a peanut butter sandwich. Bread is deficient in lysine and isoleucine, but peanut butter supplies these amino acids. On the other hand, peanut butter is low in methionine, a compound provided by bread. The traditional diets of many countries meet protein requirements through nutritional complementarity. In Latin America, beans are used to complement corn tortillas; soy foods are eaten with rice in parts of Southeast Asia and Japan. People in the Middle East combine bulgur wheat with chickpeas or eat hummus, a sauce of sesame seeds, and chickpeas, with pita bread. In India, lentils and yogurt are eaten with unleavened bread.

Livestock, especially beef cattle, also benefit from complementarity. They are fed a variety of grains with a complete set of amino acids to incorporate ultimately into steaks and hamburger. However, the second law of thermodynamics applies to beef cattle as well as to everything else. There is a loss of efficiency with each step of energy transfer, whether in electrical power plants or in cells during metabolism. Cattle are notoriously inefficient in converting the energy in their feed into meat on the hoof. It takes about seven pounds of grain to produce one pound of beef. Put into human terms, the 1.75 pounds of grain used to produce a "quarter-pounder" can provide two days of food for someone on a vegetarian diet. Other animals are more efficient than cattle in converting grain to meat. Hogs require six pounds of grain per pound of meat, turkeys need four, and chickens even less, only three pounds. It is obviously much more efficient to get food energy directly from grains, rather than through secondary or tertiary sources further along the food chain. On the other hand, one should keep in mind that pasture land used to graze cattle is often unsuitable for growing crops. Moreover, much of the food consumed by animals would be indigestible or unpalatable to humans.

A postscript to the protein story is provided by the unusual case of aspartame, a sweet dipeptide. Because of the great American preoccupation and battle with excess calories and excess pounds, artificial sweeteners have become a billion dollar business. Gram for gram, these compounds are much sweeter than sugar, but they have little if any nutritive value. Hence, they are non-fattening. The principal use (75%) of artifical sweeteners is in soft drinks. Currently the most widely used artificial sweetener is aspartame, the principal ingredient in NutraSweet and Equal. Somewhat surprisingly, the compound is related to proteins. Aspartame is a dipeptide made from aspartic acid and a slightly modified phenylalanine. The molecular structure of aspartame is given in the following diagram.

Americans drink an annual average of about 53 gallons of soft drink per person, up nearly 50% from the amount in 1985.

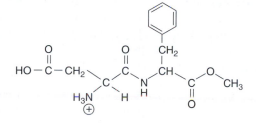

Alone, neither of the two amino acids in aspartame tastes sweet. Yet, the compound that results from their chemical combination is about 200 times sweeter than sucrose. The fact that sucrose and aspartame are chemically and structurally very different invites speculation about the molecular features that convey sweetness. But this digression will be long enough without taking up the issue of sweetness. For whatever reason, aspartame is sufficiently sweet to be used by millions of people worldwide. A few cases of adverse side effects have been attributed to aspartame, but exhaustive reviews have failed to show an unequivocal and direct connection between the symptoms and the sweetener. For the vast majority of consumers, aspartame is a safe alternative to sugar. There is, however, one group of people who definitely should not use aspartame. The warning on packets of artificial sweeteners and products containing aspartame is explicit: "Phenylketonurics: Contains Phenylalanine."

This is a case where "one man's meat is another man's poison." Phenylalanine is an essential amino acid that is converted in the body to tyrosine, another amino acid. Individuals with phenylketonuria, a genetically transmitted disease, lack the enzyme that catalyzes this transformation. Consequently, the conversion of dietary phenylalanine to tyrosine is blocked and the phenylalanine concentration rises. To compensate for the elevated phenylalanine, the body converts it to phenylpyruvic acid, excreting large quantities of this acid in the urine. Phenylpyruvic acid is termed a "keto" acid because of its molecular structure; hence the disease is known as phenyl*keto*nuria or PKU. People with the disease are called phenylketonurics.

Excess phenylpyruvic acid causes severe mental retardation. Therefore, the urine of newborn babies is tested for this compound, using special test paper placed in the diaper. Infants diagnosed with PKU must be placed on a diet severely limited in phenylalanine. This means avoiding excess phenylalanine from milk, meats, and other sources rich in protein. Commercial food products are available for such diets, their composition adjusted to the age of the user. Because phenylalanine is an essential amino acid, a minimum amount of it must still be available, even in phenylketonurics. Supplemental tyrosine may also be needed to compensate for the absence of the normal conversion of phenylalanine to tyrosine. A phenylalanine-restricted diet is recommended for phenylketonurics at least through adolescence. Adult phenylketonurics must also limit their phenylalanine intake, and hence curtail their use of aspartame.

In April 1998 the FDA approved another artificial sweetener, sucralose, a sucrose derivative that is 600 times sweeter than sucrose. In addition to normal "table-top" use where it is added directly to foods by consumers, sucralose's other approved uses include putting it into baked goods, non-alcoholic beverages, chewing gum, frostings, sweet sauces and syrups, and frozen dairy desserts.

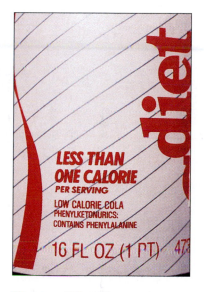

Warning: "Phenylketonurics: Contains Phenylalanine" on a diet beverage can.

11.15 Consider This: How Sweet They Are!

Consider this information about the sugar content of different food products.

Food Product	Sugar	Calories	Serving Size
Altoids, Peppermint	2 g	10	3 pieces (2 g)
Chicken in a Biskit Crackers	2 g	160	12 crackers (30 g)
Critic's Choice Tomato Ketchup	3 g	15	1 tbsp (13 g)
Delmonte Pineapple Cup	13 g	50	Individual cup (113 g)
Dr Pepper Soft Drink	40 g	150	1.5 cups
French Vanilla Coffee Mate	5 g	40	1 tbsp (15 mL)
Hostess Twinkies	14 g	150	1.5 oz
LifeSavers, WintOgreen	15 g	60	4 mints (16 g)
Ocean Spray Cranberry Juice	25 g	140	1 cup
Snickers Bar	29 g	200	2.1 oz
Sunkist Orange Soda	52 g	190	1.5 cups
Wheatables Crackers	4 g	130	13 crackers (29 g)

a. Examine this list of twelve food products. Which item has the highest ratio of grams of sugar to the number of Calories (g sugar/Cal) in one serving?

b. Does the sugar content of any of these foods surprise you? Explain your response.

c. Do you expect that the specific sugars in Dr Pepper are the same sugars found in Sunkist orange soda? In cranberry juice? In the pineapple cup? Why or why not?

d. There are 16 g of total carbohydrates listed on the label for WintOgreen Lifesavers, of which 15 g are sugars. What type of compounds do you think account for the other 1 g of carbohydrates?

11.8 Energy: The Driving Force

If you need a refresher on chemical energetics, see Section 4.4.

All of the energy needed to run the complex chemical, mechanical, and electrical system called the human body comes from carbohydrates, fats, and proteins. This energy initially arrives on Earth in the form of sunlight, which is absorbed by green plants during photosynthesis. Under the influence of a catalyst called chlorophyll, carbon dioxide and water are combined to form glucose. In the process, the Sun's energy is stored in chemical bonds of the sugar.

$$\text{Energy (from sunshine)} + 6\ CO_2 + 6\ H_2O \longrightarrow C_6H_{12}O_6 + 6\ O_2$$

During metabolism the photosynthetic process is reversed, the food is converted into simpler substances, and the stored energy is released.

$$C_6H_{12}O_6 + 6\ O_2 \longrightarrow 6\ CO_2 + 6\ H_2O + \text{Energy (from metabolism)}$$

The breaking of chemical bonds in glucose and oxygen molecules requires the absorption of energy. But, more energy is released in the exothermic reactions in which carbon dioxide and water are formed. Thus, there is a net release of energy. This energy balance is schematically represented in Figure 11.10.

Energy provided by food we eat is used to drive the chemical reactions that constitute the processes of life. The most obvious example of an energy-requiring process is muscular motion, including the beating of the heart. But most of the energy released by metabolism goes to maintain differences in ionic concentrations across cell membranes. The natural tendency is for diffusion to move substances from regions of higher con-

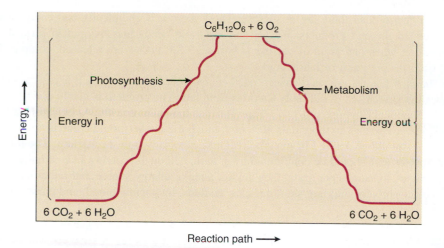

Figure 11.10
Energy from photosynthesis and metabolism.

Chemistry for Health-Related Sciences by Sears/Stanitski, © 1976. Reprinted by permission of Prentice-Hall, Inc., Upper Saddle River, NJ.

centration to those of lower concentration. Energy is required to prevent this from happening. The proper concentration differences that are essential for nerve action and other physiological functions are maintained at great energetic expense. In short, spontaneous reactions furnish the energy to non-spontaneous reactions so that they can occur. As an analogy, consider an automobile storage battery. The battery produces electrical energy spontaneously because of chemical reactions in the battery. These spontaneous processes provide energy that can be used to permit non-spontaneous processes to take place, for example, starting the car or making the headlights and horn work.

In addition to having a supply of sufficient energy, the body must have some way of regulating the rate at which the energy is released. Without such control, wild temperature fluctuations and high inefficiency could result. Again, the automobile provides an analogy. Dropping a lighted match into the fuel tank would burn all the gasoline (and the car as well.) This is a drastic but not particularly effective way to move a car. Under normal operating conditions, just enough fuel is delivered to the ignition system to supply the automobile with the energy it needs without raising the temperature of the car and its occupants beyond reason. In this way, by releasing a little energy at a time the efficiency of the process is enhanced. So it is with the body. The conversion of foods ultimately into carbon dioxide and water occurs over many small steps, each one involving enzymes, enzyme regulators, and hormones. As a result, energy is released gradually, as needed, and body temperature is maintained within normal limits.

The chief sources of this energy in a well-balanced diet are carbohydrates and fats. When metabolized, carbohydrates provide about 4 kcal/g, fats release about 9 kcal/g. A kilocalorie (kcal) is identical to a dietary Calorie, written with a capital C. Package labels and nutritional tables that state how much energy we get from our meals typically use Calories, as in "one chocolate chip cookie provides 50 Calories." We will do the same. Keep in mind that whatever units are used, on a gram-for-gram basis, fats provide about 2.5 times as much energy as carbohydrates.

The reason for this dramatic difference is implicit in the chemical composition of these two types of material. Compare the formulas of a fatty acid, lauric acid, $C_{12}H_{24}O_2$, with that of sucrose (table sugar), $C_{12}H_{22}O_{11}$. Both compounds have the same number of carbon atoms per molecule and very nearly the same number of hydrogen atoms. When the fatty acid or the sugar burn, their carbon and hydrogen atoms combine with added oxygen to form CO_2 and H_2O. But more oxygen is required to burn a gram of lauric acid, $C_{12}H_{24}O_2$, than a gram of sucrose, $C_{12}H_{22}O_{11}$. This is evident from the equations for the two reactions.

$$C_{12}H_{24}O_2 + 17\ O_2 \longrightarrow 12\ CO_2 + 12\ H_2O \qquad (11.4)$$
$$C_{12}H_{22}O_{11} + 12\ O_2 \longrightarrow 12\ CO_2 + 11\ H_2O \qquad (11.5)$$

In the jargon of chemistry, the sugar is more "oxygenated" or more "oxidized" than the fatty acid. There are more C—H bonds in the fatty acid to "burn" to CO_2 and H_2O and

Each heartbeat uses one Joule of energy.

1 dietary Calorie = 1 Calorie = 1 kcal = 1000 calories

Energy released:
Carbohydrates = 4 Cal/g;
Fats = 9 Cal/g

Oxygenated fuels were discussed in Section 4.10.

release energy than there are in sucrose. Thus, sucrose is chemically and energetically "closer" to the end products of CO_2 and H_2O and needs less oxygen to form them. Therefore, when one gram of sucrose is burned, 3.8 Calories are released, compared to 8.8 Calories per gram of lauric acid.

The fact that fats are such concentrated energy sources means that it is easy to get an unhealthy percentage of our daily Calories from fats. The problem is nicely illustrated by 11.16 Sceptical Chymist and 11.17 Your Turn. Nutritionists and the American Heart Association advise that no more than 30% of your caloric intake should come from fats, and 55–60% should be derived from carbohydrates, especially polysaccharides. The remainder, 10% or less, should be contributed by protein. Although proteins, like carbohydrates, yield about 4 Calories per gram, they are not a major energy source, but rather a store of molecular parts for building skin, muscles, tendons, ligaments, blood, and enzymes.

11.16 The Sceptical Chymist: Low-Fat Cheese

A popular brand of low-fat shredded cheddar cheese advertises that it provides 1.5 g of fat with 15 Calories from total fat per serving. There are 50 Calories per serving and of the total fat, 1.0 g is saturated fat. A serving is defined as 1/4 cup or 28 g. Is this a "low-fat" cheese? Defend your decision with some calculations. Remember that the dietary recommendation is that no more than 30% of Calories should come from fat.

11.17 Your Turn

An alternative to potato chips is bagel chips. The label on one popular brand of bagel chips lists 130 Calories per serving, with 35 Calories from fat. The total fat is listed as 4 g per serving, with 1 g being saturated fat, 1 g being polyunsaturated fat, and 2 g being monounsaturated fat.

a. What percentage of daily value of calories based on a 2,000 Calorie diet is provided by one servings of these chips?
b. Potato chips have 150 Calories in a 28 g serving, 10 g of which is fat. Are bagel chips a healthier alternative to potato chips in your opinion? Give reasons for your answer.

11.9 Energy: How Much is Needed?

The current unhealthy level of overeating and obesity in the United States and the occasional reports of anorexia and bulimia prompt the question that serves as the title of this section. The answer is vague: "It depends." The number of Calories your diet should supply each day depends on your level of exercise or activity, the state of your health, your sex, age, body size, and a few other factors. You can probably identify yourself in Table 11.5, which summarizes the daily food energy intakes that have been recommended for Americans.

Some generalizations can be drawn from Table 11.5. Most men require more Calories per day than women do, but the difference is not totally due to differences in body weight. The column showing the recommended number of Calories per kilogram of body weight indicates that the magnitude of this indicator decreases with age. Growing children need a proportionally large energy intake to fuel their high level of activity and to provide raw material for building muscle and bone. Therefore, children are particularly susceptible to undernourishment and malnutrition. Mortality rates among infants and young children are disproportionately high in famine-stricken countries. It was reported in the early 1990s that the average energy intake in Somalia had dropped to 200 Calories daily, far below the required minimum for children or adults.

The minimum amount of energy required daily, the **basal metabolism rate or BMR,** is the amount necessary to support basic body functions—to keep the heart beating, the

Table 11.5 **Recommended Daily Energy Intake (United States)**

Age (yrs)	Avg. weight (kg)	Avg. weight (lb)	Avg. height (in)	Avg. Cal/kg	Avg. Cal/day
0.5–1.0	9	20	28	98	850
4–6	20	44	44	90	1800
7–10	28	62	52	70	2000
Males					
15–18	66	145	69	45	3000
19–24	72	160	70	40	2900
25–50	79	174	70	37	2900
51+	77	170	68	30	2300
Females					
15–18	55	120	64	40	2200
19–24	58	128	65	38	2200
25–50	63	138	64	36	2200
51+	65	143	63	30	1900

lungs inhaling and exhaling, the brain active, the blood circulating, all major organs working, and body temperature maintained at 37°C. This corresponds to approximately one Calorie per kilogram (2.2 lb) of body weight per hour, although it varies with size and age. The BMR is experimentally determined in a resting state, and the quantity of energy used in digestion is eliminated by having the subject fast for 12 hours before the measurement is made. To individualize this, consider a 20-year old female weighing 55 kg (121 lb). If her body has a minimum requirement of 1 Cal/(kg hr), her daily basal metabolism rate will be 1 Cal/(kg hr) $\times$ 55 kg $\times$ 24 hr/day or about 1300 Cal/day. According to Table 11.5, the recommended daily energy intake for a woman of this age and weight is 2200 Cal. This means that 59% of the energy derived from this food goes just to keep her body systems going.

Where the rest of it goes depends on what she does. The law of conservation of energy decrees that the energy must go somewhere. If she "burns off" the extra Calories in exercise and activity, none will be stored as added fat and glycogen. But if the excess energy is not expended, it will accumulate in chemical form. Putting it more crassly, "those who indulge, bulge," unless they work and play hard.

Some indication of how hard and how long we have to work or play to use up dietary Calories is given in Tables 11.6 and 11.7. The former reports the energy expenditures

Your basal metabolism rate is approximately $\frac{1 \text{ Cal}}{\text{kg hr}}$.

$\frac{1300 \text{ Cal}}{2200 \text{ Cal}} \times 100 = 59\%$

Table 11.6 **Energy Expenditure (Cal/min) for Various Activities in Relation to Body Mass (lbs)**

Activity (Cal/min)	120 lbs (Cal/min)	150 lbs (Cal/min)	180 lbs (Cal/min)	200 lbs
Aerobics (vigorous)	7	9	11	12
Basketball (vigorous)	10	13	15	17
Bicycling (11 mph)	6	7	9	10
Golf (carrying clubs)	5	6	7	8
Jogging (10 min/mile)	9	11	14	15
Rollerblading (12 mi/hr)	10	12	13	14
Running (7 min/mile)	12	15	18	20
Studying	1.3	1.7	1.9	2.0
Swimming (fast)	9	11	13	14
Volleyball	5	6	7	8
Walking (20 min/mile)	3	4	5	6

Table 11.7 How Much Exercise Must I Do If I Eat This Cookie? Calories and Minutes of Exercise for a 150-Pound Person

Food	Calories	Walk (min)	Run (min)
Apple	125	31	8
Beer (regular) 8 oz	100	25	7
Chocolate chip cookie	50	12	3
Hamburger	350	88	23
Ice cream, 4 oz	175	44	12
Pizza, cheese, 1 slice	180	44	12
Potato chips, 1 oz	108	27	7

for various activities as a function of body weight. Table 11.7 quantifies exercise in readily recognizable units such as beers, hamburgers, and potato chips.

11.18 Your Turn

First calculate your BMR. Then select from Table 11.6 the activities you do in a typical day. Calculate the supplemental energy you need for these activities. Then add your BMR and your supplemental energy needs to determine the total number of Calories you require per day. How does this result compare with the recommended energy intake for your age and sex (see Table 11.5)?

In many parts of the world, a major nutritional problem is not how to get rid of excess Calories, but how to get enough of them. This dietary discrepancy is evident from Table 11.8, which reports the average Dietary Energy Supplies (DES) in Calories per person per day for different global regions. Although the DES has generally increased since 1969, in each of the three time intervals listed the average DES for developed nations was significantly greater than that for developing countries. In 1986–88, the average daily individual energy intake in the developed countries was 44% higher than that in less industrially advanced nations. DES values for North America are conspicuously high—well above the recommended daily energy intake. Another way to compare the data of Table 11.8 is by calculating the DES in various parts of the world as a percentage of the North American level. Again for 1986–88, the DES for Africa was only 58% of that for North America, the Far East was 61% of the North American value, Latin America 75%, and the Near East 80%.

Table 11.8 Average Dietary Energy Supplies (DES) in Calories Per Person Per Day

Region	1969–71	1979–81	1986–88
Developed Countries (averages)	3186	3300	3389
North America	3371	3487	3626
Western Europe	3233	3371	3445
Developing Countries (averages)	2158	2317	2352
Africa	2046	2148	2119
Latin America	2514	2675	2732
Near East	2399	2794	2914
Far East	2049	2185	2220
World (averages)	2435	2587	2671

(Source: Data from *The State of Food and Agriculture,* 1990; Rome: *Food and Agriculture Organization of the United Nations,* 1991.)

11.19 Your Turn

A 150-lb person consumes a meal consisting of two hamburgers, 3 oz of potato chips, 8 oz of ice cream, and a 12-oz beer. Calculate the number of Calories in the meal and the number of minutes the person would have to run to "work off" the meal.

Ans.   1524 Cal, 102 min running (from Table 11.6) or 139 min jogging (from Table 11.6).

11.10 Vitamins: The Other Essentials

Your daily diet should supply an adequate number of Calories, but Calories alone are not enough. You have already read about the essential fatty acids and amino acids that must be ingested for good health, and you are well aware that a balanced diet must also provide certain vitamins and minerals. Unfortunately, many popular foods that are high in sugars and fats are lacking in these essential micronutrients. It is thus possible that a person can be overfed with excess Calories but malnourished through a diet lacking adequate vitamins and minerals.

A detailed understanding of the role of vitamins and minerals is of relatively recent origin. Over the ages, humans learned that if certain foods were lacking, illness often resulted, but the correlation between diet and health was often accidental and anecdotal. More systematic studies began early in this century, with the discovery of "Vitamine B_1" (thiamine). The particular designation, B_1, was the label on the test tube in which the sample was collected. The general term "vitamin" was chosen because the compound, which is *vit*al for life, is chemically classed as an *amine*. The final "e" disappeared with the discovery that not all vitamins are amines. Today **vitamins** are defined by their properties: they are essential in the diet, although required in very small amounts; they all are organic molecules with a wide range of physiological functions; and they generally are not used as a source of energy, although some of them help break down macronutrients.

Vitamins are often classified on the basis of solubilities; they either dissolve in water or in fat. Vitamins A, D, E, and K dissolve in fat, but not in water, because they are nonpolar molecules. For example, the molecular structure of vitamin A, shown below, is based almost exclusively on carbon and hydrogen atoms. Thus, it is similar to the hydrocarbons derived from petroleum. Vitamins that are not fat soluble are soluble in water because their polar molecules contain several —OH groups, which form hydrogen bonds with water molecules. Vitamin C is a case in point.

The relationship between molecular structure and solubility was discussed in Section 5.13.

Hydrogen bonds were discussed in Section 5.9.

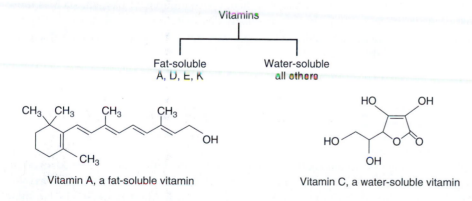

Vitamin A, a fat-soluble vitamin

Vitamin C, a water-soluble vitamin

11.20 Your Turn

Examine the molecular structures given for vitamin A and vitamin C. Use solubility and molecular structure relationships described in Section 5.13 under the general principle that "like dissolves like" to explain why vitamin A is fat soluble, not water soluble, and vitamin C has the opposite solubility behavior.

These solubility differences among vitamins have significant implications for nutrition and health. Because of their fat solubility, vitamins A, D, E, and K are stored in cells rich in lipids where they are available on biological demand. This means that the fat-soluble vitamins need not be taken daily. It also means that these vitamins can build up to toxic levels if taken far in excess of normal requirements. For example, high doses of vitamin A can result in fatigue, headache, dizziness, blurred vision, dry skin, nausea, and liver damage. Vitamin D toxicity occurs at just four to five times its RDA, making vitamin D the most toxic vitamin. Such high levels of the vitamin are reached using vitamin supplements, not through a normal diet. Cardiac and kidney damage can result. Water-soluble vitamins, by contrast, are not generally stored; any unused excess is excreted in urine. Thus, they must be consumed frequently and in small doses. Unfortunately, when taken in very large doses, water-soluble vitamins can also accumulate until they reach toxic levels, although such cases are rare. For example, there are reports that vitamin B_6, taken at 10 to 30 times the recommended dose per day for extended periods, results in nerve damage, including paralysis. Even higher doses of vitamin B_6 supplements, up to 1000 times the recommended dosage, have been consumed to alleviate the symptoms of pre-menstrual syndrome (PMS), again causing abnormal neurological symptoms. For most people, a balanced diet should provide all the necessary vitamins and minerals in appropriate amounts, making vitamin supplements unnecessary.

Even a brief review of various essential vitamins and minerals is well beyond the scope of this book, but a few observations might be of interest. For example, niacin or nicotinic acid illustrates the way in which vitamins, especially members of the B family, act as coenzymes. **Coenzymes** are generally small molecules that work in conjunction with enzymes to enhance the enzymes' activity. **Niacin** plays an essential role in energy transfer during glucose and fat metabolism. The synthesis of niacin in the body requires the essential amino acid tryptophan. Thus, a diet deficient in tryptophan may lead to niacin deficiency. Such a deficiency causes pellagra, a condition involving a darkening and flaking of the skin, as well as behavioral aberrations.

Vitamin C (ascorbic acid) must also be supplied in the diet, typically via citrus fruits and green vegetables. An insufficient supply of the vitamin leads to scurvy, a disease in which collagen, an important structural protein, is broken down. The link between citrus fruits and scurvy was discovered more than 200 years ago when it was found that feeding British sailors limes or lime juice on long sea voyages prevented the disease. Ascorbic acid is also required for the uptake, use, and storage of iron, important in the prevention of anemia. The claims that high doses of vitamin C can prevent colds and ward off certain cancers remain largely unsubstantiated.

The last vitamin in this brief overview is **vitamin E,** important in the maintenance of cell membranes and as protection against high concentrations of oxygen, such as those that occur in the lungs. Vitamin E is so widely distributed in foods that it is difficult to create a diet deficient in it, although people who eat very little fat may need supplements. Vitamin E deficiency in humans has been linked with nocturnal cramping in the calves and fibrocystic breast disease.

> This practice also led to British sailors being called "limeys."

11.11 Minerals: Macro and Micro

An adequate supply of a number of **minerals** (ions or inorganic compounds) is also essential for good health. Table 11.1 lists calcium, phosphorus, chlorine, potassium, sulfur, sodium, and magnesium among the major elements in the body. These seven **macrominerals,** although not nearly as abundant as oxygen, carbon, hydrogen, or nitrogen, are nevertheless necessary for life. The adult Recommended Daily Allowances (RDAs) for these macrominerals typically range from one to two grams. The body requires lesser amounts of iron, copper, zinc, and fluorine—the so-called **microminerals**. Trace minerals, including iodine, selenium, vanadium, chromium, manganese, cobalt, nickel, molybdenum, and tin are usually measured in micrograms (1×10^{-6} g). Arsenic, cadmium, and even lead, which are generally classified as toxic, are needed in very small amounts. Although the total amount of trace elements in the body is only about 25–30 grams, their slight amounts belie the disproportionate importance they have in good health.

A multivitamin type tablet can provide vitamins and minerals in addition to those found in a balanced diet.

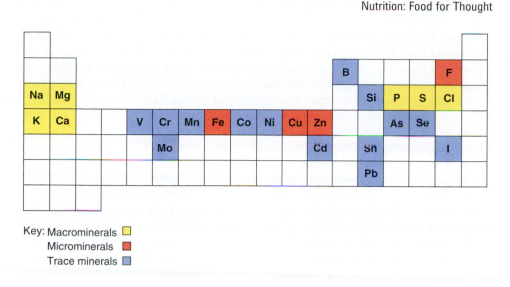

Key: Macrominerals ▢
     Microminerals ▢
     Trace minerals ▢

Figure 11.11
Periodic table indicating macrominerals, microminerals, and some trace minerals necessary for human life. Boron, silicon, arsenic, lead, and cadmium are essential in animals and likely essential in humans.

The essential minerals are highlighted in the periodic table outline that appears as Figure 11.11. The metallic elements exist in the body as cations, for example, Ca^{2+} (calcium), Mg^{2+} (magnesium), K^+ (potassium), and Na^+ (sodium). The nonmetals typically are present as anions, thus chlorine is found as Cl^- and phosphorus appears in the phosphate ion, PO_4^{3-}. The physiological functions of minerals are widely diverse.

Calcium is the most abundant mineral in the body. Along with phosphorus and smaller amounts of fluorine, it is a major constituent of bones and teeth. Blood clotting, muscle contraction, and transmission of nerve impulses also require Ca^{2+} ions.

Sodium is also essential for life, but not in the relatively excessive amounts supplied by the diets of most Americans. Physicians recommend a maximum of 1.2 grams of sodium (Na^+) per day. This corresponds to 3 grams of salt (NaCl) and is twice the estimated minimum requirement. Most Americans exceed the recommended daily sodium intake, sometimes by three to fourfold. The major culprits are processed foods and fast foods, which are very heavily salted for flavor. The major concern with excess dietary sodium is its correlation with high blood pressure (hypertension).

Oranges, bananas, tomatoes, and potatoes help supply the recommended daily requirement of 2 grams of **potassium**, a mineral that is essential for the transmission of nerve impulses and intracellular enzyme activity. Potassium and sodium are elements in the first column of the periodic table. Neutral atoms of the two elements each have one electron in the outer level. These electrons are readily lost to form K^+ and Na^+ ions. Because sodium and potassium ions have similar chemical properties, their physiological functions are also closely related. In intracellular fluid (the liquid within cells), the concentration of potassium ions is considerably greater than that of sodium ions. The reverse situation holds in the lymph and blood serum outside the cells. There the concentration of potassium ions is low and that of sodium ions is high. We have already noted the large amount of energy that must be expended to properly maintain these essential concentration gradients. The relative concentrations of K^+ and Na^+ are especially important for the rhythmic beating of the heart. Individuals who take diuretics to control high blood pressure commonly also take potassium supplements to replace potassium excreted in the urine. However, such supplements should be taken only under a physician's directions because of the potential danger that they could dramatically alter the potassium/sodium balance and lead to cardiac complications.

In most instances, the microminerals and trace elements have very specific biological functions and may be incorporated in only a relatively small number of biomolecules. **Iron**, for example, is an essential part of hemoglobin, the protein that transports oxygen in the blood, and of myoglobin, which is used for temporary oxygen storage in muscle. There are four Fe^{2+} ions in a hemoglobin molecule, and each of them binds reversibly to an O_2 molecule. Insufficient iron in the diet causes iron-deficiency anemia, a condition in which the red blood cells are low in hemoglobin and correspondingly can carry less oxygen. Symptoms include fatigue, listlessness, and decreased resistance to

Macrominerals are needed dietarily per day in amounts greater than 100 mg (0.100 g) or are present in the body in amounts greater than 0.01% of body weight.

The average U.S. daily diet contains about nine grams of NaCl (3.6 g Na^+). Some common dietary sources of sodium (mg Na^+ per serving): processed macaroni and cheese (1086); canned chicken noodle soup (1100); baked ham (950); hot dog (640); potato chips (200); white bread slice (114). Recall that 1 g = 1000 mg.

The Latin word for salt is *sal*. Salt was so highly valued in Roman times that soldiers were paid in *sal*, thereby forming the root for the modern word salary.

infection. Iron-deficiency anemia is a major problem in developed as well as developing nations. Iron deficiencies have been estimated as high as 20% in the United States, particularly among post-puberty women. On the other hand, too much iron in the diet can cause gastrointestinal distress and contribute to cirrhosis of the liver. Children have been fatally poisoned by ingesting iron supplement tablets. To be utilized by the body, iron must be absorbed as Fe^{2+} ions, not in the Fe^{3+} form or simply as elemental iron. Therefore, it is a bit surprising that some iron-fortified cereals contain metallic iron dust that can be removed by a magnet from a slurry of the cereal and water. Foods naturally rich in iron include liver and spinach.

Iodine is another element with a specific biological function. Most of the body's iodine is concentrated in the thyroid gland, where it is incorporated into thyroxine, a hormone that regulates basal metabolism rate. Excess thyroxine is associated with hyperthyroidism or Graves disease, in which basal metabolism is accelerated to an unhealthy level, rather like a racing engine. On the other hand, a thyroxine deficiency, sometimes caused by insufficient dietary iodine, slows metabolism and results in tiredness and listlessness. Both hyper-and hypothyroidism can lead to goiter, an enlargement of the thyroid gland. One way to help prevent goiter is by consuming adequate amounts of iodine. Seafood is a rich source of the element, but it is also provided by iodized salt, normal sodium chloride to which 0.02% of potassium iodide (KI) has been added. The tendency of the thyroid gland to concentrate iodine is key to the use of radioactive iodine-131 as a treatment for an overactive thyroid and to the risks of accidental exposure to this isotope.

It is not enough that a society have an adequate food supply to furnish the necessary vitamins, minerals, and macronutrients. To be useful, the food also must be kept uncontaminated until it is consumed. In the next section, we consider food preservation methods, including food irradiation.

11.12 Food Preservation

Over the centuries, people have used a variety of methods to preserve foods. Heavily salting foods or storing them in concentrated sugar syrups were the traditional methods used before modern refrigeration. These two methods create a salt or sugar concentration very much greater than that in any contaminating microorganisms such as bacteria, yeasts, and molds. At such high concentrations of salt or sugar, osmosis causes water to leave the cells of the microorganisms, killing them by rupturing their cell membranes. Heat is also used to kill microorganisms, as in home canning or pasteurization. Modern refrigeration retards, but does not ultimately prevent, spoilage.

To supplement these methods, substances are added to foods to reduce spoilage and extend useful shelf-life. As increasing numbers of consumers turn to packaged foods for meals rather than cooking "from scratch," adequate shelf-life of packaged products assumes greater importance. Such products carry warning labels like "Best if used by (date)" or ones that give a specific expiration date.

Anti-oxidants are one type of such additives, compounds that prevent packaged, processed foods from becoming rancid due to oxidation of oil or fats, which form harmful free radicals. If you examine the label of any such processed foods (dry cereals, potato chips and other so-called junk food snacks), you are likely to see the letters BHT or BHA. These stand for *butylated hydroxytoluene* and *butylated hydroxy anisole*, respectively, the two most common anti-oxidant food additives.

> Osmosis was discussed in Section 5.18.

> Refrigeration lowers the temperature so that the rates of reactions in offending microorganisms in foods are slowed, thus retarding spoilage.

Snack foods are preserved using BHT or BHA.

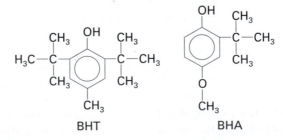

BHT   BHA

Anti-oxidants such as BHT and BHA act by preventing the build-up of free radicals, which are molecular fragments formed when fats and oils react with oxygen from the air in the food package.

$$\text{Fat (or oil)} + \text{oxygen} \longrightarrow \text{Free radical} \cdot + \text{other products}$$

A free radical has an unpaired electron (designated by a dot), which makes the species highly reactive. BHT, BHA, and other anti-oxidants scavenge the unpaired electron from the free radical to form a stable radical species. This prevents further oxidation of the fat, thus preventing rancidity.

A modern method of food preservation uses radiation. Food, of course, is irradiated in a number of circumstances, but not usually to preserve it. It's a matter of what part of the electromagnetic spectrum is used (Section 2.4). It should be rather obvious that radiation is essential for food production; visible light from the sun drives photosynthesis to give us fruits and vegetables. We use longer wavelength IR radiation from a stove, or even longer wavelength microwaves to cook food or warm up leftovers. Irradiating foods to preserve them is an entirely different case because it uses short-wavelength, high-energy gamma radiation to kill microorganisms. Such radiation is *ionizing* radiation, in contrast to that from visible, infrared, or microwave radiation, which are non-ionizing.

Classified as a food additive by Congress in 1958, food irradiation was approved by the FDA in 1963. It has been used to preserve food for astronauts to take with them as they circle the Earth. The method is used in more than 30 countries and is especially prevalent in Europe, Mexico, and Canada. It has the enthusiastic endorsement of the Food and Agricultural Organization of the United Nations. Irradiated foods even have their own international logo. Yet irradiated foods are controversial, especially in the United States. Why the controversy?

The irradiation procedure is relatively simple in the 160 such facilities worldwide. The material to be irradiated is placed on a conveyer belt that moves past a tight beam of high-energy gamma radiation generated by a cobalt-60 or cesium-137 source. The source and the irradiation facility are enclosed and shielded so that extraneous radiation does not escape. Over 40 different foods have been approved internationally for preservation by irradiation. Yet, only a small number of irradiated foods have been approved by the United States, including potatoes and strawberries for domestic consumption, and fish, shrimp, and grapefruit for export (Figure 11.12).

Those opposed to food irradiation question whether irradiated foods are safe to eat. The effectiveness as well as the need for this technology are questioned, including the proliferation of radioactive material for a possibly unneeded commercial application. Even so, critics and proponents agree that irradiating foods to preserve them does not make them radioactive. But the irradiated foods do have the normal background radiation that all foods naturally possess (Section 7.8).

The most serious charge brought by critics concerns the formation of possible "unique radiolytic products" (URPs) generated by gamma radiation breaking chemical bonds. Recall from Section 2.4 that with its short wavelength, gamma radiation has sufficient energy to break chemical bonds and create free radicals or ions. Gamma radiation has much more energy than microwave, infrared, visible, or even ultraviolet radiation. Because most foods contain a high percentage of water, the gamma radiation is absorbed by water to form irradiation products in extremely small amounts, which then react with other components of food to form stable products. It must be kept in mind that cooking food also causes chemical changes in the food, changes that are many times greater than those from gamma irradiation. Nearly five decades of research suggests that the byproducts of irradiation are the same chemical substances formed by conventional cooking or other preservation methods. Studies based on animal feeding research have repeatedly demonstrated no toxic effects from irradiated foods. The World Health Organization, the Food and Agricultural Organization of the United Nations, and the U.S. FDA all have concluded that food irradiation is safe when proper procedures and practices are used.

The need for food preservation is serious and should be kept in perspective. Worldwide, food spoilage and contamination is a significant problem, claiming up to 50% of

Free radicals were discussed in Sections 2.13 and 9.4 in relation to atmospheric pollutants and polymerization, respectively.

The international label for irradiated food.

Figure 11.12
Strawberries preserved by irradiation.

a food crop in some parts of the world, including many developing nations. Closer to home, we are not immune to such contamination in our food supply. Outbreaks of food poisoning in the United States occur periodically from chicken tainted with bacteria, commonly *Salmonella,* due to inadequate treatment in processing plants. Food contaminated with this organism has been linked to about 4000 deaths in the United States alone. The symptoms of food poisoning—abdominal pain, diarrhea, nausea, and vomiting—mimic those of short-term flu. Thus, food poisoning is often mistaken as being caused by flu. It is estimated that almost half of the raw chicken sold in the United States is contaminated with *Salmonella.* With proper handling and cooking, even chicken that is contaminated with bacteria can be prepared so as to prevent illness. Irradiation of chicken meat would lower the threat of accidental poisoning by *Salmonella.*

Food irradiation can be viewed in terms of risk-benefit; does the benefit outweigh the risk? Some would respond that irradiating strawberries to keep them fresh for a few days longer is a misuse or misapplication of a suspect technology. People are not likely to become ill or die from eating strawberries that are a bit past their peak. On the other hand, trichinosis, a serious disease, can occur by eating pork contaminated with the *Trichinella spiralis* parasite. Low-level gamma irradiation of pork kills the parasite, making the pork safe to eat. Although irradiated pork and chicken have been approved by the FDA, pork and chicken processing firms are wary that consumers will not buy these irradiated products. This is in spite of the fact that in countries where humans have consumed irradiated foods for years, including poultry and seafood, no adverse effects have been observed. Apparently the U.S. chicken processing companies feel that the risks (and costs) do not outweigh the benefits (at least to them).

Refrigerated chicken has a shelf-life of three days; after gamma irradiation, chicken can have a three-week refrigerated shelf-life.

11.21 Consider This: Food Irradiation...Thanks or No Thanks?

Food irradiation remains controversial. The Foundation for Food Irradiation Education, a web site under development in 1999, claimed that the web provides "a unique opportunity to communicate the facts about food irradiation to journalists, educators, food company executives and the general public." Indeed, the web can link a host of constituents with differing viewpoints on a topic such as food irradiation. Use the web to prepare a position paper on whether food should be irradiated. The paper can be written from the standpoint of a food company executive, a manufacturer of irradiation equipment, a government official, or a consumer activist. Be sure to cite your sources. Later, you may wish to join with others to stage a class debate about the issues involved.

11.13 Feeding A Hungry World

Even when food preservation methods are available, there must still be sufficient food produced to feed the population. To meet growing populations, global grain and cereal production has almost doubled over the past quarter century, and supplies of vegetables, fruits, milk, meat, and fish have also increased. It is estimated that at current production levels there is enough food to provide for 6.1 billion people—the projected world population at the turn of the twenty-first century. Nevertheless, many in the world go hungry; 500 million of our fellow human beings are undernourished, more than at any time in human history. Clearly, the world's food supply is not equally distributed among its inhabitants. Piles of corn and wheat rot in the American Midwest or on docks around the world, while half a billion men, women, and children go to bed hungry.

Economic, political, and social, as well as agricultural reasons account for this inequity. Some are geographic fates—certain areas of the world simply do not have enough arable land and adequate soil to produce sufficient food for their people. This is especially true in areas in which population growth is exploding. In some areas, pro-

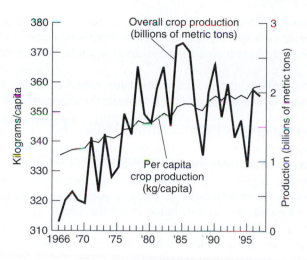

Figure 11.13

World grain harvests have risen, but because of population increases, per capita production has not changed significantly.

(Reprinted with permission from *Science*, Vol. 283, January 15, 1999. Source; FAOSTAT. Copyright © 1999 American Association for the Advancement of Science.)

longed droughts reduce crop yields, and episodic floods in other locales wash away crops. The fertilizers needed to supplement mineral-depleted soils may not be available because of costs, and beasts of burden, to say nothing of tractors, may be too expensive for farmers in some regions.

A nation that cannot feed itself must import food, which means that it must have something to sell. Consequently, economics and international trade dictate the nutrition equation. Under these conditions, some countries face the difficult choice between supporting domestic agriculture to achieve self-sufficiency or investing in manufacturing goods for export to establish a favorable trade balance. And, civil strife in some countries, created by political and military actions, blocks the flow of food and other agricultural products.

The disparity in food production is particularly great between developed and developing countries. The developed nations, including the United States and Canada, supply over 50% of the world's food, but have only 20% of its population. Food production per capita is one of the most meaningful ways of looking at the data, because it corrects for differences in population. In general, overall per capita food production has increased more rapidly in developed countries over the past two decades than it has in most developing countries, even though the latter started at much lower levels. One mitigating factor is the higher rate of population growth in developing regions (Figure 11.13) In many developing countries, food production has generally not kept pace with growing populations. The exception has been Asia, where the rate of increase in per capita food production has been greater than that even in the developed world. In stark contrast, Africa has experienced a long-term, continuing decline in per capita food production. One major reason for the increase in Asian crop yields has been greater use of fertilizers and pesticides. The application of both has often been criticized as being harmful to the environment, but the fact remains that millions have been saved from starvation, thanks to fertilizers and pesticides.

 Urea, $(NH_2)_2CO$, is a major fertilizer used worldwide. It provides nitrogen to soil by decomposing to ammonia and carbon dioxide when acted on by urease, an enzyme in soils. The ammonia is then taken up by plants.

$$(NH_2)_2CO + H_2O \xrightarrow{\text{urease}} 2\,NH_3 + CO_2 \qquad (11.6)$$

However, the efficiency of urea as a fertilizer is typically reduced because of the direct loss of ammonia through evaporation, in excess of 30%, before it can be taken up by plant roots. To overcome this inefficiency, the IMC-Agrico Company, a 1997 entrant in the Presidential Green Chemistry Challenge Awards Program, developed AGROTRAIN," a formulation containing a compound that is converted into a urease inhibitor. Spread on a field, the AGROTRAIN-linked product produces the urease inhibitor, which reduces the rate at which urease decomposes urea so that ammonia is released more

Enzymes and enzyme inhibition are discussed in Section 10.5.

slowly and efficiently. The higher efficiency is important, especially in no-till applications, an environmentally friendly approach where there is little or no disturbance of topsoil. This method reduces soil erosion and requires much less energy for application of the fertilizer.

An even more striking, and generally less controversial contribution to world agriculture has been the **Green Revolution.** A fundamental component of this enterprise has been the development of high-yield grains, principally wheat, rice, and corn, that were genetically modified to grow best in particular regions. These new varieties mature faster, permitting more harvests per year, so that the same amount of cultivated land can produce more crops. Since 1960, the Green Revolution has helped world grain harvests to more than double. Billions of people in India, Asia, and Africa have benefited from the practice. But the Green Revolution is neither a panacea nor the ultimate answer. In spite of its successes, the Green Revolution is not universally applicable, and has not been without costs. It works best in areas where water for irrigation is abundant, where money is available for supplemental fertilizers such as ammonia, urea, or nitrates, and where technological understanding and application exist.

Researchers estimate that within twenty years, global demand for the world's three most important crops—rice, maize (a type of corn), and wheat—will have increased by 40%, simply to keep pace with global food requirements. Genetic engineering and other applications of biotechnology now hold out promise for a second Green Revolution to meet such demands. In the next chapter we turn to more closely examine genetic engineering—its methods, accomplishments, and limitations.

Conclusion

This chapter began at the breakfast table with a taste for cereal and continued by considering why we eat cereal and many different kinds of foods. Even though our individual tastes vary, our biological needs are much the same. We need carbohydrates as our primary energy source; fats for cell walls, synthesis, and lubrication; proteins to build muscle and create the enzymes that catalyze the wonderful chemistry of life; and vitamins and minerals to help make that chemistry happen. People with too much to eat, like most Americans, seem preoccupied with food, although with too little regard for what they eat. The hungry and the starving think of little else beyond how to feed themselves. Chemistry is only part of the solution to one of the great challenges of our time— how to meet all individual dietary needs, regardless of region or wealth.

Chapter Summary

Having studied this chapter, you should be able to:
- Recognize the frequency and regional occurrence of malnutrition and undernourishment (11.1)
- Understand the physiological functions of food (11.1)
- Describe the distribution of water, carbohydrates, fats, and proteins in the human body and some typical foods (11.1)
- Identify the major elements found in the human body (11.1)
- Recognize and use the chemical composition and molecular structure of carbohydrates (11.2)
- Differentiate among the structures and properties of sugars, starch, and cellulose (11.2)
- Describe the symptoms and cause of lactose intolerance (11.2)
- Recognize and use the chemical composition and molecular structure of fats and oils or triglycerides (11.3)
- Identify sources of saturated and unsaturated fats and their significance in the diet (11.4)
- Differentiate among saturated, monounsaturated, and polyunsaturated fatty acids and fats (11.4)
- Discuss sources of cholesterol and its significance in the diet (11.5)
- Describe Olestra as a non-metabolizable fat; its benefits and costs (11.5)
- Give the general molecular structure of amino acids (11.6)
- Identify and use the chemical composition and molecular structure of proteins (11.6)
- Discuss the importance of essential amino acids and their dietary significance (11.7)
- Describe the symptoms and cause of phenylketonuria (11.7)

- Explain carbohydrates, fats, and proteins as energy sources (11.8)
- Discuss typical recommended daily energy intakes (11.9)
- Relate energy expenditures in various activities (11.9)
- Differentiate among international variations in dietary energy supplies (11.9)
- Discuss the effects of selected vitamins on human health (11.10)
- Describe the effects of selected minerals on human health (11.11)

- Identify and use basal metabolism rate (BMR) (11.9)
- Differentiate between fat-soluble and water-soluble vitamins (11.10)
- Dicuss the necessity of macrominerals, microminerals, and trace minerals for human health (11.11)
- Discuss various methods of food preservation, including the advantages and disadvantages of food irradiation (11.12)
- Describe various strategies for feeding the world's growing population (11.13)

Questions

Emphasizing Essentials

1. Food provides four fundamental types of materials to keep our bodies functioning. What are those types of materials?

2. Is it possible for a person to be malnourished even when eating a sufficient number of Calories every day to meet metabolic needs? Explain.

3. Use the information in Figure 11.1 to answer these questions for the time interval from 1990–92 to 1994–96.
   a. Which areas of the world showed a decrease in the proportion of the population that was undernourished?
   b. Which areas of the world showed an increase in the proportion of the population that was undernourished?
   c. Estimate the percentage decrease in the proportion of undernourished people in Latin America and the Caribbean.
   d. What additional information is needed to determine the change in the *number* of undernourished people in Latin America and the Caribbean during the same interval?

4. a. What are macronutrients and what role do they play in keeping us healthy?
   b. Name the three major classes of macronutrients.

5. Consider this chart.

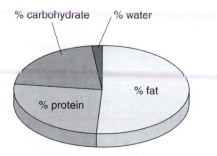

Based on the percentages of protein, carbohydrate, water, and fat given, is this graph more likely a representation of steak, peanut butter, or chocolate chip cookies? Justify your choice based on the relative percentages of the components shown.

6. Answer each of these questions about the common foods shown in Figure 11.2.
   a. Identify the top three foods that are good sources of carbohydrates and arrange them in order of decreasing percentage of carbohydrates.

   b. Identify the top three foods that are good sources of protein and arrange them in order of decreasing percentage of protein.
   c. Which of these foods should be avoided if you are controlling dietary intake of fat? Identify the top three and arrange them in order of decreasing percentage of fat.

7. Answer each of these questions about the common foods shown in Figure 11.2.
   a. Which food has the highest protein-to-fat ratio? What is that ratio?
   b. Which food has the highest fat-to-protein ratio? What is that ratio?

8. An 18-ounce steak is the manager's special at a local restaurant. Use the information in Figure 11.2 to calculate the number of ounces of protein, of fat, and of water that the customer eating this entire steak would consume.

9. Water is not considered as a macronutrient, but it clearly is essential in maintaining health. What are some of the roles that water plays in our bodies? *Hint:* You may want to refer to Chapter 5.

10. Examine the data in Table 11.1 and explain why hydrogen ranks first in atomic abundance in the human body, but third behind oxygen and carbon in terms of mass percent.

11. Use the information in Table 11.1 to answer these questions.
   a. What is the ratio of the relative abundance of potassium to sodium in the human body?
   b. What is the ratio of grams of potassium to grams of sodium in the human body?
   c. Are these elements included in the composition of the human body shown in Figure 11.3? Why or why not?

12. a. Consider the composition of the human body shown in Figure 11.3. What are the principal elements that make up water, proteins, carbohydrates, and fats? Which elements are in common among these major components of the human body?
   b. Compare your answers to part **a** with the relative abundance of the elements in the body given in Table 11.1. Is there a correlation? Explain the correlation between your lists and the relative abundances of the elements in the body.

13. This figure is a schematic diagram of the "food pyramid." What foods are in each section of the pyramid and how many servings a day should you consume?

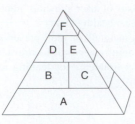

14. A large piece of sausage pizza would fall into several food groups. Identify each group and name the part of the pizza that is responsible for representing that particular food group.

15. Fructose, $C_6H_{12}O_6$, is a carbohydrate.
    a. Rewrite the formula for fructose to emphasize the original meaning of the term "carbohydrate."
    b. Write a structural formula for one of the isomers of fructose.
    c. Do you expect the different isomers of fructose to all have the same sweetness? Explain why or why not.

16. Fructose and glucose both have the formula, $C_6H_{12}O_6$. How do their structural formulas differ?

17. State what is meant by each term, and give an example of a substance that fits that term.
    a. monosaccharide   c. polysaccharide
    b. disaccharide

18. What problems can arise from regularly consuming excess dietary servings of carbohydrates?

19. Use the lock-and-key model discussed in Section 10.6 to offer a possible explanation why individuals who suffer from lactose intolerance can digest other sugars such as sucrose and maltose, but not lactose.

20. a. What are the similarities between fats and oils?
    b. What are the differences between fats and oils?

21. From the entries in Figure 11.7, identify the fat or oil with the highest percentage of each type of fat.
    a. polyunsaturated fat   c. total unsaturated fat
    b. monounsaturated fat   d. saturated fat

22. The label of a popular brand of soft margarine lists "partially hydrogenated" soybean oil as an ingredient. What does "partially hydrogenated" mean? Why doesn't the label simply say soybean oil, rather than partially hydrogenated soybean oil?

23. The text describes substitutes that have been developed for fat (Olestra) and for sugar (NutraSweet). Why have there not been attempts to develop a comparable substitute for protein?

24. What is the nutritional significance of the elements shaded on this periodic table?

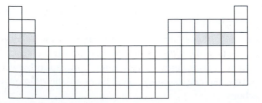

25. Why is it safer to take large doses of vitamin C than it is to take large doses of vitamin D?

Concentrating on Concepts

26. Explain to a friend why it is impossible to go on a highly advertised "all organic, chemical-free" diet.

27. Your friend wants to cut food costs and has learned that peanut butter is a good protein source. What additional information should your friend consider before making the decision to make peanut butter the major dietary protein source? *Hint:* There is relevant information in Figure 11.2.

28. a. What percentage of the total number of elements in the periodic table are utilized by the human body to produce proteins, carbohydrates, and fats?
    b. What relationship is there between the type of bonds these elements can form and what makes them so prevalent in the human body?

29. When the USDA made the decision to change dietary recommendations, they also changed the way the information was visually displayed to consumers. Rather than using a restructured pie chart, the food pyramid was introduced. Is the food pyramid a better communication tool for dietary recommendations than the pie chart used previously? Why or why not?

30. According to one USDA study, nearly 40% of the food that the average American eats each day consists of milk or dairy products. Would such a diet be possible and still meet the guidelines of the food pyramid?

31. In people who exhibit lactose intolerance, lactase, the enzyme that normally catalyzes the lactose breakdown, is either missing or is present at levels too low to support normal enzymatic activity. How does this inability to break down lactose parallel our ability to metabolize starch, but not cellulose?

32. For each statement, indicate whether it is always true, may be true, or cannot be true. Justify your answers by explaining your reasoning.
    a. Plant oils are lower in saturated fat than are animal fats.
    b. Lard is more healthful than butterfat.
    c. There is no need to include fats in our diets because our bodies can manufacture fats from other substances we eat.

33. Experimental evidence suggests that some physiological effects of saturated fats, compared to unsaturated fats, may be caused by differences in folding or wrapping of the molecules. The hydrocarbon chains in saturated fatty acids can fold or wrap more tightly than those of unsaturated or polyunsaturated fatty acids.
    a. Explain why saturated fatty acid molecules are able to fold more tightly than molecules of unsaturated or polyunsaturated fatty acids. *Hint:* If you have a model set available, make suitable molecular models to help you see the effect single or double bonds can have on the ease of folding.
    b. Explain why the extent of molecular folding influences the melting points of stearic, oleic, linoleic, and linolenic acids. See Table 11.2 for melting point values.

34. Some people prefer to use non-dairy creamer rather than real cream or milk. Some, but not all non-dairy creamers, use coconut oil derivatives to replace the butterfat in cream. Should a person trying to reduce dietary saturated fats by using non-dairy creamer use non-dairy creamers such as these? Why or why not?

35. The reaction of free radicals and oxidizing agents with unsaturated and polyunsaturated fats in the body has been suggested as a cause of premature aging. What is the chemical basis for this assertion? *Hint:* You might find it helpful to consider the mechanism of addition polymerization in Chapter 9.

36. Why is it more difficult for a person to control her or his cholesterol level than to control her or his fat intake? What steps are effective in minimizing cholesterol in the blood?

37. a. Which are the "good" lipoproteins — LDL or HDL?

   b. What function do the "good" lipoproteins perform?

   c. If a person has an LDL reading of 100 mg/100 mL and an HDL reading of 150 mg/100 mL, does this person meet current guidelines from the AHA?

38. Consider the structure for riboflavin, one of the B vitamins found in leafy green vegetables, milk, and eggs.

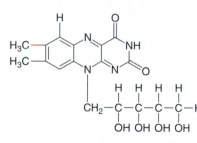

Why is it safer to take large doses of vitamin B than it is to take large doses of vitamin D?

39. American diets depend heavily on bread and other wheat products. A slice of whole wheat bread (36 g) contains approximately 1.5 g of fat (with 0 g saturated fat), 17 g of carbohydrate (with about 1 g of sugar), and 3 g of protein.

   a. Calculate the total calorie content in a slice of this bread.

   b. Calculate the percent calories from fat.

   c. Do you consider bread a highly nutritious food? Explain your reasoning.

40. What is your opinion about food preservation by irradiation? Are there some cases in which you feel irradiation is justified as a way to ensure better quality food to the consumer? Explain your position and be prepared to defend it.

Exploring Extensions

41. 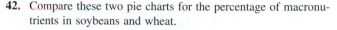 Figure 11.1 pictures the proportion of undernourished people in different areas of the world, and how that proportion has changed during the 1990s. The figure does not show how these trends have changed over a longer period, or how the total *numbers* of undernourished people have changed with time. Focus on any one region of the world, and find the necessary information to speak to these two points. Then devise a visual way to represent these data.

42. Compare these two pie charts for the percentage of macronutrients in soybeans and wheat.

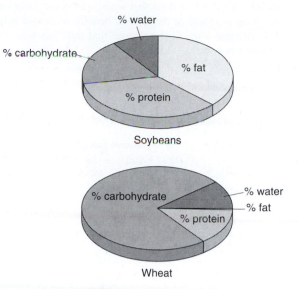

   a. Use these charts to help explain why the World Health Organization has helped develop several soy-based, rather than wheat-based, food products for distribution in parts of the world where protein deficiency is a major problem.

   b. Suggest some cultural reasons why soy might be preferable to wheat for some areas of the world.

43. The Sceptical Chymist finds the statement that the composition of the human body is " . . . roughly similar to the stuff we stuff into it" an idea hard to believe, but is willing to try to justify this statement, at least for the macronutrients. Compare the information found in Figures 11.2 and 11.3. Does it give you adequate information to decide whether the " . . . roughly similar to the stuff we stuff into it" statement is reasonable, assuming you eat only the foods shown in Figure 11.2? Why or why not?

44. How does the elemental composition of the human body compare to the elemental composition of the Earth's crust? How does the human body's composition compare to the elemental composition of the universe? Table 11.1 gives the values for the human body. Research the composition of the Earth's crust and that of the universe, listing your references. Then comment on the comparative values for the first five elements listed in order of mass abundance in each of the three circumstances—the human body, the Earth's crust, and the universe.

45. The food guide pyramid gives a range of servings for each food group. What factors do you think determine the number of servings you should eat?

46. The food guide pyramid gives a range of servings for each food group. To use this information, the consumer must know what constitutes a reasonable serving size. Investigate what constitutes reasonable serving sizes for one of the food groups, and then prepare a poster with your results to share with others who are investigating the reasonable serving sizes for other food groups. Were you surprised by any of the serving sizes? Which ones?

47. ◈ Not everyone considers milk as nature's "perfect food." Compare and contrast the viewpoints of the dairy industry with activist coalitions that work against the dairy industry. What are some of the specific benefits attributed to milk, and what are some of the reasons that milk has been called "nature's not-so perfect food"?

48. Here is the label information from a popular brand of canned chicken noodle soup.

Serving Size: 1/2 cup (4 oz; 120 g)
Servings per container: about 2.5
Amount per serving
Calories 75   Calories from Fat 25
                %Daily Value

Total Fat 2.5 g	4%
Saturated Fat 1.5 g	8%
Cholesterol 20 mg	7%
Sodium 970 mg	40%
Total Carbohydrates 9 g	3%
Dietary Fiber 1 g	4%
Sugars 1 g	
Protein 4 g	
Vitamin A	15%
Vitamin C	2%
Calcium	2%
Iron	4%

a. Analyze this information to see if the soup conforms to dietary recommendations of the AHA.

b. Is the serving size recommended on the label adequate? Explain.

c. What effect would changing the serving size have on your answer to part **a**?

49. ◈ Examine Figure 11.13, which gives the world grain harvests from 1966 to 1997.

a. Write a paragraph summarizing the information displayed by the graph.

b. Has this information changed since the text version of the graph? Why or why not?

50. Every month, a certain consumer-advocate organization presents an "Unnatural Living Award" to a person, product, or institution that demonstrates an unnatural ability to provide an unnatural product to the American people. What are the criteria by which you would make your nomination for this award? What do you consider would be a good candidate to receive this award? Explain your reasons for suggesting this candidate.

Genetic Engineering and the Chemistry of Heredity

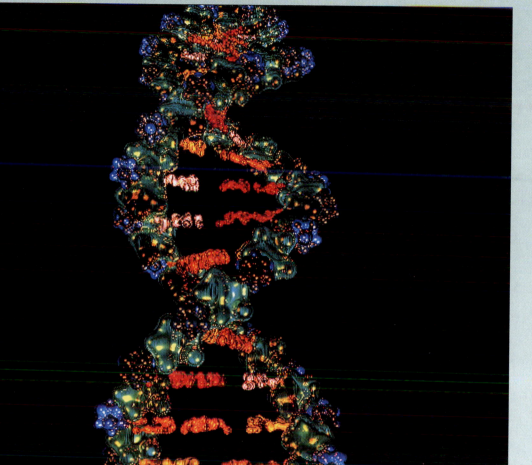

This is an artistic stylized illustration of DNA (deoxyribonucleic acid), the so-called "molecule of life." The DNA molecule is a double-stranded helix of deoxyribose and phosphate units (green and blue) bridged by nitrogen-containing bases (orange and yellow) in particular arrangements

SCIENTIST REPORTS FIRST CLONING EVER OF ADULT MAMMAL
FEAT IS SHOCK TO EXPERTS
In Creating Lamb, Researcher Sees Benefits for Medicine, but Others Fear Abuse

In a feat that may be the one bit of genetic engineering that has been antici-pated and dreaded more than any other, researchers in Britain are reporting that they have cloned an adult mammal for the first time.

The group, led by Dr. Ian Wilmut, a 52-year old embryologist at the Roslin In-stitute in Edinburgh, created a lamb using DNA from an adult sheep . . .

HEADLINE AND BEGIN-NING OF FIRST-PAGE STORY FROM FEBRUARY 23, 1997 *NEW YORK TIMES*

"What this [the technique used to create Dolly] will mostly be used for is to produce more health care products. It will enable us to study genetic diseases for which there is presently no cure and track down the mechanisms that are involved. The next step is to use the cells in culture in the lab and target genetic changes into that culture . . . I am not actually sure it is such an incredible breakthrough [cloning mammals from adult cells] but there is this aura about it that makes people jump."

Dr. Ian Wilmut, Roslin Institute researcher, as reported by Reuters News Service

"It's unbelievable. It basically means there are no limits. It means all of science fiction is true. They said it could never be done and now here it is, done before the year 2000."

Dr. Lee Silver, Princeton University biologist *New York Times* February 23, 1997

Dolly at seven months, the first mammal cloned from an adult cell.

An un-nucleated cell is one whose nucleus has been removed.

All three of the references above are about the birth of Dolly, a lamb cloned from the cell of an *adult* ewe. Using genetic engineering techniques, the DNA of one sheep's mammary tissue was transferred to an unfertilized, un-nucleated egg of another sheep. The fused new cell was implanted into a sheep's uterus and developed into Dolly. For some, this breakthrough exemplifies cutting edge research that opens up a grand new era of agricultural and pharmaceutical developments. Others see a more sinister side, a Pandora's box that should never have been opened.

To understand cloning and how Dolly came to be requires learning about DNA: three of the most important letters of the late twentieth century. Those letters have been in the news with growing frequency over the past decade. Even before Dolly, public awareness of DNA rose through the O. J. Simpson murder trial, and by being mentioned regularly in connection with new sources of pharmaceutical drugs, the diagnosis and treatment of a wide range of diseases, and prospects for the creation of new life forms. Manipulating the molecules of heredity is an activity filled with potential and fraught with peril. It is thus appropriate that this book should end with a chapter devoted to the chemistry of heredity and genetic engineering.

Chapter Overview

During the past 50 years, biology has been transformed by the application of chemical knowledge and methodology. In this chapter we first examine our understanding of how genetic information is transmitted and used. We begin with an introduction to the molecular basis of heredity—deoxyribonucleic acid (DNA) and its component chemical parts. These parts—four nitrogen-containing bases, a sugar (deoxyribose), and phosphate groups—are combined into a double helix. That structure and the story of its remarkable discovery are recounted next. Thanks to superb chemical cryptographers, DNA has been decoded. Today we know the molecular code in which genetic instructions are written. This code directs the synthesis of proteins and determines the sequence in which the constituent amino acids are combined.

The remainder of the chapter addresses some of the many applications of genetic engineering. First you will encounter the recombinant DNA techniques that have made it possible to use bacteria to produce proteins such as human insulin and growth hormones. A whole generation of new drugs and vaccines has also been created through molecular manipulation. New methods of diagnosing diseases are presented next, followed by a discussion of gene therapy, in which normal genes are introduced into patients lacking them. It is now possible to take a unique genetic fingerprint of any one of us. Such information is valuable in solving crimes, identifying human remains, and constructing genetic family trees. Some very old DNA has already been isolated and studied, raising the question of whether a Dino-Disneyland is possible. Some unusual inter-species genetic combinations also surface. We revisit the making of Dolly and issues pertaining to that achievement. Then we devote some attention to the massive project to map all the genes in the human species. The chapter ends with a hope and a warning as we look to a future filled with the benefits and risks of cloning and genetic engineering.

12.1 The Chemistry of Heredity

The human body is the world's most complicated chemical "factory." Thousands of chemical reactions involving an even greater number of chemicals occur each second. Some compounds are decomposed and others are synthesized; energy is released, transformed, and used; and chemical signals are transferred and processed. But in spite of the dazzling complexity of these processes, the last half century has seen a phenomenal increase in our knowledge of the chemistry of life. Indeed, in many respects, biology has become a chemical science. Today, much biological research is focused on molecules, not cells or organisms. This research has lead to an understanding, at the molecular level, of the very basis of life itself.

The practical manifestations of this intellectual achievement are manifold. In 1900, average life expectancy at birth in the United States was under 50 years; today it is a bit over 76 (Figure 12.1). Reasons for this dramatic increase are many: better nutrition, improved sanitation, advances in public health, more accurate medical diagnoses, new medical procedures, and numerous new medicines, drugs, and vaccines. Chemistry has contributed to all of these innovations, and it is an integral part of the latest revolution in health care—biotechnology and molecular engineering. There seems little doubt that genetic engineering will profoundly affect human life in the twenty-first century.

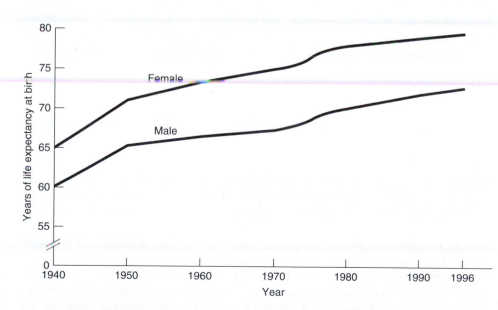

Figure 12.1

Average life expectancy at birth in the United States since 1940.

(Source: Data from Centers for Disease Control and Prevention, National Center for Health Statistics, 1996.)

Some of the most significant advances in biochemistry have involved our rapidly growing knowledge of the molecular basis of heredity. To put this into a personal perspective, you contain about 10 million million (10×10^{12}) cells that have nuclei. Each of these cell nuclei contains a complete set of the genetic instructions that make you what you are—at least biologically. The genetic instruction is organized in 23 pairs of chromosomes and approximately 100,000 genes, each of which conveys one or more hereditary traits. This is the **human genome,** the totality of human hereditary information in molecular form. This information is uniquely yours unless you happen to have an identical twin, as does one of the authors of this book. (We think we're working with Conrad, but it might be Carl.)

Human blood cells do not have nuclei.

Your special book of life is written in a molecular code on a tightly coiled thread, one invisible to the unaided eye. This thread is **deoxyribonucleic acid (DNA),** the molecule that carries genetic information in all species. Unraveled, the DNA in *each* of your cells is about two meters (roughly two yards) long. If all of the DNA in all 10 million million of your cells were placed end to end, the resulting ribbon would stretch from here to the Sun and back more than 60 times! But, you will soon discover that this astronomical figure is far from the most astounding feature of this amazing molecule.

It is approximately 93 million miles from Earth to the Sun.

12.2 The Sceptical Chymist: Stretching DNA

Sometimes authors get carried away with their rhetoric. Check the correctness of the claim that the DNA in an adult human being would stretch from the Earth to the Sun over 60 times. The other necessary information is in the paragraphs above and the unit conversion factors in Appendix 1.

The molecular structure of deoxyribonucleic acid dictates the way DNA encodes genetic information. A strand of DNA consists of three types of fundamental chemical units, repeated thousands of times. The units are **nitrogen-containing bases**, the sugar **deoxyribose**, and **phosphate groups**. All are illustrated in Figure 12.2. Two of the bases, adenine (symbolized A) and guanine (G), are built on a six-membered ring fused to a five-membered ring. Carbon and nitrogen atoms make up the rings. Cytosine (C) and thymine (T) each contain six-membered molecular rings consisting of four carbon atoms and two nitrogen atoms. These four compounds are bases because they react with water to form basic solutions. H^+ ions are transferred from H_2O molecules to nitrogen atoms of the nitrogen-containing bases, creating OH^- ions in solution, forming a basic (alkaline) solution.

H^+ and OH^- ions are discussed in Sections 6.1 and 6.2.

$$H_2O + \text{N-base}(aq) \longrightarrow {}^+\text{HN-base}(aq) + OH^-(aq) \qquad (12.1)$$

Deoxyribose is a monosaccharide (a "single" sugar) with the formula $C_5H_{10}O_4$. Figure 12.2 reveals that the deoxyribose molecule is a five-membered ring formed by four carbon atoms and one oxygen atom. The phosphate group can be represented as PO_4^{3-}, but, depending on the pH, an H^+ ion can be attached to one or more of the O^-s. The ultimate situation is H_3PO_4 or phosphoric acid. It is, in fact, the ionizable hydrogen atoms on the phosphate groups that make nucleic acids acidic.

A **nucleotide** is a combination of a base, a deoxyribose molecule, and a phosphate group. Figure 12.3 indicates how these units are linked in a nucleotide called adenosine phosphate. A covalent bond exists between one of the ring nitrogen atoms of the

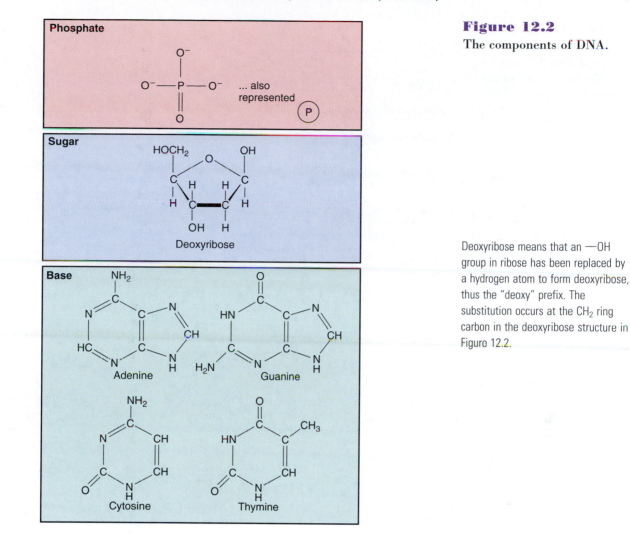

Figure 12.2
The components of DNA.

Deoxyribose means that an —OH group in ribose has been replaced by a hydrogen atom to form deoxyribose, thus the "deoxy" prefix. The substitution occurs at the CH₂ ring carbon in the deoxyribose structure in Figure 12.2.

adenine molecule and one of the ring carbons in deoxyribose. Another covalent bond connects the deoxyribose sugar molecule to the phosphate group. The other three bases form similar nucleotides of deoxyribose joined covalently to a base and to a phosphate group. Although there are other possible molecular sites for linking the base, sugar, and phosphate units, the arrangement pictured in Figure 12.3 is found in DNA.

A typical DNA molecule consists of thousands of nucleotides covalently bonded in a long chain. Consequently, a segment of DNA may have a molecular mass in the millions. The phosphate groups link the individual nucleotides. Note that in Figure 12.3, one —OH group on the deoxyribose ring remains unreacted. The phosphate group of

Sugars are discussed in Section 11.2.

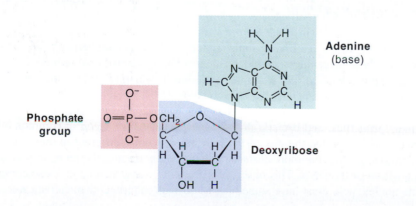

Figure 12.3
Molecular structure of a nucleotide, adenine phosphate. Adenine, a nitrogen base, is highlighted in green, the deoxyribose molecule is highlighted in blue, and the phosphate group is highlighted in pink.

Figure 12.4

The molecular structure of a segment of deoxyribonucleic acid. The bases present are guanine (G), cytosine (C), adenine (A), and thymine (T).

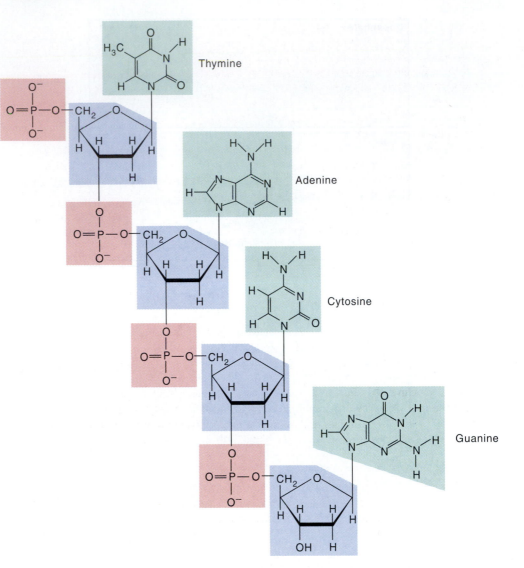

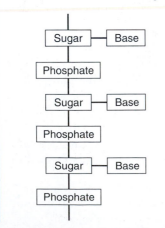

If you studied Chapter 9 you will recognize DNA as a polymer of nucleotide monomers.

A schematic of the —sugar— phosphate—sugar—phosphate— chain with bases linked to the sugar.

another nucleotide can react with this —OH, eliminating an H_2O molecule and connecting the two nucleotides, as shown in Figure 12.4 for four nucleotides linked in this manner to form a segment of DNA. This alternating chain of deoxyribose-phosphate-deoxyribose-phosphate units runs the length of the nucleic acid molecule, like the vertical rails of a ladder. Attached to each of the deoxyribose rings is one of the four possible bases.

The specific bases and their sequence in a strand of DNA turn out to have great significance. Some of the early clues to the structure of DNA and the mechanism by which it conveys genetic information came as a result of the research of Erwin Chargaff in the 1940s and 1950s. Chargaff and his co-workers were able to determine the percentage of the four bases present in DNA from a variety of species. They found that the relative amounts of the bases in a DNA sample are identical for all members of the same species. Moreover, these percentages are independent of the age, nutritional state, or environment of the organism studied. For example, according to Chargaff's data, the DNA from all members of our species *Homo sapiens* contains 31.0% adenine, 31.5% thymine, 19.1% guanine, and 18.4% cytosine. Table 12.1 also contains such findings for other species. Humans, fruit flies, and bacteria do not seem to have very much in common, and it is perhaps reassuring that the mix of the four bases is quite different in the three species. But it turns out that the more closely related the species are, the more similar the base composition of the DNA. This observation certainly suggests that the base composition of the nucleic acid must have something to do with inherited characteristics.

Table 12.1 The Base Compositions of DNA for Various Species

Species	Adenine	Thymine	Guanine	Cytosine
Homo sapiens (human)	31.0	31.5	19.1	18.4
Drosophila melanogaster (fruit fly)	27.3	27.6	22.5	22.5
Zea mays (corn)	25.6	25.3	24.5	24.6
Neurospora crassa (mold)	23.0	23.3	27.1	26.6
Escherichia coli (bacterium)	24.6	24.3	25.5	25.6
Bacillus subtilis (bacterium)	28.4	29.0	21.0	21.6

Note that the percentages of adenine and thymine are consistently similar, as are the percentages of cytosine and guanine.

(From I. Edward Alcamo, *DNA Technology: The Awesome Skill.* Copyright © 1996 The McGraw-Hill Companies, Inc. All Rights Reserved. Reprinted by permission.)

A more careful examination of Table 12.1 discloses that the DNA of *Homo sapiens* and the DNA of *Escherichia coli* (*E. coli*), a form of bacteria that inhabits the human intestine, do exhibit a very important common characteristic. They obey the same compositional regularity, now called Chargaff's rules. *In every species, the percent of adenine almost exactly equals the percent of thymine. Similarly, the percent of guanine is essentially identical to the percent of cytosine.* Put more simply: *A = T and G = C.* Such a correlation can hardly be coincidental; it must be based on biochemical form and function at the molecular level. As soon as Chargaff's rules were announced, the conclusion seemed obvious: the nitrogen-containing DNA bases somehow come in pairs. Adenine always appears to be associated with thymine, and guanine is consistently matched with cytosine.

12.2 The Double Helix of DNA

What was not so obvious was how the paired bases were part of the overall molecular structure of DNA. Therefore, scientists set out to determine the way in which nucleotides are incorporated into the DNA molecule. The most fruitful experimental strategy was X-ray diffraction, a technique that had been known since early in the twentieth century. In **X-ray diffraction,** a beam of X-rays is directed at a crystal. The X-rays strike the atoms in the crystal, interact with their electrons, and bounce off the atoms. Stated a bit more precisely, the X-rays are diffracted or scattered by the atoms. The crucial point is that the X-rays are only scattered at certain angles, which are related to the distance between atoms (Figure 12.5).

In the instruments used during the 1950s, the scattered X-ray beams struck and exposed photographic film. The resulting spots correspond to the angles of diffraction, and they represent a two-dimensional map of a three-dimensional structure. Figure 12.5 is such a map; it is the X-ray diffraction pattern of a DNA fiber obtained in late 1952 by the British crystallographer Rosalind Franklin. To the uninitiated, the photograph does not appear to contain much useful information, but the correct interpretation of these spots led to the determination of the structure of DNA.

The scientists responsible for this revolution were James D. Watson, a 24-year-old American, and Francis H.C. Crick, a loquacious and supremely self-confident Cambridge University biophysicist. Watson and Crick concluded that the X-shaped pattern in Franklin's diffraction photograph was consistent with a repeating helical arrangement of atoms, similar to a loosely coiled spring. Moreover, the spacing of the large smudges at the top and bottom of the figure was evidence of a regular repeat distance of 0.34 nanometers (1 nm = 1×10^{-9} m) within a DNA molecule.

With these clues, Crick and Watson set out to combine the molecular pieces. They did so on a large scale, using the highly accurate metal scale models shown in Figure 12.6. After a variety of attempts, they finally came up with a structure that agreed with

X-rays have short wavelengths and high energy (Section 2.4).

Rosalind Franklin's work contributed significantly to breaking the DNA code.

Figure 12.5

Photograph of the X-ray diffraction pattern of a fiber of DNA, obtained in 1952 by Rosalind Franklin. The X-pattern suggests a helical structure and the spacing of the large smudges at the top and bottom of the photograph correspond to a regular spacing of 0.34 nanometers.

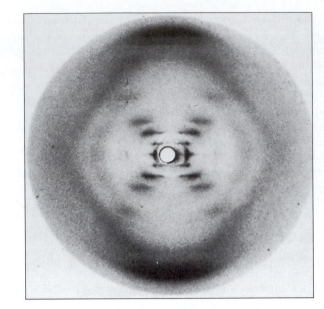

Hydrogen bonds, which help to determine many of the special properties of water, are discussed in Section 5.9.

the data. A major breakthrough was the recognition that adenine and thymine molecules fit together almost perfectly, like pieces in a jigsaw puzzle. Moreover, these two bases can be connected with two hydrogen bonds (Figure 12.7). Similarly, cytosine and guanine align by forming three hydrogen bonds. Adenine and thymine are said to be **complementary**, as are cytosine and guanine. This base pairing is the molecular basis underlying Chargaff's rules: A = T and C = G.

In the model of DNA developed by Watson and Crick, the hydrogen bonds between the complementary bases help hold together two strands of a double helix. Perhaps an even better metaphor is a spiral staircase. The steps of this molecular staircase are the bases, always paired A with T and C with G. One of the members of each base pair belongs to one strand of a helix, the other to the matching complementary helical strand, thus creating a double helix. Recall that the bases are connected to the deoxyribose rings, which in turn are linked by phosphate groups. Thus, the deoxyribose and phosphate units are in effect the stair rails to which the steps are attached. The Anglo-American team concluded that the base pairs are parallel to each other, perpendicular to the

Figure 12.6

James Watson (left) and Francis Crick (right) demonstrating their model of DNA in 1952.

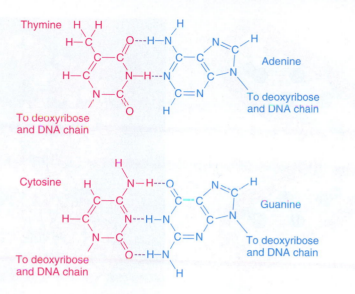

Figure 12.7
Base pairing of adenine with thymine and cytosine with guanine in DNA. Hydrogen bonds are indicated as dashed lines.

axis of the DNA fiber, and separated by 0.34 nm, the repeat distance calculated from the diffraction pattern. In addition, Franklin's results also suggested another repeat distance of 3.4 nm. Watson and Crick took this to be the length of a complete helical turn consisting of ten base pairs.

12.3 Your Turn

Identify the base sequences that are complementary to each of these sequences.

a. ATACCTGC
b. GATCCTA

Ans. **a.** TATGGACG; **b.** CTAGGAT

12.4 Your Turn

The distance between bases in a molecule of DNA is 0.34 nm.

a. Calculate the length (in centimeters) of the shortest human chromosome, which consists of 50,000,000 base pairs.
b. Mark off that length on your paper. If this is the length of the unstretched DNA molecule, what does this imply about the organization of DNA in the chromosome?

Ans.

a. $\dfrac{0.34 \text{ nm}}{\text{pair}} \times \dfrac{1 \text{ m}}{10^9 \text{ nm}} \times \dfrac{10^2 \text{ cm}}{1 \text{ m}} \times \dfrac{50,000,000 \text{ pairs}}{\text{chromosome}} = 1.7$ cm per chromosome

b. This is a very small distance, about two-thirds of an inch. The only way that this many base pairs could fit into such a small space is to have them tightly packed in a spiral.

The overall structure of DNA is represented by Figure 12.8. The space-filling model in which individual atoms are represented by spheres (Figure 12.8b) is too complicated to be of much help. In the simplified drawing on the left, the alternating sugar (S) and phosphate (P) groups are represented by two twisting ribbons. The four bases are attached to this backbone and paired in the A to T and C to G fashion described earlier.

Figure 12.8

The molecular structure of DNA. (*a*) A schematic representation in which
P = phosphate group,
S = sugar,
A = adenine,
C = cytosine,
G = guanine, and
T = thymine.

(*b*) A space-filling model in which atoms are represented by spheres.

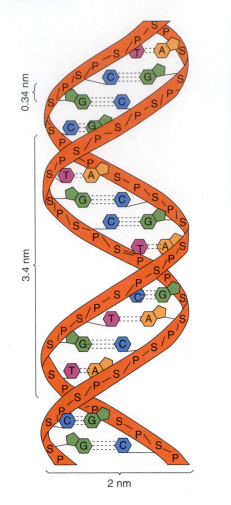

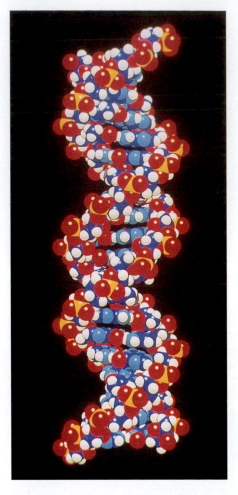

12.5 Your Turn

The DNA in each human cell consists of three billion base pairs. Calculate the length of this DNA. Does this length agree with that given in Section 12.1? Why or why not?

12.6 Your Turn

The structural features of a DNA molecule are more apparent if you can look at a three-dimensional (3-D) representation. At the *Chemistry in Context* website, with the help of Chime, you can view and rotate several different molecular structures.

a. How is a 3-D structure of DNA shown at that site similar to the one shown in your text? How is it different?
b. Look carefully at the Chime structures of the four bases that compose DNA. What are their structural common features? How do they differ?
c. Look again at the large DNA molecule. Can you spot the bases? How are they aligned?

Watson and Crick's research paper, "Molecular Structure of Nucleic Acids: A Structure for Deoxyribose Nucleic Acid," appeared in *Nature* on April 24, 1953. It is only one page long, but it is undoubtedly the most important paper to appear in that prestigious journal since the announcement of nuclear fission 15 years earlier by Meitner and Frisch in an equally short communication. The Watson-Crick paper is written with the customary passionless detachment of contemporary scientific prose. Even the most significant statement in the communication is delivered with typical British understatement: "It has not escaped our notice that the specific pairing we have postulated immediately suggests a possible copying mechanism for the genetic material."

The history of science is full of examples of how a single discovery can release a flood of related research. So it was with the discovery of the structure of DNA. Scientists immediately set out to discover the molecular details of how DNA is replicated, how it encodes genetic information, and how that information is translated into physiological characteristics. **Replication**, the process by which copies of DNA are made, is now well understood, and it is diagrammed in Figure 12.9.

Before a cell divides, the double helix partially unwinds at a rate of about 10,000 turns per minute. This results in a region of separated complementary single strands of

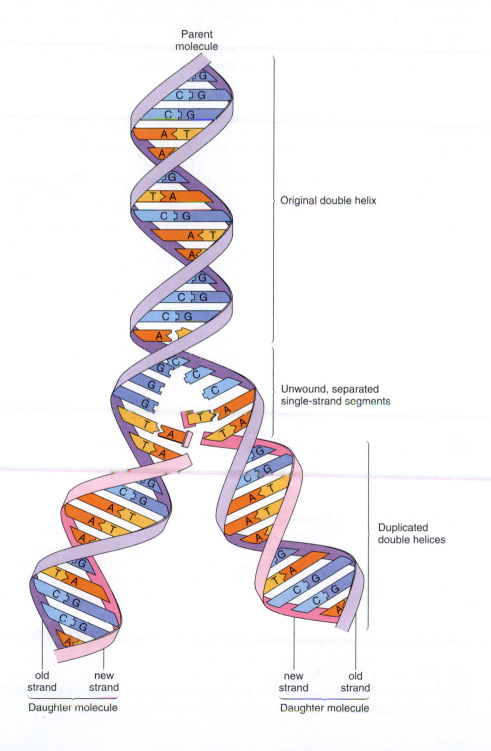

Parent molecule

Original double helix

Unwound, separated single-strand segments

Duplicated double helices

old strand   new strand

Daughter molecule

new strand   old strand

Daughter molecule

Figure 12.9

Diagram of DNA replication. The original double helix (top of figure) partially unwinds and the two complementary strands separate (middle). Each of the strands serves as a template for the synthesis of a complementary strand (bottom). As a result, the original DNA molecule is copied.

DNA, as pictured in the middle portion of Figure 12.9. Individual nucleotides in the cell are selectively hydrogen-bonded to these two single strands that serve as templates: A to T, T to A, C to G, and G to C. Held in these positions, the nucleotides are bonded together (polymerized) by the action of an enzyme. Every minute, about 90,000 nucleotides are added to the growing chain. By this mechanism, each strand of the original DNA generates a complementary copy of itself. The original template strand and its newly synthesized complement coil about each other to form a new double helix, a daughter molecule identical to the first. Similarly, the other separated strand of the original molecule twines around its new partner, another new daughter molecule. Thus, where there was only one double helix, there are now two, represented at the bottom of Figure 12.9. As the nucleus splits and the cell divides into two daughter cells, one complete set of chromosomes is incorporated into each of them. This process is repeated again and again, so that each of the 2×10^{12} nucleated cells in a newborn baby contains all the genetic information first assembled from parental DNA when the sperm combined with the ovum.

In 1968, James Watson published a highly personal account of the research that led to the determination of the DNA structure. This very readable book, entitled *The Double Helix,* is recommended to anyone who still doubts that scientists are flesh and blood with feet of clay. In the book, Watson candidly describes the process of scientific discovery: the competition and ambition, the lucky guesses and the blind alleys. His account does not always conform to a textbook definition of the scientific method, but then neither does scientific research. Two decades later, Crick followed with *What Mad Pursuit,* his own reminiscences of those heady days in Cambridge.

It is noteworthy that neither Watson nor Crick was an expert in the field of genetics when they began their research on DNA. Moreover, they did few experiments themselves. Instead, they drew on the work of experts such as Erwin Chargaff, Rosalind Franklin and her crystallographer colleague Maurice Wilkins, and the American chemist Linus Pauling. At the time, all of these scientists were better known and more highly regarded than Francis Crick and his young American collaborator. But Watson and Crick seem to have brought a fresh point of view to the problem of DNA structure, and hence they saw what more experienced and better informed scientists missed. There are those who argue that Rosalind Franklin's very significant contribution of crystallographic data was not sufficiently acknowledged, either in 1953 or in *The Double Helix.* But there are few if any who would quarrel with the decision to award Watson, Crick, and Wilkins the 1962 Nobel Prize in Physiology or Medicine. By that time, Franklin had died of cancer, and the Nobel Prize is not awarded posthumously.

12.7 Consider This: Discovering Rosalind Franklin

In 1958, Rosalind Franklin died an untimely death from ovarian cancer at age 37. Thus, she did not live long enough to add her own account to the history that was written (and rewritten) about the discovery of DNA. Because her work was long minimized or ignored, some historians now assert that *both* DNA and Rosalind Franklin were discovered.

To set the record straight, several excellent biographies of Franklin are now available. You can find reviews of these books as well as other accounts of her life on the web by searching for Rosalind Franklin. What were her contributions to the structure of DNA? Why was her work not given its full credit during her time? What questions would you ask her if you could interview her?

12.8 Your Turn

What role did knowledge of Chargaff's rules play in the discovery of the DNA double helix?

12.3 Cracking the Chemical Code

The discovery of the molecular code in which the genetic information is written is arguably history's most amazing example of cryptography. Key to the code is the sequence of bases in the DNA. The three billion base pairs repeated in every human cell provide the blueprint for producing one human being. Although these specifications are carried in DNA, they are expressed in proteins. Proteins are everywhere in the body: in skin, muscle, hair, blood, and the thousands of enzymes that regulate the chemistry of life. It follows that, by directing the synthesis of proteins, DNA can dictate the characteristics of the organism.

Chapter 10 and especially Chapter 11 contain a good deal of information about proteins. They are large molecules formed by the combination of individual **amino acids**. The 20 amino acids that commonly occur in proteins can be represented by the following general formula:

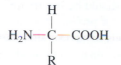

The amino group is $-NH_2$, the acid group is $-COOH$, and R represents a side chain that is different in each of the 20 amino acids. When the amino acids combine, the $-COOH$ group of one of them reacts with the $-NH_2$ group of another, forming what is known as a peptide bond and eliminating an H_2O molecule. A **protein** is thus a long chain of amino acid residues, as these structural units are called once they have been joined together.

For more information about amino acids and proteins, see Section 11.6.

The biochemists who set out to decipher the genetic code concentrated on translating base language into amino acid language. They assumed that somehow the order of bases in DNA determines the order of amino acids in a protein. The hypothesis that the code is related to the sequence of base pairs is the only reasonable one. The phosphate and deoxyribose units are identical in all DNA. Therefore, they could not supply the variability in the DNA structure to account for the individuality among species, and variation is essential in genetic material. Only the base pairs provide the opportunity for variability in the structure of DNA.

It was obvious at the outset that the code could not be a simple one-to-one correlation between bases and amino acids. There are only four bases in DNA. If each base corresponded to an individual amino acid, DNA could encode for only four amino acids. But 20 amino acids appear in our proteins. Therefore, the DNA code must consist of at least 20 distinct code "words," each word representing a different amino acid. And the words must be made up of only four letters—A, T, C, and G—or, more accurately, the bases corresponding to those letters.

Some simple statistics can help us determine the minimum length of these code words. To find out how many words of a given length can be made from an alphabet of known size, one raises the number of letters available to a power corresponding to the number of letters per word,

$$\text{number of words} = (\text{number of letters available})^{\text{number of letters per word}}$$

Thus, four letters could be used to make 4^2 or 16 different two-letter words. Similarly, DNA bases read in pairs (akin to two letters per word) could encode for only 16 amino acids. Again, this vocabulary is too limited to provide a unique representation for each of the 20 amino acids. So we repeat the calculation, this time assuming that the code is based on three sequential base pairs, or, if you prefer, that we are dealing with three-letter words. Now the number of different combinations is 4^3 or $4 \times 4 \times 4 = 64$. This system provides more than enough capacity to do the job.

12.9 Your Turn

Suppose the DNA code used four bases instead of three. How many different four-base sequences would result?

Ans: $4 \times 4 \times 4 \times 4 = 256$ different four-base sequences

Obviously, more than mathematical reasoning was required to prove the molecular basis of genetics. Once again, Francis Crick was a leader in this research. His work clearly established that the genetic code is written in groupings of three DNA bases, called **codons**. And today, thanks to Marshall Nirenberg, Har Gobind Khorana, and others, this triplet-base code has been cracked and specific amino acids have been related to particular codons.

No Rosetta Stone was available to aid these scientists in their efforts at translation. Instead, they relied on elegant and imaginative experiments that ultimately yielded a genetic dictionary. If you were to use the letters A, T, C, and G in a game of Scrabble, you could generate 64 different three-letter combinations. A few, CAT, TAG, and ACT, for example, make sense. Most are like AGC, TCT, and GGG and are meaningless—at least in English. Nature does far better than that; 61 of the 64 possible triplet codons specify amino acids. Thus, the codon sequence GTA in a DNA molecule signals that a molecule of the amino acid histidine should be incorporated into the protein, AAA codes for phenylalanine, and GGC stands for proline. The three-base sequences that do not correspond to amino acids are signals to start or stop the synthesis of the protein chain.

Because there are more codons than amino acids, there is redundancy built into the code. Some amino acids are represented by more than one codon. Leucine, serine, and arginine have six codons each. On the other hand, tryptophan and methionine are each represented by only a single codon. Significantly, the code is identical in all living things. The instructions to make Albert Einstein, bacteria, or trees are written in the same molecular language.

12.10 Your Turn

Suggest some advantages of a genetic code in which several codons represent the same amino acid.

The amount of information carried by your deoxyribonucleic acid is truly phenomenal. The DNA in each of your cell nuclei consists of approximately 1×10^9 (one billion) triplet codons. You have just read that each triplet is at least potentially capable of encoding one of the 20 amino acids found in human protein. If each codon could be assigned a letter of the English alphabet, rather than an amino acid, your DNA could encode 1×10^9 letters or about 2×10^8 five-letter words. These words would fill 400,000 pages of 500 words each, or 1000 volumes of 400 pages each. And you carry that library in two meters of a helical thread, invisible to all but electron microscopists. The miniaturization of this information to the molecular level puts to shame the most sophisticated supercomputers.

Unfortunately, scientific fact interferes a bit with the hyperbole of the previous paragraph. It has been determined that less than 2% of human DNA actually constitutes unique gene sequences. The human genome contains multiple copies of some genes that code for frequently used proteins. Moreover, there are many copies of DNA sequences that are too short to function as genes. For example, there are millions of copies of sequences consisting of only 5–10 base pairs. But the presence of this "junk" DNA in no way detracts from the wonder of molecular genetics or from its challenge. It merely gives scientists something else to study.

12.11 Your Turn

There are about three billion base pairs in the human genome, but only about 2% of this DNA consists of unique genes. The number of genes is estimated at 100,000. Use this information to calculate the average number of base pairs per gene.

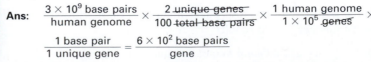

$$\textbf{Ans:} \quad \frac{3 \times 10^9 \text{ base pairs}}{\text{human genome}} \times \frac{2 \text{ unique genes}}{100 \text{ total base pairs}} \times \frac{1 \text{ human genome}}{1 \times 10^5 \text{ genes}} \times$$

$$\frac{1 \text{ base pair}}{1 \text{ unique gene}} = \frac{6 \times 10^2 \text{ base pairs}}{\text{gene}}$$

12.4 Protein Structure and Synthesis

The mechanism by which DNA directs protein synthesis is known in great detail—too much detail for this text. For our purposes, it is sufficient to recognize that the transfer of information and matter is extremely complicated. Given this complexity, it is amazing that errors in protein synthesis are very rare. Consider, for example, chymotrypsin. This protein, an enzyme that catalyzes the digestion of other proteins, consists of 243 amino acid residues. Chymotrypsin is just one of the proteins that can be made from 20 different amino acids by using 243 amino acid residues. Statistically, 20^{243} different protein molecules could be formed. Expressed relative to the more familiar base 10, this number corresponds to 1.4×10^{316}, a number larger than the estimated number of atoms in the universe! Each member of this immense group of molecules would have its own unique **primary structure**, the identity and sequence of the amino acids present. One and only one primary structure is the biologically correct form of chymotrypsin with the desired enzymatic properties. The fact that the body unfailingly (or almost unfailingly) synthesizes this particular protein out of 1.4×10^{316} possibilities is evidence of a molecular blueprint and a cellular assembly line of almost incomprehensible specificity and accuracy. And of course similar considerations apply to each of the proteins in the entire organism.

All enzymes are proteins, but not all proteins are enzymes.

After the amino acids are strung together in the correct sequence, the resulting protein chain should be able to twist and turn into an infinity of shapes. Surprisingly, it does not. Rather, the protein molecule assumes a characteristic shape that is generally influenced by variables such as temperature and pH. Once again, X-ray diffraction provides a means of determining this structure. For chymotrypsin, the result is pictured in Figure 12.10. What looks like a jumble of videotape is the carefully ordered backbone of the protein molecule. The figure shows helical segments and parallel chains that constitute the intermediate level of molecular organization, called the **secondary structure**. The overall shape or conformation of the molecule is termed its **tertiary structure**. Evidence suggests that the three-dimensional conformation of a protein molecule is stabilized by the interaction of various functional groups. Hydrogen bonds are particularly important in stabilizing secondary structural subunits that occur in many proteins.

The catalytic activity of chymotrypsin and any other enzyme is evidence of the reliable regularity of the tertiary structure of the protein. For an enzyme to carry out its chemistry, functional groups on certain amino acid residues must come close enough to form an active site. The **active site** is the region of the enzyme molecule where its catalytic effect occurs. Sometimes the amino acids involved are adjacent, in other cases they are widely separated in the protein chain, but close together in the tertiary structure.

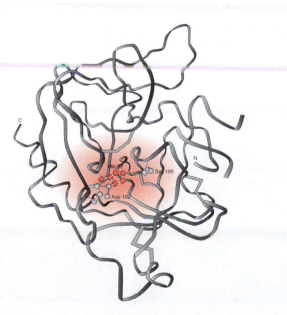

Figure 12.10

Tertiary structure of chymotrypsin. The tape represents the polymerized amino acid chain and the active site is shown in color.

Figure 12.10 indicates that the active site in chymotrypsin consists of three amino acids that would be far apart if the protein were unwound. These groups help hold the **substrate**, the molecule or molecules whose reaction is catalyzed by the enzyme. In the case of chymotrypsin, the active site of the enzyme catalyzes the breaking of peptide bonds in the substrate, another protein. In some other enzymes, the active site promotes the formation of chemical bonds. In all cases, the orientation of the active site and the conformation of the rest of the enzyme molecule are of critical importance. The fact that a newly synthesized protein molecule automatically assumes the enzymatically active shape almost suggests that the chain of amino acid residues possesses some sort of molecular memory. In fact, the favored tertiary structure is the most energetically stable conformation.

Sometimes a very subtle change in the primary structure of a protein can have a profound effect on its properties. A much studied example is provided by hemoglobin, the blood protein that transports oxygen, and the condition called **sickle-cell anemia**. When an individual with a genetic tendency toward sickle-cell disease is subjected to conditions that involve high oxygen demand, some red blood cells distort into rigid sickle or crescent shapes (Figure 12.11). Because these cells lose their normal deformability, they cannot pass through tiny openings in the spleen and other organs. Some of the sickled cells are destroyed and anemia results. Other sickled cells can clog organs so badly that the blood supply to them is reduced.

The property of sickling has been traced to a minor change in the amino acid composition of human hemoglobin. There are 574 amino acid residues in a hemoglobin molecule. The only difference between normal hemoglobin and hemoglobin S in persons with the sickle-cell trait, is in two of these amino acids. In hemoglobin S, two of the residues that should be glutamic acid residues are replaced with valine. Apparently this substitution is sufficient to cause the hemoglobin to convert to the abnormal form at low oxygen concentration.

Sickle-cell anemia is hereditary; the error in the amino acid sequence reflects a corresponding error in a DNA codon. Normally, mutations detrimental to a species are eliminated by natural selection. Perhaps sickle-cell trait has survived because it may also convey some benefit. A clue to what the benefit might be comes from studying the carriers of the gene for hemoglobin S. The gene is most common in people native to Africa and other tropical and subtropical regions and in their descendants. The fact that these are also areas with the highest incidence of malaria has led to speculation that an individual whose hemoglobin has a tendency to sickle may be protected against malaria. Specific mechanisms have been proposed to account for this protection. If the hypothesis is correct, it is an interesting example of how a genetic trait that originally had survival advantage can become a detriment in a different environment. Of course, the fact that sickle-cell disease is a genetic disease at least raises the possibility that genetic engineering may some day eliminate it.

Scientists have not yet synthesized hemoglobin, but they have made simpler proteins in laboratory glassware. The first successful effort was in 1968, when two groups of scientists, one at Rockefeller University and the other at the pharmaceutical firm of Merck, Sharpe and Dohme, independently prepared bovine ribonuclease A. Ribonuclease is an enzyme that catalyzes the cleavage of ribonucleic acid (RNA). Once the 124 amino acids

The mode of action of most enzymes can be explained by the lock-and-key model described in Section 10.6.

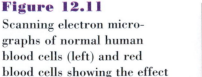

Figure 12.11

Scanning electron micrographs of normal human blood cells (left) and red blood cells showing the effect of sickle-cell disease (right).

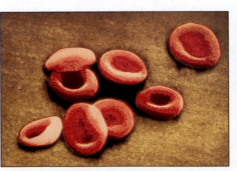

that make up bovine ribonuclease A were linked in the correct sequence, the resulting molecule possessed the catalytic activity of the naturally produced enzyme. This is additional evidence that the secondary and tertiary structure of the protein is determined by the primary structure.

The laboratory synthesis of bovine ribonuclease A was a great scientific achievement, richly deserving of the Nobel Prize it received. What a cow does in a minute or two required about 18 months of actual work, plus some 50 years of preliminary research. Although this research proved that the method worked, the direct laboratory or industrial synthesis of proteins is seldom carried out. There are easier ways of doing it. Today we can replace both the cow and the chemist with bacteria, thanks to recombinant DNA technologies. It is to that topic that we now turn.

12.5 Recombinant DNA: Therapeutic Proteins and Other Useful Substances From Other Organisms

Mythology is full of fanciful creatures: the sphinx with the head of a woman and the body of a lion; the griffin, which is half-lion and half-eagle; and the chimera with a lion's head, a goat's body, and a serpent's tail. In 1973 two American scientists, Herbert Boyer and Stanley Cohen, created another unnatural hybrid. They introduced a gene for manufacturing a protein from the African clawed toad into a common bacterium, *E. coli*. Upon replication, the *E. coli* bacteria produced the toad's protein. To be sure, humans have been manipulating the gene pool for thousands of years. We have created mules by crossbreeding horses and donkeys, dogs as diverse as Chihuahuas and Saint Bernards, and fruits and vegetables that never existed in nature. All of these were done by selective breeding. Boyer and Cohen created their chemical fantasy in laboratory glassware through the manipulations of genetic engineering.

To illustrate the technique, consider a real response to a very real need. **Insulin** is a small protein consisting of 51 amino acids. It is produced by the pancreas, and it influences many metabolic processes. Most familiar is its role in reducing the level of glucose in blood by promoting the entry of that sugar into muscle and fat cells. People who suffer from a common type of diabetes have an insufficient supply of insulin, and hence elevated levels of blood sugar. Left untreated, the disease can result in poor blood circulation, especially to the arms and legs, blindness, kidney failure, and early death. But, diabetes can be controlled by diet and by insulin injections.

Before 1982, all insulin used by diabetics was isolated from the pancreas glands of cows and pigs, collected in slaughterhouses. It turns out that the insulin produced by cattle and hogs is not identical to human insulin. Bovine insulin differs from the human hormone in three out of 51 amino acids; porcine and human insulin differ in only one. These differences are slight, but sufficient to undermine the effectiveness of bovine and porcine insulin in some human diabetics. For many years, there seemed to be no hope of obtaining enough human insulin to meet the need. Although insulin has been synthesized in the laboratory, the process is far too complex for industrial adaptation. However, since 1982 the lowly bacterium, *E. coli,* has been tricked into making human insulin.

This unlikely bit of inter-species cooperation is a consequence of using **recombinant DNA techniques** to introduce the gene for human insulin into this simple organism (Figure 12.12). Bacteria contain rings of DNA called **plasmids**. These rings can be removed and cut by the action of special enzymes. Meanwhile, the gene for human insulin is either prepared synthetically or isolated from human tissue. This human DNA is inserted into the plasmid ring by other enzymes. The result is inter-species recombinant DNA. The modified plasmids (also called vectors) are then reintroduced into the bacterial "host." Once inside the cell, the biochemistry of the bacterium takes over (Figure 12.12). Every 20 minutes, the *E. coli* population doubles, and soon there are millions of copies or clones of the "guest" (human) DNA.

Clones are a collection of cells or molecules identical to an original cell or molecule. It is possible to harvest the cloned DNA, but in the insulin example we are interested in

Figure 12.12

A general schematic of genetic engineering, showing the two major end products: cloned genes on the left and proteins on the right.

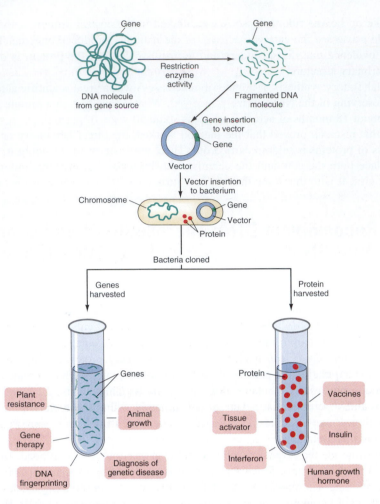

a supply of the protein, not its gene. Therefore, the bacteria are allowed to synthesize the proteins encoded in the recombinant DNA. Although the *E. coli* has no use for human insulin, it generates it in sufficient quantities to harvest, purify, and distribute to diabetics. Today, the cost of bacterially produced insulin is less than that of insulin isolated from animal pancreas. Currently, more than three million Americans use genetically engineered insulin to treat their diabetes.

Figure 12.12 is a representation of the recombinant DNA techniques just described. The actual operations are a good deal more complicated than the figure suggests, and many details have been omitted. A variety of vectors and host organisms have been used in molecular engineering. Other bacterial species are sometimes used instead of *E. coli*, and yeasts and fungi are often employed.

12.12 Consider This: Where Does Insulin Come From?

If a person must take insulin daily, how would that person know the source of the insulin? Does the source make a difference in how the insulin acts in the body? Use the resources of the web and/or consult a pharmacist to answer these questions.

Another success for genetic engineering has been the synthesis of **human growth hormone (HGH)**. This protein, produced by the pituitary gland, stimulates body growth by promoting protein synthesis and the use of fat as an energy source. Children with insufficient HGH fail to reach normal size. If the condition is diagnosed early, injections of the hormone over eight to ten years can prevent dwarfism. Formerly, a year's HGH therapy for one person required the pituitary glands from about 80 human cadavers. That source is no longer used, thanks to the production of human growth hormone in bacteria. However, the cost of treatment, even with cloned HGH, can be as high as $20,000 per year.

12.6 Engineering New Drugs and Vaccines

You have just read about two examples of replacement therapy, in which an insufficient natural supply of an essential protein is augmented with a genetically engineered supplement. Similar biochemical methods are also being used to create new drugs or larger supplies of already known drugs. The gene coding for the drug is introduced into a host organism, which then synthesizes the desired product. This is currently one of the most rapidly growing applications of recombinant DNA technology.

Other products of biotechnology appear to be effective against viruses. **Viruses are simple, infectious, almost living biochemical species.** We say "almost living" because viruses do not have the necessary biochemical machinery to carry out metabolism or to reproduce by themselves. A typical virus consists of nucleic acid and protein. It is essentially inert, and it can survive for years without losing its infectious potential. When it invades a cell, the virus takes charge, forcing the cell to make more viral nucleic acid and protein. In effect, the virus does naturally what genetic engineers accomplish with recombinant DNA techniques. Because they are so simple, viruses are notoriously difficult to combat or defend against. Thus, pneumonia, "strep," or other bacterial infections, though potentially far more dangerous than a common cold, are much easier to treat than a cold. Pneumonia and strep are caused by bacteria, which can be destroyed by penicillin and other antibiotics. But colds are caused by viruses, and about all one can do is to treat the symptoms.

Genetically engineered **interferons** may help change all that. Interferons are nature's way of providing protection against viruses. Over 20 distinct naturally occurring interferons have been identified. All of them are proteins, and some also contain carbohydrate portions. The mechanism by which these molecules defend against viral invasion is not fully understood, but it has been exploited. Thus far, genetically engineered interferons show promise against hepatitis, herpes zoster (shingles), a type of multiple sclerosis, and a variety of cancers including some forms of leukemia, malignant melanoma, multiple myeloma, and certain kidney cancers. The use of an interferon nasal spray to control the common cold may still be years away because of the high costs and biochemical complexity associated with cloning these proteins.

For the treatment of many diseases, the ultimate goal is not just the development of a drug to treat it, but the creation of a **vaccine** to prevent contracting the disease. Vaccines work by mobilizing the body's own defense mechanism. The idea is to expose the body to a molecule or organism closely related to the virus or bacterium that causes the disease. The immune system responds to this stimulus by generating **antibodies** against it. These antibodies remain in the body, where they offer protection against subsequent infection by the virus or bacterium. Of course it is important that the vaccine does not itself cause the disease. Therefore, vaccines are typically made from bacteria or viruses that have been killed or weakened, or from fragments or subunits of the virulent invaders. The latter approach is the preferable one, because there is no danger that the disease will be transmitted in the process of vaccination. Fortunately, it is here that genetic engineering is most promising. The DNA encoding for a characteristic but noninfectious part of a virus—for example, its protein coat—can be introduced into plasmids. The bacteria will consequently produce this particular protein. The protein is then isolated, concentrated, and used as a vaccine that carries essentially no risk of infection. This technique has been employed to synthesize a vaccine against hepatitis B. Because hepatitis is transmitted by blood, health care professionals are often vaccinated against the disease. If and when a vaccine is developed against HIV, it may be through similar technology.

12.7 Diagnosis Through DNA

Until very recently, the most sensitive methods of diagnosing disease were based on the detection of enzymes or antibodies in an infected organism. These proteins are generated by the host organism in response to the invasion. This means that infection is often well established before a positive test can be obtained. Fortunately, genetic engineering has enabled diagnosticians to identify the DNA of the infectious agent, even at early stages and a low concentration.

Such sensitivity is possible only because of the development of two important techniques: DNA probes and the polymerase chain reaction. *DNA probes are engineered so that they are complementary to some segment of the infecting viral or bacterial DNA* (the target). The probes are single-stranded DNA, with from 10 to over 10,000 bases, and they are usually labeled with a radioactive isotope. These radioactive probe molecules are introduced into a sample of biological fluid or cytoplasm that is suspected of containing an infectious agent. The test is carried out at a temperature and pH at which the DNA double helix of the infectious agent separates into single strands. If the probe encounters a strand with a complementary segment of bases, hydrogen bonds form between A and T and C and G and the probe sticks to the target molecule. Because the probe is radioactive, it can easily be traced using a radiation detector. The radioactivity level in a certain fraction indicates that the infectious DNA is indeed present.

Successful early diagnosis involves detecting the infecting virus or bacterium before the disease is well established. Even the most sensitive DNA probes will not work if the concentration of the invading DNA is too low. Here the **polymerase chain reaction (PCR)** has proved to be of great utility. *This technology makes it possible to start with a single segment of DNA and make millions or even billions of copies of it in a few hours.* The PCR process, which won its inventor, Kary Mullis, a Nobel Prize in 1993, is diagrammed in Figure 12.13. At the top of the figure is the double-stranded DNA molecule that the researcher is seeking to copy. In the step marked *a,* the sample is heated to about 95°C to break the hydrogen bonds between the two DNA strands, causing them to separate and unwind. Then the mixture is cooled to about 65°C and short segments of "primer" DNA are attached to each of the separated strands (step *b*). These **primers** are synthetic single-stranded nucleotides that bracket and identify the section of DNA to be copied. The unwound DNA strands and the primers are mixed with the enzyme DNA polymerase, plus an ample supply of the four free nucleotides to be incorporated into the DNA chain. Starting at the primer, the polymerase enzyme directs the attachment of complementary nucleotides along the strands of DNA. In this way, *each of the two original single DNA strands is copied to form two new DNA molecules (step c).*

Figure 12.13 shows that at the end of the first cycle there are two double-stranded DNA molecules where there formerly was one. The entire cycle is then repeated and at the end of each cycle there are twice the number of DNA molecules. The amount of DNA doubles exponentially so that starting with one double strand of DNA, the total number of double strands after *n* complete cycles will be 2n. As the following chart indicates, such exponential growth is rather impressive.

> DNA probes were used to identify *Helicobacter pylori* as the causative agent of gastric ulcers and also the likely cause of stomach cancer.

> Radioactivity is discussed in Sections 7.7–7.9.

> Exponential doubling is no small change. Consider being given a penny on the first day of a month, two pennies the second day, four pennies the third day, and so on, doubling the number of pennies from the previous day. At the end of thirty days, you would have over $10 million! Finding someone to give you that kind of money is the real trick.

n = number of cycles	number of DNA molecules
1	$2^1 = 2$
2	$2^2 = 4$
3	$2^3 = 8$
4	$2^4 = 16$
5	$2^5 = 32$
10	$2^{10} = 1024$
100	$2^{100} = 1.27 \times 10^{30}$

The polymerase chain reaction is very rapid; only one to two minutes are needed for a complete cycle. Thus, 100 cycles would require just two to three hours. But the reac-

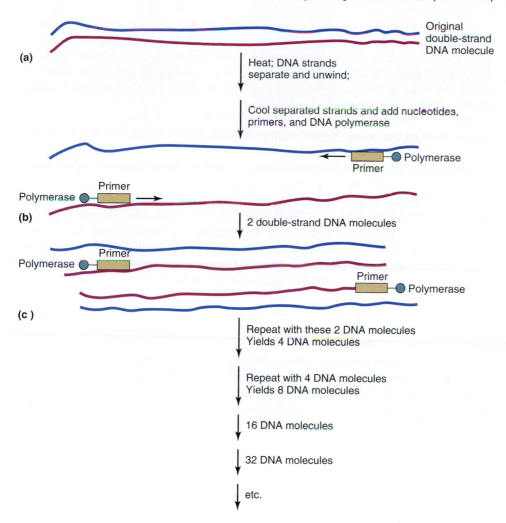

Figure 12.13

Diagram of the polymerase chain reaction.

tion would run out of starting materials long before then. A little arithmetic (12.13 The Sceptical Chymist) shows that 1.27×10^{30} double-stranded DNA molecules, each of 100 base pairs, would weigh over 139,000 tons!

12.13 The Sceptical Chymist: 140,000 Tons of Base Pairs

It is a good idea to check the assertion that 1.27×10^{30} double-stranded DNA molecules of 100 base pairs each would weigh almost 140,000 tons.

Hint: Start by using Avogadro's number to determine how many moles of DNA are represented by this large number of base pairs. Assume an average mass of 300 g per mole of nucleotide.

12.14 Your Turn

A technician starts with a single DNA molecule. How many cycles of PCR must take place to form each of these numbers of DNA molecules?

a. 16 DNA molecules
b. 256 DNA molecules
c. 1.0×10^6 DNA molecules

Ans: a. 4 cycles;   **b.** 8 cycles;   **c.** 20 cycles

PCR technology has proved to be indispensable in any procedure where a small sample of DNA must be dramatically amplified to get a sufficiently large supply for subsequent studies. Thus, it is used in developing "DNA fingerprints" in criminal cases, in studying archeological remains, and, of course, in diagnosing disease. In the latter instance, the DNA in specimens thought to contain infecting or defective DNA is multiplied by PCR so that DNA probes can be used.

The early diagnosis of HIV infection is one of the achievements of this new technology. The DNA of human immunodeficiency virus can be detected several weeks before antibodies to HIV build up to the point at which they can be identified. Similarly, recombinant DNA methods have been used to speed up the diagnosis of tuberculosis. DNA probes and PCR have also been used to identify a number of hereditary diseases. Defective genes have been identified for cystic fibrosis, Huntington's disease, some forms of Alzheimer's disease, and amyotrophic lateral sclerosis (ALS), better known as Lou Gehrig's disease, after the great New York Yankee first baseman who died of it in 1942. Scientists identified the altered gene responsible for the disease in 1993. As in sickle-cell anemia, the mutation is slight and subtle. A single amino acid is altered in superoxide dismutase, an enzyme that eliminates free radicals, highly reactive chemical species with unpaired electrons. Free radicals appear to accumulate in Parkinson's disease, Alzheimer's disease, and in normal aging. In the case of ALS, their build-up results in the destruction of motor nerves. ALS gives little early warning. Early diagnosis, based on detection of the altered gene, might permit preventative treatment.

The "might" in the previous sentence indicates a major problem in the diagnosis of hereditary diseases and defects. In many cases, the ability of modern science to respond to the effects of a defective gene has not equaled the ability to detect the gene. The tendency to develop genetic diseases is programmed in the DNA, and it may or may not be stimulated by infection. One can well ask what is the advantage of knowing that an individual is the carrier of one of these genes when there is no way to prevent or even to treat its deleterious effects. Is it helpful to know that you have a high probability of acquiring ALS or Alzheimer's disease when there is nothing you can do about it but wait?

To be sure, such knowledge could be useful in case new therapies are developed. Moreover, some carriers of inheritable diseases choose not to have children or to use *in vitro* (literally "in glass") fertilization and genetic screening. The latter approach was recently taken by a married couple, both of whom were carriers of the gene for cystic fibrosis. Five ova taken from the woman were fertilized with her husband's sperm, and the resulting embryos were analyzed for the cystic fibrosis gene at the eight-cell stage. An embryo with no or only one copy of the defective gene, neither of which would result in the disease, was implanted into the woman's uterus, and a healthy baby was born.

Genetic screening is more frequently used on embryos and fetuses that are further developed. If there is a likelihood of the parents passing on a defective gene, they sometimes request **amniocentesis**. In this procedure a sample of the amnionic fluid is withdrawn from the mother's uterus. This fluid contains fetal cells that are then analyzed for their genetic makeup. By this means, the defective genes for Down's syndrome and other hereditary conditions can be detected. If they are present, the parents face a difficult decision with a significant ethical component: whether to abort the fetus. Such painful choices could be avoided if science were to develop ways of actually changing the DNA of the fetus or even of children or adults. We now consider this prospect.

> Free radicals are also involved in stratospheric ozone depletion (Sections 2.12 and 2.15), in addition polymerization (Section 9.7), and in cooking foods (Section 11.12).

12.8 Gene Therapy

Medical researchers estimate that about 2000 hereditary diseases and genetic defects are caused by errors in a single gene. These conditions seem to be ideal candidates for **gene therapy**, which involves introducing normal genes into patients lacking them. Simply stated, cells are taken from a patient, altered by the introduction of normal genes, and then returned to the patient. If all goes well, the imported genes function normally.

The first successful application of gene therapy to a human being was in 1990, when the technique was used to treat a four-year-old girl suffering from **severe combined**

immunodeficiency disease (SCID). This is a very serious, and fortunately very rare, condition. Because of a genetic defect, a specific enzyme is not synthesized. The absence of this enzyme leads to the destruction of the white blood cells that protect the body against infection. Children suffering from SCID have essentially no functioning immune system, and the slightest infection can prove fatal. In the past, some children with SCID have survived for a few years, but only by living in sterile isolation chambers.

Today, several victims of the disease are enjoying relatively normal lives, thanks to their new genes. In the procedure followed with the first patient, the gene that encodes for the missing enzyme was identified and isolated from other sources. Special viruses were used to introduce copies of this gene into cells that had been removed from the patient. These modified cells were then reintroduced into the girl's body. The new genes have continued to function well, producing the previously absent enzyme. As a result, the concentration of white blood cells has increased significantly and the girl's antibody-producing defense mechanism is working (Figure 12.14).

Critics have warned that the introduction of the external DNA and its virus vector carries some risk of infection. However, the treatment appears to be more effective than the alternative, bone marrow transplants. Such transplants can succeed only if the donor and the patient are closely related. Even then, there is a danger that the transplanted cells may be rejected by the recipient. In gene therapy, the patient's own cells are slightly altered and returned to the body. Thus, there is no chance of rejection.

The body's own defenses are enhanced in the gene therapy approach to certain cancers. For example, **malignant melanoma** can be treated with engineered DNA that encodes for an anti-cancer agent called tumor necrosis factor. The DNA is incorporated into tumor-infiltrating white blood cells that have been removed from the patient. These cells are specifically cultured to attack the cancerous melanoma cells and are then transfused back into the bloodstream. Another approach has been used with certain brain tumors. A sample of the tumor is removed and its DNA is modified so that the tumor becomes particularly susceptible to anti-cancer drugs. The modified tissue is then returned to the body and the appropriate drug is administered. The cancer cells carrying the engineered "suicide genes" are destroyed.

Patients looking for a genetic cure to a currently incurable hereditary disease should realize that the biochemical procedures described above are difficult, complex, and time-consuming. So are the associated political procedures. Before gene therapy can be used on a human subject in the United States, the protocol must receive the approval of a committee at the researcher's hospital or other home institution, the Recombinant DNA Advisory Committee of the National Institutes of Health (NIH) and its Human Gene Therapy Subcommittee, and the Food and Drug Administration (FDA). As in the case of ordinary drug approval (Chapter 11), this approval process can take years. For patients suffering from some of these diseases, a year is literally an eternity. In 1993, in response to this situation, Bernadine Healy, then director of NIH, proposed that the

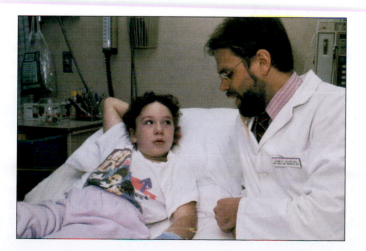

Figure 12.14

Curing disease through genetic engineering. One of two young girls who were the first humans "cured" of a hereditary disorder by transferring into their bodies healthy versions of the gene they lacked. The transfer was successfully carried out in 1990, and the girls remain healthy.

Recombinant DNA Advisory Committee institute an accelerated approval procedure when the gene therapy was intended for critically ill patients. The Committee granted greater authority and flexibility to NIH officials in these circumstances, but in general gene therapy remains highly regulated and highly experimental. Although gene therapy holds great promise for humans, it cannot be categorized as a general success to date. A statement issued recently by NIH indicates "Human gene therapy has fallen short of expectations. Between 1992 and 1997, 106 clinical trials of experimental gene therapies were done in more than 597 patients. The results indicate that clinical efficacy has not been definitely demonstrated at this time in any gene therapy protocol."

12.15 Consider This: Gene Therapy

There are promising developments in using gene therapy for treating serious human conditions, but there is not universal success.

a. Why do you think that experimental protocols with gene therapy have shown mixed results? Explain some of the factors that you feel may influence outcomes.

b. Given the mixed results, under what conditions would you consider gene therapy a valuable tool? Explain the reasons for your criteria.

12.9 Genetic Fingerprinting

A woman has been sexually assaulted. She cannot identify her attacker, who was masked. The police have two suspects, but neither man was seen in the neighborhood on the night of the attack. There seems to be little evidence, except for several drops of semen on the victim's clothing and in her vaginal canal. That may be sufficient. Thanks to **DNA fingerprinting**, it may prove possible to identify the attacker with a probability approaching certainty.

DNA fingerprinting is based on the fact that each of us (again, with the exception of the Stanitski brothers and other identical twins) has his or her own unique DNA. It is not surprising that the really important genes, those that encode for insulin, hemoglobin, chymotrypsin, and most other proteins, are identical in almost all of us. Here, as we have already noted, mutations are rare. We differ primarily in the "junk" DNA that makes up about 98% of the three billion base pairs in each human cell nucleus. Therefore, it is to this apparently nonessential DNA that forensic scientists look when they seek to determine "who dun it."

The authors recognize that jurors typically turn off when expert witnesses go into great detail about the biochemistry of DNA fingerprinting. For that reason we will try to be brief without sacrificing accuracy. In short, the technique is based on the fact that every individual appears to have a unique set of the DNA segments that serve as the spacers or punctuation marks between genes. Consider the semen sample found on the victim's clothing. Even if it is a very tiny spot, it contains more than enough DNA for a reliable genetic fingerprint; 1×10^{-9} g of DNA is sufficient. First the DNA is extracted, and then it is multiplied many times over by PCR. These copies are exposed to the action of enzymes that cut the DNA strands before and after the spacer segments just mentioned.

The spacer fragments are then subjected to **electrophoresis**, a method of separating molecules based on their rate of movement in an electric field. In the technique used in DNA fingerprinting, samples are applied to a strip of a polysaccharide gel, and electrophoresis is carried out in this medium. Because the phosphate groups of the DNA are negatively charged, the fragments migrate toward the positive electrode or pole. The speed at which a DNA segment moves depends on the magnitude of its electrical charge and its size or molecular mass. Shorter strands of DNA, consisting of fewer base pairs, will move faster than longer strands, which encounter more resistance from the gel.

It is necessary to see and measure how far the DNA fragments have traveled in a fixed period. This is done by using radioactive markers that can be detected because they expose photographic film. The fingerprint thus consists of a film with black smudges or bars, each one corresponding to the distance migrated by a particular segment of DNA. The heaviest and longest segments are closest to the point of application, the lightest and shortest ones are the farthest away.

In most criminal cases, the electrophoresis pattern produced from the evidence collected at the crime scene is compared to the electrophoresis pattern made by DNA obtained from the suspect or suspects. Figure 12.15 is the electrophoretic evidence in the sexual assault case we are investigating. Each spot indicates how far DNA segments of specific size migrated during the electrophoresis experiment. Rows 1, 5, 8, and 9 are reference markers of DNA exhibiting a known range of molecular masses. Spots at the far left represent the longest, heaviest segments of DNA; spots at the far right represent the lightest, shortest segments. Row 3 is DNA from a semen sample found on the victim's clothing, and row 6 is semen DNA obtained by swabbing her vaginal canal shortly after the attack. Not surprisingly, the positions of the spots in these two rows are identical. To avoid possible misidentification, it is also important to include a sample of the woman's own DNA, which gives the pattern in row 7. Now look at rows 2 and 4. Row 2 is characteristic of the DNA in a blood sample obtained from suspect A; row 4 is from a blood sample from suspect B. B's genetic fingerprint matches the DNA from the semen samples in rows 3 and 6.

In the case represented by Figure 12.15, the evidence strongly indicates that suspect A is innocent and that suspect B is probably guilty, but caution is required. A matching DNA fingerprint does not *absolutely* prove the guilt of a suspect. It is possible that DNA from two individuals might yield the same electrophoresis patterns, but it is highly unlikely. To increase the odds of accurate identification, comparisons of the sort just described are typically done on DNA from three or more different chromosomes. As the number of determinations increases, so does the improbability of finding any two individuals with identical DNA fingerprints.

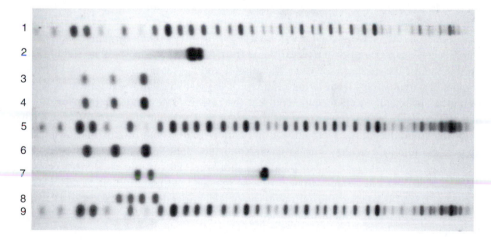

Figure 12.15

DNA fingerprints in a sexual assault case. The various rows represent the electrophoretic migration pattern of DNA segments.
Rows 1, 5, 8, 9: Reference markers of a mixture of DNA segments of known length and mass.
Row 2: DNA from a sample of blood from suspect A.
Row 3: DNA from a semen sample found on the victim's clothing.
Row 4: DNA from a sample of blood from suspect B.
Row 6: DNA obtained by swabbing the victim's vaginal canal.
Row 7: DNA from a sample of the victim's blood.

To see how this works, suppose that the statistical data base indicates that one person in 100 will exhibit a particular DNA pattern obtained from chromosome number 1. This would mean that there is a 1 in 100 chance that a suspect with that pattern is *not* the source of the sample. But also assume that the frequency of the observed pattern from another chromosome, say number 8, is 1 in 1000; and that the frequency of the observed pattern from chromosome number 12 is again 1 in 1000. Probabilities are multiplicative, therefore,

$$\text{Total probability} = 1/100 \times 1/1000 \times 1/1000 = 1/100{,}000{,}000$$

In this example, the odds that any two individuals would have the same genetic fingerprints, are 1 in 100 million. Conversely, if the DNA from a suspect matches a sample from this crime scene, chances are 99,999,999 out of 100,000,000 that the person *was* the source of the original sample.

Such probabilities can be very convincing, especially if the suspect has a motive and can be otherwise placed at the scene of the crime. However, some juries have been skeptical of DNA data. In the O. J. Simpson murder trial, defense attorneys suggested that the blood samples taken from the scene of the crime had been contaminated or planted. Apparently the doubts raised in the jurors' minds were enough to outweigh the evidence of the DNA fingerprints.

12.16 Your Turn

Suppose that a DNA fingerprint is based on three different chromosomal segments. The frequency of a match is determined to be 1 in 10 based on the first segment, 1 in 100 based on the second segment, and 1 in 1000 based on the third segment. What is the probability of a match in all three tests?

Ans: 1 in 10^6 or 1 in a million

Criminal investigation is not the only use of DNA fingerprinting. It is also a powerful tool in genetic identification. For example, it is used routinely to prove or disprove paternity, because it is far more specific than blood typing. DNA can even be extracted from bone, and the base sequence can be used to identify human remains. Such studies were recently carried out on bones exhumed from a mass grave near Yekaterinburg, Russia, the site of the massacre of Czar Nicholas II and his family in 1918. The DNA was compared with samples from living relatives of the last Russian royal family and found to match sufficiently well to conclude that the remains were indeed those of the Romanovs. In 1998, DNA fingerprinting was used to determine the identity of a soldier who was killed during the Vietnam war and buried in the Tomb of the Unknowns at Arlington National Cemetery. The tests resolved the identity of the entombed soldier, which had been narrowed to two possible candidates. The profile of DNA taken from the bones of First Lieutenant Michael Blassie matched sufficiently that of a DNA sample taken from his mother, thus allowing proper identification.

DNA analysis is not merely confined to the living and the recently deceased. Researchers have cloned and investigated DNA samples obtained from a 2400-year-old Egyptian mummy and some even older human remains. Scientists have interpreted the results to gain information about the relationship of ancient peoples, their migration routes, and their diseases.

12.17 Consider This: Lincoln's DNA

Some researchers have speculated that Abraham Lincoln suffered from a genetic condition known as Marfan's syndrome, which causes a person to grow tall and gangly. DNA fingerprints could answer the question. Do you support exhuming Lincoln's body from the Oak Ridge Cemetery in Springfield, IL? Give some reasons to explain your decision.

The current DNA age record is held by a bee and a termite that lived about 30 million years ago. Since that time, the insects had been entombed and protected in amber, which is solidified plant resin. In 1992, researchers released their perfectly preserved bodies, extracted their DNA, and subjected it to a number of studies. This research yielded important information about the evolutionary connection of these ancient organisms to modern species. But could 30-million-year-old DNA yield more? Could it be cloned into a living fossil? That is the premise of *Jurassic Park.* In the science fiction film and the novel, the chief interest is not in the fossilized insects themselves, but in the dinosaur blood they contain. The blood is the source of the DNA that is cloned and introduced into crocodile egg cells, where it replicates until it creates modern copies of long-extinct creatures. "Could this happen in real life?" That question is posed by I. Edward Alcamo in *DNA Technology: The Awesome Skill,* a book that is a useful source of information and illustrations for this chapter. Professor Alcamo's answer may be mildly reassuring for those who would rather not encounter a *Tyrannosaurus rex:* "Possibly. But you would need an entire set of dinosaur chromosomes, and only a minuscule fragment has been obtained up to now. And that's only the first of a thousand problems that must be solved."

Science fiction books and films to the contrary, dinosaurs and humans never coexisted. Fossil records are clear that dinosaurs were long gone before the first humans appeared on Earth.

12.18 Consider This: Science Fiction Success

One of the reasons that science fiction is successful is that it starts with a known scientific principle and extends, elaborates, and sometimes embroiders it. Perhaps you can be as successful as Michael Crichton was with *Jurassic Park.* Start by identifying a scientific principle from this or another chapter, and then writing a one- or two-page outline for a story based on that principle. Be sure to identify the chemical concepts you plan to include and any pseudo-science that you might employ for the sake of the story.

12.10 Mixing Genes: Improving on Nature(?)

One of the more sensational dimensions of genetic engineering has been the creation of higher plants and animals that share the genes of another species, so-called **transgenic organisms**. Inserting foreign DNA becomes progressively more difficult as one moves up the evolutionary ladder from bacteria through plants to animals. Nevertheless, some of the most spectacular successes of recombinant DNA technology have involved modifications of agricultural crops. Researchers have sometimes resorted to a "shotgun approach" to introduce DNA into plant cells. Millions of microscopic tungsten spheres are coated with the DNA to be inserted into the host, and these projectiles are fired into a group of cells. Some of the DNA finds its way into the plant chromosomes, but it is largely a hit-or-miss proposition. The use of plasmid carriers, similar to those used with bacteria (Section 12.5) is generally more dependable.

In spite of some rather formidable difficulties, altering the genetic makeup of plants by genetic engineering is faster and more reliable than relying on traditional crossbreeding. Moreover, some of the species that have been genetically combined are so dramatically different that interbreeding is impossible. This is clearly evident from Table 12.2, a listing of some recent transgenic plant experiments. Chickens and potatoes are very strange bedfellows indeed!

Among the most promising combinations have been those that confer on the host plant built-in pest resistance. Plants have been engineered to produce their own internal insecticides. For example, bacterial DNA that produces a compound that is toxic to various caterpillars has been inserted into the genome of cotton and corn plants. When cotton bollworms or corn borers attempt to eat the plants, they consume the toxin and die. Insects that do not feed on the genetically altered cotton or corn are unaffected, though they might well be killed with a conventional insecticide spray. Similar advances that reduce agricultural dependence on pesticides and herbicides are likely to be welcomed by environmentalists. Some herbicides are "broad spectrum" and unselective in their

Table 12.2 Transgenic Plant Experiments

Source of Genes	Transgenic Plant	Objective of Experiment
Chicken	Potato	Increased disease resistance
Giant silk moth		Increased disease resistance
Greater waxmoth		Reduced bruising damage
Virus		Increased disease resistance
Bacteria		Herbicide tolerance
Wheat	Corn	Reduced insect damage
Firefly		Introduction of marker genes
Bacteria		Herbicide tolerance
Flounder	Tomato	Reduced freezing damage
Virus		Increased disease resistance
Bacteria		Reduced insect damage
Chinese hamster	Tobacco	Increased sterol production
Bean, Pea	Rice	New storage proteins
Bacteria		Reduced insect damage
Virus	Melon, Cucumber, Squash	Increased disease resistance
Brazil nut	Sunflower	Introduction of new storage proteins
Bacteria	Alfalfa	Production of oral vaccine against cholera
Tobacco	Lettuce, Cucumber	Increased disease resistance
Synthetic	Sweet potato	Enhanced storage proteins with high essential amino acids content

(From I. Edward Alcamo, *DNA Technology: The Awesome Skill.* Copyright © 1996 The McGraw-Hill Companies, Inc. All Rights Reserved. Reprinted by permission.)

actions, killing many crops as well as weeds. To protect against this, a segment of DNA from *E. coli* inserted into soybeans, corn, cotton, and other crops, makes the plants resistant to certain herbicides used for weed control. Thus, the weeds are killed without damaging the crop. To cope with limited water supplies for irrigation in some areas, genetic engineering has been proposed for developing plants whose underside leaf openings would close more readily, thus controlling water loss from the plants.

During 1998, transgenic plants were grown on nearly 69 million acres worldwide (Table 12.3). For example, in 1998, 75% of cotton harvested in Alabama was from plants genetically engineered to resist insect pests. Plants genetically engineered to be herbicide or disease resistant produced one-half of the world's soybeans and one-third of its corn. By-products of these plants are found in breakfast cereals, soft drinks, candies, cooking oil, and other food products.

Someday, genetic engineering may be able to make a significant contribution to better nutrition. Researchers are investigating ways of incorporating the DNA of nitrogen-fixing bacteria into wheat, rice, and corn. These bacteria are present in the root systems of soybeans, alfalfa, and other legumes. Thanks to the bacteria, these plants can absorb N_2 directly from the atmosphere and use it in biochemical reactions. Other plants also need nitrogen, but they can use it only in soluble form from mineral sources in the ground and water. Hence, nitrogen-containing fertilizers are applied to help these plants grow well. But the need for fertilizers would be greatly reduced if all food crops were able to use atmospheric nitrogen directly.

The application of green chemistry principles to fertilizers was described in Section 11.13.

Table 12.3 Millions of Acres of Transgenic Crops Planted in 1998

Country	Millions of Acres
United States	50.7
Argentina	10.6
Canada	6.9
Australia	0.25
Mexico	0.25
Total	**68.7**

Researchers also have developed sweet potatoes with enhanced protein content by inserting a gene coded for storage of a protein with a high content of essential amino acids. Such an improvement is important in many poorer tropical countries where high-quality protein sources are expensive, but sweet potatoes are a dietary staple food. A great benefit of these genetically altered plants is that the newly acquired trait is conserved in the seeds and passed on to the next generation.

The grocery store of tomorrow may also contain other products of rearranged genes. For example, it may prove possible to genetically induce cows to give human milk for newborn babies, lactose-free milk for those suffering from lactose intolerance, naturally iron-enriched milk for anemics, casein-rich milk for cheese-makers, and skim milk for weight watchers—not all from the same cow, of course. Bovine growth hormone (BGH) is now produced in bacteria and injected into cattle to increase milk and beef production. The FDA has ruled that this practice presents no hazards to human health because BGH is biologically inactive in humans. However, critics have correctly noted that cows exposed to elevated BGH levels are more prone to infectious disease and consequently are administered higher levels of antibiotics than normal. Some of these drugs can be carried over into the milk.

12.19 Consider This: Where in the World is BGH?

In 1985, the U.S. FDA approved the use of BGH in cows for boosting milk production. Although regulatory agencies around the world have reached the same conclusion, the debate over whether BGH should be used this way is far from over.

Search the web for "bovine growth hormone" (or its synonyms "bovine somatotropin" or "BST") to check the current situation and report the arguments on both sides. If you can, document the situation in countries other than the United States.

12.20 Consider This: Turning Over a New Leaf

The New Leaf Superior potato has been genetically engineered to produce its own insecticide. This gives the potato the ability to resist attack by potato beetles, a destructive insect that causes significant damage to potato crops. You may have already eaten some of these potatoes without knowing it, for they do not have to be labeled as a food produced through biotechnology.

a. Identify some of the benefits and risks associated with genetically engineered potatoes.
b. Would you knowingly eat potato chips made from New Leaf potatoes? Explain the reasons for your opinion.

By using a cloned growth hormone, salmon and rainbow trout grow to sufficient size to be sold in supermarkets in one year rather than the two or three years normally required. Pork with only 10–20% of the usual amount of fat, and sheep whose fleece can be pulled off like a sweater during shearing season are also on the genetic engineering drawing boards. Genetic engineers are also exploring the possibilities of using plants to produce edible vaccines or insulin, plastics, and even a naturally grown blend of cotton and polyester. Extracting plastics from plants on a commercial scale would conserve significant quantities of petroleum, the conventional current feedstock for all commercial plastics.

"Molecular pharming" is the term coined to describe the practice of using domesticated animals to produce drugs and other medically significant substances. Transgenic beef and dairy cattle, hogs, and fish have been developed by transferring foreign DNA into the nuclei of eggs cells or embryos. For example, transgenic pigs grown from pig embryos injected with human hemoglobin genes produce human hemoglobin, which may be used as a blood substitute. Genetic engineering, however, is more than simply a collection of techniques. There is a human face to it among those who have benefited medically from it. Before recombinant DNA, there was no hepatitis C vaccine, insufficient erythropoieten, a protein used to stimulate red blood cell growth in dialysis patients, and a lack of tissue plasminogen activator (TPA) to dissolve blood clots in cardiac patients.

Using transgenic plants and animals to make compounds currently produced mainly in vats and vials is a far-reaching concept from the standpoint of economics and resource management. In the October 23, 1998 issue of *Science,* DuPont board chair Jack Krol commented about this matter: "In the 20th century, chemical companies made most of their products with nonliving systems. In the next century, we will make many of them with living systems." In support of that premise, DuPont has research underway to use microbes and plants to produce a wide range of compounds, from chiral drugs to plastics. Jerry Caulder, an agricultural genetic engineering entrepreneur, is also "bullish" on this approach. In the same issue of *Science,* he asked: "Organisms are the best chemists in the world. Why not use them to produce the chemical feedstocks you want rather than using petroleum?"

Used in almost all of the 200,000 kidney dialysis treatments given annually, erythropoieten stimulates red blood cell growth, thus reducing anemia and the need for blood transfusions.

Section 4.8 deals with petroleum; Section 9.1 discusses polymers derived from petroleum.

12.21 Consider This: Acceptance of the New Wave of Agriculture

Biotech foods have not been universally accepted, particularly in Europe. Organic chemists in the United States and in Europe will be heavily involved in the research to use microbes and plants to produce a wide range of compounds. What advice would you offer your friend, who is a chemistry major from Europe, about the "hot" areas of research? Would your advice be different if the friend were a chemistry major from the United States?

12.11 Hello Dolly: Cloning Mammals

Although easily fleeced sheep are not yet available, a sheep did make headlines in 1984, an ironically appropriate year, given what was reported. Researchers cloned a sheep by a technique called **nuclear transfer**. Using a very thin, hollow needle, nuclei were removed from the cells of sheep embryos and transferred into unfertilized sheep eggs from which the nuclei had been removed. The inserted nuclei gave the eggs a complete set of genes. The eggs were then implanted into a sheep's uterus, carried full term, and a lamb was born, the first mammal produced by cloning. Soon after, by using nuclear transfer, embryonic cells were cloned into cows, rabbits, goats, and rhesus monkeys.

In 1997, headlines worldwide, such as in the *New York Times,* heralded the birth of another lamb, Dolly. But Dolly was no ordinary lamb, even for a cloned one. She was the product of the first cloning of a mammal using nuclei from *adult* cells, long thought to be an extremely difficult, if not impossible, process, the grist of science fiction. That barrier fell in July 1996 with the birth of Dolly at the Roslin Institute, near Edinburgh, Scotland. She has a unique pedigree—a mammal who is an exact genetic copy, a clone, of an *adult,* but without a father. Ian Wilmut, the surrogate "father" of Dolly, used nuclei

Dolly is the first mammal cloned from a mature animal, but she was not the first cloned mammal. That mammal was cloned from embryonic cells, not adult ones.

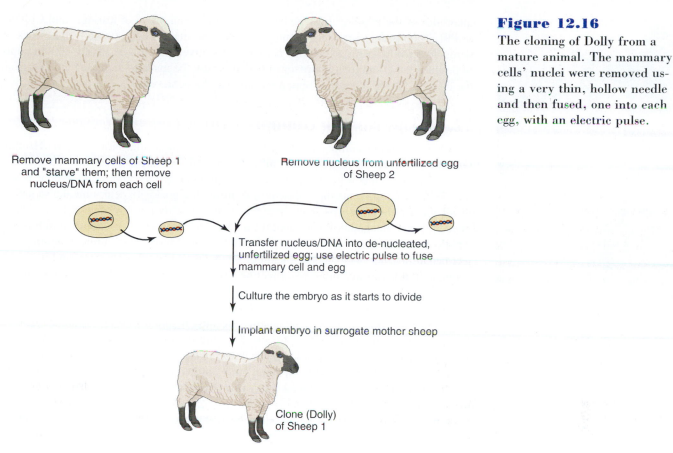

Figure 12.16
The cloning of Dolly from a mature animal. The mammary cells' nuclei were removed using a very thin, hollow needle and then fused, one into each egg, with an electric pulse.

Remove mammary cells of Sheep 1 and "starve" them; then remove nucleus/DNA from each cell

Remove nucleus from unfertilized egg of Sheep 2

Transfer nucleus/DNA into de-nucleated, unfertilized egg; use electric pulse to fuse mammary cell and egg

Culture the embryo as it starts to divide

Implant embryo in surrogate mother sheep

Clone (Dolly) of Sheep 1

removed from mammary glands of a 6-year old ewe and transferred them, one into each unfertilized sheep egg from which the nucleus had been removed. These "fertilized" eggs were then implanted into the uterus of a sheep, culminating in the birth of Dolly in July 1996. This process was similar, but not identical, to earlier clonings of mammals (Figure 12.16).

It took 277 such transfers to create Dolly. Until she came along, conventional wisdom had it that adult cells lacked the workable versions of all the genes necessary to create an entire organism. To circumvent this possible difficulty, Wilmut and his co-workers deprived the mammary cells of nutrients for five days before extracting their nuclei, forcing the cells out of their normal growth pattern and into a resting stage. Speculation is that this procedure might have increased the likelihood that once implanted into an egg, the chromosomes could be reprogrammed to establish the growth of an entire organism. Since Dolly, other researchers have used nuclear transfer with adult tissue to clone mice and cows, thus demonstrating the applicability of the technique to other mammals.

It should be kept in mind that only the nucleus of the adult ewe's donor cell was transferred to the egg. Thus, Dolly shares most, *but not all*, of her genes with her nuclear donor. Several dozen genes are outside the nucleus in her cell's mitochondria. These genes came from the recipient egg, not the donor nucleus.

The goal of Wilmut's work, supported by PPL Therapeutics, is to develop a cloning technique to produce animals on a large scale that can be used to create pharmaceuticals in their milk. The target drugs are ones that are difficult and expensive to obtain by conventional methods. PPL Therapeutics has created animals that provide a protein in commercially useful quantities that could possibly be used to treat cystic fibrosis. The protein inhibits the enzyme elastase that breaks down connective tissue and is found in excess in the lungs of patients with the disease. This research has produced sufficient

Dolly's cells contain the DNA of only one parent, unlike identical twins (natural clones), who share DNA from both parents.

Mitochondria, sometimes called the "powerhouse of the cell" are structures that are vital to energy production. They are within a cell, but not in its nucleus.

quantities of the inhibitor for it to be in the clinical trial phase of testing. Alan Colman of PPL Therapeutics sees such cloning as a way "to make a cell into an instant flock or herd." But Dolly seems to be a longshot success, given that it took nearly 300 attempts to produce her. Therefore, improvements will need to be made to increase the yield from the cloning attempts before Colman's vision can be achieved.

12.12 The Human Genome Project

The most ambitious component of current research in molecular biology is the **Human Genome Project**, an effort to map all the genes in the human organism. This means identifying all 100,000 genes found on the 46 human chromosomes and, where possible, discovering the traits they convey. In addition, the goal is to determine the sequence of all three billion base pairs in the entire genome. This massive enterprise has been likened to the Manhattan Project that resulted in the creation of the atomic bomb or the Apollo program that sent American astronauts to the Moon. The Human Genome Project began in 1989, with James Watson of DNA fame as its first director, and it is estimated that at least three more years will be required to complete it. The cost for this international venture will be about $3 billion, or one dollar per base pair.

Often the reason why scientists undertake a research project is not unlike the reason why mountain climbers climb a mountain: "Because it's there." But the Human Genome Project involves more than the spirit of adventure or the thrill of discovery. The more we know about our genetic makeup, the more likely we will be to diagnose and cure disease, understand human development, trace our evolutionary roots, recreate our family tree, and utilize these discoveries for, one hopes, the benefit of our species and our fellow species.

You may well wonder whose DNA has been selected for this unprecedented scrutiny. In fact, most of the DNA being analyzed in the Human Genome Project happens to come from members of over 60 multigenerational French families whose lineage is well documented. But it does not really matter; any one of us could serve as a DNA donor and a representative of *Homo sapiens*. In spite of our apparent differences and our long history of disputes based on those differences, the DNA of all humans is remarkably similar. It differs from individual to individual by about 0.1% of the base sequences. Within that tiny fraction resides our genetic uniqueness. Biology and chemistry provide irrefutable evidence of a lesson we as nations and as individuals have been slow to learn: we are all brothers and sisters.

The Human Genome Project appears to be up and running, with new genes being identified daily. But determining the sequence of the DNA base pairs is difficult and time-consuming. The smallest human chromosome contains 50,000,000 base pairs. Given those numbers, researchers are attempting to develop an automated base sequencer that is both fast and accurate. The accuracy is very important, because a single missed base will throw off all the subsequent base assignments, just as a skipped button hole is transmitted down the length of a shirt. The target error rate is one in 10,000 bases, a 99.99% level of accuracy. As of July 1999, 11% of the genome had been sequenced, with the sequencing rate roughly doubling each year.

Unfortunately, the spirit of cooperation has already been strained as the project has progressed. Some researchers have seen the potential for financial gain in patenting genes they have identified. The first patent application created a controversy that involved both scientists and the broader public. The NIH are funding, supervising, and actually doing much of the research, and NIH applied for some of the first patents. Spokespersons for NIH argued that by securing patents on some of the genes, they would be protecting public investment in the project. Critics retorted that because government sources are providing most of the funds, the results should belong to the public. Others reasoned that free and open exchange of information is essential in such a complex international project, and that patents would inhibit this communication. James Watson called the idea of patenting genes "sheer lunacy." In 1992, the U.S. Patent Office ruled that gene fragments could not be patented unless they had some known function. But patent applications have been filed in other countries, and industrial firms in-

Dolly and the lamb she gave birth to by conventional means.

Reading off bases at the rate of 10 per second nonstop would require 9.5 years for you to get through all three billion base pairs in your genome. In computer terms, if base pairs were considered as bytes, you would need three gigabytes (3×10^9 bytes) of computer storage to accommodate the entire human genome.

U.S. patents have been issued for genetically engineered organisms, but not DNA sequences.

volved in the project are interested in patent protection. Therefore, the patent problem remains unresolved.

We have already mentioned some of the potential benefits of the Human Genome Project, but the enterprise is not without its critics. There are those who point out that a genetic map ("being caught with your genes down") represents the ultimate invasion of privacy. Information about an individual's genetic makeup might be used by insurance companies to discriminate against those with a hereditary tendency toward certain diseases, by businesses to refuse to hire people who may be genetically at risk, or by ruthless governments to identify the "genetically inferior."

12.22 Consider This: The Human Genome Project

In 1999, a government report asserted, "Sequence completion by the end of 2003 is a major challenge, but within reach and well worth the risks and effort." How well are we progressing to meet the goal of complete sequencing? Check one of the national web sites on the Human Genome Project to find out the current status of the project. Note also any new developments posted on the site, especially any applications to gene therapy.

12.13 The New Prometheus(?)

Nature is an indispensable aid and ally in medicinal and biological chemistry. Much of our success has come from understanding and imitating natural processes. For centuries, animal breeders, agricultural researchers, and observant farmers have brought about genetic transformations in animals and plants by selective breeding. Recent applications of chemical methods to biological systems have greatly increased our capacity to effect such changes. Molecular engineering has made it possible to create nucleic acids, proteins, enzymes, hormones, drugs, and other biologically important molecules that do not exist in nature. The new molecules can be designed to be more efficient catalysts than their naturally occurring counterparts, more effective and less toxic drugs for treating a wide range of diseases, or modified hormones that actually work better than the original. It is not at all fanciful to imagine a whole range of enzymes, engineered to consume environmentally hazardous wastes that are impervious to naturally occurring enzymes. Or creating an enzyme that far surpasses the one that catalyzes photosynthesis, one of the most inefficient of all natural enzymes. As Mark Twain suggested: "Predictions are extremely difficult, especially those about the future." Yet, it is likely that AIDS and at least some forms of cancer may some day become as infrequent as polio or smallpox are now.

Even more tantalizing is the possibility of eradicating certain genetic defects. Our growing knowledge of the human genome, coupled with our understanding of the chemistry of genetics, holds the promise of altering our inheritance. Prospects include the elimination of sickle-cell anemia, diabetes, hemophilia, phenylketonuria, and dozens of other serious hereditary traits.

The next logical step would seem to be the creation of new organisms. Scientists have already cloned "new and improved" mice, tomatoes, and other animals and vegetables. Mammals have been cloned from adult cells. As this book went into final production in 1999, South Korean researchers claimed to have cloned a four-cell human embryo that was an identical genetic copy of a 30-year-old woman. They reported injecting the DNA from one of her ovarian cells into her egg from which the nucleus had been removed. Although the researchers stopped the experiment at the four-cell stage, in accordance with the South Korean ban on more fully developed human embryos, their work at least raises the spectre of cloning human beings. Other researchers remain skeptical about the report, noting that the first four cell divisions occur spontaneously in human embryos. It is only when the 16-cell stage is reached that genes begin to direct further development. Ian Wilmut of the Roslin Institute, responding to the cloning report, said that the experiment did not provide the kind of information needed to determine whether human embryos can be cloned. In 1998, Richard Seeds, a Ph.D. physicist turned biologist, announced plans to launch a human cloning clinic in Chicago.

Human cloning has been dealt with in fiction such as Huxley's *Brave New World*—using eugenics to create different classes of people, and in films like Woody Allen's *Sleeper*—attempts to clone a dead dictator, and *The Boys From Brazil*—cloning Nazis. How should we view the possibility of cloning human beings, perhaps even "new and improved" ones? The question goes straight to our identity as a species and as individuals. Are we ready to cross this genetic Rubicon? With the possible exception of the South Korean case, cloning has been done only with somatic (non-reproductive) cells; ova and sperm (germ cells) have not been used in cloning or even in cases of gene therapy. This means that although the genetic makeup of the individual may be slightly altered by the gene therapy, her or his reproductive cells (germ cells) are not changed. Any therapeutic effect benefits only the individual and not his or her descendants. One could reason that the best way to treat a genetic disease or disability is to remove the defective gene from the gene pool by intentionally altering the suspect DNA in the sperm or ova. But one could also argue that *Homo sapiens* is not yet sufficiently wise to assume such god-like power. Our species has had some tragic experiences in this century with political leaders who used less subtle methods of "genetic cleansing."

12.23 Consider This: A Cloning Clinic?

Plans to develop a human cloning clinic were announced in January 1998. What are the risks and the benefits associated with such a clinic? Has the clinic been completed at this time?

And cloning through genetic engineering raises not only the possibility for humans to duplicate themselves by unconventional means, but also the chilling potential to design a master race or to subjugate or eliminate "defectives" through genetic manipulation. Hence, there is an intentional irony in the title of this final section. Prometheus was the demigod who stole fire and the flame of learning from the gods and brought these incomparable gifts to humanity. "The New Prometheus" is the subtitle of *Frankenstein*, Mary Shelley's classic study of scientific knowledge run amok. Using our ever-growing knowledge of the chemistry of heredity wisely, and well, will surely be one of the greatest challenges of the twenty-first century. As Ian Wilmut carefully points out about his discovery: "We are aware that there is potential for misuse, and we have provided information to ethicists and the Human Embryology Authority [of Britain]. We believe that it is important that society decides how to use this technology and make sure it prohibits what it wants to prohibit. It would be desperately sad if people started using this sort of technology with people." (*New York Times* February 23, 1997, p. 22)

12.24 Consider This: Send in the Clones

The late Isaac Asimov, noted science fiction writer and biochemist, co-authored this verse, called "The Misunderstood Clone"

> Oh, give me a clone
> Of my own flesh and bone
> With its Y chromosome changed to an X.
> And when it is grown
> Then my own little clone
> Will be of the opposite sex.

(From *The Sun Shines Bright* by Isaac Asimov. Copyright © 1981 by Nightfall, Inc. Used by permission of Doubleday, a division of Random House, Inc., and Ralph Vicinanza Agency.)

Evaluate whether the verse is actually describing gene therapy or cloning. Support your answer with an explanation of the two different processes.

12.25 Consider This: Cloning Humans: For Good or for Evil?

James D. Watson of DNA fame had this to say about manipulating germ cells to create superpersons. "When they are finally attempted, germ-line genetic manipulations will probably be done to change a death sentence into a life verdict—by creating children who are resistant to a deadly virus, for example, much the same way we can already protect plants from viruses by inserting antiviral DNA segments into their genomes."[1]

Draft a statement supporting or opposing Watson's position on this issue. Your argument should be grounded in the scientific principles learned in this chapter.

[1]*Time*, "All for the Good," January 11, 1999, vol. 153, No. 1.

Conclusion

The last chapter of this book, like almost all those that preceded it, ends with a dilemma: how can we balance the great potential benefits of modern chemical sciences and technology and the risks that seem inevitably to be part of the Faustian bargain that brought us knowledge? Throughout this text, the authors have occasionally looked, with myopic professorial vision, into the cloudy crystal ball of the future. It is in the nature of science that we cannot confidently predict what new discoveries will be made by tomorrow's chemists. Nor can we know the applications of those discoveries, good or bad. Such uncertainty is one of the delights of our discipline. A chemist must learn to live with ambiguity, indeed, to thrive on it, in the search to better understand the nature of atoms and their intricate combinations, in all their various guises.

But all citizens of this planet must at least develop a tolerance for ambiguity and a willingness to take reasonable risks, especially considering that life itself is a biological, intellectual, and emotional risk. Of course, we all seek to maximize benefits, but we must recognize that individual gain must sometimes be sacrificed for the benefit of society. We live in multiple contexts—the context of our families and friends, our towns and cities, our states, our countries, our special planet. We have responsibilities to all. *You*, the readers of this book, will help create the context of the future. We wish you well.

Faust is a literary figure, an old philosopher who sells his soul to the devil in exchange for knowledge and power.

Chapter Summary

Having studied this chapter, you should be able to:

- Understand the chemical composition of deoxyribonucleic acid (DNA): a polymer of nitrogen-containing bases, deoxyribose, and phosphate groups (12.1)
- Recognize the utility of Chargaff's rules: A = T, G = C (12.1)
- Interpret evidence for the double helical structure of DNA and its base pairing (12.2)
- Understand DNA replication (12.2)
- Describe the genetic code: codons of three bases corresponding to specific amino acids (12.3)
- Relate to the amount of information encoded in the human genome (12.3)
- Discuss the primary, secondary, and tertiary structure of proteins (12.4)
- Describe the molecular basis of sickle-cell anemia (12.4)
- Understand the production of human proteins (insulin and human growth hormone) in bacteria (12.5)

- Recognize the need to establish priorities for the use of limited supplies of drugs produced by recombinant DNA technology (12.5)
- Recognize recombinant DNA techniques: insertion of foreign DNA into bacteria (12.5)
- Relate how new drugs to treat hepatitis, herpes, cancer, heart attack, strokes, AIDS, etc., and new vaccines are developed via genetic engineering (12.6)
- Understand the genetic diagnosis of ALS, Alzheimer's disease, and other conditions (12.7)
- Understand the connection between DNA probes and the polymerase chain reaction and the factors related to appropriate responses to genetically diagnosed hereditary diseases (12.7)
- Describe gene therapy as a treatment for severe combined immunodeficiency disease, malignant melanoma, and other diseases (12.8)

- Link medical and ethical issues in approval of human gene therapy (12.8)
- Discuss DNA fingerprinting for identification and evolutionary and anthropological studies and the technical and legal issues associated with DNA fingerprinting (12.9)
- Refute the possibility and wisdom of cloning long-extinct animals (12.9)
- Describe the laboratory techniques in DNA fingerprinting (12.9)
- Describe transgenic organisms and their uses: pest-resistant and nitrogen-fixing plants; animals with genes to produce human proteins and other modified biochemicals (12.10)

- Discuss ethical issues associated with transgenic organisms (12.10)
- Understand mammalian cloning using nuclear transfer (12.11)
- Describe the role of nuclear transfer in the cloning of mammalian cells (12.11)
- Relate to the Human Genome Project: its aims, significance, and ethical implications (12.12)
- Debate issues associated with the prudent and ethical applications of mammalian cloning and genetic engineering (12.13)

Questions

Emphasizing Essentials

1. The letters DNA have been called three of the most important letters of the late twentieth century. What do they literally stand for?

2. **a.** Use Figure 12.1 to determine the percent increase in human life expectancy at birth in the United States between 1940 and 1980 for males and for females.

   b. Are these percentages the same as those calculated in 12.1 Your Turn? Why or why not?

3. Consider the structures in Figure 12.2.

   a. What functional groups are in adenine?

   b. What functional groups are in deoxyribose?

   c. What functional groups are in the phosphate group?

4. Consider the structural formula of deoxyribose given in Figure 12.2.

   a. Why is deoxyribose classed as a monosaccharide?

   b. What is the molecular formula for deoxyribose?

   c. Why isn't deoxyribose an acid in aqueous solution?

5. Equation 12.1 shows the general case for a nitrogen-containing base reacting with water. Use the structural formula of thymine in Figure 12.2 to write an equation showing how thymine could react with water to generate hydroxide ions.

6. **a.** What three types of units must be present in a nucleotide?

   b. What type of bonding holds these units together in a nucleotide?

7. Table 12.1 lists the base composition of DNA for various species. The four bases are adenine, cytosine, guanine, and thymine. What relationships exists among these bases, no matter what the species?

8. **a.** What happens experimentally during X-ray diffraction?

   b. The first X-ray diffraction patterns were of simple salts such as sodium chloride. X-ray diffraction studies of nucleic acids and proteins did not come until much later. Suggest why.

9. Figure 12.7 shows the pairing of nucleotide bases in DNA.

   a. What type of *intramolecular* bonding occurs within each base?

   b. What type of *intermolecular* bonding holds the base pairs together?

10. Identify the base sequence that is complementary to each of these sequences.

    a. ATGGCAT   **b.** TATCTAG

11. Given that the distance between adjacent bases is 0.34 nm, how many base pairs are present in a chromosome that is 3.0 cm long? $1 \text{ m} = 10^2 \text{ cm}$; $1 \text{ m} = 10^9 \text{ nm}$.

12. During cell division, as many as 90,000 nucleotides per minute can be added to the growing DNA chain.

    a. The shortest human chromosome contains 50 million bases. What is the minimum time required to form a strand of this chromosome?

    b. Determine the length of this chromosome (in cm) that would be formed in one minute if the distance between bases is 0.34 nm. $1 \text{ m} = 10^2 \text{ cm}$; $1 \text{ m} = 10^9 \text{ nm}$.

13. Twenty amino acids commonly occur in proteins.

    a. What is the *general* structural formula for all of these amino acids?

    b. What functional groups are present in all amino acids?

14. The text states that if you were to use the letters A, T, C, and G in a game of Scrabble, you could generate 64 different three-letter combinations. (Your Scrabble opponent would surely challenge some of these combinations!) Nature pairs the four bases A, T, C, and G, and uses the pairs to encode for amino acids. Write down all of the possible paired combinations of A, T, C, and G to find the maximum number of amino acids that can be encoded from these four bases. *Hint:* In nature, unlike Scrabble, a letter can be used more than once in forming a pair.

15. What is a codon and what is its role in the genetic code?

16. Only 61 of the possible 64 triplet codons specify certain amino acids. What is the function of the other three codons?

17. Describe what is meant by the primary, secondary, and tertiary structure of a protein. Is one of these more important than the others? Why or why not?

18. Explain how an error in the primary structure of a protein in hemoglobin causes sickle-cell anemia.

19. How can "recombinant DNA" technology be used to overcome a shortage of insulin for use by diabetics?

20. **a.** What basic molecular building blocks are found in viruses?

    b. Why are viruses referred to as "...almost living..." biochemical species?

c. Why is it not possible to treat a viral infection with antibiotics?

21. The letters DNA have been called three of the most important letters of the late twentieth century. A long-shot candidate for this honor might be PCR.

   a. What do those letters stand for?

   b. What problem did PCR technology successfully address?

   c. Name some examples of the success of PCR technology for diagnosis.

22. What is the minimum number of PCR cycles that would be needed to convert two DNA molecules to:

   a. 5000 DNA molecules?   c. 500,000 molecules?

   b. 50,000 molecules?

23. How does electrophoresis separate different molecules?

24. What is meant by the term "molecular pharming"?

25. a. What is the human genome project?

   b. Why is it a significant step in understanding the genetic basis of humans?

   c. What progress has been made in mapping the human genome?

Concentrating on Concepts

26. What is meant by the term "cloning"?

27. Life expectancy in the United States has increased dramatically during the twentieth century, as shown in Figure 12.1. Is a similar increase possible during the twenty-first century? Explain your answer.

28. Compare this representation with Figure 12.4.

Base 1   Base 2   Base 3

—Phosphate—Sugar—Phosphate—Sugar—Phosphate—Sugar—

Both show a segment of deoxyribonucleic acid. Discuss the strengths and weakness of each representation in conveying similar information.

29. Consider Chargaff's discovery that there are equal percentages of adenine and thymine and of cytosine and guanine in DNA. Was his discovery as important to understanding the nature of DNA as was Crick and Watson's discovery of the double helix? Give reasons for your answer.

30. Use Figure 12.7 to help explain why stable base pairing does *not* occur between adenosine and cytosine, thymine and guanine, adenine and guanine, and thymine and cytosine.

31. Errors sometimes occur in the base sequence of a strand of DNA. But, not all of these errors result in the incorporation of an incorrect amino acid in a protein for which the DNA codes. Explain how this happens, and why it is advantageous.

32. Human insulin and human growth hormone have both been made through the use of recombinant DNA technology. Which do you believe is a more significant use of this technology and why? Discuss what factors have influenced your opinion.

33. DNA probes are being used for diagnosis. Describe the principle of using DNA probes to identify *Helicobacter pylori*, the cause of gastric ulcers and possibly stomach cancer.

34. a. The first successful application of gene therapy to a human being was in 1990. What is meant by "gene therapy" and what was the disease treated in that landmark case?

   b. Has gene therapy been used successfully for treating other diseases as well?

   c. Will it work for all diseases?

35. Lou Gehrig's disease is caused by the alteration of a single amino acid in the enzyme superoxide dismutase. How many base pairs are responsible for specifying this amino acid in the gene that codes for the protein? What is the minimum number of base pairs that would have to be changed to produce the disease?

36. Do you favor the patenting of genes? What are the advantages and disadvantages of this approach?

37. How widely available is the New Leaf Potato? If you wanted to plant this in your garden, would you be able to obtain these potato plants? What are the reasons that you would want to obtain this plant?

38. Consider the information in Table 12.3. Use a pictorial representation of this information to convey the differences among the countries in using transgenic crops.

39. Consider the idea of mixing genes as an improvement on nature.

   a. What are transgenic organisms?

   b. Why is the alteration of the genetic makeup of plants by genetic engineering preferred to traditional crossbreeding methods?

40. Consider some of the successful transgenic plant experiments given in Table 12.2. What generalizations can be drawn between the source of the genes, the transgenic plant, and the objectives of the experiment?

Exploring Extensions

41. Dolly's birth surprised genetic engineering experts and shocked members of the media who had the responsibility of reporting the birth and the method. Locate one or two different early reports about the cloning of Dolly. Evaluate these reports for their scientific accuracy, and for what they reveal about the opinion of experts at that time.

42. Use the following information to act as a Sceptical Chymist in checking the correctness of the claim that the DNA in an adult human would stretch from the Earth to the moon and back more than a million times.

Useful Information	
Distance, Earth to moon	3.8×10^5 km
Adult human	1×10^{13} DNA-nucleated cells
Length, one stretched human DNA thread	2 m

43. Consider the structural formula of deoxyribose shown in Figure 12.2. The prefix "deoxy" means without oxygen; the —OH group is replaced by a hydrogen atom. In the specific case of deoxyribose, the —OH group replaced by a hydrogen was bonded to the only carbon in the ring that bonds to two hydrogens in deoxyribose. Use this information to draw the

structural formula of ribose side-by-side with the structural formula of deoxyribose.

44. The text states that the more closely species are related, the more similar the DNA base compositions are in those species. The Sceptical Chymist is having trouble believing this, particularly if it means a close relationship between a fruit fly and a bacterium. Use the information in Table 12.1 to determine whether it supports the generalization about similar base pairs in similar species.

45. Perhaps you have learned some memory aids when taking music lessons (Every Good Boy Does Fine), memorizing the names of the Great Lakes (HOMES), or learning about oxidation and reduction (OIL RIG). One of the authors learned "All-Together, Go-California" as the mnemonic to remember the correct base pairings in DNA.

   a. What is the relationship in this mnemonic to DNA base pairing?

   b. Design a different memory aid that will help you remember such base pairings.

46. Of the major players in the discovery of DNA's structure, only Rosalind Franklin had a degree in chemistry. What was her background and experience that enabled her to make significant contributions? Did her contributions receive adequate credit and recognition? Write a short report on the results of your findings, giving references to web information or other citations.

47. Classify the objectives of the experiments in Table 12.2 into different categories and rate the relative importance of each of these categories.

48. Transgenic plants have not been widely accepted in all countries. What are some of the reasons for their rejection in some European markets?

49. Genetic diseases are also called inborn errors of metabolism. You may be familiar with some of these diseases, such as hemophilia, PKU, Tay-Sachs, or sickle-cell anemia. One that does not get much attention is a condition known as Niemann-Pick disease. Find out what inborn metabolic error causes this condition, how many children are born with this disease in the United States each year, what treatments are available, and whether a cure is possible. Write a report to be discussed with your classmates.

50. Remains of a soldier shot down over Vietnam in 1972 and buried in the Tomb of the Unknowns in 1984 were identified in 1998 by the Armed Forces DNA Identification Laboratory (see Section 12.9). The confirmatory tests were based on matching mitochondrial DNA (mtDNA) from the soldier's bones with mtDNA from his mother.

   a. Why were these tests not done before the remains were placed in the Tomb?

   b. Find more information about the specificities of test results when using nuclear DNA rather than mtDNA.

   c. As of this writing, scientists at the National Institute of Standards & Technology are preparing an mtDNA standard for forensic laboratories to measure the accuracy of their results. Has that standard been issued?

Measure for Measure: Conversion Factors and Constants

Metric Prefixes

deci (d) $1/10$ $= 10^{-1}$
centi (c) $1/100$ $= 10^{-2}$
milli (m) $1/1000$ $= 10^{-3}$
micro (μ) $1/10^6$ $= 10^{-6}$
nano (n) $1/10^9$ $= 10^{-9}$

deka (da) $10 = 10^1$
hecto (h) $100 = 10^2$
kilo (k) $1000 = 10^3$
mega (M) 10^6
giga (G) 10^9

Length

1 centimeter (cm) = 0.394 inch (in)
1 meter (m) = 39.4 in = 3.24 feet (ft) = 1.08 yard (yd)
1 kilometer (km) = 0.621 miles (mi)
1 in = 2.54 cm = 0.0833 ft
1 ft = 30.5 cm = 0.305 m = 12 in
1 yd = 91.44 cm = 0.9144 m = 3 ft = 36 in
1 mi = 1.61 km

Volume

1 cubic centimeter (cm^3) = 1 milliliter (mL)
1 liter (L) = 1000 mL = 1000 cm^3 = 1.057 quarts (qt)
1 qt = 0.946 L
1 gallon (gal) = 4 qt = 3.78 L

Mass

1 gram (g) = 0.0353 ounce (oz) = 0.00220 pound (lb)
1 kilogram (kg) = 1000 g = 2.20 lb
1 metric ton (mt) = 1000 kg = 2200 lb = 1.10 ton (t)
1 lb = 454 g = 0.454 kg
1 ton (t) = 2000 lb = 908 kg = 0.908 mt

Time

1 year (yr) = 365.24 days (d)   1 day = 24 hours (hr or h)
1 hr = 60 minutes (min)   1 min = 60 seconds (s)

Energy

1 joule (J) = 0.239 calorie (cal)
1 cal = 4.184 joule (J)
1 kilocalorie (kcal) = 1 dietary Calorie (Cal)
        = 4184 J = 4.184 kilojoule (kJ)
1 kilowatt-hour (kWh) = 3,600,000 J = 3.60×10^6 J

Constants

Speed of light (c) = 3.00×10^8 m/s
Planck's constant (h) = 6.63×10^{-34} J $\cdot$ s
Avogadro's number (N_A) = 6.02×10^{23} objects per mole
Atomic mass unit (m) = 1.66×10^{-24} g

The Power of Exponents

Scientific or exponential notation provides a compact and convenient way of writing very large and very small numbers. The idea is to make use of positive and negative powers of 10. Positive exponents are used to represent large numbers. The exponent, written as a superscript, indicates how many times 10 is multiplied by itself. For example:

$$10^1 = 10$$
$$10^2 = 10 \times 10 = 100$$
$$10^3 = 10 \times 10 \times 10 = 1000$$

Note that the positive exponent is equal to the number of zeros between the 1 and the decimal point. Thus, 10^6 corresponds to 1 followed by six zeros or 1,000,000. This same rule applies to 10^0, which equals 1. One billion, 1,000,000,000, can be written as 10^9.

When 10 is raised to a negative exponent, the number is always less than 1. This is because a negative exponent implies a reciprocal, that is, 1 over 10 raised to the corresponding positive exponent. For example:

$$10^{-1} = 1/10^1 = 1/10 = 0.1$$
$$10^{-2} = 1/10^2 = 1/100 = 0.01$$
$$10^{-3} = 1/10^3 = 1/1000 = 0.001$$

It follows that the larger the negative exponent, the smaller the number. The negative exponent is always one more than the number of zeros between the decimal point and the 1. Thus, 10^{-4} is equal to 0.0001. Conversely, 0.000001 in scientific notation is 10^{-6}.

Of course, most of the quantities and constants used in chemistry are not simple whole number powers of ten. For example, Avogadro's number is 6.02×10^{23} or 6.02 multiplied by a number equal to 1 followed by 23 zeros. Written out, this corresponds to $6.02 \times 100,000,000,000,000,000,000,000$ or 602,000,000,000,000,000,000,000. Switching to very small numbers, a wavelength at which carbon dioxide absorbs infrared radiation is 4.257×10^{-6} m. This number is the same as 4.257×0.000001 or 0.000004257.

Your Turn

Express the following numbers in scientific notation.

a. 10,000   **b.** 430   **c.** 9876.54
d. 0.000001   **e.** 0.007   **f.** 0.05339

Express the following numbers in conventional decimal notation.

a. 1×10^6 **b.** 3.123×10^6 **c.** 25×10^5
d. 1×10^{-5} **e.** 6.023×10^{-7} **f.** 1.723×10^{-16}

Clearing the Logjam

You may have encountered logarithms in mathematics courses but wondered if you would ever use them. In fact, logarithms (or "logs" for short) are extremely useful in many areas of science. The essential idea is that they make it much easier to deal with very large *ranges* of numbers, for example, moving by powers of 10 from 0.0001 to 1,000,000.

It is certainly likely that you have met logarithmic scales without knowing it. The Richter scale for magnitudes of earthquakes is one example. On this scale, an earthquake of magnitude 6 is 10 times more powerful than one of magnitude 5. An earthquake of magnitude 8 would be 100 times more powerful than one of magnitude 6. Another example is the decibel scale. Each increase of 10 units represents a tenfold increase in sound level. Thus, a normal conversation at 1 meter (dB 60) is 10 times louder than quiet music (dB 50). Loud music (dB 70) and extremely loud music (dB 80) are 10 times and 100 times as loud, respectively, as a normal conversation.

A simple exercise using a pocket calculator can be a good way to learn about logs. You will need a calculator that "does" logs and preferably has a "scientific notation" option. Start by finding the logarithm of 10. Simply enter 10 and press the "log" button. The answer should be 1. Next find the log of 100 and then the log of 1000. Write down the answers. What pattern do you see? (The pattern may be more obvious if you recall that 100 can be written as 10^2 and 1000 is the same as 10^3.) Predict the log of 10,000 and then check it out. Then try the log of 0.1 or 10^{-1} and log of 0.01 (10^{-2}). Predict the log of 0.0001 and check it out.

So far so good, but we have only been considering whole-number powers of 10. It would be helpful to be able to obtain the logarithm of any number. Once again, your handy little electronic wizard comes to the rescue. Try calculating the logs of 20 and 200, then 50 (5×10^1) and 500 (5×10^2). Predict the log of 5×10^3 or 5000. Now for something slightly more tricky: the log of 0.05. Try to make sense of the answer. Finally, try the log of 2473 and the log of 0.000404. In each case, does the answer seem to be in the right ballpark? If you see any other interesting relationships, you may want to experiment further.

In Chapter 6, the concept of pH is introduced as a quantitative way to describe the acidity of a substance. A pH value is simply a special case of a logarithmic relationship. It is defined as the negative of the logarithm of the H^+ concentration, expressed as molarity. Expressing this relationship as a mathematical equation: $pH = -\log (M_{H^+})$. The negative sign indicates an inverse relationship: as the H^+ concentration diminishes, the pH increases. Let's apply the equation by using it to calculate the pH of a beverage with a hydrogen ion concentration, M_{H^+}, of 0.000546 mole/liter. We first set up the mathematical equation and substitute the hydrogen ion concentration into it.

$$pH = -\log (M_{H^+}) = -\log (5.46 \times 10^{-4}\ M)$$

Next, we take the negative logarithm of the H^+ concentration by entering it into a calculator and pressing the log button, then the "minus" key. This gives 3.26 as the pH of the beverage. Apply the pH relationship to calculate the pH of milk with a hydrogen ion concentration of $2.20 \times 10^{-7}\ M$.

If we can convert hydrogen ion concentration into pH, how do we go in the reverse direction, that is, how to convert pH into a hydrogen ion concentration? Your calculator can do this for you if it has a button labelled "10^x." (Alternatively, it may use two buttons: first "Inv" and then "log.") To demonstrate the procedure, suppose you wish to find the hydrogen ion concentration of human blood with a pH of 7.40. Proceed as follows: enter 7.40 and hit 10^x (or follow whatever steps are appropriate for your calculator). The display should give the hydrogen ion concentration as $3.98 \times 10^{-8}\ M$. Now apply the same procedure to calculate the H^+ concentration of an acid rain with a pH of 3.6.

Test yourself further by determining: the H^+ concentration of tomato juice with a pH of 4.8; the pH of milk of magnesia with an H^+ concentration of $3.16 \times 10^{-11}\ M$.

Answers to Your Turn Questions Not Answered in Text

Chapter 1

1.11 **d.** element   **e.** compound   **f.** mixture

1.12 **c.** hydrogen, oxygen   **d.** carbon, hydrogen, oxygen

1.13 **c.** hydrogen chloride   **d.** sodium sulfide

1.14 **c.** calcium sulfide   **d.** lithium nitride

1.15 **b.** A molecule of the compound represented by the formula SO_2 consists of one atom of the element sulfur combined with two atoms of the element oxygen.

     c. A molecule of the compound represented by the formula SO_3 consists of one atom of the element sulfur combined with three atoms of the element oxygen.

     d. A molecule of the compound represented by the formula O_3 consists of three atoms of the element oxygen.

1.16 **c.** sulfur trioxide   **d.** trioxygen (commonly called ozone)

1.17 **b.** Balanced equation: $N_2 + 2\ O_2 \longrightarrow 2\ NO_2$

     Sphere Equation:

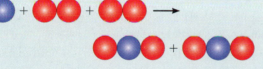

1.18 **b.** $2\ C_4H_{10} + 13\ O_2 \longrightarrow 8\ CO_2 + 10\ H_2O$

1.19 In Equation 1.6, there are 16 C, 36 H, and 34 O on each side of the equation.

1.26 $\dfrac{2 \times 10^{17}\ \text{molecules}}{5 \times 10^9\ \text{people}} = \dfrac{4 \times 10^7\ \text{molecules}}{\text{person}}$

        $= \dfrac{40{,}000{,}000\ \text{molecules}}{\text{person}}$

Chapter 2

2.2 **c.** 79 protons, 79 electrons   **d.** 24 protons, 24 electrons

2.3 **c.** 5 (Group 5A)   **d.** 2 (Group 2A)

2.4 Oxygen, sulfur, selenium, and tellurium are all in Group 6A. Each atom has 6 outer electrons.

2.6 **b.** There are 7 outer electrons per atom of iodine, for a total of 14 outer electrons in I_2. This is the Lewis structure.

$$:\!\overset{..}{\underset{..}{I}}:\overset{..}{\underset{..}{I}}\!:$$

2.7 **b.** In dichlorodifluoromethane, CCl_2F_2, the carbon has 4 outer electrons, each of the 2 chlorine atoms has 7 outer

electrons, and each of the two fluorine atoms has 7 electrons. The total number of outer electrons is 32. This is the Lewis structure.

$$\begin{array}{c} :\overset{..}{F}: \\ :Cl\!:\!C\!:\!Cl: \\ :\overset{..}{F}: \end{array}$$

2.8 **b.** In sulfur dioxide, SO_2, the sulfur has 6 outer electrons and each of the two oxygen atoms has 6 outer electrons. The total number of outer electrons is 18. This is the Lewis structure.

$$:\overset{..}{O}::S:\overset{..}{\underset{..}{O}}: \longleftrightarrow :\overset{..}{\underset{..}{O}}:S::\overset{..}{O}:$$

Chapter 3

3.7 **b.** tetrahedral,   [structure with C central, Cl top, F left, F right, Cl bottom] 109.5°   **c.** bent, $H\!-\!\overset{..}{\underset{..}{S}}:$ 109.5° H

3.8 **a.** bent, $:\overset{..}{O}\!-\!S:$ 109.5° $:\overset{..}{\underset{..}{O}}:$ **c.** triangular planar, [structure with S central, O double bond top, O left, O right] 120°

3.10 **a.** Au-197 has 79 protons, 118 neutrons, and 79 electrons.

     b. Au-198 has 79 protons, 119 neutrons, and 79 electrons. Only the number of neutrons has changed.

3.11 Given that there are only two isotopes, the other isotope must have a smaller number of neutrons because the weighted average of the two isotopes is just under 7. The other isotope is Li-6, which makes up 7.5% of all lithium atoms.

3.12 **b.** 1×10^{-10} g   **c.** 1×10^{-7} g

3.13 **b.** 32.0 g/mole NO   **c.** 137.5 g/mole $CFCl_3$

3.14 **b.** 0.636 g N/g N_2O; 63.6% N

3.15 **b.** 71.1 million metric tons

Chapter 4

4.5 **c.** 390 kJ

     d. 3.2×10^3 blocks

4.12 The ultraviolet radiation absorbed by O_2, which has greater energy than the radiation absorbed by O_3, has a shorter wavelength. Energy is inversely proportional to wavelength.

4.17 18% oxygen in MTBE

4.23 b. 1.4×10^{12} g and 1.4×10^6 metric tons

4.27 b. 1.8×10^3 lb CO_2 **c.** 31 lb SO_2

Chapter 5

5.6 a. soluble
   b. soluble
   c. partially soluble
   d. soluble
   e. insoluble
   f. soluble (pure aspirin)

5.7 a. 975 mg RDA or approximately 1000 mg
   b. 25 bottles

5.8 a. 16 ppb; 1.6×10^{-2} ppm
   b. over standard of 15 ppb

5.9 a. 7.5×10^{-1} mole of NaCl in 500 mL of 1.5 M NaCl and 7.5×10^{-2} mole of NaCl in 500 mL of 0.15 M NaCl
   b. To calculate molarity, divide moles of solute by liters of solution. 0.5 moles of NaCl in 250 mL of solution is 2 M NaCl; 0.6 moles of NaCl in 200 mL of solution is 3 M NaCl, so the second solution is more concentrated.

5.10 a. H—F
   b. O—H
   c. N—O

5.14 b. Mg^{2+}, the Lewis structures are: $\cdot Mg \cdot$ and Mg^{2+}

   c. O^{2-}, the Lewis structures are: $\cdot \ddot{O} \cdot$ and $\left[\ddot{\underset{\cdot\cdot}{O}} \colon \right]^{2-}$

   d. Al^{3+}, the Lewis structures are: $\cdot Al \cdot$ and Al^{3+}

5.15 b. KF
   c. Li_2O
   d. $SrBr_2$

5.16 c. sodium hydrogen carbonate or sodium bicarbonate
   d. calcium carbonate
   e. magnesium phosphate

5.17 b. Li_2CO_3
   c. KNO_3
   d. $BaSO_4$

5.19 b. soluble
   c. insoluble
   d. insoluble

5.20

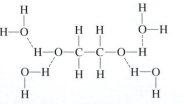

5.26 a. $Ca(HCO_3)_2$ **b.** $MgSO_4$ **c.** $MgCl_2$

5.29 a. 20 ppb = 20 μg/L; higher concentration.
   0.003 mg/L = 3 μg/L = 3 ppb
   b. 20 ppb > 15 ppb standard; 3 ppb < 15 ppb standard

5.30 b. 5.5 ppm
   c. out of range; dilute sample

Chapter 6

6.2 a. $HI(aq) \longrightarrow H^+(aq) + I^-(aq)$
   b. $HNO_3(aq) \longrightarrow H^+(aq) + NO_3^-(aq)$
   c. $H_2SO_4(aq) \longrightarrow H^+(aq) + HSO_4^-(aq)$
   d. $H_3PO_4(aq) \longrightarrow H^+(aq) + H_2PO_4^-(aq)$

6.3 a. $KOH(aq) \longrightarrow K^+(aq) + OH^-(aq)$
   b. $LiOH(aq) \longrightarrow Li^+(aq) + OH^-(aq)$
   c. $Ca(OH)_2(aq) \longrightarrow Ca^{2+}(aq) + 2\ OH^-(aq)$

6.4 b. $H_2SO_4(aq) + 2\ NaOH(aq) \longrightarrow Na_2SO_4(aq) + 2\ H_2O(aq)$
   $2\ H^+(aq) + SO_4^{2-}(aq) + 2\ Na^+(aq) + 2\ OH^-(aq) \longrightarrow$
   $\qquad\qquad 2\ Na^+(aq) + SO_4^{2-}(aq) + 2\ H_2O(aq)$
   $2\ H^+(aq) + 2\ OH^-(aq) \longrightarrow 2\ H_2O(aq)$
   $H^+(aq) + OH^-(aq) \longrightarrow H_2O(aq)$
   c. $2\ H_3PO_4(aq) + 3\ Mg(OH)_2(aq) \longrightarrow$
   $\qquad\qquad\qquad Mg_3(PO_4)_2(aq) + 6\ H_2O(aq)$
   $6\ H^+(aq) + 2\ PO_4^{3-}(aq) + 3\ Mg^{2+}(aq) + 6\ OH^-(aq)$
   $\qquad \longrightarrow 3\ Mg^{2+}(aq) + 2\ PO_4^{3-}(aq) + 6\ H_2O(aq)$
   $6\ H^+(aq) + 6\ OH^-(aq) \longrightarrow 6\ H_2O(aq)$
   $H^+(aq) + OH^-(aq) \longrightarrow H_2O(aq)$

6.5 a. acidic
   b. basic
   c. basic

6.7 pure water (7.0), club soda (4.8), tomatoes (4.2), Coca-Cola (3.1), apples (3.0), vinegar (2.5), lemons (2.3)

6.10 c. 3×10^4 tons SO_2

6.13 a. $Ca(OH)_2(aq) + 2\ HCl(aq) \longrightarrow CaCl_2(aq) + 2\ H_2O(l)$
   b. $CaO(s) + H_2O(l) \longrightarrow Ca(OH)_2(aq)$

Chapter 7

7.4 b. $_{0}^{1}n + _{92}^{235}U \longrightarrow [_{92}^{236}U] \longrightarrow _{52}^{137}Te + _{40}^{97}Zr + 2\ (_{0}^{1}n)$

7.5 $_{0}^{1}n + _{92}^{235}U \longrightarrow [_{92}^{236}U] \longrightarrow _{54}^{143}Xe + _{38}^{90}Sr + 3\ (_{0}^{1}n)$

7.6 1.00×10^{14} J/day; 1.11 g

7.10 b. $_{86}^{222}Rn \longrightarrow _{84}^{218}Po + _{2}^{4}He$
   c. $_{53}^{131}I \longrightarrow _{54}^{131}Xe + _{-1}^{0}e$

7.14 a. $_{88}^{226}Ra \longrightarrow _{86}^{222}Rn + _{2}^{4}He$
   b. Four half-lives, or 15.2 days would have to elapse.

Chapter 8

8.3 The difference in the energy, (285 kJ − 242 kJ) or 43 kJ, represents the amount of energy needed to change liquid water to gaseous water.

8.14 The element gallium, Ga, is in Group 3A, and has 3 outer electrons. The element arsenic, As, is in Group 5A, and has

5 outer electrons. When they bond in a similar array to germanium, as shown in Figure 8.13, they will each have a share in a stable octet of electrons.

8.16 $2 \left({}^{3}_{2}\text{He}\right) \longrightarrow {}^{4}_{2}\text{He} + 2 \left({}^{1}_{1}\text{H}\right)$

Chapter 9

9.4

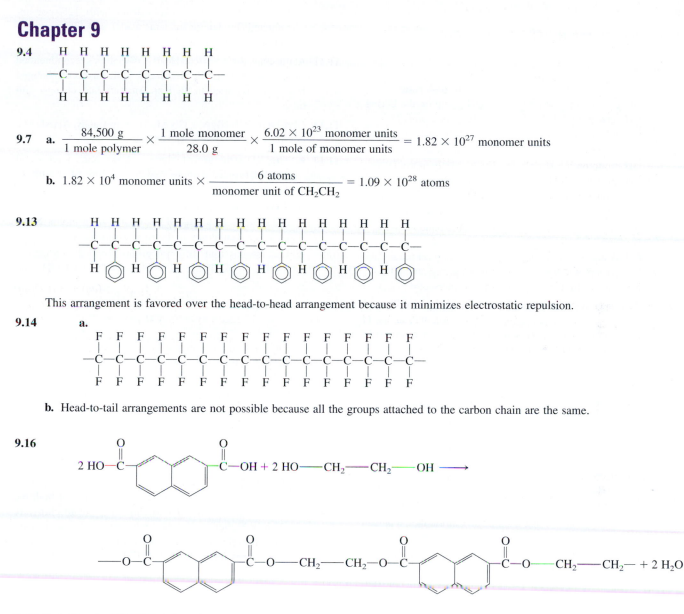

9.7 **a.** $\dfrac{84{,}500 \text{ g}}{1 \text{ mole polymer}} \times \dfrac{1 \text{ mole monomer}}{28.0 \text{ g}} \times \dfrac{6.02 \times 10^{23} \text{ monomer units}}{1 \text{ mole of monomer units}} = 1.82 \times 10^{27} \text{ monomer units}$

b. $1.82 \times 10^{4} \text{ monomer units} \times \dfrac{6 \text{ atoms}}{\text{monomer unit of } CH_2CH_2} = 1.09 \times 10^{28} \text{ atoms}$

9.13

This arrangement is favored over the head-to-head arrangement because it minimizes electrostatic repulsion.

9.14 **a.**

b. Head-to-tail arrangements are not possible because all the groups attached to the carbon chain are the same.

9.16

9.21 Propylene is CH_2CHCH_3 or C_3H_6.
$2500 \, C_3H_6 + 11{,}250 \, O_2 \longrightarrow 7500 \, CO_2 + 7{,}500 \, H_2O$

Chapter 10

10.3 *Note:* The hydrogen atoms have been omitted to make the linkage of the carbon atoms clear.

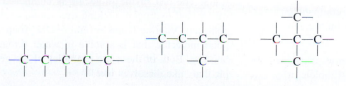

10.4 All three molecules have a benzene ring with functional groups attached that can hydrogen bond with water. Aspirin and ibuprofen also have a carboxylic acid group, —COOH.

10.7 There is only one carbon that has four different groups attached, making it the chiral carbon.

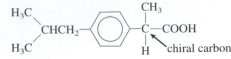

10.8 a.   **estradiol**
—OH group on the D ring
—CH₃ group on C/D ring intersection

    testosterone
—OH group on the D ring
—CH₃ group on C/D ring intersection

b.   **estradiol**
—OH group on the D ring
—CH₃ group on C/D ring intersection

    progesterone
C=O group on the D ring
—CH₃ group on C/D ring intersection

c.   **cholic acid**
no double bonds in any ring
—OH group on the A ring
—CH₃ group on C/D ring intersection

    cholesterol
no double bonds in A, C, or D rings
—OH group on the A ring
—CH₃ group on C/D ring intersection

10.12 Androstenedione and testosterone have the same structural features, except that androstenedione has a C=O on the D ring. In the D ring of testosterone there is a —C—O—H in place of the C=O in androstenedione.

Chapter 11

11.3 The number of atoms present is proportional to the number of moles of each element for every 100 g of body weight. Oxygen has a higher atomic mass than hydrogen, so even though it is about 2.5 times more abundant, oxygen has a smaller number of moles present per 100 g of body weight, when compared with the same values for hydrogen.

11.5 d. It is recommended that for a 2,500 Calorie diet, a person should consume less than 80 g of total fat, 300 mg of cholesterol, 2,400 mg of sodium, 3,500 mg of potassium, and 375 g total carbohydrate. A diet of Ko-Ko Crunchies cereal certainly meets the requirement of a low-fat, zero cholesterol diet, but is higher in carbohydrates and sodium than recommended. It is below the recommended level of potassium intake and this cereal is not a significant source of protein.

11.6 Salivary enzymes break down the complex carbohydrates in unsweetened crackers into simple sugars, which are responsible for the sweet taste.

11.7 Enzymes catalyze breaking bonds during digestion. There must be a close "lock-and-key" fit between the enzyme and the molecule substrate upon which it acts. The molecular architecture of human enzymes apparently is not compatible with the beta linkage between glucose units in cellulose, but the enzymes fit the alpha linkage in starches.

11.9 a. Canola oil is 6% saturated (likely capric, lauric, myristic, or palmitic acids), 36% polyunsaturated (linoleic or linolenic acids), and 58% monounsaturated fat (oleic acid).

b. Olive oil is 14% saturated (likely capric, lauric, myristic, or palmitic acids), 9% polyunsaturated (linoleic or linolenic acids), and 77% monounsaturated fat (oleic acid).

c. Lard is 41% saturated (likely stearic acid), 12% polyunsaturated (linoleic or linolenic acids), and 47% monounsaturated fat (oleic acid).

11.11 At this time, the American Heart Association's recommended values for cholesterol, HDL, and LDL have not changed from those reported in the text. Answers may change with time.

11.13 GlyGlyGly, GlyGlyAla, GlyAlaAla, GlyAlaGly, AlaAlaAla, AlaAlaGly, AlaGlyGly, AlaGlyAla

11.14 a. The 3-D structure of glycine shows bond angles and generally gives a more realistic picture of the three-dimensional molecule.

b. Some amino acids contain sulfur atoms in their structure, and some contain rings of five or six carbon atoms. Some have nitrogen atoms in the ring, another difference.

c. Leucine has four different groups bonded to a central carbon atom. They are —H, —NH₂, —COOH, and

$$-CH_2-\underset{\underset{CH_3}{|}}{CH}-CH_3$$

This molecule is optically active; the chiral carbon is indicated in this structure.

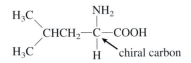

11.17 a. $\dfrac{130 \text{ Calories}}{2000 \text{ Calories}} \times 100 = 6.5\%$

b. Assuming that the serving size is the same for both the bagel chips and the potato chips, the potato chips are slightly higher in Calories per serving. Remembering that fats release 9 Cal/g when digested, a serving of bagel chips releases 4 g × 9 Cal/g = 36 Cal. The percentage of daily caloric intake from this fat is only 1.8% based on a 2,000 Calorie diet. A serving of potato chips releases 10 g × 9 Cal/g = 90 Cal, which is 4.5% of the caloric intake based on a 2,000 Calorie diet. It appears that these bagel chips are slightly lower in fat and Calories per serving than these particular potato chips. Both types of chips are snack food, and not meant to be a major component of the diet.

11.18 Different individuals will calculate different answers, resulting in different comparisons.

11.20 Vitamin A has only one —OH that can hydrogen bond with water. The rest of the molecule is composed of nonpolar carbon and hydrogen, making this vitamin soluble in nonpolar fats. Vitamin C has several —OH groups, making it a polar molecule that is soluble in water through hydrogen bonding. Both of these are examples of the general principle that "like dissolves like."

Chapter 12

12.1 a. Approximately a 21% increase for females; about an 18% increase for males.

b. Factors that have been proposed are that women have better diets, and hormonal differences that offer some protection against heart attacks before menopause.

12.5 Section 12.1 notes that the length is 2 meters, but this is for the combined length of both DNA strands.

12.6 a. The 3-D structure of DNA gives a better representation of the actual arrangement of atoms in this complex molecule, making the text figure more understandable.

b. The four bases all have a six-membered ring containing two nitrogen atoms. Adenine, guanine, and cytosine have an —NH$_2$ group attached to the six-membered ring. The six-membered ring of adenine and of guanine is also connected to a five-membered ring containing two nitrogen atoms.

c. The bases are always paired in the same way: two hydrogen bonds connect thymine to adenine, and cytosine is paired with guanine by three hydrogen bonds.

12.8 Chargaff's rules pointed to the complementarity of the base pairs, leading to the conclusion that adenine, and thymine must occur together, as do guanine and cytosine.

12.10 Redundancy in the code means that a mistake in the amino acid produced by the codon and introduced into a protein may not necessarily result in a mistake in the amino acid introduced into the protein. Multiple codons for certain amino acids may also speed up protein synthesis.

Answers to Selected End-of-Chapter Questions

Chapter 1

1. $\dfrac{1 \text{ L}}{1 \text{ breath}} \times \dfrac{15 \text{ breaths}}{1 \text{ minute}} \times \dfrac{60 \text{ minutes}}{1 \text{ hour}} \times \dfrac{8 \text{ hours}}{1 \text{ working day}} = 7200 \text{ L}$

4. $9000 \text{ ppm} \times \dfrac{100 \text{ parts per hundred}}{1,000,000 \text{ ppm}} = 0.9 \text{ parts per hundred}$
or 0.9%

6. $50,000 \text{ ppm} \times \dfrac{100 \text{ parts per hundred}}{1,000,000 \text{ ppm}} = 5 \text{ parts per hundred}$
or 5%

8. The percentage is calculated by comparing the difference in the two readings to the standard. $\dfrac{(0.15 - 0.12)}{0.12} \times 100 = 25\%$ above the standard

10. **a.** 85,000 g
  b. 10,000,000 gallons

11. **a.** 7.2×10^7 cigarettes
  b. 1.5×10^3 °C

13. **a.** Group 7A
  b. fluorine, chlorine, bromine, iodine, and astatine

16. **a.** potassium oxide
  b. aluminum chloride

18. **a.** One atom of the element carbon, two atoms of the element hydrogen, and one atom of the element oxygen are combined to form one molecule of formaldehyde.
  b. Two atoms of the element hydrogen are combined with two atoms of the element oxygen to form one molecule of hydrogen peroxide.

19. **a.** $N_2(g) + O_2(g) \longrightarrow 2NO(g)$

20. **a.**

24. **a.** platinum = Pt, palladium = Pd, rhodium = Rh
  b. All three metals are in Group 8B on the periodic table. Platinum is directly under palladium, and rhodium is just to the left of palladium.
  c. These metals are solids at the temperature of the exhaust gases, so they must have relatively high melting points. Also, they do not undergo permanent chemical change in catalyzing the reaction of CO to CO_2 in the exhaust stream.

32. **a.** The order of increasing length is 5.0×10^{-3} m, 1 m, and 3.0×10^2 m.
  b. If 1 meter is 1 year, then 3.0×10^2 m is equivalent to 300 years. 5.0×10^{-3} m is equivalent to 0.0050 years. This is about 1.8 days.

36. Sample **a.** represents a compound because two different atoms are joined.
  Sample **b.** represents a mixture because two different types of atoms are shown.

42. Formaldehyde can be released from cigarette smoke and from synthetic materials such as foam insulation, and from the adhesives used in dying and gluing carpet pads, carpets, and laminated building materials. The air indoors is often not well circulated, leading to an accumulation of formaldehyde and other pollutants. Efforts to make homes airtight, leading to greater energy efficiency, have led in some cases to making problems of indoor air pollution worse, rather than better.

Chapter 2

2. **a.** Yes, diamond and graphite are two distinct forms of the same element, carbon.
  b. No, water and hydrogen peroxide cannot be allotropes because they are compounds, not elements.
  c. Yes, white phosphorus and red phosphorus are allotropes because they are two distinct forms of the same element. Red phosphorus forms when several P_4 molecules combine into a chain structure.

6. **a.** 8 protons, 10 neutrons, 8 electrons
  b. 16 protons, 19 neutrons, 16 electrons
  c. 92 protons, 146 neutrons, 92 electrons

9. **a.** $Ca\!:$
  b. $\cdot \overset{\displaystyle \cdot}{\underset{\displaystyle \cdot}{N}} \cdot$
  c. $:\overset{\displaystyle \cdot \cdot}{\underset{\displaystyle \cdot \cdot}{Cl}} \cdot$
  d. $He:$

11. **a.** 2(5) = 10 outer electrons $\;:N:::N:\;$ and $\;:N\!\equiv\!N:$
  b. 1 + 4 + 5 = 10 outer electrons $H\!:\!C:::N:$ and
    $H\!-\!C\!\equiv\!N:$
  c. 2(5) + 6 = 16 outer electrons $\;:\overset{\cdot\cdot}{N}::\overset{}{N}::\overset{\cdot\cdot}{O}:$ and
    $:\overset{\cdot\cdot}{N}\!=\!N\!=\!\overset{\cdot\cdot}{O}:$

13. a. $5 \text{ cm} \times \dfrac{1 \text{ m}}{10^2 \text{ cm}} = 5 \times 10^{-2}$ m. This is in the microwave region of the spectrum.

14. $c \times \nu \cdot \lambda$ and $\nu = \dfrac{c}{\lambda}$; $c = 3.0 \times 10^8$ m/s

    a. $\nu = \dfrac{3.0 \times 10^8 \text{ m/s}}{5 \times 10^{-2} \text{ m}} = 6 \times 10^9 \text{ s}^{-1}$

15. $E = h \cdot \nu$ and $h = 6.63 \times 10^{-34}$ J·s

    a. $E = 6.63 \times 10^{-34} \text{ J} \cdot \text{s} \times 6 \times 10^9 \times \text{s}^{-1} = 4 \times 10^{-24}$ J

20. a. Cl, 7 outer electrons, $\overset{\cdots}{\underset{\cdots}{:}}\text{Cl}\cdot$

    NO, 11 outer electrons, $\cdot\text{N}::\overset{\cdots}{\text{O}}:$

    ClO, 13 outer electrons, $\cdot\text{Cl}:\overset{\cdots}{\text{O}}:$

    HO, 7 outer electrons, $\text{H}:\overset{\cdots}{\text{O}}\cdot$

  b. Each of these species has less than an octet of outer electrons on one non-hydrogen atom. Their reactivity is based on trying to attain an octet of outer electrons for each atom capable of holding an octet.

23. a. Methane, CH_4, has $4 + 4(1) = 8$ outer electrons.

$$\begin{array}{c} \text{H} \\ | \\ \text{H} - \text{C} - \text{H} \\ | \\ \text{H} \end{array}$$

    Ethane, C_2H_6, has $2(4) + 6(1) = 14$ outer electrons.

$$\begin{array}{c} \text{H} \quad \text{H} \\ | \quad\quad | \\ \text{H} - \text{C} - \text{C} - \text{H} \\ | \quad\quad | \\ \text{H} \quad \text{H} \end{array}$$

  b. There are 14 different CFCs that can be formed from methane.

28. SO_2 has a Lewis structure identical to ozone. This should not be surprising because sulfur is in the same family as oxygen, so it has the same number of outer electrons. The difference is that the electrons in the sulfur atom in SO_2 are at a further distance from the nucleus than is the case for the oxygen atoms in O_3. This difference does not show in the Lewis structures. Here are the resonance structures.

$:\overset{\cdots}{\text{O}}:\overset{\cdots}{\text{O}}::\overset{\cdots}{\text{O}}: \longleftrightarrow :\overset{\cdots}{\text{O}}::\overset{\cdots}{\text{O}}:\overset{\cdots}{\text{O}}:$ and $:\overset{\cdots}{\text{O}}:\overset{\cdots}{\text{S}}::\overset{\cdots}{\text{O}}: \longleftrightarrow :\overset{\cdots}{\text{O}}::\overset{\cdots}{\text{S}}:\overset{\cdots}{\text{O}}:$

35. UV-C radiation is extremely dangerous, but it is completely absorbed by normal oxygen, O_2, as well as by ozone, O_3, before it can reach the surface of the earth.

39. a. There are $4 + 3(7) + 6 = 31$ outer electrons available.

$$\begin{array}{c} :\overset{\cdots}{\text{F}}: \\ | \\ :\overset{\cdots}{\text{F}}:\overset{}{\text{C}}:\overset{\cdots}{\text{O}}\cdot \\ | \\ :\overset{\cdots}{\text{F}}: \end{array}$$

  b. This free radical is quite reactive in the troposphere, so it does not last long enough to reach the stratosphere.

43. O_2, O_3, and N_2 all have an even number of electrons in their Lewis structures. N_3 would have 15 electrons. Molecules with odd numbers of electrons are generally more reactive than those in which all electrons are present in pairs.

48. a. For CFC-12, add 90 and get 102. The compound contains 1 carbon, no hydrogens, and 2 fluorines. Therefore, there must be 2 chlorines. The formula is CF_2Cl_2.

  b. There is 1 carbon, no hydrogens, and no fluorine atoms in CCl_4. The code number plus 90 must be 100 and 100 minus 90 gives 10. This is CFC-10.

  c. $22 + 90 = 112$. This means 1 carbon, 1 hydrogen, and 2 fluorines. This coincides with the formula, which is CHF_2Cl_2.

Chapter 3

2. a. $6 CO_2(g) + 6 H_2O(l) \xrightarrow[\text{sunlight}]{\text{chlorophyll}} C_6H_{12}O_6(aq) + 6 O_2(g)$

  b. The number of atoms of each element on either side is the same. $C = 6$, $O = 18$, $H = 12$.

  c. The number of molecules is not the same on either side of the equation. There are 12 on the left, but only 7 on the right. The large molecule glucose has formed on the right hand side of the equation, using 24 atoms per molecule.

4. a. The concentration of CO_2 in the atmosphere at the present time is about 280 ppm. 20,000 years ago, the concentration was about 190 ppm, far lower. However, 120,000 years ago the concentration was about 270 ppm, close to today's levels.

  b. The average temperature at present is somewhat above the 1950–1980 average temperature of the atmosphere. 20,000 years ago, the average temperature was lower by about 9°C. However, 120,000 years ago the average temperature was lower by only about 1°C.

  c. Although there appears to be a *correlation* between average temperature and CO_2 concentration, this Figure does not prove *causation* of either factor by the other.

7. These are the two Lewis structures. $\text{H}-\text{H}$ and $\text{H}-\overset{\cdots}{\text{O}}-\text{H}$.

  For H_2 with only two atoms, the atoms can only be arranged in a straight line. For H_2O, even though the Lewis structure has been *shown* as a straight line, this does not mean that the molecule is linear. In fact, the bent structure of water is so well known that the Lewis structure is often written in this manner.

$$\begin{array}{c} \text{H} \diagdown \overset{\cdots}{\text{O}}: \\ | \\ \text{H} \end{array}$$

9. a. $4 + 2(1) + 2(7) = 20$ outer electrons; the Lewis structure

is $\begin{array}{c} :\overset{\cdots}{\text{Cl}}: \\ | \\ \text{H}-\text{C}-\text{H} \\ | \\ :\overset{\cdots}{\text{Cl}}: \end{array}$ and the shape is tetrahedral.

  b. $4 + 6 = 10$ outer electrons. The Lewis structure is $:\text{C}\equiv\text{O}:$ and the shape is linear. (This could have been predicted from the fact that there are just two atoms.)

  c. $1 + 4 + 5 = 10$ outer electrons. The Lewis structure is $\text{H}-\text{C}\equiv\text{N}:$ The shape is linear.

  d. $5 + 3(1) = 8$ outer electrons. The Lewis structure is

$$\begin{array}{c} \text{H}-\text{P}-\text{H} \\ | \\ \text{H} \end{array}$$ The shape is triangular pyramidal also called trigonal pyramidal.

16. $C_6H_{12}O_6(aq) \xrightarrow{\text{yeast}} 2 C_2H_5OH(aq) + 2 CO_2(g)$

19. The weighted average for silver is 107.870 g/mole.

$107 \times 0.52 + x \times 0.48 = 107.870$

$0.48x = 107.870 - 55.64$

$x = 109$; The other isotope of silver is silver-109. Interestingly, silver-108 is not a stable isotope.

21. **a.** $2(1) + 16 = 18$ g/mole

b. $12 + 2(19) + 2(35.5) = 121$ g/mole

c. $12 + 16 = 28$ g/mole

26. **a.** $\dfrac{358 \text{ ppm} - 280 \text{ ppm}}{280 \text{ ppm}} \times 100 = 28\%$ increase in the concentration of CO_2

b. $\dfrac{1.7 \text{ ppm} - 0.70 \text{ ppm}}{0.7 \text{ ppm}} \times 100 = 140\%$ increase in the concentration of CH_4

$\dfrac{0.31 \text{ ppm} - 0.28 \text{ ppm}}{0.28 \text{ ppm}} \times 100 = 11\%$ increase in the concentration of N_2O

29. Drilled ocean cores can be analyzed for the number of types of microorganisms present. Another correlating piece of evidence is the changing alignment of magnetic fields in particles in the sediment over time. Another possibility is to analyze the deuterium-to-hydrogen ratio in ice cores.

33. In BF_3, there are $3 + 3(7) = 24$ outer electrons. This is the Lewis structure. $:\overset{..}{\underset{..}{F}}-B-\overset{..}{\underset{..}{F}}:$

$\overset{|}{\underset{..}{:}}\overset{}{F}:$

In NH_3, there are $5 + 3(1) = 8$ outer electrons. This is the Lewis structure. $H-\overset{..}{N}-H$

$\overset{|}{H}$

Note that in BF_3, there are just three pairs of electrons around the central boron. To achieve maximum separation of these three electron pairs, which will mutually repel, a fluorine-to-boron-to-fluorine bond angle of 120° is predicted. This is why BF_3 has a triangular planar geometry. In NH_3, however, there are four pairs of electrons—three bonded pairs and one lone pair. This determines the triangular pyramidal geometry. Although the predicted hydrogen-to-nitrogen-to-hydrogen bond angle is 109.5°, the repulsion of the lone pair with the bonded pairs closes the angle somewhat so that the experimentally measured angle is 103°.

37. Living systems < oceans < the Earth

This order is chosen because living things have finite lifetimes that are relatively short. The oceans can capture carbon from carbon dioxide to form carbonates, which are stable for a relatively long time. The Earth, however, contains fossil fuels. They can keep a carbon atom in place for millions of years before the fuel is mined and burned.

42. **a.** It does *not* prove that global warming theories are correct. It is a short-range observation that is consistent with global warming trends that are predicted, but it may not predict large-scale or long-range changes.

b. It does *not* prove that global warming theories are correct. Although it is not definite proof, it is important because it provides one more piece of experimental evidence for the predicted effects of global warming on biological species.

46. **a.** 20 electrons are required if 3 atoms join. This is the general Lewis structure. $:\overset{..}{\underset{..}{X}}-\overset{..}{\underset{..}{Y}}-\overset{..}{\underset{..}{Z}}:$

b. This molecule will be bent. There are four pairs of electrons around the central Y atom. However, two of the pairs are bonding pairs, and there are two lone pairs. Repulsion between the two lone pairs and their repulsion of the bonding pairs is predicted to cause the bond angle to be less than 109.5°.

c. There are new possibilities if double and triple bonds are allowed.

Number of outer electrons	Lewis Structure	Shape	Predicted bond angle
18	$:\overset{..}{\underset{..}{X}}-\overset{..}{Y}=\overset{..}{\underset{..}{Z}}:$	bent	less than 120°
16	$:\overset{..}{X}=Y=\overset{..}{Z}:$	linear	180°
16	$:\overset{..}{\underset{..}{X}}-Y\equiv Z:$	linear	180°

Chapter 4

2. **a.** Estimating values from the graph, domestic oil production accounted for approximately 60% in 1970, 30% in 1990, and is predicted to be about 18% in 2010.

4. The temperature is the same, 70°C, in each container. Temperature is commonly measured in °C for scientific work, or in °F for household applications. The heat content of the water is not the same. Heat depends on both the temperature and the size of the sample. There is twice the heat in the water in Container 1 compared with the water in Container 2 because there is twice the mass of water present.

7. $\dfrac{260,000 \text{ kcal}}{1 \text{ day}} \times \dfrac{365 \text{ days}}{1 \text{ year}} = \dfrac{9.5 \times 10^7 \text{ kcal}}{1 \text{ year}}$

This value can be related to each of the energy sources.

a. $\dfrac{9.5 \times 10^7 \text{ kcal}}{1 \text{ year}} \times \dfrac{1 \text{ year}}{65 \text{ barrels}} = \dfrac{1.5 \times 10^6 \text{ kcal}}{\text{barrel}}$

10. **a.** $2 C_2H_6 + 7 O_2 \longrightarrow 4 CO_2 + 6 H_2O$

b. $2 \; H-\overset{\overset{\displaystyle H}{|}}{C}-\overset{\overset{\displaystyle H}{|}}{\underset{\underset{\displaystyle H}{|}}{C}}-H + 7 :\overset{..}{O}=\overset{..}{O}:$

$\longrightarrow 4 :\overset{..}{O}=C=\overset{..}{O}: + 6 \; H-\overset{..}{\underset{|}{O}}:$
$\overset{|}{H}$

c.

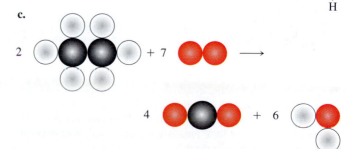

11. $\dfrac{52.0\,\text{kJ}}{1\,\text{g}\,C_2H_6} \times \dfrac{30.1\,\text{g}\,C_2H_6}{1\,\text{mole}\,C_2H_6} = \dfrac{1560\,\text{kJ}}{\text{mole}\,C_2H_6}$

14. **a.** Exothermic; a charcoal briquette releases heat as it burns.

    b. Endothermic; liquid water gains the necessary heat for evaporation from your skin, and your skin feels cool.

15. Bonds broken in the reactants

2 carbon-to-oxygen triple bonds	= 2(1072 kJ)	= 2144 kJ
1 oxygen-to-oxygen double bond	= 1(494 kJ)	= 494 kJ

Total energy *absorbed* in
    breaking bonds   2638 kJ

Bonds formed in the products

4 carbon-to-oxygen double bonds	= 4(799 kJ)	= 3196 kJ

Total energy *released* in
    forming bonds   3196 kJ

Net energy change is (+2638 kJ)
    + (−3196 kJ) = −558 kJ

Notice that the overall energy change has a negative sign, characteristic of an exothermic reaction.

18. **a.** $2\,C_5H_{12}(g) + 11\,O_2(g) \longrightarrow 10\,CO(g) + 12\,H_2O(l)$
$8(346\,\text{kJ}) + 24(411\,\text{kJ}) + 11(494\,\text{kJ}) \longrightarrow$
$10(1072\,\text{kJ}) + 24(459\,\text{kJ})$
The net energy change is −3670 kJ and the reaction is highly exothermic.

22. There are two different isomers.

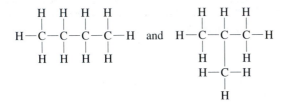

28. All of these refer to the same idea, which is that energy is not consumed during a chemical reaction. Energy can be transformed, but the total energy is constant during a chemical reaction.

30. **a.** The C—F bond requires 485 kJ/mole, the C—Cl bond requires 327 kJ/mole, and the C—Br bond requires 285 kJ/mole to break the bond. The C—Br bond is the least energetic, and the bromine free radical can interact with ozone even more effectively than the chlorine free radical.

    b. This is the structure for C_2F_4HCl.

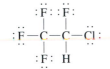

The C—F bond requires 485 kJ/mole, the C—Cl bond requires 327 kJ/mole, and the C—H bond requires 411 kJ/mole to break the bond. This means that the chlorine free radical forms with the least energy into the system. Although this seems contradictory in a replacement for Halons, this compound's lifetime in the atmosphere is significantly shorter than the Halons, making it less effective in depleting ozone.

39. Entropy values increase with increasing complexity and randomness. Diamond is a highly ordered crystalline structure, and so has a low entropy. Methanol is a liquid, and the molecules of methanol have greater complexity and more opportunity for random movement.

Chapter 5

1. **a.** 3×10^9 gallons now $\times \dfrac{1\,\text{gallon 20 years ago}}{10\,\text{gallons now}} =$
3×10^8 gallons 20 years ago

    b.
$\dfrac{(3 \times 10^9\,\text{gallons now}) - (3 \times 10^8\,\text{gallons 20 years ago})}{3 \times 10^8\,\text{gallons 20 years ago}}$
$\times 100 = 900\%$ growth

3. The drinking water supply is limited by the amount of available fresh water. Figure 5.4 shows 97.4% is salt water, which means only 2.6% is fresh water. Further restricting the amount of fresh water is the fact that 90% of fresh water is frozen in Antarctica, and that 80% of fresh water in the United States is used for irrigating crops and cooling electric power plants. The water available for drinking is only 0.26 L or 260 mL.

500 L total $\times \dfrac{2.6\,\text{parts fresh}}{100\,\text{parts total}} \times \dfrac{10\,\text{parts not frozen}}{100\,\text{parts fresh}}$
$\times \dfrac{20\,\text{parts available}}{100\,\text{parts not frozen}} = 0.26\,\text{L available}$

5. 55 mg of Ca^{2+} per liter of bottled water is the same as 55 ppm.
$\dfrac{55\,\text{mg}\,Ca^{2+}}{1\,\text{L}\,H_2O} \times \dfrac{1\,\text{g}}{10^3\,\text{mg}} \times \dfrac{1\,\text{L}}{10^3\,\text{mL}} \times \dfrac{1\,\text{mL}\,H_2O}{1\,\text{g}\,H_2O} =$
$\dfrac{55\,\text{g}\,Ca^{2+}}{10^6\,\text{g}\,H_2O}$ or 55 ppm Ca^{2+}

10. Water is a very good solvent for other polar substances and for many ionic substances.

14. The arrow points to a hydrogen bond, an example of an *inter*molecular force, which is a force *between* water molecules and not *within* each water molecule.

18. **a.** $250\,\text{g}\,H_2O \times \dfrac{1\,\text{cal}}{\text{g} \cdot {}^\circ\text{C}} \times (100^\circ\text{C} - 15^\circ\text{C}) = +2.1 \times 10^4\,\text{cal}$
of heat absorbed

    b. $500\,\text{g}\,H_2O \times \dfrac{1\,\text{cal}}{\text{g} \cdot {}^\circ\text{C}} \times (55^\circ\text{C} - 95^\circ\text{C}) = -2.0 \times 10^4\,\text{cal}$
of heat released

21. **a.** Na_2S sodium sulfide

    b. Al_2O_3 aluminum oxide

28. These values are not easy to compare, because the units are not the same. There are 100 mg of magnesium per tablet, which is usually swallowed with a minimum of water. The concentration of magnesium reaching the body is probably far greater from the multivitamin tablet than it is from this brand of bottled water.

30. **a.** Electronegativity values generally increase from left to right across a period (until the 8A group is reached) and from bottom to top within any group. This means that element #2 is predicted to have the greatest electronegativity value.

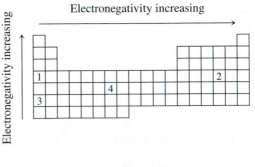

33. a. The Lewis structure for ethanol is

$$H-\overset{\overset{\displaystyle H}{|}}{\underset{\underset{\displaystyle H}{|}}{C}}-\overset{\overset{\displaystyle H}{|}}{\underset{\underset{\displaystyle H}{|}}{C}}-\overset{..}{\underset{..}{O}}-H$$

This is a polar molecule, and it will exhibit hydrogen bonding.

b. The cube sinks because, as is the case for most substances, the density of the solid phase is greater than the density of the liquid phase. Despite the fact that hydrogen bonding between molecules takes place, no regular lattice pattern such as exists in ice forms in the solid ethanol.

36. With the exception of contaminants that are known carcinogens, MCLG and MCL values are usually very close to the same.

Chapter 6

1. a. Acids have a sour taste, react with carbonates, and impart a characteristic color to acid-base indicators.

3. a. $HBr(aq) \longrightarrow H^+(aq) + Br^-(aq)$

b. $H_3C_6H_5O_7(aq) \longrightarrow H^+(aq) + H_2C_6H_5O_7^-(aq)$

4. a. Bases taste bitter and have a slippery feel in water.

b. Bases release hydroxide ions, OH^-, in aqueous solution.

6. a. $RbOH(s) \longrightarrow Rb^+(aq) + OH^-(aq)$

7. a. $KOH(aq) + HNO_3(aq) \longrightarrow KNO_3(aq) + H_2O(aq)$
$$K^+(aq) + OH^-(aq) + H^+(aq) + NO_3^-(aq) \longrightarrow$$
$$K^+(aq) + NO_3^-(aq) + H_2O(aq)$$
$$OH^-(aq) + H^+(aq) \longrightarrow H_2O(aq)$$

9. a. Basic; the concentration of the $OH^-(aq)$ is greater than the concentration of the $H^+(aq)$

12. $S(s) + O_2(g) \longrightarrow SO_2(g)$
$SO_2(g) + H_2O(l) \longrightarrow H_2SO_3(aq)$
$H_2SO_3(aq) \longrightarrow H^+(aq) + HSO_3^-(aq)$

15. a.
$$\frac{14 \text{ g N}}{(135 \times 12)(96 \times 1)(9 \times 16)(1 \times 14)(1 \times 32) \text{ g coal}} \times$$
$100 = 0.73\%N$

b. $3 \text{ tons coal} \times \dfrac{0.73 \text{ parts N}}{100 \text{ parts coal}} = 0.02 \text{ tons nitrogen}$

19. The Figure shows that about 1×10^6 tons of SO_2 come from transportation, and about 7×10^6 tons of NO_x come from transportation. Together, this is about 8×10^6 tons of combined pollutants compared to a total of about 40×10^6 tons of SO_2 and NO_x.
% combined SO_2 and NO_x $= \dfrac{8 \times 10^6 \text{ tons}}{40 \times 10^6 \text{ tons}} \times 100 =$ 20% from transportation
The Figure shows that about 16×10^6 tons of SO_2 come from electric power production, and about 11×10^6 tons of NO_x come from electric power production. Together, this is about 27×10^6 tons of combined pollutants.

% combined SO_2 and NO_x $= \dfrac{27 \times 10^6 \text{ tons}}{40 \times 10^6 \text{ tons}} \times 100 =$

68% from electric power production

Chapter 7

1. *E* represents energy, *m* represents mass lost in a nuclear transformation, and *c* represents the speed of light.

3. a. 94 protons

b. 52 neutrons

c. 92 protons and 146 neutrons

5.

Isotope	Number of Protons	Number of Neutrons
C-12	6	6
Ca-40	20	20
Zn-64	30	34
Sn-119	50	69

For the lighter elements with an even number of protons, there is a matching number of neutrons in a stable isotope. From these data, an atomic number of 20 seems to be the limit for this case. For heavier elements with an even number of protons, the number of neutrons must be greater than the number of protons to create a stable nucleus. The lower limit given in the data is atomic number equal to 30 or above. No generalization can be reached for elements with an odd number of protons or for the elements with atomic numbers 11–29.

8. a. Because there are only two stable isotopes, there must be a higher percentage of B-11. In fact, there is about 80% of B-11 and only 20% of B-10.

10. a. $^2_1H + ^2_1H \longrightarrow [^4_2He] \longrightarrow ^1_0n + ^3_2He$

b. $^{238}_{92}U + ^{14}_7N \longrightarrow [^{252}_{99}Es] \longrightarrow ^{247}_{99}Es + 5\,^1_0n$

12. a. The sum of the masses of the reactants is 5.02838 g and the sum for the products is 5.00878 g. This means that the mass difference, 0.0196 g, has been converted to energy following Einstein's equation, $E = mc^2$.

b. $E = mc^2$; $E = 1.76 \times 10^{12}$ J

14. **A** = control rod assembly, **B** = cooling water out of the core, **C** = control rod, **D** = cooling water into the core, **E** = fuel rod

17. There are two advantages to using water rather than graphite in U.S. reactors as the moderating material. Water has a higher heat capacity than graphite, and when water gets hot, it does not burn the way the graphite did in the Chernobyl reactor.

21. There is no difference in what is represented. Each of these represents an alpha particle, which is the same as a helium nucleus.

23. a. $^{131}_{53}I \longrightarrow ^{131}_{54}Xe + ^0_{-1}e$

b. $^{238}_{92}U \longrightarrow ^{234}_{90}Th + ^4_2He$

25. 2 half-lives = $(1/2)^2 = 1/4$; 4 half-lives = $(1/2)^4 = 1/16$; 6 half-lives = $(1/2)^6 = 1/64$

Chapter 8

1. a. Both equations represent the combustion of hydrogen gas. Equation 8.1 represents twice the number of moles of reactants and products compared with equation 8.2.

b. The amount of energy released in the reaction shown in equation 8.1 will be double the amount of energy released in the reaction represented by equation 8.2.

3. a. To check this equation, $H_2(g) + \frac{1}{2} O_2(g) \longrightarrow H_2O(l) + energy$, first consider the Lewis structures for reactants and products.

$$H-H + \tfrac{1}{2} \ :\overset{..}{O}=\overset{..}{O}: \longrightarrow H-\overset{..}{\underset{\underset{\displaystyle H}{|}}{O}}$$

Energy needed to break bonds:
$432 \text{ kJ} + \frac{1}{2}(494 \text{ kJ}) = +679 \text{ kJ}$
Energy released as new bonds form: $2(-459 \text{ kJ}) = -918 \text{ kJ}$
Overall, according to this calculation, the reaction releases 239 kJ of energy.

b. Average bond energies are based on bonds within molecules in the gaseous state. In the given chemical equation, the H_2O formed is present as a liquid rather than as a gas. Additional energy is released when gaseous water condenses to the liquid state, so the stated value of 286 kJ is greater than the 239 kJ calculated in part **a.**

4. No, the order will not be the same; it will be inverted. In each case, the heat of combustion per mole is found by multiplying the heat of combustion per gram by the molar mass. For example, for octane, 46 kJ/g $\times$ 114 g/mole = 5244 or 5200 kJ/mole. For methane, 54 kJ/g16 g/mole = 864 kJ/mole, which rounds to 860 kJ/mole. For hydrogen, 143 kJ/g $\times$ 2.0 g/mole = 286 kJ/mole, which rounds to 290 kJ/mole.

6. This must be the process of *reduction*; electrons must be *gained* to convert oxygen gas to combined oxygen in water. This is the reduction half-reaction.
$\frac{1}{2} O_2(g) + 2 H^+(aq) + 2e^- \longrightarrow H_2O(l)$

8. 1.8 amps $\times$ 1.5 volts $\times$ 4.0 hours $\times$ 60 minutes/hr $\times$ 60 seconds/minute = 3.9×10^4 amp $\cdot$ volt
Because 1 amp $\cdot$ volt = 1 J, this is also 3.9×10^4 J

11. a. A fuel cell is an electrochemical cell in which a chemical reaction corresponding to combustion takes place, but the energy is released as electricity rather than as heat and light.

b. $2 H_2(g) + 4 OH^-(aq) \longrightarrow 4 H_2O(l) + 4e^-$ This is the process of oxidation.

c. $O_2(g) + 4e^- + 2 H_2O \longrightarrow 4 OH^-(aq)$ This is the process of reduction.

14. a. $Li \longrightarrow Li^+ + e$ half-reaction of oxidation
$I_2 + 2e^- \longrightarrow 2 I^-$ half-reaction of reduction

b. $2 Li + I_2 \longrightarrow 2 LiI$ overall reaction in the cell

c. Oxidation occurs at the anode, so this is the half-reaction.
$Li \longrightarrow Li^+ + e^-$
Reduction occurs at the cathode, so this is the half-reaction.
$I_2 + 2e^- \longrightarrow 2 I^-$

16. a. This is the balanced equation. $2 Mg(s) + SiCl_4(l) \longrightarrow 2 MgCl_2(l) + Si(s)$
Each magnesium atom loses 2 electrons and there are 2 magnesium atoms. Silicon in $SiCl_4$ picks up the four electrons.

b. The silicon in $SiCl_4$ gains 4 electrons to form silicon atoms, which is reduction.

18. In current usage, the term "hybrid car" refers to the combination of a gasoline engine together with a nickel–metal hydride battery, an electric motor, and an electric generator. Other hybrids using fuel cells will continue to be developed.

20. Note that there are 8 electrons around each silicon atom, but there are 9 electrons around the central atom. Each carbon atom has 4 outer electrons, so the central atom in the Figure must have 5 outer electrons. This is consistent with arsenic, which is in Group 5A. The additional electron forms an *n*-type silicon semiconductor.

22. Fusion and fission are both nuclear processes. In the case of fusion, small nuclei join to form larger nuclei. In the case of fission, larger nuclei are split to form medium-weight nuclei. This is a sample equation for the process of fission.
$^{235}_{92}U + ^1_0n \longrightarrow [^{236}_{92}U] \longrightarrow ^{87}_{35}Br + ^{146}_{57}La + 3\ ^1_0n$
This is a sample equation for the process of fusion.
$^2_1H + ^2_1H \longrightarrow [^4_2He] \longrightarrow ^1_0n + ^3_2He$

24. The difference in mass is $2(2.01355) - (3.01550 + 1.00728) = 0.00432g$. Using $E = mc^2$, and recalling that $1 \text{ J} = \text{kg} \cdot \text{m}^2/\text{s}^2$, the energy released is 3.89×10^{11} J.

Chapter 9

1. Plastics are synthetic polymers. This means that plastics are polymers, but not all polymers are plastics.

6. a. 330 lbs/person

b. 160 lbs/person

c. 150 % change

d. 110 % change

8.

346 kJ *released* in forming
2 moles of half-bonds; equivalent
in energy to 1 mole of covalent C-C bonds

602 kJ must be *added* to break one mole of this bond; 1204 kJ for 2 moles of monomers

346 kJ are *released* in forming one mole of each bond; 1038 kJ for 3 moles of bonds

Altogether, note that 1204 kJ of energy were added to the system and (1038 kJ + 346 kJ) = 1384 kJ were released from the system. That means that 180 kJ of energy are released from the system, making the reaction exothermic.

10. a. 1430 monomer units

b. 2860 carbon atoms

13. There are *three* pairs of electrons around each carbon in the monomer, making the geometry around the carbon trigonal planar and the Cl—C—H bond angle 120°. Each carbon in the polymer has *four* electron pairs around the carbon, making the geometry tetrahedral, and the bond angle 109° in the polymer.

15. The bottle on the left is likely made of flexible, low-density branched polyethylene. The one on the right is likely made of rigid, high-density, linear polyethylene. The structures of LDPE and HDPE, found in Figure 9.10, can be used to help explain this difference in properties at a molecular level. The low-density polyethylene is highly branched, preventing close interactions between the chains and allowing the plastic to be softer and more easily deformed. The high-density polyethylene chains are linear. The chains can more closely approach each other, creating opportunity for interactions. The bottle on the right is more rigid than the one on the left.

17. All except #1, PETE, share a common structural feature. They have the same basic structure as the ethylene molecule, but one of the hydrogen atoms has been replaced with a different atom or group of atoms.

20. This is the head-to-head, tail-to-tail arrangement of PVC formed from three monomer units. Note the carbon containing two hydrogen atoms attached to another carbon containing two hydrogen atoms, and the carbon containing two chlorine atoms attached to another carbon containing two chlorine atoms.

22. The CO_2 most likely replaces CFCs that were formerly used for this process. Most of the CO_2 used comes from existing commercial and natural sources, so no additional CO_2 is contributed to global warming. CFCs are implicated in the depletion of the ozone layer and can no longer be used for blowing Styrofoam. For several years, pentane, C_5H_{12}, has been used to replace CFCs, but pentane is flammable, and CO_2 is not.

24. a. This is the Lewis structure.

$$H-\overset{\overset{\displaystyle H}{|}}{C}=\overset{\overset{\displaystyle H}{|}}{C}-C\equiv N:$$

b. When Acrilan fibers burn, one of the products is the poisonous gas hydrogen cyanide, HCN.

29. a. Yes. Each division represents 10 billion pounds.

b. No. The years spaced along the x-axis of this representation are not equally spaced. It starts out with each bar being 5 years apart, but after 1960, the spacing of the bars becomes first 7 years, then 4 years, then 3 years, and continues to vary until 1985. After that, each bar is spaced 3 years apart.

30. There are several other features of polymer chains that can have an influence on the properties of the polymer formed. These include:
1. the length of the chain (the number of monomer units)
2. the three-dimensional arrangement of the chains
3. the branching of the chain
4. the chemical composition of the monomer units
5. the bonding between chains
6. the orientation of monomer units within the chain

36. The "Big Six" polymers are almost completely nonpolar molecules and therefore do not dissolve in polar water molecules. The generalization, developed in Chapter 5, is that "like dissolves like." Some of the "Big Six" dissolve or soften in hydrocarbons or chlorinated hydrocarbons because these nonpolar solvents interact with the nonpolar polymeric chains.

38. A monomer must have a double bond in its structure. Although double bonds in rings may be present, the double bond used for addition polymerization must be along the chain. That double bond must be accessible to attach by a free radical, resulting in a single carbon-to-carbon bond that is left along the chain. As that reaction repeats itself, the addition polymer grows. An example is the formation of PVC.

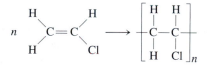

Chapter 10

1. a. An antipyretic drug reduces fever.
   b. An analgesic drug reduces pain.

c. An anti-inflammatory drug reduces inflammation, which is redness, heat, swelling, and pain caused by irritation, injury, or infection.

d. Yes. Aspirin is an example of a drug with all three properties.

3. The condensed formulas are $CH_3CH_2CH_2CH_2CH_3$ (or $CH_3(CH_2)_3CH_3$), $CH_3CH_2CH(CH_3)CH_3$, and $CH_3C(CH_3)_2CH_3$.

5. There are four possible isomers with the formula C_4H_9OH. Here is the structural formula for each isomer (H omitted on Cs).

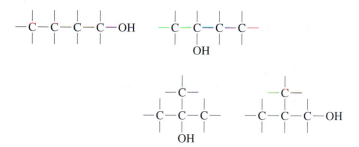

7. Alcohols, aldehydes, and acids all have examples with one carbon atom. The others require more than one carbon atom by the nature of their functional groups.
   a. alcohol   An example is methanol, CH_3OH.
   b. aldehyde   An example is formaldehyde, CH_2O.
   c. carboxylic acid   An example is formic acid, HCO_2H.
   d. ester   This does not have a possible one-carbon example. The carbonyl group must have a carbon-containing group on either side, one bonded to the carbon and the other to the singly bonded oxygen of the carbonyl group.
   e. ether   This does not have a possible one-carbon example. The central oxygen atom must have a carbon-containing group on either side. If one position or both positions are occupied by hydrogen atoms, the compound is an alcohol.
   f. ketone   This does not have a possible one-carbon example. There must be a carbon-containing group on either side of the carbon that is double bonded to an oxygen. If one or both positions are occupied by hydrogen atoms, the compound is an aldehyde.

11. a. amine group ($-NH_2$) and carboxylic acid group ($-COOH$)
    b. two amide groups ($-\overset{\overset{\displaystyle O}{\|}}{C}-\overset{\overset{\displaystyle H}{|}}{N}-$)
    c. amide group ($-\overset{\overset{\displaystyle O}{\|}}{C}-\overset{\overset{\displaystyle H}{|}}{N}-$), carboxylic acid group ($-COOH$); also sulfide ($-S-$) and amine ($-N-$) in ring.

13. There are several nonpolar groups, so it is likely to be very soluble in the lipids that form the cell walls. Only the amine group is polar, helping the molecule to dissolve in the fluid inside and outside the cells.

17. "Superaspirins" are a new class of medicines that are selective in preferentially blocking the COX-2 enzyme that makes prostaglandins associated with inflammation, pain, and fever. These "superaspirins" do not have any effect on the COX-1

enzyme, which means there will be fewer side effects such as stomach irritation and kidney disfunction.

20. **a.** This cannot exist in chiral forms. There are two groups that are the same, —CH₃.

   b. This can exist in chiral forms, because the four groups attached to the central carbon are all different.

   c. This cannot exist in chiral forms. There are two groups that are the same, CH₃.

24. **a.** Although technically not an anabolic steroid, androstenedione is converted to testosterone when ingested and testosterone is a banned anabolic steroid.

   b. Androstenedione is not an anabolic steroid, and therefore does not fall under the policy of being banned.

28. **a.** These are the possibilities.

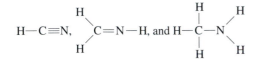

   b. These are the possibilities.

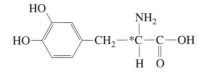

35. L-dopa is effective because the molecule fits in the receptor site, but the nonsuperimposable mirror image of D-dopa does not. This is the structure of L-dopa, with the chiral carbon atom marked by a star. Note that there are four different groups attached to the starred carbon atom.

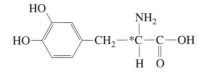

Chapter 11

1. The four fundamental types of materials provided by food are water, energy sources, raw materials, and metabolic regulators.

3. **a.** Sub-Saharan Africa, east and southeast Asia, Latin America, and all developing regions showed a decrease in the proportion of the population that was undernourished.

   b. The Near East and North Africa and South Asia showed an increase in the proportion of the population that was undernourished.

   c. Estimating from the graph, this is the percentage change.
   $$\frac{15\% - 12\%}{15\%} \times 100 = 20\%$$

   d. To determine the change in the number of undernourished people in Latin America and the Caribbean during the same interval, it would be necessary to know the total population during both periods. Multiplying the percentage by the population would give the number of undernourished people in each time interval, and then the change in the number of

undernourished people can be calculated by subtraction.

$$\frac{\text{(Number of undernourished in '90–'92)(Number of undernourished in '94–'96)}}{\text{(Number of undernourished in '90–'92)}}$$
$$\times 100 = \% \text{ change}$$

5. There is too much carbohydrate for steak, and too much protein for chocolate chip cookies. The pie chart is likely a representation of peanut butter. (See Figure 11.2 for confirmation.)

10. Although hydrogen atoms are far more abundant than oxygen or carbon atoms, hydrogen has a far smaller mass than either oxygen or carbon.

13. **A.** 6–11 servings of bread, cereal, grain, and pasta

   B. 3–5 servings of vegetables

   C. 2–4 servings of fruits

   D. 2–3 servings of milk, yogurt, and cheese

   E. 2–3 servings of meat, poultry, fish, dry beans, eggs, and nuts

   F. fats, oils, and sweets used sparingly

16. The structure of glucose is based on a six-membered ring of five carbons and one oxygen. The structure of fructose is based on a five-membered ring of four carbons and one oxygen. Glucose has one —CH₂OH side chain, and fructose has two.

20. **a.** Fats and oils are both composed of nonpolar hydrocarbon chains. Edible fats and oils both contain some oxygen. Most fats and oils are triglycerides, which are esters of three fatty acid molecules and one glycerol molecule. Both oils and fats feel greasy and are insoluble in water.

   b. Oils tend to contain more highly unsaturated fatty acids and smaller fatty acids than fats. If the triglycerides are solid at room temperature, the material is termed a fat. If the triglycerides are liquid at room temperature, the material is termed an oil.

27. According to Figure 11.2, peanut butter is 26% protein, which makes it a very good source of protein. However, it is also 52% fat, which is quite high if one needs to limit fat, in the diet. It is, however, relatively low in saturated fat, if not hydrogenated.

30. It would be unlikely to meet the guidelines. Only 2–3 servings per day are recommended for milk, yogurt, and cheese; 6–11 servings of bread, cereal, grains, and pasta are required, 3–5 servings of vegetables, 2–4 servings of fruit, and 2–3 servings of meat, poultry, fish, dry beans, eggs, and nuts. If 40% came from milk or dairy products, there is not enough room in the diet for the other food groups to be well represented.

Chapter 12

1. DNA stands for **d**eoxyribo**n**ucleic **a**cid.

3. **a.** The base adenine has the amino group, —NH₂.

   b. The sugar deoxyribose has several —OH groups.

   c. There are no functional groups *in* the phosphate, but the phosphate itself is a functional group. Do not mistake the doubly bonded oxygen atom for a ketone, for there is no bond to a carbon.

6. **a.** A nucleotide must contain a base, a deoxyribose unit, and a phosphate group linked together.

b. Covalent bonding holds the groups together.

8. **a.** A beam of X-rays, which have high energy and short wavelengths, is directed at a target. The X-rays are then diffracted at certain angles, which are related to the distance between atoms.

    b. Ions in a salt like sodium chloride have a very regular structure that is easily determined by X-ray studies. Atoms in nucleic acids and proteins do not show the same well-known patterns of crystalline regularity, making the interpretation of the X-ray diffraction pattern more difficult.

13. **a.** This is the general formula for an amino acid, in which R represents a side chain that is different in each of the 20 amino acids.

$$\text{H}_2\text{N}-\overset{\overset{\displaystyle \text{H}}{|}}{\underset{\underset{\displaystyle \text{R}}{|}}{\text{C}}}-\text{COOH}$$

    b. The functional groups are —COOH, which is the carboxylic acid group, and the —NH$_2$ group, the amine group.

15. A codon is a grouping of three DNA bases. Codons are used to signal that a molecule of a certain amino acid should be incorporated into a protein.

18. There is only a minor change in the amino acid composition of human hemoglobin that leads to sickle-cell anemia. In hemoglobin S, two of the residues that should be glutamic acid are replaced by valine. This seemingly innocuous change has rather drastic results for the person with this genetic disease.

21. **a.** PCR stands for **p**olymerase **c**hain **r**eaction.

    b. Using DNA for diagnostic probes is not effective if the concentration of the invading DNA is too low. PCR technology made it possible to start with a single segment of DNA and make millions or billions of copies of it in a relatively short time.

    c. Amplification of DNA has made possible early diagnosis of HIV infection, for example, as well as the diagnosis of several genetic diseases.

24. This is the term coined to describe the practice of using domesticated animals to produce drugs and other medically significant substances.

29. The discovery that %A = %T and that %C = %G provided the basis for asking *why* this pattern was observed. Chargaff's contribution was in finding the data that suggested that the bases were paired. Crick and Watson took this information a step further to discover both *how* and *why* they were paired, and the influence the pairing had on the structure of DNA.

Glossary

The numbers indicate the pages where these terms are defined and explained in context.

A

absolute (Kelvin) temperature scale Zero on the Kelvin scale is absolute zero or $-273°C$, the lowest possible temperature *169*

acid a substance that releases hydrogen ions, H^+, usually in aqueous solution *232*

acid anhydride a compound, typically an oxide of a nonmetallic element, that reacts with water to generate an acid *240*

acid deposition the process by which acid is deposited through precipitation, fog, or airborne particles *237*

acid neutralizing capacity (ANC) the capacity of a lake to resist change in pH when acids are added to it *249*

acid precipitation (*see acid deposition*) *236*

acid rain (*see acid deposition*)

acidic solution an aqueous solution in which the H^+ ion concentration is greater than the OH^- concentration; a solution with a pH less than 7.0 *234*

activation energy the energy necessary to initiate a chemical reaction *153*

active (receptor) site the region of an enzyme molecule where the catalytic activity occurs *388, 469*

addition polymerization a process in which monomeric molecules combine to form a polymer without the elimination of any atoms *351*

aerosol a form of liquid in which the droplets are so small that they stay suspended in the air rather than settling *28*

alcohol an organic compound bearing an —OH functional group with the generic formula ROH *384*

alkali a term applied to some bases *233*

allotropes two forms of the same element that differ in their molecular or crystal structure, and hence in their properties *47*

alpha particle a particle given off during radioactive decay consisting of two protons and two neutrons; it has a mass of 4 amu and a charge of $+2$ *280*

amine a basic organic compound with the generic formula RNH_2 *384*

amino acid a compound containing a carboxylic acid group (—COOH), a basic amino group (—NH_2), and a characteristic identifying group; amino acids polymerize to form proteins *434, 467*

amino acid residues amino acids that have been incorporated into a peptide chain *434*

amniocentesis the procedure in which a sample of the amniotic fluid is withdrawn from the mother's uterus *476*

ampere a unit of electrical current *312*

anabolic steroid a compound that promotes muscle growth, but can have serious negative side effects *399*

analytical chemistry the branch of chemistry concerned with developing and applying techniques to detect the presence of various chemical components in a sample and to determine the concentration of those components *109*

androgens male sex hormones *395*

anion a negatively charged ion *198*

anode the electrode at which oxidation occurs *312*

antagonist a molecule that occupies the active site of an enzyme, but exhibits no activity *397*

antibody a protective biological agent generated by the body in response to infection *473*

aqueous solutions solutions in which water is the solvent *188*

aquifer a large natural underground reservoir *186*

atmospheric pressure the force with which the atmosphere presses down on a given surface area *14*

atom the smallest unit of an element that can exist as a stable independent entity *18*

atomic mass the mass of an atom expressed relative to a value of exactly 12 for carbon-12 *110, 111*

atomic mass unit (amu) a unit used to express the mass of individual atoms and molecules, equal to 1.66×10^{-24} g

atomic number the number of protons in an atomic nucleus, equal to the number of electrons in an electrically neutral atom *48*

atomic weight (*see atomic mass*) *111*

Avogadro's number the number of objects in one mole, 6.02×10^{23} *112*

B

basal metabolism rate (BMR) the number of Calories necessary to support basic body functions *440*

base a substance that releases hydroxide ions, OH^-, usually in aqueous solution *233*

basic solution an aqueous solution in which the OH^- concentration is greater than the H^+ concentration; a solution with a pH greater than 7.0 *234*

beta particle an electron released during radioactive decay; it has a mass of 1/1838 amu and an electrical charge of -1 *280*

biochemistry the branch of chemistry that deals with the chemistry of living things *109*

biomass materials produced by biological processes *165*

biotechnology technology based on the manipulation and alteration of biological materials, especially genetic material *420*

bond energy the amount of energy that must be absorbed to break a specific chemical bond, usually expressed in kJ/mole of bonds *149*

breeder reactor a fission reactor that converts U-238 to fissionable Pu-239 while it generates energy *279*

buckminsterfullerenes (fullerenes) C_{60} and related compounds *340*

C

calibration graph a graph of the absorbances versus the concentrations of several solutions of known concentration *216*

calorie the amount of heat necessary to raise the temperature of exactly one gram of water by one degree Celsius; 4.184 J *142*

Calorie the amount of heat necessary to raise the temperature of exactly one kilogram of water by one degree Celsius; 1000 cal; used in nutrition *439*

carbohydrate a compound containing carbon, hydrogen, and oxygen, the latter two in the same 2:1 atom ratio as found in water *423*

carbon cycle the cyclic process by which carbon and its compounds circulate through the animal, vegetable, and mineral kingdoms *108*

carboxylic acid an acidic organic compound with the generic formula RCOOH *384*

catalyst a chemical substance that participates in a chemical reaction and influences its speed without undergoing permanent change *77, 386*

cathode the electrode at which reduction occurs *312*

cation a positively charged ion *198*

Chapman cycle the set of four related reactions that represents the natural steady-state formation and destruction of ozone in the stratosphere *67*

Chargaff's rules the generalization that in DNA from all species, the percent of adenosine equals the percent thymine and the percent guanine equals the percent cytosine *460*

chemical change (*see chemical reaction*) *22*

chemical equation a representation of a chemical reaction using chemical symbols and formulas *22*

chemical formula a representation of the elementary composition of a chemical compound *19*

chemical reaction a process in which substances described as reactants are transformed into different substances called products *22*

chemical symbols one- or two-letter symbols that represent the chemical elements *16*

chiral isomers two forms of a compound, with the same formula and the same number and elementary identity of atoms, whose molecules are non-identical mirror images of each other; also known as optical isomers *390*

chlorination disinfection of water supplies with chlorine gas, sodium hypochlorite, or calcium hypochlorite *208*

chlorofluorocarbon a compound composed of the elements chlorine, fluorine, and carbon *46, 76*

chromosomes thread-like strands within cell nuclei that are the repository of genetic information in the form of DNA molecules *458*

clone an identical copy of a molecule, cell, or organism *471*

codon a sequence of three nitrogen-containing bases in a DNA molecule that encodes for a specific amino acid during protein synthesis *468*

coenzyme a substance, generally consisting of small molecules, working in conjunction with an enzyme to enhance the enzyme's activity *444*

combustion burning; the rapid combination of oxygen with a flammable material, accompanied by the evolution of heat energy *22, 145*

complementary bases the DNA base pairs: adenine with thymine and guanine with cytosine *462*

compound a pure substance made up of two or more elements in a fixed, characteristic chemical combination and composition *18*

concentration the ratio of amount of substance (solute) to amount of water (solvent or solution) *190*

condensation reaction a process in which monomeric molecules combine to form polymers by the elimination of small molecules such as H_2O *357*

conservation of energy, law of (first law of thermodynamics) energy is neither created nor destroyed; the energy of the universe is constant *143*

conservation of matter and mass, law of in a chemical reaction, matter and mass are conserved; the mass of the reactants converted equals the mass of products formed *23*

copolymer a polymer consisting of two or more different monomeric units *355*

covalent bond a chemical bond created when two atoms share electrons (usually an even number) *204*

covalent compound a compound consisting of molecules that are in turn made up of covalently bonded atoms; a molecular compound *204*

cracking breaking down of large molecules in petroleum into smaller ones in the gasoline range *161*

current the rate at which electrons flow; measured in amperes *312*

D

daughter the isotope formed by the radioactive decay of the "parent" isotope *281*

density mass per unit volume; usually expressed in grams per cubic centimeter or grams per milliliter *195*

deoxyribonucleic acid (DNA) the compound constituting the genetic material of all living things *458*

desalination any process that removes ions from salty water, such as sea or brackish waters *223*

dipeptide a compound composed of two amino acid units *434*

distillation a purification or separation process in which a solution is heated to the boiling point and the vapors are condensed and collected *158, 223*

DNA fingerprinting the technique of DNA matching that can be used to identify the individual source of a DNA sample *478*

DNA probes relatively short segments of single-stranded DNA used to specifically bind to other DNA *474*

double bond a covalent bond consisting of two pairs of electrons shared between two atoms *54*

double helix description of the molecular structure of deoxyribonucleic acid (DNA) *461*

E

efficiency the fraction of heat energy that is converted to work in a power plant *169*

electrochemical cell (battery) a device that converts the energy released in a spontaneous chemical reaction into electrical energy *315*

electrode an electrical conductor that serves as the site of chemical reaction in an electrochemical or electrolytic cell *312*

electrolysis the electrical decomposition of a compound into its constituent elements *309*

electrolyte a solution that conducts electricity or a compound that will conduct electricity when dissolved in water *198*

electrolytic cell a device in which applied electrical energy is used to bring about a nonspontaneous reaction *315*

electromagnetic spectrum the entire range of radiant energy, including X-ray, gamma, ultraviolet, visible, infrared, microwave, and radio radiation *57*

electron a subatomic particle with a mass of 1/1838 amu and a charge of −1 unit that is of great importance in atomic structure and chemical reactivity *48*

electronegativity a measure of the attraction of an atom for the electrons that constitute a covalent bond *193*

electrophoresis a method of separating molecules based on their rate of movement in an electric field; the speed at which a molecule travels depends on its size (mass) and electric charge *478*

element a substance that cannot be broken down into simpler stuff by any chemical means *16*

endothermic absorbing heat *147*

energy the capacity to do work *141*

entropy a measure of randomness in position or energy *172*

enzyme a biochemical catalyst, a protein that influences the rate and direction of a chemical reaction *424*

essential amino acid an amino acid that cannot be synthesized by the body and must be supplied in the food eaten *436*

essential fatty acid a fatty acid that cannot be synthesized by the body and must be supplied in the food eaten *429*

ester an organic compound with the generic formula RCOOR′; formed by the condensation reaction of a carboxylic acid and an alcohol *355*

estrogens female sex hormones *395*

exothermic releasing heat *145*

exposure the amount of a substance encountered *10*

F

fat a triglyceride; a compound made from fatty acids and glycerol *425*

fatty acid an acidic compound with a long hydrocarbon chain; a component of fats and oils *426*

first law of thermodynamics (law of conservation of energy) energy is neither created nor destroyed; the energy of the universe is constant *143*

fission (nuclear) a reaction in which a large atomic nucleus, such as uranium-235, splits when struck by a neutron to form two smaller fragments and release large quantities of energy 270

fraction a component separated from bulk crude oil (petroleum) by fractional distillation based on the fraction's boiling point 158

free radical an unstable chemical species with an unpaired electron 70

frequency in wave motion, the number of waves passing a fixed point in one second 57

fuel cell a cell in which a fuel such as hydrogen is allowed to react with oxygen under controlled conditions and the energy is liberated as electricity 311

functional groups groupings of atoms that confer characteristic properties on the molecule and the compound 355, 383

fusion (nuclear) a reaction in which nuclei of light atoms combine to form heavier nuclei and release large quantities of energy 328

G

gamma rays short-wavelength, high-energy electromagnetic radiation released during radioactive decay 58, 281

gas chromatography an analytical method that uses the differential absorption of components in a mixture carried by a gas as they move down a packed column; a detector indicates the emergence of each component from the mixture as it leaves the column 218

gene therapy the introduction of normal genes into patients lacking them 476

generic drug a drug that is equivalent to a pioneer drug, but not able to be marketed until the patent protection of the pioneer drug has run out (20 years) 409

genetic engineering the manipulation and alteration of genetic material (DNA) for a wide variety of purposes 450

global greenhouse effect the return of 84% of the energy radiated from the surface of the Earth 98

green chemistry designing chemical products and processes that reduce or eliminate the use and/or generation of hazardous substances 27

greenhouse effect the process by which atmospheric gases such as CO_2, CH_4, and H_2O trap and return a major portion of the heat (infrared radiation) radiated by the Earth 98

green revolution the development of high-yield grains through genetic modification 450

groundwater water pumped from wells that have been drilled into aquifers 186

H

half-life the time required for the level of radioactivity to fall to one-half of its initial value 286

hard water water containing a significant concentration of magnesium or calcium ions 210

heat the form of energy that flows from a hotter to a colder body 141

heat of combustion the quantity of heat released when a fuel is burned; variously expressed in cal/g, cal/mole, J/g, or J/mole 146

heavy metal a member of a rather ill-defined group of metallic elements with high densities and large atomic masses 214

high-density lipoprotein (HDL) a combination of lipid and protein that transports cholesterol from dead or dying cells back to the liver; the HDL's density depends on the ratio of lipid to protein 431

high-level nuclear waste (HLW) waste typically from spent fuel taken from commercial nuclear reactors and from nuclear weapons production 288

hormones substances produced by the body's endocrine glands that can have a wide range of physiological functions, including serving as "chemical messengers" 386

human genome the totality of human genetic information 486

Human Genome Project an international effort to map all the genes in the human organism and determine the DNA base sequence 458, 486

hybrid car an automobile that uses a gasoline engine and an electric motor as alternative methods of propulsion 320

hydrocarbon a compound of hydrogen and carbon 25, 158

hydrogen bond a relatively weak electrostatic attraction between a hydrogen atom bearing a net positive charge and a nitrogen, oxygen, or fluorine atom bearing a net negative charge; hydrogen bonds exist between some molecules and within some molecules 194

hydronium ion the H_3O^+ ion that is responsible for acidic properties in solution 232

I

inertial confinement a process being tested as a way of inducing nuclear fusion 330

infrared radiation heat radiation; the region of the electromagnetic spectrum that is adjacent to the red end of the visible spectrum and is characterized by wavelengths longer than red light 58

inorganic chemistry the branch of chemistry that deals primarily with chemicals of a mineral origin 109

interferons naturally occurring protein-based molecules that provide protection against viruses 473

ion an electrically charged atom or a group of covalently bonded atoms; can be positive (cation) or negative (anion) 198

ion exchange a process in which ions are interchanged, usually between a solution and a solid 212, 223

ionic bond a chemical bond created by the electrostatic attraction between oppositely charged ions 198

ionic compound a compound consisting of positively and negatively charged ions 198

isomers different compounds with the same formula and the same number and elementary identity of atoms; isomers differ in molecular structure, that is, the way in which the constituent atoms are arranged 161, 382

isotopes two (or more) forms of the same element whose atoms differ in number of neutrons and therefore in atomic mass 51

J

joule a unit of energy corresponding to $kg \cdot m^2/s^2$ 142

K

Kelvin scale the absolute temperature scale whose zero corresponds to $-273°C$, the lowest possible temperature 169

L

lethal dose-50 (LD_{50}) the minimum dose required to kill 50% of test animals 405

Lewis structure a representation of molecular structure based on the octet rule that uses dots to represent electrons 51

lipids fats, oils, and related compounds 425

lipoproteins compounds consisting of lipid and protein portions 431

low-density lipoprotein (LDL) a combination of lipid and protein that transports cholesterol from the liver to peripheral tissues; the LDL's density depends on the ratio of lipid to protein 431

low-level nuclear waste (LLW) nuclear waste contaminated with relatively small quantities of radioactive materials; other than high-level nuclear waste 293

M

macromolecule a molecule with large molecular size and a high molar mass; term often applied to polymers 340

macronutrients the major classes of compounds required for nutrition; carbohydrates, fats, and proteins 419

magnetic containment a method using strong magnetic and electrical fields to confine a beam of nuclei during fusion 330

malnutrition a condition caused by a diet lacking in the proper mix of nutrients, even though enough calories are eaten daily *418*

mass a measure of the quantity of matter in a body, often expressed in grams or kilograms and measured by weighing the object with a balance *18*

mass number the sum of the number of protons and the number of neutrons in any atomic nucleus *51, 271*

Maximum Contaminant Level (MCL) the legal limit of a contaminant, expressed in ppm or ppb *206*

Maximum Contaminant Level Goal (MCLG) the level, in ppm or ppb, at which a person weighing 70 kg (154 lb) could drink 2 L of water containing the contaminant every day for 70 years without suffering any ill effects *206*

mesosphere the region of the atmosphere above an altitude of 50 kilometers *13*

microminerals the microminerals are zinc, copper, iron, and fluorine *444*

microwave radiation electromagnetic radiation with relatively long wavelengths, low frequencies, and low-energy photons; stimulates molecular rotations; used in microwave ovens and in radar *58*

millirem (mrem) one thousandth of a rem *284*

minerals in general, inorganic chemical substances; more specifically, the inorganic chemical substances required for healthy nutrition and body function; classified as macrominerals, microminerals, and trace minerals *444*

mixture a physical combination of two or more substances (elements or compounds) present in variable amounts *15*

molar mass the mass of one Avogadro's number (one mole) of atoms, molecules, or whatever particles are specified; usually expressed in grams *113*

molarity (M) the number of moles of solute present in 1 L of solution *191*

mole one Avogadro's number of anything; 6.02×10^{23} atoms, molecules, electrons, etc. *112*

molecular compound a compound consisting of molecules; a covalent compound *204*

molecular mass the mass of a molecule expressed relative to a value of exactly 12 for carbon-12 *113*

molecular pharming the practice of using domesticated animals to produce drugs and other medically significant substances *484*

molecular weight (*see molecular mass*)

molecule a combination of a fixed number of atoms, held together by chemical bonds in a certain geometric arrangement *19*

monomer a small molecule that combines with other monomers to yield a polymer *340*

monounsaturated having one carbon-carbon double bond per molecule; usually applied to fats or fatty acids *427*

N

n-type semiconductor a material that will not normally conduct electricity well, but will do so under certain conditions through the movement of electrons *324*

neutral solution an aqueous solution containing equal concentrations of H^+ and OH^- ions; a solution with a pH of 7.0 *234*

neutralization the chemical reaction of an acid and a base *233*

neutron a subatomic particle with a mass of 1 amu and no electrical charge *48*

nitrogen balance a state in which the body excretes as much nitrogen as it ingests *435*

nonelectrolyte a substance that does not conduct electricity, either by itself or in solution *198*

nonspontaneous a process that will not occur by itself, but only if energy is supplied from some external source *172*

nuclear transfer the use of a very thin, hollow needle to remove a cell's nucleus to transfer it to an unfertilized egg from which the nucleus has been removed *484*

nucleotide the repeating unit of DNA, consisting of a nitrogen-containing base, a deoxyribose sugar, and a phosphate group *458*

nucleus (atomic) the center of an atom *48*

O

octet rule a generalization that in most stable molecules, all atoms except hydrogen will share in eight outer electrons *52*

oils triglycerides that are liquid at room temperature *425*

optical isomers (*see chiral isomers*) *390*

organic chemistry the branch of chemistry that deals primarily with compounds of carbon *109, 380*

osmosis the natural tendency for a solvent to move through a membrane from a region of higher solvent concentration to a region of lower solvent concentration *224*

outer electrons the electrons in the outer energy levels of an atom; the outer electrons are chiefly responsible for the chemical properties of that particular element *49*

oxidation a process in which an atom, ion, or molecule loses one or more electrons *311*

oxygenated gasoline gasoline blended with oxygen-containing compounds such as MTBE, ethanol, and methanol *163*

P

p-type semiconductor a material that will not normally conduct electricity well, but will do so under certain conditions through the

movement of positively charged "holes" *324*

parent the isotope undergoing radioactive decay *281*

parts per million (ppm) a measure of concentration that can be expressed in units of mass or in numbers of atoms, molecules, and/or ions *7, 190*

peptide bond the molecular linkage bonding amino acids in proteins and monomers in nylon *357, 434*

periodic table an organization of the elements in order of increasing atomic number and grouped according to similar chemical properties and similar electron arrangements *16*

pH a number, typically between 0 and 14, that indicates the acidity of a solution; also, the negative logarithm of the hydrogen ion concentration when the concentration is expressed as moles of H^+ ion per liter of solution; $pH = -\log(M_{H^+})$ *235, App. 3*

phenyl group a common molecular fragment based on a hexagon of six carbon atoms (the benzene ring); $-C_6H_5$ *353*

photon a "particle" of radiant energy *59*

photosynthesis the process by which green plants use sunlight to power the conversion of carbon dioxide and water into sugars such as glucose plus oxygen *96*

photovoltaic cell a device that converts radiant energy into electricity *307, 322*

physical chemistry the branch of chemistry that seeks to elucidate the structure of matter and discover the general principles governing its transformation *109*

pioneer drug the first version of a drug to be marketed under a brand name *408*

Planck's constant the proportionality constant relating the energy of a photon to the frequency of radiation; 6.63×10^{-34} joule second *59*

plasma a gas-like form of matter consisting of electrons and positively charged ions *330*

plasmid a ring of bacterial DNA; used as a vector for introducing new genes in recombinant DNA research and technology *471*

polar covalent bond a covalent bond in which the electrons are not equally shared, but displaced toward the more electronegative atom *193*

polyamide a polymer formed by the condensation reaction of amino acids (in which case the polyamide is a protein) or of diacids and diamines (in which case the polyamide is nylon) *356*

polyatomic ion a group of covalently bonded atoms bearing a positive or negative electrical charge *200*

polyester a polymer formed by the condensation reaction of diacid and dialcohol monomers *355*

polymer a substance consisting of long macromolecular chains   *340*

polymerase chain reaction (PCR) a technique for rapidly making many copies of a segment of DNA   *474*

polyunsaturated having more than one carbon-carbon double bond per molecule; usually applied to fats and oils   *427*

positron a subatomic particle with the mass of an electron, but with a charge of +1   *328*

primary structure the identity and sequence of the amino acids present in a protein molecule   *469*

primers single-stranded nucleotides that bracket and identify the section of DNA to be copied in the polymerase chain reaction   *474*

products the substances formed from reactants as a result of a chemical reaction   *22*

prostaglandins a group of hormone-like compounds produced by the body, where they cause a variety of responses including fever, swelling, and pain; inhibited by aspirin   *386*

proteins polymers made up of various amino acids as the monomeric units; essential components of the body and the diet   *467*

proton a subatomic particle with a mass of 1 amu and a charge of +1   *48*

Q

quantized separated into discrete energy levels, as, for example, the electronic energy levels in an atom   *59*

quantum mechanics a branch of physical science that treats the structure of atoms and molecules by constructing theoretical models   *49*

R

radiation absorbed dose (rad) the absorption of 0.01 joule of radiant energy per kilogram of tissue   *283*

radioactivity the phenomenon in which certain unstable atomic nuclei emit radiation and thereby undergo structural transformation   *280*

reactants the starting materials in a chemical reaction that are transformed into products during the reaction   *22*

receptor site a site on a cell or molecule where a hormone or other biologically active molecule can bind   *388*

recombinant DNA DNA that has incorporated into it DNA from another organism   *471*

reduction a process in which an atom, ion, or molecule gains one or more electrons   *311*

reformulated gasolines (RFG) oxygenated gasolines that contain a lower percentage of certain volatile hydrocarbons   *139, 165*

rem (*see roentgen equivalent mammal*)

replication the process by which copies of DNA are made   *465*

resonance a representation of electron transfer between two (or more) electron distributions within a molecule; the actual electronic distribution is intermediate between the extremes of the resonance forms   *55*

reverse osmosis a process for water purification in which water is forced through a membrane and ions and other contaminants are filtered out   *224*

risk assessment the process of analyzing and balancing the risks and benefits associated with some particular course of action   *10*

roentgen equivalent mammal (rem) a unit of radiation exposure equal to the number of rads absorbed multiplied by a constant, n, characteristic of the type of radiation   *284*

S

saturated having only single carbon-carbon bonds; usually applied to fats and oils   *427*

scientific notation a system for writing numbers as the product of a number, usually with one digit to the left of the decimal point, and 10 raised to the appropriate power or exponent, for example, 6.02×10^{23} *11*

second law of thermodynamics it is impossible to completely convert heat into work without making some other changes in the universe; heat will not of itself flow from a colder to a hotter body; the entropy of the universe is increasing   *171*

secondary structure helices, parallel chains, and other localized structural features in the overall structure of a protein molecule   *469*

semiconductors materials that do not normally conduct electricity well, but will do so under certain conditions   *322*

significant figures the number of numerals that correctly represents the accuracy with which an experimental quantity is known   *36*

single bond a covalent bond consisting of one pair of electrons shared between two atoms   *52*

solute a component that dissolves in a solvent to form a solution   *188*

solution a homogenous mixture at the atomic, molecular, and/or ionic level, consisting of a solute (or solutes) dissolved in a solvent   *188*

solvent in a solution, the component present in the largest concentration, usually the liquid component in which the solute dissolves   *188*

source reduction decreasing the amount of plastic waste generated by reducing the quantity of plastics produced and used   *361, 367*

specific heat the quantity of heat energy that must be absorbed to increase the temperature of one gram of a substance by one degree Celsius   *196*

spectrophotometer an instrument in which light of a desired wavelength is passed through a sample into a detector where the light is converted into an electrical signal   *216*

spent fuel material remaining in fuel rods after they have been removed from a nuclear reactor   *279*

spontaneous a process that can occur by itself, though it may be necessary to initiate the reaction   *172*

steady state a condition in which a dynamic system is in balance so that there is no net change in the concentration of at least some of the participants in the reactions   *67*

steroids a class of ubiquitous and diverse organic compounds that contain three six-membered carbon rings and one five-membered ring   *392*

stratosphere the region of the atmosphere between altitudes of 15 and 50 kilometers above sea level; location of the ozone layer   *13*

substrate the species upon which an enzyme acts   *470*

surface water lakes, rivers, and reservoirs   *186*

T

temperature a property that determines the direction of heat flow; when two bodies are in contact, heat always flows from the object at the higher temperature to that at the lower temperature   *141*

teratogen a substance that gives rise to birth defects   *403*

tertiary structure the overall three-dimensional structure of a protein molecule   *469*

tetrahedron a regular figure with four identical sides, each one an equilateral triangle; the four corners of the tetrahedron correspond to the locus of the four electron pairs (bonding and/or nonbonding) around an atom that obeys the octet rule   *102*

thermal energy energy characterized by the random motion of molecules   *171*

titration the reaction of a reagent with a measured volume of a solution of known concentration   *210*

tokamak a donut-shaped reaction vessel in which a plasma of light isotopes is contained by magnetic fields, with the object of inducing nuclear fusion   *330*

toxicity the intrinsic hazard of a substance   *10*

trace minerals dietary minerals generally needed in microgram quantities; iodine, selenium, vanadium, chromium, manganese, cobalt, nickel, molybdenum, and tin are the trace minerals required by humans   *444*

transgenic organism an organism having the genes of more than one species   *481*

triglyceride an ester composed of three fatty acids and glycerol; fats and oils are typically triglycerides   *425, 426*

triple bond a covalent bond consisting of three pairs of electrons shared between two atoms *55*

troposphere the part of the atmosphere that lies directly on the surface of the Earth *13*

U

ultraviolet the region of the electromagnetic spectrum that is adjacent to the violet end of the visible spectrum and is characterized by wavelengths shorter than violet light *58*

undernourishment a condition in which the daily intake of food is insufficient to supply the body's energy requirements *418*

unsaturated having at least one carbon-carbon double bond per molecule; often applied to fats and oils *427*

V

vaccine a biological agent that produces or increases immunity to a particular disease *473*

vector a molecular or cellular component, for example a plasmid, that is used to import foreign DNA into a host cell *471*

virus a biochemical species consisting of nucleic acid and protein that can be replicated in a host cell, where it often causes disease *473*

vitamins organic compounds that serve a variety of functions essential to life, often by promoting metabolic processes *443*

volt a unit of electrical potential *312*

voltage a measure of difference in electrical potential *312*

W

wavelength in wave motion, the distance between successive peaks *56*

weight a measure of the attraction of gravity on an object, proportional to mass *18*

work work is done when movement occurs against a restraining force; equal to the force multiplied by the distance over which the motion occurs *141*

X

X-radiation electromagnetic radiation with short wavelengths, high frequencies, and high-energy photons; used in medical diagnosis and therapy and in determining crystal structures; can damage biological tissue *58*

X-ray diffraction a procedure for determining crystal and molecular structure by measuring the pattern formed when X-rays are scattered by the constituent atoms *461*

Z

zeolite a clay-like mineral made up of aluminum, silicon, and oxygen; often used as a water softener or catalyst *212*

Credits

Text and Line Art

Chapter 1

Fig. 1.9: From M. Lippmann, "Lead and Human Health: Background and Recent Findings" in *Environmental Research*, 51:1–24, 1990. Copyright © 1990 Academic Press, Inc., Orlando, FL. Reprinted by permission; **Fig. 1.11:** Government of Canada-1991. *The State of Canada's Environment, Environment Canada.* Reproduced with the permission of the Minister of Public Works and the Government Services Canada, 1999.

Chapter 2

Fig. 2.3: From Muhammad Iqbal, *An Introduction to Solar Radiation*, 1983. Copyright © 1983 Academic Press, Inc., Orlando, FL. Reprinted by permission; **Fig. 2.5:** Reprinted by permission of John E. Frederick, University of Chicago; **Fig. 2.6:** Reprinted by permission of John E. Frederick, University of Chicago; **Fig. 2.8:** From *Scientific American*, Vol. 275 No. 1, July 1996. Copyright © Laurie Grace. Reprinted by permission; **Fig. 2.9:** Copyright © 1998 by The New York Times. Reprinted by permission; **Fig. 2.16:** From R.D. Sojkov, *The Changing Ozone Layer.* Copyright © 1995, World Meteorological Association. Reprinted by permission; **Fig. 2.17:** From World Meteorological Association web site: www.wmo.ch/web/arep/nhoz.html. Reprinted by permission; **Fig. 2.18:** From Joseph P. Glas, "The Phaseout of CFCs: The End of One Era and the Beginning of Another." Reprinted by permission.

Chapter 3

Opening text pp. 93–94: Reprinted courtesy of The Boston Globe; **Fig. 3.5:** From Chang, *Chemistry*, 6th edition. Copyright © 1998 The McGraw-Hill Companies, Inc. All Rights Reserved. Reprinted by permission; **Fig. 3.7:** Courtesy of Dr. Jeffrey A. Draves, University of Central Arkansas; **Fig. 3.11:** From Tom Wigley, et al., "Implications of Proposed CO2 Emissions Limitations," John T. Houghton (ed.), in *World Meteorological Association*, October 1997, figure 4. Reprinted by permission; **Fig. 3.12:** From "Climate Change and Risk Due to Global Warming" from UNEP/WMO document entitled Common Questions About Global Change. Reprinted by permission of World Meteorological Association.

Chapter 4

Fig. 4.1a: From *The Chemical World: Concepts and Applications*, Second edition by Moore, Stanitski, et al., Copyright © 1998 Saunders College Publishing, reproduced by permission of the publisher; **Fig. 4.1b:** From *Scientific American*, March 1998. Copyright © Laurie Grace. Reprinted by permission; **Fig. 4.6:** *Chemistry: The Central Science* 5/E by Brown/LeMay/Bursten. © 1985. Reprinted by permission of Prentice-Hall, Inc., Upper Saddle River, NJ; **Fig. 4.12:** Reprinted with permission from *Chemistry in the Community (ChemCom)*, 1988. Copyright © 1988 American Chemical Society; **Fig. 4.14:** From *The Chemical World: Concepts and Applications*, Second edition by Moore, Stanitski, et al., copyright © 1998 Saunders College Publishing, reproduced by permission of the publisher; **Fig. 4.16:** From *Chemistry: Imagination and Implication,* by A. Truman Schwartz, copyright © 1973 by Harcourt, Inc., reproduced by permission of the publisher.

Chapter 5

Fig. 5.8: From Jacqueline I. Kroschwitz, et al., *Chemistry: A First Course.* Copyright © 1995 The McGraw-Hill Companies, Inc. All Rights Reserved. Reprinted by permission; **Fig. 5.10:** From Jacqueline I. Kroschwitz, et al., *Chemistry: A First Course.* Copyright © 1995 The McGraw-Hill Companies, Inc. All Rights Reserved. Reprinted by permission; **Fig. 5.21:** © International Bottled Water Association. Reprinted by permission.

Chapter 6

Fig. 6.3: From National Atmospheric Deposition Program (NRSP-3)/National Trends Network. (1998). NADP Program Office, Illinois State Water Survey, 2204 Griffith Dr., Champaign, IL 61820. Reprinted by permission.

Chapter 7

Fig. 7.2: From Martin S. Silberberg, *Chemistry: The Molecular Nature of Matter and Change*, 2nd edition. Copyright © 2000, The McGraw-Hill Companies, Inc. All Rights Reserved. Reprinted by permission. **Fig. 7.4:** From Northern States Power Nuclear Plant, Monticello, MN. Reprinted by permission; **Fig. 7.9:** From Martin S. Silberberg, *Chemistry: The Molecular Nature of Matter and Change,* 2nd edition. Copyright © 2000, The McGraw-Hill Companies, Inc. All Rights Reserved. Reprinted by permission **Fig. 7.11:** From *The Nuclear Waste Primer,* 1985. Copyright © 1985 The Lyons Press. Reprinted by permission.

Chapter 8

Fig. 8.8: From Chang, *Chemistry*, 6th edition. Copyright © 1998 The McGraw-Hill Companies, Inc. All Rights Reserved. Reprinted by permission.

Chapter 9

Definition p. 339: © Merriam-Webster's Collegiate ® Dictionary, Tenth Edition. Reprinted by permission of Merriam-Webster, Inc.; **Fig. 9.3:** Adapted with permission, from *Chemistry* by J.W. Moore, W.G. Davies, and R.W. Collins, The McGraw-Hill Companies, 1978; **Fig. 9.20:** Reprinted by permission of NAPCOR (National Association for PET Container Resources).

Chapter 12

Text p. 455: Copyright © 1997 by The New York Times. Reprinted by permission. **Fig. 12.2:** From Robert H. Tamarin, *Principles of Genetics*, 4th edition. Copyright © 1993 The McGraw-Hill Companies, Inc. All Rights Reserved. Reprinted by permission; **Fig. 12.7:** From *Chemistry: Imagination and Implication,* by A. Truman Schwartz, copyright © 1973 by Harcourt, Inc., reproduced by permission of the publisher; **Fig. 12.8a:** From Sylvia S. Mader, *Biology*, 6th edition. Copyright © 1998 The McGraw-Hill Companies, Inc. All Rights Reserved. Reprinted by permission; **Fig. 12.9:** From Sylvia S. Mader, *Biology,* 6th edition. Copyright © 1998 The McGraw-Hill Companies, Inc. All Rights Reserved. Reprinted by permission. **Fig. 12.10:** From B.S. Hartley and D.M. Shotten, in P.D. Boyer, (ed.), *The Enzymes*, 3/e, 1971, Vol. 3, Copyright © Academic Press, Inc., Orlando, FL. Reprinted by permission; **Fig. 12.12:** From I. Edward Alcamo, *DNA Technology: The Awesome Skill.* Copyright © 1996 The McGraw-Hill Companies, Inc. All Rights Reserved. Reprinted by permission.

Photographs

Chapter 1

Opener: NASA; **p. 2 (left):** Tony Stone Images; **p. 2 (right):** NASA; **p. 3:** Garry D. McMichael/Photo Researchers, Inc; **p. 19:** Science VU/IBM/Visuals Unlimited; **p. 23:** © The McGraw-Hill Companies, Inc./Bob Coyle, photographer; **p. 29:** Courtesy of Corning Incorporated; **p. 35:** © The McGraw-Hill Companies, Inc./Ken Karp, photographer; **p. 37:** Sheila Terry/SPL/Photo Researchers, Inc; **p. 38:** SIU/Visuals Unlimited

Chapter 2

Opener: NASA; **p. 46:** Felicia Martinez/Photo Edit; **p. 50:** The McGraw-Hill Companies, Inc./Stephen Frisch, photographer; **p. 54:** Donald Clegg; **p. 66:** David Young Wolff/Photo Edit; **p. 74:** © The McGraw-Hill Companies, Inc./Stephen Frisch, photographer; **p. 85:** Courtesy of Pyrolcol Technologies; **p. 86:** Courtesy of Whirlpool, Inc.

Chapter 3

Opener: Geospace/SPL/Photo Researchers, Inc; **p. 94:** AP Photo/Katsumi Kasahara; **p. 102:** Michael Newman/Photo Edit; **p. 103 (top):** © The McGraw-Hill Companies, Inc./Stephen Frisch, photographer; **p. 103 (bottom):** Richard Megna/Fundamental Photographs; **p. 114:** Conrad Stanitski; **p. 119:** Courtesy of Robert Collier, Oregon State University, Corvallis, OR; **p. 131:** NASA

Chapter 4

Opener: Bill Ross/Tony Stone Images; **p. 140 (top):** Roy David Farris/Visuals Unlimited; **p. 140 (bottom):** Providence Management Communications; **p. 142:** Charles Winters; **p. 145:** Steve Lehman/Tony Stone Images; **p. 148:** Courtesy of Mercedes-Benz; **p. 152:** Corbis-Bettmann; **p. 153:** Charles D. Winters/Photo Researchers; **p. 156:** Jeff Smith; **p. 158:** Werner H. Muller/Peter Arnold; **p. 162:** Michael Newman/Photo Edit; **p. 163:** AP/Wide World Photos; **p. 165:** Bob Daemmrich/Stock Boston; **p. 167:** Truman Schwartz; **p. 174:** H. Schwarzbach/Peter Arnold

Chapter 5

Opener: Michael Newman/Photo Edit; **p. 185:** David Young Wolff/Photo Edit; **p. 186 (top):** Michael Newman/Photo Edit; **p. 186 (bottom):** D. Cavagnaro/Visuals Unlimited; **p. 197:** Tom Pantages; **p. 205:** Jerry Mason/SPL/Photo Researchers; **p. 209:** Lawrence Migdale/Photo Researchers, Inc; **p. 211:** Conrad Stanitski; **p. 212:** Courtesy Betz Corp; **p. 213:** © The McGraw-Hill Companies, Inc./Bob Coyle, photographer; **p. 224:** PUR ® Drinking Water Systems, a division of Recovery Engineering, Inc.

Chapter 6

Opener: Rafael Macia/SPL/Photo Researchers, Inc.; **p. 232:** © The McGraw-Hill Companies,

Inc./Stephen Frisch, photographer; **p. 233:** Felicia Martinez/Photo Edit; **p. 235, 241:** Conrad Stanitski; **p. 243:** PhotoDisc; **p. 246 (top left):** NYC Parks Photo Archive/Fundamental Photographs; **p. 246 (top right):** Kristen Brochmann/Fundamental Photographs; **p. 246 (bottom):** A.J. Copley/Visuals Unlimited; **p. 248:** Jo Prater/Visuals Unlimited; **p. 251:** M. Edwards/Peter Arnold; **p. 252, 253:** Wil Stratton; **p. 259:** Mark Gibson/Visuals Unlimited

Chapter 7

Opener: Sylvain Coffie/Tony Stone Images; **p. 266 (top):** Associated Press; **p. 266 (bottom):** John Edwards/Tony Stone Images; **p. 269:** Rob Crandall/The Image Works; **p. 275:** Joe Sohm/The Image Works; **p. 277:** Igor Kostin/Sygma; **p. 278:** Corbis-Bettmann; **p. 279:** SPL/Photo Researchers, Inc.; **p. 280 (top):** Michael Koch/Associated Press; **p. 280 (bottom):** AIP Emilio Segre Visual Archives/W.F. Meggers Collection; **p. 287:** Photo Researchers, Inc. **p. 288:** Vladimir Syomin; **p. 289:** Visuals Unlimited: **p. 291:** U.S. Department of Energy/Photo Researchers, Inc.; **p. 293:** U.S. Department of Energy/Photo Researchers, Inc.

Chapter 8

Opener: Martin Bond/SPL/Photo Researchers; **p. 306:** Associated Press; **p. 310:** NASA; **p. 313:** Courtesy of Georgetown University; **p. 314:** Courtesy of Mercedes-Benz; **p. 315:** Tony Freeman/PhotoEdit; **p. 319:** Mark Hopkins Hotel/Associated Press; **p. 321 (left):** Associated Press; **p. 321 (right):** Atsushi Tuskada/Associated Press; **p. 322 (top):** Truman Schwartz; **322 (bottom):** Tony Freeman/PhotoEdit; **p. 323:** Westinghouse/Visuals Unlimited; **p. 326 (top):** Ken Lucas/Visuals Unlimited; **p. 326 (bottom):** A.J. Copley/Visuals Unlimited; **p. 327 (both):** Associated Press

Chapter 9

Opener: Courtesy of Lonaz, Inc.; **p. 339:** Paul Mozell/Stock Boston; **p. 320:** Jess Stock/Tony Stone Images; **p. 341 (left):** American Chemical Society, photo by Rudy Baum; **p. 341 (right):** George Dillon/Stock Boston; **p. 342 (Fig. 9.3, top row, left to right):** Michael Newman/Photo Edit; Amy C. Etra/Photo Edit; Deborah Davis/Photo Edit; C.J. Allen/Stock Boston; **p. 342 (Fig. 9.3, bottom row, left to right):** Michael Newman/Photo Edit; Olear/Photo Edit; Peter Byron/Photo Edit; **p. 342 (Fig. 9.4 top row, left to right):** Ron Sherman/Stock Boston; Spencer Grant/Photo Edit; Bill Bachman/Photo Edit, **p. 342 (Fig. 9.4 bottom row, left to right):** Myrleen Ferguson/Photo Edit; Rudi Vonbriel/Photo Edit; Myrleen Ferguson/Photo

Edit; **p. 344 (top):** Elizabeth Zuckerman/Photo Edit; **p. 344 (middle):** David Young Wolff/Photo Edit; **p. 344 (bottom):** Michael Newman/Photo Edit; **p. 347:** Bill Aron/Photo Edit; **p. 351 (left and middle):** Tony Freeman/Photo Edit; **p. 351 (right):** James Shaffer/PhotoEdit; **p. 353 (both):** Bill Aron/Photo Edit; **p. 354 (top and middle):** Michael Newman/Photo Edit; **p. 354 (bottom):** David Young Wolff/Photo Edit; **p. 356 (top):** Tom McCarthy/Photo Edit; **p. 356 (bottom):** Myrleen Ferguson/Photo Edit; **p. 357:** Courtesy of Dupont; **p. 363:** The Garbage Project, The University of Arizona; **p. 364 (top):** Courtesy of ACS, photo by Mike Ciesielski; **p. 364 (bottom):** Gayna Hoffman/Stock Boston; **p. 370:** Martha G. Clarke

Chapter 10

Opener: Larry Kolvood/The Image Works; **p. 372:** Courtesy of the Alexander Fleming Laboratory Museum, St. Mary's Hospital, Paddington, London; **p. 376:** Terry Wild Studio; **p. 380:** Corbis/Bettmann; **p. 382 (both):** © The McGraw-Hill Companies, Inc./Bob Coyle, photographer, **p. 385:** Felicia Martinez/Photo Edit; **p. 387:** Dr. Lawrence Marnett, Department of Biochemistry, Vanderbilt University Medical Center; **p. 390 (top):** Courtesy of Intercardia; **p. 390 (bottom):** Courtesy of Stephen J. Cato, V.P. of Chemical Design; **p. 403:** Corbis/Bettmann; **p. 409:** Michael Newman/Photo Edit

Chapter 11

Opener: Matthew McVay/Stock Boston; **p. 423, p. 425, p. 428:** Michael Newman/Photo Edit; **p. 430 (left):** Lund/Custom Medical Stock Photos; **p. 430 (right):** Roseman/Custom Medical Stock Photos; **p. 432:** Spencer Grant/Stock Boston; **p. 433:** Michael Newman/Photo Edit; **p. 437:** Leonard Lessin/Peter Arnold, Inc.; **p. 444:** Bernd Wittich/Visuals Unlimited; **p. 447:** Tony Freeman/Photo Edit

Chapter 12

Opener: Douglas Struthers/Tony Stone Images; **p. 456:** Paul Clements/AP Photo; **p. 461:** Oesper Collection in the History of Chemistry, University of Cincinnati; **p. 462 (top):** Courtesy of the Biophysics Department, Kings College, London; **p. 462 (bottom):** Corbis/Bettmann; **p. 464:** Nelson Max/LLNL/Peter Arnold, Inc; **p. 470 (both):** Bill Longcore/Photo Researchers, Inc.; **p. 477:** Courtesy Dr. Ken Culver, photo by John Crawford, National Institutes of Health; **p. 479:** Courtesy of Lifecodes Corporation; **p. 486:** APTV via Media/AP Photos

Index